Lecture Notes in Networks and Systems

Volume 368

The series "Lecture Notes in Networks and Systems" publishes the latest developments in Networks and Systems—quickly, informally and with high quality. Original research reported in proceedings and post-proceedings represents the core of LNNS.

Volumes published in LNNS embrace all aspects and subfields of, as well as new challenges in, Networks and Systems.

The series contains proceedings and edited volumes in systems and networks, spanning the areas of Cyber-Physical Systems, Autonomous Systems, Sensor Networks, Control Systems, Energy Systems, Automotive Systems, Biological Systems, Vehicular Networking and Connected Vehicles, Aerospace Systems, Automation, Manufacturing, Smart Grids, Nonlinear Systems, Power Systems, Robotics, Social Systems, Economic Systems and other. Of particular value to both the contributors and the readership are the short publication timeframe and the world-wide distribution and exposure which enable both a wide and rapid dissemination of research output.

The series covers the theory, applications, and perspectives on the state of the art and future developments relevant to systems and networks, decision making, control, complex processes and related areas, as embedded in the fields of interdisciplinary and applied sciences, engineering, computer science, physics, economics, social, and life sciences, as well as the paradigms and methodologies behind them.

Indexed by SCOPUS, INSPEC, WTI Frankfurt eG, zbMATH, SCImago.

All books published in the series are submitted for consideration in Web of Science.

More information about this series at https://link.springer.com/bookseries/15179

Elena G. Popkova
Editor

Imitation Market Modeling in Digital Economy: Game Theoretic Approaches

Springer

Editor
Elena G. Popkova
Moscow State Institute of International Relations
Moscow, Russia

ISSN 2367-3370 ISSN 2367-3389 (electronic)
Lecture Notes in Networks and Systems
ISBN 978-3-030-93243-5 ISBN 978-3-030-93244-2 (eBook)
https://doi.org/10.1007/978-3-030-93244-2

This Springer imprint is published by the registered company Springer Nature Switzerland AG
The registered company address is: Gewerbestrasse 11, 6330 Cham, Switzerland

Imitation Modeling as A Scientific and Methodological Approach to Studying the Digital Economy Markets (Introduction)

Transitioning to the digital technological mode is a new trend of the modern market economy. Like any change, this new trend has increased the uncertainty of the perspectives of markets' development and markets' economic subjects. This leads to increased complexity and reduced effectiveness of managing markets and market agents in the conditions of a high level of uncertainty. Thus, an important scientific and practical problem is the reduction of uncertainty in the digital economy markets.

The existing scientific and methodological approach to reducing the uncertainty of economic processes and systems, which is actively used in the market economy and is connected to the creation and application of conceptual models, cannot be used in the conditions of digitalization. This is because the digital economy is completely new, changes dynamically, and envisages a multi-vector development of economic systems (with the foundation on different technologies). The conceptual model is based on a reconsideration of the past experience of the market economy through an expert's lens—i.e., it has a certain share of subjectivism; it is static—i.e., envisages the permanence of economic processes and systems; it allows comparing alternative scenarios from the positions of only one economic subject (the one that uses the model).

As an alternative, this book offers a new—game—scientific and methodological approach to reducing the uncertainty of economic processes and systems, which envisages their imitation modeling. The advantages of the new approach make it especially suitable for studying the digital economy markets. Thus, the imitation model obtained with the help of the game approach has the following properties:

- it is based on scientific logic, official statistics, and mathematical tools, due to which it is objective and very precise;
- it is dynamic; i.e., it allows taking into account the possible changes of economic processes and systems based on the modeled logical connections and regularities;

– it allows comparing alternative scenarios from the positions of all economic subjects, considering each scenario as a conditional game (e.g., a "prisoner's dilemma"), and considering the costs and winnings of each participant of the game.

To solve this problem, the book sets four research questions (RQ), the answer to each of which allows receiving the imitation modeling of the digital economy markets.

RQ1: What is the influence of globalization and digitalization on the modern economy and sectorial markets? The existing views on globalization and digitalization are contradictory. When focusing on advantages, it is possible to see that freedom of international trade increases availability of resources, and digitalization ensures the growth of resource efficiency of economic subjects and economic systems; opens new opportunities for improving positions in the world arena for developing countries, in a new—digital—form of competition and competitiveness.

The focus on drawbacks shows that globalization hinders the achievement of national and entrepreneurial interests, subjecting them to international tendencies. After decades of the intensive international integration and domination of the philosophy of free trade, the interest in moderate protectionism for the protection of national markets from the negative influence of the global economy is returning. Digitalization opens the opportunities for developing countries only formally; in reality, it strengthens the positions of developed countries in international relations and world trade. Digital technologies are unpredictable—their implementation might not satisfy the expectations of growth of business processes' effectiveness and might require unexpected and unaffordable—for economic subjects and economic systems—expenditures.

Due to the contradiction of globalization and digitalization, their influence on the modern economy and sectorial markets cannot be correctly described by conceptual models, since they would focus either on the advantages or drawbacks of this influence. Imitation models are necessary; they allow for the objective reflection of this influence with the systemic consideration of the pros and cons of globalization and digitalization.

RQ2: What is the experience and perspectives of improving the use of the leading technologies in the digital economy markets? Digital technologies are diverse, and they are implemented and used without any system. However, there are three obvious alternative scenarios of the development of the digital economy under the impact of the Fourth Industrial Revolution. The first scenario envisages a transition to Industry 4.0 and dissemination of the leading technologies—robots, artificial intelligence, the Internet of Things, etc—in the market. This scenario is most typical for progressive (hi-tech) markets with sufficient investments.

According to the second scenario, the leading technologies are not available for the market, but digitalization is profitable for market agents and becomes very widespread. Standard digital technologies—mobile communication, Internet, etc—are implemented. The third scenario means a lack of perspectives of the participation in the Fourth Industrial Revolution or unavailability of digital technologies for market participants—e.g., due to the limited influence of the mechanism of competition of the deficit of investments. In this case, digital deprivation or conscious and voluntary refusal (protest against) from digitalization takes place with the preservation of the pre-digital technological mode.

Each of the described scenarios requires a deep elaboration in view of precise and reliable causal connections of the change of economic subjects and economic systems, which could be determined by imitation modeling. Each scenario is connected to its conditional winners and losers; the systemic consideration of their interests and consequences for them requires the use of game theory.

RQ3: What is the manner of implementation, what is the role, and how is it possible to optimize business management and business integration in the digital economy? On the one hand, integration of entrepreneurial structures allows them to join efforts and resources, distribute risk, and increase the chances for a successful transition to the digital technological mode. On the other hand, integration in the conditions of the digital economy acquires a new—electronic—form and requires reconsideration and revision of the existing business practices. Imitation modeling and game theory are especially useful in this.

RQ4: How to ensure (in the most effective way) economic security and sustainable development of markets and companies in the digital economy? New—digital—risks, which emerge and grow in the conditions of the Fourth Industrial Revolution, have to be controlled based on the progressive practices of management. Together with this, the sustainable development goals receive new impulses and face new barriers on the path of their practical implementation. The very notion and approach to the achievement of economic security and sustainable development require a certain revision in view of the modern digital and global context. Imitation modeling allows for the fullest and the most correct answer to this question as well.

The goal of this book is to use imitation modeling as a scientific and methodological approach to studying the digital economy markets. This goal is achieved consecutively in the four thematic parts of the book. The first part is devoted to imitation modeling of the impact of globalization and digitalization on the modern economy and sectoral markets. It studies the transformation of the added value chains and managerial decisions under the influence of automatization and considers the trends of digital transformation of the market economy with the focus on the experience and case examples of Russia and the EAEU.

In the second part, the use of the leading technologies in the digital economy markets is studied from the positions of game theory. The authors take into account the leading technologies of digital transport, systems of information security, alternate reality in e-commerce, mobile technologies, etc.

The third part considers business management and business integration in the digital economy with the help of imitation models. The authors elaborate on the specifics and prospects of networkization of business in the digital economy, the issues of improving logistics, development of the regional economy, the use of cluster mechanisms for the development of digital entrepreneurship, etc.

In the last, fourth, part of the book, economic security and sustainable development of markets and companies in the digital economy are studied with the help of game theory. Main attention is paid to the issues of modeling of the labor market, efficiency and sustainable employment, business risks management in the conditions of the digital economy, and organizational and legal regulation of economic security and sustainable development with reference to the experience of Russia and countries of the EU.

This book is aimed primarily at scholars who study the digital economy markets. They will find in this book the scientific and methodological recommendations and developments on the use of game theory, as well as imitation models of the digital economy markets. To a certain extent, this book could be of interest for the subjects of state and corporate management of the digital economy markets. In this book, they will find answers to the question on the most effective management with the use of imitation models and game theory.

Contents

The Influence of Globalization and Digitalization on the Modern Economy and Sectoral Markets

Impact of Digitalization in Customs on Increasing the Speed and Reliability of the Supply Chain

Anastasia A. Vorona[1(✉)], Anastasia G. Getman[1], and Valentina Yu. Dianova[2]

[1] The North-West Institute of Management of the Presidential Academy of National Economy and Public Administration, St. Petersburg, Russia
getman-ag@ranepa.ru

[2] Plekhanov Russian University of Economics, Saratov, Russia

Abstract. In current conditions, the development of economy and economic processes require all structures of the system to use advanced technology to accelerate and simplify the performed actions. In this regard, the issue of implementing digital technology in all government agencies and business structures becomes particularly relevant. In turn, this requires financial and managerial costs for the optimal and quality work on the implementation of the digital benchmark. The Federal Customs Service is an executive authority that plays a significant role in the country's economy. It is one of the main administrators of the country's budget revenues. Moreover, it provides fiscal, regulatory, and controlling functions in foreign economic activity. This study presents the main results of the ongoing measures on digitalization in customs. The research methodology includes the analysis of works on the development and digitalization of the Federal Customs Service. From a scientific point of view, the research can help significantly contribute to the development of customs theory. The paper identifies the key areas of development of customs processes for the creation of intellectual customs. The authors identify the dependence of customs operations on digitalization and its impact on the speed of the supply chain across the customs border.

Keywords: Federal Customs Service · Customs administration · Customs authorities · Digitalization · Electronic declaration center · Supply chains · Reliability · Speed

JEL Codes: M10

1 Introduction

Nowadays, the world economy sees global processes, the understanding of which requires the consideration of a whole set of areas of interaction between different spheres of life. One of these processes is digitalization, which is now seen as an integral part of the development of the world.

E. G. Popkova (Ed.): ISC 2020, LNNS 368, pp. 3–11, 2022.
https://doi.org/10.1007/978-3-030-93244-2_1

The use of promising and innovative technologies becomes the key to the success of any activity. The digitalization of Russia's economy necessitates the transition of all spheres of social and economic activity to digital technology. Customs is one of the essential areas. At the same time, the Federal Customs Service of Russia (FCS) is one of the fastest-growing government agencies, which, in turn, dictates the need to apply advanced digital technologies in the daily activities of customs authorities and ensure the acceleration and simplification of work.

The paper aims to analyze the impact of the implementation of digitalization in customs on the reliability of supply chains in current conditions.

2 Materials and Methods

The research used the methods of theoretical generalization and comparison, analysis and synthesis, and methods of statistical analysis.

The issues of the development of customs authorities are discussed by Russian and international authors, such as V. Y. Dianova [7], V. V. Makrusev [14], Yu. V. Malevich [15], A. Kh. Tursunzoda [19], and K. V. Yakovlev [22].

Digital benchmarks in the activities of the Federal Customs Service are discussed in the works of O. P. Matveeva, R. P. Meshechkina [16], T. V. Luzina, E. A. Molchanov [13], A. A. Vorona, S. N. Gamidullaeva, L. A. Kopteva, and Yu. A. Maksimov [23].

M. S. Beck, K.V. Petelina, S. P. Kalmykov [3], A. A. Seleznev [18], and others study the issues of digitalization of the commodity supply chain. Our review of secondary literature demonstrates the increased interest in the digitalization of cargo movement across the customs border.

3 Results

Modern trends dictate the need for widespread use of digital technology in public administration, business, and everyday life. The active use of these technologies entails the need to create conditions that will provide a comfortable environment for all participants and ensure the protection of transmitted information. The measures taken by the government are aimed at developing and facilitating the interaction of all participants in the economic system. The most striking example of such transformations in the field of informatization processes is a single portal for providing state and municipal services in digital form. This technology provides contactless interaction between citizens and officials of state and municipal authorities for obtaining various types of public services, including customs services (inclusion of intellectual property in the customs register, obtaining a preliminary classification decision, etc.). By the end of 2020, the Public Services Portal of the Russian Federation (Gosuslugi) provided 228 million services and processed 78 million payments to the amount of 76 billion rubles. By the end of the year, the service had 126 million registered users. Of them, 24 million users joined the service just in 2020 [17].

Transformations caused by the introduction of digital technologies influenced almost all priority sectors of the economy, including foreign trade and customs.

Originally, digitalization referred to the conversion of information into a digital form, which, in most cases, led to lower costs and the emergence of new opportunities [12].

In retrospect, the first use of the term digitalization dates back to 1971 and refers to the "increasingly widespread use of digital technology" [20].

Nowadays, the term digitalization represents the introduction of digital technology into various spheres of life and society. Since the first mention of digitalization, there have been significant changes, primarily in the technology itself.

In this context, the digitalization of customs and near-customs structures also plays a special role. The main focus of digitalization is to improve the efficiency of custom processes. Improving the efficiency of customs processes is seen in reducing the time of customs operations, increasing the efficiency of customs control, and ensuring the completeness and timeliness of the transfer of customs duties to the country's budget, which will undoubtedly ensure national and economic security.

Let us look at the foreign trade statistics of the Russian Federation for January-November 2018–2020 (Table 1).

Table 1. General results of foreign trade of the Russian Federation for January–November 2018–2020, million $

Indicator	2018 (January–November)	2019 (January–November)	2020 (January–November)	Growth rate 2019 to 2018, %	Growth rate 2020 to 2019, %	Growth rate 2020 to 2018, %	Growth rate 2019 to 2018, million $	Growth rate 2020 to 2019, million $	Growth rate 2020 to 2018, million $
Turnover	629,179	610,938	511,358	97.10	83.70	81.27	−18,241	−99,58	−117,821
Export	410,088	386,416	302,02	94.23	78.16	73.65	−23,672	−84,396	−108,068
Import	219,091	224,523	209,339	102.48	93.24	95.55	5,432	−15,184	−9,752
Balance	190,997	161,894	92,682	84.76	57.25	48.53	−29,103	−69,212	−98,315

Source Compiled by the authors based on [10]

The data presented in Table 1 shows a significant decrease in trade flows in January-November 2020, compared to the same periods in 2019 and 2018. Trade turnover decreased by 16.3%, exports decreased by 21.84%, and imports decreased by 6.76% compared to 2019. Compared to 2018, the year 2020 saw an even more significant decrease in turnover (18.73%). It is worth noting that this circumstance is mainly caused by the COVID-19 spread, which led to the destabilization of international supply chains [4]. In general, we can say that the COVID-19 pandemic had a negative impact on Russia's foreign trade since there was a decrease in foreign trade turnover, which led to a reduction in customs payments administered to the federal budget. Next, let us provide statistics on customs payments, which is presented on the website of the Federal Customs Service of Russia (Fig. 1).

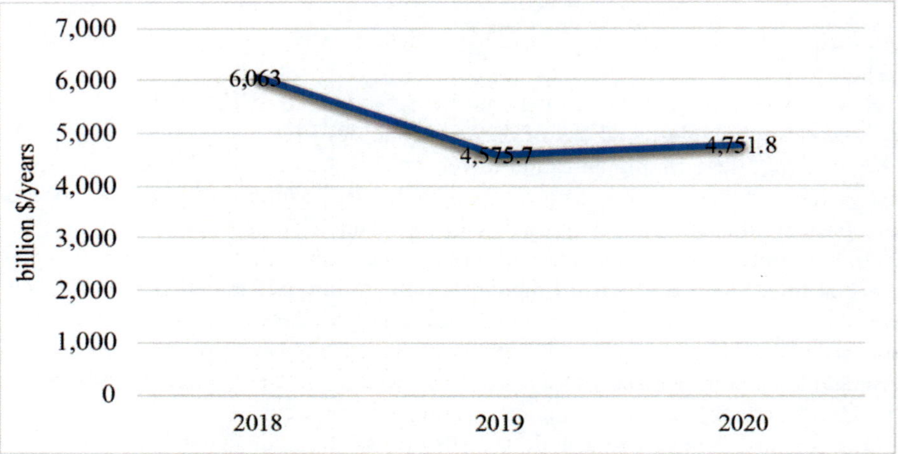

Fig. 1. Administration of customs payments by the FCS to the federal budget for 2018–2020. *Source* Compiled by the authors

Figure 1 shows that the administration of customs payments did not decrease in 2020 compared to 2019 due to the decrease in trade turnover and the measures taken by the government to combat the spread of COVID-19. Customs borders were closed, which led to increased prices for goods and services due to reduced competition in national markets of many countries. However, we can observe a sharp decline (by 21.63%) in the transfer of payments to the budget compared to 2018.

Due to restrictive measures, customs authorities and participants in foreign economic activity (FEA) faced the problem of interaction in the passage of goods across the customs border. This has led to an even greater need for digital technology. According to the head of the Federal Customs Service of Russia V. I. Bulavin, transparency of the customs sphere directly depends on the density of technologies in the customs sphere and the level of trust between all participants of foreign economic activity [8].

The Concept of creating a network of electronic customs houses and electronic declaration centers (EDC) has been implemented to simplify the interaction between customs authorities and participants in foreign economic activities. Sixteen electronic declaration centers have already been created and are in operation; approximately 90% of all declarations of goods are drawn up in these centers. The goal of the FCS for 2021 is to bring the level of declaration processing in the EDC to 95% of the entire declaration array. The digitalization of declaration processing allows to speed up this process without reducing quality.

Digitalization also plays a significant role for the business community engaged in foreign economic activities. In this case, it minimizes the logistics costs associated with implementing customs procedures and, above all, the reduction of time costs, which directly affects the competitiveness of business structures. To provide the best assessment of competitiveness, it is necessary to make observations over a long period and observe the dynamics of indicator values. However, most consulting firms nowadays study a company's business processes only once [6].

One of the most discussed and applied methodologies for assessing the optimal conditions for international trade is the Doing Business ranking.

A. A. Vorona describes the areas of application of the Doing Business ranking [21]. According to 2019 data, Russia ranks 28th in the ease of doing business in the Doing Business 2020 report [24], which is 12 points higher than the 2017 position and 3 points higher than the 2019 position.

It is worth noting that the achieved results and the strengthening of Russia's position in this rating are mainly due to the digitalization of the customs sector. There is a direct correlation between the speed of customs operations and the speed of all stages in the supply chain. The use of digital technology by customs authorities has reduced the time required to complete customs operations for importing and exporting goods. This statement is graphically presented in Table 2, which contains information about the time costs at the checkpoints on the state border of Russia.

Table 2. Speed of customs operations in export-import flows in Q3 2013–2020, min.

Operation name	2013 Q3	2014 Q3	2015 Q3	2016 Q3	2017 Q3	2018 Q3	2019 Q3	2020 Q3
Time of operation by customs authorities of the Russian Federation related to state control at road checkpoints (minutes) for goods subject to veterinary, phytosanitary, and sanitary-quarantine control	47.5	37.1	31.8	29.62	29.86	28.14	28.83	27.52
Time of operation by customs authorities of the Russian Federation related to state control at road checkpoints (minutes) for other goods for which no risks of violations have been identified	29.8	27.2	20.7	17.61	18.26	17.91	17.25	19.93

Source [9]

According to Table 2, over the past seven years, there has been a reduction in time for customs authorities to perform state control operations by 12.87 min. Accordingly, the obtained data indicate the acceleration of processes, which, in turn, has a positive impact on the speed of goods transportation.

It is also possible to highlight such an aspect as the reliability of the supply chain. In his dissertation, P. A. Bochkarev indicates the following characteristics of supply reliability: market sensitivity, virtuality, process matching, and network approach [5]. However, potential risks may arise due to the specifics of the product itself [11]. In this regard, the authors indicate the need to develop mathematical tools to assess the speed and reliability of the supply chain. It is also necessary to determine the role of customs authorities in this. This is the task for our future studies.

Accordingly, electronic interaction between customs authorities and participants in foreign economic activity, through creating a network of electronic customs and EDC, will lead to a more reliable and high-quality process of moving goods across the border.

Thus, Russia is steadily improving its position by reducing the time for customs operations. One of the factors affecting the timing of customs operations is the digitalization of processes, namely, the use of information customs technology in customs control.

In this regard, the "Strategy of development of customs service of the Russian Federation until 2030" (Strategy 2030) was adopted by order of the Government of the Russian Federation of May 23, 2020, No. 1388-r [1]. Strategy 2030 formulated 23 target benchmarks.

The strategy for the development of the customs service is aimed at creating an intellectual custom by 2030 [2]. It provides for the implementation of several fundamental areas of development of customs processes. Certain results were achieved, and strategic objectives for the next ten years were set as a result of the measures taken by the government (Fig. 2).

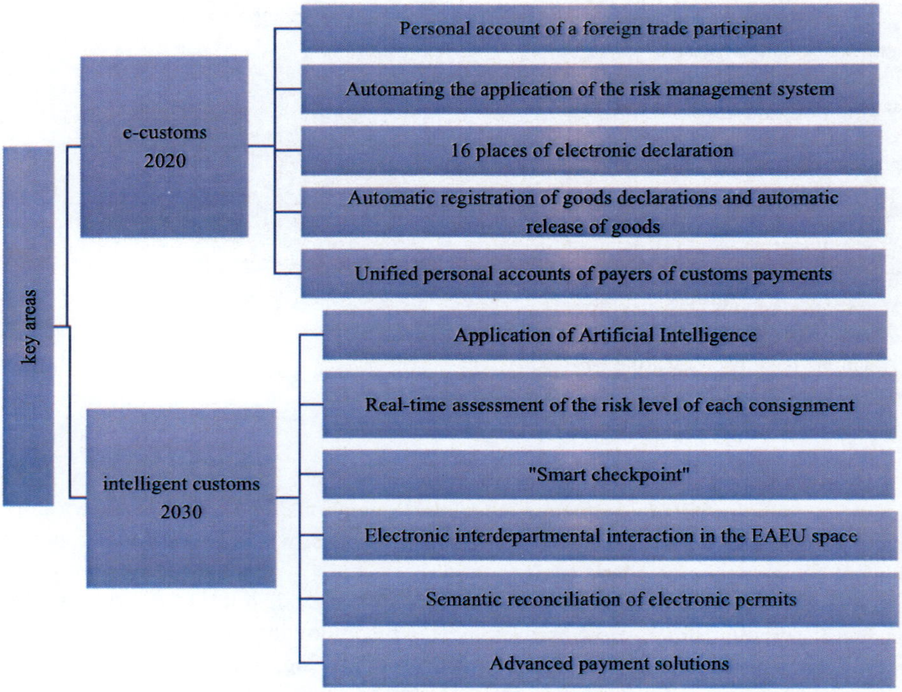

Fig. 2. Key areas of the development of customs processes for the creation of intellectual customs. *Source* Compiled by the authors

During the transition from e-customs, the creation of which was completed in 2020, to intelligent customs, significant projects are planned through the active introduction of a digital platform and elements of artificial intelligence in custom activities, which makes it possible to store, adapt, optimize, and transmit a huge array of incoming information. This system will also allow making sound and correct management decisions in the shortest possible time intervals.

For the period up to 2030, it is essential to build effective work on the construction, reconstruction, and modernization of customs checkpoints on the state border of the Russian Federation. This direction of customs policy is one of the most important since it directly relates to increasing the level of protection of state interests of the country.

However, it is now clear that digital technology is not enough for a qualitatively new functioning of the entire customs system. Thus, it is necessary to introduce technological solutions that can simulate the thought processes and human functions and achieve positive results. Further implementation of the strategic objectives defined by Strategy 2030 depends on the quality of the planned activities to implement the targets. It is now becoming a priority task for the FCS of Russia.

4 Conclusion

The Federal Customs Service has already taken a big step toward the digitalization of customs technology. Customs control processes that depend on the results of intelligent analysis of unformalized data and verbal information present the greatest challenges for digitization and automation. Our study indicates that the measures taken by the Federal Customs Service of Russia on digitalization and transformation are aimed at the further development of the agency. The implementation of strategic targets should improve the business climate in Russia. Bonafide participants in foreign economic activities will have the opportunity to optimize the supply chain by simplifying customs procedures and improving the quality of service through the automation of processes. The authors identified the need to develop mathematical tools to assess the speed and reliability of the supply chain and determine the role of customs authorities in this process.

References

1. Government of the Russian Federation. Order “Strategy for the Development of the Customs Service of the Russian Federation Until 2030” (23 May 2020, No. 1388-r) (2020). http://www.consultant.ru/document/cons_doc_LAW_353557/
2. Presidential Executive Office. Decree “On the Development of Artificial Intelligence in the Russian Federation” (10 Oct 2019, No. 490) (2019). http://www.consultant.ru/document/cons_doc_LAW_335184/
3. Baranov, D.N.: The essence and content for “The digital economy” category. Bull. Moscow Univ. Named S. Yu. Witte. Ser. 1 Econ. Manag. **2**(25), 15–23 (2018). https://doi.org/10.21777/2587-554X-2018-2-15-23

4. Beck, M.S., Petelina, K.V., Kalmykov, S.P.: Digital transport corridors as one of the initiatives in the framework of the digital agenda of the EAEU 2025. In: Kalmykov, S. P. (ed.) Actual Issues of State Regulation of Foreign Economic Activity: Domestic and Foreign Experience, pp. 6–10. Siberian Transport University, Novosibirsk, Russia (2020)
5. Bochkarev, P.A.: Management of Reliability of Supply Chains in Logistics of Supply (Dissertation of Candidate of Economics). St. Petersburg, Russia (2015). https://www.hse.ru/data/2015/07/22/1102695943/Диссертация_20Бочкарев.pdf. Accessed 24 Dec 2020
6. Boyko, I.V., Getman, A.G.: Global supply chains: new trends in the context of the Coronavirus pandemic. Admin. Consult. **11**(143), 42–48 (2020). https://doi.org/10.22394/1726-1139-2020-11-42-48
7. Dianova, V.: Arrangement of favorable conditions for foreign trade activity within the EAEU by improvement of customs administration. Custom. Bus. **2**, 18–21 (2019)
8. Federal Customs Service of the Russian Federation. All Planned Electronic Declaration Centers have been Established in the Russian Federation-comments by Vladimir Bulavin at the Opening Ceremony of the Far Eastern Electronic Customs (4 Sept 2020). https://customs.gov.ru/press/aktual-no/document/249388. Accessed 23 Dec 2020
9. Federal Customs Service of the Russian Federation. Information on the Performance of the Main Indicators of the Customs Authorities of the Russian Federation, 2013–2020 (n.d.). https://customs.gov.ru/folder/89130
10. Federal Customs Service of the Russian Federation. Results of Foreign Trade with All Countries, 2018–2020 (n.d.). https://customs.gov.ru/statistic/vneshn-torg/vneshn-torg-countries
11. Getman, A.G.: Actual issues of risk management in the international chains of deliveries of the goods containing intellectual property Items. Admin. Consult. **10**, 64–69 (2018). https://doi.org/10.22394/1726-1139-2018-10-64-69
12. Khalin, V.G., Chernova, G.V.: Digitalization and its impact on the Russian economy and society: advantages, challenges, threats and risks. Admin. Consult. **10**(118), 46–63 (2018). https://doi.org/10.22394/1726-1139-2018-10-46-63
13. Luzina, T.V., Molchanov, E.A.: Regulation of custom operations in the context of digitization of the economics. Bus. Educ. Right **3**(48), 131–139 (2019). https://doi.org/10.25683/VOLBI.2019.48.349
14. Makrusev, V.V.: Topical issues of the customs administration institution development within the digital economy. In: Niyazova, M.V. (ed.) Pacific Russia on the Economic and Political Map of the World, pp. 99–103. Vladivostok Branch of the Russian Customs Academy, Vladivostok (2019)
15. Malevich, Yu.V.: Improving customs administration of global supply chains: analysis of results. In: Shcherbakov, V.V. (Ed.). The Logistics Potential of St. Petersburg in the Formation of an Innovative Economy, pp. 179–182. St. Petersburg, Russia (2016)
16. Meshechkina, R.P.: Digitalization as a factor of increasing the efficiency of customs administration. Bull. Russ. Custom. Acad. **3**(52), 33–39 (2020)
17. Public Services Portal of the Russian Federation. Public Services in Figures for 2020 (2020, Dec 30). https://www.gosuslugi.ru/help/news/2020_12_30_results_of_the_year
18. Seleznev, A.A.: On the issue of digital logistics. In: Gamidullayev, S.N. (Ed.), Customs Readings-2019: Science and Education in the Context of the Formation of an Innovative Economy, pp. 104–107. V. B. Bobkov Saint-Petersburg Branch of the Russian Customs Academy, St. Petersburg (2019)
19. Tursunzoda, A.Kh.: Improvement of customs administration measures in the Republic of Tajikistan. Sci. Notes St. Petersburg Branch Russ. Custom. Acad. Named After V. B. Bobkov **4**(76), 67–69 (2020)

20. Volynchikov, I.B., Timchenko, T.N.: Provisions for the formation of a maritime shipping company's competitiveness management system. Int. J. Econ. Bus. Admin. **7**(Special Issue 1), 93–100 (2019). https://doi.org/10.35808/ijeba/255
21. Vorona, A.A.: Recent trends of the score in "Doing Business" and Russia's position in it. Sci. Lett. Russ. Custom. Acad. St. Petesrburg Branch named after Vladimir Bobkov **4**(68), 51–55 (2018)
22. Vorona, A.A., Yakovlev, K.V.: Directions of customs administration in the activities of the Federal Customs Service and improvement of customs control after the release of goods. Sci. Lett. Russ. Custom. Acad. St.-Petesrburg Branch Named After Vladimir Bobkov **1**(73), 9–12 (2020)
23. Vorona, A.A., Gamidullaev, S.N., Kopteva, L.A., Maksimov, Yu.A.: Economic and Information Security in the Conditions of the Activity of Electronic Declaration Centers. V. B. Bobkov Saint-Petersburg Branch of the Russian Customs Academy, St. Petersburg (2020).
24. World Bank. Doing Business 2020. Comparing Business Regulation in Economies. World Bank, Washington, DC (2020). https://openknowledge.worldbank.org/bitstream/handle/10986/32436/9781464814402.pdf

Transformation of Customs Administration and the Impact of Automation on Decision-Making by Customs Authorities

Maya V. Boikova[1], Anastasia A. Vorona[2](✉), Dmitry V. Gubarev[3], Yuri A. Maksimov[4], and Tatiana N. Timchenko[3]

[1] Russian Customs Academy, Moscow, Russia
[2] The North-West Institute of Management of the Presidential Academy of National Economy and Public Administration, St. Petersburg, Russia
[3] Admiral Ushakov State Maritime University, Novorossiysk, Russia
[4] St. Petersburg Branch of the Russian Customs Academy named after V. B. Bobkov, St. Petersburg, Russia

Abstract. The authors focus on current trends in the development of customs authorities and the informatization of the administration of customs services. The paper considers customs reforms aimed at introducing elements of artificial intelligence into the processes of dispatching and decision-making on the release of goods during an electronic declaration. The authors briefly analyze the problems faced by customs authorities during the ongoing reforms and automation of decision-making. Some directions for overcoming these problems are proposed. The proposed algorithm for applying the decision support system in electronic declaration centers demonstrates the critical perception of the decision-making system without the participation of officials in the customs authorities.

Keywords: Federal Customs Service · Customs administration · Customs authorities · Digitalization · Automation · Decision-making · Electronic declaration centers · Economic security

JEL Codes: M10

1 Introduction

Nowadays, the use of digital technology in practice is a prerequisite for the effective functioning of entities and organizational structures. Decision-making largely depends on the quality of automation since advanced technology speeds up and simplifies all procedures. However, the seeming simplicity of digitalization forces organizations to spend considerable resources and make considerable efforts to establish a digitalization mechanism within the system.

E. G. Popkova (Ed.): ISC 2020, LNNS 368, pp. 12–20, 2022.
https://doi.org/10.1007/978-3-030-93244-2_2

Thus, it seems relevant to consider the transformation of customs administration caused by the need to adapt to changing global trends and the widespread digitalization of all processes.

The paper aims to analyze the transformational processes occurring within the Federal Customs Service and determine the impact of digitalization on the quality and effectiveness of the decisions made.

2 Materials and Methods

V. Karpovich [5], A. O. Ponomarev [11], and others study the digitalization of business and public administration. Of particular relevance are the issues of customs administration disclosed in the works of S. N. Garmonnikov [3]; R. P. Meshechkina and I. A. Aleynikova [9]; and A. A. Vorona and K. V. Yakovlev [14]. O. P. Matveeva, R. P. Meshechkina [8], A. A. Vorona [10], T. V. Luzina, and E. A. Molchanov [7] study digital benchmarks in the activities of the Federal Customs Service. The transformation of customs authorities towards the implementation of the concept of transition to electronic customs and electronic declaration centers is disclosed in the monograph by A. A. Vorona, S. N. Gamidullaev, L. A. Kopteva, and Yu. A Maksimov [15]. The reviewed works testify to the relevance of the issue under consideration and require further consideration.

3 Results

In current conditions, the Federal Customs Service of Russia is an executive authority focused on achieving results. In this regard, a variety of reforms affecting the quality and speed of processes are actively pursued. The transformation of customs administration in a fast and convenient high-tech process through a full-scale digitalization and automation of customs authorities is one of the priorities carried out following the instructions of the President and the Government of the Russian Federation on a large-scale reform of customs authorities. The Strategy 2030 [1] defines a reasonably significant list of tasks for their improvement.

The concentration of the entire declaration file in electronic declaration centers (EDCs) led to a higher level of administration of customs authorities, ensured the efficiency and effectiveness of customs control, provided the collection of customs duties, and created the unity of law enforcement. The reform allowed the Federal Customs Service of Russia to significantly reduce customs clearances, namely, 16 EDCs instead of 600 customs posts, which previously issued customs declarations.

There are currently nine electronic customs offices and 16 electronic customs. They have become the main places of customs clearance. Their effectiveness is proved by the positive results. Namely, 97.5% of declarations were declared electronically at the end of 2020, with more than 1,000,000 automatically issued goods declarations (GDs). As of November 26, 2020, almost 3.4 million and more than 964,000 GDs were registered in a fully automated mode. This figure exceeds the same figures for 2019 (2.8 million automatically registered GDs and 643,000 released GDs in 2019) [2]. Impressive results testify to the effectiveness of the implemented concept of customs administration.

In terms of structural changes that have occurred in the customs authorities, it should be noted that the division of functionality into electronic and non-electronic components became possible due to the transfer of customs operations in electronic form. Electronic declarations, preliminary informing, and remote release have formed the basis, which, as new software is developed and introduced, will be filled with new content in the form of customs operations, the execution of which will be possible in electronic form. It seems reasonable to assume that the functionality of electronic declaration centers will increase over time. Customs posts of actual control (CPAC) will focus exclusively on the manual functions of customs control, which cannot be performed without human participation.

Based on the fact that electronic customs interacts with the outside world (participants in foreign economic activities, other customs authorities, state controlling bodies, etc.) exclusively in electronic form, we believe it is legitimate to conclude that electronic customs, as a structural unit, interacts with the information flow formed in connection with the movement across the customs border of a particular shipment and in connection with the information environment formed and existing in the information storage of customs authorities and containing information on previously displaced consignments, decisions made by the customs authorities for each consignment, the results of customs control of goods, etc. Customs operations are performed in relation to this flow of information. Each operation successively changes the status of specific goods moved across the customs border: arriving, placed under the customs procedure of customs transit, temporary storage, declared, released (conditionally released) by the customs authority, and exported.

To outline the main directions of customs administration, the "Strategy for the development of the customs Service of the Russian Federation until 2030" was adopted [1]. Its adoption outlined the development of a new digital vector of development of the Federal Customs Service of Russia. This new digital vector will make customs not just electronic but also intelligent. For this purpose, it is planned to use artificial intelligence (AI). For example, the checkpoints will be a unified information system receiving data from inspection and weight-size complexes, radiation control systems, and other means of customs control.

In current conditions, the actions performed by the EDC in relation to the incoming declaration array are becoming more and more automated due to the introduction of technologies such as automatic dispatch of goods declaration, automatic registration of goods declaration, and automatic release of goods declaration. The Federal Customs Service of Russia has already established and optimized the registration and release of goods declaration. It also improved the risk management system and categorization of the participants in foreign trade. It is planned to automate 50% of all work processes by 2024 and 100% of processes by 2030.

However, any modernization entails internal and external problems. In the Federal Customs Service of Russia, these problems are failures in the operation of digital services. Thus, participants in foreign trade indicate problematic moments associated with the incomplete display of customs receipts and failures in the formation of reports on the expenditure of funds when working with their personal accounts. Additionally, the users note the problem of the inability to establish feedback with the issuing inspector. Although the Federal Customs Service of Russia has set the task of eliminating contacts between inspectors and declarants to minimize corruption risks, bona fide participants note the complete lack of feedback. After receiving a denial of automatic release, participants in foreign economic activities are forced to determine the reasons for the denial on their own, which potentially complicates the process of making adjustments.

The electronic declaration has a considerable advantage over its paper counterpart, which is becoming a thing of the past. However, such an information system is not yet complete and requires perfection. In the current environment, digitalization has provided a completely transparent process of customs formalities, eliminating the corruption component. On the other hand, it has almost completely limited the possibility of personal contact between a participant in foreign economic activity (FEA) and the releasing inspector. This sometimes leads to delays in release due to the lack of awareness of participants in FEA and the inability to obtain timely information about the reasons for the delay in making decisions on the goods declaration.

The next area of customs administration was the introduction of automatic dispatching of goods declarations. The system of dispatching customs declarations between 16 electronic declaration centers was launched on February 1, 2021.

Dispatching is an algorithm of automatic distribution of customs declarations submitted by the participants in FEA. Dispatching in customs is designed to distribute declarations to the EDCs in a more homogeneous way. This was done to reduce the workload of the EDCs and customs officials. Despite the elaborate system of dispatching and the high level of technical support of this procedure, participants in FEA have encountered several problems related to the automatic distribution of declarations.

The main problem is the information support of the participants in FEA. According to the participants in FEA, the introduction of the dispatch of customs declarations was an unexpected measure about which there was not enough advance information (even though the dispatch project had been under development since 2018).

This fact poses serious reputational problems for the Federal Customs Service of Russia. According to some participants in FEA, the time for a complete inspection of goods has increased from 1–2 days to 5–7 days. This increase is due to the increased workload of inspectors at customs posts and the shortage of space in temporary storage warehouses.

The problem is exacerbated by the fact that the consequences of idle vehicles, increased time for customs control, and freight delivery time significantly impact the performance of foreign trade participants (particularly, it decreases profits and additional costs).

The Federal Customs Service of Russia has defined directions (Fig. 1) to develop automatic and automated decision-making technologies further.

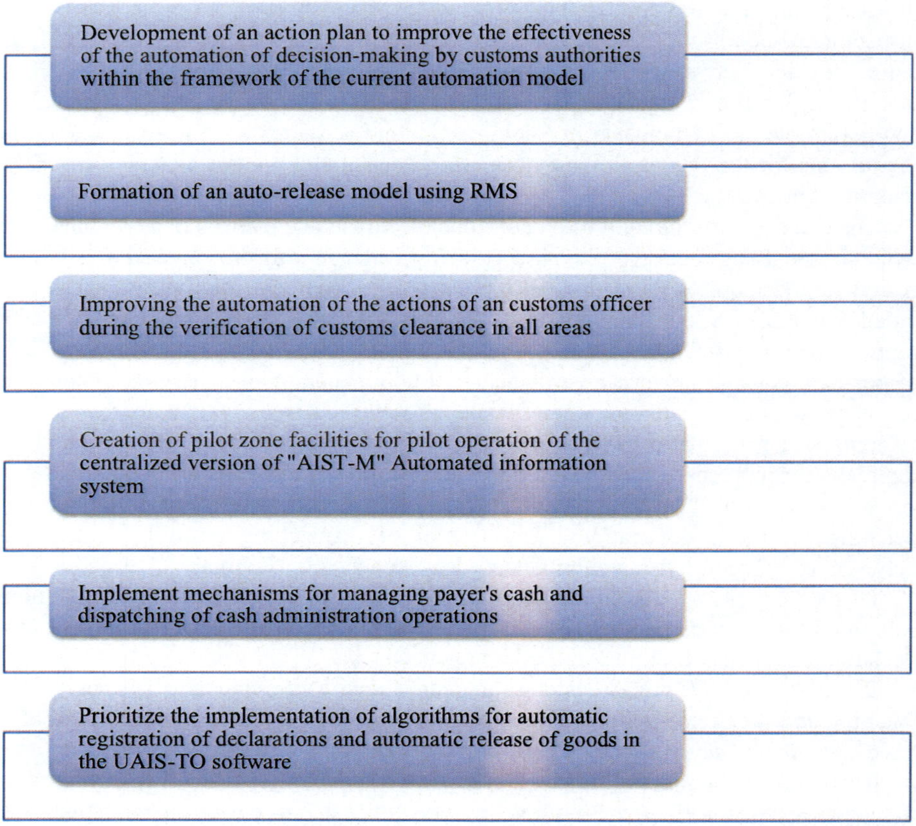

Fig. 1. Directions for the development of technologies of automatic and automated decision-making of the Federal Customs Service of Russia. *Source* Compiled by the authors

However, the authors are critical of the complete automation of the processing of the declaration file. Full automation entails external problems associated with the insufficient development of the information environment to work with paperless media, the lack of training of professional staff, and the uncoordinated system of information exchange between the EAEU member states and interagency exchange with other structural bodies (Ministry of Internal Affairs, Federal Security Service, and Tax Service).

Another problem is errors in filling out the declaration. This problem has been occurring for many years regardless of the type of declaration: on paper/electronic media. If errors are detected in the electronic declaration, an electronic notification is sent to the declarant. Such notification includes a list of errors detected during format-logic control. Errors must be corrected, and a corrected electronic declaration must be re-submitted. The re-submitted electronic declaration is assigned a unique registration number if it successfully passed the control of documents and there are no conditions for refusal of registration. At this stage, there is a potential need for justification of errors in filling out the declaration by the inspector of the EDC. At this stage, the quality and speed of the decision depend on the issuance of the GDs.

In this regard, there is a need to define how a decision is made during the implementation of activities by customs authorities within the EDC. According to T. K. Kravchenko and D. V. Isaev [6], decision-making is the process of choosing one of the alternative solutions (or just alternatives). Thus, there appears a need to systematize the approach to understanding the mechanism of decision-making, subject to the application of digital technology in EDCs. Thus, there arises the need to justify the conceptual apparatus. The authors propose the introduction of two definitions (Fig. 2).

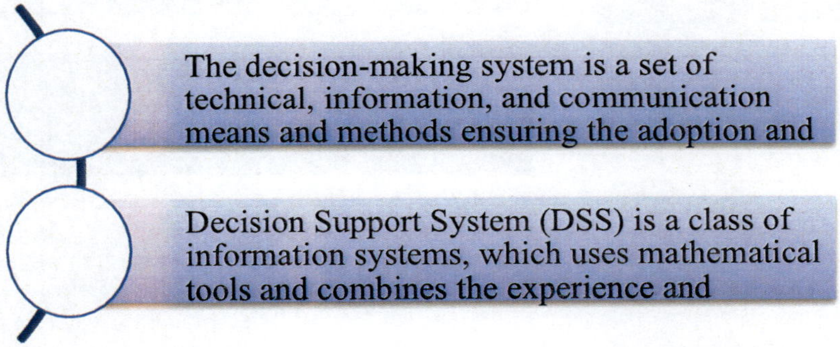

Fig. 2. Proposed conceptual apparatus. *Source* Compiled by the authors

Decision Support System (DSS) is a class of information systems, which uses mathematical tools and combines the experience and informalized knowledge of the decision-maker.

DSS is also a computer system, which allows one to make evidence-based decisions of the organizational plan by collecting and analyzing large amounts of information [12].

Such systems also allow making decisions and justifying the reasons for the decision made [4].

They are also "human-machine" systems allowing decision-makers to use available data and objective and subjective models to make decisions.

The considered definitions show that the created system of electronic customs and EDCs uses a decision-making support system. It can be schematically represented as an algorithm (Fig. 3).

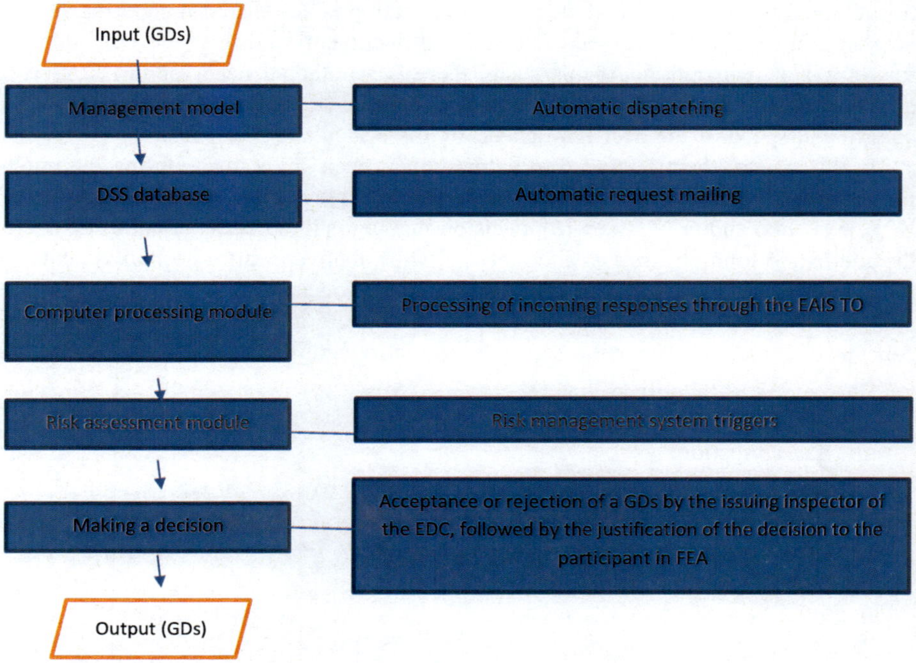

Fig. 3. Algorithm of the application of decision support system in the EDC. *Source* Compiled by the authors

The presented algorithm shows the need for the issuing inspector of the CED in the decision-making system. At the stage of triggering the risk management system (RMS), the system cannot conduct an independent analysis. Thus, the system requires the experience and qualification of the officer, who, based on the presented data, can assess the degree of severity of the situation and choose the most appropriate solution. This provision is necessary to ensure the protection of the economic interests of the government and participants in FEA. Thus, the implementation of DSS ensures compliance with economic security criteria and allows decisions to be made directly by the operator rather than the digital system. In this case, any country seeking to ensure its political independence and economic security regulates foreign economic relations [13].

On the other hand, if there are problems at the stage of triggering the RMS, the digital system automatically rejects the incoming declaration without explaining the reasons for the rejection, which leads to the re-submission of the declaration. To determine the reasons for the rejection, participants in FEA have to make an inquiry and wait for an answer, which leads to loss of time and money from the downtime of the cargo.

4 Conclusion

The digitalization of customs regulation goes far beyond the control of international trade activities and should include the widest application of information and communication technology. The application of information and communication technology will open up the following target horizons for transforming the system of customs administration:

- Use of big data to form statistics, administrate budget revenues, and conduct performance audits and regulatory and legal regulations in the field of customs;
- Application of innovative ways to obtain and process data used for customs purposes;
- Implementation of various technologies related to AI in customs control of goods and vehicles.

The main direction of the development of customs administration is seen in creating the optimal balance between the interests of the state and business in the implementation of customs activities. Strengthening the traceability system will make the process of monitoring goods from the moment of their production to the moment of their consumption completely transparent. Created automatic algorithms are designed to support bona fide participants in FEA, which is also a positive direction for businesses and the government.

References

1. Government of the Russian Federation. Order "Strategy for the Development of the Customs Service of the Russian Federation Until 2030" (23 May 2020, No. 1388-r). http://www.consultant.ru/document/cons_doc_LAW_353557/
2. Federal Customs Service of the Russian Federation. Main Results of Activity for 11 Months of 2020 (2020, Dec 25). https://customs.gov.ru/press/federal/document/261656. Accessed 15 Jan 2021
3. Garmonnikov, S.N.: On the need to improve the definitions of the concepts "customs business" and "customs administration" in the Customs Union within the framework of the EurAsEC. Customs Affair **2**, 8–12 (2013)
4. Golovko, V., Kroshchanka, A., Kovalev, M., Taberko, V., Ivaniuk, D.: Implementation of an intelligent decision support system to accompany the manufacturing process. Open Seman. Technol. Intell. Syst. **4**, 175–182 (2020)
5. Karpovich, Y.: From digitalization of business to digitalization of public administration. In: Rudenko, M.N., Subbotina, Yu.D. (eds.) Forecasting of Innovative Development of the National Economy within the Framework of Rational Nature Management, pp. 48–51. Perm State University, Perm (2020)
6. Kravchenko, T.K., Isaev, D.V.: Decision Support Systems: A Textbook and a Workshop for Universities. Yurayt Publishing House, Moscow (2020)
7. Luzina, T.V., Molchanov, E.A.: Regulation of custom operations in the context of digitization of the economics. Bus. Educ. Right **3**(48), 131–139 (2019). https://doi.org/10.25683/VOLBI.2019.48.349

8. Meshechkina, R.P.: Digitalization as a factor of increasing the efficiency of customs administration. Bull. Russ. Custom. Acad. **3**(52), 33–39 (2020)
9. Meshechkina, R.P., Aleynikov, I.A.: Improving customs administration as a factor in the development of state customs services. Bull. Belgorod Univ. Cooper. Econ. Law **3**(55), 157–161 (2015)
10. Meshechkina, R.P., Vorona, A.A.: Digitalization of customs processes as a basis for improving the Federal Customs Service activities. Bull. Belgorod Univ. Cooper. Econ. Law **1**(86), 26–36 (2021). https://doi.org/10.21295/2223-5639-2021-1-26-36
11. Ponomarev, A.O.: The essence of digitalization, ways, stages and goals of digitalization implementation in public service institutions. In: Stepanyan, T.M. (Ed.), Institutions, Forms and Methods in State and Municipal Administration, Moscow, pp. 65–67 (2020)
12. Rajabov, K.Y.: Adaptation of decision support systems in the management system of the regional agro-industrial complex. In: Ismailov, T.A. (ed.) Fundamental and Applied Problems of Mathematics, Computer Science in Modern Science: Theory and Practice of Actual Research, pp. 96–100. Dagestan State Technical University, Makhachkala (2016)
13. Timchenko, T.N., Tonkonog, V.V., Pogarskaya, A.S., Golovan, T.V., Konfino, K.V.: Drug turnover in the Russian Federation: customs aspect. Probl. Soc. Hygiene Public Health Hist. Med. **28**(S), 803–809 (2020). https://doi.org/10.32687/0869-866X-2020-28-s1-803-809
14. Vorona, A.A., Yakovlev, K.V.: Improving customs administration as a factor of development of customs authorities at the present stage. Sci. Lett. Russ. Custom. Acad. St.-Petersburg Branch Named After Vladimir Bobkov **1**(69), 17–21 (2019)
15. Vorona, A.A., Gamidullaev, S.N., Kopteva, L.A., Maksimov, Yu.A.: Economic and information security in the conditions of the activity of electronic declaration centers. V. B. Bobkov Saint-Petersburg Branch of the Russian Customs Academy, St. Petersburg (2020)

Current Issues in Regulating Commodity Flows Through the Territory of Russia and Its Borders

Galina P. Semenova(✉), Lyudmila V. Bobrova, Igor A. Salnikov, Vladislav I. Salnikov, and Larisa G. Vorobyova

Russian Customs Academy, Moscow, Russia

Abstract. Currently, Russian economic growth is weak, as evidenced by economic reports of the regions for several branches of the economy. The provision of industries with material goods is unstable. Countries use their own rules and legislation to provide their economies with products, goods, and services. The ratio of import-export and domestic production varies depending on the decisions made by the national leadership. The issues of regulating import and export flows serve as a complex economic problem faced by many economists. Most of such problems depend on various, often contradictory, factors resulting in significant economic disruptions or negative processes. These issues lie at the intersection of economics and mathematics. However, most of them are easily solved using unique economic-mathematical methods. This study analyzes the current state of commodity flow regulations in Russia to facilitate decision-making on providing the population with quality goods. Moreover, we focused on developing a methodology for limiting quantitative regulation of commodity flows using not only commodity balances. Since the proposed solutions are costly and require financial support from the state, we propose using artificial intelligence methods, particularly the game theory methodology, to solve the problem at the initial stage.

Keywords: Export-import operations · Regulation of commodity flows · Economic security

JEL Codes: C44 · C73 · O24 · O32

1 Introduction

Currently, Russia has no monopoly on external trade – export-import enterprises are free to conclude foreign trade transactions based on their financial turnover and market needs [5, 7]. This causes several problems:

1) The problem with accounting and analyzing the total amount of goods imported into Russia and produced domestically;
2) The problem of introducing a single standard of quality of products, goods, and services, adapted to the imported products;
3) The problem of a unified system of quantity limits for import and export of goods;

E. G. Popkova (Ed.): ISC 2020, LNNS 368, pp. 21–27, 2022.
https://doi.org/10.1007/978-3-030-93244-2_3

4) The problem of justifying shipping volumes using the cost approach methodology. This, in turn, causes the problem of justifying rates for transporting goods through the country.

In this study, we aim to solve these problems by developing a methodology for limiting quantitative regulation of commodity flows using not only commodity balances.

2 Materials and Methods

Currently, the government maintains statistical records of commodity turnover but cannot fully establish the volume of goods on regional markets [4]. Accounting for the volume of products, goods, and services in the regional and national turnover will require a comprehensive adjustment of management accounting and analysis, both at the level of economic entities and regional departments. The following areas should be changed:

1) Restructuring information flows that collect quantitative parameter and introduce methods of quantitative accounting at regional statistics centers;
2) Adjusting the system of automated accounting for commodity flows at the national level;
3) Using economic-mathematical methods in the decision-making of management accounting for commodity flows (MACF) in regions;
4) Economic justification of rates to determine the profitability of shipping through the national territory.

At the first stage of identifying the ways of complex redefining of MACF, we must establish the risks of the following directions:

1) Information flow restructuring is based on collecting new parameters of MACF and regrouping the existing ones. The recurrent technological risks of this approach include the following:
   - Delays in the development of a set of intra-departmental and intra-administrative accounting documents at the sectoral and departmental level of enterprises and in the creation of regional reports;
   - Delays in the work of IT specialists that collect and link data parameters;
   - Delays in the development of additional analysis techniques that should be integrated into the existing MACF system.
2) Changing the system of automated accounting will require centralizing regional accounting systems, which will cause additional delays for the development of interconnected artificial intelligence (AI) subsystems for analyzing quantitative parameters and evaluating the qualitative ones;
3) Using economic-mathematical methods for MACF, taking into account the quantitative relations between business entities in local and regional markets, involves developing relevant projects, their terms of reference, program modules, as well as pilot implementation and trial operation. Like any other project, these

measures require appropriate funding and paying the salaries of professional workers [2].

4) A comprehensive approach to the economic justification of rates for shipping through Russian territory would require the joint work of AI subsystems and economists. Such joint activities would use marketing situation analysis to create different shipment rate plans. The main issue of using AI for establishing rates of commodity flow is calculating an economically justified minimum rate with intrinsic norms and standards, providing the projected profits for a given transit situation. Here, great importance is given to methods of calculating rates with several constraints depending on the need for the shipping contract, its profitability, and the profit-to-cost ratio of shipping. This approach uses AI for multivariate decision-making, which takes time to develop.

The algorithm for comprehensive adjustment of MACF is a multifunctional system with complex interconnections of its components. The functions of the algorithm must be consistent, non-contradictory, and compatible. Since the adjustment requires modernizing the management system, the general management algorithm should follow the functional-algorithmic reference model with a hierarchical structure. The structure should be detailed down to the operations level and support future re-engineering as the "bottom-up" parameter management system develops.

In economics, various market participants (firms, associations, and ministries) often need to coordinate their activities, even if their interests do not align. In such cases, game theory helps determine the optimal solution for the participants who have to reconcile their behavior in a conflict of interest. In practice, game theory is increasingly being applied to optimize economic decisions and to various scholarly activities. In this way, game theory becomes an essential tool for increasing the efficiency of planning and management decisions [1, 3].

3 Results

Below we will define the algorithm for comprehensive adjustment of MACF by representing it as a modernization of a management system or as a process of giving qualitatively new and modern features to the existing system. The latter should be done by introducing new digital resources and technologies allowing to set and achieve fundamentally new goals at enterprises.

The main task of the MACF adjustment algorithm is developing relevant projects, their terms of reference, program modules, as well as pilot implementation and trial operation. These stages require appropriate funding and staff.

Staff plays a paramount role in solving set tasks, meaning that a strict and flexible system for workflow organization and incentives in the management system must be developed. Economic, management, and engineering-technical decisions are made at different levels. The task of decision-makers is to analyze all relevant factors, risks, and threats of a given situation, as well as to develop and implement the final version of a decision. This task is facilitated by the use of intelligent systems [8].

The work of experts and decision-makers results in innovative projects for comprehensive adjustment of MACF, scientific recommendations on improving the efficiency of regulating trade flows, expert opinions in the analysis of complex abnormal cases. These results are primarily used by people who make strategic decisions or are responsible for selecting and implementing promising and commercially viable proposals at the highest management level.

Let us examine closely the ways of optimizing choosing between different projects that the experts developed. Taking the respective projects (e.g., Project 1, Project 2, <…>, Project 5) as the foundation, the strategies of managing commodity flows are developed (e.g., S1, S2, <…>, S5). Then, the efficiency of each project is established using a multiple regression model:

$$Y_{Pij} = b_{0(ij)} + b_{1(ij)}X_{1ij} + b_{2(ij)}X_{2ij} + \ldots + b_{n(ij)}X_{nij} \quad (1)$$

In Formula (1):

- Y_{Pij} – forecasted income generated by Project i and management strategy S_j;
- The X_{1ij}, X_{2ij}, …, X_{nij} set of independent variables accounts for different market factors (cost of goods, inflation rate, advertising costs, risk chance, etc.) associated with Project i and management strategy S_j.

Fluctuations in the values of factor variables must be established through market research and statistical data collection.

Regression coefficients are calculated based on the collected statistical material via the standard method of least squares or using the Regression procedure in the Excel Analysis ToolPak.

We will examine the potential of game theory for the formalization of knowledge management and decision-making processes in selecting the optimal project.

International trade is often mistakenly viewed as a zero-sum game. For example, an average American might think that when China gets richer through trade with the United States, America suffers a loss. However, international trade is a nonzero-sum game – a game in which both players can win (and usually do). From the point of economic theory, nonzero-sum games are the most interesting [6].

One can calculate the projected income from different projects by placing them in the payoff matrix. In this matrix, the rows will be the projects, and the columns will be the management strategies. The intersection of the rows and columns would contain the income, calculated by Formula (1). The income would depend on the project number, strategy number, and the current economic environment.

Depending on the specific values of the model parameters, different types of equilibria can be identified. The equilibrium type is crucial because it predetermines the managerial activity in the decision-making process.

1. If the equilibrium in the pure strategies is evident, one can choose the specific projects corresponding to a management strategy;
2. If there is no equilibrium in pure strategies, a decision must be made to use mixed strategies. This situation means that one should not be limited to a single project or a single management strategy. The management scheme becomes more complex;

therefore, it makes sense to implement a combination of several projects or several strategies.

In analyzing the different variants of the current economic environment and assessing risks, one could use the simulation method since it allows to "play out" the situation for different sets of initial factors.

Then, the algorithm of working with the model for reproducing an equilibrium situation takes the following form:

- Based on the market research and collected statistics, determine the range of variation of the main factor variables that make up the model;
- Using random number generation, establish the values of model parameters;
- Calculate the projected income for each project;
- Input the results into the payoff matrix;
- Apply well-known criteria and models (Wald's, Laplace's, Hurwicz's) to the payoff matrix and determine the most profitable project.

The first three steps of the mixed strategies algorithm are the same (determining the range of variation of the main factor variables, getting values of model parameters, calculating projected income for each project). Nevertheless, after inputting the results into the payoff matrix, one needs to calculate the probabilities for selecting each project and determine the proportion of usage for each project in a particular management strategy.

In this situation, scenario analysis can also be used. Scenario analysis involves creating technologies that provide a higher probability of producing an effective solution in those situations where it is possible and a higher probability of reducing losses in those situations where losses are unavoidable. A major advantage of this method is that the scenarios allow (with some degree of reliability) the specialists to describe possible development trends, the relationship between the existing factors, and to create a set of states that the situation may reach under specific influences.

Therefore, the corrected algorithm for MACF adjustment (that accounts for mixed strategies) should consist of the following steps:

- Based on marketing research data, develop possible projects for MACF adjustment and created the main management strategies S_i;
- Create several arrays of predicted values of independent variables X (market factors - the cost of goods, inflation, advertising costs, risk chance, etc.). Each predictive array will correspond to a particular management strategy;
- Calculate the parameters of multiple regression for Formula (1), which evaluating the effectiveness of each project;
- Using simulation modeling, calculate the forecasted values of commodity flow parameters for each strategy and each project;
- Fill out the payoff matrix, where the strategies C_i make up the rows, and the arrays of factor variables make up the columns. Calculated values of the interaction of factor variables, obtained by simulation modeling and the multiple regression equation, serve as elements of this payment matrix;
- Determine whether pure strategies provide a solution. If there is such a solution, carry out several steps to select the best management strategy:

1. Carry out maximax evaluation of management strategies, determining the maximum possible effect of each project and strategy.
2. Carry out maximin evaluation of management strategies, determining the minimal guaranteed effect of each project and strategy (usually calculated using Wald's model).
3. Choose the optimal (compromise) management strategy (can be easily computed using Hurwicz's criterion, the pessimism-optimism index for which is calculated taking into account the estimates obtained in steps 1 and 2).;

- If pure solutions do not provide a solution, analyze the dependence between project profitability and cumulative influence independent factors. Then, calculate the probability with which each strategy should be used. Based on the calculated probabilities, a "hybrid" strategy is developed (i.e., joint implementation of several projects).

A variant of the game model can also be used as the database for the main matrix of valid projects and their combinations, taking into account the economic efficiency of the projects. In this matrix, along with the estimated forecast parameters of profitability, one could include the costs for acquiring computing equipment, other resources, and salaries of professional staff. Using this matrix of valid projects, management could decide whether to select a particular project or not.

The proposed algorithm of comprehensive MACF adjustment (based on AI, game theory, and scenario method) can be used for optimizing other economic processes.

4 Conclusion

Accounting for the volume of products, goods, and services in the regional and national circulation will require a comprehensive adjustment of MACF, both at the level of economic entities and regional bodies.

All proposed directions of restructuring are costly and require appropriate financial support, both from the national budget and public-private partnerships. Time potential remains the main risk, but it can be eliminated by restructuring the set tasks and framing a "product" result of examining each stage and each planned situation. The proposed approach requires creating a national system project, where sets of subsystems are developed by different organizations and agencies (possibly on a tender basis), which would reduce the cost and time of developing and testing solutions. We propose modeling the task via game theory to solve the problem at the initial stage.

Customs authorities exercise quantitative control over commodity flows according to the relevant departmental legislation, which can always be improved depending on changes in external factors. We draw attention to the methodological approach for evaluating the ratio between the growth of costs and the throughput capacity of transportation channels. This approach helps determine the practical viability of transit shipment amounts, taking into account the costs of maintaining, servicing, and replacing transportation channels.

References

1. Arbnor, I., Bjerke, B.: Methodology for Creating Business Knowledge, 3rd edn. SAGE Publications, Los Angeles, London, New Dehli (1997). https://doi.org/10.4135/9780857024473
2. Bobrova, L.V., Luchina, N.A.: Business Modeling: Textbook for Graduate Students. National Open Institute, St. Petersburg (2019)
3. Bobrova, L.V., Luchina, N.A.: Simulation Modeling in Economics: Textbook for Graduate Students. National Open Institute, St. Petersburg (2020)
4. Gurova, T.I., Konev, I.P.: Some approaches to use of game theory in economics. Vestnik RIAT **1**, 38–43 (2019)
5. Kravtsova, D.S.: Game Theory in Economics and International Relations. Molodoy Ucheny **17**(255), 1–4 (2019)
6. Nagolova, A.D., Dolgopolova, A.F.: Practical application of game theory in solving economic problems. Mod. Sci. Inten. Technol. **5–2**, 166–168 (2014)
7. Sliva, I.I.: Application of game theory to solve economic problems. Izvestiya MGTU "MAMI" **1**(15), 154–162 (2013). https://cyberleninka.ru/article/n/primenenie-metoda-teorii-igr-dlya-resheniya-ekonomicheskih-zadach
8. Smirnov, A.V., Rakhmanova, I.O.: Information Support of Business Processes: Textbook, 2nd edn. Peter the Great St. Petersburg Polytechnic University, St. Petersburg (2012)

Problems and Prospects of the Russian Economy Emerging from a Quasi-static State Against the Background of Global Trends

Alexander E. Illarionov[1(✉)], Alexander I. Novikov[1], Ekaterina P. Garina[2], Natalia S. Andryashina[2], and Elena V. Romanovskaya[2]

[1] Russian Presidential Academy of National Economy and Public Administration (Vladimir Branch), Vladimir, Russia
illarionov_a@vlad.ranepa.ru

[2] Minin Nizhny Novgorod State Pedagogical University, Nizhny Novgorod, Russia
natali_andr@bk.ru, alenarom@list.ru

Abstract. The reaction and aggravation of the Russian economy's problems to changes in the surrounding biological and natural processes against the background of global development trends is justified. It shows the growth of chaotic processes in the system of public relations, which inevitably leads to aggravation of relations and wars for vital space and resources, the growth of man-made and other disasters like pandemics and, in general, leads to crisis phenomena in the economy and human vulnerability in everyday life. Tasks were set and solved to find analogies for the influence of changes in the biological system, including those caused by coronavirus COVID-19 on the formation of social relations.

A characteristic of the quasi-static state of the Russian economy is given, due to low growth rates, falling incomes of the population and a high proportion of the poor. The withdrawal of Russia from the quasi-static state implies the need to assess and realize the potential for growth of the Russian economy, taking into account socio-economic and natural-environmental processes. It turns out that the existing potential can provide the necessary level of development of the Russian economy and its exit from the quasi-static process, provided that sharp polarization in income is eliminated, that is, the causes of antagonization of the social sphere are eliminated.

It is proved a reconfiguration of the economic mechanism of the Russian economy is required. Factors were identified and criteria were proposed that could affect Russia exit from the quasi-static state.

Keywords: Quasi-static state · Economy · Development · Directions

JEL Codes: F15 · H56

E. G. Popkova (Ed.): ISC 2020, LNNS 368, pp. 28–36, 2022.
https://doi.org/10.1007/978-3-030-93244-2_4

1 Introduction

Due to COVID-19 pandemic the International Monetary Fund (IMF) made changes to the forecast concerning a condition of world economy in 2020, having assumed fall of world GDP by 4,9% [15]. The situation is similar in Russia. The budget for 2021 includes GDP growth of only 3,3%, which does not cover the fall of the 2020 economy – 3,9% [4]. In accordance with the forecast and in 2022, Russia will not yet be able to overcome the consequences of the pandemic. The pandemic has exacerbated problems primarily in trade, social, environmental and quality of life. Humanity has faced an unprecedented problem up to this time, namely the negative influence of one (biological) system on other systems of social order in general and economic development in particular.

The aim of this research is to formulate an adequate economic response to the consequences of the epidemic and other modern challenges in relation to Russia against the background of global trends.

The following tasks have been formulated and achieved:

- search of analogies of changes' impact in a biological system, natural processes on formation of the public relations including caused by COVID-19 coronavirus;
- assessment of the economy state through the lens of entropy growth;
- characterization of the quasi-static state of the Russian economy;
- the potential identification of the Russian economy and mechanisms for its mobilization.

2 Methodology

Ensuring the reliability of research and solving problems to increase the sustainability of economic development involves taking into account the actions of the laws of entropy.

The withdrawal of Russia from the quasi-static state implies the need to assess and realize the potential for growth of the Russian economy taking into account natural processes and solve environmental problems.

Within the framework of behavioral economics, it is planned to make adjustments to the meaning of human existence and formulate a system of value guidelines based on the principles of harmony.

3 Results

Currently, the entire world community is affected by the pandemic on the economy and other social processes. No similar reaction has previously been noted.

In what direction should methods of «treating» the economy and increasing its resilience from the negative impact of the pandemic be sought?

As part of the solution of this problem, we will expand the conceptual apparatus of entropy formation and propose to take it beyond the substrate-functional features, namely, the effect of the virus on the growth (reduction) of entropy. Is the virus a living organism? The virus is essentially a complex molecule. In the world of science, it is believed that there is no generally accepted phenomenon of life, but a living cell is a generally recognized form of life. Let us turn to the hypothesis of the famous American scientist Evgeniy Kunin (a native of Russia) that «the basis of everything alive is a cell, also arose from viruses several billion years ago» [8].

To eliminate the formation of molecules in the form of viruses, it is necessary to achieve a decrease in internal energy - entropy. According to the Austrian physicist E. Schrödinger, «living organisms kind of feed on negative entropy (non-entropy), extracting it from the environment and thereby increasing the increase in positive entropy in it» [14].

Modern man compared with progenitors has increased the production of negative entropy (energy) many times, in addition, the population is growing. Therefore, the production of negative entropy is increasing on the planet, and in order to eliminate it (recycle), the need for living creatures that feed on it increases. Hence the conclusion that if a person downplayed the activity of the influenza virus, then a free niche appeared, which has free energy for the emergence of a new virus, the so-called COVID-19. Entropy growth (reduction) is an integral component of the life process.

From the economic development's point of view, it is important to reduce entropy in the system of public relations, primarily related to the production, circulation and consumption of goods, that is, it is important to ensure the harmonious development of society.

An analysis of the development of the Russian economy over the past 30 years (1990–2020) indicates there have been no fundamental changes in the Russian economy, quantitative growth is extremely insignificant, and for a number of qualitative indicators the situation has worsened, that is, there is a process of «trampling in place». According to A. Aganbegyan, «Russia is only adjacent to the lower group of developed countries, and its economy does not yet meet the maturity criterion» [16].

This is evidenced by indicators (ratings) reflecting Russia place in the world:

- GDP per capita at nominal value - 62nd place [25];
- human development index - 49th place (2016);
- life expectancy in Russia - 103rd place (data from the World Health Organization) [6].

Despite low indicators regarding the quality of life of the population, Russia remains among the leaders in oil production (2nd–3rd place), gas (1st–2nd place), gold production (3rd place), production of aluminum, titanium, nickel and other non-ferrous metals (2nd–5th places). Russia share in food production increased, in particular, in wheat, rye, barley, buckwheat - 1 place. Russia share in the arms market is great - 2nd place, by the number of mobile phones - 5th place, etc.

The theoretical justifications and study of the experience of other countries show that in order to ensure the qualitative development of the country, on the example of Russia, it is required, according to A. Aganbegyan [1]:

- reduction of tax burden on business to 30% or less, in Russia it exceeds 40%;

– increase of investments up to 30% [3, p. 40]. In Russia in 2017, the Investment/GDP indicator in terms of fixed capital was 21,8%, in China the investment rate (Investment/GDP) in recent years amounted to 44–48%, in India – 30–39% [9];
– a decrease in the share of public spending (the experience of countries catching up with 10–13%), in Russia this share exceeds 20%.

The reconfiguration of the economic mechanism did not happen, and at present it is important to understand the state of the Russian economy and what should be done?

The development of the Russian economy can be characterized as a quasi-static state. This concept has been transferred by us from the course of physics, the section of thermodynamics.

Quasi-static process (quasi-equilibrium process) (from lat. Quasi - as it were, like Greek. Statikos - stopping, related to equilibrium, status - stationary). In a thermodynamic system, this means an infinitely slow transition of it from one equilibrium state to another.

The quasi-static state of the Russian economy (very slow development) is confirmed by the following statistics:

a) the low growth rate of the economy and its lagging behind the global average;
b) real incomes of the population lag behind inflation, which led to a high level of poverty (according to Rosstat, about 20 million people);
c) problems with employment of the population. At present, especially in connection with the pandemic, there are serious problems in terms of employment, as Russian President Vladimir Putin spoke about at the G20 (G20) summit in November 2020, calling «mass unemployment and rising poverty the main risks to the world» [17]. In Russia, of the 82 million able-bodied population in 2018, only 44 million were considered employees of organizations and enterprises, about 10 million were state employees, military personnel, and employees of law enforcement agencies. Following these statistics, about 18 million people fall on the self-employed, but for this category there is a guarantee problem. Taking advantage of legal loopholes, many employers of small enterprises in the context of the pandemic transfer employees to the status of self-employed [18];
d) there was a drop in life energy and a change in the vector of its dynamics from the active to the passive side. And in a pandemic, doctors, teachers and other social workers are forced to work in 1, 5–2 shifts [10, 24]. That is, there are all the signs of a quasi-static process not only in the economy, but also in the life of the country [13, 19, 21].

The UN in terms of achieving the goals of sustainable development and improving the quality of life proclaimed 17 Goals and 169 Goals (2015). «A person in the UN-proclaimed coordinate system is considered an active element to increase the sustainability of socio-economic and natural systems and decision-making on their transformation» [10, 24].

For Russia, in terms of developing and solving practical problems of exit of the quasi-static situation, it is necessary:

1. Solving the problem of national preservation. The mechanisms for achieving this goal can be found in more detail in the collective monograph «People's Preservation and/or Preservation of Human Health: Social, Economic, Legal, Medical and Environmental Aspects» [5].
2. Solving problems of social reproduction. We have established that in connection with the decrease in real incomes of the population in Russia since the beginning of the pandemic (March 2020), the aggregate consumer and investment demand has decreased. For this reason, the reproduction process, including the reproduction of labor resources, has been disrupted. And the underutilization of benefits explains the increase in the number of poor. This circumstance negatively affected the activity of economic entities and provoked in certain areas the transition of the Russian economy from an inflation state to a deflation phase. This can lead to self-isolation of resources and the emergence of elements with impaired reproductive functions, which, despite the measures taken, prevent the restoration of normal life in rural areas to date.
3. In biological systems, immunity is a measure of counteraction to coronavirus. And in economic systems, in order to overcome the quasi-static situation, the questions of the response and sensitivity of socio-economic systems to changes in macroeconomic parameters and institutional changes are of fundamental importance. Oddly enough, the topics of increasing the sensitivity of systems and accelerating their reaction in the economy are poorly studied.

In the development of this topic, we will touch on a number of publications on the issues of achieving the stability of systems, increasing their sensitivity and response, published in Russia and published in domestic scientific publications. Significant ones include:

a) studies of M. Vukobratovich and R. Tomovich [2];
b) the works of A. Novikov and A. Astakhin on the development of increasing the sensitivity and reaction of socio-economic systems [11, p. 134–139]. They argue that «against the background of a multi-level crisis unfolding in the world, there is an increase in chaotic processes in the system of public relations, which inevitably leads to aggravation of relations and wars for vital space and resources, the growth of man-made and other disasters like pandemics, and in general leads to human vulnerability in everyday life».

Based on the theory of optimal functioning and some prerequisites of an axiomatic nature, they identified 6 key factors generating negative entropy:

F_1 - man-made; F_2 - natural;
F_3 - environmental; F_4 - biological;
F_6 - anthropogenic; F_5 - socio-economic;
Formally, this can be written:
$F = y\ (F_1; F_2; F_3; F_4; F_6; F_5)$.

To determine the potential of the Russian economy, we will give world trends and forecast scenarios for the development of national economies of the main economic blocs, see Table 1.

Table 1. Annual GDP growth rate [8] (in %, at purchasing power parity, at constant prices in 2013)

	1991–2000	2001–2010	2011–2013	2014–2020	2021–2030	2031–2035
Whole world	3,0	3,8	3,5	3,8	3,7	3,7
Developed countries	2,7	1,7	1,4	2,3	2,6	3,2
USA	3,4	1,6	2,0	2,7	2,8	2,8
EU	2,1	1,5	0,5	1,7	2,3	2,3
Developing countries and countries in transition	3,4	6,2	5,4	4,9	4,5	4,3
China	10,4	10,5	8,3	6,5	5,0	4,0
India	5,6	7,5	6,2	5,9	5,5	5,3
Russia	−3,9	4,8	3,0	2,5	3,5	4,0

Source Compiled by the authors based on [7]

It follows from the table that the growth rate of the economy will be determined by China, India, developing countries, the role of Russia will increase.

Russia has the necessary resources to protect security, including the necessary capital, but in our opinion there are no effective mechanisms for its use [12, pp. 126–134]. Currently, major savings include [24]:

- National Welfare Fund (as of 01.11. 2020 amounted to $167,63 billion or 13,298 trillion rubles, 11,7% GDP) [20];
- net assets of the banking sector amounted to RUB 2020 trillion in January 88.6 [21];
- Central Bank's gold and foreign exchange reserves ($433 billion) [22];
- capital of Russian oligarchs (200 families have a total capital of 485 billion dollars) [22];
- monetary savings of Russians (despite the decrease in savings during the pandemic, their share by 01.01. 2021 g amounted to about $389 billion) [22]. At the beginning of the pandemic, almost 40% of Russians had accumulations, experts at the "Perspective" scientific and technical center found out [23].

The savings in the country, according to A. Aganbegyan, can provoke processes of «warming up» [1].

Mechanisms for mobilizing the resources of the Russian economy.

Economic, social, demographic, environmental and political factors, as well as elements of the behavioural economy that determine the confidence of the people and determine the sustainability of the system, are important in order to overcome the quasi-static situation of Russia [12, pp. 126–134].

To do this, we propose a descriptive model of Russia exit from a quasi-static position, taking into account the influence of the above-mentioned factors.

Factors influencing a country's exit from a quasi-static state will be in series, and indicators will be separate criteria that determine the impact of these factors on the stability of the system (Table 2).

Table 2. Factors and criteria which can lead Russia out of a quasi-static state

Symbols	Title of factors	Criteria
FE	Socio-economic	FE_1 GDP per capita (purchasing power parity)
		FE_2. Investment in% of GDP
		FE_3. Share of innovative products in total output
		FE_4. Inflation rate in%
		FS_1 Proportion of the poor
FD	Demographic	FD_1. Birth rate
		FD_2. Death rate
		FD_3. Average life expectancy
		FD_4. Population ageing rate (proportion of persons over 65)
FEc	Ecological	FEc_1. Environmental costs, % of GDP
		FEc_2. Income from environmental payments in% of GDP
B	Behavioural	FB_1 Level of public confidence in the authorities
		FB_2. Crime rate (number of crimes per 100 thousand people)
		FB_3. Income ratio of richest to poorest in decimal terms

Source Developed and compiled by the authors

Formally, this can be written:

$$\begin{aligned} SS = & (FE_1 + FE_2 + FE_3 + FE_4 + FS_1) + (FD_1 + FD_2 + FD_3 + FD_4) \\ & + (FEc_1 + FEc_2) + (FB_1 + FB_2 + FB_3). \end{aligned}$$

4 Conclusion

1. The necessary level of development of economy of Russia and her exit from quasi-static process on condition of elimination of sharp polarization in income, that is elimination of the reasons of antagonization of the social sphere can provide the available potential of resources.
2. In the conditions of a failure of a course of democratic liberalization and formation of an ideological vacuum for Russia it is important to offer himself to the whole world as leader of traditional national interests with a claim for formation of the Russian civilization.
3. For an exit of the Russian Federation from a quasi-static state it is required:

   - to raise quality indicators of growth of economy of Russia, to reduce the level of the poor population;
   - improvement of investment and innovative policy will allow to achieve increase in cost efficiency of production;
   - activization of measures of social support will allow to reduce tension of demographic character;
   - more large-scale and drastic measures on realization of environmental policy are required;

- technological and technical innovations will promote increase in stability of functioning of key branches of the national economy;
- stimulation of measures of development of civil society and creating favorable conditions of material and power, information character will allow to increase trust of the population to public authorities.

References

1. Aganbegyan, A.G.: On overcoming stagnation, recession and achieving five percent growth. Econ. Revi. Russ. **2**(60), 17–24 (2019). https://e-v-r.ru/wp-content/uploads/2019/05/%D0%92%D0%B5%D1%81%D1%8C-%D0%B6%D1%83%D1%80%D0%BD%D0%B0%D0%BB-%E2%84%96-2-2019-%D0%B3
2. Vukobratovich, M., Tomovich, R.: General Theory of Sensitivity/Per. from English. In: Tsypkin, J.Z. (ed.) M.: Soviet Radio, 240 p. (1972). https://cat.libnvkz.ru/CGI/irbis64r_14/cgiirbis_64.exe?LNG=&I21DBN=KNIGY&P21DBN=KNIGY&S21STN=1&S21REF=&S21FMT=&C21COM=S&S21CNR=&S21P01=0&S21P02=0&S21P03=I=&S21STR=32.95%2F%D0%A2%2056-171690
3. Davies, K.: China Investment Policy/OECD Working Papers on International Investment, Jan 2013, p. 40
4. Zamakhina, T.: Budget Adopted//Russian Newspaper, 27 Nov 2020. №. 268(8322) (2020). https://rg.ru/gazeta/2020/11/27.ht
5. Illarionov, A.E., Kartukhin, V.Yu., Zhambrovsky, V.M., Novikov, A.I., Togunov, I.A.: People's preservation and (or) preservation of people's health: social, economic, legal, medical and environmental aspects. In: Novikova, A.I.: Collective Monograph/General Edition. Vladimir Branch of RANEPA, Vladimir, 324 p. (2019). https://elibrary.ru/item.asp?id=38488475
6. World Health Organization. Life Expectancy Data by Country (2016)
7. Joint Research of the Institute of IMEMO named after E. M. Primakov and the Strategic Forecasting Group of the Atlantic Council. Global system at a turning point: paths to a new normality: per. from English. In: Dynkin, A., Barrows, M. (eds). M.: IMEMO RAS, p. 32 (2016). https://www.imemo.ru/files/File/ru/publ/2016/2016_008
8. Medvedev, Yu.: Direct speech. Professor Konstantin Severinov: Is it possible to make protection against future viruses in advance. Virus is a Human Friend. Russian Newspaper, 25 Nov 2020, No. 265 (8319) (2020). https://rg.ru/2020/11/24/mozhno-li-zaranee-sozdat-zashchitu-ot-budushchih-virusov.html
9. Mirkin, Ya.M., Zhukova, T.V., Bakhtaraeva, K.B, Komova, A.V.: Mechanisms for stimulating ultra-fast growth: world practice. In: Mirkina, Y.M. (ed.). M.: Master, 2018, p. 94 (2018). http://www.mirkin.ru/_docs/book097
10. Novikov, A.I., Illarionov, A.E., Togunov, I.A.: Research and Analysis of Socio-Environmental-Economic Policy Through the Lens of State Social Standards: National and Regional Aspects: Collective Monograph, p. 294. Vladimir Branch of RANEPA, Vladimir (2017). https://elibrary.ru/item.asp?id=32696997
11. Novikov, A.I., Astakhin, A.S.: Reducing the level of impact of destabilizing factors on the public security system in the light of the "rapid response" policy. Manag. Consult. St. Petersburg **9**(105), 134–139 (2017). https://www.elibrary.ru/item.asp?id=30543144

12. Novikov, A.I., Astakhin, A.S.: Ensuring national security: problems and solutions. Theor. Econ. **6**(42), 126–134 (2017). https://cyberleninka.ru/article/n/obespechenie-natsionalnoy-bezopasnosti-problemy-i-puti-resheniya
13. Decree of the President of the Russian Federation of 21.07.2020 No. 474. On National Development Goals of the Russian Federation for the Period until 2030. Russian Newspaper. № 159(8213) (2020). https://rg.ru/2020/07/22/ukaz-dok.html. Accessed 20 Nov 2020
14. Schredinger, E.: What is life? Physical aspect of a living cell. Regular and Chaotic Dynamics. NIC, Moscow, Izhevsk (2002). https://elar.urfu.ru/bitstream/10995/1743/1/1213213
15. Electronic Resource. https://www.rbc.ru/economics/24/06/2020/5ef35b8a9a79474a0c6a16f6. Accessed 20 Nov 2020
16. Electronic Resource. https://aif.ru/money/opinion/gde_mesto_ekonomiki_rossii_v_mirovyh_reytingah. Accessed 20 Nov 2020
17. Electronic Resource. https://iz.ru/1090365/2020-11-21/putin-nazval-rost-bezrabotitcy-i-bednosti-glavnymi-riskami-dlia-mira. Accessed 20 Nov 2020
18. Electronic Resource. http://www.soloby.ru/21599/%D1%87%D0%B8%D1%81%D0%BB%D0%B5%D0%BD%D0%BD%D0%BE%D1%81%D1%82%D1%8C-%D1%82%D1%80%D1%83%D0%B4%. Accessed 20 Dec 2020
19. Electronic Resource. https://www.elibrary.ru/contents.asp?id=36580777. Accessed 20 Dec 2020
20. Electronic Resource. https://minfin.gov.ru/ru/document/?id_4=27068-obem_fonda_natsionalnogo_blagosostoyaniya. Accessed 20 Dec 2020
21. Electronic Resource. https://cbr.ru/Collection/Collection/File/27385/razv_bs_20_01.pdf. Accessed 20 Dec 2020
22. Electronic Resource. https://inosmi.ru/politic/20180426/242084824.html. Accessed 20 Dec 2020
23. Electronic Resource. https://aif.ru/money/mymoney/skolko_deneg_u_rossiyan. Accessed 20 Dec 2020
24. Electronic Resource. https://otr-online.ru/programmy/segodnya-v-rossii/rekordnaya-ubyl-naseleniya-v-chyom-realnye-prichiny-padeniya-roz. Accessed 20 Dec 2020
25. Electronic Resource. https://zen.yandex.ru/media/economics_channel/epoha-putina-mesto-rossii-v-mire-po-urovniu-vvp-5ec617916acfae70773ce43d. Accessed 15 Jan 2021

Marketing of Customs Services: Directions of Development in the Context of Digitalization

Yulia V. Malevich(✉), Irina A. Plastunyak, Aleksandr V. Dmitriev, Ekaterina M. Ksenofontova, and Anna A. Belostotskaya

St. Petersburg State University of Economics, St. Petersburg, Russia
{malevich.yu,plastounjak.i,dmitriev.a, belostockaya.a}@unecon.ru, koka007@list.ru

Abstract. The article analyzes the problems of the marketing of customs services improving in the context of modern digital technologies development. The importance of marketing as an effective tool for increasing the efficiency of interaction between customs authorities and participants in foreign economic activity is substantiated.

Keywords: Marketing · Customs services · Digitalization · Foreign economic activity

JEL Code: L86 · M16 · M31 · O32 · F15

1 Introduction

Digitalization processes in the Eurasian Economic Union (EAEU) contribute to the formation and development of the customs services market. Marketing as an interactive and targeted concept of impact on consumers is an effective tool for solving problems of interaction between customs authorities and participants in foreign economic activity (FEA), which allows effective process management of providing customs services, including using modern digital technologies.

It should be noted that the application of marketing concepts in the customs administration from a methodological point of view provides an opportunity to consider the state control departments' activities not only as a regulatory structure but also as subjects providing complex customs services within the framework of the functioning of the global foreign trade market. At the same time, the marketing of customs services can be characterized as a conceptual tool for the interaction of the Federal Customs Service with participants in foreign economic activity in the following context:

- solution of national tasks contributing to the maintenance of economic growth of Russia;
- increasing the social well-being of society with unconditional compliance with the requirements of regulatory legal acts in the field of customs administration;

E. G. Popkova (Ed.): ISC 2020, LNNS 368, pp. 37–45, 2022.
https://doi.org/10.1007/978-3-030-93244-2_5

– satisfaction of business needs, facilitating world trade, by creating favorable conditions for doing business.

2 Methodology

A significant number of domestic and foreign scientific works are devoted to finding the optimal balance between government regulation and the creation of comfortable conditions for foreign economic activity. In recent years, researchers have been focusing, among other things, on transformation in all areas related to the transition to a new technological order, digitalization of the economy, and the introduction of intelligent technologies [2].

Analysis of works in available sources allows us to state the indisputable relevance of the application and the need for further development and implementation of tools in such a scientific direction as marketing in the field of customs. So, in the work [5], the authors considered the concept of service-oriented administration, on the basis of which the directions of improving the tools for marketing customs services are determined. The conditions are highlighted and the necessary solutions are considered to achieve a qualitatively new level of customs administration in the field of foreign economic activity. In the resulting part, the authors managed to outline the main concepts of strategic management and development of the customs system based on the application of the concept of marketing customs services [12].

The implementation of marketing concepts and models into the activities of the customs administrations of various countries, as already mentioned, is considered by many scientists to be an effective tool. It should be noted that it comes not only for the customs system as a whole but also for its elements: customs operations, customs control, administration of customs payments, etc. As an illustration, let us refer to the work [10], in which the authors consider post-release customs control (post-audit) as an independent management function that directly affects both the quality of customs services and business processes.

World integration processes, globalization trends have become preconditions for the redistribution of priorities and shift of emphasis. Cross-border trade, including electronic trade, is actively developing; approaches to cooperation between customs administrations of neighboring states and integration associations are changing. One of the goals of international customs cooperation is the increasing efficiency of cross-border trade by reducing barriers and establishing partnerships. A similar opinion is shared by the authors of [13], noting that cooperation and partnership is an important tool that promotes cross-border integration and stimulates the development of border regions, and the exchange of experience and knowledge makes it possible to define better the needs in the provision of customs services, thereby influencing economic development of cooperating regions, as well as increasing the efficiency of provided customs services. And in this regard, the issue of data exchange is also relevant, the methods of which are diverse in the era of the digital economy.

In the era of the digital economy, the communication processes of marketing take over - in international exchange, as noted by the authors of [11]. In our opinion, digitalization can achieve other effects as well. Agreeing with the author of work [3], considering the issues of electronic cross-border payments and stating the change in the international monetary landscape due to the introduction of digital currencies, which will contribute to the integration of banking systems in the future, we consider that the above points can also be attributed to indirect marketing effects, which in the result will contribute not only to increasing the transparency of international transactions, facilitating international trade but also to the positive image of government agencies, including customs.

Paying sufficient attention to the communicative function, we believe that modern customs technologies developed and implemented as one of the elements of the strategy by the customs authorities of the Russian Federation contribute to the achievement of effective results. So, in the article [8], the best practice of introducing modern technologies, primarily customs, in the seaports of the Mediterranean Sea is considered. Based on the research, the authors identified those that should be implemented into the activities of an “ideal” port that meets all the requirements of foreign economic activity participants. These technologies should be aimed at improving the performance of any port.

Without pleading with the role of customs administrations, we believe that balanced development of the system of customs services is possible only with a similar response from the business, for which marketing tools are becoming key. A similar opinion is shared by the authors of [7], who substantiate the relationship between global enterprise strategies and their value chains in foreign trade as part of the development of a typology of four areas: managed cross-border activities, optimization of distribution networks, supplier orientation, and strategic joint evolution. This integrative typology provides an analytical framework for understanding the relationship between the value chain management methodology and the dynamics of the firm’s strategy in the field of foreign economic activity, which allows the development of effective enterprise policies, including the context of digitalization, as well as global challenges such as the coronavirus pandemic.

In the work [9], the research is devoted to the influence of external and internal factors on the marketing strategy of a logistics operator in a developing country. It is highlighted that the most important external factor is the logistics infrastructure, while the legal environment is noted by the respondents as the least significant factor. At the same time, the authors note that this may be due to the lack of transparency of the legal system.

3 Results

In the context of the logic of this study, we formulate a scientific idea of improving the complex of state customs services in the Russian Federation based on marketing tools.

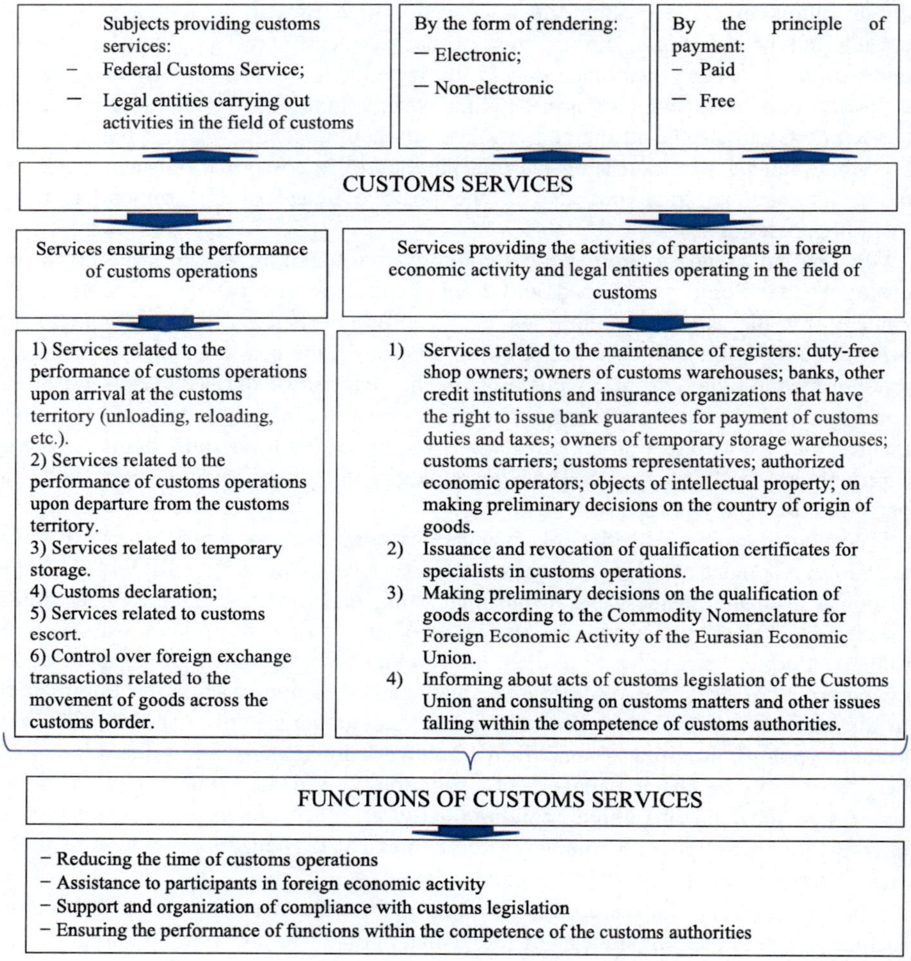

Fig. 1. Consolidated structure of customs services. *Source* Developed and compiled by the authors

The development of public services provided by customs authorities (further - customs services) is a priority area for improving the legal support for the activities of customs authorities, as defined in the "Strategy for the development of the customs service of the Russian Federation until 2030" (Fig. 1).

On the other hand, the development of "electronic provision of customs service" is included in the concept of "electronic government" development, and will contribute to a positive perception of information technologies by the business community.

Today, according to the "Global E-Government Development Index", which is maintained by the United Nations (UN) and which is a comprehensive indicator that assesses the readiness and capabilities of national government agencies in the use of information and communication technologies (ICT) for the provision of public services to citizens Russian Federation as of 02.02.2021 takes 36th place (ICT 0.8244). Denmark (ICT 0.9758) and South Korea (ICT 0.9560) occupy the first lines of the rating [14].

Currently, the FCS provides some government services in the field of customs. Some statistic indicators are presented in Figs. 2 and 3.

This research demonstrates, on the one hand, the need for services increase, and on the other hand, the insufficient volume of services provided in electronic form, while in modern society, the digitalization of consumer behavior is traced.

Thus, there are many problems associated with the provision of customs services to participants in foreign economic activity, the solution of which can be facilitated by the development of customs marketing tools based, among other things, on the use of digital technologies [4].

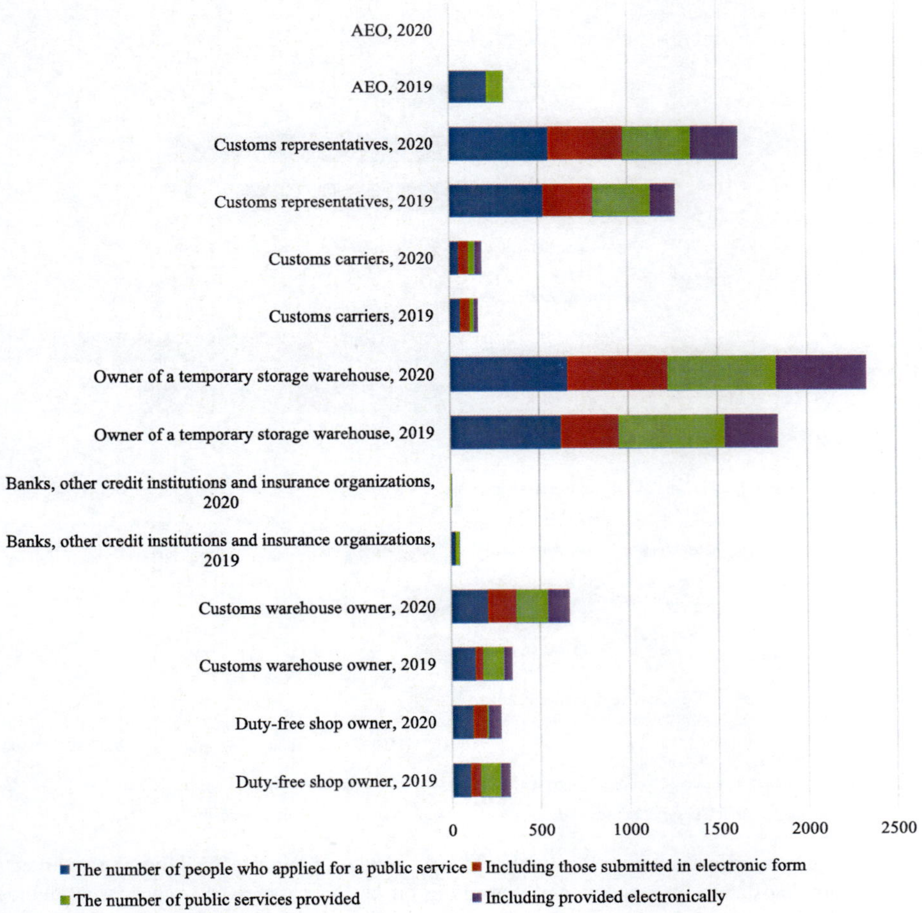

Fig. 2. Information about the number of state services provided for maintaining registers provided by the FCS of Russia for 2019–2020. *Source* Compiled by the authors based on [15]

Digitalization leads to a transformation of the infrastructure of state authorities, namely, to a reduction in the provision of services in a traditional format and a significant increase in electronic interaction through information channels and, in the short term, blockchain platforms. Of course, traditional interaction cannot be completely ruled out, but the shift in emphasis towards remote channels is visible.

In general, the main consequence of digitalization is the active formation of an omnichannel system for the provision of public services, including all possible and, most importantly, convenient for the consumer, methods of processing and issuing applications for public services and their payment, united by a single information system, where the electronic catalog of public services acts as a single "showcase" that provides information on the availability and openness of public authorities.

We believe that the development of the customs services complex should be based not only on modern technologies but also on increasing business confidence in the

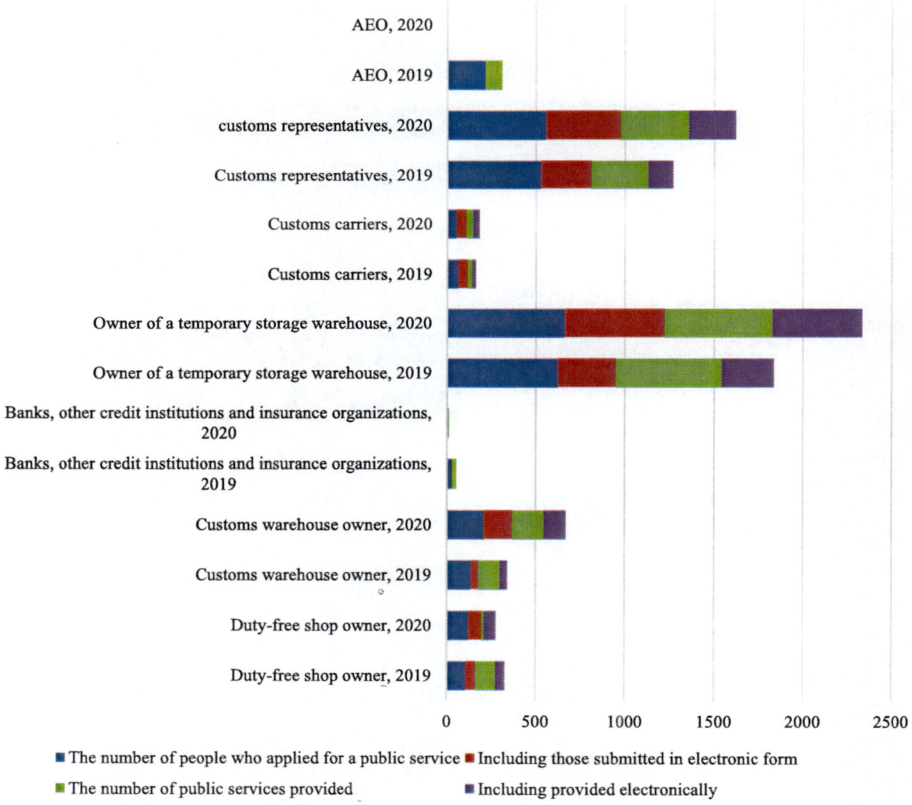

Fig. 3. Quantity information on the number of state services provided for informing about acts of customs legislation on customs and consulting on customs issues and other issues falling within the competence of customs authorities and on making preliminary decisions on qualifying goods according to the CNFEU EAEU, provided by the Federal Customs Service of Russia for 2019–2020. *Source* Compiled by the authors based on [15]

customs service. For the research, we have performed the decomposition of customs services and elements of participants' trust in foreign economic activity to the customs authorities, presented in Fig. 4.

The key principle of modern marketing is to ensure the company's transparency for society. Becoming more and more open to public authorities is the foundation of trust on which effective and long-term relationships are built. Confidence in customs authorities on the part of foreign trade participants is based on such components as quality and safety, openness (transparency), visibility, and social contribution [6].

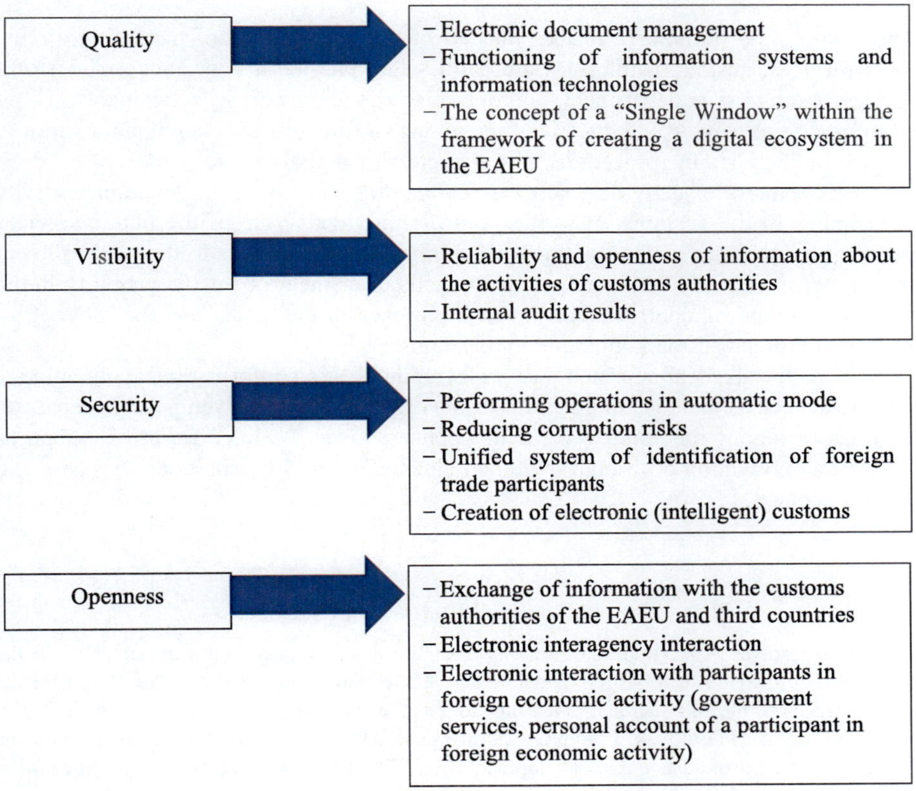

Fig. 4. Decomposition of customs services and elements of confidence of the foreign trade participant in relation to the customs authorities. *Source* Developed and compiled by the authors

Confidence can be achieved through the high quality of services provided by the customs authorities, ensuring their safety in consumption. The second important component of trust is the openness of customs authorities for foreign trade participants. The more public authorities open their internal environment and disclose their business processes, technologies, and other aspects of their activities, the more foreign trade participants will trust the state customs authorities [1].

The openness of the customs service is significantly influenced by the public communications system. Interaction of the FCS of Russia with the business community is carried out through the Public Advisory Council on Customs Policy under the FCS and advisory councils under the regional customs administrations (RCA) and customs. Hence the PR value increases. But it is worth mentioning that public PR differs significantly in nature, goals, and objectives from business PR. An important mechanism in customs PR is feedback modeling.

4 Conclusion

Thus, within the framework of the concept for the development of marketing of customs services, customs authorities are entities that provide a wide range of customs services, both to state departments and to foreign economic activity participants, whose main goal is to support the effective functioning of foreign trade activities to simplify and accelerate customs procedures in the context of digitalization.

The openness of activities increases the confidence of foreign economic activity participants in the customs authorities and the provided value of the public services perception, increases their assessment, which, in turn, contributes to the establishment of long-term mutually beneficial relations. Another element of trust is visibility in the information environment. Visibility means presence in the media and the Internet, as well as personalized marketing communications.

It can be talked about such ways of communicating information about relative openness, such as the level of openness of the organization at a given point in time, and information about the improvement in openness over a while. Ensuring openness affects the motivation of foreign trade participants in two different areas - process and client openness.

References

1. Belostotskaya, A.A.: Impact of activity openness on sustainable consumption. In: In the Collection: Science, Research, and Practice. Collection of Selected Articles Based on the Materials of the International Scientific Conference, St. Petersburg, pp. 163–166 (2020)
2. Dmitriev, A., Plastunyak, I.: Digital platforms for managing transport and logistics systems in the context of sustainable development. E3S Web Conf. **208**, 01007 (2020). https://doi.org/10.1051/e3sconf/202020801007
3. He, D.: Digitalization of cross-border payments. China Econ. J. (2021). https://doi.org/10.1080/17538963.2020.1870272
4. Konnikov, E.A., Konnikova, O.A., Bogacheva, T.V., Yuldasheva, O.U.: The study of the conformity of marketing solutions to the concept of sustainable consumption. Innov. Manag. Educ. Excell. Thr. Vis. **2020**(I–XI), 5226–5232 (2018)
5. Kopteva, L.A., Efimova, I.A.S.: Improving the customs marketing toolkit. Russ. Sci. Goals Object. (2018). https://doi.org/10.18411/sr-10-12-2018-59

6. Malevich, Y.V., Plastunyak, I.A., Aitova, K.A.: Current issues of development of modern transport-logistics and customs technologies in the conditions of economic integration. In: 3rd International Scientific and Practical Conference, DEFIN 2020 (2020). https://doi.org/10.1145/3388984.3390642
7. Pananond, P., Gereffi, G., Pedersen, T.: An integrative typology of global strategy and global value chains: the management and organization of cross-border activities. Spec. Iss. Glob. Strat. Glob. Val. Chains **10**(3), 421–443 (2020). https://doi.org/10.1002/gsj.1388
8. Morales-Fusco, P., Saurí, S., Lekka, A.M., Karousos, I.: Assessing customs performance in the mediterranean ports. KPI selection and best practices identification as part of the MEDNET project. Transp. Res. Procedia **18**, 374–383 (2016). https://doi.org/10.1016/j.trpro.2016.12.049
9. Van Hong, P., Nguyen, T.-T.: Factors affecting marketing strategy of logistics business – cse of Vietnam. Asian J. Shipp. Log. **36**(4), 224–234 (2020). https://doi.org/10.1016/j.ajsl.2020.03.004
10. Selyukov, M., Kostin, A., Shalygina, N., Evdokimov, S., Moskalenko, O., Solodilov, D.: Improvement of customs and tax authorities interaction in the process of carrying out customs control after the release of goods. In: Proceedings of the 8th International Conference on Contemporary Problems in the Development of Economic, Financial and Credit Systems (2020). https://doi.org/10.2991/aebmr.k.201215.011
11. Smirnov, E.N.: Directions of development and regulation of cross-border data flows in international trade. E-Management **3**(1), 16–26 (2020). https://doi.org/10.26425/2658-3445-2020-1-16-26
12. Winterhalter, J., Andreas, N.: How to tax data in the context of international taxation. Trilemma of data-based taxation approaches and its solution through digital documentation (2020). https://doi.org/10.13140/RG.2.2.15136.00000
13. Zabielska, I.: Cooperation, partnership and integration in the cross-border area: the role of borders and cross-border cooperation. Econ. Law **19**(3), 569 (2020). https://doi.org/10.12775/EiP.2020.038
14. Global E-Government Development Index (2021). https://gtmarket.ru/ratings/e-government-development-index
15. Information on the number of public services provided by the FCS (2020). https://customs.gov.ru/svedeniya-o-kolichestve-predostavlennyx-gosuslug

Trends in Digital Transformation of the Russian Insurance Market

Anna S. Obukhova[1], Tatiana S. Kolmykova[1](✉), Nadezhda A. Serebryakova[2], Aleksandr Yu. Goncharov[3], and Elena O. Astapenko[4]

[1] Southwest State University, Kursk, Russia
obuhova_anna@inbox.ru
[2] Voronezh State University Engineering Technology, Voronezh, Russia
[3] Voronezh State Medical University N.N. Burdenko, Voronezh, Russia
[4] Yugorsk State University, Khanty-Mansiysk, Russia
kgtu_fk@list.ru

Abstract. The modern insurance market is characterized by an active spread of digitalization. The use of digital technologies leads to a reduction in the operating cycle due to the expansion of functions for analysis and data processing. Digitalization processes are accompanied by the emergence of new insurance products and services, as a result, the infrastructure develops, and new principles of communication with customers and partners appear. There is a process of development of new business models, which directly depends on the perception of new technologies by consumers.

The purpose of this study is to consider the main trends in the digitalization of the Russian insurance market and also to substantiate the practical significance of the introduction of modern digital technologies.

The objectives of the study are to determine the main directions for the development of digital technologies in the Russian insurance market.

The main results of the study are obtained based on the application of statistical, monographic methods, the method of scientific abstraction, economic analysis, and sample observation.

This study makes it possible to identify the main directions of digitalization of the insurance market, to evaluate the practical aspects of using digital services to do business and organize economic activities. The digitalization of the insurance market also allows simplifying all processes of interaction between the consumer and the insurance company: from choosing and issuing a policy to the moment of receiving compensation for an insured event. The use of modern information and communication technologies in the insurance market makes the search and application of new models for organizing business processes, management, and ways of interacting with customers more important.

Keywords: Innovation management insurance · Insurance market · Digitalization of the insurance market · Digital sales channels

JEL Codes: G22

E. G. Popkova (Ed.): ISC 2020, LNNS 368, pp. 46–53, 2022.
https://doi.org/10.1007/978-3-030-93244-2_6

1 Introduction

A digital system is a globally competitive infrastructure that transmits, stores, and processes data based on the use of modern information and communication technologies [8, 10]. Digitalization creates fundamentally new conditions for doing business, guides companies along the path of creating new business models based on networkization and the spread of innovative ecosystems [7, 15]. Researchers note that the introduction of new digital technologies is especially important for the financial market and the insurance market [1, 4, 11, 14].

The potential for the use of digital technologies in the insurance industry is due to the possibilities of expanding channels of communication with policyholders, the creation of new forms of online marketing and sales [9]. Expanding the range of using digital technologies creates not only new opportunities for interaction with consumers, but also leads to the restructuring of business processes within the companies, and determines their transformation into more flexible, adaptive structures. The processes of digital modernization of insurance organizations take place on the external circuit (at the level of interaction with the market and customers) and in the internal environment (change in business operations, management, and control).

The introduction of digital technologies and the changes accompanying this process have exacerbated the competition between insurance companies. According to the authors, the current trends in competition between insurance companies are presented as follows (Fig. 1).

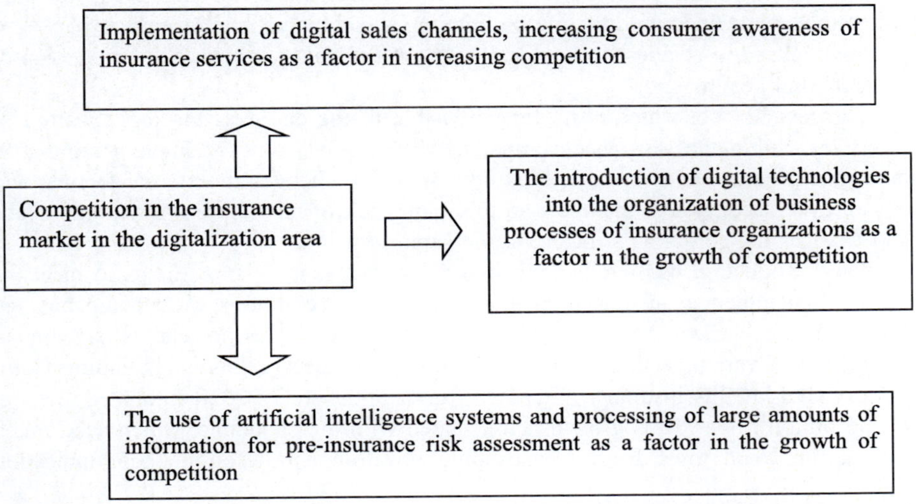

Fig. 1. Directions of competition among insurers in the digitalization area. *Source* compiled by the authors

The competition between insurers in the area of digitalization is associated with the emergence of new sales channels via the Internet, and mobile applications, which has led to the replacement of mono-channel communication with multi-channel communication. Omni channel and "multi touch" have also emerged.

Today's customers expect an insurance company the possibility to interact using the channel that is preferred at any given time and all stages of the buying process. The use of digital technologies in the work of insurers affects the behavioral characteristics of consumers, which affect the choice of a package of insurance services, and the insurance organization.

2 Materials and Methods

The digitalization of the insurance market is aimed at using new models of business processes that contribute to the development of interaction with customers through the use of information technology. Digitalization has simplified the mechanisms of interaction with customers, from choosing and issuing an insurance policy to receiving compensation as a result of an insured event.

Interaction of a client with an insurance organization can be presented in two perspectives. The first one, visible to the client, is carried out through a personal account. In the client's account, the functions of storing online policies are implemented, it's possible to place contracts drawn up offline, view and download the policy in PDF format, save the cost calculation with the ability to continue processing in a mobile application, or online store. The most convenient is the mobile application, which includes a wide product line. Simple types include vehicle insurance, packaged products related to accident insurance. The more difficult complex is the protection of property and health.

The second one, which is no longer visible to the client, is the digitalization of processes within the insurance company, which allows quickly implementing new insurance products and services, flexibly respond to changes in external factors, and ensure cybersecurity. All this leads to the joint work of mixed cross-functional teams and changes the corporate structure at the same time.

The introduction of digitalization into the activities of insurers helps to meet the needs of customers in insurance coverage. In modern conditions, most companies are transferring all client services online - from policy sales to claims settlement. According to various estimates, the volume of insurance premiums excluding Compulsory civil liability insurance (CCLI) in Russia at the end of 2020 amounted to about 30–35 billion rubles. This growth is associated with the pandemic, which accelerated the current trend towards the outstripping development of online communication channels with consumers of insurance products and services. According to the Expert RA agency, 80% of the surveyed insurers are using new IT products, 50% are implementing online services for the website, mobile application, and personal account (Fig. 2).

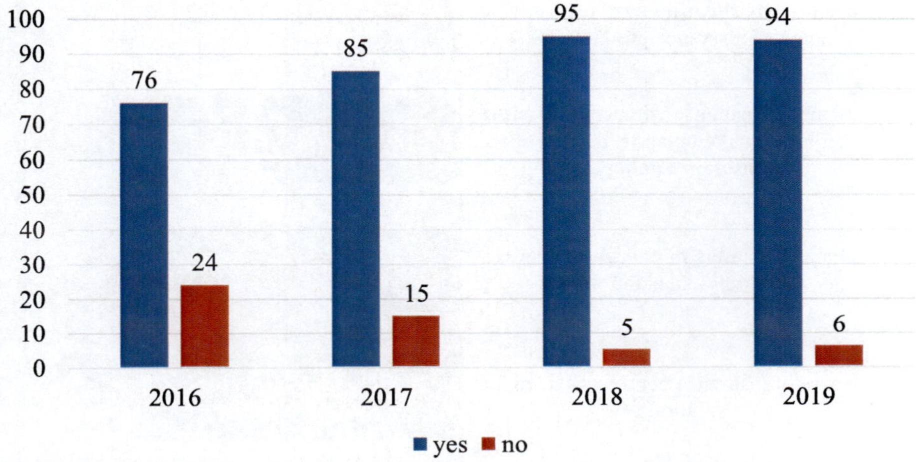

Fig. 2. Insurers' use of IT solutions related to the use of the Internet. *Source* rating agency "Expert RA" https://www.raexpert.ru

More than 85% of insurance companies use online sales. The online sales channel for insurance services is the most widespread among digital sales channels. It's the process of buying an insurance policy online, including choosing an insurance service, paying, and obtaining an insurance policy. At the same time, it should be noted that such channels are promising in the future because the number of active Internet users in the world is growing. This has already accounted for more than three billion people, i.e. about 42% of the world's population now [2].

Most insurers that use online services to sell policies provide customers with a wide range of voluntary insurance products. These include such common products as voluntary medical insurance, property insurance for individuals, motor third party liability insurance, insurance services against accidents and illnesses, and others.

In recent years, trends have been appearing in the development of online sales not only through the websites of insurance companies but also through the websites of partners (banks, microfinance organizations, Internet aggregators, online travel agencies, and other organizations) [12, 13]. The dynamics of indicators of the use of Internet insurance in Russia is presented in Fig. 3.

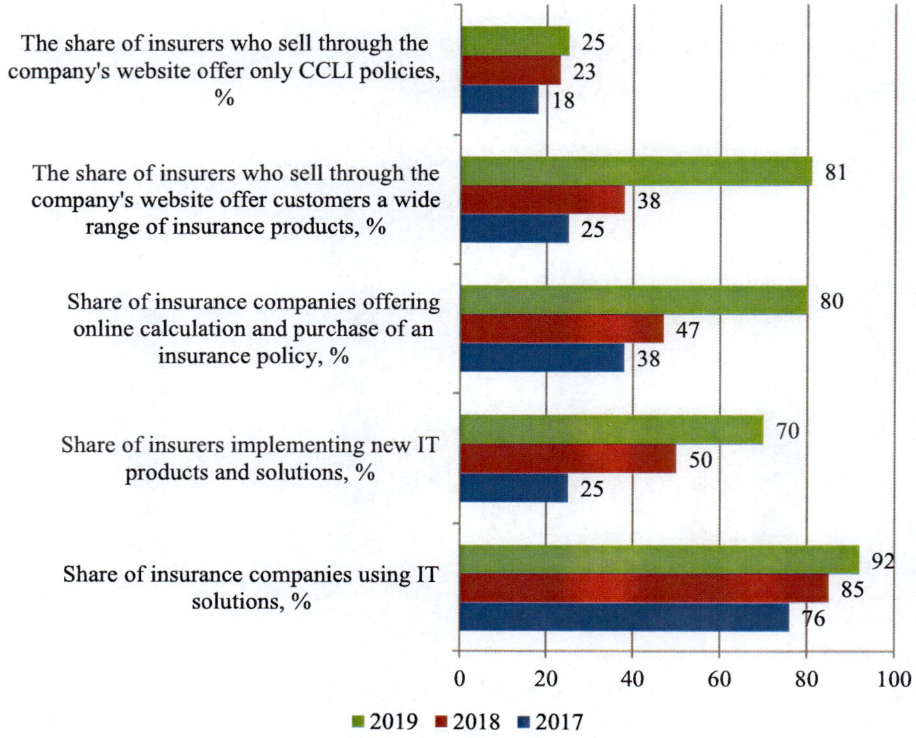

Fig. 3. Change in indicators of Internet insurance in Russia, %. *Source* rating agency "Expert RA" https://www.raexpert.ru

Analysis shows that the Internet plays a significant role in the insurance sector in terms of efficient and quick use of information, as it becomes more accessible to many people, and it's easier to use [3, 5, 6].

Another area of digitalization in the insurance market is the introduction of chatbots. Their use makes it possible to organize consulting and customer support, as well as the sale of insurance services through digital channels, websites, mobile applications, voice terminals, and social networks. Chatbots are widely used in contact centers and help staff to work with clients. This allows reducing the waiting time on the line or providing the customer with an alternative self-service channel.

Using the chatbot software also allows analyzing the client's voice or written request and providing the most appropriate response. If the request isn't correct or not clear, the employee refuses or offers to try again. The chatbot workflow is presented in Fig. 4.

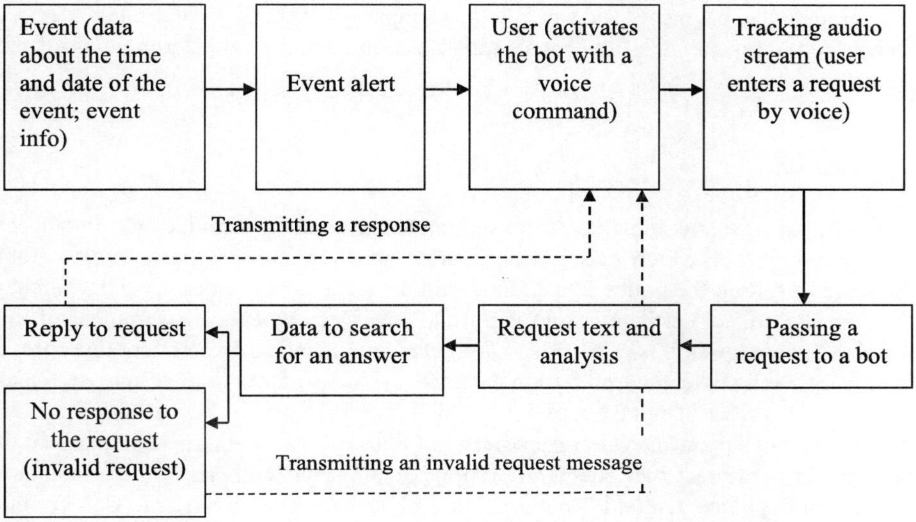

Fig. 4. Chatbot workflow. *Source* compiled by the authors

Chatbots are used to optimize the internal processes of the insurance company: the time required to search for information is saved, the number of direct communications is reduced.

The impact of digitalization also affects the active automation of labor in insurance companies. Thus, operations related to the control of regular contributions were always performed manually by managers, and it was laborious, time-consuming, and the cost of an error, given the scale of the company, was quite significant. If the client didn't have time to pay off his debts on time, he was forced to solve the problem personally in one of the company's offices. Taking into account the workload of employees, this further postponed the timing of payments. These processes have become automated with the introduction of digital technologies. In addition, the use of decision-making systems using artificial intelligence technologies, chatbots, robotization of the preparation and processing of insurance contracts makes it possible to form an individual offer for each client in terms of tariff, insurance conditions, and additional services.

Among Russian companies, IJSC "VSK" is actively introducing digital technologies. "VSK" entrusted the solution to the digitalization problem to the international company Bell Integrator, which has experience in digital business transformation and owns technologies of artificial intelligence, augmented reality, blockchain, Internet of Things, and other innovations. Bell Integrator has combined insurance accounting, account management, and SMS customer interaction to create a one-stop debt management solution that is scalable. The created robot, on command, unloads data from the Oracle database, checks for contracts in the accounting system and, if the documents aren't in the database, creates and deletes them after the expiration date. This robot, using a web service, sends SMS notifications to customers with links to pay the bill through the payment gateway on the company's website. If the client doesn't receive funds for payment, the robot initiates the process of early termination of the contract.

3 Results

Currently, insurance companies are observing the process of introducing a document recognition system, which makes it possible to abandon paper carriers, starting from the stage of selling the policy and ending with the settlement of losses without forcing the client to collect certificates and confirmations. Decree of the Government of the Russian Federation of 03.06.2019 No. 710 "On Conducting an Experiment to Improve the Quality and Coherence of Data Contained in State Information Resources" launched a pilot project "Digital Profile", in which insurance organizations take part with the Ministry of Digital Development and the Central Bank of Russia. The purpose of this project is the fact that clients of insurance companies authorize on the insurer's website through the Unified Identification and Authentication System (UIAS), and directly receive the opportunity to use reliable digital profile information.

In the development of the legislative process, the State Duma of the Russian Federation adopted federal laws No. 211-Federal Law of July 20, 2020, On "Financial Transactions Using a Financial Platform" and No. 212-Federal Law of July 20, 2020 "On Amendments to Certain Legislative Acts of the Russian Federation on the Issues of Financial Transactions using the Financial Platform". The marketplace is developing now and its functions according to the "Plug & Play" principle and unites the following financial market participants:

- Financial organizations - banks, insurance companies, mutual funds, issuers of corporate and government bonds;
- Financial platforms through which interaction between clients and financial institutions are carried out;
- Aggregator sites are internet resources and mobile applications. Their use allows the client to compare financial products and making a choice.

4 Conclusions/Recommendations

The development of the insurance market in Russia based on digital technologies has significant prospects. The spread of internet access and mobile devices will expand for the foreseeable future. The existing IT infrastructure will expand and new digital platforms will emerge, the insurance companies will become active participants in which. These changes provide insurance companies with the opportunity to enter a new era of digital insurance: the time for servicing policyholders will be significantly reduced, the possibility of fraud will be minimized or eliminated, operating costs will be reduced, and the opportunities for customers in insurance coverage will expand.

References

1. Bohnert, A., Fritzsche, A., Gregor, S.: Digital agendas in the insurance industry: the importance of comprehensive approaches. Geneva Pap. Risk Insur. Issu. Pract. **44**(1), 1–19 (2019)
2. Cappiello, A.: Technology and the Insurance Industry: Re-configuring the Competitive Landscape. Springer, Cham (2018). https://doi.org/10.1007/978-3-319-74712-5
3. Kazarenkova, N.P., Svetovtseva, T.A.: Transformation of the banking system of Russia under the influence of digitalization of the economy. Bull. Southwest State Univ. Ser. Econ. Soc. Manag. **4**(29), 188–195 (2018)
4. Kazarenkova, N.P., Kolmykova, T.S.: Modern growth points of the Russian banking sector and their impact on the economic development of the country. J. Appl. Econ. Sci. **12**(4), 985–994 (2017)
5. Kolmykova, T., Merzlyakova, E., Preobrazhensky, B., Serebryakova, N.: Theoretical approaches to the study of the socially-oriented infrastructure of the region, education excellence and innovation management through vision 2020. In: Proceedings of the 33rd International Business Information Management Association Conference, IBIMA, pp. 1006–1011 (2020)
6. Kolmykova, T.S.: Place and role of a human in the system of robotized circular reproduction in the regional economy. Lect. Notes Netw. Syst. **111**, 265–273 (2020)
7. Kolmykova, T.S., Ostimuk, O.V., Obukhova, A.S.: Role of institutional investors in the resource provision of the region's economic growth. J. Adv. Res. Law Econ. **8**(8), 2439–2447 (2017)
8. Kvashnina, D.V., Ershova, I.G.: State regulation of innovative policy of the region in the framework of the digital economy. Bull. North Caucasus Federal Univ. **2**(77), 60–65 (2020)
9. Moşteanu, N.R., Faccia, A., Cavaliere, L.P.L.: Digitalization and green economy-changes of business perspectives. In: Proceedings of the 2020 4th International Conference on Cloud and Big Data Computing, pp. 108–112 (2020)
10. Obukhova, A., Merzlyakova, E., Ershova, I., Karakulina, K.: Introduction of digital technologies in the enterprise. E3S Web Conf. **159**, 04004 (2020)
11. Obukhova, A., Merzlyakova, E., Sokolov, A., Chernykh, Ya.: Technology and prospects of digital economy development, vision 2025: education excellence and management of innovations through sustainable economic competitive advantage. In: Proceedings of the 34th International Business Information Management Association Conference, IBIMA, pp. 13648–13655 (2019)
12. Obukhova, A.S., Pavlova, O.A., Chernykh, Ya.V.: Research trends and prospects for development of the digital economy. Region Syst. Econ. Manag. **3**(46), 23–30 (2019)
13. Prosvetova, A.A., Dozhdeva, E.E.: Insurance activities in the digital economy of Russia. In: Ashmarina, S.I., Horák, J., Vrbka, J., Šuleř, P. (eds.) IES 2020. LNNS, vol. 160, pp. 736–741. Springer, Cham (2021). https://doi.org/10.1007/978-3-030-60929-0_95
14. Sitnikova, E.V., Kolmykova, T.S., Astapenko, E.O., Grivachev, E.A.: An assessment of commercial banks' financial sustainability. In: Bogoviz, A.V., Ragulina, Y.V. (eds.) ISCI 2019. LNNS, vol. 115, pp. 370–378. Springer, Cham (2020). https://doi.org/10.1007/978-3-030-40749-0_44
15. Sviridova, S., Shkarupeta, E., Dorokhova, O.: The use of neural networks and a genetic algorithm for modeling the innovative environment of enterprises. In: Topical Problems of Green Architecture, Civil and Environmental Engineering, E3S Web of Conferences, TPACEE 2019, pp. 10–45 (2020)

Prospects for the Use of Neural Networks to Analyze the Indicators of Export of Goods and Services of the Regions of the Russian Federation in the Medium Term

Anna A. Abrosimova(✉), Uryi V. Trifonov, Olga U. Angelova, Alevtina V. Kulikova, and Pavel S. Shalabaev

Lobachevsky State University, Nizhny Novgorod, Russia

Abstract. The main purpose of the article is the forecasting the indicators of export of goods and services to the world economy using neural networks and testing the method on the example of a specific region of the Russian Federation.

For data analysis, the authors use not the regression analysis, which is traditional for such purposes, but a neural network (implementation takes place using the Deductor program). The input and output parameters for the program are determined, an analytical table is compiled, predicted values are built. A feature of the analysis presented by the authors is that a forecast was made for the period, for which the data are already known to test the neural network. This allows evaluating how accurate the values produced by the program are.

This article shows that the use of neural networks for export analysis is possible and realizable. The article describes the main subtleties and limitations of the possible application of the methodology. Also, the main directions of using this analysis and the possibilities of its application in practice are described. The obtained data are corresponded and close to reality. This indicates the accuracy of the constructed neural network.

This method of forecasting exports and other indicators can be used by enterprises to adjust business plans and conduct SWOT and PEST analysis.

Keywords: Export · Neural network · Indicators · Analysis · Model

JEL Codes: C55 · C61 · E27

1 Introduction

Export is a fundamental factor of the economy, which requires regular monitoring and analysis. The Russian Federation is trying to strengthen its position in the world market by concluding new trade agreements with foreign enterprises, optimizing the import and export performance of various products. Therefore, the relevance of competent leadership, as well as analysis, accounting, and audit of exports is increasing every year [5, 6].

E. G. Popkova (Ed.): ISC 2020, LNNS 368, pp. 54–63, 2022.
https://doi.org/10.1007/978-3-030-93244-2_7

Analysis of the dynamics of export indicators of the Russian Federation shows that there is a decrease in the share of exports in the structure of foreign trade turnover of the Russian Federation by 1.9% - from 60.2% (in 2019, with a total turnover of $490.5 billion) to 58.3% (according to data for January–September 2020, when exports amounted to $237.8 billion with a foreign trade turnover of $408.1 billion) [3]. This suggests that additional study of export indicators is necessary to identify reduction factors. It is relevant to study the impact of the 2020 pandemic on the situation in Russia in general and on indicators of foreign trade in particular.

The export structure as a driver of economic growth and the factors affecting it are fairly popular research topics in the modern economy. Some scientists consider the export of individual countries as a research base, such scientists as [3, 5, 6, 19] analyzed the main indicators and dynamics of change. In [20], the structure of Turkey's exports is analyzed, the features of exports of countries with developing economies are determined and the impact of exports of high-tech products on economic growth in the country is analyzed. In [10], a study of the structure of Macedonian exports is carried out; a positive causal relationship is revealed between the increase in the export of goods of high complexity and the country's economic growth at the beginning of the 21st century. At the same time, the author states the fact that the structure of the country's export is highly concentrated. Similar studies were carried out in Canada [3], Pakistan [1], Morocco [11], and based on data from the BRICS countries [16].

Mironova concludes that raw materials prevail in the structure of exports in most regions of Russia based on the analysis of the commodity structure of Russia's exports to the CIS countries and far abroad, as well as exports of federal districts of the Russian Federation. The article [13] emphasizes the importance of developing a strategy for the development of regional exports to increase the share of high-tech products, especially in mechanical engineering.

The authors note that the values of the export indicators of the Arkhangelsk Region are more dependent on the export activity of the timber industry complex (TIC) in their work [14]. While analyzing the dynamics of the export of TIC products in this region, they state the growth of exports of the main types of products, the expansion of export partner countries, the predominance of low-processed products in the export structure.

Other researchers have focused on analyzing the factors affecting a country's export structure. So, the influence of financial institutions on the dynamics of exports is shown in [17]. The role of financial integration in export diversification based on statistical data from 96 countries is analyzed in [9]. The use of various production factors in the European Union countries and their impact on the export structure are analyzed in [12].

Studies in which the authors attempt to assess the impact of exports and their structure on economic growth and develop recommendations that will allow countries with different levels of development to ensure economic growth, which is especially important in the context of the current global recession, are interesting. In [4], it is clearly shown that a stronger effect on economic growth is exerted not by a change in the structure of exports as a whole, but by a change in that part of it where raw materials are taken into account for countries with economies in transition based on an analysis of data from 11 countries over 20 years. The author's recommendations suggest the implementation of a balanced export strategy, in which revenues from the export of raw materials should be directed to support national producers of high-tech goods and domestic demand for them that should increase the rate of economic growth. The IMF edition [5] contains recommendations for optimizing the structure of a country's export through its diversification, developed using machine learning methods.

In the scientific research [3], the authors propose their effective methodology for determining the share of non-resource exports and argue that economic growth of both individual regions and the country as a whole can be achieved only by shifting the export structure to the side that requires comprehensive state support of exporting enterprises.

Recently, researchers have begun to actively apply machine learning methods to analyze economic processes. One of the most common ML methods is neural networks. Neural networks are used to analyze and predict data in economics [2, 14, 15, 19, 21], including for stock market data [18], in agricultural sciences [8], and concerning indicators of the energy industry to predict oil and electricity prices [7]. At the same time, the features of subject areas, unstructured data sources, a significant number of influencing factors impose certain restrictions on the use of ML methods and necessitate the involvement of experts from the subject area in the process to improve efficiency. Besides, attempts are being made to systematize and automate the application of ML methods, the choice of the most optimal one in applied areas occurs through the creation of specialized software [2].

2 Methodology

In its most general form, the management model of export and import indicators can be presented in the following form:

$F_i\left(\underline{x}_1^i, \underline{x}_2^i, \ldots, \underline{x}_{ni}^i\right) = \left(\underline{y}_1^i, \underline{y}_2^i, \ldots, \underline{y}_{mi}^i\right)$, where

$i = \underline{1, p}$

$\underline{x}_k^i \cdot \left(k = \underline{1, n_i}\right)$ is a group of indicators, called input parameters, which describes a separate aspect of the state of the economic system;

$\underline{y}_j^i \cdot \left(j = \underline{1, m_i}\right)$ – this group of criteria makes it possible to assess the state of the economic system and agents within the framework of a certain feature. The behavior of agents can also be considered here;

F_i is a function that allows calculating the values of a vector $\underline{y}_j^i$ depending on the values of the vectors $\underline{x}_1^i$, $\underline{x}_2^i$, ..., $\underline{x}_{ni}^i$.

It is necessary to optimize the number of input indicators required in the future to adequately describe the state of the economic system, conscious that an increase in their number leads to an increase in costs of all types (financial, time, organizational, including related with the collection of information) when building a model for managing the export and import of a country. At the same time, the objectives of the study, the availability of information sources, the presence of a pool of experts who are focused specifically on the specifics of the institutional environment, and many others will be the factors of justifying and choosing a set of specific indicators.

F_i ($i = \overline{1, p}$) function properties by which the specific values of the criteria are calculated are determined by the chosen approach. Figure 1 shows a model of F_i function with clear input information ($\underline{x}_1^i$, …, $\underline{x}_{ni}^i$ vector), a clear scheme and procedure for calculating output indicators, and, finally, specific criteria (output parameters) in the form of vectors ($\underline{y}_1^i$, …, $\underline{y}_{mi}^i$).

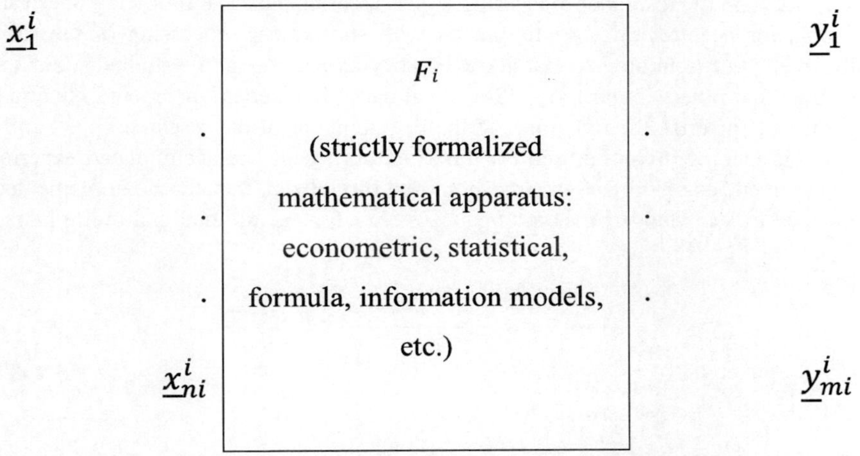

Fig. 1. F_i function as a pronounced form. Source Compiled by the authors based on [22]

The "black box" phenomenon can be used in the second approach. It will be rational in cases when the construction of the F_i function is impossible for various reasons. Figure 2 shows the dependencies describing the connection of "input" ($\underline{x}_1^i$, …, $\underline{x}_{ni}^i$) with "output" ($\underline{y}_1^i$, …, $\underline{y}_{mi}^i$).

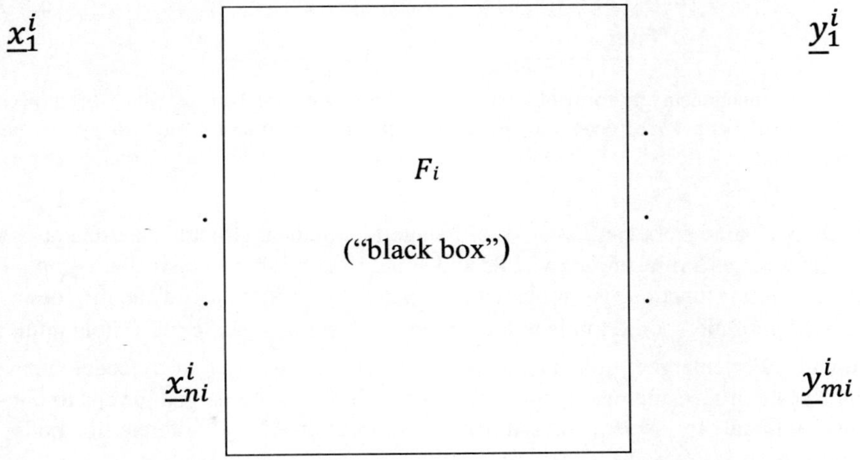

Fig. 2. F_i function as a "black box". Source Compiled by the authors based on [22]

In our opinion, the most acceptable option with this approach is the use of neural networks, which can form a structure «within themselves», that is modeled by the regularity of the investigated object, process or phenomenon, in the learning process. The procedure for applying the apparatus of neural networks presupposes a preliminary solution to the problem of machine learning of the network, which requires a significant number of input examples (samples) in a given subject area, which can be both statistical data and the results of processing expert assessments. The following ones can be the tools for peer review: questionnaires with subsequent processing of the results, polls, intelligent technologies that allow identifying patterns in the studied subject area (Big Data and other similar tools). Statistical data (time series) of ordered quantitative data that characterize the dynamics of the development of the research object and the factors influencing it can be used for the same purposes. The accumulated experience and competencies of subject experts should be formalized, that can be implemented in the format of knowledge bases in expert systems, for institutional and multiple use of their knowledge (Fig. 3).

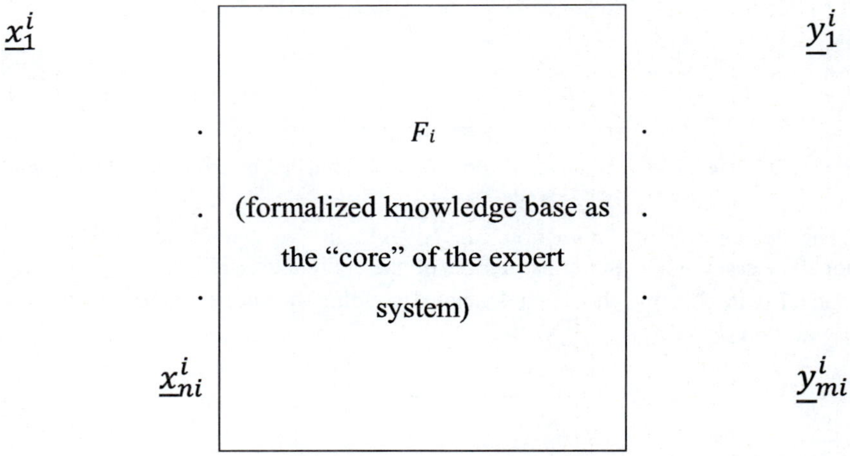

Fig. 3. F_i function in the form of a knowledge base, with the help of which the process of managing the behavior of economic agents is implemented. Source Compiled by the authors based on [22]

Output parameters can have a multi-stage hierarchical structure. Particular assessment criteria, as a rule, are indivisible and should not be split into smaller components. When the F_i function is implemented using the apparatus of neural networks, knowledge of the $\underline{y}_j^i$ criterion is expressed in the form of specific indivisible numerical estimates. For enlarged groups of output parameters, it is required to reduce them to a single scale of measurement and convolution of individual enlarged groups to obtain a single integral assessment, which can be implemented by different methods and algorithms.

Let's pay special attention to the export and import research model based on neural networks. In our case, it is possible to predict the dynamics of the output parameters affecting consumer behavior using a similar model. Tasks related to clustering (grouping) of economic agents (consumers) are another class of tasks for which neural networks are successfully used. Neural networks allow determining the most acceptable number of clusters (if necessary, and then correlate a specific economic agent with one of the clusters).

The apparatus of neural networks also makes it possible to analyze the influence of changes in the input parameters included in the initial information base on the output parameters of economic agents. One of the variants of the analysis algorithm consists in changing (without leaving the given range of changes) the components of the $\underline{x} = (x_1^1\ x_2^1, \ldots, x_{j1}^1, x_1^2, x_2^2, \ldots, x_{j2}^2, \ldots, x_1^n, x_2^n, \ldots, x_{jn}^n)$ common input vector and keeping track of how the output parameters change (or do not change). It is possible to identify the most significant input parameters by alternately changing the values of the input vector $\underline{x}$. If significant changes in the numerical input values of the input parameter practically do not affect the output, then such a parameter is insignificant and such input parameters can be ignored in the future and even excluded from the initial Infobase.

What is especially important, such an intelligent analysis allows managing economic processes without presenting and formulating in an explicit form the regularities connecting the input parameters of economic agents with the output parameters of economic agents.

We assume that the neural network will issue an export forecast, adjusting this indicator taking into account the input factors based on this data. We will use a neural network for analysis (Deductor program, which allows automating some routine operations and visualize decision-making logic). The forecasting method is a multilayer perceptron. It is expected that the application of an approach based on neural network analysis will improve the speed and quality of management decisions and provide the possibility of deep analytics.

3 Results

The data of 2016–2019 on the export of the Russian Federation officially presented by the Federal State Statistics Service were taken for analysis.

Object - Subject of the Russian Federation (region).

The input data to the neural network are:

X1 – Export volume in N-1 year, million USD
X2 – Export volume in N-2 year, million USD

The output parameter will be the export volume in 2019. We especially took the indicator of the period, which is already known, to see the accuracy of the neural network forecast (Y1 - Export volume in N year, million USD).

A fragment of the input and output data of the neural network is shown in Fig. 4.

Subject	X1	X2	Y1	Y1_OUT	Y1_ERR
5	149246,6	118381	197609	190545,581623306	0,00127766847494654
5	197609	149246,6	183541,6	190545,993454769	0,0012564042141877
5	118381	149137,5441	149246,6	149255,655110087	2,09978962972857E-9
21	26451,8	21815	27613,9	28392,6503774369	1,55304792494503E-5
21	21815	15802,8	26451,8	22786,9990834687	0,000343945072637162
21	15802,8	15726,83467	21815	18075,916397035	0,000358029381983108
52	14498,9	11949,1	20154,2	16692,4786156495	0,000306882793713934
52	20154,2	14498,9	19348,3	21115,9680942178	8,00183816456183E-5
58	13873,2	9602,5	16444,7	15821,58769295	9,94308095715577E-6
43	13106,7	9316,3	15406,3	15155,5670389626	1,60994550695222E-6
73	10574,6	8998,7	15272,7	12648,9364236934	0,000176294074366023
73	15272,7	10574,6	15188,6	16999,4637488253	8,39769054722089E-5
58	16444,7	13873,2	14873,7	18248,6850099647	0,000291697034437733
52	11949,1	12504,8	14498,9	15002,5794105516	6,49675835405826E-6
58	9602,5	10346,62645	13873,2	12103,335821892	8,02173287644376E-5
43	9316,3	11408,30635	13106,7	12227,7793652015	1,97827944604534E-5
43	15406,3	13106,7	12555,3	17468,2238074696	0,000618113708040627
73	8998,7	11583,75935	10574,6	11956,0398724171	4,8871119892064E-5
34	6844,7	5545,7	9061,3	6976,39600178745	0,000111316730398685
50	6923,9	7118,8	8559	7604,78827328555	2,33172761596137E-5
31	7001	5588,8	8439,4	7140,92228619046	4,31774825715538E-5
34	9061,3	6844,7	8272	9974,08538034907	7,419096163593E-5
11	7076	6580	7626,8	7564,98812634781	9,78435347238505E-8
31	8439,4	7001	7411,1	9278,64841283181	8,9316573826171E-5

Fig. 4. Training data array for a neural network with output indicators. Source Developed and compiled by the authors

The neural network organized for research has the following characteristics:

- Number of layers – 2;
- The number of neurons: 1 layer – 8, 2 layer – 4.

The neural network graph shows the connections built by the program within the framework of these indicators and is presented below in Fig. 5.

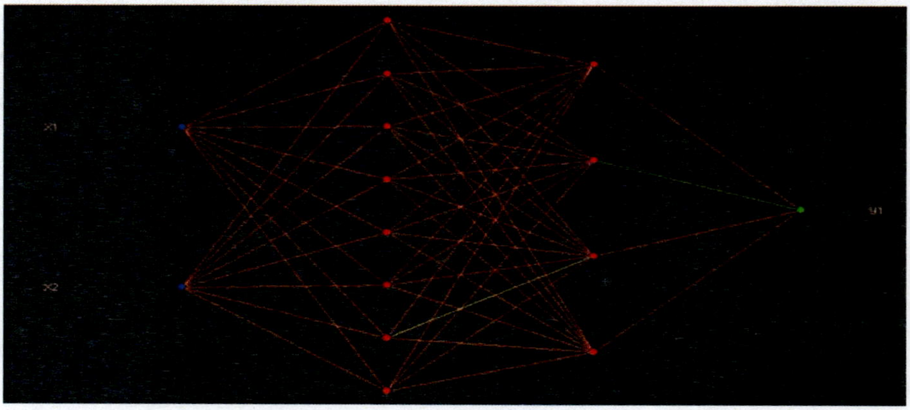

Fig. 5. Neural network graph. Source Developed and compiled by the authors

The scatter diagram illustrated in Fig. 6 shows that the trained neural network meets the requirements since there are practically no indicators outside the scattering boundaries.

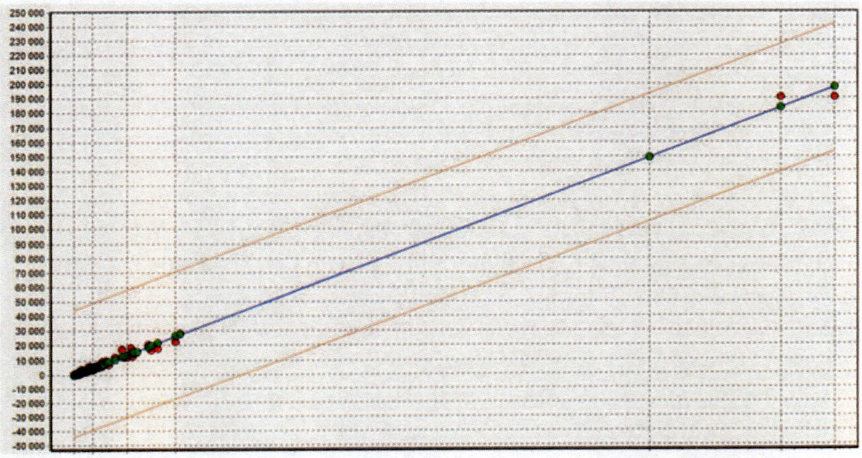

Fig. 6. Scatter plot. Source Developed and compiled by the authors

The results show that the forecast of the neural network is close to reality and the deviation obtained as a result of the analysis does not exceed 10%. This indicates the accuracy of the constructed neural network and its possibility of using it in practice.

4 Conclusion

We presented the data obtained by the neural network in the form of Table 1.

Table 1. Prediction results using a neural network, million USD

Region	Real export data for 2018–2019	Neural network predicted data for 2020
Moscow	X1 – 183,541.6 X2 – 197,609	У1 – 190,544.794
Saint Petersburg	X1 – 27,613.9 X2 – 26,451.8	У1 – 29,233.8
Nizhny Novgorod	X1 – 4,873.6 X2 – 5,486	У1 – 5,269.4

Source Developed and compiled by the authors

Let's compare this forecast with real export indicators by regions for 2020 (data are known for 9 months of 2020) (Table 2).

Table 2. The result of comparing the forecast and actual indicators, million USD

Region	Real export data for 9 months of 2019	Neural network predicted data for 2020
Moscow	96,040.0	190,544.794
Saint Petersburg	15,591.9	29,233.8
Nizhny Novgorod	3,173.7	5,269.4

Source Developed and compiled by the authors

As you can see, the forecast made by the neural network differs from the real numbers. This is because there was a sharp economic downturn in 2020 caused by the coronavirus pandemic. This had a significant impact on all economic indicators of the country. Table 3 reflects information, from the analysis of which it can be concluded that the total volume of exports for 2020 decreased by 20.7% compared to 2019.

Table 3. General results of foreign trade of the Russian Federation

	January-December 2019, million USD	January-December 2020, million USD	Growth rate, %
Turnover	674,278.6	571,913.1	84.8
Export	426,317.3	338,183.6	79.3
Import	247,961.3	233,729.5	94.3
Balance	178,356.1	104,454.1	–

Source Developed and compiled by the authors

It can be concluded that the neural network has trained quite correctly, the obtained values are close to real values, considering the overall decrease in the volume of exports, as well as the analysis of data for 9 months of 2020.

The data obtained as a result of such analysis can be used by small and medium-sized businesses to adjust business plans and assess external environmental factors. In particular, they can be used to carry out PEST and SWOT analysis. Such a study can also be carried out on the indicators of imports and other data of international trade.

References

1. Ahmed, R.R., Ghauri, S.P., Vveinhardt, J., Streimikiene, D.: An empirical analysis of export, import, and inflation: a case of Pakistan. ESPERA **21**(3) (2018)
2. Athey, S.: The impact of machine learning on economics. In: The Economics of Artificial Intelligence: An Agenda, pp. 507–547. University of Chicago Press, Chicago (2018)
3. Baleevskikh, A.S.: Comparative analysis of trends in foreign trade between Russia and the Perm Territory in the current period. Bull. Perm Natl. Res. Polytech. Univ. Socio-econ. Sci. **1**, 152–164 (2020)

4. Belkania, D.: Export structure and economic performance in transition economies. Eur. Res. Stud. J. **XXIII**(1) (2020). https://ssrn.com/abstract=3580924. Accessed 25 Feb 2020
5. Che, N.X.: Intelligent Export Diversification: An Export Recommendation System with Machine Learning (2020/175) (2020)
6. Gunçavdı, O., Kayam, S.S.: Unravelling the structure of Turkish exports: impediments and policy. J. Policy Mod. **39**(2), 307–323 (2017). https://doi.org/10.1016/j.jpolmod.2016.09.001
7. Ghoddusi, H., Creamer, G.G., Rafizadeh, N.: Machine learning in energy economics and finance: a review. Energy Econ. **81**, 709–727 (2019). https://doi.org/10.1016/j.eneco.2019.05.006
8. Storm, H., Baylis, K., Heckelei, T.: Machine learning in agricultural and applied economics. Eur. Rev. Agricult. Econ. **47**(3), 849–892 (2020). https://doi.org/10.1093/erae/jbz033
9. Jangam, B.P., Akram, V.: Does financial integration drive export diversification? Evidence from a cross-country analysis. J. Financ. Econ. Policy **13**(1), 45–61 (2020). https://doi.org/10.1108/JFEP-06-2019-0110
10. Lazarov, D.: Empirical analysis of export performance and economic growth: the case of Macedonia. Int. J. Trade Glob. Mark. **12**(3–4), 381–393 (2019)
11. Lotfi, B., Karim, M.: Export diversification and economic growth in Morocco: an econometric analysis. Appl. Econ. Finan. **4**(6), 27–35 (2017)
12. Maciejewski, M., Wach, K.: What determines export structure in the EU countries? The use of gravity model in international trade based on the panel data for the years 1995–2015. J. Int. Stud. **12**(1), 151–167 (2019). https://doi.org/10.14254/2071-8330.2019/12-1/10
13. Mironova, O.A.: The need to optimize the commodity structure of Russian exports at the macro and meso levels. Econ. Law State **1**(9), 19–28 (2020)
14. Myakshin, V.N., Petrov, V.N., Pesyakova, T.N.: Trends in the development of foreign economic relations of the regional timber industry complex (on the example of the Arkhangelsk region). Bull. Perm Univ. Ser. Econ. **15**(1), 110–130 (2020)
15. Puzina, N.V., An, Y.N., Simonova, N.Y.: The place of Russia in the world oil market. Bull. Siberian Inst. Bus. Inf. Technol. **4**(32), 60–69 (2019)
16. Chang, R., Yang, L.: Export structure decomposition and competitiveness measurement of BRICS exports in global value Chain. J. Chongqing Jiaotong Univ. (Soc. Sc. Edn.) **06** (2018)
17. Minetti, R., Mulabdic, A., Ruta, M., Chun Zhu, S.: Financial structures, banking regulations, and export dynamics. J. Bank. Fin. **124**, 106056 (2021). https://doi.org/10.1016/j.jbankfin.2021.106056
18. Ruf, J., Wang, W.: Neural networks for option pricing and hedging: a literature review. J. Comput. Fin. Forthcom. (2020). https://ssrn.com/abstract=3673153. Accessed 25 Feb 2020
19. Sadriev, A.R., Mustafina, O.N.: Study of the influence of the technological development of the Russian economy on the export structure. In: Babkin, A.V. (ed.). Clustering of the Digital Economy: Theory and Practice: Monograph, St. Petersburg, pp. 29–51(2020). https://doi.org/10.18720/IEP/2020.6/2
20. Sahin, B.E.: Impact of high technology export on economic growth: an analysis on Turkey. J. Bus. Econ. Finan. **8**(3), 165–172 (2019)
21. Sukharev, O.S.: Innovations: dynamics, structure, export and import strategy. Metal. Eur. **3**, 18–21 (2019)
22. Trifonov, Y.V., Brykalov, S.M., Trifonov, V.Y.: Assessment of the state of economic systems according to various criteria. Fundamental Research. LLC "Publishing House 'Academy of Natural Sciences'" **3**, 107–112 (2020)

The Model of Managing a Region as a Quasi-corporation, Taking into Account the Innovation Component

Oleg L. Goiher(✉), Ekaterina M. Gusarova, Olga S. Bugrova, Ivan M. Vinogradov, and Roman I. Kudryakov

Vladimir State University, Vladimir, Russia
r.kudryakov-vlsu@bk.ru

Abstract. Purpose: The purpose of this study is to create a methodology that allows a comprehensive assessment of the region as a system, including taking into account the innovative potential of the territory in a new interpretational model of the region as a quasi-corporation.

Design/Methodology/Approach: The article presents the main elements of the process of regional governance, describes the features of the emergence of communicative ties and assesses them in terms of their importance for the authorities and key participants in regional governance. The authors carried out a comparative assessment of the classical region and the region as a corporation in terms of functions, goals and constituent elements of the system.

Findings: It is shown that the region can be considered as a quasi-corporation and the use of corporate management methods is permissible. The authors describe the mechanism of the influence of the main forces on the processes of regional management in a static situation: society, government and business. In the model of acting forces, a special place is given to the innovative potential of the region.

Originality/Value: The described calculation results will become the basis for further research and the creation of a new classification model of regions, and will also be useful to regional authorities for a comprehensive assessment of the region's system and the development of development programs.

Keywords: Region · Quasi-corporation (corporation) · Stakeholders · Regional management · Innovation

JEL Codes: O38 · P41 · R11 · R58

1 Introduction

Modern society is a new level of development of civilization and, therefore, requires new approaches to the field of human resource management, business, the development of new approaches in interpersonal and intersubjective interaction. As a result, regional governance at the present stage of development is undergoing a shift in the basic principles of governance from closedness and bureaucratic complex mechanisms towards openness of government bodies and general accessibility for the majority of

E. G. Popkova (Ed.): ISC 2020, LNNS 368, pp. 64–72, 2022.
https://doi.org/10.1007/978-3-030-93244-2_8

participants and stakeholders within (externally) regional interaction, as well as further development of inclusive institutions of regional management.

The main idea of describing the region as a quasi-corporation is, first of all, in the abstracted representation of the subject as a kind of business unit or business project aimed at positive results of activity (in this case, production and social are key for the region), taking into account the needs of interested participants in regional management and existing restrictions (resource, time, institutional, etc.). It should be noted that the use of subsidies as a tool to eliminate regional inequality, as well as as a method of stimulating economic processes for the development of a territory did not show adequate efficiency due to various reasons: firstly, the late distribution of intergovernmental transfers; secondly, the federal center is not always able to adequately assess regional needs [1], in addition, one should not forget about the corruption component, errors in targeted financing, in addition, this approach can be attributed to demotivating, because it does not form incentives and ideas for the development of regions (in this case, it does not mean the provision of strictly guaranteed subsidies for unprotected segments of the population, as well as social benefits).

In this regard, there is a need to develop new, modern methods and management tools, the purpose of which is to achieve positive results of activities based on their own resources (labor, material, natural and financial), potentials and limitations. This approach prompts the consideration of these processes from the point of view of «business units» or, quasi-corporations, as mentioned above.

The purpose of this study is to develop a model for the functioning of the region as a quasi-corporation and endowing it with the features that will be inherent in it within this paradigm.

The object of research is the territory as an object of regional management, the subject of research is a combination of interactions between elements of regional government and key participants, as well as the features of communication links obtained as a result of such interaction.

2 Methodology

It is advisable to use a systematic approach within the framework of the description of the process of regional management, proceeding from the fact that the region is a complex system of interconnected independent elements and subsystems. The most frequently considered definitions of the region can be considered the following, for example, A.S. Marshalov and A.S. Novoselov understand a region as not only a subsystem of the country's socio-economic complex, but also a relatively independent part of it with a complete reproduction cycle, special forms of manifestation of reproduction stages and specific features of the course of social and economic processes [2].

It is necessary to recall the classical interpretation of the concept of «region» given by N.N. Nekrasov, in contrast to this definition: «A region is a large territory of a country with more or less homogeneous natural conditions, a characteristic direction of the development of productive forces based on a combination of natural resource complexes with a material and technical base, productive and social infrastructure» [3].

Both of these statements are correct, they complement each other and enrich the concept of «region», revealing it from different sides. It can be argued that both of them view the region as a complex system of connections in one way or another.

The region functions in the external and internal environment like any other open system. A distinctive feature in this case is that the object (region) can and does influence the formation of both the internal (to a greater extent) and external (to a lesser extent) environment.

Let's consider and compare the basic principles of the classical region and the region as a quasi-corporation. The basic principles of regional governance include: decentralization, partnership, subsidiarity, mobility and adaptability [4].

Next, let's consider the principles of organizing regional management for the region as a quasi-corporation, correlating the two approaches to the concepts presented (Table 1).

Table 1. Principles of organizing regional management for a region as a quasi-corporation

Principle name	Region as a unit of government	Region as a quasi-corporation
Decentralization	It is necessary to clearly define the boundaries of decentralization, because centralization is the guarantor of the preservation of state power, the state itself and prevents the emergence of undesirable destructive social phenomena	
Partnership	Poorly implemented	Should become one of the main within the framework of this concept: understanding not only mutual interests, but the interests of indirectly involved persons
Subsidiarity	Intervention by a higher authority is possible only under certain conditions and if the lower authority has shown its incompetence in resolving the issue	
	–	The source of the request for such interference must be the holder of power under the Constitution
Mobility and adaptability	Poorly implemented. Regional governments remain closed, bureaucratic structures with little flexibility	The principle should form the basis of a new paradigm. After all, the ability of commercial organizations to quickly respond to changes in the external environment strikingly distinguishes it from the authorities

Source Compiled by the authors

Of course, the considered list of principles is not exhaustive, for example, R.G. Mannapov, L.G. Akhtarieva add the principles of scientific validity; strategicness; innovativeness; resource provision of the assigned competence to it [5]. The author considers it necessary to introduce a clarifying principle – the principle of the prevalence of society, proceeding from the dominant role of the social sphere in the region only on the basis that the population is the «source» and the main resource of the region and its potential, based on the general principles of building democracy, civil society and the functioning of inclusive government institutions.

It can be argued that the region as a quasi-corporation is a multifaceted open-type system operating on the principles of openness, adaptability and partnership as a result of all of the above.

Thus, if we consider the region as a quasi-corporation to its expanded formulation, we can add features inherent in the region as a spatial socio-economic corporation:

- clear outline of borders and location in space, borders with other regions (objects)
- socio-economic goals: economic goals should become a priority for the region
- activity aimed at satisfying customer requirements is the main motive for the activity.

Further, it is necessary to proceed to the consideration of interested participants in regional management, taking into account their practical activities in conditions of natural and artificial resource constraints. This interaction can be represented in the form of the author's conceptual model of the region as a quasi-corporation (Fig. 1).

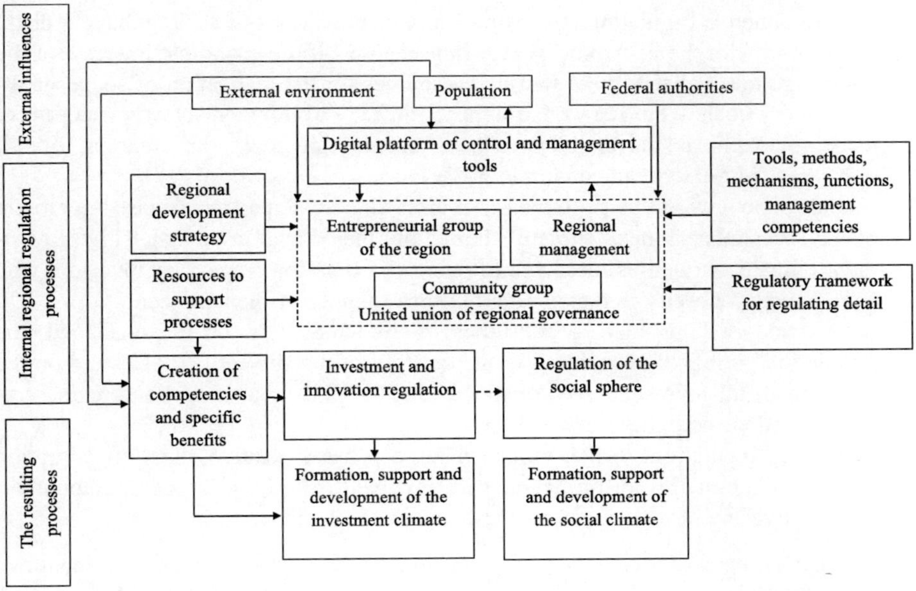

Fig. 1. Model of regional management of the region as a quasi-corporation. *Source* Compiled by the authorsObjectives of the subject and object of management

Let's describe the subject of modeling (a region as a quasi-corporation) according to the following parameters: subject and objects of management; functions of the subject and object of management; the goals of the subject and object of management; influencing factors.

Subjects and objects of management

Representation of the region as a quasi-corporation implies a shift in terms of the subjects of management: the subjects of management in this case are a public group (as a representative of the population as a «consumer» of management services), an entrepreneurial group (as representatives of the business environment and the provision of services) and regional management itself as an accumulator of optimal management decisions and their direct executor - their union will be called a single union of regional management. Objects of regional management can be considered everything that is within the boundaries of an administrative territorial unit, and not only objects belonging to municipal property, as well as objects outside of production activities affecting other spheres of life of the population, as well as the population permanently residing in the region.

Objectives of the subject and object of management

The purpose of the subject of regional management is to develop an optimal accumulated management solution or a set of measures to ensure the achievement of the goals of strategic development of the region. These include:

1. Economic goals – goals aimed at obtaining additional income from the implementation of investment programs, stimulating innovation;
2. Social goals (socio-psychological) are manifested in the social policy of the region and are aimed at regulating the socio-economic conditions of society, have a direct connection with the economic goals of the region - if the economic interests of the region are not realized, there will be no sources for the realization of social goals.
3. Investment goals - the region is interested in the development of investment projects, since this is the creation of new jobs (social goal), the creation of new industries (ultimately, an economic goal).
4. As part of achieving this goal, measures are being developed to increase the growth of the investment attractiveness of the region, the formation of its positive investment climate, as well as measures to reduce barriers for projects or investors who have already chosen the region as a place to implement their project.
5. Satisfaction of consumers – consumers for the region is, on the one hand, the population permanently residing in the territory of the subject (PPRTS) and on the other hand, all representatives of the production and non-production sphere, carrying out their activities here.
6. Improving reputation, image, rating, value, etc. has a positive effect on migration processes (the quality of the labor force involved is higher), the demographic situation, investment processes and increases tourist interest.

Ultimately, each of these goals will contribute to the achievement of the economic goals of the region.

The conscientious implementation of actions to implement the intended results of their internal strategies is the goal of regional management objects.

Functions of the subject and object of management: The subject of regional management in the model of managing a region as a quasi-corporation is represented in the form of 3 groups: a social group, an entrepreneurial group and regional management. More details on other aspects can be found in the previously presented article «External and internal environment of the region as a spatial socio-economic corporation» [6].

We will compose the distribution matrix of functions for clarity and completeness of data presentation (see Table 2). All the functions of the subjects of regional management will be divided into a classic form: forecasting, regulation, programming and control. Let's present the most important and clearly expressed subfunctions and tasks of subjects in the context of the forecasting function within the framework of generalized groups.

Table 2. Matrix of distribution of functions of regional management in the model of managing a region as a quasi-corporation

№	Function name	Regional management				Business group				Community group			
1	Forecasting												
1.1	Placement of productive forces	O		C	R	O	P	C			P	C	
1.2	Development of territory development programs	O	P	C	R		P	C	R		P	C	R
1.3	Improving the environment	O	P	C	R	O	P	C	R	O	P	C	R
1.4	Basic industries development plan	O		C	R		P	C	R		P	C	
1.5	Plans to stabilize the economic situation	O	P	C	R	O	P	C	R	O	P	C	

Source Compiled by the authors

Symbols used in the Table 2: O – performs an operation, P – provides data, information, ideas; C – consults, participates in the discussion; R– is responsible for this.

Table 2 shows that the functions, roles and responsibilities of business and society in regional governance are increasing. The organization and functioning of business and public groups should not be reduced to a formal education procedure, the work of the formed groups should not be conciliatory due to subjective reasons, the work of the three subjects should be organized on the principles of dialogue, freedom of expression and sound criticism.

The functions of management objects are expressed in independent and specific actions in the process of regional management. In particular, I.V. Afonin notes the following functions of the population in the process of regional development and management in a series of articles on the organization of public hearings [7]: firstly, the investment function - the population itself can become an investor. territory of residence; secondly, the function of making decisions – it may depend on the population which investor will get the right to implement the project and the territory; and the guiding function – the direction of the vector of development of the territory depends on the general mood of the population.

The authors can add the following functions of the population within the framework of the concept of the region as a quasi-corporation: first, the aggregation of interests – the unification of several points of view of various social groups; secondly, the representative function is realized through direct representation of one's interests in government bodies (not through elective structures).

The sphere of business (or entrepreneurship) as an object of regional management performs the following functions in the system of the region: firstly, the formative function – business creates an environment and base for civilized industrial, social, investment, market and competitive relations, and also actually forms the internal economic system; secondly, the demographic function, then the investment and production functions, and finally, the ecological function - manifests itself in caring for the environment and adherence to the principles of sustainable development [8].

The nature and scheme of interaction between government and business as a whole is the subject of many studies, in particular, the theory of stakeholders, formulated by E. Firman [9], is interesting in this respect. Further, the concept was developed by many scientists such as V.L. Tambovtsev [10], M.A. Petrov [11], M. Clarkson [12], I. M. Jawahar, G.L. McLaughlin [13], J. Post, L. Preston, S. Sachs [14] and others.

Thus, the system of regional management should be aimed at the effective implementation of the functions laid down in the subject and object, take into account their complex structure, just as business and the population, on the one hand, are an object of management, but on the other hand, they are a subject. In this situation, it is important to observe the principles of honesty and openness of one's intentions and actions, to exclude the principle of affiliation, unnecessary bureaucracy and corruption.

3 Results

The final element of the descriptive part of the model (see Fig. 1) is the factors of influence. However, we will not talk about the standard and well-known factors of the external and internal environment of the region as a quasi-corporation. In this case, the authors propose to consider the influence of the main forces in the region as a quasi-corporation and form a certain simplified model of their interaction in the conditions of the functioning of the region as a quasi-corporation with the influence of 3 dynamically changing elements: society (population) - IS, business - IB and authority IG:

$$R = (IS, IB, IG) \tag{1}$$

The state of the region will change depending on the dynamics of the constituent elements. The authors propose to take into account the following to measure the values of each coefficient in the calculation:

For IS – the dynamics of real incomes, the rate of decline/increase, unemployment

$$IS = f(\Delta ri, \Delta br, u) \tag{2}$$

For IB - GRP value, innovation index of the region, share of profitable enterprises

$$IB = f(GDP, II, B) \tag{3}$$

For IG - the number of civil servants, the governor's rating in the all-Russian rating, the investment rating of the region

$$IG = f(g_n, r_g, Ir) \quad (4)$$

Let's calculate the value of the integral indicator R for the Vladimir region from 2017 to 2019. Each component is calculated separately. Let's consider the calculation of the index using the example of IB (see Table 3).

Table 3. Calculation of the R indicator for the Vladimir region – IB

IB calculation	*Business*								
	GRP			*Regional innovation index (II)*			*Share of profitable firms*		
Specific weight	0,35			0,35			0,3		
Period	**2017**	**2018**	**2019**	**2017**	**2018**	**2019**	**2017**	**2018**	**2019**
Value	1,05	1,16	1,12	43	35	35	68,60	67,8	67,80
Grade	5	5	5	3	3	3	4	4	4
Coefficient	1,75	1,75	1,75	1,05	1,05	1,05	1,2	1,2	1,2

Source Compiled by the authors

Figure 2 shows the calculation results graphically.

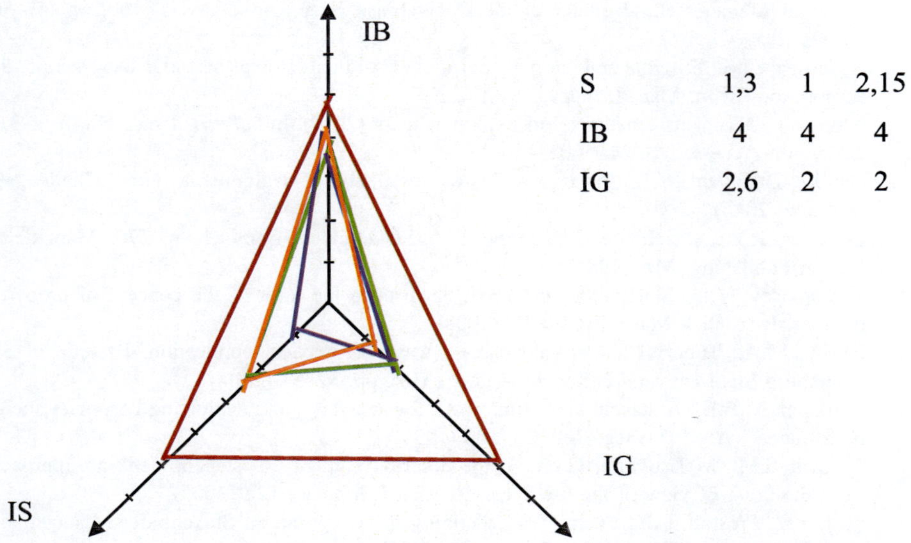

Fig. 2. System configurator of the model region as a quasi-corporation under the action of 3 forces of influence. *Source* Compiled by the authors

In its ideal position, the system should show the maximum (or close to them) values, which will mean the harmonious development of the system and the maximum manifestation of the synergistic effect from the actions of 3 systems.

The 2 components of the system showed average values in 2019. Presumably, the results for 2020 will be much worse and it is not recommended to include it in the analytical series, since it is not typical and macroeconomic factors are not identical. The vector of influence of business IB will be considered the leading one, the vector of influence of power IG must be increased to 3 units, and then the vector IS will be formed inertially.

Subsequently, this study provides regional authorities with the opportunity to take prompt action and implement corrective actions on the proposed «forces». Also, the proposed model can become the basis for a new classification system of regions.

References

1. Frumina, S.: Problems of providing subsidies from the federal budget to the subjects of the Russian Federation. Finan. Banks No. **5**, 63–64 (2016)
2. Marshalova, A.S., Novoselov, A.S.: Fundamentals of the Theory of Regional Reproduction: A Course of Lectures, Moscow, pp. 34–35 (1998)
3. Nekrasov, N.N.: Regional Economy, Moscow, pp. 45–46 (1975)
4. Gavrilov, A.I.: Regional Economy and Management: Textbook. Handbook for Universities, Moscow (2002)
5. Mannapov, R.G., Akhtarieva, L.G.: Modern Laws, Principles, Tasks and Mechanism of Regional Management. Bulletin of the Chelyabinsk State University, March, pp. 47–56 (2009)
6. Gusarova, E.M.: External and internal environment of the region as a spatial socio-economic corporation. Econ. Manag. Manag. Syst. (2021)
7. Afonin, I.: Managers-amateurs, amateurs-managers (2019). http://terraplan.ru/pdf/tip_1-31-26-29.pdf. Accessed 10 Mar 2021
8. Danilov-Danilyan, V.I., Piskunova, N.A: Sustainable Development: New Challenges, Moscow (2015)
9. Freeman, R.E.: Stakeholder Management: A Stakeholder Approach, p. 276. Marshfield, Pitman Publishing, MA (1984)
10. Tambovtsev, V.L.: Stakeholder theory of the firm in the light of the concept of property regimes. Russian J. Manag. **6**, 20–21 (2008)
11. Petrov, M.A.: Theory of interested parties: ways of practical application. Bulletin of St. Petersburg State University. Ser. 8. Issue 2 (316), pp. 52–53 (2004)
12. Clarkson, M.B.E.: A stakeholder frame-work for analyzing and evaluating corporate social performance. Acad. Manag. Rev. **20**, 106 (1995)
13. Jawahar, I.M., McLaughlin, G.L.: To the descriptive theory of stakeholders: an approach from the point of view of the life cycle, Russian J. Manag., 122 (2007)
14. Post, J.E., Preston, L.E., Sachs, S.: Redefining the Corporation: Stakeholder Management and Organizational Wealth, Stanford, p. 17 (2002)

The Problem of Providing a Highly Effective Flexible Methodology in the Management of Regional Marketing Projects and Its Solution

Elena S. Lovkova[1](✉), Tatiana N. Kashitsina[1], Nadezhda V. Kapustina[2], Navai N. O. Rustamov[3], and Aleksandra V. Sultanova[4]

[1] Vladimir State University named after Alexander and Nikolay Stoletovs, Vladimir, Russia
[2] Financial University under the Government of the Russian Federation, Moscow State University of Technologies and Management named after K.G. Razumovski, Moscow, Russia
[3] Russian New University, Moscow, Russia
irada@inbox.ru
[4] Samara State Technical University, Samara, Russia

Abstract. This article defines the relevance of using modern, universal methodologies for managing marketing projects at the regional level. It is necessary to use various management tools for the effective development of the region in modern economic conditions. This article discusses such tools as, ERP system, Kanban, PERT method, PATTERN system: The application of the Scrum methodology as the most effective methodology for planning and managing marketing projects at the regional level is considered in more detail. The essence of Scrum is described in detail, it consists in tracing the project life cycle by the owner of the product at all stages of its development, which leads to the most coordinated and effective work of performers, obtaining the necessary result and minimizing risks in communication. The main roles of Scrum and their responsibilities in the regional project management system are considered. The application of this methodology at the regional level will increase the rating of the socio-economic situation of the region, the level of quality of life of the population, the level of its investment attractiveness and attract new resources and interregional and international projects. Thus, the application of this methodology will allow achieving the goals both at the regional and state levels in general.

Keywords: Regional project management · Management tools · Marketing projects · Scrum · Regional attractiveness

JEL Codes: R01 · P0 · M310

E. G. Popkova (Ed.): ISC 2020, LNNS 368, pp. 73–79, 2022.
https://doi.org/10.1007/978-3-030-93244-2_9

1 Introduction

It is obvious that the level of country's development depends on the intensity of the development of the regional economy at the present stage of Russia's economic development. The development of regions in Russia is historically uneven, this is due to the geographical location of the regions, the presence of raw materials on their territories and the proximity of the region to the country's investment centers.

Thus, the regions are developing in different economic conditions, as a result, they have formed distinctive levels of stability, quality of life, investment attractiveness, infrastructure development, etc. In this situation, it becomes necessary to form unique competitive advantages of each individual region. Marketing design of the territory is becoming an effective tool in solving this problem. Regions can increase their level of recognition and attractiveness at the regional and global levels thanks to a competent marketing policy.

2 Methodology

The success in this direction is the clear positioning of the priority areas of the region's development, the formation of marketing development strategy and the organization of regional marketing design. At the same time, the main problems are going beyond the budget, not meeting deadlines, the same tasks. Regions are resorting to the use of project management to establish processes for organizing the work of marketing structures, now this area is more popular and relevant than ever. Competently structured project management in the region affects its economic development, investment attractiveness and image. Progress in this area does not stand still and every day more and more new approaches and methods of project management appear. The relevance of the chosen topic is due to the fact that many regional marketing projects fail because of improper development of the management system.

3 Results

It is necessary to use various management tools for the effective development of the region in modern economic conditions. Let's consider a small list of management tools used by management entities:

- ERP system is a system of organizational strategy for the integration of production activities, organization and management of human and financial resources of the project. The ERP system is designed to balance and optimize all resources (labor, material, intangible, financial, etc.) [2].
- Kanban is a management tool that allows you to keep under control projects that involve the execution of several jobs, sometimes in parallel.
  - PERT method;

    The most typical application of the above management tool is to determine the estimated time it takes to complete a job, process, project, etc.

- PATTERN system;
  This management tool is complex because it combines the use of several tools and methods in combination. Firstly, such a tool as PATTERN implies building a certain «scenario» - characteristics of the state of affairs at the moment and analysis of how everything will develop. Secondly, it is necessary to build the already mentioned management tool - a «tree of goals», where the coefficient of importance or significance is calculated in a certain way for each goal [1].

One of these methods is SCRUM, this technique is most effective for planning and managing marketing projects at the regional level.

The essence of the SCRUM methodology is to track the life cycle of the project by the owner of the product at all stages of its development, this leads to the most coordinated and effective work of performers, obtaining the necessary result and minimizing risks in communication.

There are only three roles in the SCRUM methodology (Fig. 1).

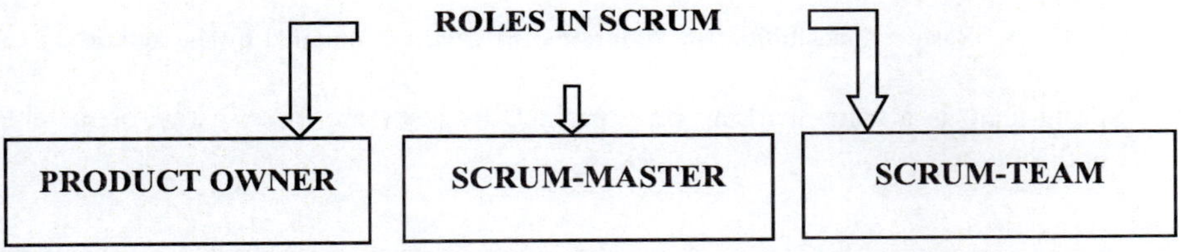

Fig. 1. «Roles in Scrum». *Source* [6]

Product owner is an individual representing the interests of the client, he can be either an official representative or a confidant of the customer. The Product Owner is directly interested in the high-performance product and is in constant communication with the Scrum-team (Fig. 2).

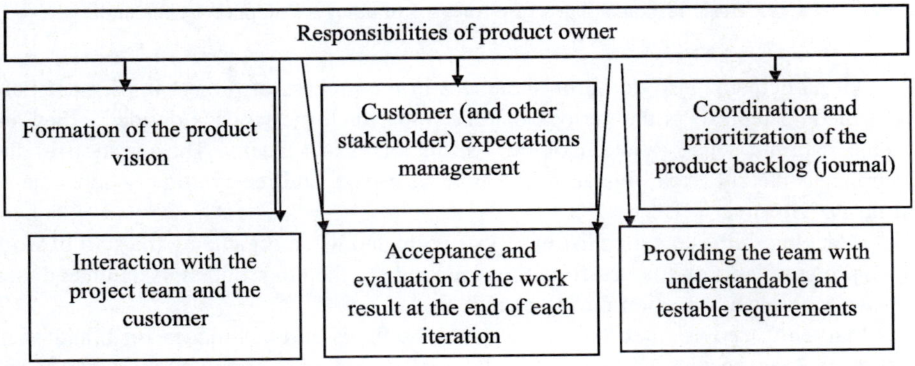

Fig. 2. Basic responsibilities of product owner. *Source* Compiled by the author

Scrum-master is a project manager who is responsible for the effectiveness and quality of work, oversees the organization and implementation of the scrum principles. He gathers quick meetings, promotes team development and minimizes conflict situations in the team as a project manager (Fig. 3).

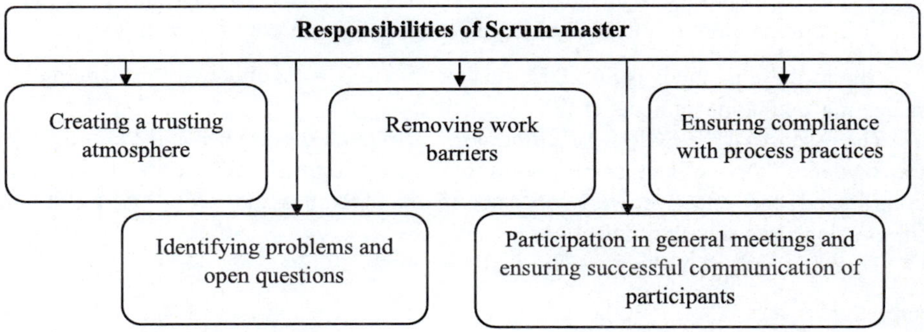

Fig. 3. Basic responsibilities of Scrum-master. *Source* Compiled by the author

Scrum-team is a team working on a project. Its tasks are to set a real achievable goal and then achieve this goal (Fig. 4).

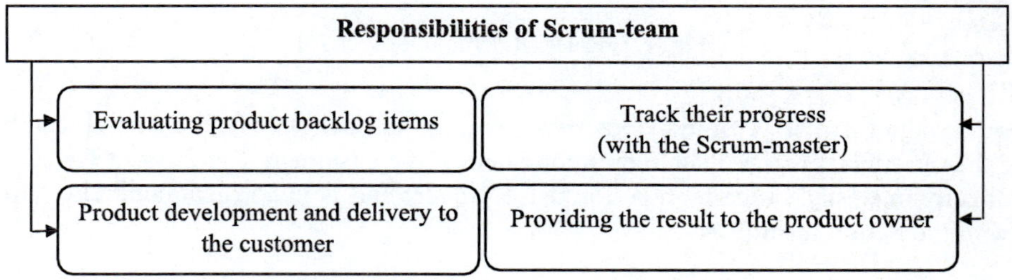

Fig. 4. Basic responsibilities of Scrum-team. *Source* Compiled by the author

All participants also do Scrum-practices in the process of project implementation. The method represents the periods of time into which the work is divided - they are called «sprints». The duration of the «sprint» is 1 - 4 weeks. The duration of the «sprint» is clearly fixed, this allows to minimize risks and receive timely adjustments from the customer [4].

«Backlog» - the so-called list of tasks - is created when organizing the start of work on a project. Tasks are moved from the «backlog» to the «in progress» section and then to the «completed» section during execution.

Marketing people often talk about ideas, methods and techniques for brand management. Many people are good at budgeting, strategizing, and creating departments. However, the difficulty lies in the implementation of the targets. Here, the most common are: constant alterations, clarifications, budget cuts and the development of new schemes, which were not even discussed at the beginning. And the plus to everything is the market, where there is a huge number of competitors with a constantly improving product. And this is where Scrum is more relevant than ever.

The marketing project team is most suitable for the scrum method, as it is a small group of 5 to 9 people. More participants will make it more difficult to exchange information. There are four events in scrum that are held regularly and therefore they are called rituals [10] (Fig. 5).

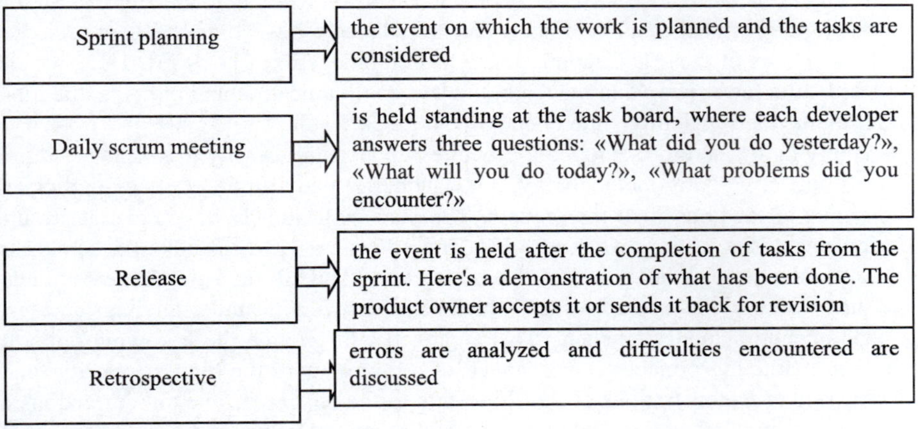

Fig. 5. Four events in scrum. *Source* [7]

Implementing a scrum methodology in any area requires a certified scrum master who will explain and help to implement the basic tools. Then the team begins to finalize them to fit their industry. One of the features of scrum is the lack of hierarchy, because everyone is a member of the group. Stakeholders involved in scrum are actually divided into three roles – product owner (sales director of the firm), Scrum-master (marketing manager) and Scrum-team (specialists from designer to web analyst).

At the first meeting, questions about the setting of tasks and the duration of the sprint are resolved, that is, into how many short stages it will be possible to divide the work with a deadline, for example, three months. Thus, if the parties, namely the product owner, the Scrum-master and the Scrum-team, come to the choice of a two-week sprint, then every two weeks the team will present the work done on tasks to the product owner, who accepts this work or sends it for revision [3]. Sometimes there is an excess or shortage of tasks, in which case the scrum master agrees with the product owner which tasks are prioritized in case of excess, and which tasks can be transferred from the next sprint in case of shortage. It should be considered that a task that is half completed in scrum is an unfulfilled task. All tasks set for one sprint must be 100% completed by the end of this sprint [9].

One of the difficulties of marketing projects lies precisely in the implementation of the plan. Competitors do not stand still and are also constantly improving, calculating moves, rebuilding. As a result, because of this, someone is left behind, and someone breaks into leaders. As a rule, this concerns changes, adjustments, minimization of budget sections or the formation of completely new sections in the project. Thus, flexibility and speed are very important in marketing projects. And that's why Scrum is chosen.

Daniil Vakhovsky, Marketing Director of Uklon, said the following when answering a question about the positive changes after the introduction of scrum in the company: «We have become very fast and flexible. We need much less time and people to launch a national promotional campaign. Now, remembering my previous jobs, it is difficult for me to imagine what 25–30 people can do in the marketing department. But there are even departments of 50 people.» He also noted that the number of the marketing department of any company can be reduced by 2–3 times without prejudice to the activities of the organization, this will even reduce fixed costs [5]. Thus, we can conclude: the fewer people in the team, the less communication and red tape, the more transparency and visibility - the effective and quick result.

Applying the scrum methodology requires understanding why it is needed. Scrum is a complete, working methodology, not a modern trendy gimmick. Each participant must give all of himself to the process. The team will already have concrete results during one sprint lasting two weeks [8]. The advantage of the method is quickness, quick planning, quick analysis and error correction. And all this is implemented under the supervision of the product owner, which helps to avoid misunderstandings and such a phenomenon as understatement. This approach will provide the following opportunities in marketing projects: quick launch of a project with the highest priority functions, regular demonstrations of developments to the customer, making corrections at the stage of implementation. All this will avoid complete failure and disappointment of the customer. This will give undoubted advantages in the implementation of a marketing project in an unstable economic market environment, where there are thousands of competitors with a constantly changing product.

4 Conclusion

Thus, it is advisable to further develop regional marketing design to achieve the following goals:

1. Formation and increase of the region's place in the ratings of investment and business attractiveness of regions;
2. Attracting new investment resources and projects to the region's economy;
3. Stimulation and implementation of own marketing projects in other regions;
4. Development of interregional network organizational and marketing structures and projects;
5. Development of mass media and communication at the regional and interregional levels;
6. Increase in the level of political and economic stability in the region, the quality of life of the population, security, tourist and investment attractiveness, the level of development of social and innovative infrastructure.

References

1. Erlygina, E.G., Vasilyeva, A.D., Polikarpova, A.S.: Features of project management in crisis. Krasnoyarsk Sci. **9**(1–4), 96–102 (2020)
2. Lovkova, E.S., Mishacheva, E.S.: Modern management methods used by small and medium-sized businesses in the regions of Russia. Krasnoyarsk Sci. **10**(2–2), 102–107 (2021)
3. Polzunova, N.N., Polzunov, I.S., Sukhareva, T.A.: Problems of implementation of project management in small business. Econ. Manag. Prob. Solut. **2**(4(100)), 48–53 (2020)
4. Kostygova, L., Polzunova, N.: Improving the organization of the recycling process. In: 19th International Multidisciplinary Scientific Geo Conference SGEM Conference Proceedings. pp. 407–414 (2019)
5. Polzunova, N., Savelev, I., Nikiforova, S., Ushakov, S.: State support for small enterprises in the countries of the European union and the Russian federation. In: Popkova, E.G., Sergi, B. S. (eds.) ISC 2019. LNNS, vol. 87, pp. 261–269. Springer, Cham (2020). https://doi.org/10.1007/978-3-030-29586-8_31
6. SCRUM in Marketing: Why is it Needed? (Article) [Electronic Resource] (2020). https://njt.ru/local/templates/navigator_new/include/download_file.php?file=92219
7. Scrum: a revolutionary method of project management/D. Sutherland, - M.: Publishing house MIF (2016)
8. Scrum in marketing on the example of the Uklon service (Article) [Electronic resource] (2017). https://l-a-b-a.com/blog/show/104. Accessed 9 Apr 2021
9. SCRUM (Article) [Electronic resource]. https://ru.wikipedia.org/wiki/SCRUM#SCRUM. Accessed 9 Apr 2021
10. SCRUM: effective method of project management (Article) [Electronic resource]. https://4brain.ru/blog/scrum/. Accessed 9 Apr 2021

Potential of Operations Research and System Analysis in Improving the Quality of State Support for Regional Science and Innovation on the Example of the Krasnoyarsk Regional Science Foundation

Irina A. Panteleeva[1(✉)], Kristina V. Parfenteva[2], Yury V. Shan'ko[3], Vladimir I. Byvshev[1], and Vadim G. Demin[1]

[1] Siberian Federal University, Krasnoyarsk, Russia
bobbyz@bk.ru
[2] Krasnoyarsk Regional Fund for Support of Scientific and Technical Activities, Krasnoyarsk, Russia
[3] Institute of Computational Modelling of SB RAS, Krasnoyarsk, Russia

Abstract. Development institutes act as a major driving factor in the development of innovations. Therefore, they must use effective and progressive methods of work. In this study, we aim to develop a mechanism for compiling the list of projects to be reviewed by the expert council for further financial support. The mechanism is based on the methods of system analysis and operations research. The study was conducted on the example of project applications to the contests of the Krasnoyarsk Regional Science Foundation.

We constructed a logistic regression model using a binary linear classifier based on the contest application review. As a result, we determined the lower limit for the inclusion of a project in the short-list. By comparing the resulting project criteria with the actual results of the review, we received a condensed list of projects that eliminates the influence of the human factor on the contest result.

This model is universal and can be scaled up to other development institutions if adjusted as necessary.

Keywords: Operations research · Systems analysis · Logistic regression · Development institutions · Scientific expertise

JEL codes: C44 · C63 · O38 · O43

1 Introduction

Socio-economic development of countries and regions is moving rapidly; innovations are becoming increasingly widespread both in the economy and in everyday life. One cannot possibly imagine the functioning of the basic economic sectors without the scientific and technological achievements that only yesterday felt like the realm of fiction. Economic development is impossible without global integration. But, integration poses the question

E. G. Popkova (Ed.): ISC 2020, LNNS 368, pp. 80–87, 2022.
https://doi.org/10.1007/978-3-030-93244-2_10

of national and regional interests, which again rely on science, technology, and innovation. Research and development (R&D) institutions should promote the development at the federal and regional level, according to Federal law No. 127-FZ "On science and state scientific-technical policy" dated August 23, 1996. The development of science, technology, engineering, and innovation relies on goal-setting and the differentiation of mechanisms, methods, and ways of providing financial support.

The efficiency of development institutions at the federal level became a pressing issue at the end of 2020, contributing to their subsequent reform and reorganization. The Government of the Russian Federation approved key performance indicators linked to the national and industry-specific development goals of the state, aimed at promoting the development of economic branches. However, these indicators are impossible to achieve without a well-built system of management and process control for R&D institutions [5]. The smooth operation of R&D institutions at both the regional and federal scales may benefit from the use of operations research (OR) methodology in their business processes.

OR is a relatively young science; its boundaries and content are still vaguely defined. The term OR generally refers to the application of quantitative mathematical methods to justify managerial decisions. OR starts with the use of a mathematical method of systems analysis to substantiate solutions and decisions.

The available secondary literature on operations research and systems analysis is mainly theoretical in nature. However, these analytic tools are extensively used in Western companies, as well as in manufacturing companies globally, to maximize profits, reduce losses, or improve output. Their use in R&D institutions is a novelty; nevertheless, some federal development institutions already employ this methodology.

One example of working with mathematical-statistical models and OR is the Russian Foundation for Basic Research, where these tools are used at the stage of project review and evaluation (Multiple criteria approach to R&D project efficiency developed by Alexei B. Petrovsky) [2, 3].

This approach uses methods of group verbal analysis of decisions (ARAMIS, KLAVA-I, KLAVA-NI, MASKA, PAKS, and PAKS-M). It allows solving problems of collective multi-criteria choice, taking into account different opinions of many participants, the diversity and mismatch of their goals, and ways of expressing their preferences. The methods are based on the use of multiset theory, which is a new mathematical apparatus that allows working with qualitative data without converting it into a numerical format.

One example is the use of mathematical modeling methods to assess the progress in scientific topics within a competitive environment created by a system of scientific performance ratings [4]. The model features a population-based structure that relies on a solution to evolutionary optimality problems. Optimality is a theorem on the extreme properties of Malthusian functions calculated at stable equilibrium positions for components that have non-zero values in these equilibrium positions. The model allows identifying the prospects of scientific development (i.e., determining which topics will become primary or secondary).

Furthermore, OR and system analysis methods are applied in information security [1]. The complex protection of digital communication facilities is largely determined by the cost of acquiring protection hardware, which represents a significant portion of the total cost. The developed model considers various options for replacing hardware while minimizing costs and damage from possible system intrusion or unauthorized access to information and maximizing the protection of communication hardware. The model provides a rationale for the strategic decision-making in the selection, operation, and replacement of hardware. The advantage of mathematical models for inventory management lies in their diversity, which determines the applicability of these models to the problem of information security.

Another example is the use of the methodological approach on the level of governmental financial management level. This approach can be used to choose a science-based solution to the interaction between optimization mechanisms of tax regimes and the subsidies for agricultural producers [6]. In this case, the model of production possibility frontier was taken as the basis. The developed model takes into account the impact of taxes and subsidies on the motivation of agricultural producers and helps determine which sectors should be supported first to maximize the socio-economic effect for both enterprises and the government. The model was used to determine the optimal size of subsidies, the level of the tax burden, and the main directions of financial support for agricultural producers.

Below, we will examine the prospects of applying OR methodology on the example of the Krasnoyarsk Regional Science Foundation.

In 2019, Krasnoyarsk Regional Science Foundation implemented the model of project contest for applied scientific-technical, sociological, and humanities research. This measure aimed to improve the mechanisms of support for applied research and ensure the competitiveness of the Krasnoyarsk Territory (Fig. 1). The contest is co-financed by the applicants. Before the contest, the Foundation invites federal and local authorities to submit proposals for the topics relevant to the socio-economic development of the Krasnoyarsk Territory. After the list of contest topics is formed, the Foundation collects applications according to the above list.

Application review includes an independent evaluation of each project proposal by three experts, as well as feedback or in-person meeting between a representative of authorities, who proposed the topic, with the relevant expert commission. The Presidium of the Expert Council takes into account the opinion of the authorities in evaluating project applications. The contest winners cooperate with supervisors from authorities in the implementation of the project. As a result, the project output is implemented by regional ministries and departments. Over the two years that the contest has been held, a total of 38 projects worth more than 67 mln. RUB were supported.

In 2020, prior to the launch of the competition for 2021, the Foundation decided to design a procedure for determining the threshold value of expert reviews. If a project is evaluated higher than the threshold, it is included in the list of projects recommended to be supported. The procedure uses mathematical modeling methods to objectively determine the winning applications and effectively distribute budget funding.

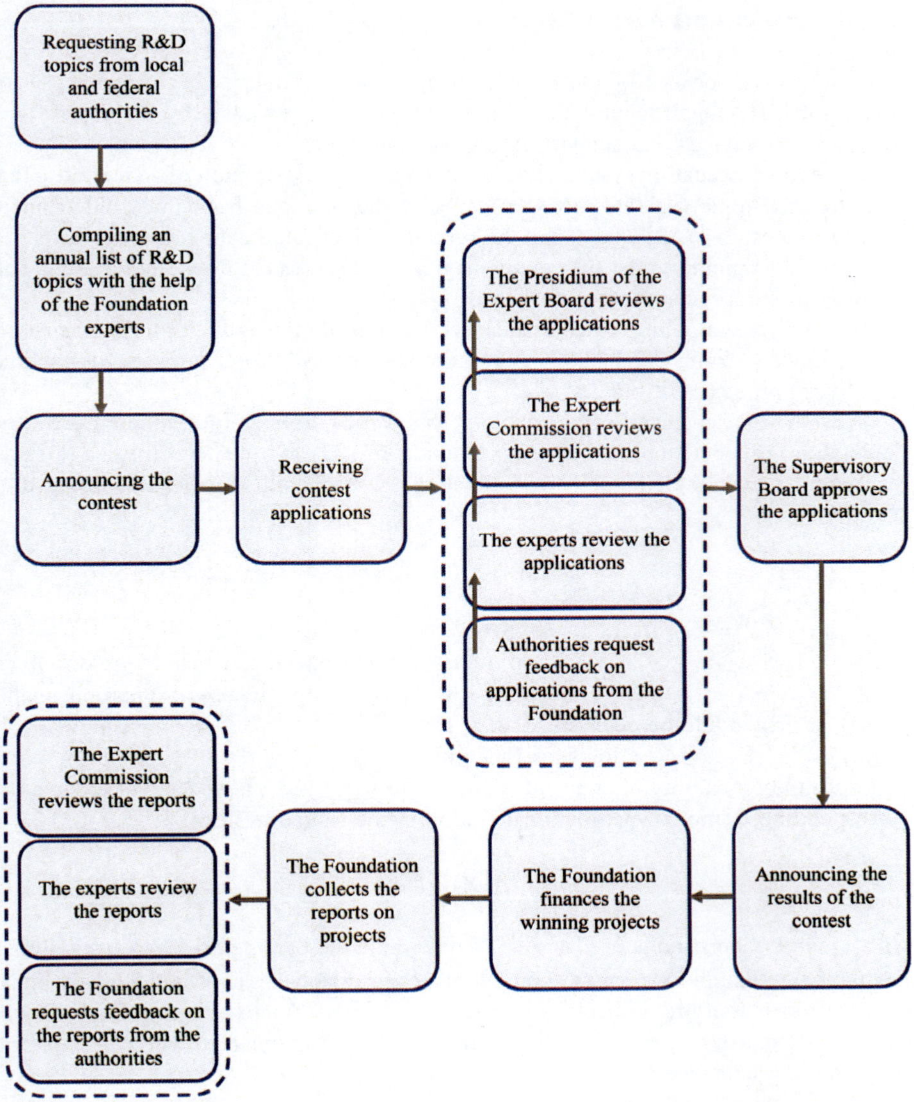

Fig. 1. Project selection model within the framework of the contest for applied scientific-technical, sociological, and humanities R&D projects, aimed at creating products and technologies for the competitiveness of the Krasnoyarsk Territory. *Source* Compiled by the authors

2 Materials and Methods

The initial data consists of the sample of applications submitted to the contests of Krasnoyarsk Regional Science Foundation, which were reviewed by the experts and were deemed advisable to support by regional authorities.

The expert evaluation system includes three blocks: the opinion of an authority that submitted the topic (values: "+" / "–" / "±"), the opinions from three independent experts (values: "+" / "–" / "±") and the sum of points from the three experts (values: 0 to 360). If the opinion of the authority is negative, all scores are multiplied by zero, and the project is not recommended for support.

Based on the sampling structure analysis, we must derive a rule for forming a short-list of projects, from which the Expert Commission will select projects for funding according to the contest budget.

We introduce a dependent variable y that varies from 0 (not supported by the Foundation) to 1 (supported by the Foundation). Now, we construct a logistic regression model (logit model) for estimating the probability of y being equal to 1 (y = 1).

$$\mathbb{P}\{y = 1|X\} = f(z),$$

where $z = A^T X = a_0 + a_1 x_1 + \ldots + a_N x_N$;

$X = (1, x_1, x_2, \ldots, x_N)$ – vector-columns of independent variables (regressors);

$A^T = (a_0, a_1, a_2, \ldots, a_N)$ – a vector-row of model parameters (regression coefficients), which should be determined; $N \in \mathbb{N}$;

$f(z) = 1/(1 + e^{-z})$ – a logistics function.

Since the probability of an event is $\mathbb{P}\{y = 0|x\} = 1 - f(z)$, we can write down the corresponding Bernoulli distribution function in the following form:

$$\mathbb{P}\{y|X\} = f\left(A^T X\right)^y \left(1 - f\left(A^T X\right)\right)^{1-y} \quad (1)$$

with a distribution parameter $f\left(A^T X\right)$. To determine the parameters of A^T, we will use the maximum likelihood method. Using logistic regression, we construct a binary linear classifier for classifying projects with a set of attributes X into a class, which we will call the short-list of projects. All algorithms were implemented via the statistical programming language R.

3 Results

To build a classifier, we examine a training sample of M examples – pairs set $\left\{\left(X^{(j)}, y^{(j)}\right)\right\}_{j=1}^{M}$, that describe the specific values of the independent variables $X^{(j)}$ and the corresponding values of the dependent variable $y^{(j)} \in \{0, 1\}$. We shall look at the following values:

- $x_1 \in \{0, 1, 2, 3\}$ – the number of experts who had an overall positive ("+") opinion on supporting the project;
- $x_2 \in \{0, 1, 2, 3\}$ – the number of experts who had a somewhat positive ("±") opinion on supporting the project;
- $x_3 \in \{0, 1\}, x_3 = 1$, if an authority approved the project and $x_3 = 0$ otherwise;
- $x_4 = x_1 + x_3, x_4 \in \{0, 1, \ldots, 4\}$;
- $x_5 \in \mathbb{R}$ – the standardized value of total points that the project received from the three experts. The standardized value is obtained by subtracting the average of points from their sum for the entire sample and dividing the resulting difference by the standard deviation of the sum for the entire sample.

In this study, we examined three different sets of regressors: $X_1 = \{x_1, x_2, x_5\}$, $X_2 = \{x_1, x_2, x_3, x_5\}$, $X_3 = \{x_2, x_4, x_5\}$. Then, the maximum likelihood method for finding the parameters of the logistic regression A^T involves maximizing the value of the likelihood function on a subset $\mathcal{P}$ of the training sample, which includes projects that an authority approved:

$$\widehat{A}^T = \underset{\mathrm{A^T}}{\operatorname{argmax}} \prod_{\left(X^{(j)}, y^{(j)}\right) \in \mathcal{P}} \mathbb{P}\left\{y = y^{(j)} | X = X^{(j)}\right\}. \tag{2}$$

After we find a set of parameters $\widehat{A}^T$ of the logistic regression model, the values of $f(A^T X)$ can be interpreted as the probability of the project making it to the short-list. To develop the classifier, we define the threshold value $\widehat{p}$ of the probability of the project reaching the short-list as the minimum value of the logistic function on the subset $\mathcal{P}$ of the training sample: $\widehat{p} = \min_{\left(X^{(j)}, y^{(j)}\right) \in \mathcal{P}} f(\widehat{A}^T X^{(j)})$.

The classifier takes the form of the following rule: *a project gets on the short-list if the value of its logistic function is not lower than the threshold value*:

$$f(A^T X) \geq \widehat{p}. \tag{3}$$

The classifier was trained on a sample of 49 projects corresponding to a single 2019 competition. For each set X_k, $k = 1, 2, 3$ of regressors we calculated threshold values $\widehat{p}_1 = 0.347$, $\widehat{p}_2 = 0.342$, $\widehat{p}_1 = 0.346$. The classifier was tested on a set of 120 projects from the 2019–2020 competitions. The classifier generates a small number of false positives – it put some projects that were not funded on the short-list. However, having some false positives is beneficial in this case since it creates a small reserve of projects that qualify for funding and require the attention of the expert committee. There also were some false negatives (5 projects, 4.2%) – the classifier failed to include some of the financed projects on the short-list. This, in turn, is unacceptable. The error is caused by the small training sample. As the data on contests increases, we should refine the logistic regression parameters according to the new samples and investigate the method of choosing the threshold value $\widehat{p}$.

The logistic model and classifier allowed us to specify the following threshold values of a project making it to the short-list:

- Expert opinions: "+ + −";
- Sum of expert points: 236;
- The opinion of the authority: "+" (Fig. 2).

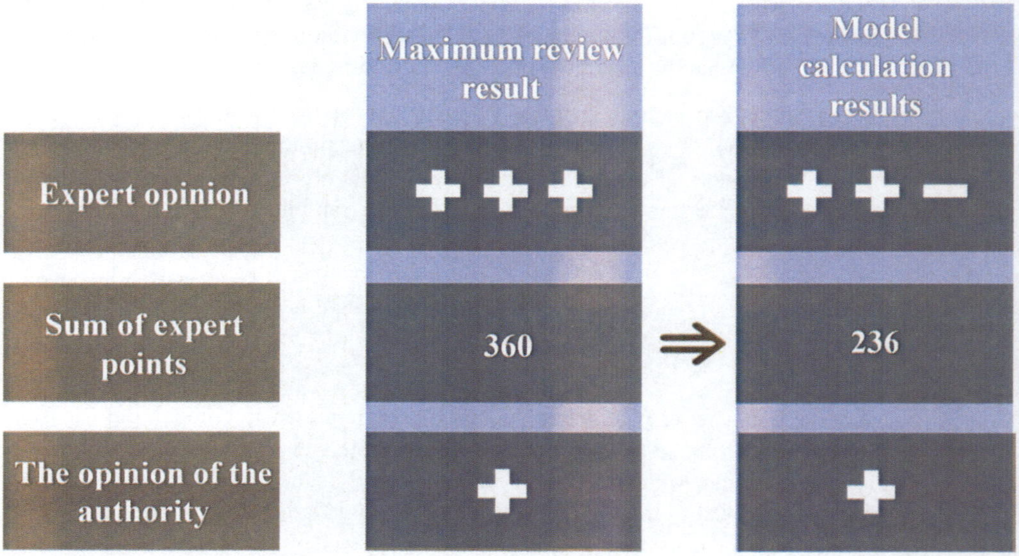

Fig. 2. Calculation results of the project selection model. *Source* Compiled by the authors

Having compared the obtained criteria for project selection with the actual review results, we received a list of 30 projects instead of 49. This substantially simplifies the expert analysis work and eliminates the influence of the human factor. The model will be further tested and improved.

4 Conclusion

Methods of OR and system analysis are widely used in various domains of life, especially in manufacturing, where the criteria of optimality are more pronounced. However, in R&D institutions, the methods of mathematical modeling are rarely used at the moment.

The developed model of choosing projects according to the results of expert reviews allows to objectively determine the winning project. This, in conjunction with a rigorous and independent review of the reports, will subsequently increase the efficiency of financing and supporting the most promising and productive research, ultimately improving the socio-economic situation in the region and the country.

Moreover, the developed model is universal and scalable, enabling its future application in other innovative regional institutions.

Acknowledgment. The reported study was funded by RFBR, project number 20-010-00482.

References

1. Klimenko, I.S.: Research of operations in complex protection information and communication objects. News of the Tula state university. Techn. Sci. **2**, 369–375 (2021)
2. Petrovsky, A.B.: Research projects expertise as a collective multicriteria choice. RFBR J. **3**(99), 34–43 (2018). https://doi.org/10.22204/2410-4639-2018-099-03-34-43
3. Petrovsky, A.B., Royzenson, G.V., Tikhonov, I.P., Balyshev, A.V.: Multiple criteria approach to R&D project efficiency. Herald of the National Technical University "KhPI". Ser. Inf. Sci. Modell. **43**, 138–148 (2009)
4. Razzhevaikin, V.N.: Mathematical model of scientific themes competition. Operat. Res. Mod. Syst. Solut. **6**, 23–33 (2020). https://doi.org/10.14357/ORMSS20200103
5. Stupina, A.A., Antamoshkina, O.I., Ruiga, I.R., Korpacheva, L.N., Kovzunova, E.S.: Building the strategy for innovative development of industrial enterprises based on network planning methods. IOP Conf. Ser. Mater. Sci. Eng. **1047**, 012039 (2021). https://doi.org/10.1088/1757-899X/1047/1/012039
6. Svetlov, N.M., Kazakevich, I.A.: Improving the methodology of state regulation of Belarusian agriculture. Izvestiya of Timiryazev Agric. Acad. **4**, 148–161 (2013)

Controlled Energy Transition: Economic War Among Global Stakeholders

Alexander G. Simonov(✉) and Sergey N. Lavrov

RUDN University, Moscow, Russia
simonov_ag@rudn.ru

Abstract. This article is dedicated to the mid-term and long-term prospects of the European gas market transformation. The energy market is observed as an integrated system of corporate and governmental interests. The forced energy transition is contemplated as a game-changing undertaking, causing landslide effects for all parties involved: households, companies, states. The contradictory goals of alleged environmental and economic efficiency are considered as factors of an inherent dilemma for the future of the EU gas market. The regulatory and political intervention in different industries (coal, nuclear, gas, etc.) due to cost and benefit trade-offs causes a joint reaction of the entire complex. A potential contradiction between declared goals and probable consequences is spotted.

Authors attempt to provide an insight scenario analysis of key economic alterations for different types of consumers and suppliers. An emphasis is put upon the roles of major players – Russian and US suppliers – and the transformation of market structure in general. The results of the study may be practically applied in terms of elaborating business strategies on the EU market within the span of 10–20 years.

Keywords: European gas market · Energy regulation · Hydrogen · Economic and environmental efficiency · Economic development · Chinese and European approach · Sanctions · Competitive strategy · Development goals · Carbon neutrality

JEL codes: F63 · F64 · O24 · Q47

1 Introduction

According to a classical approach, world energetics has lived through several transformations [11]. The first one took place in mid-XIX with massive industrialization (accompanied by worldwide development of steam engines) and switch from biomass to coal. The second one happened in early and mid-XX with an extensive spread of internal combustion engines and the increasing role of oil (the tendency was significantly catalyzed by WWII with many battles fought with armored vehicles and for oil fields). The third energy transition (utilization of natural gas) mainly started in the postwar period and is going on. Unlike biomass and coal, initial technologies of natural gas extraction and, especially, transmission and consumption required large-scale, usually cross-border infrastructure. The fourth energy transition is emerging and shall

E. G. Popkova (Ed.): ISC 2020, LNNS 368, pp. 88–97, 2022.
https://doi.org/10.1007/978-3-030-93244-2_11

be based on the proliferation of renewable energy sources, including "clean hydrogen", according to the vision of the EC (2020). The expansion of the fourth transition may significantly affect and transform the business of leading oil and gas companies [6].

The main problem lying in-between the third and the fourth transition is the way it will develop: evolutionary or revolutionary, governed by economic reasons or politically imposed. The share of natural gas in primary consumption is relatively high only in developed and/or extracting countries (see Fig. 1).

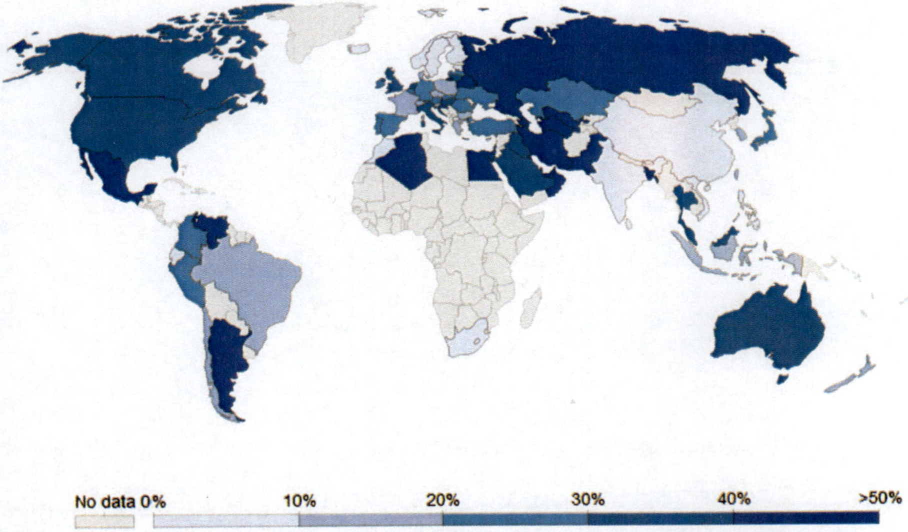

Fig. 1. Share of primary energy from natural gas, 2019. *Source* BP [13]

The authors hypothetically assume that the third energy transition in the European Union is not finished, on the contrary, it will accelerate; the declared "green" energy strategy is aimed at devaluing the US and Russian gas imports and creating ground for European business to participate in profit distribution.

2 Analytical Paradigm and Methodology

Although natural gas became energy carrier #2 (after oil) in total energy supply (TES) in the EU in 1996 (see Fig. 2), it has surpassed coal in power generation only in 2018, with nuclear power still holding leadership (see Fig. 3).

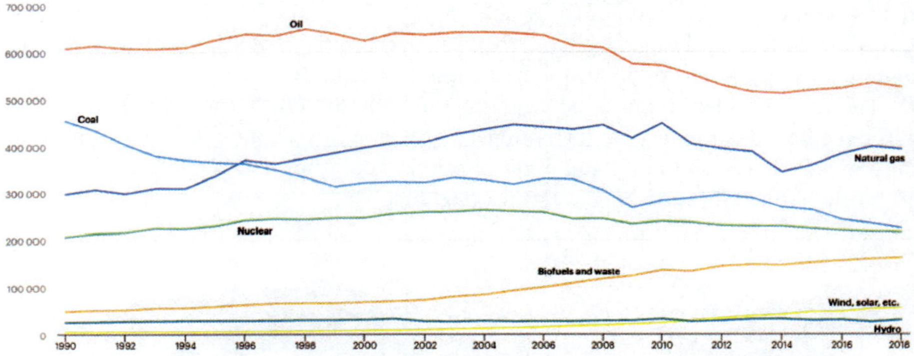

Fig. 2. Total energy supply (TES) by source, EU-28 (ktoe). *Source* IEA [19]

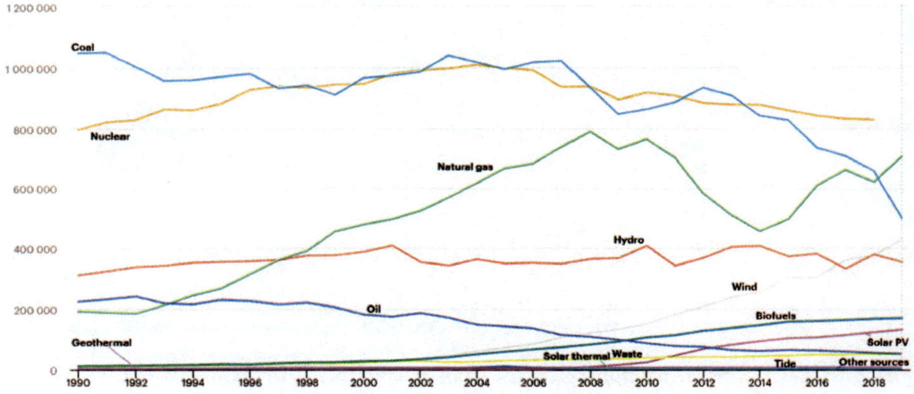

Fig. 3. Electricity generation by source, EU-28 1990–2019 (GWh). *Source* IEA [19]

In 2020, the European Commission has adopted a new hydrogen strategy [17], which may cause a landslide effect (in the long run) for all parties acting on the EU gas market. The claimed goal of renewable hydrogen supply is 1 million tons by 2025 and 10 million tons by 2030. These volumes correspond to an equivalent of 0,5 and 4.7 bln cubic meters of natural gas respectively.

However, implementing a new hydrogen strategy inevitably affects the existing energy system in terms of infrastructure, cost structure, long-term investment strategies of energy suppliers, and, finally, their customers.

To be sustainable European energy policy shall form a comprehensive approach towards other energy sources as it seems impossible to shift from all energy carriers to hydrogen at once. A sort of road map shall be developed, therefore providing priority prescriptions regarding sources being substituted as well as steps and timelines being observed.

Nuclear Energy. Despite several acts (national and European) being adopted the absolute value of energy produced by nuclear stations even slightly exceeds the level of 1990. Although deadlines were drawn for several power plants, it is clear that (a) there is no adequate direct substitution in terms of stable energy supply as almost all renewable are susceptible to high output volatility and (b) the enforcement of foreclosures will cause significant governmental expenditures due to compensation for respective costs and lost profit: e.g. Germany's Constitutional Court ruled Eon, RWE and Vattenfall, are entitled to "appropriate compensation" for the state decision to rush the shutdown of their reactors [14].

Oil. Oil (and oil products) is currently the main energy source for passenger and commercial transport. There are several technologies to overthrow it: LNG/CNG vehicles, electric vehicles (EV), and hydrogen power cells vehicles. Only gas vehicles have been massively introduced in transport activities and the provision of public services (e.g. Italian program of CNG transport).

Coal and Natural Gas. As for now, the toughest source competition takes place between coal and natural gas with market players being economically forced to switch from one to the other influenced by market price and cost of emissions. A phenomenon and a respective term of coal-to-gas switching (and vice versa) are widely acknowledged [16]. The substitution of coal with natural gas is also driven by technological development. Combined cycle plants have 60% heat efficiency compared to 40% of coal-fired ones ten years ago [20]; as for now, this gap has increased by another 2–3 p.p.

Renewables. Wind and solar power supplies currently account for 3.4% of TES. Rapid growth (in relative values) has been a result of the low base effect. Moreover, solar and wind energy cannot provide stable supplies due to natural reasons. The energy mix shall contain a stable fraction to cover volatile consumption.

So it may be concluded that ongoing technological development and increasing costs of emission will put further pressure upon coal generation and political decisions have already stalled further development of nuclear projects (even in case of turnback the industry will require decades to recover). Renewables cannot securely replace traditional energy carriers. In the midterm, only gas can provide energy sector redemption for the reasons described below.

The cost comparison (Fig. 4) indicates that natural gas is a frontrunner among fuels. Only wind plants being subsidized can challenge it. However, there is no substitution for natural gas as a stable energy source. Without a stable energy supply, the energy mix can fail to satisfy peaking energy demand, wind power is subjected to extreme volatility and therefore cannot establish an energy basis.

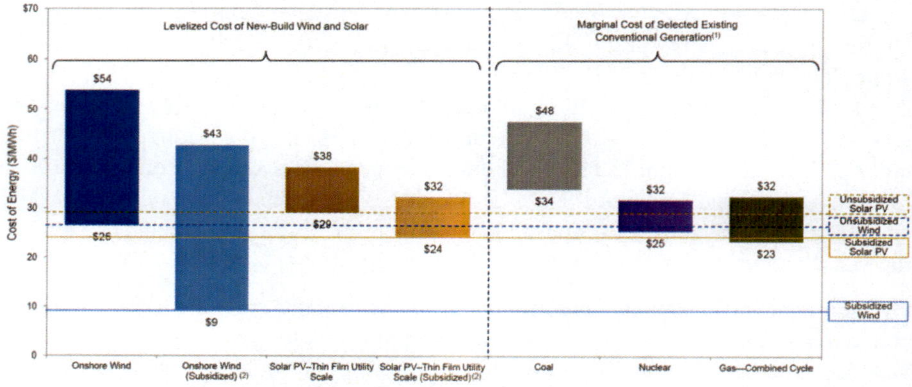

Fig. 4. Levelized Cost of Energy (LCOE), $/MWh. *Source* Lazard [5]

All in all the situation may be analyzed from the point of Schumpeter's creative destruction approach [9]. To stay competitive existing players/technologies/product shall upkeep their efficiency against new (and potential as well) entrants [8].

To push out previous leaders the candidates shall possess decisive competitive advantages. The substitution of coal looks rather economically and environmentally reasonable. The substitution of nuclear power seems to be less economically, but more politically and socially stipulated (although there are tools to assess risks of meltdown and costs of recovery).

3 Results and Conclusions

The declared hydrogen strategy looks rather premature from the points of energy security, economic and environmental efficiency. The prospects of its implementation for separate sectors (Fig. 5) shall be observed. The substitution of natural gas appears to be neither economically nor environmentally efficient.

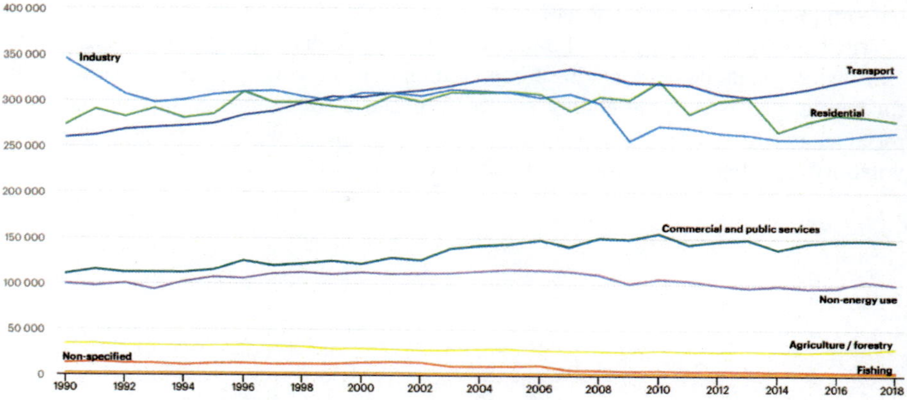

Fig. 5. Total final consumption by sector in EU-28, ktoe. *Source* IEA [19]

Transport Sector. Transport consumes 328 Mtoe in the EU, which is 29% of total final consumption (TFC). Despite several initiatives, most cars a fueled with oil products. There are many alternative technologies, which include CNG/LNG vehicles, EV, and hydrogen cells vehicles. The efficiency of hydrogen cell vehicles is twice lower than that of EV (instead of direct plug-in charge additional losses occur: electricity is further used for electrolysis, produced hydrogen requires compression, transportation, power is generated again from hydrogen) [7].

Residential Sector. Residential consumption (24% of TFC) is widely scattered. Large cities and small villages require stable power and heat supplies (especially in winter). The existing infrastructure is designed to provide network deliveries (electricity, gas); some households are still heated with coal or fuel oil. Producing hydrogen via electrolysis by households seems senseless (electricity may be used directly; besides, there is a need for additional security precautions). Shipping hydrogen to households is either not viable: it will require vast investments into new infrastructure, while the existing electricity grid may be used for power deliveries.

Industrial Sector. Probably, this sector is most suitable (compared to others) for hydrogen transition. To avoid expedient investment costs for infrastructure development it shall be reasonable to organize industrial clusters with local hydrogen production, which shall minimize distribution challenges. It is expected that new undertakings will be first in line to establish these clusters. Existing industrial sites with developed infrastructure and supply chains will probably take years to switch for new energy carriers if ever (subject to economic and regulatory conditions).

However, the EU hydrogen strategy contains a few inherent contradictions, which inter alia include definitions and perception of "clean" hydrogen technologies, the expected pace of transition, the volume of substitution, the level of environmental effects of new conventional technologies and hazard of hydrogen equipment, and, finally, the economic outcome for European industrial and household consumers as well as the unpredicted response of key suppliers.

A number of the most vivid drawbacks of European strategy shall be named. The following aspects of hydrogen energetics currently prevent massive transition and corresponding ban of fossil fuels:

1. Hydrogen is produced via the electrolysis of water or dissociation of hydrocarbons. The former requires energy to be first produced, the latter is still dependant on hydrocarbons.
2. Electrolysis hydrogen is susceptible to power volatility. Neither renewable (i.e. solar and wind) can currently guarantee stable supplies.
3. Electrolysis and subsequent combustion of hydrogen incur a 50% loss of efficiency (direct electricity consumption seems more prudent).
4. Hydrogen has the lowest density, it can penetrate and leak from metal, plastic, and glass pipes and tanks. Moreover, hydrogen causes the embrittlement of high-pressure steel pipelines [3].
5. Hydrogen is extremely explosible in a wide concentration range.
6. Despite very high calorific value by weight, due to low-density hydrogen has very low calorific value by volume (ca. 2,7 less than natural gas).

7. Low density, hence, low liquefaction temperature, causes high energy losses in the case of liquefying.

Another approach to the declared hydrogen strategy may be established from the point of cost and benefit analysis in terms of national competitiveness. The forced transition substantially transforms supply, hence, value chains. European endeavors may actually target two goals other than declared: first, imposing additional regulatory requirements on energy supplies may be aimed at diminishing the value of the imported energy commodities (primarily from the USA and Russia); second, due to significant hydrogen transportation difficulties the production is to be mainly based in the EU, causing a necessity of additional level in the supply chain – electrolyzer facilities, which are expected to be produced by European companies [12]. The latter reason is crucial as a multi-billion domestic industry with millions of new jobs may be created.

The process of reforming the EU gas market influences the global competitiveness of other international participants, including Russian and US companies, as well as host countries [4]. Thus a range of corresponding counteractions shall be contemplated. For instance, US companies practically do not participate in infrastructural projects in the European Union directly, their main business approach is delivering LNG to regas terminals; moreover, in the case of higher gas prices on the Asian markets, the tankers are headed for higher profit margins [10]. European orientation towards spot market development has also stalled many infrastructure projects, developed by Russian investors, which is conditioned by unbundling regulation, third-party access provisions as well as long-term uncertainty. According to International Energy Agency [18], unlike Qatar and Australia, the volumes of Russian and US supplies will increase dramatically for the next 5 years (Fig. 6). The final destination of these shipments will be stipulated to regional market prices and relevant costs.

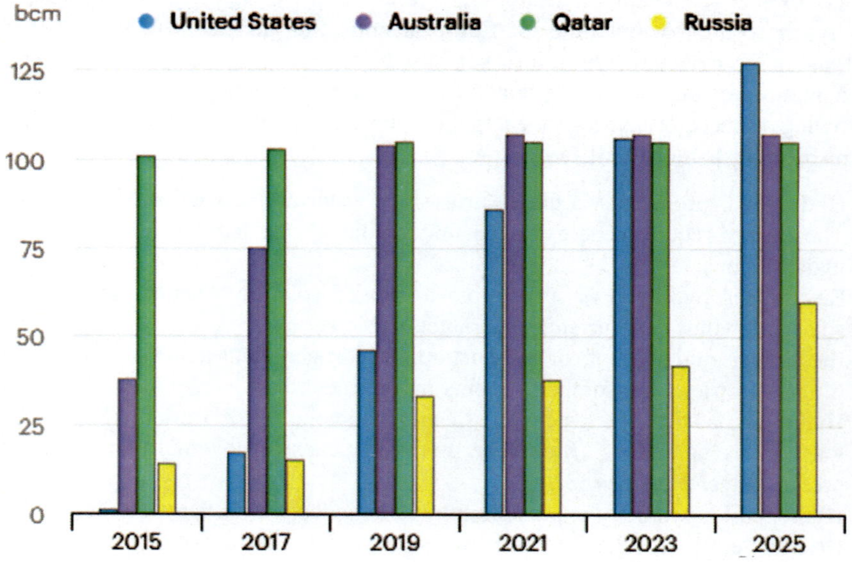

Fig. 6. LNG exports for selected countries (bcm). *Source* IEA [19]

However, despite heavy dependency on energy imports, the EU may regain a strong stand via declared environmental agenda and create its own model of energetics, based upon regulatory and market mechanisms of promoting "clean" energy. Granting hydrogen energy supplies a top dispatch priority accompanied by imposing additional taxes and tariffs on fossil fuels inevitably turns natural gas into a raw product and extends the supply and value chain. This approach may cause a significant redistribution of profit margin in case of the low market power of importers. However, the number of suppliers is rather limited and the outcome of such an approach will significantly depend upon their strategy on their handling of the new rules. Anyway, the situation will not be stable as other regional markets exist and there is a steady trend of increasing LNG production and transportation capacities by major producers. Alternatives of the EU gas market will be monitored.

Another way of developing and handling the energy (in particular gas) market and relevant infrastructure projects can be observed in China [1]. Unlike the European trend towards decentralization and increasing uncertainty, the Chinese midstream is being centralized and uniformly regulated. The latter is secured via integration of main pipelines within a national provider (PipeChina), obliged to grant equal access to third parties. Chinese approach can potentially minimize transaction costs of doing energy business in China and curb long-term uncertainty for suppliers. The establishment of a national transportation provider may be regarded as an alternative to the European Third Energy Package, causing far fewer interruption threats. The forecast by IEA has already predicted the rise of the Chinese and Indian LNG market, while Europe is losing its positions (Fig. 7).

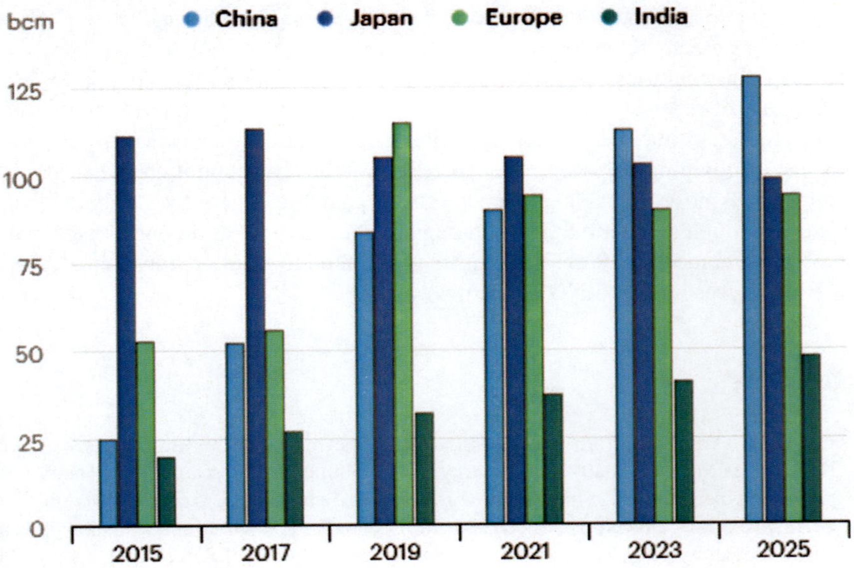

Fig. 7. LNG imports for selected countries (bcm). *Source* IEA [19]

The dynamic standoff among global gas suppliers (primarily the USA and Russia) and key regional markets (European and Asian) will shape the future of the world energy economy. The ongoing third energy transition boosts regional gas markets, providing for increasing interdependency of regional ones, which may ultimately form a global market. The increasing role of natural gas in energy mixes of the largest economies (USA, Russia, China, the EU, etc.) will lead to emerging new competition models aimed at enhancing the efficiency of national markets and regulatory systems. Several features of the new global economy have already appeared. Due to the large scope and economies of scale in oil and gas business, the number of companies is relatively low, many of them form a significant pillar for national economies, both in terms of budget incomes and sustainable economic growth and development in general (hundred thousands of well-paid jobs, raw materials, and energy procurement for other industries, technological development, macroeconomic stability, etc.). This fact inevitably causes a solid business and state relationship, resulting in escalation of business rivalry into intergovernmental rivalry and even economic war.

The European gas market has been mostly buyer's market for this decade which was the outcome of natural reasons: several acting and potential suppliers, while market regulation being enforced by European institutions. Moreover, the sellers have been actively competing (up to state involvement and sanctions exchange).

The rise of China and its stake on clean energy (including primarily coal-to-gas switch in power generation) transforms the world gas market into a global oligopsony-oligopoly with significant states' intervention. Therefore some scenarios may be predicted and tackled. The ongoing sanctions (the USA-Russia, the EU-Russia, the USA-China) may significantly fragment the world gas market into several parts: the European one will be limited for gas imports from Russia (in case of banning Nord Stream 2 and deterioration of Ukraine gas system), which will cause high energy prices and weakening national competitiveness due to cost increase for other industries. The stake on promoting green technologies may help only in case of global acceptance, in this case, European countries can make up for the lack of domestic fuels deficit by leading development of renewable technologies and global success of the relevant enterprises. In China situation may be just the opposite: Russian gas pipelines are to be built at a rapid pace, providing for the cost leadership of Russian supplies against US LNG. However, the most probable prospects encompass the emerging of the global gas market with tough competition both among suppliers as well as among consumers. The market will embrace long-term deals (primarily linked to pipeline flows) as well as spot market deals (secured by LNG capacities).

References

1. Downs, E., Yan, S.: Reform is in the Pipelines: PipeChina and the Restructuring of China's Natural Gas Market. (2020). www.energypolicy.columbia.edu/research/commentary/reform-pipelines-pipechina-and-restructuring-china-s-natural-gas-market. Accessed 10 Apr 2021
2. Granovskii, M., Dincer, I., Rosen, M.A.: Economic and environmental comparison of conventional, hybrid, electric and hydrogen fuel cell vehicles. J. Power Sour. **159**(2), 1186–1193 (2006). https://doi.org/10.1016/j.jpowsour.2005.11.086

3. Hafsi, Z., Mishra, M., Elaoud, S.: Hydrogen embrittlement of steel pipelines during transients. Proc. Struct. Integ. **13**, 210–217 (2018). https://doi.org/10.1016/j.prostr.2018.12.035
4. Karpova, N.S., Lavrov S.N., Simonov A.G.: Mezhdunarodnye Gazovye Proekty Rossii: Evropeĭskiĭ Al′ians I Strategicheskie Al′ternativy [International Gas Projects of Russia: European Alliance and Strategic Alternatives]. Moscow, TEIS (2014). ISBN 9785721813504
5. Lazard: Levelized Cost of Energy and Levelized Cost of Storage – 2020 (2020). www.lazard.com/perspective/levelized-cost-of-energy-and-levelized-cost-of-storage-2020/
6. Mastepanov, A.M.: Energy transition: what should the oil and gas world get ready for. Prob. Econ. Manag. Oil Gas Comp. **10**, 5–14 (2019)
7. Morris, J.: Why Hydrogen Will Never Be the Future of Electric Cars. Forbes (2020). www.forbes.com/sites/jamesmorris/2020/07/04/why-hydrogen-will-never-be-the-future-of-electric-cars/?sh=693ccc6312fa
8. Reinert, E.S., Reinert, H.: Creative destruction in economics. New Nietzsche Stud. **9**(3), 1–23 (2015)
9. Schumpeter, J.A.: [1942]. Capitalism, Socialism and Democracy. London, Routledge (1994)
10. Sheldrick, A.: U.S. LNG Export Bonanza Reshapes Energy Map in Asia. Reuters (2021). www.reuters.com/business/energy/us-lng-export-bonanza-reshapes-energy-map-asia-2021-04-02/. Accessed 10 Apr 2021
11. Labussière, O., Nadaï, A. (eds.): Energy transitions. ECE, Springer, Cham (2018). https://doi.org/10.1007/978-3-319-77025-3
12. Yodwong, B., Guilbert, D., Phattanasak, M., Kaewmanee, W., Hinaje, M., Vitale, G.: AC-DC converters for electrolyzer applications: state of the art and future challenges. Electronics **9**(6), 912 (2020). https://doi.org/10.3390/electronics9060912
13. BP. Statistical Review of World Energy, 2020 (2021). https://www.bp.com/content/dam/bp/business-sites/en/global/corporate/pdfs/energy-economics/statistical-review/bp-stats-review-2020-full-report.pdf
14. Deutsche, W.: German Utilities Win Compensation for Nuclear Phaseout. (2016). www.dw.com/en/german-utilities-win-compensation-for-nuclear-phaseout/a-36639314. Accessed 10 Apr 2021
15. European Commission, 2018: A Clean Planet for All. A European strategic long-term vision for a prosperous, modern, competitive, and climate neutral economy, COM 773 (2018)
16. European Commission, 2019: Market Reports Show Further Drop in Coal-Fired Power Plants in Third Quarter (2019). ec.europa.eu/info/news/market-reports-show-further-drop-coal-fired-power-plants-third-quarter-2019-dec-23_en. Accessed 10 Apr 2021
17. European Commission: Communication from the Commission to the European Parliament, the Council, the European Economic and Social Committee and the Committee of the Regions. A hydrogen strategy for a climate-neutral Europe (2020). COM/2020/301 final
18. International Energy Agency: Gas 2020. Analysing the Impact of the Covid-19 Pandemic on Global Natural Gas Markets (2020). https://webstore.iea.org/download/direct/3005. Accessed 10 Apr 2021
19. International Energy Agency: Data & Statistics (2021) www.iea.org/data-and-statistics/data-browser. Accessed 27 June 2021
20. National Petroleum Council US: Facing the Hard Truths about Energy – a Comprehensive View to 2030 of Global Oil and Natural Gas (2007)

Antitrust Regulation in the Context of Digitalization

Nizami V. Abdullayev(✉) and Andrey A. Antonov

Vladimir State University named after Alexander and Nikolay Stoletovs, Vladimir, Russia

Abstract. The purpose of the chapter is to study the changes in antitrust regulation caused by the digitalization of the economy. The authors draw attention to the fact that for digital markets, the usual indicators of monopoly power, which could be used by a state market monitoring body, do not work. Indicators such as innovative technologies, intellectual property, and Big data are now coming to the fore.

The authors note that in the context of the digitalization of the economic space and the development of markets for digital services, regulations should be revised and clarified, in accordance with which markets are monitored and a policy to protect competition is implemented.

Keywords: Monopoly · Cartel · Digital platform · Network effect · Antitrust regulation

JEL Codes: E65 · K21

1 Introduction

The introduction of digital technologies into economic processes has led to a change in many areas of human activity. We are witnessing the emergence of innovative technologies that affect all spheres of human society.

Predicting the consequences is one of the most important tasks at this stage of the development of a new industry. It is important to understand what society may face to successfully bring about a technological revolution. The emerging phenomenon has become a frequently discussed topic. There is an active accumulation of scientific material, thanks to which each person can be involved in the process of changing society.

In 2017, a program for the development and implementation of digital technologies was developed and approved. The digital economy began to cover more and more spheres of public life. In turn, the introduction of innovative technologies and the digitalization of society entail improvements that begin with a simplification of a person's daily life and end with an increase in the quality of life of the population. However, digital technologies also contribute to the emergence of new types of problems. These problems include, first, the emergence of fundamentally new factors of market monopolization.

E. G. Popkova (Ed.): ISC 2020, LNNS 368, pp. 98–102, 2022.
https://doi.org/10.1007/978-3-030-93244-2_12

2 Materials and Method

There are many definitions of monopoly in the economic literature. They reflect a variety of views and methodological approaches to the analysis of various aspects of market monopolization.

Soviet-era dictionaries either equate a monopoly with a large company with a significant market share or view it as the owner of a unique license to manufacture or sell a product or to exclusively use something. In modern conditions, these criteria for monopolization are insufficient.

McMillan's Dictionary defines a monopoly as a firm that is the only supplier of a homogeneous product that has no substitutes but has multiple buyers [5–10]. At the same time, the importance of such a formal feature of monopoly as the inelasticity of market demand, which coincides with the value of demand for the firm, is emphasized. Similar definitions are presented in textbooks on economics.

In the Economic Encyclopedia, edited by Abalkin's monopoly is defined as the exclusive right of any person to produce, purchase, or trade-in goods, as an enterprise or group of enterprises that occupy a dominant position in an industry, national or world market, as well as a form of the market controlled by one or more enterprises [3].

Monopoly expresses a certain type of economic relations, relations of the interaction of an economic entity with the economic space. However, the monopoly is not necessarily a large firm, although it is more likely to take advantage of monopoly advantages. The monopoly position can also be caused by the narrowness of the market, or by the fact of a closed economic space. The reasons for the occurrence of such situations are quite diverse. This is the factor of the territory, the availability of licenses for a certain type of activity, technological innovations, the scale of production, etc.

Thus, the main property of a monopoly is the presence of monopoly power, which is realized as the possibility of using control over the parameters of the economic space for the realization of one's economic interests. That is, the bearer of a monopoly is a certain economic entity - a company pursuing its economic interest. It is implemented for the long-term goals of the company. Ultimately, the firm's economic interest is driven by the desire to reduce business risks, profitability, and growth.

The components of the external environment of the firm are its various counterparties - suppliers of various resources, intermediaries, end users, as well as competitors, the local community, government agencies. The specificity of monopoly relations with the environment is since an economic entity has economic power, which allows it to establish control over key factors of the external environment in its economic interests.

Thus, within the framework of our research, we will understand monopolies as an economic entity with monopoly power, which allows them to control the parameters of economic space to realize their economic interests. It should be noted that the digitalization of the economic space significantly changes both the monopoly power itself and the degree of its impact on market entities.

3 Results

The provision of digital services and the sale of digital products has led to the fact that transaction costs tend to zero, and the boundaries of markets are gradually disappearing. Digital services can be obtained instantly anywhere in the world, it is enough to have access to the INTERNET. According to the Ministry of Digital Development of Russia, the number of fixed broadband Internet subscribers in the Republic of Ingushetia is 1.51 per 100 people, in the Republic of Dagestan - 3.18 and the Chechen Republic - 4.78 in comparison with the indicator of 37.48 of the leaders in indicator - Novosibirsk region [8].

In the context of the active use of digital technologies, the tools for exercising monopoly power are changing. These are innovative technologies, intellectual property, and big data [11]. Therefore, to implement an effective antitrust policy, it is important to introduce into the legislation norms that prevent monopoly pressure on digital markets.

First, the rules that do not allow the use of cartel agreements when trading on electronic platforms. So, according to the FAS Russia for 2019, 87% of the initiated cases of anti-competitive agreements (320 cases) consisted of bid-rigging (cartels and collusion with the customer (organizer) of the bidding) [4].

Auction robots are used to conduct electronic trading. Their legal use allows significant savings in the efforts and time of bidders when determining and automatically offering contract prices.

However, auction robots, when used illegally, can facilitate cartel agreements. For example, suppliers of medical supplies for public hospitals Valiria LLC and Egamed LLC used auction robots built into the interface of the electronic trading platform of Sberbank-AST CJSC and configured to maintain prices at the auctions. The cartel members pre-configured the auction robots so that they had the same price reduction limit, as well as the same price proposals submitted with a difference of 0.003 s. In 2018, the FAS Russia found the companies guilty of entering a cartel at an auction aimed at maintaining prices.

Secondly, market participants can use new software products that allow them to adapt their business strategies to market changes in a short time. On the one hand, the use of special online platforms allows the company to operate more efficiently in the market and literally anticipate the actions of competitors. On the other hand, the availability of effective tools for collecting market information even from completely open sources allows you to quickly adapt your prices to the prices of competitors, which, is a copy of a cartel agreement and gives a similar market effect - the establishment and maintenance of monopoly high or monopoly low prices.

The use of price algorithms is not illegal until it leads to a violation of the Federal Law "On Protection of Competition" No. 135-FZ of July 26, 2006. This happens when the manufacturer, based on the information received, has the opportunity to dictate retail prices to resellers for certain types of products or certain models of complex technical products.

Third, in the context of the digitalization of the economy, network effects have a significant impact on markets. Large network firms that own databases can have a significant impact not only on markets and prices but also on consumers of products, shaping their taste preferences and creating the effect of "digital slavery". Having complete information about consumer preferences, their income, marital status, etc. allows you to keep consumers in a certain market by creating anti-competitive barriers.

The Bayer/Monsanto merger was an example where the FAS was forced to consider new market power factors. This transaction led not only to a change in the structure of the seed market but also to the formation of a new oligopoly market for agricultural technologies.

The agricultural technology market is a socially significant market that determines the food security of the state, is global, characterized by a radical transformation of the agro-industrial complex (high-tech solutions protected by patents, IT technologies, "big data" are involved in the transaction) [1]. Both companies owned new information technologies, innovations, digital platforms, algorithms, and technologies that allowed them to influence market conditions, create barriers to market entry for other players and dictate conditions for the further development of the agro-industrial complex [11].

Fourth, the use of digital technologies can lead to the creation of "network effects" in that the bargaining power of the parties to a business combination significantly enhances their bargaining power. Network effects mean the dependence of the consumer value of a product on the number of users of the same group (direct network effect) or a change in the value of a product for one group of users with a decrease or increase in the number of users in another group (indirect network effect) However, digitalization has given the development of platforms a new round and markets with network effects are no longer a special case. An example is a deal to merge the main taxi aggregators in the Russian market Yandex/Uber [2].

Digital markets have become an essential part of economic life and are growing in importance. At the same time, the influence of industry leaders is growing for example, in 2020, the top 5 most expensive brands according to Forbes belonged to digital giants. At the same time, as the practice of antitrust cases has shown, these corporations sometimes use their influence to hinder the activities of competitors. The European Commissioner for Competition in 2016 in the case against Google noted: "Google has developed many innovative products that have changed lives. But that doesn't give Google the right to deny other companies competition and innovation." Thus, antitrust regulators are faced with the task of preventing the restriction of competition in digital markets without prejudice to innovation processes.

4 Conclusion

Thus, digital technologies are significantly changing the approaches to antitrust regulation and the activities of antitrust authorities.

First, conventional approaches are not always applicable to the analysis of digital markets; therefore, the antitrust regulator faces the task of updating regulations and research methods carried out as part of antitrust investigations. Especially important. To ensure that the activities of the antimonopoly body do not worsen the investment climate and create additional institutional risks for both national investors and large world players.

Traditional criteria for defining market boundaries may be untenable for markets with digital platforms: the interchangeability of services is ambiguous, and it is not possible to conduct a "hypothetical monopolist test" for free services.

Second, for digital markets that are very flexible and fluid, using only quantitative monopolization criteria (such as market share) does not always provide an objective assessment of the degree of market monopolization. A high share in digital markets does not always mean market power. Therefore, antitrust authorities should pay increased attention to the company's ability to dictate its conditions to the market.

Third, there can be positive and negative network effects in digital markets. Their presence or absence can significantly affect the conclusion about the dominance of the company, which means that Order No. 220 requires revision and the inclusion of recommendations for assessing network effects and the degree of their influence on the market.

Fourth, the issue of the antitrust authority's access to the big data of digital giants remains problematic. In particular, the possibility of applying a special antitrust regime to the leaders of the digital industry aimed at limiting their market power is being considered in European countries. Particular attention should be paid to studying the positive experience of antitrust regulators in the control of mergers and acquisitions.

References

1. Consideration of the Bayer-Monsanto transaction revealed some problems, the solution of which requires a change in the existing legislation (2018). https://fas.gov.ru/news/24965
2. The decision of the FAS Russia in case No. AG/82029/17 (2017)
3. Economic encyclopedia. In: L.I. Abalkin (ch. ed.). Moscow, 444 (1999)
4. Federal Antimonopoly Service: Report on the state of competition in the Russian Federation for 2019 (2020). https://fas.gov.ru/documents/687048
5. Frank, R.H. (1991). Microeconomics and Behavior. New York, 370–371
6. Hayman, D.N.: Modern Microeconomics: Analysis and Application, vol. 2, 3 (1992)
7. Laidler, D., Estrin, S.: Introduction to Microeconomics. 3rd edn. New York, 224–225 (1989)
8. Mankiw, N.G.: Principles of Economics. Saint Petersburg, 317 (1999)
9. McConnell, K.R., Bru, S.L.: Economics **2**, 92 (1992)
10. Samuelson, P.: Economics **2**(90), 92 (1992)
11. Tarkhova, K.V., Alifirov, V.I., Gorokhova, O.N.: Evolution of Antitrust Regulation in Russia in the Digital Age (2020). https://doi.org/10.38044/2686-9136-2020-1-4-38-55
12. The Ministry of Digital Development of Russia. https://digital.gov.ru/ru/pages/statistika-otrasli/#section-674

Policy Review of Global Gas Market Players: A Combined State and MNE Approach Towards New Balance of Efficiency and Security

Alexander G. Simonov(✉), Yulia A. Konovalova, Arina A. Tinkova, and Nataliya V. Dyuzheva

Peoples' Friendship University of Russia (RUDN University), Moscow, Russia
{simonov_ag, tinkova-aa}@rudn.ru,
konovalova_yua@pfur.ru

Abstract. The article is dedicated to the analysis and prediction of mid-term trends in the globalizing natural gas market. A special emphasis is put upon the problem of efficient energy procurement in terms of national economies and international affairs. The problem is allegedly created by emerging global trends as a result of contradicting interests of key market players. The issue becomes even more topical as erosion of international rules takes place (sanctions, WTO institutional crisis, etc.). The authors suggest observing corporate and government interests as an interdependent economic complex, not as separate issues. Such an approach provides for a broader view of international strategic developments and for internalizing some variables, which have been external for many commodity market researchers beforehand. For example, this approach enables us to comprehend state policy towards green energy proliferation and carbon footprint downsizing not as a declared transcendental value, but as an economic tool for enhancing national competitiveness and competitiveness of domestic energy and manufacturing industries.

Keywords: Regional gas markets · Energy regulation · Hydrogen · Energy security · Economic development · National competitiveness · Development goals · Carbon neutrality · Fuels competition

JEL Codes: F63 · F64 · O24 · Q47

1 Introduction

Due to the ongoing global third energy transition, the role and value of natural gas are to be increasing further on, while a full-scale switch to renewable sources is not to be expected until at least 2050 even in OECD countries [8]. Increasing dependency of global energy upon natural gas, caused by the overall growth of energy demand accompanied by phase-out of nuclear generation and decommissioning of coal plants, creates new challenges and opportunities for international companies and policymakers.

E. G. Popkova (Ed.): ISC 2020, LNNS 368, pp. 103–113, 2022.
https://doi.org/10.1007/978-3-030-93244-2_13

As far as the power supply is a business with economies of scale only a few major players act regionally/globally, primarily in a B2B segment.

Most countries with a market economy have a multi-level structure of delivering energy to final consumers. On average it includes three stages: energy source production (or imports)/power generation, power distribution/retail, in some cases these levels may be merged or split. The differences in energy procurement structure among countries and the varying market power of producers, intermediaries, and final consumers create the basis for a wide variety of national and regional terms and conditions of energy markets. It has been observed that even in countries with heavy dependency on gas imports the margin of distributors (usually domestic companies) may be equal to or sometimes exceed the importer's price. For instance in Germany (Fig. 1) a sharp decrease in import prices (purchase price of suppliers) was accompanied by almost even price for private households [2]. Therefore exploring the market structure and existing system of supply contracts is inevitable on the way to a complex comprehension of the natural gas business.

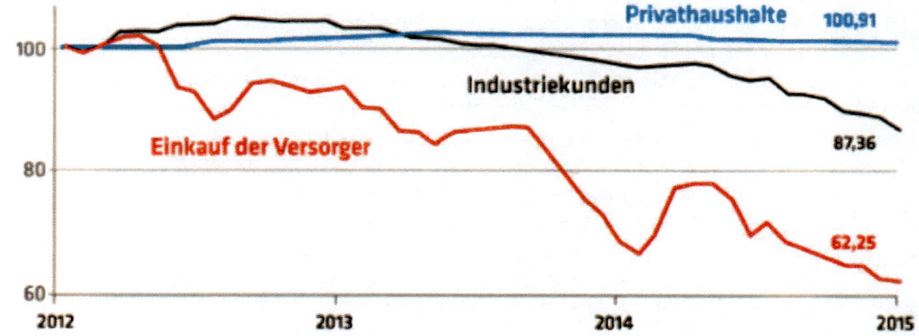

Fig. 1. Gas price development in Germany (January 2012 – 100%): households (blue), industrial customers (black), suppliers' purchase price (red). *Source* Bundestag Economic Service, Germany [2]

Another crucial factor, affecting the market structure and, hence, profit distribution, is local legislation and its enforcement. National (supranational in case of the EU) regulation may pose a significant challenge for one company and provide opportunities for another. Due to a significant link between national companies' wellbeing and state macroeconomic stability (energy security, jobs, the balance of payment, added value, etc.), many governments may experience a conflict of interests with economic reasons prevailing upon declared values and legislative framework.

A failure (de jure or de facto) of many international institutions governing or facilitating international activities on state and corporate levels has resulted in increasing state and business interaction and junction [7].

The integral and overlapping national interests and corporate goals in combination with globalizing energy (gas) markets resulted in blurring lines between market competition and geopolitical rivalry with large energy MNEs being both key economic players and conductors of national policies.

The complex interdependence between national policy framework and leading domestic enterprises makes it almost senseless to explore separate markets or single political or legislative steps. A complex and far broader view on national and global competitiveness is required. Therefore establishing a comprehensive system of indicators and factors may give additional impulse to the theoretical and applied development of natural gas economics.

To perceive and explore national policy frameworks and business strategies of different MNEs it can be worth investigating a reliable indicator of real state intentions – governmental expenditure. State budgets for hydrogen R&D a presented below (Fig. 2). Hydrogen is declared as a new energy carrier with zero greenhouse gas emissions. In limited appliances hydrogen may be a substitute for natural gas. It is worth mentioning that both fuels are gaseous under normal conditions and some technologies enable to receive H_2 from water with the help of electric power (electrolysis), other technologies provide for gaining hydrogen from hydrocarbons (pyrolysis of natural gas gives the highest specific hydrogen output). To some extent, hydrogen and natural gas may be considered substitutes.

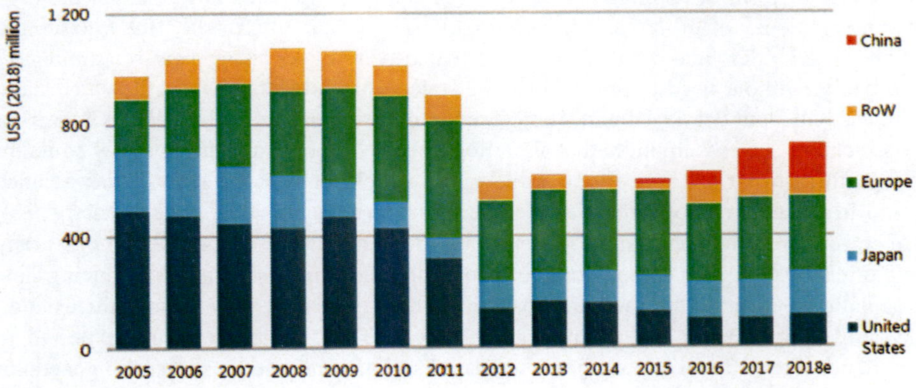

Fig. 2. Government RD&D budgets for hydrogen and fuel cells. *Source* IEA [6]

Government researches of hydrogen-related technologies are located mainly in four origins – the US, Japan, Europe, and recently China. The rest of the world spends less than 10% of the total budget. It is highly important to highlight several landslide changes over the last two decades, which reflects underlying intentions of global players:

1. The share (and absolute value) of the USA had been steadily decreasing from 60% to 40% in 2005–2011. In 2012 US hydrogen research budget plummeted (cut in half). Currently, the United States spend less than Europe, Japan, and even China.
2. Japanese budget dynamics is opposite to the US one: there was a stiff decrease in 2005–2011, but there has been steady growth since 2012; currently, Japan is a No. 2 (after Europe) by state investments into hydrogen associated researches.

3. Despite some fluctuations, Europe has the most stable approach towards the hydrogen topic. Increasing state budgets has resulted in becoming the world's leader (ousting the US and Japan).
4. China, being a freshman in the hydrogen field, puts great stress upon this issue: the country invested more than the USA after only two years of observed (by IEA) engagement.
5. The final, most important for the current investigation, the conclusion is that the largest importers of natural gas engage in the development of hydrogen technologies: the USA switch to exporting natural gas was accompanied by cutting budgets for hydrogen R&D.

1.1 Analytical Paradigm and Methodology

But why approach towards green energy in general and hydrogen, in particular, is so important? According to arising global agenda of minimizing carbon footprint and promoting renewable traditional energy, industries are to face a severe challenge. Production, distribution, and utilization of fossil fuels may bump upon significant obstacles in form of additional taxes and regulatory barriers. This will affect the economic efficiency of oil & gas producing nations and their enterprises. But who are the main stakeholders and what may be the real drivers of lobbying new requirements? Probably, without need to prove, it can be stated, that subjects, levying new taxes and duties, will doubtlessly benefit, therefore. A more complicated hypothesis requiring justification is an assumption, that alteration of cost system can result in a new structure of profit distribution, hence, new build-up of supply chains and overall risk-balance ratio in the majority of industries (as energy underlies almost all business activities). According to some conclusions of the Coase theorem changes in transaction costs may significantly influence the outcome of economic allocation [4]. Thus introducing new regulations, affecting transactional costs in the energy market, may reshape the existing system of the international natural gas business. Thus governments are expected to influence international approaches towards natural gas and energy sources in general to achieve the best conditions (from the individual point).

Revealing potential impacts on natural gas market rules and respective reactions of key players may establish a ground to building a variety of most probable scenarios of further market developments.

Regarding prospects of the natural gas market in general, it shall be noted that as a matter of fact necessity to hold at least one stable source in the energy mix for fixing peaks and fluctuations in combination with preclosure of nuclear plants and decommissioning of coal generation leaves only natural gas as a keystone energy source [8]. The economic value of natural gas can be demonstrated by the share of primary energy supplied on its basis: the highest levels of primary energy, obtained from natural gas are reached in two types of countries, which are either OECD countries or gas extracting ones.

The balance of demand and supply will further alternate, leading to new shocks for regional markets. Unlike previous decades the most rapid and extensive growth of natural gas consumption will be located in large Asian economies (primarily China in the future decade, later accompanied by India). The demand lost in 2020 will be probably recovered already this year (Fig. 3).

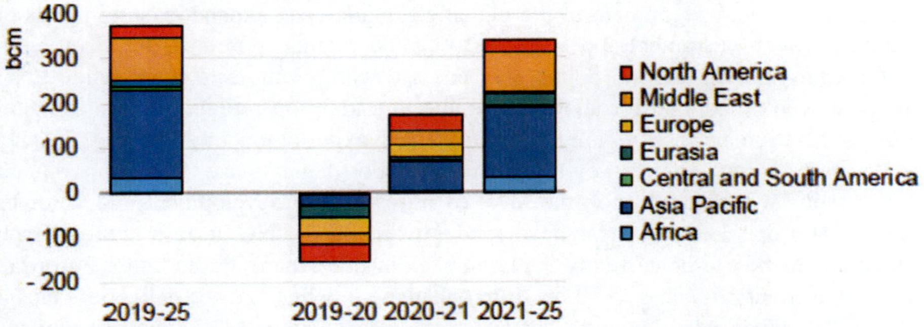

Fig. 3. Regional breakdown of demand growth until 2025, bcm. *Source* IEA, URL: https://www.iea.org/reports/gas-2020/2021-2025-rebound-and-beyond

The supply side of the market will also live through some significant evolutionary changes (Fig. 4). The domestic production in Europe will keep shrinking at a rapid pace; later on, extraction may deteriorate in Southern America. In the Pacific region production will be increasing, but incremental growth will become even lower (i.e. in about two decades production capacities may reach plateau). North America may demonstrate a similar growth pattern (but the volumes supplied are to be two times larger). Probably, the only region with increasingly growing production may become Eurasia (in particular, Russia).

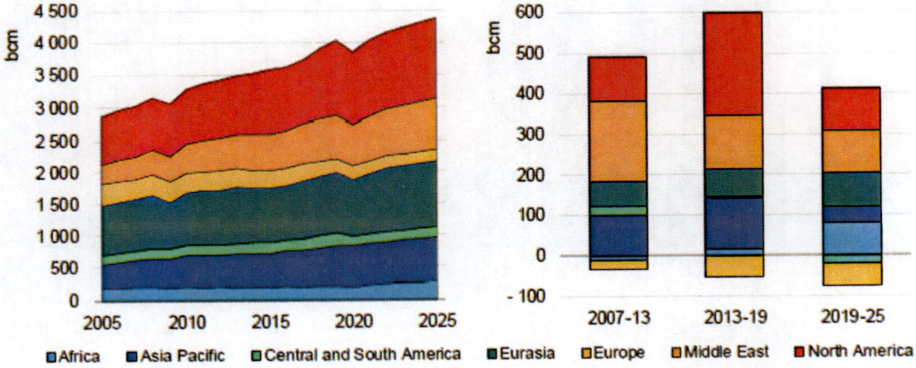

Fig. 4. Global natural gas supply per region 2005–2025, bcm. *Source* IEA, URL: https://www.iea.org/reports/gas-2020/2021-2025-rebound-and-beyond

LNG is a product, which can finally globalize the natural gas market. Increasing LNG production and transportation capacities with cost optimization will balance regional markets. It is important to notice, that LNG will complement pipeline shipments: in case of price differences among regional markets, LNG will flow to the customers, ready to pay the higher price, which finally shall cause a globally balanced gas market (the only differences may occur due to transportation costs or short-term fluctuations). The development of the global gas market can expand existing trends of dealing between customers and suppliers: pipelines, being more specific assets with fixed starting and destination points, will be used within long-term deals, while LNG shipments can be applied for short- and midterm trading operations (or for long-term term cooperation in case of joint projects). LNG exports and imports (Fig. 5) may evolve in different ways for key market players. For demand size, LNG imports and their reasons will depend upon the level of infrastructure development and volumes required. Europe has the most developed pipeline and LNG infrastructure, which causes this market to be currently the largest (alongside Japan), in the future European growing necessity in energy will result in balancing pipeline and see deliveries. Japan, being a developed island country has no other options except LNG (pipeline projects with Russia are only being discussed without any reasonable prospects of success), which may slightly decrease imports via implementing energy-saving technologies. China and slightly afterwards India will become main frontrunners of demand for LNG, which is caused by two main factors – growing economies and deficit of existing pipeline infrastructure projects, linking them with key suppliers (existing Russia-China pipelines are considered to be insufficient to meet growing consumption).

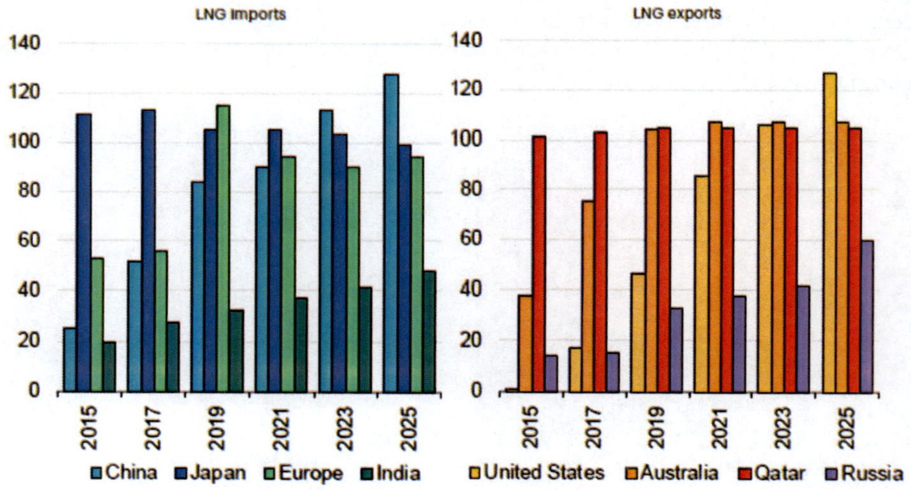

Fig. 5. LNG exports and imports by some countries, bcm. *Source* IEA. URL: https://www.iea.org/reports/gas-2020/2021-2025-rebound-and-beyond

The supply side of LNG segment is also not homogeneous: exporters like Qatar (a typical representative of Middle East exporters) and Australia have no other sales channels due to geographical reasons, they have reached or near achievement limit values. The main influence in the short- and midterm horizon will be exercised by the USA and in the mid- and long-term by Russia.

2 Results and Conclusions

2.1 Australia

Australia is a gas reach country with a developed gas industry. Stipulated by geographic location the country has no options of exporting natural gas other than LNG deliveries. Due to its proximity to the Asian market, Australia is primarily focused on it. Shipments to the European market are not executed or not reported. Like Russia and the USA Australia can enjoy the synergy of further developing its gas industry through combining exporting valuable energy sources to foreign markets with the development of the domestic economy based on the relatively low cost of energy. It is not expected, that Australia will promote "green" agenda, as it benefits from the existing energy market condition.

2.2 China

China has followed a complex of measures, similar to a Third Energy Package of the European Union but with some details, making this approach more specific and development-oriented. The main difference in market consequences is preferred conditions for end consumers: European unbundling has caused emerging of additional stages in distribution chains, which may cause a price increase for final customers (Fig. 1), while the Chinese approach (creating a single midstream operator – Pipe-China) on the contrary facilitated interaction and bargaining between supply and consumption, leading market structure to perfect competition (Fig. 6).

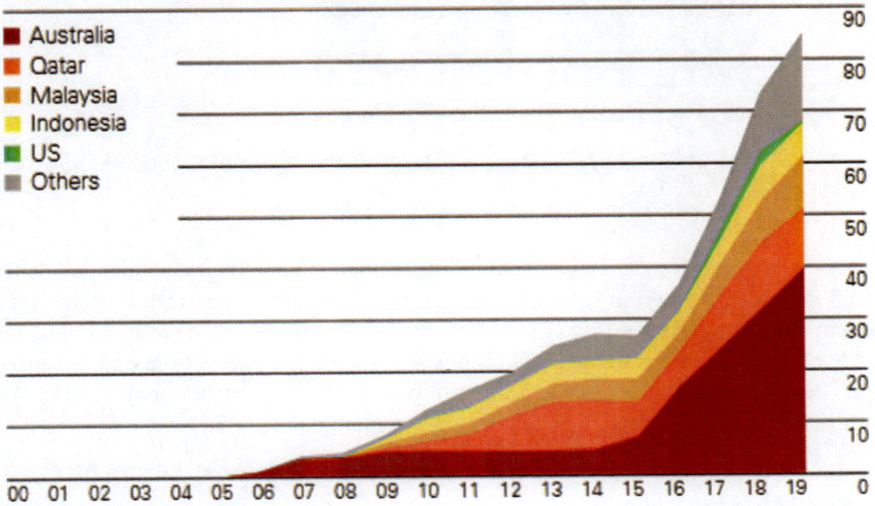

Fig. 6. China: LNG imports by source, bcm. *Source* BP Statistical Review of World Energy 2020 [3]

Another way to saturate the domestic market and create competition among suppliers is Chinese policy towards the development of infrastructure (pipelines from Russia and Turkmenistan in combination with regas terminals for LNG deliveries). However, it shall be noted that China has started to invest in green technologies (Fig. 2). Probably it shall be regarded as plan B for outbalancing dependency on natural gas imports and as an attempt to seize the existing problem of significant environmental pollution.

2.3 Eu

The European Union represents one of the most significant and key regional gas markets; it has a well-developed infrastructure, rapidly growing demand, and shrinking domestic production (Fig. 7).

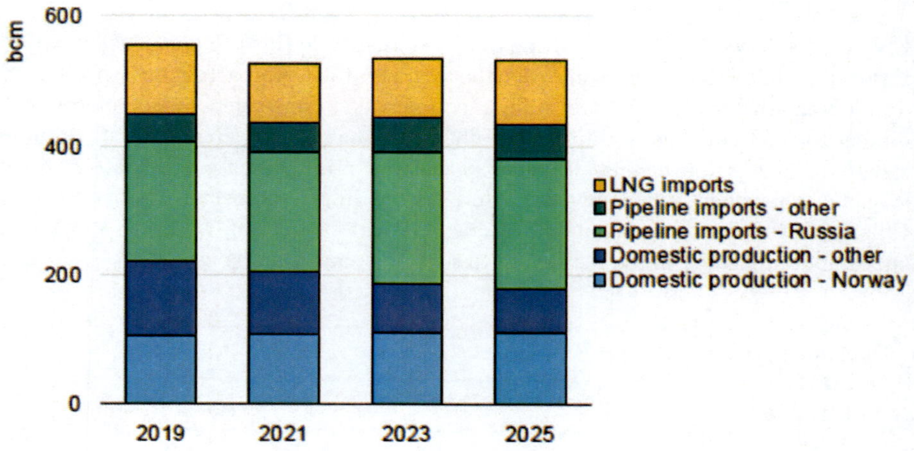

Fig. 7. Evolution of European natural gas supply, 2019–25. *Source* IEA, URL: https://www.iea.org/reports/gas-2020/2021-2025-rebound-and-beyond

As it was mentioned above unbundling, creating new stages within the delivery chain has caused unwilling results: the end price consumers may grow higher for the benefit of intermediaries (Fig. 1). It may be beneficial for the country in case intermediaries are presented by national companies, which may be not the case for the free open economy. In this respect a forced promotion of green agenda may be considered as a tool to devalue supplies of natural gas, enabling:

a) to obtain a better import price (via levying additional taxes and duties for emission permits), and.
b) to create a new industry with domestic companies, developing and providing green technologies and offering green equipment (e.g. electrolyzers, wind and solar generators, etc.).

2.4 Japan

For a long time, Japan has been a key Asian consumer of natural gas and a market with the highest prices. The prospects of this market remain stable, as this developed country is accustomed to paying high (relative to other markets) prices for natural gas. The concerns about nuclear power may become a factor of a further increase in gas consumption; however, implementing energy-saving technologies may neutralize this trend. Nevertheless, high prices in combination with the development of new LNG projects in main gas producing countries create a sort of hedge for further price growth: LNG flows will first target Japan as the most profitable market, causing prices to remain stable. In terms of the global gas market Japan will hold a medium position between "big" country (having an influence on price as a virtual upper limit) and "small" country (having relatively low price influence as domestic consumption is going to be flat or even slightly declining – Fig. 5 – as well as loosing dynamic against new regional benchmarks – China and India later on).

2.5 Qatar

Qatar, like other Middle East oil & gas exporters, is going to further develop LNG capacities. However, due to relatively low global political power the country will have to consider threats of substitution products or green policies, jeopardizing its profits. In this respect, Middle East economies will probably have to invest current fossil fuels incomes into development industries, which shall form a prospective economic basis.

2.6 Russia

Despite a focus on the manufacturing and tertiary sector, being declared for several decades, the country is heavily dependent upon oil & gas exports. On one hand, Russia relies upon a stable and rather ecological energy source – natural gas, on the other there is a challenge of the globalizing natural gas market and rigid competition with other players. For this reason, export routes are being diversified (pipeline projects to China, new liquefaction terminals in Sakhalin and Yamal). However, to secure the risk of devaluing natural gas by the joint policy of the key customers' investments into renewable are performed; some hydrogen technologies are being developed (primarily technological processes of methane pyrolysis with carbon capture). In case of pure economic rivalry, Russia can become a global energy market leader, having the least transportation costs to the existing core market (the European Union) and the emerging core market (Asia Pacific). The position of the country on the new global market will depend heavily upon the model of interaction between the USA and Russia in terms of economic and political rivalry, as both countries have significant influence in their regions and can influence global gas prices.

2.7 USA

The United States of America had been a country, importing energy sources for a long period. Extraction of natural stood near-zero values until 2000 when the development of non-conventional gas fields started and relative technologies (e.g. horizontal drilling and fracking) were widely applied. However, only in 2017, the exports of natural gas exceeded imports (Fig. 8). Currently, the USA are to become a world-leading natural gas supplier (domestic production surpassed national consumption in 2017) (Fig. 9).

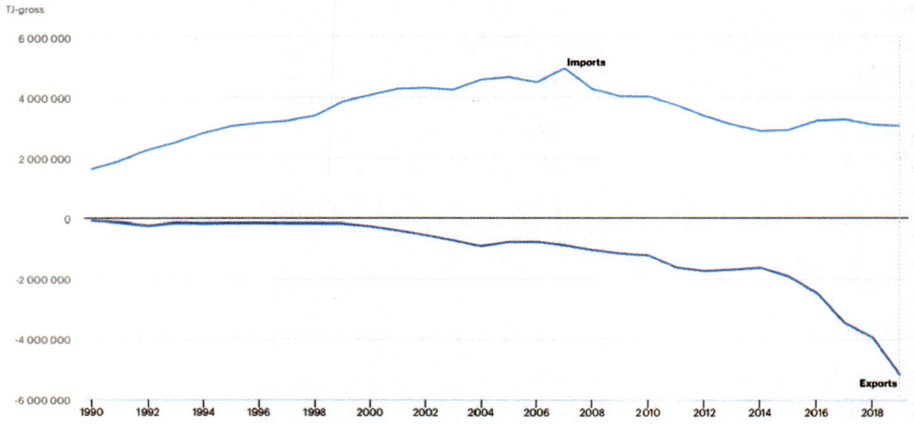

Fig. 8. Exports VS Imports of natural gas by the US. *Source* IEA, URL: https://www.iea.org/data-and-statistics/data-browser?country=USA&fuel=Natural%20gas&indicator=NatGasImportsExports

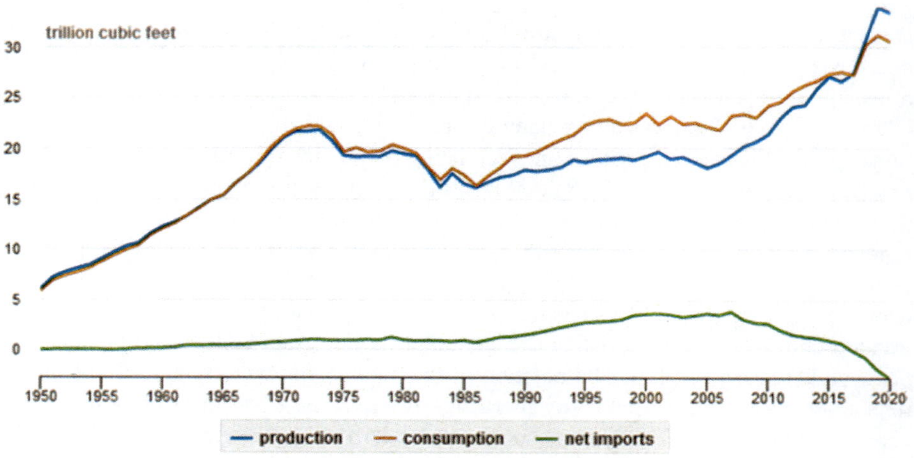

Fig. 9. US production, consumption and net imports. *Source* US Energy Information Administration, 2021, URL: https://www.eia.gov/energyexplained/natural-gas/where-our-natural-gas-comes-from.php

US natural gas industry according to American Petroleum Institute [1] was expected to create 1 million additional manufacturing jobs. This heavy rise in demand for the labour force is caused by the development of non-conventional gas fields and applying rather labour-intensive technologies. Despite almost triple growth of US proved reserves over the last two decades (4.5 bcm in 1999 against 12.9 bcm in 2018) the extraction capacity upkeep constantly requires some new drillings. The US abstention to enter environmental treaties may be linked with the unwillingness to take on obligations, which may burden the development of job-creating industries. The US energy policy would rather target export capacities of existing and potential rivals, as well as launching new markets for US exports.

Acknowledgement. This paper has been supported by the RUDN University Strategic Academic Leadership Program and Grant of RUDN University No. 060119–0-000.

References

1. American Petroleum Institute. How Many Jobs Has the Oil and Natural Gas Industry Created? (2012). https://www.api.org/oil-and-natural-gas/energy-primers/hydraulic-fracturing/how-many-jobs-has-the-oil-and-natural-gas-industry-created
2. Bundestag. Gaspreise Und Endverbraucher Sachstand [Gas Prices and Final Consumer State of Affairs]. WD 5 - 3000 - 104/16 (2017). https://www.bundestag.de/resource/blob/487666/2b587e6276bef604426c7bfdeb2b6006/WD-5-104-16-pdf-data.pdf
3. British Petroleum. BP Statistical Review of World Energy 2020 (2020). https://www.bp.com/content/dam/bp/business-sites/en/global/corporate/pdfs/energy-economics/statistical-review/bp-stats-review-2020-full-report.pdf
4. Coase, R.H.: The nature of the firm. Economica [Economy] **4**(16), 386–405 (1937). https://doi.org/10.1111/j.1468-0335.1937.tb00002.x
5. Downs, E., Ya, S.: Reform is in the Pipelines: PipeChina and the Restructuring of China's Natural Gas Market", 16 September 2020 (2020). www.energypolicy.columbia.edu/research/commentary/reform-pipelines-pipechina-and-restructuring-china-s-natural-gas-market
6. IEA. The Future of Hydrogen. Report prepared by the IEA for the G20, Japan. Seizing today's opportunities (2019). https://www.iea.org/reports/the-future-of-hydrogen
7. Klinova, M.: State and business interaction: survey of economics and policy through time. Econ. Strat. **4**(138), 186–197 (2016)
8. Lavrov, S., Simonov, A.: Gas industry a factor of international competitiveness. RUDN J. Econ. **29**(1), 164–172 (2021)
9. Zhukov, S.: Globalizatsiia Rynka Prirodnogo Gaza: Vozmozhnosti I Vyzovy Dlia Rossii [Globalization of Natural Gas Market: Opportunities and Challenges for Russia]. IMEMO RAS, Moscow (2010). ISBN 9785953502672

Digital Leadership in An Uncertain Technological Environment

Julia A. Kovalchuk[1,2], Liliya S. Valinurova[3](✉), Marina Yu Telegina[4], and Igor M. Stepnov[1,4]

[1] MGIMO-University, Moscow, Russia
[2] Moscow Aviation Institute, Moscow, Russia
[3] Bashkir State University, Ufa, Russia
[4] Financial University under the Government of the Russian Federation, Moscow, Russia

Abstract. The paper investigates the impact of leadership in digital solutions on reducing technological uncertainty. Advanced technology allowed forming digital leadership and global superiority of companies based on information processing. During the research, the authors use system analysis, which allows for a multidisciplinary comparison of subjective judgments and objective development of reality, taking into account the reflection of society on the diffusion of advanced technology. The authors identify several most common significant problems of most companies striving for leadership. The research question posed in this paper is: "Do new digital leaders reduce future technological uncertainty, or are they an operational or strategic response to the current opportunities and challenges of the current digital agenda?" The conducted study of the success of digital companies has shown that it is impossible to ignore the leadership of their leaders, who provide technological superiority. Two forms of leadership (tech vanguards and kinetic leaders) are given the closest attention. The authors find that Russian companies identify success mostly with kinetic leadership, which is more operational in nature. The leading companies were from the banking, information and communication, and high-tech industries. The authors conclude that subsequent research should present more measurable relationships between tech vanguards, kinetic leadership, the overall success of a company, and reduced technological uncertainty.

Keywords: Digital technology · Uncertainty · Leadership · Digital leadership · Technological superiority

JEL Codes: O33 · O47 · M21

1 Introduction

Historically, the continuous development of technology has not bypassed any sphere of society, taking priority in providing large-scale superiority in the business environment. Advanced technology allows for the formation of digital leadership and ensures the global superiority of companies based on information processing. Digital solutions

E. G. Popkova (Ed.): ISC 2020, LNNS 368, pp. 114–122, 2022.
https://doi.org/10.1007/978-3-030-93244-2_14

have become a source of new changes in society. The digital world is beginning to dictate its terms of development.

Undoubtedly, the introduction of technological innovation can provide a new level of business, directly impacting all production sectors. In this environment, leadership (including digital leadership) becomes increasingly relevant on the road to global technological superiority, implying leadership in the digital sphere and all areas of business development, including those not directly related to the development of technology.

The central research assumption is that a leader with high adaptation and motivation can multiply the chances of a company to function more effectively in a constantly changing environment and significantly accelerate the achievement of technological superiority in the industry. The presence of the leader in the industry can shape the long-term development agenda of society. In this regard, it is necessary to determine whether leadership in management and leadership in technology can be separated at this point in time.

The paper aims to investigate the distinctive features of the leaders and highlight the features of the leading companies in the technology race in the context of uncertainty. The emergence of such phenomena as tech vanguards and kinetic leaders allows us to pose the following research question: "Do the new digital leaders reduce future technological uncertainty, or are they an operational or strategic response to the current opportunities and challenges of the current digital agenda?".

2 Materials and Methods

To answer the questions posed in this research, the authors use several research approaches and methods, including a comparative analysis of the discussion available in secondary literature related to technological uncertainty.

The predominant research methodology used in this research is system analysis, with its multidisciplinary nature and the ability to compare subjective judgments and the objective development of reality.

The analysis relies on the multifaceted nature of judgments about technological uncertainty, taking into account that most judgments about the future are of an expert nature based on past experience and that they do not always reflect the new information picture of the world. During the analysis, it was essential to consider different approaches to implementing technology in the economic world and society's reflection on the diffusion of advanced technology.

The subjective component of system analysis allowed the authors to identify the position of researchers regarding the technological future. The techniques and methods of system analysis in relation to objective reality also allowed to establish the unity of the positions of researchers, collecting them in the concept of leadership. Two types of leadership are given the most attention – tech vanguards and kinetic leaders.

3 Results

The introduction of digital technological innovations in various spheres of society has given rise to a new style of leadership, which can inspire companies to bright, extraordinary, and large changes in market behavior in an environment of increasing uncertainty.

3.1 Technological Uncertainty

As a radical tool for transforming society, digitalization is marked with a high degree of uncertainty about the future characteristics of society. Contemporary secondary literature on the subject often relies on futuristic projections. Thus, there are claims [15] of a shift in the focus of research toward the popularization of scientific and technological progress [1, 4, 6, 16].

Due to such influences on public and business opinion, a technology popular nowadays may turn out to be an outsider in the future. F. J. Milliken describes uncertainty as an individual's perceived inability to accurately predict something due to the lack of sufficient information, classifying such uncertainty as the uncertainty of state, effect, and response [18].

Simultaneously, technological uncertainty has become a basic condition for the dynamics of economic systems. Uncertainty greatly impacts the innovative decisions and actions of entrepreneurs [17].

There are at least four types of uncertainty in the current digital race:

- Uncertainty of new digital technology;
- Uncertainty of success in markets;
- Uncertainty of strategies (business models);
- Uncertainty of public acceptance [21].

Therefore, uncertainty is the basis of many decisions and the content of discussions. According to the survey [10], it is essential that businesses self-assess their readiness for technological reforms, for example, in six dimensions (technology, personnel, strategy, leadership, process, and innovation). In another opinion [3], many decisions on technology assessment are still based on expert judgment. However, we agree that it is necessary to consider uncertainty from two perspectives – from the perceived market uncertainty and from the capabilities of the firm, in which leaders play an essential role [12].

3.2 Leadership

When considering digital leadership, it is necessary to rely on the balance of management and leadership in management. R. Straub [22] notes that P. Drucker was "ahead of the game" in defining management, seeing management as the art of getting things done, and leadership as the art of determining what needs to be done from the dry technocratic discipline of management studies. According to J. P. Kotter [14], leadership has several features. Since "leadership" and "management" in his understanding (similar to P. Drucker) are not identical, these features help distinguish these concepts. According to the views of J. P. Kotter, the purpose of management is to bring clarity and predictability and ensure consistency, which contributes to the achievement of the goals set by interested parties. J. P. Kotter indicates that leadership is a unity of three characteristic components:

- Future-orientation – forming the necessary strategies to facilitate the effective achievement of goals in the future, which is essential during technological renewal;
- Aligning communications according to the vision of future results provided by the digital environment;
- Moving in the right direction – a forward-looking vision of the necessary resources and opportunities for development.

P. Horner-Long and R. Schoenberg consider two oppositions in understanding the essence of leadership [11]. According to P. Horner-Long, the main distinguishing feature of all leaders is a set of qualities and forms of behavior that are unique to them. On the other hand, R. Schoenberg believes that the position of a leader pushes him or her to adopt a certain style and behavioral traits appropriate to his or her developmental environment.

Several researchers support R. Schoenberg and note that in case of uncertainty, it is necessary to consider situational factors affecting the organization's position in the market [9].

Based on the above, we can identify several most pressing problems of most companies seeking leadership in the digital world and the possibilities of overcoming them:

- Low level of interaction inside and outside the company;
- Lack of a clear vision;
- Limits of the previously achieved growth;
- Presence of disagreements within the company;
- Lack of control over the implementation of planned goals.

The solution to the problems mentioned above continues to be relevant in the digitalization of the company's activities.

3.3 Leadership of Companies

When the leader establishes a balanced communication system in the company, providing seamless communication between the parties interested, it is possible to increase companies' effectiveness significantly. A particular company or specific technology can achieve leadership. When the incentives of an organization and its employees are aligned, and conditions to track progress and monitor deviations from the set development goals are in place.

In any business, a successful decision is always followed by the consolidation of the "formula for success" – the chosen business strategy. Even though leading companies strive to get the most out of a profitable idea, there is no one right way on the road to retaining leadership positions through endless improvement and expansion of the scale of success. When the market experiences various changes (especially technological ones), the business model developed earlier becomes irrelevant and fails.

Nevertheless, large companies are often confronted with problems that they cannot overcome due to the significant changes brought by technological development. As a rule, under the influence of competitors' attacks, they realize that they cannot hold the leading position. Consequently, there comes a decline in the performance of leading companies, the level of their sales and profits are steadily declining [23].

It is worth noting that the use of traditional methods and approaches to managing the development and achievement of leading positions in the market is inappropriate for knowledge-intensive industries due to the unconventional nature of their functioning. This fact is caused by the destruction of hierarchical and organizational foundations occurring due to expanding the technological spectrum of possibilities to adapt and systematize information for the convenience of its subsequent processing [2]. Thus, we can conclude that the leader is the main link in the innovation process and the initiator of promoting technological development strategies through effective communication, high resourcefulness, and adaptability [8].

3.4 Kinetic Leadership

A. Haddud and D. McAllen believe that companies digitize workplaces everywhere, regardless of size, industry, or other features [7]. In studies [5] and later [20], the authors discussed the concept of implementation and subsequent development of digital technology with their comprehensive application.

The Deloitte study "The kinetic leader: Boldly reinventing the enterprise" [24] notes the expansion of the circle of participants directly involved in process management. There appear new professions (e.g., IT director or technical director). Nowadays, significant attention is given to the participation of general, financial, and operational directors and leaders in strategic and innovative development, as well as the participation of R&D managers in the company's technological development. It is worth noting that this research covers more than 1,300 participants from about 70 countries and more than 20 industry sectors. Meanwhile, the increase in respondents who are not digitally connected is 18% (22% in 2018 and 40% in 2020), which indicates a growing scale of technology development despite their lack of direct digital affiliation.

In 2021, one of the effects of technological innovations is saving the time of the executives through the automation of business processes, the introduction of digital financial solutions in the practices of corporate management based on the use of big data, predictive analytics, and artificial intelligence, performing routine and one-time tasks in less time intervals, etc. [24].

Consequently, the introduction of technology in different production areas has a range of advantages. Nevertheless, the problem is that leaders often cannot use the right technology in certain situations. This requires using the right technology at the right time and in the right way. This should subsequently contribute to better results in the decision-making and management of the human resource.

That is why so much attention is given to the emergence of a new generation of leaders – kinetic leaders. A kinetic leader is a powerful agent of changes responsible for driving business technology strategy. A kinetic leader focuses not only on launching technological transformation but also on executing large-scale organizational change to achieve advanced business goals and make the organization more sustainable and agile in the business environment.

Given the complexity and ambiguity of the world economy, dictated by the COVID-19 crisis, kinetic leaders have a unique opportunity to assert themselves and prove their worth. After all, they can offer strategies for overcoming the crisis, take the lead in launching transformational processes, and help businesses with adaptation and the subsequent formation of the results of post-industrial and post-pandemic civilization.

3.5 Tech Vanguards

In the era of transition to the Sixth Industrial Revolution, the opinion that the impact of traditional methods and ways of management is not enough for the qualitative advancement of the company has become absolutely stable, especially in the context of the COVID-19 pandemic. To achieve leadership, companies now need to respond to the challenges dictated by COVID-19 in a qualitative and timely manner. Moreover, they need to thoroughly engage in the processes of "healing" companies and create a set of procedures aimed at developing the latest tools to adapt to the possible changes in development standards of the existing business environment.

The introduction of the latest digital capabilities in manufacturing, the powerful communication skills of leaders within the company, focused development, and the instant change in strategies in response to new and emerging uncertainties create new ways of doing business. Companies with the traits mentioned above are considered to be tech vanguards.

Tech vanguards are growth-oriented organizations that have leaders who are highly interested in the development of these organizations. These organizations usually outperform other organizations in the market on three parameters: market performance, clarity of vision, and maturity of technological functioning [13].

Such organizations are customer-focused. They aim to engage customers in the process of innovative development. When competitors prefer to cut costs, tech vanguards seek to increase investment and put more and more money to attract and encourage customers to invest in the organization's development.

With the threat of a crisis, most organizations in the market can focus on more thorough cost-cutting and saving resources, which seems logical at first glance. However, the result of these aspirations will inevitably be the loss of the organization's significance and withdrawal from the market.

The Deloitte research [13] shows the following:

- Tech vanguards allocate 5% more to innovation than traditional organizations (20% and 15%, respectively);
- Tech vanguards spend 0.6% more of their income on technology (4.8% and 4.2%, respectively).
- When estimating the possible increase in revenues for technology development in monetary terms, tech vanguards are twice as likely to increase the budget for technology development (24% and 11%, respectively).

Table 1 presents data based on a sample of the top 20 Russian companies by capitalization level in the leading sectors of the Russian economy: banks and financial services, oil and gas production and refining, metallurgy, mining, information technology, trade, telecommunications, multibranch holdings, construction and real estate, and transport [19]. The selected companies were evaluated by experts. Based on the data presented, we can conclude that the capitalization of companies using high-tech solutions continues to grow despite the crisis conditions. This means that investors should expect increased interest in Yandex, Magnit, Rostelecom, and other companies.

Table 1. The level of capitalization of the leading sectors of the Russian economy

No.	Sector	Company	Capitalization		Digital leader/ecosystem	Tech Vanguard	Kinetic Leader
			End of 2020, mln. $	Change for the year, %			
1	Banking and financial services	Sberbank	79504	−10.2	+/+	–	+
2		Tinkoff Bank	6557	53.0	+/+	+	+
3	Oil and gas production and refining	Gazprom	68012	−30.7	+/−	+	–
4		Rosneft	62534	−18.9	−/−	+	–
5	Metallurgy	Norilsk Nickel	50604	2.7	−/−	+	+
6		Novolipetsk Steel	16941	22.1	−/−	–	+
7	Mining	Polus	27738	81.7	−/−	–	–
8		Polymetal	11142	49.6	−/−	+	–
9	IT	Yandex	22122	54.9	+/+	+	–
10		Mail.ru Group	5928	20.8	−/+	–	–
11	Retail	X5 Retail Group	9809	4.7	−/+	–	–
12		Magnit	7828	38.5	−/−	–	–
13	Telecommunications	MTS	8916	−14.0	−/+	–	+
14		Rostelecom	4299	31.5	−/+	–	+
15	Multi-sectoral holding company	En+ Group	6693	4.8	−/+	+	–
16		AFK Sistema PAO	3772	58.7	−/+	–	+
17	Construction and real estate	PIK Group	5338	25.4	−/−	–	–
18		LSR Group	1211	−4.7	−/+	–	–
19	Transport	Aeroflot	2363	26.8	−/−	–	–
20		Novorossiysk Commercial Sea Port	2108	−24.4	–	–	–

Source Compiled, systematized, and evaluated by the authors based on [19]

In many organizations and even industries, crisis conditions lead to a steady decline in financial and technological capabilities. However, the most stable companies can overcome the pressure of the crisis due to their high ability to adapt and quickly change their strategic views, choose new directions of development, and redistribute funds. Consequently, technology leaders may have a much greater ability to function in a crisis.

4 Conclusion

Digital technology is implemented based on digital business models, using the information to achieve performance and success. At its core, technological superiority is not just the possession of the best and most advanced technology with the best prospects and the best ways to commercialize and create value with the consumer but also the implementation of sound changes in management models aimed at leadership.

In our opinion, the emergence of such new phenomena as tech vanguards and kinetic leaders responds to the challenges of the modern economy and ensures the success of companies. Nevertheless, it does not reduce technological uncertainty.

Still, the findings based on the capitalization of Russian companies show that kinetic leadership reduces near-term uncertainty to a greater extent than tech vanguards, including the ability to set barriers.

References

1. Barrat, J.: Our Final Invention: Artificial Intelligence and the End of the Human Era. Thomas Dunne Books/St. Martin's Press, New York (2013)
2. Cortellazzo, L., Bruni, E., Zampieri, R.: The role of leadership in a digitalized world: a review. Front. Psychol.**10**, 1938 (2019). https://doi.org/10.3389/fpsyg.2019.01938
3. Dissel, M.C., Phaal, R., Farrukh, C.J., Probert, D.R.: Value road mapping. Res. Technol. Manage.**52**(6), 45–53 (2009). https://doi.org/10.1080/08956308.2009.11657599
4. Megatech, F.D.: Technology and Society 2050 in the Forecasts of Scientists and Writers. Bombora, Moscow (2018)
5. Gerth, A.B., Peppard, J.: The dynamics of CIO derailment: how CIOs come undone and how to avoid it. Bus. Horiz.**59**, 61–70 (2016). https://doi.org/10.1016/j.bushor.2015.09.001
6. Gordon, R.: The Rise and Fall of American Growth: the U.S. Standard of Living Since the Civil War. Princeton University Press, Princeton (2017)
7. Haddud, A., McAllen, D.: Digital workplace management: exploring aspects related to culture, innovation, and leadership. In: Proceedings of the PICMET 2018: Portland International Conference on Management of Engineering and Technology, pp. 1–6, Honolulu (2018)
8. Harvard Business Review Analytic Services. Operationalizing digital transformation: New insights into making digital work (2017). https://hbr.org/sponsored/2017/05/operationalizing-digital-transformation-new-insights-into-making-digital-transformation-work. Accessed 15 Mar 2021
9. Heinz, U., Baga, T., Gebert, D., Kearney, E.: Leadership and cooperation as success factors in innovative R&D projects on electronic platforms. Team Perform. Manage.**12**, 66–76 (2006). https://doi.org/10.1108/13527590610674077
10. Hizam-Hanafiah, M., Soomro, A.M., Abdullah, N.L.: Industry 4.0 readiness models: a systematic literature review of model dimensions. Information **11**(7), 364 (2020). https://doi.org/10.3390/info11070364
11. Horner-Long, P., Schoenberg, R.: Does e-business require different leadership characteristics? An empirical investigation. Eur. Manage. J.**20**, 611–619 (2002). https://doi.org/10.1016/S0263-2373(02)00112-3

12. Hoskisson, R.E., Busenitz, L.W.: Market uncertainty and learning distance in corporate entrepreneurship entry mode choice. In: Hitt, M.A., Ireland, R.D., Camp, S.M., Sexton, D.L. (eds.) Strategic entrepreneurship: Creating a new mindset, pp. 151–172. Blackwell, Oxford (2002)
13. Kark, K., Phillips, A.N., Briggs, B., Lillie, M., Tweardy, J., Buchholz, S.: Deloitte insights. The kinetic leader: Boldly reinventing the enterprise (2020). https://www2.deloitte.com/us/en/insights/topics/leadership/global-technology-leadership-study.html. Accessed 13 Mar 2021
14. Kotter, J.P.: A force for change: how leadership differs from management. In: Management and leadership, pp. 3–18. Simon and Schuster, New York (2008)
15. Kovalchuk, J., Stepnov, I.: Research of the technological future's uncertainty. MGIMO Rev. Int. Relat.**12**(6), 214–224 (2019)
16. Leonghard, G.: Technology vs. Humanity: the Coming Clash Between Man and Machine. Publishing House AST, Moscow (2018)
17. Meijer, I.: Uncertainty and Entrepreneurial Action. The Role of Uncertainty in the Development of Emerging Energy Technologies. Utrecht University, Utrecht (2008)
18. Milliken, F.J.: Three types of perceived uncertainty about the environment: state, effect, and response uncertainty. Acad. Manage. Rev.**12**, 133–143 (1987). https://doi.org/10.2307/257999
19. RIArating: Top 100 largest Russian companies in Russia – Rating 2021, 4 February 2021. https://riarating.ru/infografika/20210204/630194238.html. Accessed 16 Mar 2021
20. Schwarzmüller, T., Brosi, P., Duman, D., Welpe, I.M.: How does the digital transformation affect organizations? Key themes of change in work design and leadership. Manage. Rev. **29**(2), 114–138 (2018). https://doi.org/10.5771/0935-9915-2018-2-114
21. Stepnov, I.: The uncertainty of the technological future. In: Stepnov, I. (ed.) Technology and Business Strategy, pp. 19–37. Springer, Cham (2021). https://doi.org/10.1007/978-3-030-63974-7_2
22. Straub, R.: The power of ecosystems. Glob. Focus **13**(2), 28–32 (2019)
23. Sull, D.: Why good companies go bad (1999). https://hbr.org/1999/07/why-good-companies-go-bad. Accessed 15 Mar 2021
24. Valinurova, L.S., Leibert, T.B., Khalikova, E.A.: Innovative approaches to corporate financial management in business systems. In: Popkova, E.G., Sergi, B.S. (eds.) ISC 2019. LNNS, vol. 91, pp. 83–95. Springer, Cham (2020). https://doi.org/10.1007/978-3-030-32015-7_11

Impact of the COVID-19 Pandemic on Medical Devices Market and Medical Services Market of the Eurasian Economic Union Member States (EAEU)

Irina N. Belova(✉), Maria I. Dyatlova, Anna E. Zhukova, and Elena A. Egorycheva

Peoples' Friendship University of Russia, (RUDN University), Moscow, Russia
{belova_in, egorycheva-ea}@rudn.ru

Abstract. The spread of coronavirus infection has become a trigger mechanism for global economic shocks for almost all industries, especially for the medical sector. The health care system, pharmaceutical industry enterprises, the market of medical services, and manufacturers of medical devices are known to have experienced the greatest burden.

Despite all the contradictions, such as control and restrictive measures, disagreements on models of combating coronavirus, restrictions related to the border closure, transport links, etc., the EAEU countries (Armenia, Belarus, Kazakhstan, Kyrgyzstan, and Russia) have shown an example of timely response and counteraction to the spread of COVID-19 pandemic. At the same time, Russia acted as a guarantor and basis for the economic well-being of the entire Eurasian economic entity.

Based on all abovementioned, there is a necessity to consider detailed current changes in the markets of medical devices and medical services in the Eurasian Economic Union countries and, in particular, in the Russian Federation, what is considered to be the object of the research.

The latest data on the market of medical devices and services, legislative acts, theoretical and methodological developments in the field of evolutionary and institutional economics were used in the research.

Keywords: Medical device market · Medical services market · Coronavirus pandemic · COVID-19 · Medical device manufacturing · Dynamics and volume of the medical device market

JEL Codes: F20 · F29 · E44 · L10

1 Introduction

The coronavirus pandemic is known to have had a significant impact on health systems around the world. The high demand for medical services and products has exceeded the capacity of some countries to provide them. The difficulties encountered during the pandemic, such as border closure and restrictions on the mobility of the population,

E. G. Popkova (Ed.): ISC 2020, LNNS 368, pp. 123–131, 2022.
https://doi.org/10.1007/978-3-030-93244-2_15

particularly affected the supply of components, the reduction and reorientation of production; led to changes not only in the volume and nature of the production of medical devices but also in the demand for certain medical services.

Thus, the demand for such medical devices (MD) as X-ray diagnostic equipment, personal protective equipment, blood transfusion equipment, defibrillators began to prevail, and in medical services (MS), a special increase was observed in laboratory diagnostics.

The COVID-19 pandemic has contributed to the development and activation of processes, both at the state and private enterprises levels. The introduction of new technologies and changes like production was the driver of adaptive changes in the markets of medical devices and medical services. To protect the domestic market of MD in Russia, several legislative acts were adopted to protect the domestic market and support local producers, such as the introduction of a simplified procedure for registering MD, the Federal Law (FL) on restricting access to public procurement for foreign manufacturers, and the introduction of plans to expand the representation of Russian manufacturers in public procurement.

According to some experts, despite the consequences of the pandemic, the market of medical devices and services will not only recover but also grow by 5–6% by the end of 2021 [5]. Nevertheless, the experience of the first stages of the fight against the COVID-19 spread showed the necessity to increase the state budget not only in the field of health but also for the development and subsequently increase the production of competitive medical equipment and medical services to meet the population needs. Such conclusions and recommendations on prospects of the MI and MS market were obtained by reviewing the main draft laws and the latest analytical data in this area, which were studied in detail later in the study.

2 Methodology

In the research, such methods as systematic and structural data analysis, economic-statistical, comparative methods, and logical analysis were used. To conduct the research, the authors used reports of scientists and analytical agencies specializing in the markets of medical products and services, as well as the health sector in general.

3 Results

In connection with the COVID-19 pandemic, the global medical device market decreased by 2.3% in 2020 compared to 2019 and amounted to 411.4 billion US dollars [4].

At the same time, the coronavirus pandemic has led to an unprecedented demand for masks, gloves, emergency vehicles, products laboratory diagnostics, etc. The dynamics of the production industry leaders proved the change of the vector needs in medical devices.

Thus, the leading international companies BD (Becton Dickinson) (medical equipment, devices, reagents), Siemens Healthineers (equipment for medical imaging and diagnostics), Thermo Fisher Scientific (special diagnostics) in 2020 received a profit from sales from 5 to 14%. Pfizer (laboratory diagnostics, pharmacy, vaccines) became the leader in sales with a revenue of $42 billion. At the same time, Johnson & Johnson (prostheses, implants), Philips (remote diagnostics technologies) suffered serious losses [8].

The Russian medical services market in the period from 2016 to 2019 was characterized by a gradual growth at an average of 3.8% per year (calculated in the number of admissions) [10]. The main positive dynamics was observed both in the sphere of compulsory health insurance (CHI) (expansion of state programs) and in the commercial sectors. But in 2020, the coronavirus pandemic made its adjustments in the medical services market. So in 2020, the volume of the medical service market decreased by 10.8%, compared to 2019. Among the main reasons for the decline there are: in case of the pandemic- the suspension of planned medical care and the number of appointments; a decrease in outpatient appointments; in case of population real income decrease - the refusal of "planned" visits; limiting the number of visits to medical institutions due to the risk of COVID-19 infection. According to some experts, the number of admissions is expected to increase by 12.8% by 2021 compared to 2020. Since 2021, no restrictions were imposed on the work of private and public medical institutions. The main reason for the increase is the resumption of medical examinations of the population, mass vaccination of the population against COVID-19, additional examinations, and tests during the recovery period after coronavirus infection.

It is worth noting that during the quarantine period, a special growth was observed in the laboratory diagnostics sector, which became the main driver of the commercial medicine market. This is primarily due to the high speed of service delivery, the ability to receive services at home, and the convenience of obtaining test results.

At the end of 2020, there were changes not only in the structure or volume of the Russian medical devices market but also in the field of regulation - the adoption of a list of regulatory legal acts.

Speaking generally about the industry, the COVID-19 pandemic contributed to the introduction of new technologies and became a driver of adaptive changes in technological processes. They allowed developing a strategy for reducing production costs. Both the State and the package of measures taken at the interstate levels play an important role in coordinating the activities of enterprises and solving social problems.

At the same time, the pandemic has set the most difficult tasks for states related to the speed of the virus spread, the systematic impact on all aspects of the population's activities, the increased burden on the medical sphere, the fall in GDP, the decline in demand for hydrocarbons, the increased importance of statehood, etc.

The EAEU is regarded to be an example of a timely response and counteraction to the spread of COVID-19. In 2020, in terms of GDP, the EAEU rose to 5th place in the ranking of the world's leading economies. Nevertheless, the EAEU was faced with some problems during the pandemic:

- contradictions related to control and restrictive measures. In particular, the EAEU member states expected significant assistance, for example, with the supply of test systems;
- disagreements on the fight measures against coronavirus (Belarus- light, Russia, Kazakhstan, Armenia, and Kyrgyzstan-quarantine-restrictive);

- restrictions related to the closure of borders and transport links;
- the outflow of migrants from Russia – problems with labour force;
- sanctions that restrict the introduction of high technologies in production, etc. [11].

As a result, on March 16, 2020, Armenia suggested duties on medical products used to combat coronavirus to be at zero levels. The Kyrgyzstan representatives proposed to create a working group to take operational measures in the context of the pandemic. It should be noted that on March 19, 2020, the coordinating Council under the Eurasian Commission for the prevention of the spread of Coronavirus Infection was established. Its main objective was to adopt emergency response measures to stabilize the social and economic situations in the EAEU countries. According to the Order of the Eurasian Intergovernmental Council No.6 "On measures taken within the framework of the Eurasian Economic Union aimed at ensuring economic stability in the context of the development of the COVID-19 coronavirus pandemic" [2], the following changes were indicated:

- permission for the sales of medicines used to counter the spread of the coronavirus infection COVID-19 in the territory of the EAEU of, as well as registered and manufactured in the EAEU member states and intended for the markets of third countries;
- zeroing or reducing for a certain period of customs duties on the import of materials and components for the designated industries in connection with the social and economic significance;
- strengthening the role and importance of science, primarily in the fields of medicine, biology, and social relations;
- "Explosive" digitalization of communications with virtuality both at the internal and external state levels;
- strengthening health systems, conducting joint research in the development of means of prevention, diagnosis and treatment of infections, improving the algorithms of interaction in responding to epidemics;
- prevention of the coronavirus spread while maintaining the principle of free movement of goods, including socially important goods, food, medical equipment, personal protective equipment [8];
- strengthening of economic cooperation between the countries, including reducing the impact of negative factors, the ability to act in conditions of uncertainty in the world markets;
- maintaining the stability of the functioning of domestic markets, including the exemption from payment of duties and the ban on the export of certain goods from the EAEU.
- support for citizens, businesses and system-forming enterprises to overcome the negative consequences, including tax deferrals, credit and rental holidays, postponement of insurance premiums, a moratorium on fines and bankruptcy, etc.

In November 2020, the Decision of the Council of the Eurasian Economic Commission (EEC) No. 86 of October 01, 2020, was published, according to which the duty-free regime for the import of materials and components for the production of some medicines will be extended until March 31, 2021. Also, since October 1, 2020, the ban on the export of certain types of medical devices, protective equipment, and personal protective equipment, materials established by Decision No. 41 of the EEC Board of March 24, 2020 [3] has ceased to apply.

To protect the domestic market of medical devices in Russia, several legislative acts have been adopted that restrict the access of foreign manufacturers to public procurement. According to Federal Law No. 249-FZ of 31.07.2020 "On Amendments to the Federal Law "On the Contract System in the Field of Procurement of Goods, Works, and Services for State and Municipal Needs" a minimum share of purchases from the countries of the Eurasian Union was introduced. A list of products with certain characteristics and conditions for setting the maximum price will be formed, as well as suppliers will be identified by the Government of the Russian Federation, necessary for the implementation of the program of the minimum share of purchases from the countries of the Eurasian Union.

A decision was made to enter into the Unified Information System (UIS) information on the minimum share of purchases made from manufacturers of the EAEU countries. The project was developed by the Ministry of Industry and Trade of the Russian Federation according to which the minimum share of public procurement is established in the period from 2021 to 2023, the specified share of MD will change depending on the year.

Also in April 2020, the "Third Extra" rule extended to medical devices such as prosthetics, medical furniture, orthopaedic products, and accompanying parts. The "Third Extra" rule implies that in public procurement (including MD), applications coming from foreign countries will be rejected if at least two proposals were submitted from EAEU suppliers. Previously, the list of medical devices included such products as adhesive dressings, ultrasound scanning devices, artificial lung ventilation devices, etc. At the moment, the medical devices, which is covered by Resolution No. 102, includes more than 140 types of medical equipment.

A ban was also imposed on the purchase of foreign medical masks until the end of 2021, since according to the indicators of the Ministry of Industry and Trade of the Russian Federation, in the first 6 months of 2020, more than 80% of medical masks were imported from Southeast Asian countries. At the same time, there was an active production of masks in Russia, which increased by 20 times and amounted to about 21 million masks per day, when the demand of state institutions is 10.8–11 million masks per day. According to Government Decree No. 299 of March 18, 2020, a procedure was introduced for the accelerated registration of MD with a low degree of the potential risk of use (medical masks, insulating suits and dressing gowns, gloves and shoe covers, respirators).

By the decree of June 2020 (No. 804), the list of medical devices that can be registered under the simplified procedure was expanded to 363 types of MI. The new list includes consumables for laboratory diagnostics, medical furniture, equipment, ventilator valves, pulse oximeters and other devices.

The President of the Russian Federation gave instructions to carry out large-scale digitalization and more active introduction of technologies in medicine.

According to the Roadmap, in November 2020, a plan for the transition of the budget direction of healthcare to work with artificial intelligence (AI) was approved. Thus, each institution should have more than five medical devices and services with built-in AI. A separate industry platform is also planned to be launched. This platform will combine all the data of patients from the Federal Integrated Electronic Medical Card of the Unified State Information System in the Field of Healthcare and a decree has been issued to simplify the registration of software, including with AI, in the case of a medical device.

In connection with the aggravation of the epidemiological situation in 2020, the Russian Federation Government has developed some rules for determining the maximum selling prices for medical devices in a pandemic or emergency.

In addition, the Council of the Eurasian Economic Commission exempted from import duties goods, which are necessary to prevent the spread of COVID-19, except for goods, which are produced in sufficient quantities on the territory of the EAEU. The Consortium of Developers and Manufacturers of Medical Equipment was formed to localize the main types of medical equipment on the territory of the Russian Federation in connection with the threats and risks of dependence on external suppliers during the pandemic. Manufactured devices will be entered into the register to provide additional features after the product is released to the market.

The changes in the framework of the EAEU are worth noting as well. From January 2021, registration for medical devices that cannot be issued under the simplified procedure must be registered following the requirements of the EAEU, in which the procedure includes mandatory inspection of production. September 2020 is considered to be the start of registration under the new rules when the first EAEU registration certificate for a medical device was received.

In October 2020, the Ministry of Industry and Trade proposed a draft law on amendments to the program "Development of the pharmaceutical and medical industry", according to which the volume of consumption of Russian-produced medicines by 2024 in monetary terms should be at least 30%, namely 180 billion rubles. At the same time, the share of institutions specializing in technological innovations in the pharmaceutical and medical industries will have to be at least 56%.

The rules for granting grants to budget institutions in 2021 were also approved; they will look like subsidies from the federal budget for the implementation of projects aimed at the development of medicines and medicines. The allocated budget for this area will amount to 2.6 billion rubles. The maximum amount of one grant was designated as 260 million rubles; the winner is selected as a result of a competition held by the Ministry of Industry and Trade of Russia. To participate in the competition and receive a grant, it is necessary to prepare a business plan, a feasibility study (feasibility study), and a schedule for its implementation.

In addition to these opportunities, companies will have the opportunity to conclude contracts with the state and launch mass production of products that are competitive in the market. At the same time, the state provides tax incentives and other preferences for the investor.

In 2021, a project is being launched to update equipment in clinics at the first level of health care; the list of re-equipment consists of 121 items. From 2021 to 2025, 88 thousand units of medical equipment and 19 thousand units of vehicles will be purchased for regional programs.

The state authorities, against the background of the spread of the COVID-19 coronavirus infection in the Russian Federation, took active measures to provide health care with all the necessary means to protect, prevent and treat the population during the pandemic.

To compensate for the costs of treatment and testing of the population for COVID-19, 47.5 billion rubles were allocated from the reserve fund for regional medical institutions for compensation.

According to previously received data, one of the main directions of the development of the health system will remain the program of modernization of primary health care, which was postponed to 2021; the list of updated medical devices included more than 121 items under this program.

One of the main components in the development of the health system can be considered a plan for economic recovery after the COVID-19 coronavirus pandemic. The cost of implementing this plan amounted to 6.4 trillion rubles. If we consider the funds already provided for economic recovery, then the budget has already been spent by more than half.

Based on the experience of the current pandemic, it is planned to create a regulated system for conducting laboratory diagnostics of infectious diseases, improve and re-equip hospitals and clinics, create a federal scientific centre specializing in infectious diseases, modernize the service for the control of infectious diseases, maintain and create reserves of medicines and medicines, improve the methods of building new medical organizations.

The year 2020 showed that public authorities could adopt laws in a short time to meet the needs of public health during the fight against the pandemic. These changes primarily affected the MI circulation system.

Regarding the EEU member states, Kyrgyzstan is currently in the most difficult economic situation. In 2020, the International Monetary Fund, the Asian Development Bank, the Eurasian Development Bank, the German Development Bank, the World Development Bank, the Islamic Development Bank, and the EU assisted Kyrgyzstan in the amount of $456 million (RUB 33.6 billion) to support and implement measures to combat coronavirus [6].

Armenia is also currently in a rather difficult political situation. The main direction of the government's actions is to vaccinate the population. For these purposes, Russia will provide a batch of Sputnik-V vaccines at a reduced price in February 2021.

Kazakhstan has developed its own KazCOVID-in vaccine and launched mass production of the Russian Sputnik V vaccine [7].

In 2021 production of the Sputnik V vaccine will begin in Belarus. Russia has provided a technology developed by its scientists [1].

4 Conclusion

The coronavirus pandemic has seriously affected the EAEU medical devices and medical services market. First of all, the structure of MD consumption has changed. The most popular were some types of low-tech MD (masks, gloves, etc.) and high-tech MD (resuscitation equipment, laboratory diagnostic products, vaccines), as well as diagnostic and rehabilitation services after recovery.

At the same time, the EAEU member states confirmed the effectiveness of the union's actions, which allow each of the participants to get out of the global economic crisis with the least losses due to the preservation of the internal regional market.

Russia has shown itself as a leader in determining the strategy of the union's actions in the context of the pandemic and has become an economic guarantor of the implementation of the decisions taken. The most significant achievement of Russia can be considered as the organization of the health system during the pandemic, the development of test systems and vaccines against the COVID-19 pandemic.

Based on all abovementioned, it is necessary to conclude that the medical devices and medical services markets are currently in the recovery stage. Thus, organizations and states that were able to "resist" during the pandemic and adapt to constantly changing conditions formed a strategy for overcoming crises, and also gave an impetus for further development.

Acknowledgement. This paper has been supported by the RUDN University Strategic Academic Leadership Program and Grant of RUDN University No. 060119–0-000.

References

1. Belta. EAEU countries to jointly produce a vaccine against COVID-19 – EEC (2021). https://www.belta.by/economics/view/strany-eaes-budut-sovmestno-proizvodit-vaktsinu-ot-covid-19-eek-427673-2021. Accessed 17 Mar 2021
2. Eurasian Intergovernmental Council. Order "On measures taken within the framework of the Eurasian Economic Union to ensure economic stability in the context of the development of the COVID-19 coronavirus infection pandemic" (2020). http://docs.cntd.ru/document/564670858. Accessed 29 Mar 2021
3. Eurasian Intergovernmental Council. Review of key measures and decisions of the ECE (2020). http://www.eurasiancommission.org/ru/covid-19/Pages/measures.aspx. Accessed 7 Apr 2021
4. Medidex. Health financing in a pandemic (2020). https://meditex.ru/news_all/. Accessed 17 Apr 2021
5. Medidex. Russian market of medical devices. Results of 2020. https://www.meditex.ru/news_all/RossiyskiyrynokmeditsinskikhizdeliyItogi2020goda/. Accessed 07 Apr 2021
6. Myseldon. International donors have promised to give almost half a billion dollars to Kyrgyzstan (2020). https://news.myseldon.com/ru/news/index/228816751. Accessed 7 Apr 2021
7. Pharmaceutical Review of Kazakhstan. Exports of Kazakhstani medical products slowed down during the pandemic (2020). https://pharm.reviews/analitika/item/5749-eksport-kazakhstanskikh-meditsinskikh-tovarov-vo-vremya-pandemii-zamedlilsya. Accessed 7 Mar 2021
8. Politicalanalyst. The virus continues to make adjustments (2020). https://politanalitika.ru/posts/2020/04/15/virus-prodolzhaet-vnosit-korrektivy.html. Accessed 7 Apr 2021
9. Proclinical. Who are the top 10 medical device companies in the world? (2020). https://www.proclinical.com/blogs/2020-9/who-are-the-top-10-medical-device-companies-in-the-world. Accessed 7 Apr 2021

10. Ria-news. In 2020, the volume of the medical services market in Russia decreased by 10.8% and amounted to 1.93 billion admissions (2021). https://ria-news.ru/economics/v-2020-g-obem-rynka-meditsinskih-uslug-v-rossii-sokratilsya-na-10-8-i-sostavil-1-93-mlrd-priemov. Accessed 7 Apr 2021
11. Ritm Evrasii [Rhythm of Eurasia]. Eurasian Union: coronavirus test (2020). https://www.ritmeurasia.org/news–2020-03-24–evrazijskij-sojuz-ispytanie-koronavirusom-48100. Accessed 29 Mar 2021

Integration of Netherlands into Global Value Chains: View from the Perspectives of Domestic Value Added Decomposition

Ekaterina M. Zharkova and Natalia A. Volgina(✉)

Peoples' Friendship University of Russia, Moscow, Russia

Abstract. Over the past decades, many economies have integrated into global value chains (GVCs). These processes were elapsing unevenly, with large country and sectoral differences. Some metrics have been developed in the research literature that made it possible to assess and compare the degree and depth of chain engagement. This paper aims to identify the specifics of the integration of the Netherlands into GVCs in terms of such an indicator as the share of Dutch domestic value-added (DVA) incorporated into third country exports (DVX). The decomposition of the country's gross exports developed based on a broader input-output approach, served as a key methodology for assessing the participation of the Netherlands in GVCs. The UNCTAD-Eora GVC Database was chosen as a statistical base of the study. Based on calculations and analysis, the authors came to some conclusions. A special feature of the Netherlands is a high share of foreign added value (FVA) in gross exports, and, accordingly, a low share of DVA. At the same time, from 2008 to 2018, the value of DVA of the Netherlands showed an upward trend, while the share of DVA in gross value added exports also increased, which indicates that DVA is growing faster than FVA. During the same period, the dynamics of the DVX showed mixed results: in terms of value, it increased, but as a percentage of gross Dutch exports it slightly decreased, which, in our opinion, indicates the loss (albeit not significant) of the competitiveness of the Netherlands in the downward chain of added cost. Most of the downstream links in the chain in terms of DVX are involved in such sectors of the Dutch economy as the chemical industry, the oil and gas industry, the retail and wholesale trade, as well as agriculture, fishing and hunting. Geographically, the Netherlands' DVX was focused on the countries of the European continent, which confirms the conclusions about the emerging regional GVC hub in Europe.

Keywords: Value-added trade · GVCs · Netherlands · Exports decomposition · Foreign value added · Domestic value-added incorporated in the export of third countries (DVX)

JEL Codes: F14 · F23 · F60

E. G. Popkova (Ed.): ISC 2020, LNNS 368, pp. 132–141, 2022.
https://doi.org/10.1007/978-3-030-93244-2_16

1 Introduction

In recent decades, the economies of many countries, including the Dutch one, have integrated into global value chains (GVCs). These processes were elapsing unevenly, with large country and sectoral differences. Some metrics have been developed in the research literature that made it possible to assess and compare the degree and depth of incorporation into GVCs. Scientists [1] were pioneers in this area, often referred to as HIY, they proposed a vertical specialization (VS 1) index, where a starting point for research where imported intermediate products, that is, foreign value added (FVA) in the gross exports of countries. Note, that the second component of gross exports is a domestic value-added (DVA).

The main goal of this paper is to identify the specifics of the Netherlands' integration into GVCs in terms of such indicators as the share of Dutch domestic value-added incorporated into third country exports (DVX).

2 Methodology

The decomposition of the country's gross exports developed based on a broader input-output approach, served as a key methodology for assessing the Netherlands' participation in GVCs. This direction of analysis is based on the work of [2]; this approach is often referred to as the KPWW method. The idea behind the decomposition of exports in the KPWW approach is that gross national exports are split into two parts: FVA and DVA. At the same time, the DVA in exports is divided into the value of goods that became final consumption goods and the value of goods that acted as intermediate goods and began to "move" further along the global value chain (often this part of domestic value-added is called "vertical specialization 2" or indirect value-added export). Namely, this part of the DVA is the subject of our analysis. A very small share of exports (less than 1%) is exports of intermediate goods that are returned to the exporting country.

The main source of statistics for the study is the UNCTAD-Eora GVC database [6], which publishes data for the period 1990–2018, and it offers a breakdown of gross value-added trade flows by industry (based on I-O tables) up to 2017.

3 Results

3.1 Analysis of Netherlands' DVX Trade Dynamics

The Netherlands is one of the most GVC-integrated countries in the world. In 2018, total Dutch value-added (VA) exports amounted to $822 billion; at the same time, 198 billion dollars or 24% of total Dutch VA exports are included in the export of third countries (DVX). The global average of this index is 28%. For comparison, in the United States, the share of DVA incorporated into the export of third countries is 34%, for the EU countries it averages 29%, for China - 32%. Although the Netherlands indicator is below the EU average, it nevertheless exceeds the share of German DVA incorporated into the exports of third countries [6]. As can be seen from Fig. 1, over the past 10 years the dynamics of the Dutch value-added, incorporated into the exports of third countries was marked by a sharp decline - by 21% in 2009, which was associated with the global financial crisis. A sharp decline of 17% was also observed in 2015, which was associated with the global decline in international trade and FDI [9, p. 22].

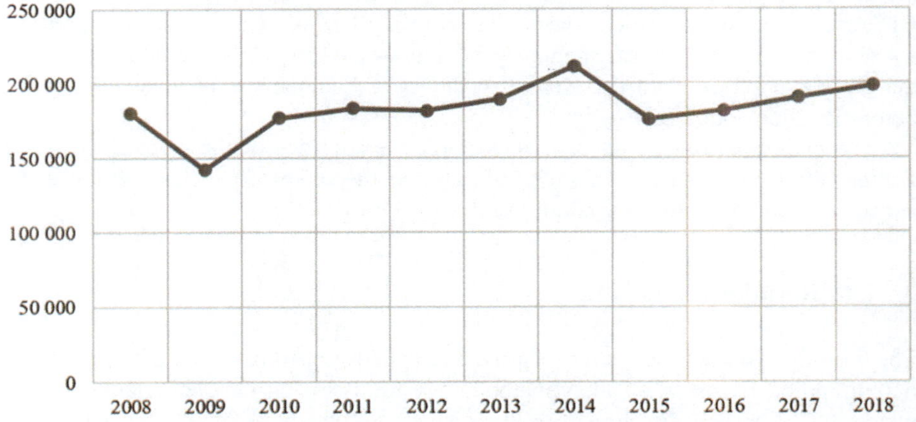

Fig. 1. Netherlands DVX dynamics of in million dollars 2008–2018. *Source* Compiled by the authors based on [6].

If we do not take into account these separate downfalls in dynamics, then the share of Dutch value-added, incorporated into the exports of the rest of the world, in the total export of VA has been gradually growing in the last 5 years with an average annual rate of 6%. However, many analysts, in particular, WTO experts, predict a decline in international trade by 9.2% in 2020 [11] due to the global sanitary and epidemiological crisis, which is likely to have a negative impact on the export of the Dutch VA, as well as other countries. Also, after the economic recession caused by the pandemic, we can observe a trend towards the increase companies' independence from external suppliers; manufacturers began to rely less on foreign supplies. This defragmentation strategy can have a strong negative effect on the economies of the countries most integrated into GVCs, but this effect is still difficult to fully assess.

Based on the data in Table 1, the Dutch VA can be called one of the most demanded in the world, since the Netherlands is in the top 10 countries in terms of the volume of VA incorporated into the export of third countries. According to this indicator, in 2018, the Netherlands took 8th place; at the same time, the country was far outstripped by such giants as China, the USA and Germany, as well as Great Britain, France, Japan and Italy. This is, largely, because these economies are a priori larger in scale.

Table 1. Top 10 countries by DVX volume

		DVX 2008, mln dollars	DVX 2018, mln dollars	VA total exports 2018, mln dollars	%% change in DVX 2018 to 2008	%% DVX in total VA exports 2018
1	China	409,113	683,158	2,155,368	67%	32%
2	USA	55,196	601,613	1,794,314	8%	34%
3	Germany	587,517	529,980	2,200,550	−10%	24%
4	UK	252,742	298,704	816,526	18%	37%
5	France	287,869	294,925	948,325	2%	31%
6	Japan	313,682	287,363	1,059,263	−8%	27%
7	Italy	228,671	225,052	908,020	−2%	25%
8	**Netherlands**	**180,306**	**198,277**	**822,017**	**10%**	**24%**
9	Russia	155,709	192,681	385,486	24%	50%
10	Belgium	142,193	157,797	674,891	11%	23%

Source Compiled by the authors based on [6]

In terms of the geographical distribution of Dutch value-added, as of 2018, the regionalization effect in Dutch value-added trade is visible, i.e. the Netherlands is more focused on trade partnerships with countries of the European region - the European Union. As shown in Table 2, among 10 countries with the highest quantitative indicator of Dutch VA in their exports, nine countries are from the European Union.

Table 2. Top 10 countries with the highest quantitative indicator of Dutch VA in their exports 2008 and 2018

	Top-10 countries exporting Dutch DVX	Netherlands DVX 2008, mln dollars	Top-10 countries exporting Dutch DVX	Netherlands DVX 2018, mln dollars
1	Germany	42,862	Germany	48,532
2	Belgium	39,077	Belgium	38,964
3	France	12,644	France	13,203
4	UK	11,770	Italy	13,187
5	Italy	11,171	UK	10,334
6	Spain	4,613	Spain	5,380
7	Sweden	4,551	Sweden	5,082
8	Denmark	4,347	Switzerland	4,340
9	Switzerland	3,502	China	3,940
10	Austria	3,358	Denmark	3,647

Source Compiled by the authors based on [6]

If we analyze the geographical structure of Dutch DVX use from the qualitative point of view (the share of Netherlands content in the gross exports of third countries) then the list of top-10 partners changes somewhat. The Netherlands participates to the greatest extent in Belgium's exports - 6.5% of Belgium's gross exports come from the Netherlands. The Netherlands is also involved in the creation of exports of Lithuania (4.3%), Luxembourg (3.7%) and some other countries listed in Table 3. At the same time, the orientation of the Netherlands remains predominantly towards the European region, which confirms one of the GVC development scenarios described in the World Investment Report 2020 - strengthening of GVCs regionalization, which will lead to the creation of shorter supply chains and increased focus on local markets [10, p. 13].

Table 3. Top-10 countries with the highest use of Dutch VA in their exports, 2008 and 2018

	Top-10 countries exporting Dutch VA input	Netherlands DVX 2008, in %% of total VA export	Top-10 countries exporting Dutch VA input	Netherlands DVX 2018, in %% of total VA export
1	Belgium	6.8%	Belgium	6.5%
2	Lithuania	4.2%	Lithuania	4.2%
3	Aruba	3.3%	Luxemburg	3.7%
4	Luxemburg	3.3%	Aruba	3.3%
5	Denmark	2.6%	Denmark	2.0%
6	Germany	2.0%	Germany	1.9%
7	Sweden	1.9%	Suriname	1.7%
8	UK	1.7%	Sweden	1.7%
9	Hungary	1.5%	Greece	1.6%
10	Suriname	1.5%	Italy	1.5%

Source Compiled by the authors based on [6]

3.2 Netherlands As a Value-Added Supplier for Third Countries Sectoral Exports

According to the data from Fig. 2, in 2017, the most significant added value created in the Netherlands was imported and then incorporated into their exports by the following sectors of third countries: re-exports and re-imports (23.3% of all Dutch DVX), machinery and equipment (17.7%), chemical products (10.8%) and vehicle manufacturing (8.4%).

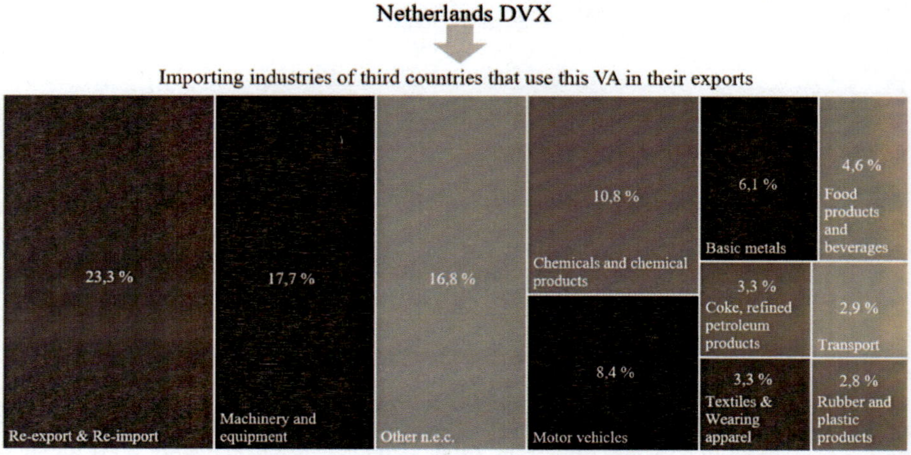

Fig. 2. Top-10 industries of third countries importing the Dutch VA for their own exports 2017 in % of total Dutch DVX. *Source* Compiled by the authors based on [6].

According to Table 4, it can be concluded that the demand for Dutch VA in 2017 increased (compared to 2008) in almost all top 10 importing industries of third countries. The demand for Dutch VA increased to a greater extent from the machinery and equipment industry (an increase of 8.7%) and the basic metals industry (an increase of 3.5%). However, the demand for the Dutch VA from such industries as oil refining and coal industry (coke), plastics and rubber production, as well as from transport services decreased markedly.

Table 4. Dynamics of top-10 industries of third countries importing the Dutch VA for their exports, 2008 to 2017

The main industries of third countries that incorporate Dutch VA in their export	mln dollars 2008	mln dollars 2017	%% of the increase
Re-export & Re-import	41,821	42,234	1.0%
Machinery and equipment	29,583	32,155	8.7%
Chemicals and chemical products	19,439	19,563	0.6%
Motor vehicles	14,832	15,234	2.7%
Basic metal	10,749	11,123	3.5%
Food products and beverages	8,377	8,423	0.5%
Coke, refined petroleum products	6,491	6,017	7.3%
Textile & Wearing apparel	5,818	5,930	1.9%
Transport	5,693	5,238	8.0%
Rubber and plastic products	5,058	5,054	0.1%

Source Compiled by the authors based on [6]

For example, 12.5% of Belgium's re-exports are Dutch VA, 16.8% of Belgium's machinery and equipment industry exports also belong to the Netherlands (for the UK this figure is 9.5%). If to mention the chemical industry, 13.8% of this industry exports in Lithuania rely on a VA produced in the Netherlands; in the case of the automotive industry, 11.7% of the industry's exports in Belgium likewise come from the Netherlands (10.3% in the UK). Dutch value added is also demanded production and export of the metallurgical and food industries in the UK: they account for 14.4% and 11.3% of the Dutch VA, respectively.

Let us consider the main industries and product groups of Dutch top 5 partners in VA trade, that use Dutch VA in their exports. Thus, the main goods in Germany, which were produced with the help of the Dutch VA (excluding re-exports) are chemical goods. The main products of Belgium and France, which were produced with the help of the Dutch VA (excluding re-exports) were chemicals, chemical products, and man-made fibres. Italy uses the Dutch VA for the production and export of machinery and equipment. Great Britain needs the Netherlands' VA (not counting re-imports) for vehicles production (1.8%).

3.3 Most Demanded Dutch DVX by Industry

As shown in Fig. 3, the most demanded Dutch VA, which is then being incorporated into third-country exports comes from the chemical industry (11.3% of the total Dutch VA incorporated into third-country exports originates from this industry), crude oil production and natural gas (9.9% of the Dutch DVX), wholesale and retail trade (7.5%), other commercial activities (5.5%), agriculture, fishing and hunting (5.4%).

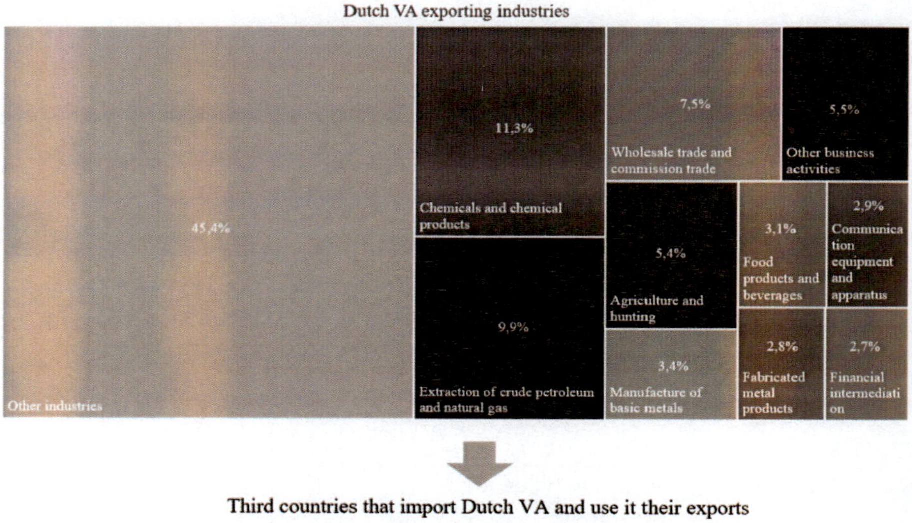

Fig. 3. Top-10 Dutch industries creating VA incorporated into third countries exports in 2017 as a percentage of total Dutch DVX. *Source* Compiled by the authors based on [6].

Compared to the base year of 2008 (Table 5), in 2017, the integration into the GVCs of such industries as metallurgy, oil and gas production, chemical, as well as the production of electrical equipment and communications decreased, at the same time as the integration into the GVCs of wholesale and retail trade services, financial services, services of other commercial activities increased. There is also a noticeable trend towards increased integration into the GVCs of Dutch agriculture, as well as the food and finished metal goods industries. Thus, there is a strong trend towards the strengthening of Dutch downward links in the GVCs, this conclusion is confirmed by the data from World Investment Report-2013 and World Investment Report-2018: in 2010, the ratio of Dutch downward links in the GVC was 76% [8, p. 132], and in 2017 it was 78% [9, p. 25].

Table 5. Dynamics of top-10 Dutch industries that create VA incorporated into the exports of third countries, 2008 compared to 2017

Main Dutch industries that export VA incorporated in exports of third countries	mln dollars 2008	mln dollars 2017	%% of the increase
Chemicals and chemical products	20,721,994	20,539,035	−0.9%
Extraction of crude petroleum and natural gas	18,438,769	18,042,354	−2.1%
Wholesale trade and commission trade	13,152,129	13,696,909	4.1%
Other business activities	9,530,714	9,988,825	4.8%
Agriculture and hunting	9,645,864	9,831,248	1.9%
Manufacture of basic metal	6,525,838	6,261,033	−4.1%
Food products and beverages	5,486,376	5,578,080	1.7%
Communication equipment and apparatus	5,451,506	5,256,047	−3.6%
Fabricated metal product	4,982,773	5,049,990	1.3%
Financial intermediation	4,828,760	49,090,949	1.7%

Source Compiled by the authors based on [6]

It is noteworthy that for all the top 10 importers of Dutch VA, the main donor industries for VA are the chemical industry and wholesale and retail trade services, whose share in the aggregate DVX for these countries averages 11.4% and 7.7%, respectively. For example, 15% of all Dutch value-added in China's exports comes from the chemical industry, for France this share is 14.4%, and for Spain, it is 14.0%. Wholesale and retail trade in the Netherlands creates the added value of Italy's exports to the greatest extent - 9.5% of the total Netherlands' DVX, Spain - 8.6%, Denmark - 8.3%. Also, for four countries from the list of the Netherlands' top 10 VA trade partners (Germany, Belgium, Italy, and France), the source of export is partially the Dutch crude oil, and natural gas industry, for example, 21.7% of the total Netherlands' VA incorporated in export to Germany occurs precisely from this industry. When examining the structure of the main industries supplying the Dutch VA to third countries, one cannot ignore the phenomenon of agriculture in the Netherlands: as on average 7.9% of the country's VA incorporated into the exports of Denmark, Spain and the UK come from the agricultural sector of the Netherlands.

4 Results and Discussion

So, by authors' calculations, Dutch most integrated into GVCs industries are the chemical industry, the oil industry, and agriculture. This situation is a direct reflection of the dynamic development of these industries in the Dutch national economy.

According to the Netherlands Foreign Investment Agency (NFIA), the Netherlands is one of the 5 largest chemical clusters in the world, accounting for 30–40% of the total chemical production in Europe, the Netherlands is a homeland for more than 2,000 chemical companies and 19 (out of 25) the most influential chemical MNEs of the world [5]. The Netherlands is also the "geographic heart" of Europe, and its location provides proximity to the majority of European consumption markets [5].

The Netherlands is also the second-largest supplier of agricultural products in the world after the United States, where 15 of the 20 largest agrifood MNEs in the world are concentrated [4]. Dutch agriculture has a "phenomenal" character, it not only fully satisfies the domestic needs of the country but also "feeds" many countries of the world, being one of the largest suppliers of vegetables, seeds and flower bulbs [7].

The Netherlands is actively developing alternative sources of energy production - windmills, solar panels, etc. The country, like other European countries, is pursuing a "green policy" to protect the ecosystem and reject classic "polluting" energy sources. Nevertheless, a significant part of the country's economic income comes from the oil sector, the leaders of which are MNEs such as NAM and Royal Dutch Shell. However, there is a noticeable trend towards a relative decrease in the role of this Dutch sector of the economy in the GVC.

Re-exports play an important role in the Dutch economy. The high share of re-exports is explained by some factors: the Netherlands is the centre of European transit trade, where the seaport of Rotterdam (and partly the port of Amsterdam) serves as the most important centre for customs clearance of export-import cargo.

The role of services in the Dutch foreign trade should be noted separately. In 2017, 43.4% of the total Dutch VA incorporated in third-country exports originated from the services sector, which indicates a significant share of downward GVCs participation (in 2008, this share was 41.9%). According to the generally accepted methodology, downward GVCs participation is an indicator of the country's reverse participation in GVCs, calculated based on the decomposition of the production of the final goods and showing the share of final products produced in the country's sector within the GVCs framework [3, p. 10]. The most demanded types of services supplied by the Netherlands and exported consequently by third countries are wholesale and retail trade, other commercial activities, financial intermediation, transport services, as well as telecommunications and postal services.

5 Conclusion

Due to the transit nature of the Dutch economy, the country relies heavily on the FVA, but the Dutch VA is also involved in the creation of overseas exports. And Netherlands' industries mostly involved in GVC in terms of DVX (or downstream links in the chain) are the chemical industry, the oil and gas industry, the retail and wholesale sector, and agriculture, fishing and hunting, with a noticeable dynamic in the relative decline in the Dutch DVX from the first two industries and the relative increase in the Dutch DVX from the latter two.

In absolute terms, there is a steady positive trend in Dutch DVX, but it is difficult to make an unambiguous conclusion about a decrease or increase in the Dutch downward ties in GVCs, since in some industries they are decreasing trends (for example, oil and gas production, production of chemical goods), and in others vice versa increasing trends (for example agriculture and service sector).

Also, from 2008 to 2018, the share of DVA in the gross exports of the Netherlands' VA increased, which indicates a possible strengthening of the country's role in the production of final consumer goods and the strengthening of the Netherlands' downward links in GVCs. The sectoral analysis also suggests that the Netherlands is reducing downstream participation in GVCs through the export of extractive industry products, and, conversely, increasing integration into GVCs through the services sector, as well as through agriculture, which to some extent confirms one of the described in World Investment Report-2020 trends - diversification, or intensification of participation in GVCs by strengthening the service sector and manufacturing industries [10, xiii].

Acknowledgement. This paper has been supported by the RUDN University Strategic Academic Leadership Program and Grant of RUDN University No. 060119–0-000.

References

1. Hummels, D., Ishii, J., Yi, K.-M.: The nature and growth of vertical specialization in world trade. J. Int. Econ. **54**(1), 75–96 (2001)
2. Koopman, R., Powers, W., Wang, Z., Wei, S.-J.: Give credit to where credit is due: tracing value added in global production chains. In: NBER Working Papers Series, vol.16426, p. 38 (2010). https://www.nber.org/system/files/working_papers/w16426/w16426.pdf. Accessed 25 May 2021
3. Li, X., Meng, B., Wang, Z.: Recent patterns of global production and GVC participation. In: Global Value Chain Development Report 2019. Technical Innovation, Supply Chain Trade, and Workers in a Globalized World, pp. 9–43. WTO (2019)
4. Netherlands Foreign Investment Agency (NFIA): Key industries – Agrifood. https://investinholland.com/doing-business-here/industries/agrifood/. Accessed 9 May 2021
5. Netherlands Foreign Investment Agency (NFIA): Key industries – Chemicals. https://investinholland.com/doing-business-here/industries/chemicals/. Accessed 9 May 2021
6. The Eora Global Supply Chain Database - UNCTAD-Eora Global Value Chain Database. https://www.worldmrio.com/unctadgvc/. Accessed 11 Mar 2021
7. Viviano, F.: This tiny country feeds the world. National Geographic (2017). https://www.nationalgeographic.com/magazine/2017/09/holland-agriculture-sustainable-farming/. Accessed 9 May 2021
8. World Investment Report 2013: Global value chains: investment and trade for development. https://unctad.org/system/files/official-document/wir2013_en.pdf. Accessed 28 May 2021
9. World Investment Report 2018: Global Value Chains: Investment and new industrial policies. https://unctad.org/system/files/official-document/wir2018_en.pdf. Accessed 6 May 2021
10. World Investment Report 2020: International production beyond the pandemic. https://unctad.org/system/files/official-document/wir2020_en.pdf. Accessed 9 May 2021
11. WTO News: Trade shows signs of a rebound from COVID-19, recovery still uncertain, 6 October 2020. https://www.wto.org/english/news_e/pres20_e/pr862_e.htm. Accessed 25 Jan 2021

Check for updates

Global Labor Market and Challenges of Digitalization

Marina V. Safronchuk[1](✉), Natalia V. Ivanitskaya[2], and Amirzhan K. Baibulov[3]

[1] MGIMO-University, Moscow, Russia
[2] National Research University "Higher School of Economics", Moscow, Russia
[3] K. Zhubanov Aktobe State University, Aktobe, Kazakhstan

Abstract. The present article examines the world labor market under the influence of increasing globalization and technologies of the Fourth Industrial Revolution, which has made the beginning of the digital transformation of the world economy. Such general alterations caused by technological changes in general as the growth of labor productivity, increasing labor mobility and competition, reducing transaction costs in the labor market are reflected in the research. Specific tendencies and new segments in the labour market have been identified. The key challenges - the substitution of a person for cars, Artificial Intelligence, the growth of structural unemployment, and inequality of incomes as a result of a digital gap, new requirements to the human capital are at the focus of attention. The state should reduce risks in the labor market by ensuring a comprehensive and systematic solution to these problems through industrial, innovative, and workforce policies, special training programs, and improving digital legislation.

Keywords: Globalization · Digitalization · Labor market · Labor demand and supply · Unemployment · Economic policy · Human capital · Labor productivity

JEL Codes: E24 · F620 · F630 · J2 · J3 · J42 · J48

1 Introduction

The digital era began since the Third Industrial Revolution, which leads to the automation of production, networked economy, e-commerce, having promoted the modern stage of globalization. Even at that time, major changes began to appear in the production, consumption, and distribution of the resources in the labor and capital markets. New conditions, arising from the participation of almost all markets in the global Internet, appeared. Evenatthattime, markets began to transform under the influence of network technologies when the conclusion of transactions became possible outside the borders and in real-time mode, that is global. The main changes are the transparency and the availability of information, the growth of resources mobility, the increase of the international labor and capital flows, the speeding up and the simplifying of deals, the reduction of transaction costs. However, the Fourth Industrial Revolution gave the world some fundamentally new digital technologies such as

E. G. Popkova (Ed.): ISC 2020, LNNS 368, pp. 142–150, 2022.
https://doi.org/10.1007/978-3-030-93244-2_17

Artificial Intelligence and blockchains, mining farms and cryptocurrency, big data and data centers, cloud storage of information arrays, 3D and 4-D printing (additive technologies), the Internet of things and for things, nano-chips, smart enterprises, and smart cities. Digitalization of the world economy has brought the transformation of global markets to a new level and new quality. By digital transformation is meant not only the implementation of technologies of the Fourth Industrial Revolution but also the creation of a digital product. The transition of employing companies to the digital model, the expansion of digital capital, have seriously affected the labor market.

The relevance of the paper is that the labor market is the most important object of macroeconomic researches. It does not only sensitively reflect the slightest changes in the economy, but also according to some indicators appears to be the carrier of the advancing information, and performs a predictive function. It gathers and processes the information from all markets and gives back the information signals, external for other markets. Not without reason, the Federal Reserve System of the USA makes decisions to change the key interest rate based on the analysis of labor market. To regulate labor market means to regulate the whole economy.

The tasks of research are to find out what challenges of the digital economy are, what processes happen in the economy and in the labor market, to determine whether these challenges can be met or adapted and what role the state should play.

2 Methodology

To identify changes in the global labor market under the conditions of digitalization, the analysis method was used in order to trace the characteristic features for each part of the labor services market, the analogue methods, the classification of labor market segments, and synthesis. The tools and the methods of microeconomic analysis (for instance, analysis of individual labor supply or positions of an employer) were applied as part of the study. The methods of macroeconomic analysis were used to identify structural imbalances and structural disparities, proposals for state regulation, and investment promotion. Data extrapolation method, causal analysis, normative and positive analysis were used.

In this paper statistics from international databases and research findings using software and building of econometric models, data from international expert communities and organizations, and materials from the World Economic Forum were used. The object of the study is the digitalization of the world economy, its changes as a consequence of the assimilation of technologies of the Fourth Industrial Revolution; the subject is the global labor market. The research studies its changes and challenges, new tendencies, and changes in response to the challenges and drivers of the digital age.

The study methodology is as follows. First, the essence of digitalization and the basic technologies of the digital revolution, their impact on economic relations, the efficiency of resources, methods and forms of business, competitive conditions, etc. are considered. Then the following points are examined: how this affects the labor market, how demand from digitalizing companies changes, its structure. What problems and challenges appear, how the labor market responds to them from the supply side, how this affects macroeconomic indicators and state regulation.

3 Results

The world public is alarmed for the growth of digital technologies, especially robotics with Artificial Intelligence, rapidly displaces people from the labor market. Robot Vulnerability Index measuring the degree of disruption of workplaces in industrial production was developed. According to Oxford Institute for Economic Policy (Oxford Economics) report "How robots change the world" (2019), the number of industrial robots will grow from 2.25 million to 20 million by 2030 and global employment in production will decrease by 8.5% [4]. China will lose the most of all - 12.5 million individual working places; the European Union will lose 3 thousand, the USA - 1.3 million and South Korea - 780 thousand. The rest of the countries (except for Japan) will lose 3000 employees in the industry only [4]. Digital technologies create a powerful effect of replacing labor with capital in other areas such as medical diagnostics, agriculture, energy (smart farms, smart enterprises, stations, etc.). Besides, international digital platforms (UBER, Amazon, etc.) in the labor market are becoming large monopsonists, and a global digital monopsony has been created.

The main problems concerning the labor market and human personality are that Artificial Intelligence being core and base technology can not only replace a person but also can impose solutions against his will and interests, eliminating him from economic life. Reduction and changes in labor demand, increase in structural unemployment, difficulties in retraining; growing digital inequality and income inequality, lack of specialists to digitalize companies, these challenges fall on the state budgets, pose new challenges to governments.

Indeed, the researchers talk about the powerful effect of replacing labor with capital, which is peculiar for all previous industrial revolutions. However, new technologies do not lead to the growth of unemployment, they reduce it [5]. For instance, over the past 100 years, the rapid development of technology has only contributed to the reduction of unemployment. Even in the depth-comparable crises of the 1929-30s (Great Depression) and in the great recession of 2008–2009, the unemployment rate varied dramatically. The fact is that the effect of displacement is overlapped by the effect of productivity, which expands production, and the effect of diversity. It is connected with the emergence of new goods and services, which create new jobs and professions. So, for instance, the development of only one Artificial Intelligence gave the world more than thirty new specialties [3].

Digital technologies do not reduce but alter the labor market seriously. The nature of labor, its parameters, and conditions change, new market segments appear. Productivity increases. For example, data processing and digital twins improve the quality of planning significantly and speed up the commissioning of industrial facilities, improve their maintenance, and reduce operational risks. The use of industrial drones has dramatically improved geologic exploration, replacing people in difficult and dangerous places of work.

The tendency noted in the research works of [2] about increasing labor mobility with a high level of human capital due to increased labor virtualization grows. This is facilitated not only by growth but also by the deepening of digital capital, which improves labor productivity.

Virtual labor collectives are increasingly created, the possibility of flexible recruitment of personnel for each specific project at the regional and international level is improved. Electronic recruitment systems and digital platforms in the labor market facilitate job search and hiring greatly, making the choice more efficient and fast, narrowly focused on a specific task. Digital selection reduces the costs of opportunistic behavior and moral hazard, as well as the overall transaction costs in the labor market, improving coordination between supply and demand. The emergence of remote flexibility in the provision of unique labor services is very important. For instance, remote surgeries through digital devices are independent of distances and boundaries. Time, haulage, and accommodation costs are saved. The possibility of obtaining remote services, especially specific and unique, increases the quality of life, and utility level of citizens. A person can flexibly combine work simultaneously in different cities and organizations.

The Fourth Industrial Revolution is distinguished by the harmonization and integration of a large number of various scientific disciplines and discoveries. Material innovation appears as a consequence of the interdependence between different technologies [6]. Accordingly, all specialties must contain a basic information technology component. Demand for a combination of IT engineering with knowledge of various industries, for related specialties in the field of biochemistry, bioengineering, nanomaterials, etc. grows. Automated design, additive technologies, engineering of materials, and synthetic biology are combined. Demand for economists, lawyers, and financiers declines.

In different market segments, according to the level of education and, correspondingly, human capital, substitution effect action with production capital (industrial robots, drones, and other devices with Artificial Intelligence) occurs unevenly. Mainly employees with medium qualifications are subject to replacement. In industry, their replacement ratio is on average 1.6 [4]. The new tendency lies in the fact that the expansion of the field of high technology causes "sagging" demand for middle-skilled workers, especially workers of monotonous jobs [10]. Such specialties, in particular, are more often replaced by Artificial Intelligence and are more susceptible to structural unemployment, and a relative decrease in wages. However, in addition to workers with high qualifications and education in the field of information technology, there is an increasing demand for workers with low or not high qualifications such as manual labor, not replaced by machines, and machine maintenance. In the graph (Fig. 1), the author presents a dynamic model of demand for labor with different levels of human capital (different qualifications of workers), reflecting the "sagging" demand and wages in the segment of middle-skilled workers. The curve in Fig. 1 reflects the dependence of wage dynamics on changes in marginal labor productivity (qualifications) as a function of human capital gain - MPL ($\Delta H/H$), where Δ w/w - real wage growth rate.

This tendency sets the government the task of increasing attention to this segment of the labor market in two directions. The first direction is to support workers from the "sagging" segment in moving to sectors with simpler work and to facilitate conditions of self-employment. Using the opportunities of e-government (for instance, digital platforms) one can assist in getting a job, starting a business, and marketing his products. The second direction is the economics of knowledge. It is necessary to strengthen investments in basic and extended education that meets the demand of

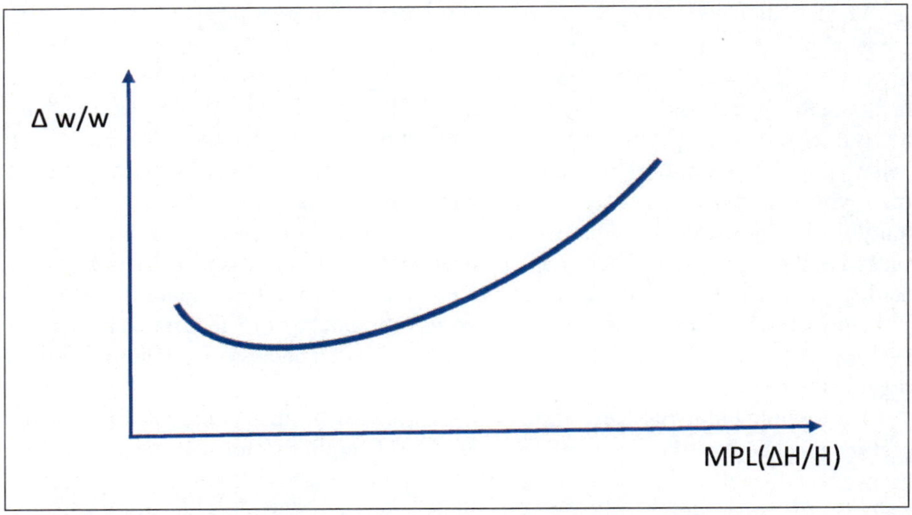

Fig. 1. The dynamic model of labor demand. *Source* Compiled by the authors.

digitalization. This is the development of online education, the creation of new educational centers, their more even and balanced territorial placement, the formation of the new architecture of education, the educational eco-environment. It is necessary to develop vocational education taking into account the needs of local enterprises and organizations, to expand "smart cities." Labor mobility and quality of life could be enhanced by the modernization and development of transport infrastructure with improved logistics of freight and passenger transport. Researches confirm a positive correlation between transport infrastructure development and the human development index, which includes human capital, especially in catching-up countries [8].

Concerning the dynamics of wages in general, there are two differently directed tendencies in the world market. On the one hand, there are practically no countries with very low-paid manpower, where transnational companies transferred their production in order to save on costs. Over time, enhancing the skills of workers in the learning process and improving living standards, in partnership with the activity of transnational companies, has led to a gradual increase in the cost of low-paid manpower in recipient countries of industrial capital. In the world labor market, thanks to the growth of mobility, wages in similar segments of different countries are equalized. Transnational companies now have incentives to replace the rising price of low-skilled labor services with machines, and it exacerbates competition and reduces wages in this segment.

Within the framework of structural and employment policies, the governments of these countries should invest more in the knowledge economy, in the development of human capital. The primal task of the state is to provide training in new specialties for digital transformation. It is required to integrate industry specialization and information technology components, combining specialties due to interaction and interpenetration of industries as follows: biochemistry, bioengineering, medicine, biology, cognitive and additive technologies, etc. There is a growing need for multidisciplinary competencies.

For instance, in e-commerce, marketing, and advertising competencies there is a need to be combined with web analytics, web design skills. It is very difficult to find training personnel. Specialties due to the rapid development of technologies and the emergence of new forms of business are constantly updated, which exacerbates personnel shortages. Private companies have a dilemma that is to dismiss and hire new workers, or to train [9]. Large corporations that can finance education, have long experience and infrastructure for training and retraining their employees, are inclined to training instead of layoffs. Medium and especially small-sized businesses, without experience and financial resources, are inclined more to hire new employees to digitalize companies [9].

However, by applying a benefits system the state can support companies of small and medium-sized businesses to train employees if appropriate, facilitating the situation in the labor market, and the burden on the educational market, on its segment of extended education.

The business has lost interest in the cheap, unqualified labour force, and under new conditions prefers to invest in individual working places with a higher level of human capital, giving a much greater return on investment. Therefore, state investment in human capital creates additional incentives for private investment in industry, regardless of the cost of credit, improves the investment climate, the investment attractiveness of the economy.

Access to investment in human capital can be facilitated by the use of smart contracts. The principle of separated responsibility in the blockchain prevents opportunistic decisions and expands the freedom of investment into effective projects, which must be taken into account in state regulation [12].

"Artificial Intelligence, quantum computing (super-powerful quantum computers) and augmented reality do not replace a person, but strengthen his professional experience, complementing with expert knowledge and the capabilities of super-powerful forecasting", said Satya Nadella, the head of Microsoft [7]. This once again emphasizes that the tasks of new technologies are not to displace a person but to improve the efficiency of human capital, to make labor more efficient and more comfortable. Artificial Intelligence is far from perfect as it does not have intuition, a reaction to non-standard situations. It's applying without human involvement can lead to emergency consequences. For instance, some accidents of the Boeing 737 Max X aircraft, failures in experiments with unmanned vehicles, etc. Artificial Intelligence experts conclude that it is more effective to use Augmented Intelligence [3].

Numerous surveys of executives of digitalizing companies have shown that they prefer to use human Artificial Intelligence, skills, and experience in approximately equal proportion [1].

Being at the stage of implementation of basic technologies, digitalization is still gaining strength. Its epicenter is knowledge-intensive services, high-tech and medium-tech industries, which even in technologically developed countries cover a small part of the workforce (for example, only 20% in the United States). Over time, this share will increase even now society changes under the influence of digitalization. Technological breaking lines generate a bifurcation of social transformations reflected in the labor market. Newmarket segmentation principles appear. In contrast to the traditional segments on employment and ability to work, industries, and professional affiliation, segments are revealed in the labor market. Segments are divided according to the specific characteristics and qualities of workers, which are naturally endowed with different groups of the population. Thanks to digital consultants, segments, corresponding to the psychological characteristics and preferences of different generations have been identified. For example, there are segments on the level of competencies and forms of remuneration, on the type of generation, on gender dominance in the sector of employment. So, in the service sector, which grows rapidly in the digital economy, women dominate. According to statistics, the efficiency of investments made in companies organized by women is higher than in startups organized by men. According to joint research by The Boston Consulting Group and Women in Energy (WoNY), over the past 5 years, women-organized startups have attracted 56% less investment, and the return on investment is 10% more than in companies organized by men [11]. Men predominate in the industrial and raw materials sectors, in which many types of work disappear. Correspondingly, women's employment and structural unemployment among men increase.

The labor market is segmented according to the principle of generations. Generations Boomers, Millennials, Gen Zens (generations X and Y) are characterized by high professional mobility and flexibility (transition from one profession and type of employment to another, the ability to combine), weariness from employment and social pressure, passive perception of reforms, choosing low incomes in exchange for calmness [9]. On the contrary, Generation Z (Centennials, Next, Indigo) born in the digital age does not divide the world into real and virtual, solves all problems with the help of applications energetically, strives to work hard, and earn money. The digital native is prone to career success and entrepreneurial business [9].

In general, the population passively perceives the problem of inequality in income and access to digital advantages. The population does not work for the competition and increase labor activity, preferring a calm life and social comfort. Most of the population is inclined to average income, and the interest of people in digitalization is not high.

Based on this, one can get an idea of the nature of the individual (in an average form) labor proposal in each of the segments considered, ranked by generation and their relation to income and wealth. These factors affect the different elasticity of individual labor supply on wages.

Such a variety (kaleidoscope) of features and preferences in the labor market allows an employer to unite workers from project to project flexibly. A calm attitude to income inequality shows that the overall level of income of workers is still growing for reasons of increasing technological efficiency and, by no means, social support of the government. The level of consumption and quality of life in each segment grows. The thinking of people is also rebuilt from a quantitative to a qualitative assessment of the well-being, and the facilitation of domestic efforts, which gives technological progress.

4 Conclusion

Information technologies of the Third Industrial Revolution, having made a breakthrough in efficiency, have exhausted their resource. Now they act as digital infrastructure, the basis for the introduction of technologies of the Fourth Industrial Revolution, which give a powerful impetus to the efficiency of not only capital but labor as well, improve working conditions and workplace safety. Productivity and efficiency increase as digital capital deepens and creative ideas are applied.

Thanks to IT technologies, labor mobility continues to grow, information in the labor market becomes more transparent and symmetrical, transactional costs are reduced, and job search and hiring efficiency are increased. Digital technologies allow discovering new segments in the labor market, combining flexibly the features of individual labor supply from different segments of the market.

The competition in the labor market grows; digitalizing companies increase the requirements to education, to the quality of human capital. Transnational companies are no longer interested in cheap work and replace it with machines. Investing in low-return human capital is not profitable. An increase in the rate of human substitution by robotics with Artificial Intelligence in industrial production with a coefficient of 1.6 can increase structural unemployment to 8.5% by 2030 in the world [4]. This challenge is very serious.

The other challenges of digitalization are the increasing imbalance of the labor market. The supply structure responds slowly to the rapidly changing demand structure. Imbalance in separate segments as follows: the demand for medium-skilled labor is interrupted, demand and incomes in the middle-class sector decrease, and demand for high qualifications, engineering, and related specialties with the IT component grows rapidly.

Due to the rapid and unpredictable development of new technologies, risks in the labor market grow; they are reinforced by the competitive undermining of digital companies and the emergence of global digital monopsonies. These and other challenges of digitalization set the government the task of forming new approaches to regulation, economic and personnel policy. It is necessary to ensure the protection of the employee through the development of labor legislation in the field of digital labor relations. As part of digitalization programs, it is necessary to increase investment in human capital, cooperate with businesses actively and flexibly to develop training in new specialties and competencies for the digital economy. It is necessary to create a new architecture of education, an eco-system of the educational environment jointly. The government should encourage small and medium-sized businesses to train and retrain employees within companies, to participate in the creation of educational platforms and flexible internal training systems. All these measures will improve the investment climate and competitiveness of the economy; reduce risks in the labor market. They will help to compensate faster for the effect of replacing labor with capital, contributing to the expansion of the effects of productivity and diversity that saved the world from rising unemployment after previous industrial revolutions.

References

1. Bruno, S., Popkova, E.: Human capital and AI in industry 4.0. Convergence and divergence in social entrepreneurship in Russia. J. Intellect. Cap. **21**(4), 565–581 (2020). https://doi.org/10.1108/JIC-09-2019-0224

2. Brynjolfsson, E., McAfee, E., Spence, M.: New world order labor, capital, and ideas in the power law economy. Foreign Aff. **93**(4), 44–53 (2014). http://www.foreignaffairs.com/articles/141531/erik-brynjolfsson-andrew-mcafee-and-michaelspence/new-world-order
3. Digital Intelligence Development International Forum, Moscow (2019). Accessed 08 Nov 2019
4. How robots change the world (2019). What automation means for jobs and productivity. Oxford economics, June 2019. http://resources.oxfordeconomics.com/how-robots-change-the-world. Accessed 8 Apr 2021
5. Pettinger, T.: Cracking Economics. Translated by Robinovich, E.L., p. 320. ACT, Moscow (2020)
6. Schwab, K.: The Fourth Industrial Revolution, 208P. World Economic Forum (2017)
7. Schwab, K., Davis, N.: Shaping the Fourth Industrial Revolution, 287P. World Economic Forum, Cologny (2019)
8. Sergeeva, M.: Transport infrastructure's effect on human development: distinguishing between different channels of influence. Econ. Gov. Probl. Decis. **2, 5**(101), 58–63 (2020)
9. Sologubova, G.: To the question on digitalization and labor market problem. Digit. Econ. **2** (2), 50–62 (2018)
10. The Future of Jobs. Report. World Economic Forum (2016). http://reports.weforum.org/future-of-jobs2016/shareable-infographics. Accessed 8 Apr 2021
11. Why Women-Owned Startups Are a Better Bet (2018). The Boston Consulting Group (BCG) report, June 2018. https://www.womeninenergy.eu/wp-content/uploads/2018/12/Women_in_Energy_in_the_CEE-E_Region_Dec2018_final.pdf
12. Zavyalova, E.B., Shumskaia, E.I., Kuzmin, M.D.: Smart contracts in the Russian transaction regulation. In: Popkova, E.G., Sergi, B.S. (eds.) ISC 2019. LNNS, vol. 129, pp. 205–212. Springer, Cham (2020). https://doi.org/10.1007/978-3-030-47945-9_22

Funding Model of the Tourism and Recreation Sector in the Regions with Limited Self-development Potential

Zhanna A. Zakharova[1(✉)], Pavel N. Zakharov[2], and Andrey A. Sherkashin[2]

[1] Vladimir Branch of the Russian Presidential Academy of National Economy and Public Administration, Vladimir, Russia
[2] Vladimir State University, Vladimir, Russia
pz@vlsu.ru

Abstract. At most Russian regions very necessary are need to solve the contradiction between reach tourist and recreation potential (a lot of picturesque, memorable places of architecture, landscape, attractions, monuments of culture, archeology, etc.) and possibilities to satisfice demand from tourist flow (internal and international tourist flow). And the most promising tool in smoothing this contradiction is the use of public-private partnership, namely a specific model of concession in the development of tourism and recreational areas of the regions. The purpose of this article is identifying the prerequisites and limitations of the self-development potential formation of tourist and recreational areas in the Russia center regions (on the materials of Vladimir, Ivanovo, Kostroma and Yaroslavl regions), and the development of a model of concession in the tourism and recreational sector in relation to the specifics of the conditions of regions with insufficient potential for self-development. The result of the research is a model of concession in the development of tourism in the Russia center regions. At the same time, it is proved that in regions with insufficient potential for self-development, the most acceptable will be those concession models that involve significant guarantees from the public authorities. In view of the limited budget resources of the Russia center regions, the most relevant source may be the resources of state financial development institutions or various funds accumulating funds on both market (rental) and non-market basis (instruments of state stimulation).

Keywords: Funding model · Concession agreement model · Tourism · Infrastructure · Regional economy

JEL Codes: R58 · Z23

1 Introduction

In the context of the turbulent development of the modern economy, caused, among other things, by the consequences of the COVID-19 pandemic, it is possible to identify the most affected industries and types of economic activity, which include the service sector in general and tourism in particular. These areas require a certain amount of

E. G. Popkova (Ed.): ISC 2020, LNNS 368, pp. 151–161, 2022.
https://doi.org/10.1007/978-3-030-93244-2_18

government support. Each country defines its own combination of tools and forms of this support. In Russia, in accordance with the Federal Law of December 8, 2019. No. 385-FZ "On the Federal Budget for 2021 and for the planning Period of 2022 and 2023" provides for financial support for the tourism sector within the framework of program measures, including subsidies for co-financing tourist infrastructure facilities, grant support, and subsidizing part of the interest rate of credit institutions that are investors in the tourism sector. In addition, from 2020, a separate measure of financial support in the form of tourist cashback is provided. It is necessary to emphasize the difficulties of access of the participants of the regional tourist market to financing at the expense of the federal budget. At the same time, funding under regional programs is very limited. Therefore, the relevance of attracting private capital to the development of the tourism sector in the region on the basis of the financial mechanism of public-private partnership is growing.

The subject matter of public-private partnerships, and, in particular, concession agreements, is today one of the most relevant and interesting for a significant number of researchers. And this is not accidentally, since the existing domestic and foreign experience in the implementation of concessions shows that with this form of interaction between the state and business a number of serious problems can be solved, both at the level of the national economy and at the regional and municipal levels.

Initially, the concept of public-private partnership implied the use of this tool in capital-intensive, socially significant projects, where state resources are limited and, at the same time, the need to realize them is the main function of the state (utilities, roads, social infrastructure). However, at present the concept involves the use of public-private partnerships, mainly concession agreements in the types of economic activities necessary to ensure sustainable socio-economic development of regions where private capital has certain objective limitations that restrain their self-development (Self-developing are the types of economic activity that, under the conditions of a modern market economy (spontaneity, uncertainty and cyclicality), independently, due to accumulation of internal resources and internal motives, react to changes in environmental factors (level of demand, etc.), ensuring their steady state in the medium term and long term periods).

Under the concession, we understand the organizational-economic form of interaction between the state and business, carried out on the basis of long-term contractual relations, as well as on the basis of the principle of complementarity (Zakharova 2016), on the one hand (state), the lack of resources (financial, labor, entrepreneurial), on the other hand (business) also the lack of resources (long-term capital, technology), insufficiency of motives, high risks of economic activity, in order to meet social needs (state, enterprises, population) in industries, types of economic activities with the lack of capacity of self-development.

That is, with this approach, the main criteria where it is advisable to use concessions are not only a shortage of resources, high risks, but also insufficient motivational mechanism for economic activity (under the current economic conditions). And the main purpose of the implementation of concessions is meeting public needs. As the established experience shows, for the effective implementation of concessions, first of all, the necessary objective prerequisites are necessary, and mainly public request for the development of a particular activity.

Undoubtedly, the peculiarities of the development of such regions as Moscow and Moscow region to the neighboring regions can be also used as a perspective and in the development of the tourist and recreational sphere.

2 Methodology

In the course of a scientific study, the authors used methods of a systematic approach, and also used both general (comparison and analogy, analysis and synthesis), and general scientific (normative and positive etc.) research methods. In addition, methods of modeling socio-economic processes were used for this research.

3 Results

The purpose of this work is to identify the factors, prerequisites and limitations of the formation of self-development potential in the tourism and recreation sector in the regions of the center of Russia (based on the materials from the Vladimir, Ivanovo, Kostroma and Yaroslavl regions), and to develop a model of a concession in the tourism and recreation sector for specific conditions of regions with insufficient self-development potential.

To achieve this goal, we tried to identify, on the basis of significant statistical material, the main objective processes taking place in the regions of central Russia, which showed the following results.

Thus, in the territories of Vladimir, Ivanovo, Yaroslavl, Kaluga, Nizhny Novgorod regions, due to the high urbanization of the population (approximately 80% of the urban population), there is significant domestic demand for recreational services (rehabilitation, recreation, treatment), as well as steady demand for visiting places of tourist-recreational sphere of these areas by the residents of Moscow and Moscow region (due to the environmental problems, high pace of life and limited territorial resource of the region). In addition, in self-sufficient regions, the improvement the quality of life of the population, the level of income in accordance with the law of elevating needs, opens up new needs, including the need for rest, which determines the priority demand for tourist and recreational services, and the requirement for their diversity (Babaev 2008). The initial desire for recovery, recovery is complemented by the desire for cultural and spiritual development, which requires the emergence of various forms of tourism (agritourism, ecological, industrial tourism, etc.).

Also due to the decline of the national currency rate in 2014, visiting tourist facilities turns out to be promising for foreign tourists as well. According to the statistical data of the Federal Agency for Tourism, it can be noted that over the period from 2014 to 2019, the number of foreign tourists placed in collective accommodation facilities almost doubled throughout Russia (Table 1). Due to the factor of the COVID-19 pandemic, the number of foreign tourists has sharply decreased (in Russia as a whole by 5 times).

Table 1. The number of foreign citizens placed in collective accommodation facilities (excluding microenterprises), thousand people

Territory	2013	2014	2015	2016	2017	2018	2019	2020
Russian Federation	4443,6	4607,9	5033,5	6092,2	8028,8	10616,8	10855,9	2159,2
Central Federal District	2235,1	2209,0	2511,0	3062,4	4434,2	6171,2	6017,8	1300,7
Vladimir region	30,1	25,3	57,6	92,7	124,6	102,5	121,9	4,3
Ivanovo region	5,6	4,5	5,1	4,6	5,6	4,8	5,2	1,44
Kaluga region	59,4	59,0	26,4	18,1	20,5	33,2	31,3	11,3
Kostroma region	6,9	7,3	4,5	7,1	7,8	9,6	7,2	1,3
Moscow region	169,6	139,5	164,7	227,9	293,0	542,2	438,6	110,5
Ryazan region	9,1	13,9	14,3	18,7	14,3	14,5	8,7	3,8
Yaroslavl region	22,2	19,6	44,1	44,8	36,0	63,7	41,8	4,4
Moscow	1822,8	1827,0	2074,5	2516,6	3790,7	5234,4	5161,4	1097,6

Source Developed and compiled by the authors according to the official website of the Federal Tourism Agency [11]

So, in the Vladimir region, the same indicator increased almost 5 times (from 25.3 thousand people in 2014 to 121.9 thousand people in 2019).

In addition, the most popular place in the Vladimir region is the city of Suzdal, which is visited annually by 1.7 million people, mainly residents of Moscow and the Moscow region as a weekend getaway. And according to the Russian travel agency Turstar, at the end of 2017, the cities of Ples and Suzdal became the most popular small cities among tourists in Russia. So in 2017, Ples (Ivanovo region) accounted for 210 tourists/resident (400 thousand tourists per year/1.8 thousand residents) and Suzdal (Vladimir region) - 175 tourists/resident (1.7 million tourists per year/9.7 thousand residents). However, the stay in Suzdal is 5,4 thousand rubles per person per day, which is clearly not available for the population with a low income level. That is, the existing proposals in these regions are not sufficient in principle (and do not sufficiently take into account the interests of different segments of the population (different income levels). Also, the use of tourist and recreational resources is limited due to the low quality of tourist services, the insufficient development of tourist and supporting infrastructure (hotels, public catering facilities, etc.), and the high level of their wear and tear. And such type of economic activity as "Activity of hotels and public catering enterprises" in the GDP structure of the Vladimir and Kostroma regions for 2019 was only 1.1%, Ivanovo-1.4%, and Yaroslavl - only 0.9% [7]. Not a high level of profitability, restrictions on the own financial resources of enterprises of the type of economic activity "Activities of hotels and public catering enterprises" do not allow for significant investment. Thus, investments in fixed assets for this type of economic activity are practically not carried out in the Ivanovo, Kostroma, and Nizhny Novgorod regions (0.0% each), and in the Yaroslavl and Vladimir regions this indicator was 0.02% and 0.01%, respectively. In absolute terms, the dynamics of investments in fixed assets aimed at the development of collective accommodation facilities (hotels) indicates a decrease in investment in this type of economic activity for the period from 2016 to 2017 in the Vladimir region by almost 5 times (from 285.8 to 57.4 million rubles). in the Ivanovo region by 3 times (from 4.1 to 1.34 million rubles), in the Ryazan region by 2.3 times (from 28.7 to 12.7 million rubles) (Table 2).

Table 2. Investments in fixed assets aimed at the development of collective accommodation facilities (hotels, other places for temporary residence) (excluding small businesses) million rubles

Territory	2013	2014	2015	2016	2017	2018	2019
Russian Federation	67449,0	80468,5	32653,5	30366,1	32645,59	42868,14	36682,9
Central Federal District	11754,0	11348,2	8391,4	9144,6	13987,29	22606,6	8245,1
Vladimir region	36,2	176,9	112,9	285,8	57,40	46,67	79,47
Ivanovo region	11,6	1,4	6,2	4,1	1,34	1,9	–
Kaluga region	118,1	79,5	729,7	62,7	121,05	72,6	42,4
Kostroma region	2,0	9,0	3,4	9,1	30,95	1,3	33,6
Moscow region	3319,6	3665,2	4316,9	2070,5	2676,87	3443,4	2801,4
Ryazan region	7,1	3,1	6,3	28,3	12,70	20,9	4,7
Yaroslavl region	38,8	242,3	429,3	1135,1	2215,92	1322,3	977,3
Moscow	7191,2	4547,7	2320,9	4369,9	8310,89	16778,6	3673,9

Source Developed and compiled by the authors according to the official website of the Federal Tourism Agency [12]

It should be noted that the maximum volume of investment in the tourism sector in Russia as a whole is observed in 2014. In the future, the volume of investment in fixed assets has unstable dynamics. In addition, you should pay attention to the fact that Moscow and the Moscow region make a significant contribution to the total investment volume. If we analyze other regions of the Central Federal District, we can distinguish a multiple lag of the Vladimir, Ivanovo and Kostroma regions from the Yaroslavl region, despite the fact that these territories are included in the famous tourist route "Golden Ring of Russia".

Summarizing the entire above, Table 3 shows the prerequisites and limitations of the formation of the potential of self-development in the tourism sector of the region.

Table 3. Prerequisites and limitations of the formation of the potential of self-development in the tourism sector of the region

№	Requirements for the formation of the potential of self-development in the tourism sector of the region	Restrictions on the formation of the potential of self-development in the tourism sector of the region
1	In the conditions of digitalization of the economy and the formation of a post-industrial society, the share of incomes that can be spent in the service sector, namely in the tourism sector, is increasing	Insufficient level of the tourism infrastructure development is holding back the growth of tourist flow
2	The sphere of tourism due to the effect of cooperation can cause a growth in business activity in related areas of the regional economy	Poor diversification of tourism types focused on educational, ecological, industrial and other types of tourism

(*continued*)

Table 3. (*continued*)

№	Requirements for the formation of the potential of self-development in the tourism sector of the region	Restrictions on the formation of the potential of self-development in the tourism sector of the region
3	High urbanization, environmental problems, positive dynamics of monetary incomes of the population result in the growth of demand for tourist and recreational services	High level of depreciation of tourist and recreational facilities, which requires significant investments
4	Significant tourist and recreational potential (the presence of picturesque, memorable places of architecture, landscape, attractions, cultural, archeological monuments, etc.) increases the investment attractiveness of the sphere, including for foreign investors	Unstable interstate relations restrain the inflow of foreign tourists
5	The depreciation of the national currency makes the tourism sector attractive for foreign tourists in terms of price	Insufficient supply for categories of citizens with different income levels

Source Developed and compiled by the authors

Focusing the attention of regional and federal authorities on the need to implement projects and pro-grams, stimulating the investment process in this area, would make it possible to activate the significant resource potential of the regions of the center of Russia, helping to satisfy both domestic public needs for recreation and the needs of foreign tourists. Due to the objective limitations on the development of this type of economic activity, in our opinion, it is advisable to attract investments in this area on the basis of pooling state and business resources in the form of concession agreements.

At the same time, the effective implementation of concession agreements requires the selection of an optimal model that takes into account not only the development of the industry or type of economic activity, but also the socio-economic development of the region (degree of self-development) and includes a different combination of state and private capital.

After examining the existing experience in implementing projects based on concessions, we note that in the regions of Russia in various types of economic activity (housing and public utilities, toll roads, health care, sports) the following models of concession are mainly used:

- model with direct payment system. This model provides for the financing of new construction or reconstruction of existing tourist infrastructure facilities at the expense of private capital and the return on investment at the expense of consumers of the service The model is implemented only in conditions of in-creased unmet market demand for project services, while the level of fees from consumers should allow the concessionaire to service the attracted funding and operate the facility without additional financial guarantee mechanisms from the state.
- the second model provides for a return on in-vestment due to operational payments paid by the state (accessibility fee). This model provides for the return of the invested investments at the expense of the municipal or regional budget. The use of this model in the financing of tourist infrastructure facilities in regions with insufficient potential for self-development is complicated by the low resource capabilities of these budgets Thus, for the time being, in the economic aspect, only developed regions can use this scheme (Moscow). Although individual industry ministries provide for such a model and, accordingly, a guarantee of reimbursement of the investor's costs at different stages of the project at the expense of development institutions (for example, at the expense of the Housing and Utilities Reform Fund) (Zakharova 2017);
- the third model involves the return of investment by a private partner at the expense of consumer payments, but under the condition of a minimum guaranteed return for a private partner from the gran-tor. That is, in case of failure to achieve certain revenue values by the concessionaire, the grantor pays direct compensation to the level approved in the con-cession agreement. The model is applied in the case of a projected growth in demand for a project service over time (with an output of design capacity in a few years). The risk of demand is shared between the state and private capital.
- the fourth model assumes as a payment mechanism the use of consumer fees and the possibility of carrying out other commercial activities (differentiation of types of activity). When private capital is given the opportunity to diversify economic activities to increase the return on investment For example, when re-constructing a museum, a manor, or a cultural proper-ty with an affordable tariff for a consumer, a private investor receives a preferential right to develop land near the objects and the possibility of organizing various types of commercial activities (hotels, parking lots, shops, etc.) demand provides high income.

Having correlated the types of economic activity, the regions and the concession models used in these activities, we made a conclusion that the models that are most appropriate in the conditions of various types of regions. Table 4 shows the criteria, models and tools for the use of concessions in regions with different potential for self-development.

Table 4. Criteria, models and tools for the use of concessions in regions with different potential for self-development

Regions	Criteria (Zakharova, Zh.A., Bogatyrev V.V. 2018)	Concession models	Tools
Self-developing regions	– regions with a significant share of sustainable activities – having an optimal set of types of economic activity – possessing a high degree of openness of the regional economy – the development of a critical mass of entrepreneurship – having a high proportion of investment in fixed assets – having a share of profitable enterprises at least 80% – having high innovative potential – having higher indicators of the quality of life of the population in terms of income and salary, in terms of the share of paid services in the structure of household expenditures as compared to the national average, which forms the individual interest of the population in living and working in the region	With direct payment	Subsidizing domestic tourism and various target programs for the development of tourism, industry development strategies
		Differentiation of activities	Preferential right to use land
With insufficient self-development potential	– having no significant differentiation of types of economic activity, which makes the region's economy less resilient to various fluctuations in the conjuncture of both the domestic and foreign markets – having a high proportion of unprofitable enterprises (30% and above) – having low indicators of the quality of life of the population in terms of income and salary, in terms of the share of paid services in the structure of household expenditures as compared with the national average – having a low share of investment in fixed assets – and etc	Minimum guarantee of profitability	Compensation from the budget of the concessionaire's shortfall in income, state support in the form of subsidies at various stages of the project in the tourism and recreation sector of the region
		Accessibility payment	Annuity payments of the constituent

Source Developed and compiled by the authors

With regard to the conditions of regions with insufficient potential for self-development (no significant differentiation of types of economic activity; a high proportion of unprofitable enterprises (30% and above), lower than the country's indicators of the quality of life of the population in terms of income, in terms of salary, household expenditures, low investment and innovation activity, etc.) the greatest prospects for use are models such as: minimum guarantee of profitability and accessibility fees (based on not having budget co-financing in them). However, taking into account the demand for the tourist and recreational sphere in the regions we are considering, a concession model with a minimum guarantee of profitability will be more suitable. At the same time, this model can also take into account some state support instruments as pre-investment (land tax, land lease), investment (subsidizing capital investments depending on the scale of the project, subsidizing interest on loans used for the construction or reconstruction of tourist facilities). - recreational infrastructure), and at the operational stage (guarantees against risk situations) of the implementation of tourism and recreation projects in the regions.

That is, in regions with insufficient self-development potential, the most acceptable will be those concession models that imply significant guarantees from the state authorities, which is reflected in budget burden of the state. Due to the limited budgetary funds of the regions of the center of Russia, the most relevant source of providing such guarantees can be resources of state financial development institutions or various funds accumulating money on both market (rental) and non-market basis (state incentive instruments).

For example, some researchers suggest that it is advisable to create non-profitable state-public financial institutions (institutions) (Babayev, Tyshina 2012) that use long-term refinancing of the Central Bank at a small percentage and commercial banks as operators that make a profit on their loan servicing costs (and not a percentage of the loan amount) will lend to various kinds of targeted infrastructure projects, monetizing, on the one hand, the real needs of the population, turning them into demand, on the other hand, by providing credit resources to business entities, in order to increase supply, satisfying increasing demand (Gritsenko 2015).

But, all the same, if we are talking about significant state guarantees (interest, guarantees against excessive risk), then we raise the question of creating appropriate funds and sources of these funds financing. Nekipelov A.D. believes that a company with 100% state participation is needed, which, like in the private sector, on a market-based, competitive basis would hand over the appropriate resources for use and for financing this company it is advisable to create a market mechanism for rent withdrawal (Nekipelov 2012). Rent is a regularly received income from “real estate” - property, land.

In the tourist and recreational sphere, the formation of rent is associated with the concept of a tourist destination. A tourist destination is an intensively produced and consumed set of tourist products, concentrated in time and space. In the traditional sense, a destination is a geographic area that is attractive for tourists, that is, it offers a certain set of services that meet the needs of tourists and satisfy their demand related to travel, overnight, meals, entertainment, etc. (Gavrilchak 2009).

The main factor ensuring the effectiveness of the practical use of the designated model is the formation of an objective specialized funds, formed on the basis of a part of the rent withdrawal from tourism revenues (tourist tax).

4 Conclusions

The central place in the system of tourist differential withdrawal is the corresponding financial mechanism, which implies the establishment of a de facto indirect raising tax in the tourist destination (tourist tax). It is advisable to accumulate these payments in a regional centralized trust fund. This fund is a source of investment in the modernization of tourist and recreational infrastructure as support for concession agreements in the development of the tourist and recreational sector of the region. The determination of the amount of this payment depends on the following factors:

– the level of the development of the tourist destination;
– the size of the tourist flow;
– the duration of stay of one tourist;
– benefits for certain categories of tourists.

The mechanism for raising tourist tax is to establish the absolute or relative tax for a tourist service unit (when paying for hotel services, when buying tickets to museums, etc.).

Thus, the use of the developed concession model implies a reduction of restrictions in the formation of the self-development potential of the tourist and recreational sphere of the regions of the center of Russia.

References

1. Babaev, B.D., Borovkova, N.V.: On the issue of demand for recreational and tourist services in the collection: innovative development of regions in the context of globalization. In: Materials of the International Scientific and Practical Conference, pp. 6–9 (2015)
2. Babaev, B.D., Tyshina, V.N.: On the concentration of society's savings by financial and credit institutions and the organization of their transfer to the real sector: theoretical and methodological aspects. Bull. N. A. Nekrasov Kostroma State Univ. **6**, 18–22 (2012)
3. Buvaltseva, V.I.: The essence and content of the process of self-development. In: All-Russian Scientific Conference "Science and education." Collection of Scientific Papers, pp. S.25–29. KemSTU, Belovo (2002)
4. Gavrilchak, N.I.: Tourist rent. Forms of recreational rent. Russ. Entrepreneurship **2**, 86–92 (2009)
5. Gritsenko, A.A.: A new paradigm of economic policy is needed. World of Changes. Spec. release, pp. 42–46 (2015)
6. Nekipelov, A.D.: Point of View, 2nd ed., Ext./M: Master: INFRA-M, 2012. -512, p. 346 (2012)
7. Rosstat, M.: Regions of Russia. Socio-economic indicators. Stat. sat./p. 1242 (2020)
8. Zakharova Zh.A.: On the principles of building a theoretical model of organizational forms by concession type. Bull. Ivanovo State Univ. Ser. Econ. **1**(27), 30-35 (2016)
9. Zakharova, Z.: State Concession policy in the housing and utilities sector. Federalism **1**(85), 119–130 (2017)
10. Zakharova, Z.A., Bogatyrev, V.V.: Interconnection between the categories of region's self-development and population's living standards. In: Popkova, E.G. (ed.) HOSMC 2017. AISC, vol. 622, pp. 440–447. Springer, Cham (2018). https://doi.org/10.1007/978-3-319-75383-6_56

11. Official website of the Federal Tourism Agency. https://tourism.gov.ru/contents/analytics/statistics/chislennost-inostrannykh-grazhdan-razmeshchennykh-v-kollektivnykh-sredstvakh-razmeshcheniya-rosstat/. Accessed 12 Apr 2021
12. Official website of the Federal Tourism Agency. https://tourism.gov.ru/contents/analytics/statistics/statisticheskie-pokazateli-kharakterizuyushchie-razvitie-turistskoy-otrasli-v-rossiyskoy-federatsii-/. Accessed 12 Apr 2021

The "Invisible Hand" of Digitalization: The Challenges of the Pandemic

Kirill D. Karpunin[1(✉)], Julia V. Ioda[2], Kristina O. Ternavshchenko[3], Zhanna A. Aksenova[4], and Tatyana G. Maglinova[5]

[1] G.R. Derzhavin Tambov State University, Tambov, Russia
[2] Lipetsk State Technical University, Lipetsk, Russia
[3] Kuban State Technological University, Krasnodar, Russia
[4] Saint-Petersburg State Forest Technical University, Saint-Petersburg, Russia
[5] Academy of Marketing and Social Information Technologies, Krasnodar, Russia

Abstract. The research aims at revealing the effect of the "invisible hand" of digitalization on the mechanisms of market self-regulation of the socio-economic system, as well as determining its capabilities to overcome the negative consequences of turbulence due to the pandemic. The authors proved that in terms of digitalization, the principle of the "invisible hand" of the market transforms its content. On the one hand, the development of the mechanism of self-regulation of market forces takes place within the framework of the procedures for agreeing on a common understanding of the essence of the development of the new markets using digital technologies and taking into account the processes of their convergence. On the other hand, the competitive advantage in the market is achieved by those entities that have access to large databases of consumer data, use targeted marketing technologies and apply new-generation digital algorithms. The authors described how a specific digital market mechanism is actively formed in the virtual space, which differs from the mechanism of the traditional market. It is determined that the effect of the "invisible hand" of digitalization leads to the creation of technological systems, automation of business processes, automation of decision-making processes and the achievement of digital convergence, generating wave processes of business activity growth. It is proved that during the pandemic, the effect of the "invisible hand" of digitalization can ensure the intensive development of the digital economy sectors against the background of a significant compression of traditional sectors.

Keywords: Digitalization · Digital economy · Invisible hand of the market · Invisible hand of digitalization · Digital convergence · Pandemic

JEL Codes: O11 · O33

1 Introduction

In terms of dynamically developing digitalization, the influence of innovative drivers, which provide competitive advantages and accelerated economic development to the economies of developed countries, is becoming more and more noticeable [21]. Digital

E. G. Popkova (Ed.): ISC 2020, LNNS 368, pp. 162–173, 2022.
https://doi.org/10.1007/978-3-030-93244-2_19

solutions and developments, technological innovations and organizational innovations are used both in the real sector of the economy and in the areas of intangible activities. At the same time, the convergence of digital technologies and processes is noteworthy. On the one hand, the convergence between data transmission services, telephony, broadcasting, and OTT services creates the basic conditions for the development of new markets using digital technologies [32]. Another area of convergence concerns the introduction of ICT systems and tools in all spheres of economic activity. In this case, the mechanism of the "invisible hand" of digitalization is activated, when the strengthening of market power is determined by access to large databases of consumer data, the use of digital technologies and artificial intelligence and generates wave processes of business activity growth. The study of the effect of the "invisible hand" of digitalization becomes especially relevant during the turbulence caused by the COVID-19 pandemic.

2 Literature Review

Smith used the term "invisible hand" of the market to describe the forces that ensure the achievement of generally useful economic results, regardless of the initial intentions and actions of individual individuals or even contrary to them. In his work "Research on the nature and causes of the wealth of nations", Smith proved that the equilibrium in the market is achieved by regulating supply and demand using market mechanisms. This happens through self-regulation between the producer and the consumer, based on meeting the needs and interests of each group [20, 30].

Digitalization leads to a radical change in the principles of the organization of production processes, and, therefore, the mechanisms for regulating supply and demand in the market [22]. On the one hand, the emerging high-tech information and network environment repeatedly expands the capabilities of producers and provides them with effective ways to promote goods and services, develop marketing communications, attract the attention of the target audience, which is becoming a scarce resource in digital conditions [17]. On the other hand, digital technologies create conditions where the consumer has a wider interactive choice of goods and services under the influence of the online community of consumers, their independent assessments and views. In addition, thanks to digital technologies, consumers can be involved in the process of creating goods and services [10, 13]. In the conditions of digitalization, there are processes of convergence between telephony, data transmission, broadcasting, OTT services [32]. It is proved that the convergence of such types is very widespread and provides a significant economic effect, primarily synergetic [4, 5, 21]. In particular, the critical role of convergence of breakthrough technological and organizational innovations is to generate wave processes of growth in production volumes and business activity [21]. In this context, researchers are increasingly using the term "invisible hand" of digitalization, implying that A. Smith's "invisible hand" of the market as a mechanism for regulating supply and demand in the market is being replaced by digital technologies and the effects they cause [3].

During the COVID-19 pandemic, there was a significant growth of the digital sector, and the use of information technologies to support various sectors of the economy and preserve the socio-economic well-being of countries [11].

3 Methodology

The purpose of the study is to reveal the effect of the "invisible hand" of digitalization on the mechanisms of market self-regulation, as well as to determine the possibilities of its use to overcome the negative consequences of turbulence due to the pandemic.

Research objectives: 1) to reveal the content of the effect of the "invisible hand" of digitalization; 2) to highlight the features of the effect of the "invisible hand" of digitalization on the mechanism of self-regulation of market forces; 3) to determine the possibilities of using the technologies of the "invisible hand" of digitalization to overcome the negative consequences of the pandemic on the socio-economic system.

Research methods: theoretical analysis and synthesis, systematization, comparative analysis, economic analysis, induction and deduction, graphical method, system approach.

4 Results

Economics says that the main principle of economic activity is to ensure the usefulness of the product to the consumer and to receive benefits from the producer. In this case, the traditional understanding of the "invisible hand" of the market consists of six main components: the market price (the value of the goods, which is determined by supply and demand); the supply of goods (the willingness to produce and offer the consumer a certain product for a certain period); free competition (the situation on the market when neither the consumer or the producer can influence the price of goods); capital flow (the ability of capital to be redirected from low-profit areas to high-profit enterprises); steady demand (demand for everyday non-replaceable goods and services, for example, public transport, food); affordable lending at reasonable interest rates.

The emergence of digital technologies has led to the formation of the foundations of the market mechanism in the virtual space, causing a strong and steady increase in the number of transactions made via the Internet. As a result, there is the formation of a new market system that carries radical changes in the current models of economic behaviour of market participants [15].

Digital technologies transform the environment of interaction between the producer and the consumer, reducing the effectiveness of the "invisible hand" of the market in terms of the organization of production and distribution of products [38]. In particular, the basis of a market economy is exclusivity as the ability of producers/sellers to force consumers to become buyers. In the digital economy, new opportunities for replication and delivery of products are emerging, and the ability of the owner of the goods to exclude competitors from their segment by simple and cheap means is disappearing. In a market environment, producers compete, performing the same operations at different costs. In the digital economy, the cost of replicating digital products is approaching zero, which eliminates differences between producers/sellers in their costs for servicing additional orders [15]. Traditionally, the market environment is characterized by transparency, which demonstrates a clear understanding of individuals' needs and choices in purchasing goods and services. By contrast, the digital environment creates conditions for the development of long-term relations between the producer and the consumer in the terms of trust [16]. Thus, the traditional scheme of interaction of economic entities, as well as the goals and mechanisms of their interaction, is being transformed.

In the conditions of digitalization, the primary subjects of the economy (stakeholders) are individuals/consumers and producers.

Digital technologies have created a more convenient market than before, where consumers can focus on exactly the products that they need. Consumers are increasingly using online technologies to select and purchase goods or services [8]. However, for certain categories of goods, they can apply traditional shopping schemes. In addition, the Internet and digital technologies are actively used by consumers to collect information about goods, services, and their qualitative characteristics when making consumer choices. They have more information about brands and are interested in a large number of brands. Thus, the relative humidity of consumers as participants in market relations increases. In addition, in the digital economy, the consumer becomes a direct participant in the process of creating consumer value [24, 25, 34]. This allows you to create a new individualized product and distribute the benefits of this activity among all participants in the interaction.

Producers in the digital economy are in a value creation matrix, where both virtual and real components are integrated. Producers use digital technologies for creating and selling goods or services (innovative technologies, smart cybernetic automation tools in business processes, smart system solutions, system integrators and aggregators), accumulate arrays of data about the individual characteristics of consumers, obtaining more complete market information.

However, interaction with consumers in the virtual space is fraught with increased competitive interaction between producers. In this case, the speed of responding to changing needs, as well as the skills of using digital technologies to retain regular customers and establish mutually beneficial long-term relationships with new ones, become the fundamental factors for achieving the competitive advantages of the producer [6, 10].

The state and consumer communities act as secondary stakeholders. However, the role of the state is also being transformed. Digital technologies contribute to the progress of creative activity and the expansion of the volume of market transactions. At the same time, the opposite process is also taking place - the increasing role and importance of the production of public goods. As a result, there is a contradiction between individual and universal intellectual property, which ensures that each individual appropriates the products of creative activity, their unlimited and free availability for each user [36]. Thus, digital technologies and creative labour undermine market exchange, and the digital economy objectively displaces the market economy with the help of the state.

Consumer communities as informal institutions in the digital economy are becoming a reflection of the personalized consumer experience [10]. In such a space, consumers who are informed, active, united in communities (usually on the Internet) and have greater market power increasingly determine the consumer choice of goods and services [25].

The interaction of economic entities is aimed at achieving the goals of digitalization: improving the properties and characteristics of consumer goods and means of production; performing work to create consumer value; launching business processes and ensuring the interests of stakeholders.

Thus, a specific digital market mechanism is actively being formed in the virtual space. It differs from the mechanism of the traditional market[15]. The subjects of the new market establish close long-term stable relationships, and the interaction between them is regulated by new market institutions formed in the virtual space. As a result of such interaction, the "invisible hand" effect of digitalization is achieved: technological systems are created; business processes are automated and automatic production is developed; automation of decision-making processes in economic systems is launched and digital convergence is achieved, generating wave processes of business activity growth [21] (Fig. 1).

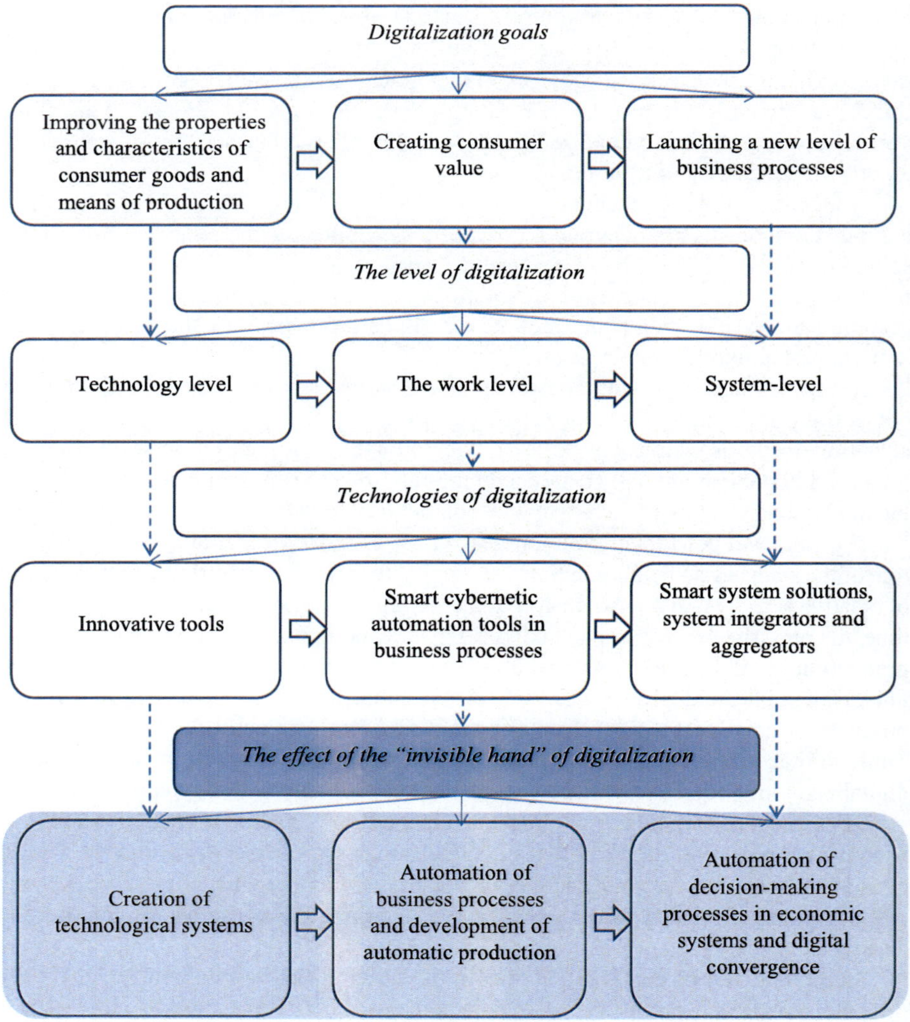

Fig. 1. The effect of the "invisible hand" of digitalization *Source* Compiled by the authors

The COVID-19 pandemic that broke out in January 2020 created not only an epidemiological threat to the life and health of the world's population but also a whole range of economic and social problems. The introduction of quarantine measures in most countries of the world, the suspension of a large number of enterprises, the prohibition of physical contact of employees at facilities have repeatedly increased the importance of digital technologies used both between enterprises and in the system of interaction between consumers and producers.

One of the technological forms of overcoming the global problem has become the technologies of remote interaction of individuals, which made it possible to organize interaction between people and industrial relations [1]. The enterprises that carried out the digital transformation at the beginning of the pandemic suffered the least financial and technological losses, as they were able to ensure the continuity of production, despite the requirements for self-isolation for employees [29].

The "invisible hand" effect worked, it ensured the intensive development of the digital economy sectors against the background of a significant "compression" of traditional sectors. In particular, examples of its activities were the intensive development of the e-commerce sector against the background of a reduction in sales of goods, the expansion of employment opportunities in online labour markets and an increase in demand for new specialities with an increase in unemployment, the accelerated development of sharing services in the context of a general reduction in rental transactions. During the pandemic, there was a significant increase in the use of e-commerce compared to the figures recorded before the COVID-19 pandemic. According to the report of Group M "E-Commerce forecast", the dynamics of the advanced development of e-commerce in 2020 has created a plateau for a new accelerated future growth [7]. For example, manufacturers of consumer packaged goods (CPG), that is, goods used daily by ordinary consumers and requiring regular replacement or replenishment, such as food, beverages, clothing, tobacco, cosmetics and household goods) note that the pandemic led to an increase in retail sales through e-commerce channels by 277% in the 2nd quarter of 2020 [7].

According to experts, in 2020, the volume of global retail e-commerce, including car sales but excluding food delivery services and delivery for consistency in each market, amounted to $ 3.9 trillion, or 17% of equivalent global retail sales. By 2027, sales in e-commerce are expected to grow to 10 trillion US dollars worldwide (Fig. 2).

Indeed, the withdrawal of consumers to self-isolation has forced them to spend more time online and use the opportunities of e-commerce to purchase essential goods. In addition, it helped to reduce the risk of contracting the virus when visiting stores. In turn, producers/sellers of goods intensified sales through websites, which helped to mitigate the overall decline in sales indicators [18, 37].

The pandemic has destabilized the situation in the labour market[23]. The transfer to remote work has become an insurmountable challenge for many employees, and mass layoffs of employees due to the bankruptcy of enterprises have deepened social inequality. In particular, according to a survey by Gartner, a third of enterprises during the pandemic abandoned permanent employees in favour of freelance (for example,

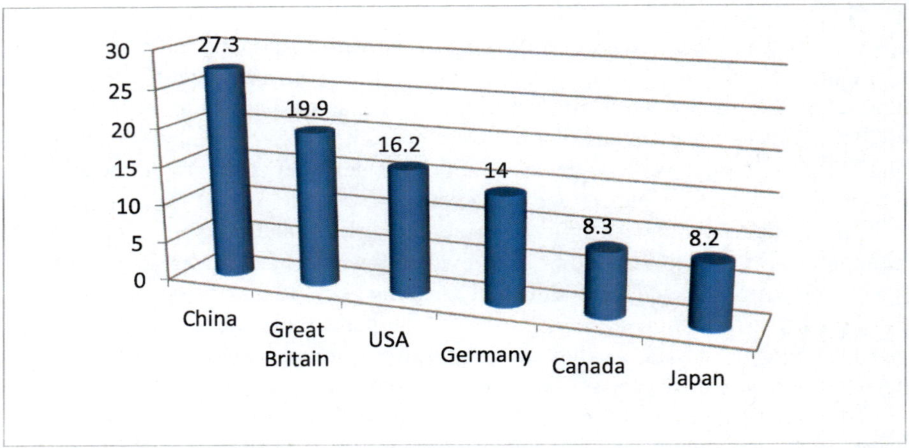

Fig. 2. The share of e-commerce in the total volume of retail trade, %, 2020 *Source* Compiled by the authors based on [7]

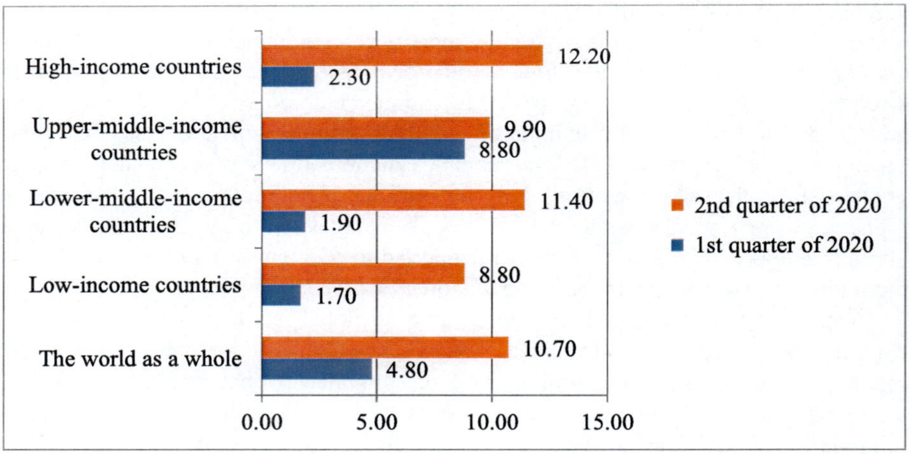

Fig. 3. Reduction in the volume of total working hours in the 1st and 2nd quarters of 2020 compared to pre-crisis indicators (4th quarter of 2019, seasonally adjusted) *Source* Compiled by the authors according to [33]

freelancers or consultants hired temporarily for a specific project). A survey by Avito Jobs and Kelly Services showed that 70% of companies operating in Russia suspended hiring employees during self-isolation, and 56% of Russian enterprises temporarily suspended their activities due to the restrictions imposed [35]. In general, there was a reduction in the volume of total working hours in all countries compared to pre-crisis indicators (Fig. 3).

According to UN estimates, in 2020, the total labour income in the world decreased by 8.3% or $3.7 trillion, which is 4.4% of global GDP [31]. Unemployment in the age group from 15 to 24 years was 8.7%. This is more than twice the average for the adult population (3.7%). The pandemic has had the greatest impact on the hotel sector and food services, where employment has declined by an average of more than 20%. Digital platforms for work came to the rescue, they transformed work processes, creating the basis for the processes of personnel selection and the resumption of production activities in difficult epidemiological conditions. The KPMG study “HR Actions in a Pandemic situation” confirmed that digital platforms successfully replace face-to-face interviews, and the remote format allows saving time for both enterprises and job seekers [35]. The International Labour Organization emphasizes that, on the one hand, the pandemic has caused an increase in the supply of labour on online platforms [9]. On the other hand, there was a reduction in the demand for labour and a significant shift in the structure of demand: the focus of attention on the part of employers shifted to specialists in such industries as information technology, marketing, R&D [9]. This area of employment opens up employment opportunities both in the offline and online labour markets. The effect of the “invisible hand” of digitalization balances supply and demand in the global labour market: demand comes mainly from developed countries, and the supply of labour comes from developing countries [37]. However, wages are a factor affecting the balance in a competitive market: their decline satisfies the demand on online platforms in the presence of numerous differentiated offers from specialists. The pandemic accelerated the onset of a long-emerging problem of modern society – automation and the release of low-skilled workers and also provoked the accelerated development of new skills and professions by the population [19].

Another area where the effect of the “invisible hand” of digitalization has become noticeable during the pandemic is the sphere of sharing. In general, this sphere is characterized by a wave growth of business activity and services, for example, aggregators of passenger transportation Uber, Gett, Yandex Taxi [21]. The pandemic affected the C2C trading segment less. The self-isolation regime stimulated the growth of conscious temporary purchases. In addition, representatives of older generations have actively joined the C2C-trade segment, making purchases using the Internet. At the same time, the geography of involvement in the sharing economy has expanded due to the formation of the habit of delivering “to the door” [14]. According to Data Insight experts, the pandemic period may bring an additional 1.6 trillion rubles to the B2C trade sector in Russia over the next 5 years [2]. The emergence of new online buyers, the isolation of the house and an increase in consumer demand for certain groups of goods, and growth in the frequency of purchases, the transition of some employees to remote work, a reduction in the effective demand of the population will contribute to the growth of this market segment with reductions in the volumes of other sectors [12]. The segment of P2P services (online freelance exchanges) was actively developing in the pre-pandemic period, in 2019 2.5 million self-employed people were freelancers on online platforms, and the total volume of transactions in the sector in 2019 amounted to 140 billion rubles [12]. With the growth of unemployment in traditional sectors of the economy, the development of online freelance exchanges becomes an effective tool in solving employment problems.

The annual growth of the car-sharing sector in 2019 was estimated at 58%. The imposed restrictions and sanitary rules had a negative impact on the work of companies in the car-sharing sector: in April-May 2020, the congestion of vehicles was recorded at the level of 60–70% compared to the same period of 2019 [26]. In June 2020, the situation began to change, and in the subsequent period, there was a gradual recovery in consumer demand. The effect of the "invisible hand" of digitalization contributed to the stabilization of the situation in this segment by changing the landscape of the car-sharing market in terms of the composition of participants and the cost and terms of service provision [12]. Rental of accommodation facilities (places in common rooms, rooms, apartments and houses) before the pandemic was a dynamically growing segment of the sharing economy in Russia. In the period from 2017 to 2019, the volume of transactions increased from 5.8 to 15.6 billion rubles [27]. However, during the pandemic, there was a decrease in domestic tourism. In addition, sanitary restrictions on movement were introduced, which negatively affected the development of the short-term rental sector of accommodation facilities (the occupancy rate decreased to 54–58%) [28]. The current situation required a change in the market situation, the format of work and service to ensure the safety of the guests' stay. The implemented changes and the effect of the "invisible hand" of digitalization are aimed at increasing the attractiveness of the sector for Russian consumers in comparison with hotel accommodation.

5 Conclusion

The authors revealed the effect of the "invisible hand" of digitalization on the mechanisms of market self-regulation. It is proved that a specific digital market mechanism is actively being formed in the virtual space, which differs from the mechanism of the traditional market. The subjects of the new market establish close long-term stable relationships, and the interaction between them is regulated by new market institutions formed in the virtual space. As a result of such interaction, the effect of the "invisible hand" of digitalization is achieved: technological systems are created; business processes are automated and automatic production is developed; automation of decision-making processes in economic systems is launched and digital convergence is achieved, generating wave processes of business activity growth. The authors determined the expediency of using the effect of the "invisible hand" of digitalization to overcome the negative consequences of the pandemic on the socio-economic system. Under the conditions of quarantine measures implemented in most countries, the suspension of a large number of enterprises, the prohibition of physical contact of employees at facilities, the importance of digital technologies used both between enterprises and in the system of interaction between consumers and producers has increased many times. It is proved that the effect of the "invisible hand" of digitalization has ensured the intensive development of the digital economy sectors against the background of a significant "compression" of traditional sectors (intensive development of the e-commerce sector against the background of a reduction in sales of goods; expansion of employment opportunities in online labour markets and an increase in demand for new specialities against the background of rising unemployment; accelerated development of sharing services against the background of a general reduction in rental transactions).

References

1. Barkhatov, V., Dyachenko, O.: Development of the digital economy of Russia in the conditions of a pandemic. Bulletin Chelyabinsk State University. Econ. Sci. **10**(444), 70, 177–182 (2020)
2. Data Insight: E-commerce 2020–2024, Data Insight forecast (2020). https://datainsight.ru/DI_eCommerce2020_2024. Accessed 20 Aug 2021
3. Dietze, I.P., Nedzvetskaya, N.: The invisible hand of the robot: man in the era of digital society. Philos. Econ. **6**(114), 109–123 (2017)
4. Dyatlov, S., Lobanov, O.: Convergence of information technologies in the Eurasian economic space. In-State and market: mechanisms and institutions of Eurasian integration in the conditions of increasing global hyper-competition: collective monograph, St. Petersburg, pp. 641–655 (2017)
5. Enikeeva, L., Sorvina, T., Torosyan, E.: Customer involvement in emerging and developed markets as marketing innovation. Int. Bus. Manag. **10**(18), 4324–4331 (2016)
6. Gordon, Y.: Partnership marketing, Peter, St. Petersburg (2014)
7. GroupM: This year next year: e-commerce forecast (2020). https://www.groupm.com/this-year-next-year-ecommerce-forecast. Accessed 20 Aug 2021
8. Gtmarket: Digital consumers prefer brands (2009). http://gtmarket.ru/laboratory/expertize/2009/2230. Accessed 20 Aug 2021
9. ILO: World Employment and Social Outlook 2021: The role of digital labour platforms in transforming the world of work (2021). https://www.ilo.org/global/research/global-reports/weso/2021/WCMS_771749/lang–en/index.htm. Accessed 20 Aug 2021
10. Karpunina, E.K., Yashin, S.N., Molchan, A.S., Lisova, E.V., Derkacheva, E.A.: The price of pleasure: the challenges and threats of the digital economy. In: Kolmykova, T., Kharchenko, E.V. (eds.) Digital Future Economic Growth, Social Adaptation, and Technological Perspectives. LNNS, vol. 111, pp. 363–373. Springer, Cham (2020). https://doi.org/10.1007/978-3-030-39797-5_37
11. Karpunina, E., Butova, L., Sobolevskaya, T., Badokina, E., Pliusnina, O.: The impact of the covid-19 pandemic on the development of Russian national economy sectors: analysis of dynamics and search for stabilization measures. In: Proceeding of the 37th IBIMA Conference, Cordoba, Spain, 1–2 April 2021, pp. 1213–1226 (2021)
12. Karpunina, E., Galieva, G., Rudenko, V., Sokolovskaya, E., Poluyanova, N.: The development of the collaborative consumption economy in Russia: regional features. In: Proceeding of the 37th IBIMA Conference, Cordoba, Spain, 1–2 April 2021, pp. 338–347 (2021)
13. Kenney, M., Zysman, J.: The Rise of the Platform Economy. Issues Sci. Technol. **32**(3) (2016). http://issues.org/32-3/the-rise-of-the-platform-economy. Accessed 20 Aug 2021
14. Kireeva, N., Zavyalov, D., Zavyalova, N., Saginova, O.: Transformation of the rank economy in the conditions of the COVID-19 pandemic. Econ. Entrep. Law **11**(2), 213–226 (2021)
15. Koblova, Yu., Murygina, N.: Features of the market mechanism in the digital economy. Bull. Saratov State Socio-Econ. Univ. **3**(72), 29–33 (2018)
16. Koblova, Yu., Popov, M., Murygina, N.: Institutes of virtual space: mechanism, patterns of formation and new threats. Bull. Saratov State Socio-Econ. Univ. **3**(67), 82–86 (2017)
17. Kozyrev, A.: Mathematical models and forecasting in the digital economy (2018). https://medium.com/cemi-ras. Accessed 20 Aug 2021

18. Lisova, E., Petrov, I., Koryakina, T., Kalombo Mulamba, V., Karpunina, K.: The well-being of Russian regions under threat: economic and social impact assessment of COVID-19. In: Proceeding of the 36th IBIMA Conference, 4–5 November 2020, Granada, Spain, pp. 7603–7616 (2020)
19. McKinsey: What the future of work will mean for jobs, skills, and wages: Jobs lost, jobs gained (2017). https://www.mckinsey.com/featured-insights/future-of-work/jobs-lost-jobs-gained-what-the-future-of-work-will-mean-for-jobs-skills-and-wages. Accessed 20 Aug 2021
20. Melnik, D.: The invisible hand of the market (2018). https://yandex.ru/search/?text=Д.+В.+Мельник+НЕВИДИМАЯ+РУКА+РЫНКА&clid=2270455&banerid=6301000000%3A5cddc5b70d2ddf00281ef889&win=385&&lr=13. Accessed 20 Aug 2021
21. Minakov, V., Shuvaev, A., Lobanov, O.: The effect of digital convergence in the economy. Bull. St. Petersburg State Univ. Econ. **2**(110), 12–18 (2018)
22. Molchan, A., Karpunina, E., Kochyan, G., Petrov, I., Velikanova, L.: Effects of digitalization: new challenges for economic security systems. In: Proceedings of the 34th IBIMA Conference, 13–14 November 2019, Madrid, Spain, pp. 6631–6639 (2019)
23. Nazarova, I., Galieva, G., Sazanova, E., Chernenko, E., Karpunina, E.: Labour market and employment problems: analysis of long-term dynamics and prospects of development in Russian regions. In: Popkova, E.G. (ed.) Imitation Market Modelling in the Digital Economy: Game-Theoretic Approaches. Lecture Notes in Networks and Systems. Springer, Cham (2021)
24. Payne, A., Storbacka, K., Frow, P.: Managing the co-creation of value. J. Acad. Mark. Sci. **36**, 83–96 (2008)
25. Prahalad, K., Ramaswamy, V.: The Future of Competition. Creating a Unique Value Together with Consumers. Publishing House Williams, Moscow (2010)
26. Quote: Will COVID destroy taxi and car-sharing: the consequences of the pandemic for the sector? (2020). https://quote.rbc.ru/news/article/5ec262799a794769190b6710. Accessed 20 Aug 2021
27. RAEC: The Economy of shared consumption in Russia 2018 (2018). https://tiarcenter.com/sharing-economy-research/. Accessed 20 Aug 2021
28. RBC: How the pandemic will change Russian hotels (2020). https://realty.rbc.ru/news/5eeb0e419a794706fe70c04d. Accessed 20 Aug 2021
29. Salygin, V., Markin, A.: The digital economy in the conditions of a pandemic. Scientific works of the Free Economic Society of Russia, vol. 223, pp. 244–250 (2020)
30. Smith, A.: Research on the nature and causes of the wealth of peoples. Eksmo, Moscow (2007)
31. Tass: The UN has calculated the loss of working time due to the pandemic (2021). https://tass.ru/ekonomika/10539553. Accessed 20 Aug 2021
32. The Federal Antimonopoly Service of the Russian Federation: A. Golomolzin on how to protect yourself from the dangers of changing the "invisible hand of the market" to "digital" (2017). https://fas.gov.ru/news/81. Accessed 20 Aug 2021
33. United Nations: Analytical note: the sphere of labour and COVID-19 (2020). https://www.un.org/sites/un2.un.org/files/sg_policy_brief_world_of_work_and_covid_19_russian.pdf. Accessed 20 Aug 2021
34. Vargo, S., Lash, R.: The development of a new dominant logic of marketing. Russian J. Manag. **4**(2), 73–106 (2006)
35. Vedomosti: How the pandemic accelerated digitalization in the labour market (2020). https://www.vedomosti.ru/partner/articles/2020/06/26/833455-pandemiya-uskorila. Accessed 20 Aug 2021
36. Voeikov, M.: State in the epoch of the digital economy. Theor. Econ. **1**, 47–50 (2019)

37. Volodina, A., Podolskaya, T.: The role of digital technologies in the recovery of the world economy after the COVID-19 pandemic. Econ. High-Tech Ind. **2**(4) (2021). https://doi.org/10.18334/evp.2.4.112188
38. Warner, M., Witzel, M.: Virtual organizations: new forms of doing business in the XXI century. Good Book [Dobraya Kniga], Moscow (2005)

Sociological Survey as a Management Tool in the Digital Economy

Marianna S. Santalova[1(✉)], Irina V. Soklakova[2], Svetlana N. Saifieva[1], Anar S. Kusainova[2], and Elvira P. Lesnikova[3]

[1] Academy of Management and Production, Moscow, Russia
[2] Moscow Economic Institute, Moscow, Russia
[3] Plekhanov Russian University of Economics, Voronezh Branch, Voronezh, Russia

Abstract. The article discusses the possibilities of increasing the efficiency and development of production activities of a furniture factory in the digital economy; the unused capacity of the company is being investigated. With the help of a sociological survey, the preferences of consumers of the factory's products are revealed and ways to expand production and its profitability are determined.

The work aims to use a sociological tool for analyzing and assessing the company's capabilities and making timely and effective management decisions in the digital economy.

The work uses research methods such as analysis (tabular, graphical), synthesis, induction and deduction, observation and collection of facts, etc. [2].

The research methodology was scientific literature, Internet sources, reporting of a furniture factory, the points of view of such authors as [1–13, 15] and others.

Keywords: Sociological survey · Analytical tool · Production efficiency · Management decisions · Production activities · Digital economy

JEL Codes: D24 · E23 · L23 · M11

1 Introduction

The use of such analytical tools as a sociological survey allows making timely management decisions in the digital economy. With the help of this tool, reserves for increasing production efficiency are identified, opportunities for enterprise development are expanded.

The object of research in this work is the furniture factory "SbkHome". The subject of the research is a sociological survey as an analytical tool, the use of which will make it possible to develop management decisions aimed at increasing the production efficiency of the furniture factory "SbkHome".

E. G. Popkova (Ed.): ISC 2020, LNNS 368, pp. 174–181, 2022.
https://doi.org/10.1007/978-3-030-93244-2_20

2 Methodology

The work uses research methods such as analysis (tabular, graphical), synthesis, induction and deduction, observation and collection of facts, etc.

The research methodology was scientific literature, Internet sources, reporting of a furniture factory, the points of view of such authors as [1–15] and others.

3 Results

In the course of the study, it was revealed that the production activities of the furniture factory "SbkHome" in recent years have been effective, however, due to unforeseen circumstances, in terms of the decrease in demand from the population against the background of the coronavirus pandemic, the efficiency of the company's production activities has slightly decreased, which is reflected in a decrease in production volumes, profits, as well as profitability, including the profitability of production of certain types of products [1].

In general, it is worth noting that all types of products manufactured by the furniture factory throughout the entire research period were profitable, that is, they brought profit to the enterprise. The most cost-effective products are furniture for children and adolescents. So, with the maximum level of profitability of the production activity of a furniture factory at 9%, the profitability of the production of furniture for children and adolescents reached the level of 20% (in 2018), and in 2020 there was a positive dynamics of the profitability of the production of this type of product (the indicator increased from 13.73% in 2019 to 16.92% in 2020). But, as the study shows, the capacity of the enterprise is not fully utilized.

In support of what has been said, we present Table 1.

Table 1. Evaluation of the effectiveness of the furniture factory "SbkHome" for the production of furniture for children and adolescents

Name of product	2018	2019	2020	Deviation 2020 -2019	
				Absolute +/-	Relative %
Children's, teenage and other furniture	651,008	890,355	893,582	3,227	100,36
Children's, teenage and other furniture	726,000	925,650	935,600	9950	101.07
Children's, teenage and other furniture	89,67	96,19	95,51	-0,68	-
Production profitability,%	20,14	13,73	16,92	3,19	-

Source Compiled by the authors

The data in Table 1 confirm the fact that the furniture factory has the potential to load additional production capacities for the production of the most profitable type of product - furniture for teenagers and children.

To identify areas for expanding the range of this type of product, a study was made of the preference of consumers of the children's furniture market in the region, using a sociological survey [13]. The idea of this sociological survey is to use this tool to research the buyers of children's furniture to identify their needs and motives when buying products [6]. The sociological survey revealed weaknesses in the assortment (product) policy of manufacturers and the direction of work on this product segment to maximize consumer satisfaction [5].

In order to study buyers when making a purchase decision, a survey was conducted in which 100 people took part [4].

At the first stage, it was revealed which furniture manufacturers consumers prefer. The distribution of their opinions is shown in Fig. 1.

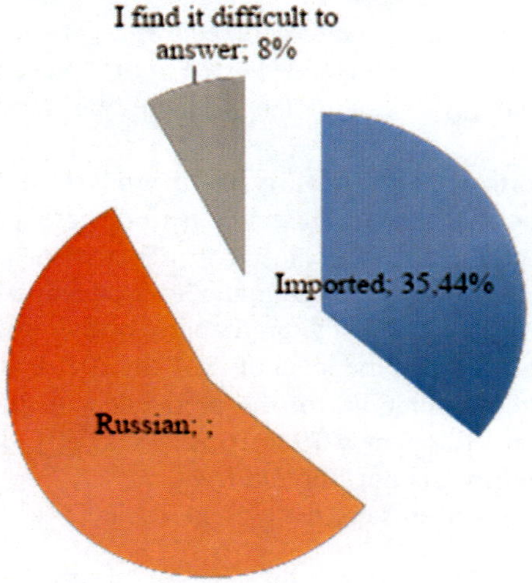

Fig. 1. Distribution of consumers concerning furniture manufacturers. *Source* compiled by the authors based on [14]

Based on the data presented in Fig. 1, we can conclude that the absolute majority of respondents prefer a Russian manufacturer (57%). The share of those consumers who would like to purchase imported furniture (36%) is also quite high, and only 8% of respondents found it difficult to answer.

In their activities, manufacturing enterprises can use various distribution channels, the main ones of which at the moment are brand stores, specialized stores, Internet stores [3]. The data indicates that the majority of buyers (44%) prefer branded stores of manufacturing enterprises, while 36% of buyers prefer specialized stores. Internet sites are also quite popular among buyers, so 19% of respondents said that they prefer to buy furniture in online stores. It must be said that one of the buyers noted that he prefers to buy used furniture for children.

Further, the respondents assessed the degree of importance of individual factors to which they pay attention when choosing children's furniture [15]. The results of this assessment are presented in Table 2.

Table 2. Assessment of the degree of importance of each factor, influencing the choice of children's furniture (1 – "absolutely unimportant", 5 – "very important")

Criteria importance	1	2	3	4	5	Total
Price level	0	0	4	22	74	100
Product quality	0	0	0	44	56	100
Safety	0	0	0	0	100	100
Product range	0	12	22	44	22	100
Appearance	0	0	30	45	25	100
Comfort	0	0	44	24	32	100

Source Compiled by the authors

The data in Table 2 indicate that none of the factors was completely unimportant for consumers who took part in a survey, conducted by a furniture factory. The most important factor can be assigned to such a factor as safety – 100, out of 100 respondents chose it as "very important". The second item is occupied by such a factor as the price level, the maximum importance of which was assigned by 74 respondents out of 100, in the third-place is the quality of the goods (56 people out of 100).

Further, using Fig. 2, let us evaluate the preferences of consumers, regarding the range of children's furniture.

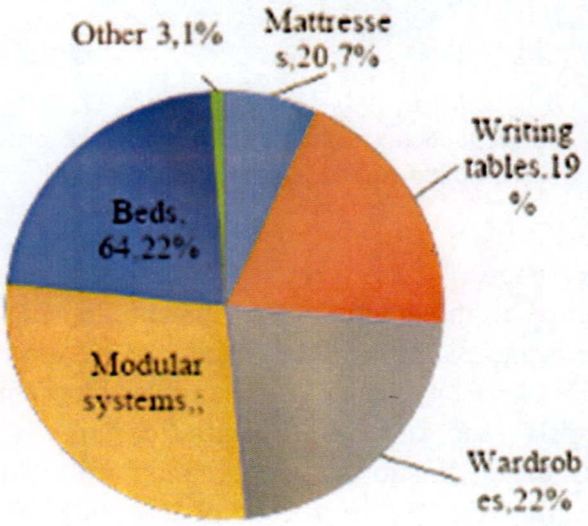

Fig. 2. Distribution of consumer preferences by the range of children's furniture. *Source* compiled by the authors based on [14]

Based on the data presented in Fig. 2, we can conclude that the survey was attended by consumers who purchased more modular systems (81%), wardrobes (64%) and beds (64%).

The next stage of the research was devoted to the question of identifying the payment methods, preferred by the respondents [7]. Thus, the majority of consumers who took part in the sociological survey prefer to pay for the purchase, including the purchase of furniture, with their funds immediately (62%), some consumers (34%) prefer to use purchase cards and instalment plans, and some (4%) prefer instalments. provided by the store.

For some consumers, the colour scheme of children's furniture is important, and for some, it is not important. We will reflect the attitude of consumers to the colour scheme of children's furniture, using Fig. 3.

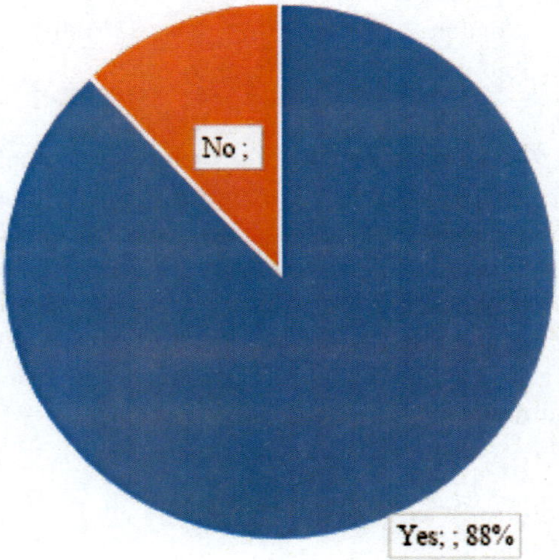

Fig. 3. Distribution of the respondents' opinions, regarding the principledness of the colour scheme of furniture. *Source* compiled by the authors based on [14]

The results of this stage of the study (Fig. 5) allow us to conclude that 88% of respondents pay attention to the colour scheme when choosing children's furniture.

Children's furniture, including furniture, produced by the investigated enterprise, may differ (have stylization) by gender (for example, blue children's furniture - for boys, and pink - for girls), and may have a neutral design.

How important is the stylization of children's furniture for consumers and its difference by gender will reveal the next stage of the study (Fig. 4).

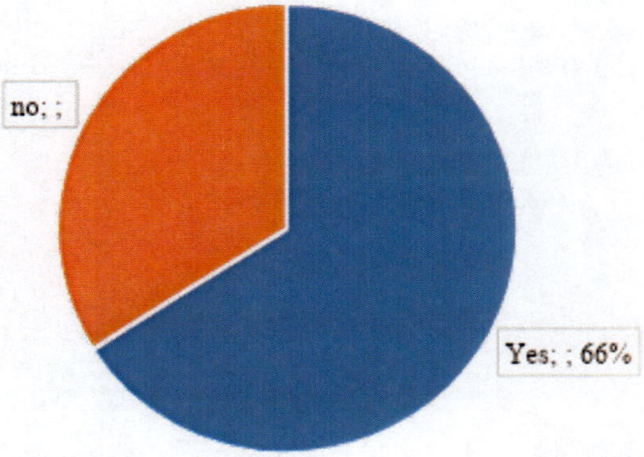

Fig. 4. Distribution of respondents' opinions regarding the stylization of the children's room. *Source* compiled by the authors based on [14]

The data in Fig. 4 indicates that the stylization of the children's room (for a girl or a boy) is important for a fairly large proportion of consumers - 66% of all respondents belong to this category. For a smaller share of respondents (34%), stylization according to the sex of the child is not important and they do not pay attention to this when choosing a children's room.

As you know, children's furniture can become outdated as the child grows up, so many consumers tend to purchase furniture for a certain period (universal), and some prefer to update it as the child grows up. What the respondents are more inclined to do will be revealed in Fig. 5.

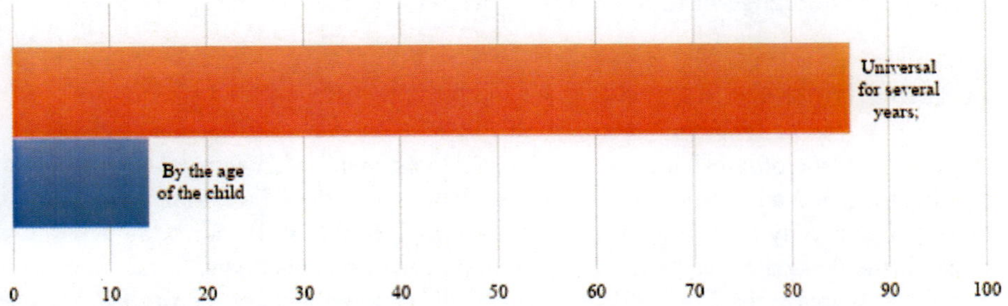

Fig. 5. Distribution of reasons of respondents regarding the duration of furniture service. *Source* compiled by the authors based on [14]

The data in Fig. 5 indicate that only 14 respondents can afford to purchase children's furniture by the age of a child, and 86 respondents prefer to purchase universal furniture for several years.

At the final stage of the study, it was revealed in which direction, according to the respondents, it is necessary to expand the range of children's furniture, that is, what kind of children's furniture is lacking for consumers in the region [10]. The distribution of consumer opinions, regarding a possible expansion of the assortment, is shown in Fig. 6.

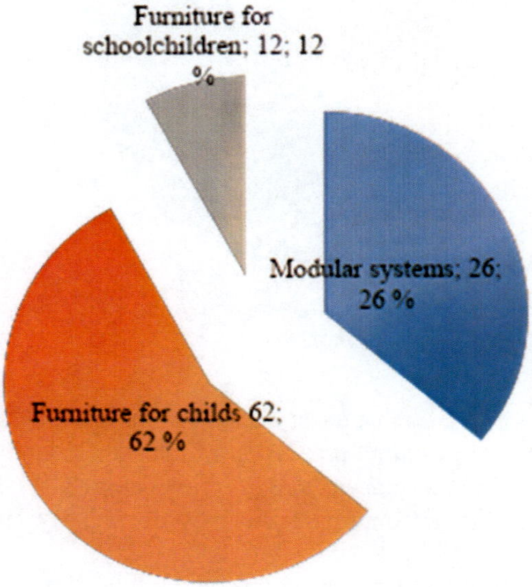

Fig. 6. Distribution of consumer opinions regarding a possible expansion of the range of children's furniture. *Source* Compiled by the authors based on [14]

Figure 6 shows that the majority of respondents would prefer to see and be able to purchase a wider range of furniture for babies (newborns).

4 Conclusion

Thus, such a sociological tool as a survey allows making an important managerial decision to improve the efficiency of the production activities of a furniture company in the digital economy [10]. Expansion of the range of furniture for newborns will allow to use of the unused capacities of the company, activate demand and attract additional customers, increase the level of sales, which means the company's profit [9].

The survey data showed that the majority of respondents would prefer to see and be able to purchase a wider range of furniture for babies (newborns).

As the calculations showed, as a result of the expansion of the range of children's furniture, the total production volume of the furniture factory "SbkHome" will be increased by 1.01%, profit from sales–by 4.84%. Finally, the effectiveness of the proposed measure is confirmed by the dynamics of the profitability indicator, which will increase by 0.28% points and amount to 7.57% [11].

Acknowledgement. The results of the study were discussed at the Academy of Management and Production, the Moscow Economic Institute, the Plekhanov Russian Economic University; introduced at the enterprise.

References

1. Akulich, V.V.: Express-analysis of the main indicators of the organization's activity. Plann. Econ. Department **10**(40), 25 (2015)
2. Akulova, E.: Methodology of sociological research. LAP Lambert Academic Publishing, Moscow (2014)
3. Dobrenkov, V.I., Kravchenko, A.I.: Methods of sociological research. Publishing House of Moscow State University, Moscow (2017)
4. Dudina, V.I.: Methodology and methods of sociological research: monograph. St. Petersburg State University St. Petersburg State University (2014)
5. Gottlieb, A.: Qualitative sociological research. LAP Lambert Academic Publishing, Moscow (2011)
6. Gottlieb, A.S.: Qualitative sociological research. Cognitive and existential horizons. Flint, Moscow (2014)
7. Kovalenko, E.M.: Assessment of the profitability of sales and factors affecting its level. Econ. Finance Control **2**, 10–15 (2018)
8. Lesnikova, E.P., Jakhongirov, J.O., Sadykova, K.V., Zakharova, T.I., Santalova, M.S.: Management of innovative working behaviour. In: Modern Global Economic System: Evolutional Development vs. Revolutionary Leap. LNNS, vol. 198, pp. 1008–106. Springer, Cham (2021)
9. Santalova, M., Didenko, S., Jakhogirov, I.J.: Need for a manager competency model. Educ. Transform. Issues **3**(3), 31–36 (2016)
10. Santalova, M.S., Soklakova, I.V., Balabanova, D.K.: The choice of the competitive strategy of the company. In: Proceedings of the International Scientific Conference "Far East Con" (ISCFEC 2020). Vladivostok, pp. 1282–1289 (2020)
11. Surat, V.I., Santalova, M.S., Soklakova, I.V., Ermilina, D.A., Zhanibekova, G.K., Lebedeva, E.V.: Strategic analysis of the organization's activities. Academy of Management and Production, Moscow (2020)
12. Tikhonova, E.V.: Methodology and methods of sociological research. Academia, Moscow (2012)
13. Vakulenko, R.Y.: Assessment of the impact of control factors on the efficiency of using the production resources of an enterprise. Directory Eng. J. **1**, 26–33 (2016)
14. Yadov, V.A.: Sociological research: methodology, program, methods, Samara (1995)
15. Zolotov, A.V., Morozova, N.A.: Labour productivity: topical problems of theory and practice. Bulletin of Nizhny Novgorod University. Publishing house Nizny Novgorod State Univ. **1**, 547–552 (2016)

Morphology of Heterogeneous Spatial Economic Systems: Concept and Structural Models

Yuri S. Kolesnikov(✉) and Natalia P. Ketova

Southern Federal University, Rostov-on-don, Russia
nketova@sfedu.ru

Abstract. The paper considers the morphology of spatial economic systems as a methodological concept that allows describing the system bases of heterogeneity of economic space and socio-economic inequality of regions. When analyzing the internal structure of the economic space of the region, the authors apply the tools of macroeconomic, territorial-economic, and organizational-economic projections presented in modern research. The authors propose a variant of the structural model of the organization of the economic space of the region, which is empirically verified based on data on the regions of southern Russia.

Keywords: Region · Economic space · Morphology · Heterogeneity · Inequality · Structural model · Economic patterns · Regional markets

JEL codes: R12 · R28 · R58

1 Introduction

As shown by many years of experience, the market and the current strategies and mechanisms of centralized economic management lead to the insurmountable divergence of regions, the growth of their socio-economic inequality, the loss of systemic homeostasis properties of the regions, and, consequently, the reduction of cohesion of the economic space of the vast Russian periphery. These internal processes receive additional impulses under the influence of market forces of globalization and the uneven concentration and centralization of production, commodity flows, resources, and economic assets. Therefore, structural interactions are the supporting framework of the region's economic space, and their specific configuration (architecture) is a fundamental factor of regional inequality. According to the research practice, the concept of the morphology of economic space (ES) of the region" can be the methodological and categorical-notional platform of research in the analysis of multistructural spatial economic systems. The idea of the morphology of the regional economic system allows us to analyze the economic structures formed by market forces and the whole totality of structural interactions and interrelations that ensure the reproduction process. This analysis requires the development of an integrated theoretical model of the polystructural organization of the ES of the region (standard model and its

E. G. Popkova (Ed.): ISC 2020, LNNS 368, pp. 182–189, 2022.
https://doi.org/10.1007/978-3-030-93244-2_21

morphology), focused on the identification of territorial, economic, and institutional factors of formation of heterogeneity of socio-economic space of regions.

The possibility of such an approach lies in the methodology of interdisciplinary synthesis, considering the peculiarities of evolution, the multi-economic nature, and institutional, environmental, and socio-economical heterogeneity of the Russian periphery. Thus, the paper aims to substantiate the heuristic value of the methodological concept of the morphology of ES of the region and develop a system of structural models of its current organization, explaining, in many respects, the multi-factorial nature of socio-economic heterogeneity and inequality in the level of regional development.

The theoretical basis of the research was the aggregate notions of heterogeneity and structural asymmetry in the development of economic space. The authors relied on theories and models of spatial economics developed in the works of leading Russian regionalists, economists, political economists, and sociologists. These works present the following:

- The morphological approach in regionalism;
- Set of concepts and structural models describing the basic natural-economic, institutional, and territorial-economic conditions and factors of socio-economic inequality of regions;
- Growth of imbalances and asymmetry in the development of regions;
- Heterogeneity of the vast Russian periphery;
- Internal mechanisms of regional divergence and problems of reproduction connectivity of regions and municipal entities of Russia.

The ideas and approaches of the leading Russian scholars such as V. N. Leksin and B. N. Porfiryev [9], G. B. Kleiner [7], V. N. Ovchinnikov [11], V. M. Odrin and V. G. Kartavov [10], and A. G. Druzhinin [2] were used directly in the subject field of this research.

2 Methodology

Based on an extensive and diverse conceptual-theoretical continuum of research, the authors used the methodology of interdisciplinary analysis of the multistructural organization of ES of the region (its morphology) focused on the analysis of the interaction of socio-economic and industrial-technological modes established in the region and operating different-scale, diversified, and multi-level economic entities and institutions [2, 7, 10, 14]. The institutional-applied part of the research used the materials of statistical compilations "Regions of Russia" and an array of data from an economic and sociological study of economic patterns in the economy of Southern Russia conducted by the authors in 2017–2018 [8, pp. 86–87].

3 Results

The concept of the morphology of the ES of a region expresses the unity and integrity of all structures whose functions are directly related to the territory and its reproductive system. Therefore, the morphology of the local ES should be understood as a systemic unity within a certain territory, which is its physical, organizational platform, and the totality of stable and ordered connections between structural elements in the process of implementation of their functions, reproducing the integrity of the ES and available for analysis. In this case, the initial typological attribute of the structurization of ES is not their complexity, size, and other qualitative and quantitative characteristics, but their purpose and functions determining all other particular parameters [9, pp. 7–8]. Thus, the identification of the structural components of ES is methodologically valid, provided that their reproductive functions and specific features are dominantly mediated by space. Based on this theoretical imperative, we can build a typical model of the morphology of EP of the region, using the methodological tools of the three research projections – macroeconomic, territorial-economic, and organizational-economic projections, which are widely and diversely represented in modern regionalistics. Simultaneously, the analysis in the subject field of these three projections is combined with the analysis of the structural forms of the spatial organization of the economy (this is a peculiarity of the author's research methodology), since it is identified within the five concepts (models) of the region established in the theory of regional studies, namely: (1) a region as a reproductive system; (2) a region as a market; (3) a region as a corporation; (4) a region as a state; (5) a region as a society.

Based on the generalization and systematization of the models of regional ES structuring contained in research practices, their grouping, and typology, the authors proposed a matrix of analysis of regional ES morphology. This matrix allows describing ES of the region by 56 structural parameters (24 macroeconomic, 14 territorial-economic, and 18 organizational-economic in the context of these five concepts of the region as an economic system). Given the limitations of the volume of the research paper, below we present only two parts of the matrix – in the context of the concept of the region as a reproductive system and the region as a market on the example of Southern Russia (Southern and North Caucasus Federal Districts. The full version of the analysis matrix of the regional ES morphology can be found in the monograph prepared with the participation of the authors [11, pp. 30–34].

The Concept of the Region as a Reproductive System.

I. Types of structural models in *macroeconomic projection*:
   1.1 Sectoral model (primary, secondary, and tertiary sectors of the economy);
   1.2 Structural and production model (production, consumer, investment, resource, energy, infrastructure, and other components of reproduction);
   1.3 Export-import oriented model;
   1.4 Pattern model (socio-economic and industrial-technological patterns);
   1.5 The component model of the national wealth of the region.
II. Types of structural models in *territorial-economic* (economic and geographic) *projection*:

2.1 The model describing the functionality of the territorial and economic subsystem (natural and economic, settlement, industrial and economic, and transport and communication);
2.2 The model describing the type of economic development of the territory (old industrial, secondary industrialization, new industrial, and pioneer development);
2.3 The model describing the dominant specializations in the region's economy (industrial, agro-industrial, agricultural, and resource-raw);
2.4 The model describing the level of sectoral heterogeneity of the regional economy (monosectoral, multisectoral, diversified, and innovative).

III. Types of structural models in *organizational and economic projection:*
3.1 Vertical-hierarchical model (global, national, regional, and local components of the economy);
3.2 Center-periphery model (centers of economic dominance, periphery, agglomerations, and support framework);
3.3 Enclave model (in the sectoral and regional dimension of the economy);
3.4 Cluster model (territorial-production complexes, clusters, free economic zones, special economic zones, etc.);
3.5 Networking model (ecosystems, network consortia, associations, etc.).

The Concept of the Region as a Market.

I. Types of structural models in *macroeconomic projection*:
1.1 The model describing the nature of market exchange circuits (open, closed, loop, and transit);
1.2 The model describing the nature of market relations in the economic sectors of the region (market, non-market, and quasi-market);
1.3 Factor model (market of production means, consumer goods, products, services, investment, etc.);
1.4 The model describing markets in categories of dynamics (mobile resources, non-mobile resources, markets growing, stagnating, and crisis);
1.5 Export-import-oriented market model.

II. Types of structural models in *territorial-economic* (economic and geographic) *projection*:
2.1 The model describing the territorial infrastructure of the regional market;
2.2 Logistic model (domestic, interregional, global, industry, network, and virtual [Internet trade] markets of the region);
2.3 "Nodal" model (points of concentration) of markets.

III. Types of structural models in *organizational-economic projection*:
3.1 The model describing the object-subject structure of the regional market;
3.2 The model describing the structure of wholesale and retail markets, as well as micro-markets;
3.3 The model describing the structure of monopolistic markets and mega-markets;
3.4 The model describing the structure of competitive markets;

3.5 The model describing the composition and institutional forms of organization of regional markets (urban, rural, corporate, local, mega-markets, fairs, online stores, etc.).

The applied implementation of the proposed methodology for constructing structural ES models as applied to the South of Russia has confirmed its heuristic value, which is illustrated by the analysis of patterns of ES models. The South of Russia historically belongs to the regions of peripheral type. It is multipattern and simultaneously reproduces and ensures the functional interaction of production and economic modes of different types of economic activity – archaic, pre-industrial, industrial, and post-industrial (segments 5 and 6 of technological modes). As an object of system morphological analysis, the ES of the South of Russia is, therefore, of particular interest in terms of relief and unambiguous heterogeneity of all its subsystems – socio-economic, industrial, technological, settlement, environmental, etc. The integrated use of statistical data and the adjustment of individual indicators, taking into account the materials of the author's study of traditional ways of life in the regions of Southern Russia [8, pp. 86–87], allowed us to identify the following structure of socio-economic patterns (Table 1).

Table 1. Structure of socio-economic patterns in the economy of Southern Russia (in % of the total)

No.	Name of the pattern	Southern federal district	North Caucasian federal district
1	Household (archaic)	16.5	31.5
2	Small-scale commodity	32.8	13.4
3	Private corporate	8.6	12.8
4	State	26.0	26.5
5	Municipal and economic	15.1	15.8
TOTAL:		100	100

Source Calculated by the authors based on [5, p. 128; 8, pp. 86–87]

The results of the calculations confirm that almost 50% of the economy of the South of Russia consists of resources, products, and economic assets of household and small-commodity patterns, which implies appropriate institutionalization of this parameter in the strategies and methods of planning the spatial development of the economy of the peripheral regions of Southern Russia. The task of identifying the production and technological deposits in the regional economy. Conceptually and methodologically, the research used a generally accepted model that includes pre-industrial, industrial, and post-industrial modes, with the allocation of the 3rd mode (machine industry technologies), the 4th mode (mass industrial production technologies), the 5th mode (information technologies), and the 6th mode (digital technologies, AI technologies, etc.) within these modes. Our calculations are based on the data of statistical compilations on the sectoral structure of GDP, the distribution of the number of enterprises by type of economic activity, the volume of ICT use, the innovative activity of enterprises and organizations, and the volume of innovative products and services. Our calculations allowed us to identify the following structure of production and technological modes (Table 2).

Table 2. Composition and share of production and technological modes in the economy of Southern Russia (compared with the Russian Federation), in %

Production and technological mode	Southern federal district	Russia*
Pre-industrial	31.0	10.0
Industrial Mode III Mode IV	41.8 20.0	30.0 50.0
Postindustrial Mode V** Mode VI	8.0 –	10.0 –
TOTAL:	100.0	100.0

Source * Data for the Russian Federation – evaluation by RAS experts [12, p. 3]. ** Compiled by the authors based on [3, p. 105; 5, p. 464, 556, 871, 944, 954];

Thus, the mode characteristic of the ES of the South of Russia demonstrates serious inertia in the evolution of socio-economic and industrial-technological modes, the enclave, and the conserved nature of some sectors of agricultural and industrial production of the service sector. This leads to a high level of heterogeneity in the development of the economic sectors of the South of Russia, which leads to the peripheralization of the economy, reduces the potential impact of development institutions and new forms of territorial organization of the economy, and hampers the diffusion of innovation to the periphery.

The empirical and applied study of the structural characteristics of the regional ES as a market showed that the market-economic structure of the ES of the South of Russia is supported by the intensive functioning of regional markets, which are local-territorial spheres of commodity circulation. Regional markets differ in location, types of represented goods, prices, conjuncture, and supply and demand. In this case, markets act as nodal elements of ES of the region, the consistent core of economic structures that form the centripetal vectors of market attraction of economic activity to the local nodes of transactional activity of economic entities.

According to the research results, in the South of Russia, the market network acts as the basis for the formation of the economic framework of the region and, above all, its agro-industrial complex (AIC). The markets of the AIC carry out the role of the supporting framework of the entire economy, forming a unifying chain of goods and services for the main part of the producers of the region. With a considerable number of small and medium agricultural producers, large farms in the South of Russia are, nevertheless, the key structures in the economy. They form the market nodes and the main commodity flows in the ES of the region since they reproduce a steady demand for agricultural machinery, equipment, seeds, fertilizers, plant protection products, feed, veterinary drugs, etc. Simultaneously, they supply the market with agricultural products and services and act as the core of regional grain, vegetable, meat, and other clusters [1, pp. 17–21].

To strengthen market positions of the agricultural sector of the regions of South Russia, it seems promising to implement deep processing of agricultural products, projects to implement predictive analytics to forecast demand in regional consumer markets for specific crop and livestock products [4, pp. 42–64], eco-oriented marketing [15, pp. 124–127], Big Data analysis, constructive projects for the agricultural sector in social networks [6], and other innovative technologies. In production processes, it is advisable to expand the differentiated application of fertilizers, the use of digital field passports, online quality control of technological processes, which has recently become widespread in crop and livestock online monitoring.

A significant role in the market structure of AIC in the region is played by the accelerated development of markets of agricultural machinery, equipment, tools, and instruments, formed mainly by the largest enterprises in the industry – GC "Rostselmash," LLC "KLAAS," JSC "Klever," and others, the end products of which has a high proportion of added value and are the result of high technological conversions.

Thus, the agro-industrial cluster of Southern Russia, including at least seven sub-industries, largely predetermines the architecture of the regional economy as a market, serves as an economic framework of reproductive relations and interactions (urban and rural areas) in the ES of the region, and has broad prospects for development.

4 Conclusion

The methodological concept of the morphology of regional ES has a heuristic function of synthesizing the system of structural relationships and interactions that have developed in the ES of the region. This system predetermines determinants, drivers, causes, and consequences of development, reproducing the structural heterogeneity and socio-economic inequality of regions. Morphological connections act as fundamental for determining the type of ES in terms of its level of development, key factors, and sources of development and growth.

The instrumental and methodological strategy of building the morphology of ES of the region is based on the matrix form of representation of the whole complex of structural links and interactions (types of structural models) identified in the ES of the region based on macroeconomic, territorial and economic (economic an geographic), and organizational and economic approaches in the subject field of each specific model (concept) of the region as a reproduction system, a market, a state, a corporation, and a society. This methodology allowed us to systematize and typologize the models of the morphology of ES of the region, expressing a set of structural relations, directly determined by the territories. Practical verification of this methodology for analyzing the ES of the region on the example of Southern Russia has confirmed its heuristic value and the validity of the specific results of the research. The full-scale deployment of the proposed strategy for the study of heterogeneity and socio-economic inequality of regions seems a promising task in the search for ways and mechanisms to develop the reproductive unity and connectivity of the ES of Russia, the importance of which is emphasized in the Strategy for Spatial Development of the Russian Federation for the period until 2025 [13, pp. 19–27]. Moreover, it is a promising area in improving the methodology and tools of regional economic policy to reduce socio-economic inequality of regions.

References

1. Analytical Center "Expert SOUTH.": Ranking of the largest companies in the Southern Federal District by sales volume in 2017. Expert SOUTH **9**, 17–21 (2018). https://expertsouth.ru/ratings/krupnejshie-kompanii/renking-krupneyshikh-kompaniy-yufo-po-obyemu-realizatsii-produktsii-v-2017-godu/. Accessed 14 June 2021
2. Druzhinin, A.G.: The most important trends in the development of Russia's socio-economic space in the post-Soviet period: globalization, regionalization, centralization, strengthening of the periphery and/or marginalization? How have they affected people's living standards and lifestyles? In: Druzhinin, A.G., Kolosov, V.A., Shuvalov, V.E. (Eds.) The space of modern Russia: Opportunities and Barriers to Development (Reflections of Geographers and Social Scientists), pp. 78–80. Vuzovskaya kniga, Moscow, Russia (2012)
3. Druzhinin, A.V., Kolesnikov, Y.S., Ovchinnikov, V.N. (Eds.): The South of Russia: institutions and strategies of economic modernization. Vuzovskaya kniga, Moscow, Russia (2014)
4. Eyal, N., Hoover, R.: Hooked: how to build habit-forming products (S. Filin Transl. from English). Mann, Ivanov and Ferber, Moscow, Russia. (Original work published 2013) (2021)
5. Federal State Statistics Service of the Russian Federation. Regions of Russia: Socio-economic indicators. Rosstat, Moscow, Russia (2018)
6. HSE University. SMM-promotion. What is social media marketing? (2020). https://marketing.hse.ru/news/416404387.html. Accessed 14 June 2021
7. Kleiner, G.B. (ed.): Mesoeconomics of transition: Markets, industries, and enterprises. Nauka, Moscow (2001)
8. Kolesnikov, Y.S., Ovchinnikov, V.N. (Eds.): Traditional economic practices of the peoples of the North Caucasus: Ethnoeconomics – XXI century. Sodeistvie–XXI vek, Rostov-on-Don, Russia (2015)
9. Leksin, V.N., Porfiryev, B.N.: Legal mechanisms in the structure of spatial development regulators (As exemplified by the Russian Arctic region). Problems Manage. Theory Practice **4**, 6–15 (2018)
10. Odrin, V.M., Kartavov, V.S.: Morphological analysis of the system. Naukova Dumka, Kiev, USSR (1977)
11. Ovchinnikova, V.N., Kolesnikova, Y.S.: Morphology of the economic space of the region. Sodeistvie–XXI vek, Rostov-on-Don, Russia (2019)
12. Radzikhovsky, L.A.: Running from time. Rossiyskaya Gazeta, **195**(7658) (2018)
13. Shamakhov, V.A., Mezhevich, N.M.: The strategy of spatial development of the Russian Federation until 2025: Economic opportunities and administrative Restrictions. Admin. Consult. **4**(124), 19–27 (2019). https://doi.org/10.22394/1726-1139-2019-4-19-27
14. Tatarkin, A.I.: Innovation-based modernization of the area of Russia. Region: Econ. Sociol., **1**(89), 6–33 (2016). https://doi.org/10.15372/REG20160101
15. Yashalova, N.N., Krylova, N.P., Levashov, E.N.: Regional environmental marketing: features and prospects. Sci. J. ITMO Univ. Series Econ. Environ. Manage. **3**, 123–131 (2020). https://doi.org/10.17586/2310-1172-2020-13-3-123-132

Organizational Processes of the Formation of Financial System in the Context of Digitalization of Economy

Tatyana A. Zhuravleva[1(✉)], Irina V. Skobliakova[1], Elena M. Semenova[1], Elena A. Zbinyakova[1], and Aleksandr V. Zakharov[2,3]

[1] Orel State University, Orel, Russia
[2] Moscow University for Industry and Finance "Synergy", Moscow, Russia
[3] Plekhanov Russian University of Economics, Moscow, Russia

Abstract. The increasing role of digital technology is transforming the country's financial system. The financial system acts as the base of end-to-end technology in the digital economy, passing through advanced tools that enhance the interaction between humans and artificial intelligence. Despite the high level of technological sophistication of the financial system, the formation of organizational processes allowing to consider the totality of risks and legal factors accumulated in the internal and external environment is a fairly significant problem in the digital economy. This condition determines the relevance of the chosen research topic. The paper considers the organizational processes of the formation of the financial system in a digital economy. The authors consider types of financial technologies in the context of the digital economy and analyze the legal specifics of the regulation of the financial system in the digital economy. Moreover, they highlight the risks of the financial system in the digital economy. The authors also form the mechanism of organizational processes affecting the development of the financial system in a digital economy. The scientific research is based on several methods such as theorizing, benchmarking of financial technologies, aximetric representation, analytical method, comparative method, grouping, diversification, method of local applicability, formation of alternative paths, data modeling, and process architecture.

Keywords: Organizational processes · Digitalization · Financial system · Digital twin · Mechanism · Legal system · Risk · Digital technology

JEL codes: E60 · G20 · G28

1 Introduction

Financial technology is a fairly important tool in digitalization policies. The introduction of financial technologies in various spheres of life allows forming a flexible system of the country's competitiveness in the international space. The development of the market for digital financial technologies has led to the need to implicate a modern

E. G. Popkova (Ed.): ISC 2020, LNNS 368, pp. 190–199, 2022.
https://doi.org/10.1007/978-3-030-93244-2_22

legal system to regulate organizational processes. Organizational processes of the formation of the financial system in the digital economy have several specific factors.

The first factor is determined by the lack of regulatory issues controlling the financial system of the Russian Federation in the digital economy. The Russian Federation has several fundamental laws in the organization and regulation of the financial system of the Russian Federation. Some of these laws are as follows:

- Federal Law "On additional measures to support the financial system of the Russian Federation" (October 13, 2008 No. 173-FZ);
- Presidential decree "On some measures to stabilize the financial system of the Russian Federation" (August 25, 1998 No. 988).

Nevertheless, the lack of declared norms on cryptocurrency, cryptocurrency exchanges, ICOs, and smart contracts create a legal gap in the system of legislative norms. To solve the identified problem, it is necessary to work on the inclusion and clear regulation of the issue and circulation of cryptocurrency and the provision of conditions and smart contracts in the Federal law "On digital financial assets, digital currency, and amendments to certain legislative acts of the Russian Federation" (July 31, 2020 No. 259-FZ).

The second factor is related to the presence of a fairly large range of financial technologies in the digital economy, which have not received proper attention within the Russian financial system. In addition to the financial technologies mentioned above, it is necessary to consider crowdfunding, crowding platforms, and financial sandboxes. The consideration of these types of financial technologies will allow for the formulation of the organizational processes of the financial system that meets the postulates of the digital economy.

The third factor highlights the risks of the digital economy. Currently, with the high concentration of transparent forces, the risk component of the financial market is short-term and fast-changing. The consideration and constant monitoring of risks allow minimizing the negative effects of the digital economy on the financial market.

The fourth factor is the formation of the mechanism of organizational processes affecting the development of the financial system in a digital economy. This is the last factor among the problems presented above. Organizational processes imply the step-by-step execution of tasks according to a plan aimed at solving the key problems stated in the system [14]. Considering organizational processes in the structure of the financial system in the digital economy, it is necessary to implicate a mechanism that allows launching and adjusting these processes in accordance with emerging conditions. This factor will allow structuring the entire aspect of organizational processes within a single system to form the financial system in a digital economy.

This research aims to consider the organizational processes of forming the financial system in a digital economy. The following research tasks are formulated to implement the research purpose:

- To consider financial technologies in the context of the digital economy;
- To analyze the legal specifics of regulating the financial system in the digital economy;

- To highlight the risks of the development of the financial system in the digital economy;
- To form the mechanism of organizational processes affecting the development of the financial system in the digital economy.

2 Materials and Methods

The methodological basis of the research lies in the elaboration of the tasks mentioned above. The consideration of financial technologies involves the use of methods such as theorizing [2], benchmarking of financial technologies [7], and aximetric representation [1]. The analysis of the legal specifics of the regulation of the financial system in the digital economy is declared through the use of analytical and comparative methods. The risks of developing the financial system in the digital economy are highlighted using the methods of grouping, diversification, and local applicability [3]. The mechanism of organizational processes affecting the development of the financial system in a digitalized economy is formed using the formation of alternative paths, data modeling, and process architecture [11].

Let us look at the most common financial technologies in the context of the digital economy. The analysis carried out by the Central Bank of Russia in the field of digitalization of the financial market has shown that the current prospects for the development of financial technology are the transformation of the banking sector from a traditional financial monopoly to a technology company and the formation of business models as tools to create banking ecosystems [15]. The most common financial technologies corresponding to the trends presented above are data analysis and mobile and cloud technologies.

According to the research results, the most promising financial technologies are cryptocurrency, blockchain, cryptocurrency exchange, and smart contracts. Cryptocurrency is a digital currency that enables the operation of automated payment systems [16]. Cryptocurrency is differentiated into more than 200 varieties [9]. The main types of cryptocurrencies are shown in Fig. 1.

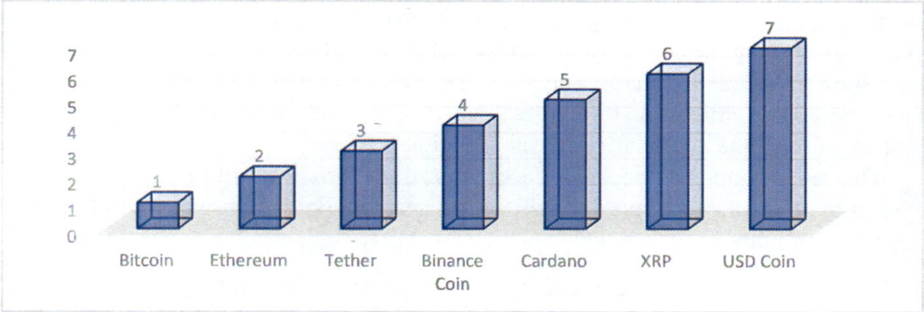

Fig. 1. Main types of cryptocurrencies. *Source* [5]

Cryptocurrency acts as an instrument of financial intermediation, which reduces the effectiveness of the monetary policy of the Central Bank of the Russian Federation in the context of digitalization. On the one hand, cryptocurrency is a type of financial instrument of the digital economy. On the other hand, it acts as a prototype of decentralized financial means, which have evolved throughout the existence of financial institutions. The main problem caused by the slowdown of the use of cryptocurrency is a weak regulatory function of the distribution of decentralized funds by the state [12]. The highlighted problem stems from the fact that the monetary circulation of cryptocurrency can be chaotic and uncontrolled by the issuing center. The formulated aspect will lead to an increase in the risks of the implementation of monetary policy. Thus, the Central Bank in the digital economy becomes not just a regulator but a risk manager to track decentralized financial means.

The main directions of cryptocurrency development are based on the redistribution of financial assets (beyond the control of the centralized regulator) and the reduction of excess liquidity in the boundaries of non-cash settlement [8]. The second direction of cryptocurrency is based on the fact that any participant registered on the cryptocurrency exchange can perform operations. In addition to private users, some companies emerge in the cryptocurrency market and decentralize this function to suit their needs. For example, Facebook released the cryptocurrency Libra, which equaled $0.39 as of July 14, 2021. Issuing the Libra cryptocurrency, Facebook tried to create a sustainable system focused on the flexible decentralization of funds (Fig. 2).

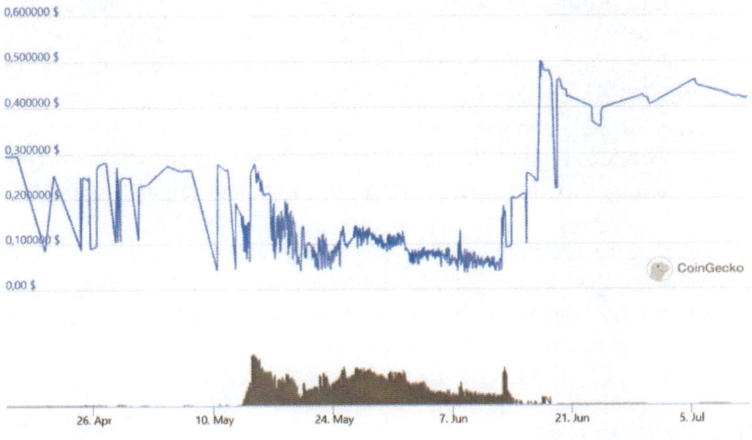

Fig. 2. Price variation of Libra cryptocurrency from April 20, 2021, to July 14, 2021. *Source* [4]

Libra does not compete with sovereign currencies since Facebook does not form an independent monetary policy and does not tie it to one of the existing currencies. To a large extent, Libra acts as a long-term investment tool.

The next tool of the financial system is blockchain technology. This technology aims to build a consistent chain of related blocks to increase the transparency of financial assets and facilitate access to financial markets [17]. Blockchain is a financial technology that allows engaging a considerable number of areas of the financial system – settlements, liquidity of assets, and anti-fraud measures. Despite cryptocurrency, it is worth mentioning tokenization – the development and issuance of tokens backed by illiquid assets – as part of the blockchain allocation. This process within the financial system is characterized by the fact that the indicated mechanism involves the sale of illiquid assets (high market entry, consideration within the venture capital market, the lack of legislation to regulate these processes).

Defining cryptocurrency and blockchain as tools of the financial system in the digital economy, we consider the main infrastructure platform – the cryptocurrency exchange. The most widespread cryptocurrency exchanges are in Japan and Singapore [6]. In the first case, the cryptocurrency exchanges are used to develop special legislation to regulate the order and organization of this type of activity. In the second case, these processes are regulated by the legislation on operators of funds [10]. Simultaneously, securities regulated based on the legislation on the organization of trading do not fit under these conditions [13]. Additionally, Russia has not worked out a legal system that fixes the process of taxation of crypto-assets. This condition should indirectly affect the taxation system of individuals participating in cryptocurrency exchange and purchasing goods using cryptocurrency.

Compared to cryptocurrency, tokens are not based on the cryptocurrency exchange. Tokens are regulated by the ICO as the process of issuance and subsequent sale of financial instruments. The advantage of the ICO is a fairly low cost of issuing tokens. ICO is an analogue of IPO. Tokens have a dual approach. On the one hand, they act as securities, and, on the other hand, they are a product and payment instrument. The first requirement can be implemented in the ICO and the second – only at the IPO. At the same time, many countries (e.g., France) do not separate tokens by type of characteristics and refer them to securities. In the Russian Federation, tokens are not considered as a tool of the financial market in the digital economy due to the lack of a legislative framework to regulate the issue and sale of tokens and the leading role of the Central Bank of the Russian Federation as the regulator of the country's financial market.

These financial technologies act as a starting point in transforming the entire financial system in the context of the digital economy. To determine the Russian realities within the framework of the studied topic, let us analyze the legal specifics of the regulation of the financial system in the digital economy.

3 Discussion

Nowadays, risks in the financial system are associated primarily with the development of the digital economy. The first risk stems from the increasing complexity of the digitalization of information infrastructure. Like the digital devices powering the financial system, the digital environment operates within a framework of limited access to the production, distribution, and exchange of data exposed to information threats from cyberattacks.

The financial system is the most vulnerable, which is associated with the introduction of the largest number of digital technologies than in other areas of the economy. Setting limits on this risk will lead to lagging behind the technological development of the country's financial system.

The second risk within the framework of the identified issues is determined by changes in the capacity of human resources in the structure of the financial system functioning in a digital economy. On the one hand, the introduction of digital technology in the financial system increases labor productivity and reduces the cost of maintaining processes. On the other hand, this circumstance forms one of the most dangerous risks of the financial system in the digital economy – the change of human resources with rising unemployment among the economically active population. The increase in unemployment and changes in the capacity of human resources is due to a reduction in the labor force related to low- and medium-skilled personnel. The digital economy involves an increase of part-time employment in the conditions of increased labor intensity and declining wages for corporate employees. The highlighted factor will lead to an increased role of automation of the financial system and the lack of highly qualified specialists in the created jobs. These changes will increase the importance of the financial environment compared with other areas of the national economy.

In the context of the increasing role of this risk, an important problem lies in the lagging of the education system in the formation of the financiers of the future. This lag is primarily caused by the ineffectiveness of the federal state educational standards, leading to the formation of a uniform education system in the Russian Federation. The minimization of this risk lies in the development of a flexible system of standardization of Russian education. The financial system concentrates on specialists with narrow competencies differentiated according to several branches of the national economy. In accordance with this fact, the model of education producing personnel for the financial system must be restructured following the transformation of the labor market. The gap between the needs of the financial system and the country must be bridged.

The decline of government control and the role of protectionist policies is related to the emergence of a financial digital global world. This risk is due to the formation of a global network financial world, which strengthens the role of decentralization of the financial system. Legal regulation at the national level weakens these processes in the context of financial transactions on world exchanges. Simultaneously, the role of international legislation and the emergence of new digital financial assets localize the problems of the development of the financial system at the national level.

4 Results

The set of identified risks and the analysis of legislation regulating the financial system in a digital economy determined the importance of forming a mechanism of organizational processes affecting the development of the financial system in a digitalized economy (Fig. 3).

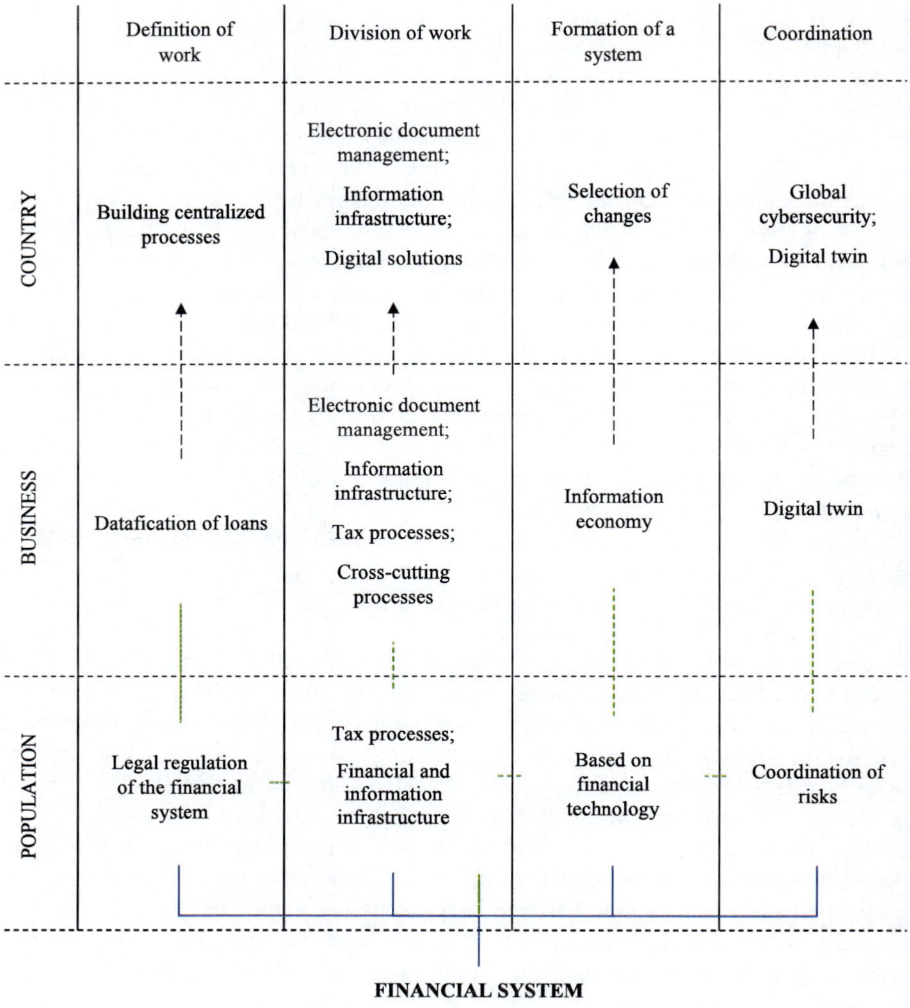

Fig. 3. The mechanism of organizational processes affecting the development of the financial system in the digital economy. *Source* Compiled by the authors

The mechanism of organizational processes affecting the development of the financial system in a digitalized economy is built based on allocating organizational processes, identifying users, revealing factors in the digital economy, and defining the relationship between financial processes. This research is based on four organizational processes that aim to build a mechanism considering the financial specification and digital environment of the economy – definition of work, division of work, formation of a system, and coordination. The users of the mechanism of organizational processes are the population, business, and government. Classical allocation of these subjects and attributing them to the users allows reflecting the most structured groups of consumers of the financial system in the digital economy.

The organizational process of the definition of work regenerates the legal regulation of the financial system at the population level. This condition is related to the problems identified in the analysis of the legal framework for developing the financial system in the digital economy. The regulation of digital financial assets used by the public is at the core of this issue. Organizational processes of the definition of work are interconnected at all levels of users of the mechanism. Accordingly, legal problems at the population level are solved together with the alignment of centralized processes at the national level. The definition of work at the business level is the most differentiated. This condition is determined by the highlighted processes – datafication of loans. After datafication, the loan is based on the disbursement of loan funds by analyzing the personal data of the business entity that is borrowing the money. This condition expands the information network of the financial system within the digital economy.

The division of work as an organizational process of the mechanism generates conditions and directions with the greatest transparency among the financial system and elements of the digital economy. Two areas are important at the population level:

- Aspects affecting the centralized connections of the financial system;
- Aspects reflecting the basics of organization and digitalization of this environment.

The first component is the tax process. The second component is aimed at the formation of financial and information infrastructure. At the business level, the division of work occurs through the implementation of electronic document management, the creation of information infrastructure, the harmonization of tax processes, and the introduction of end-to-end technologies. The government, as a user, sees the division of labor as the direction of impact on the digital financial system. These areas include electronic documentation, information infrastructure, and digital solutions.

The third group of organizational processes is the formation of the financial system. In the context of this group, the emphasis is made on highlighting the main criterion enhancing the effect of the digital economy in systems of users. The formation of the financial system occurs within this group based on the introduction of financial technology (user – population), the establishment of norms of the information economy (user – business), and the selection of changes occurring in the global financial system (user – government).

The final group of organizational processes in this mechanism is coordination. Coordination is the result of the development of the financial system in the digital economy and a monitoring point, the tracking of which allows for the transformation of the processes within the system. As a user, the population produces a fairly large number of risks reflected in the financial system. At the same time, decentralized processes do not allow to monitor these risks fully. The creation of the digital twin as an organized process of coordination is implicit in business and government. Additionally, as a user, the government requires coordination of cybersecurity processes at the global level, which helps form a holistic system of sustainable interaction of the financial component in the digital economy.

5 Conclusion

The conducted research on the organizational processes of formation of the financial system in a digital economy allowed us to formulate the following conclusions:

1. The development of financial technology is transforming the financial system in line with the creation of a modern and universal digital environment. The most common technologies and tools of the financial system are cryptocurrency and blockchain, with separate platform systems. The increasing creation of digital technologies and their chaotic introduction into the financial system weakens the role of the legal system in the regulation of the financial sector in the digital economy.
2. Currently, the key regulatory act linking the financial system with the processes of the digital economy is the Federal Law "On digital financial assets, digital currency, and amendments to certain legislative acts of the Russian Federation" (July 31, 2020 No. 259-FZ). This legal act regulates the processes of issuing, accounting, and circulation of digital financial assets, including the specifics of the operator of the information system of digital financial assets and the circulation of digital currency in the Russian Federation. Despite this fact, Russia has not yet formed a policy document to focus on the transition of the financial system to the digital economy by centralizing the main postulates of this process. Partial prerequisites of the above condition are indicated in the passport of the national project "Digital Economy of the Russian Federation" and the main directions of developing information security of the credit and financial sphere for 2019–2021.
3. The risk component of the development of the financial system in the digital economy is concentrated in the context of infrastructure and education. The research revealed the following main risks of the organizational functioning of the financial system in the digital economy:

- Increasing complexity of the system of digitalization of the information infrastructure;
- Changes in human capacity;
- Lag of the education system in the formation of financiers of the future;
- Decline of government control in the digital sphere;
- Formation of a financial digital global world.

The identified risks of organizational processes have determined the importance of forming a mechanism regulating the set of components that affect the functioning of the financial system in the digital economy.

4. The mechanism of organizational processes affecting the development of the financial system in a digitalized economy is based on a comprehensive elaboration of the functionality associated with the inclusion of users in the system of actions aimed at the maintenance of this system. The organizational processes include the definition of work, the division of work, the formation of the system, and the subsequent coordination. The users of the organizational process allow concentrating the activities on the functioning of the financial system within the groups of the population, business, and government. The final organizational process of this mechanism is coordination. It requires the implementation of digital twins at the

level of business and government (considering the measures of global cybersecurity) and monitoring risks threatening the population.

References

1. Adamyk, B., Skirka, A.: Banking system and international trade in Ukraine during the financial crisis. Int. Econ. Lett. **5**(1), 17–25 (2016). https://doi.org/10.24984/iel.2017.5.1.3
2. Bugaeva, T.N.: Transformation processes in financial system of the state. Sci. Bulletin Finance Banking Investment **3**(48), 23–31 (2019)
3. Chernenko, V.A.: A new format for the financial development of the country's economy. Econ. Vector **1**(12), 80–84 (2018)
4. CoinGecko. (n.d.). Cryptocurrency prices ranked by market capitalization. https://www.coingecko.com/ru Accessed 14 Jul 2021
5. CoinMarketCap. (n.d.). Top-100 Cryptocurrencies by market capitalization. https://coinmarketcap.com/ru/ Accessed 13 July 2021
6. Glazev, S.Y.: Problems and prospects of the Russian financial market in the context of structural changes in the world economy. Finance Theory Practice, **24**(3), 6–29 (2020). https://doi.org/10.26794/2587-5671-2020-24-3-6-29
7. Gresnigt, F., Kole, E., Franses, P.H.: Interpreting financial market crashes as earthquakes: a new early warning system for medium term crashes. J. Bank Finance **56**, 123–139 (2015). https://doi.org/10.1016/j.jbankfin.2015.03.003
8. Kai, G., Zhang, W., Jin, Z., Wang, C. Z.: Hope bifurcation and dynamic analysis of an improved financial system with two delays. Complexity, 3734125 (2020). https://doi.org/10.1155/2020/3734125
9. Khajehnasiri, A.A., Afshar Kermani, M., Ezzati, R.: Chaos in a fractional-order financial system. Int. J. Math. Oper. Res. **17**(3), 318–332 (2021). https://doi.org/10.1504/IJMOR.2020.110028
10. Kleiner, G.B., Rybachuk, M.A., Karpinskaya, V.A.: Development of ecosystems in the financial sector of Russia. Manager **11**(4), 2–15 (2020). https://doi.org/10.29141/2218-5003-2020-11-4-1
11. Kolinets, L., Uniyat, A.: Crisis measures to ensure the stability of the country's financial system. Sci. Europe **48–3**(48), 59–63 (2020)
12. Lavrushin, O.I., Sokolinskaya, N.E.: View on the financial market development challenges. Econ. Busin. Banks **12**(38), 54–74 (2019)
13. Liu, C.: Organization evolution of fuzzy system based on financial risk degree of commercial banks. Math. Problems Eng., 6698299 (2021). https://doi.org/10.1155/2021/6698299
14. Park, M.S.: The financial system and the Pasinetti theorem. Camb. J. Econ. **30**(2), 201–217 (2006). https://doi.org/10.1093/cje/bei056
15. Shkodinsky, S.V., Dudin, M.N., Usmanov, D.I.: Analysis and assessment of cyberthreats to the national financial system of Russia in the digital economy. Financial J. **13**(3), 38–53 (2021). https://doi.org/10.31107/2075-1990-2021-3-38-53
16. Szendrei, T., Varga, K.: FISS – A factor-based index of systemic stress in the financial system. Russian J. Money Finance **79**(1), 3–34 (2020). https://doi.org/10.31477/rjmf.202001.03
17. Zhuo, W., Ding, C., Xiong, Y., Peng, J.: Construction of legal system for maritime international trade financial supervision cooperation. J. Coastal Res. **103**(S1), 143–146 (2020)

Legal Consciousness of Russian Society in the Conditions of Global Digitalization

Natalia A. Ryasnyanskaya[1(✉)], Lidiya M. Balakireva[1], Natalia N. Kravchenko[1], Valery V. Meleshkin[1], and Dmitry I. Sostin[2]

[1] North Caucasian Federal University, Stavropol, Russia
[2] MIREA–Russian Technological University Branch in Stavropol, Stavropol, Russia

Abstract. The paper focuses on the current problems of modern Russian society – the formation of legal consciousness in the conditions of global digitalization. To identify the peculiarities of this process, the authors implement classical general scientific methods of cognition and special methods used in socio-political and historical-legal scientific research. To determine the peculiarities of the formation of legal consciousness of Russian society in the context of global digitalization, the authors identified the essence of this process, its prerequisites and results, and the relationship of the development of legal consciousness with digitalization. The research identified the primary goal of the political course of Russia. Nowadays, the country aims to increase the level of legal consciousness of the population in terms of introducing digital technology in everyday life. The authors conclude that the formation of Russian legal consciousness in the context of global digitalization directly depends on the values and capabilities of Russian society to create a mechanism to guarantee the education of national legal consciousness at the national level. This process significantly contributes to the public's perception of various innovations, including global digitalization itself. Currently, Russia needs to revive national values that determine the future of society, the country, and the role and place of the country in the international arena. This need is especially acute during global digitalization, which precludes the isolation of any state or region. These facts allow us to consider digitalization as one of the leading criteria for the quality of life of modern society. It is necessary to note the comprehensive coverage of this process, from the economy to the average citizen. Simultaneously, the main indicator of the effective implementation of digitalization in all spheres of society is not only the level of mastery and skills to work with it but also the result of the development of society itself, which is largely reflected in the level of legal awareness of the individual. In the world of nano-technology, informatization and computerization serve as a platform for digitalization. In contrast, digitalization forms a kind of technological environment allowing the users (regardless of gender, age, income, or education) to self-define the scope of their contacts at various levels of daily life.

Keywords: Legal consciousness · Globalization · Cyberthreat · Society · Digitalization · Digital divide · Technocratic elite

JEL code: I28

E. G. Popkova (Ed.): ISC 2020, LNNS 368, pp. 200–207, 2022.
https://doi.org/10.1007/978-3-030-93244-2_23

1 Introduction

At the current stage of development, the Russian Federation pays special attention to the formation of the legal consciousness of citizens, which implies knowledge of the law, the ability to apply it in everyday life, and acceptance and adherence to the social and moral values of society. This area is especially relevant in the conditions of global digitalization.

In the classical version, the structural elements of legal consciousness are considered from the position of legal psychology and legal ideology. Simultaneously, legal psychology is the result of experience gained in the daily life of a person. It can be seen as a societal response to ongoing transformations in various spheres of society [12]. At the same time, legal ideology is a legally established set of views, theories, and hypotheses, that allows for an objective assessment of the political situation in the country.

The main components of the mechanism of guarantees for the formation of national legal consciousness provide for its structural study in the individual, group, and universal aspects, as the main result of social development. Numerous social surveys show that children and teenagers from dysfunctional families tend to be in the risk group. Additionally, clear and sometimes insurmountable social boundaries in families, kindergartens, schools, vocational schools, colleges, universities, and society create favorable conditions for the antagonism in the social environment and produce a distorted perception of modern moral, spiritual, and legal values, and consequently, legal nihilism.

It will be difficult to break these social boundaries in the era of global digitalization because there is widespread propaganda and glorification of the bohemian life of a narrow circle of the elite of Russian society (which is sometimes far from representing the best human qualities) from TV screens, the Internet, and social networks. Every person will react differently to such information. One will ignore it, another will laugh, and a third will be inspired to illegal deeds for the sake of quickly achieving all earthly goods.

The peculiarities of this stage of global development are as follows:

- Moving most social and economic interaction to the digital environment;
- Distribution of remote forms of employment and translation of information;
- Replacement of human labor by artificial intelligence and highly automated equipment;
- Widespread introduction of cashless settlements and payments;
- Conversion of several democratic procedures into electronic form (publicly accessible databases, the possibility of remote voting, paperwork, etc.).

Digitalization is an essential strategic resource determining the vector of socio-economic and political development of society [4].

To further develop society considering the currents demands, a radical revision of the ways and types of communication is required as a critical element of digitalization and the determination of the future path of development of public legal consciousness [8]. Nowadays, digitalization has affected all spheres of society and adjusted the perception of the role and place of a person in the country and the awareness of their rights, responsibilities, and abilities to implement them in everyday life. This specific feature of digitalization in Russia allows us to talk about the relevance of considering the outlined problems.

2 Materials and Method

The methodological basis in the research of the problem of developing the legal consciousness of Russian society in the context of global digitalization is general scientific methods of knowledge and special methods used in socio-political and historical-legal scientific research [5, 6]. The methods of analysis and synthesis are used to show the features of the formation of the legal consciousness of society in the conditions of digitalization. These methods allowed defining the meaning of the terms "legal consciousness" and "digitalization" in Russian society, which reflect the complex process of formation of national legal consciousness in the conditions of conversion of a certain type of information into digital form. On the one hand, digitalization is a response to the demands of society. On the other hand, it determines the vector of development of legal consciousness in Russian society.

The method of deduction allowed the authors to identify the peculiarities of the formation of national legal consciousness and the organizational basis of digitalization under the Russian legislation based on the regulatory and legal framework. The authors used the method of induction to analyze the articles of the Constitution of the Russian Federation and Federal Laws. The main directions of the development of Russian policy were considered.

The method of abstraction allowed us to study the level of public legal consciousness in the conditions of digitalization. As a global process, digitalization has embraced all aspects of society. Nevertheless, its impact on socially important aspects, including legal consciousness, must be critically examined.

Along with general scientific methods, the authors used a block of special historical and legal methods (e.g., historical-legal and formal-legal methods, the method of comparative law, and system analysis of legal phenomena), which allowed to create a holistic picture of the development of public legal consciousness in the conditions of digitalization in Russia. The development of digitalization in Russia is associated with the spread of the Internet since the 1990s when society learned new opportunities in the field of information and communication technologies. On the one hand, there opened new opportunities for society. On the other hand, not all categories of the population could cope with the flood of freely accessible information. It is necessary to pay special attention to the fact that the Internet, high-quality technology, and the acquisition of skills in the new system have not become available to all social groups in Russia, thus creating a digital divide, which has contributed to a further strengthening of social division.

The methodological tools used in the research revealed the peculiarities of the development of the legal consciousness of Russian society in the conditions of digitalization, identified problems, and outlined ways of its further development and overcoming the digital divide.

3 Results

Nowadays, digitalization acts as a global vector of economic and social development. It aims to convert information into digital form and make it accessible to improve the efficiency of economic performance and the quality of everyday life. Compared to the initial stage of computerization, when information technology was used only for particular tasks, the modern process of digitalization involves the coverage of all aspects of society and, at the same time, the positive perception of these innovations by the society.

Digital processing, copying, storage, and transmission of various kinds of information allows digitizing various data, ranging from an address book and ending with documents of historical value. A new type of information accessibility and processing has made scientific and encyclopedic publications, updating legal information systems, and sites providing various types of services publicly available [10].

On the one hand, this process led to quick access to the information of interest. On the other hand, it contributed to the formation of the legal consciousness of people. In the life of Russian society in the 21st century, legal education, as a key factor in the formation of personal legal consciousness, is entrusted with a mission to define the social and legal values that contribute to the preservation and translation of the existing legal ideals and legal experience in everyday life [11].

Thus, we can identify the main objectives of legal education contributing to the formation of the legal consciousness of Russian society in the context of global digitalization. These objectives are as follows:

- To increase the level of legal culture through a general increase in legal awareness;
- To provide socialization of individuals through the acquisition of legal knowledge and the ability to translate it through the use of digital technology;
- To form a clear civic position in legal matters.

The solution of these problems and their implementation with the use of digital technology will allow giving an objective assessment of the Russian legal reality and determining the prospects for the development of education and upbringing and the direction of public policy in this area.

Practical developments in this area have shown that the achievement of the desired results requires continuous monitoring of the level of legal consciousness in society using digital technologies and the possibility of placing legal information in the public domain to identify the degree of its deformation in time and make prompt and appropriate decisions at the national level. Numerous studies in this area show that public legal awareness has a technological nature, which allows using different types of digital actions depending on the category of respondents, the conditions of the conduct, the tasks set, and the results expected.

In this regard, it is necessary to know the current digital capabilities of the country to respond to the situation timely. It is generally accepted to distinguish the following parameters of digitalization quality:

- Uninterrupted high-speed Internet in the country;
- Level of computer literacy of the population;

- Qualitative and quantitative indicator of public consumption of online content;
- Accessibility and quality of digitization of demanded services, availability of electronic services at the government and legal level [13].

The analysis of the digitalization of socially significant spheres in Russia showed that there are software products for cybersecurity. Simultaneously, there are opportunities for the digital transformation of all socially significant industries. All necessary prerequisites have been formed to deepen digitalization at the national level further. These prerequisites are as follows:

- Globalization of the economic sphere in the presence of a single economic space;
- Dynamic development of Internet technologies and the growth of computing power of processors;
- Ubiquity of mobile devices and the use of social media;
- Recognition of the importance of digitalization with the necessary legal and regulatory framework.

However, the accelerated transition to digital technology causes the following problems:

- Distrust of Russian society to new technologies due to the lack of adaptability of the legal framework to the current needs;
- Lack of qualified professionals in the field of information and legal sphere;
- Inaccessibility for the majority of citizens of expensive modern equipment for comfortable use of new technologies;
- Gap between the knowledge and skills gained in the digital environment and the real demands of everyday life [9].

It is necessary to highlight the leading role of the Russian government as the initiator and the main driving force of the digital transformation of all spheres of life. Currently, digitalization is one of the national development priorities, which is reflected in the policy documents issued over the last decade. Implementing the information and communication function, the government primarily cares about the quality and accessibility of socially important information, which is the primary vector of increasing the population's legal consciousness. The government guarantee of the quality and accessibility of social and political-legal information ensures the unity of all spheres of life throughout the country, the creation of an information space, and a high level of legal awareness of different social groups [3].

World practice shows that the leading role of the government in the context of global digitalization will provide a stable growth of production, create an economic and legal platform for new business projects, support the development of services, and enable young professionals to contribute to the development of the village, city, region, and country. Despite the new directions in the socio-political sphere allowing Russia to act as a worthy partner and a competitor in the international arena, a step into the world of innovation can bring the desired results when the population feels the improved quality of life and openness, support, and protection from the government. These factors are essential components of the effective development of civil society and the rule of law, regardless of the time and geographic location of the country.

As practice has shown, the era of globalization, along with technical innovations, has contributed to the emergence of several problems that have had a negative impact on the level of legal awareness of the Russian population. Liberalization and digitalization of the labor market provoked the growth of technological unemployment with a sharp fall in the social status of workers. Simultaneously, cyber-attacks, digital fraud, digital piracy, and the distribution of malicious content have become relevant in Russian daily life. In turn, these processes have led to growing public dissatisfaction with comprehensive digitalization, distrust of the pursued political course, loss of security of personal space, and the emergence of digital addiction among young people.

Global digitalization contributed to the formation of a new system of values in Russia. This system is far from the centuries-old traditional system. It was fully supported by advanced Internet users (usually young people), for whom the Internet space has become the meaning and the main reference point in life. Accordingly, in the market of educational services, mathematical and technical specialties became greatly demanded. Humanities, particularly those of legal orientation, have lost their original importance, which inevitably affected the level of thinking of young people.

The loss of critical thinking skills, logical reasoning, and analytical abilities inevitably affected the level of legal consciousness of society. Innovative processes at the global level will only yield positive results when they can coexist with the institution of national values. This should be greatly facilitated by government control to ensure the effective functioning of the digital environment. The problem of creating a platform ensuring the implementation of global technological processes remains relevant in Russia. The absence of such a platform will inevitably provoke undesirable conflicts in society, which will disturb the delicate balance of socio-political interests. Under such conditions, there emerged a new social group – the technocratic elite. This group has the necessary technical skills and material resources to make active use of the electronic social services offered by digitalization [7].

Thus, the problem of forming the legal consciousness of the population in the context of global digitalization sets the following primary task to the government:

- To ensure a gradual digitalization of socially significant areas, taking into account the current legislation to avoid violations of human and civil rights and freedoms;
- To overcome the population's distrust of the ongoing political course;
- To promote the level of national legal consciousness.

It should be emphasized that Russia has an appropriate legal framework for international cooperation in the field of digitalization. The long-term plan for the introduction of digital technology is clearly outlined in the program "On the strategy for the development of information society in the Russian Federation for 2017–2030" (May 9, 2017 No. 203) [1] and in the program "On the development of artificial intelligence in the Russian Federation" (November 10, 2019 No. 490) [2]. However, the main tasks (e.g., preserving social stability and increasing the level of national legal consciousness) still have to be solved, considering the preservation of traditional values of society and the current demands.

4 Conclusions

The formation of national legal consciousness in the context of global digitalization faces many obstacles. These obstacles are as follows:

- Social disunity created by the problem of fathers and children and the emergence of a technocratic elite;
- Distrust to the new government course due to "digital fraud" and the lack of clearly developed mechanisms of the legal protection of individual interests;
- Free-accessibility of personal data and the invasion into a person's personal space;
- Substitution of centuries-old values for the demands of the virtual world.

The new century has already been named "the era of nano-technology and global digitalization." However, society faced the problem of the coexistence of scientific discoveries with age-old national values. History shows that it is impossible to build a future on the ashes of the past. Following the direction of the development of society at the international level, to maintain social stability and build public confidence, the government needs to avoid radical breakdowns of established rules and customs and prepare a stable and mobile legal framework that can meet the demands of society and provide legal protection of the population.

References

1. Presidential Executive Office. Decree "On the strategy for the development of information society in the Russian Federation for 2017–2030". Moscow, Russia: Corpus of Legislation of the Russian Federation (2017)
2. Presidential Executive Office. Decree "On the development of artificial intelligence in the Russian Federation" (together with the "National strategy for the development of artificial intelligence for the period until 2030"). Moscow, Russia: Corpus of Legislation of the Russian Federation (2019)
3. Dobrinskaya, D.E., Martynenko, T.S.: Perspectives of the Russian information society: digital divide levels. RUDN J. Sociol. **19**(1), 108–120 (2019). https://doi.org/10.22363/2313-2272-2019-19-1-108-120
4. Drozdova, A.M., Balakireva, L.M., Vorotilina, T.V., Makarova, E.V., Meleshkin, V.V.: Legal awareness and legal culture as elements and means for the implementation of a mechanism for ensuring the legal impact. Opcion **35**(89), 922–934 (2019). https://produccioncientificaluz.org/index.php/opcion/article/view/27454
5. Kozyrev, A.N.: Digital economy and digitalization in historical retrospect. Digital Econ. **1**(1), 5–19 (2018). https://doi.org/10.34706/DE-2018-01-01
6. Kupriyanovsky, V., Sukhomlin, V., Dobrynin, A., Raikov, A., Shkurov, F., Namiot, D.: Skills in the digital economy and the challenges of the education system. Int. J. Open Inf. Technol. **5**(1), 19–25 (2017)
7. Lisenkova, A.A.: Socio-cultural challenges of the digital age. Digital Scholar Philosopher's Lab **2**(1), 173–182 (2019). https://doi.org/10.5840/dspl20192117
8. Smirnov, D.A., Tereshchenko, E.A., Botasheva, L.E., Trofimov, M.S., Melnikova, V.A., Dolgopolov, K.A.: Digital juris-prudence. Revista Inclusiones **7**(S1-1), 273-283 (2020)

9. Smirnov, D.A., Zhukov, A.P., Aparina, O.Y., Lauta, O.N., Zakalyapin, D.V.: Use of the transformational leadership model in police management. Amazonia Investiga **8**(20), 236–241 (2019). https://amazoniainvestiga.info/index.php/amazonia/article/view/90
10. Tereshchenko, E.A., Kovalev, V.V., Balakireva, L.M., Savchenko, E.V., Solovtsova, E.A.: Characteristics of legal culture study in Russian and foreign legal literature. J. Adv. Res. Law Econ. **8**(5), 1619–1628 (2017). https://journals.aserspublishing.eu/jarle/article/view/1723
11. Trofimov, M.S., Tereshchenko, E.A., Provodinaand, O.A., Tshovrebova, K.A.: From the Construction of Schools to the Digitalization of Education: Critical Issues of Education in the Caucasus in the Conditions of Integration of Educational Spaces in Russia and South Ossetia. In: Shakirova, O.G., Bashkov, O.V., Khusainov, A.A. (eds.) Current Problems and Ways of Industry Development: Equipment and Technologies. LNNS, vol. 200, pp. 1039–1045. Springer, Cham (2021). https://doi.org/10.1007/978-3-030-69421-0_115
12. Volkov, A.A., Drozdova, A.M., Kovalev, V.V., Riasnyanskaya, N.A., Chursina, V. P.: Russian legal consciousness of students in the introduction of infocommunication technologies, vol. 1. In: Drozdova, A.M. (Ed.). Moscow, Russia INFRA-M (2021)
13. Zhuzhgov, I.V., Volkov, A.A., Salnyy, A.M., Kulikova, T.B., Ryasnyanskaya, N.A.: Legal space monitoring: theoretical and legal aspects. Opcion, **35**(S20), 524–549 (2019). https://produccioncientificaluz.org/index.php/opcion/article/view/27586

Using the Leading Technologies in the Digital Economy Markets

Development of Modern Customs and Logistics Technologies in the Ecosystem of Digital Transport Corridors

Yulia V. Malevich(✉), Irina A. Plastunyak, Aleksandr V. Dmitriev, Ekaterina M. Ksenofontova, and Ksenia A. Aitova

St. Petersburg State University of Economics, St. Petersburg, Russia
{malevich.yu, plastounjak.i, dmitriev.a, ovsyannikova.k}@unecon.ru, koka007@list.ru

Abstract. The article discusses the development of modern customs and logistics technologies in the ecosystem of digital transport corridors of the Eurasian Economic Union (EAEU). The directions and methods of improving foreign economic activity through the introduction of digital services in the EAEU are investigated. The importance of digital transport corridors in the context of increasing the competitiveness of the EAEU territory as a transit route between Europe and Asia is substantiated.

Keywords: Customs and logistics technologies · Digital transport corridors · Foreign economic activity · Eurasian Economic Union

JEL Code: L86 · M15 · M16 · O32 · R40

1 Introduction

In modern economic conditions, increased attention is paid to the development of digital technologies. Various "digital ecosystems" are being formed and present a collection of digital technologies and services used by any enterprise, organization, community, or association.

The EAEU is also actively developing this area, as evidenced by the adopted Order of the Eurasian Intergovernmental Council dated 31.01.2020 No. 4 "On the Formation of an Ecosystem of Digital Transport Corridors", within which new customs and logistics technologies and services will be introduced, allowing the development of foreign economic activity both within the EAEU and beyond, as well as making the territory of the EAEU a more competitive transit route between Europe and Asia.

E. G. Popkova (Ed.): ISC 2020, LNNS 368, pp. 211–220, 2022.
https://doi.org/10.1007/978-3-030-93244-2_24

2 Methodology

Currently, in Russian and foreign studies, the authors study the problem of the development of modern customs and logistics technologies in the ecosystem of digital transport corridors (EDTC) from different sides.

The work of [4, p. 80] considers the modern project "One Belt - One Road", the Belt and Road Initiative (BRI), which allows the formation of new, alternative supply chains, to include not only developed but also underdeveloped countries in them. The main problem is the movement of goods and vehicles across the border. It is noted that the less developed the state, the more time it takes to carry out operations at the border. At the same time, on the one hand, this project provides new opportunities, but on the other, new risks appear.

Article of [2] describes the development and plans for the implementation of a cloud-based logistics platform to support and optimize cross-border transport using the China Railway Express in the context of China's recent One Belt, One Road initiative. The architecture of the authors' cross-border cloud logistics platform includes a hybrid cloud model with three layers of cloud services, including core business, value-added services such as real-time bidding, traffic information transfer, and integrated carrier service management.

In his article [8] notes that although the Belt and Road Initiative (BRI) is an ambitious infrastructure and development project that contributes to sustainable economic growth through the promotion of foreign trade in Eurasia and Africa, its potential remains underestimated. Logistics networks along the BRI routes face some problems that hinder effective operation, in particular, geopolitical and regulatory barriers, problems of organizing transport by rail and sea transport, as well as information support for the delivery of goods.

The research of [3] notes that in almost all studies, cost, time, and reliability are the main criteria for choosing a transportation option. At the same time, the cargo owner almost completely ignores such indicators as the flexibility of the transport service and additional services. Flexibility is understood as the ability of the logistics system to change depending on the needs of the shipper at any time, both before the start and during transportation, the ability to offer solutions in the event of supply chain disruptions. To solve the problem of choice, it is proposed to model based on the theory of random utility. According to the survey, recommendations for the participants were determined, including the state structure for forecasting cargo flows.

The publication of [5] notes that existing platforms (for example, UberFreight, BlackBuck, and Quicargo) allow communication between providers and consumers of transport services, but without direct control over the participants. At the same time, participants do not want to share all confidential information. The purpose of the research is to determine the performance of the system under various scenarios of information exchange between the cargo owner, carrier, and operator. For each element of the system, parameters are set, for example, for the vehicle release time, earliest start time, latest end time, capacity, start location, end location. An interactive auction procedure for multiple participants with the exchange of various information is considered.

The article of [7] discusses the creation and development of the ecosystem of digital transport corridors of the Eurasian Economic Union, pays attention to the research of the elements features of the functional and organizational structuring in the ecosystem of digital transport corridors and the definition of priority areas for the development and application of digital technologies and services in managing the transport and logistics system of the EAEU.

The research of [9] is devoted to identifying modern trends, stumbling blocks, and prospects for the joint use of transit potential by Russia and Kazakhstan in the East-West and North-South international transport corridors, the article notes that cooperation in the transit sphere can lead to an increase in the overall efficiency of economic integration within the EAEU, and also serve as a springboard for greater transcontinental connectivity.

3 Results

According to the main directions for the implementation of the digital agenda of the Eurasian Economic Union until 2025, it is planned to resolve issues and launch initiatives to implement the areas of digital transformation of economic sectors in the Eurasian Economic Union (EAEU) space, as well as the development of digital infrastructure, customs and logistics technologies and ensuring the security of digital processes. The total cost of the EDTC project is estimated at 10 billion rubles.

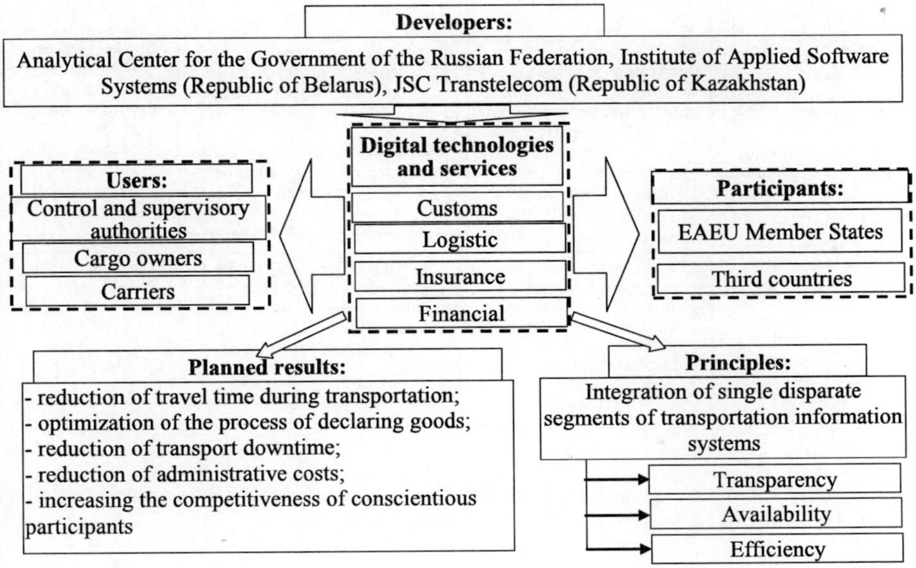

Fig. 1. Model of the digital transport corridors ecosystem. *Source*: developed and compiled by the authors

One of the priority initiatives in this context is the development of an ecosystem of digital transport corridors (EDTC), a model of which is shown in Fig. 1 [10].

According to the planned results, EDTC will allow optimizing many customs and logistics processes, which, first of all, include cargo transportation and filing of declarations.

Also, it is planned to minimize entry barriers to the international transport market and provide opportunities for bona fide participants in foreign economic activity to increase their competitiveness.

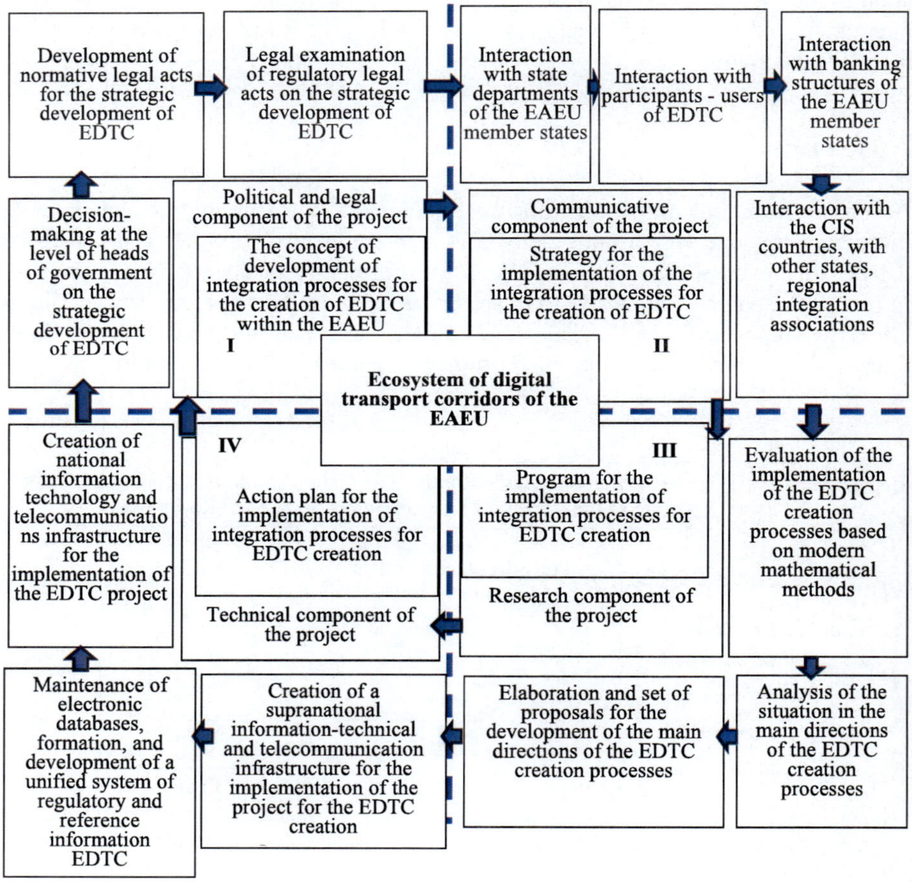

Fig. 2. Diagram of the sequential stages of decision-making to achieve the projected result of the EDTC creation. *Source*: developed and compiled by the authors

This will be achieved through the ability to choose and apply the most optimal routes and transportation conditions, to provide in electronic form a package of legally significant documents, to book a queue in advance at multilateral automobile checkpoints (MAC), as well as by increasing the efficiency of the activities of state control and supervisory bodies and improving the functionality of a single integrated information environment, including the national and supranational levels (Fig. 2) [11].

Table 1. Main Eurasian transport corridors involved in transit traffic

Name of ITC	Length of ITC, km	The nature of transportation (mode of transport)	Route of ITC
Western Europe - Western China	8,445	Automotive	St. Petersburg - Moscow - Orenburg - Aktobe - Almaty - Khorgos
Northern Sea Route	14,000	Sea	Key ports - Murmansk, Arkhangelsk, Nordvik, Ambarchik, Igarka, Dudinka, Dikson, Tiksi, Pevek, and Provideniya
Transsib	9,288	Multimodal	Berlin (Germany) - Warsaw (Poland) - Minsk (Republic of Belarus) - Moscow -Yekaterinburg - Vladivostok/Nakhodka and its branch Kurgan - Petropavlovsk (Republic of Kazakhstan)
North - South	7,200	Multimodal	St. Petersburg - the port of Mumbai (India) with the participation of Kazakhstan on the sections: the seaport of Aktau, the regions of the Urals of Russia and Aktau – Atyrau, Bereket (Turkmenistan) – Gorgan (Iran)
Trans-Caspian International Transport Route	9,400	Multimodal	China - through the territory of Kazakhstan, Azerbaijan, and Georgia - European countries through Turkey or Ukraine
Northern Corridor of the Trans-Asian Railway (TAR)	8,000	Railway	Connects Western Europe with China, the Korean Peninsula, and Japan through Russia (on the Dostyk/Khorgos – Aktogay – Sayak – Moints – Astana - Petropavlovsk (Presnogorkovskaya) section
Southern corridor (TAR)	9,000	Railway	Southeast Europe - China and Southeast Asia through Turkey, Iran, Central Asian countries (on the Dostyk/Khorgos – Aktogay – Almaty – Shu – Arys – Saryagash section)
Central corridor (TAR)	8,000	Railway	Central Asia - North-Western Europe (in the section Saryagash – Arys – Kandagach – Ozinki)

Due to the peculiarities of the geographical location of the EAEU, the bulk of the cargo transit flow will pass through the territory of Russia, the Republic of Belarus, and Kazakhstan. Despite this, in the implementation of the EDTC project, the possibilities of the transit potential of Kyrgyzstan and Armenia will also be involved (Table 1) [11].

Some factors affect the routing of transit freight traffic across the EAEU, including the lack of access to the sea in four of the five member states. This factor influences and provides prerequisites for the development of a special ecosystem of transport corridors, since a significant proportion of freight transportation by rail is carried out through sea trade ports. The main types of cargo transported in this way are coal, ore, grain, oil, and oil products.

The use of digital technologies at sea checkpoints also ensures the further development of various logistics technologies for the delivery of goods, including through the use of multimodal transportation. This opens up a wide range of opportunities for integration into the global transport and logistics system for the member states of the Eurasian Economic Union, and also contributes to the implementation of target areas for the development of the transit potential of the EAEU territory [6].

Improvement of electronic document flow at sea checkpoints improves the quality of the provision of customs and logistics services by reducing non-production, time costs associated with processing and sending paper documents between state control and supervisory bodies of participants in the foreign economic activity of the EAEU member states. Besides, it ensures the development of an information environment for multimodal technological interaction of various types of transport, cargo owners, and other participants in the transport process, and also allows the transition to electronic data exchange, accelerates, and simplifies the processes of cargo transportation [1].

These provisions are acquiring particular importance nowadays in the context of a decrease in the overall volume of cargo transportation by public transport in general across the EAEU, and in Russia in particular, where at the end of 2020 this figure amounted to 7.84 billion tons, which is 5.68% less than last year (Table 2) [12].

Table 2. Transportation of goods by mode of transport in Russia, million tons

Indicator	2018	2019	2020	2020/2019
Cargo transported, total, including:	8,157.83	8,313.77	7,841.45	−5.68%
Railway	1,293.52	1,281.44	1,246.65	−2.72%
Automotive	5,546.42	5,737.29	5,400.61	−5.87%
Sea	25.21	20.53	26.62	29.68%
Inland water	118.17	110.18	101.19	−8.16%
Air	3.19	3.17	3.19	0.73%
Pipeline	1,171.31	1,161.17	1,063.20	−8.44%

The volume of cargo transportation by sea in 2020 increased by 29.68%, and by air - by 0.73%. For the rest of the modes of transport, the decrease is observed, including on the road, which, according to the estimates of the Eurasian Economic Commission (EEC), currently accounts for about 80% of the total volume of freight traffic in the EAEU countries (excluding pipeline transport).

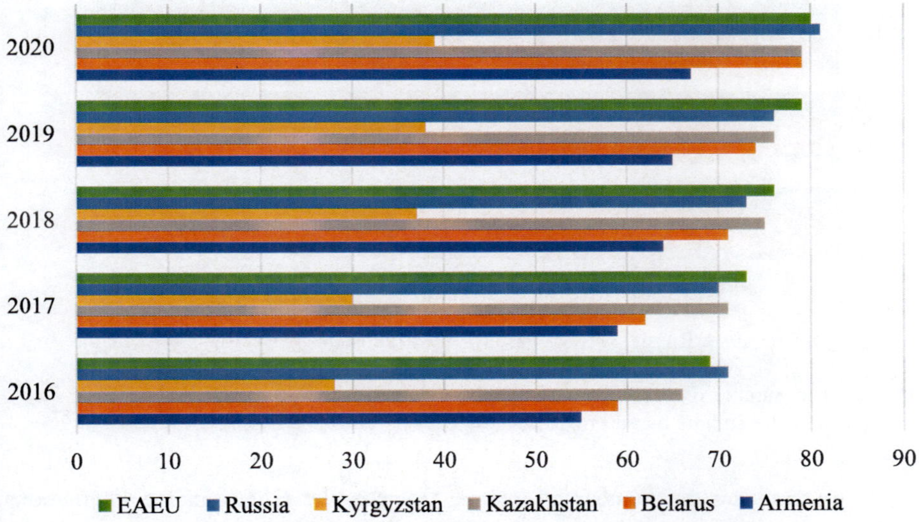

Fig. 3. The number of Internet users, %. *Source* Compiled by the authors based on [12].

According to the forecast, through the implementation of the EDTC project in terms of international road transport, starting from 2022, the annual mileage per vehicle unit will increase by an average of 20% due to a decrease in unproductive downtime. At the same time, the increase in annual revenue per car will amount to about 12 thousand euros.

However, it should be recognized that currently there are some obstacles to the successful implementation of the EDTC project, in particular, the digital inequality within the EAEU, for example in Kyrgyzstan, the number of Internet users is two times lower than the average in the EAEU (Fig. 3).

The number of subscribers of cellular networks in the EAEU member states is presented in Fig. 4.

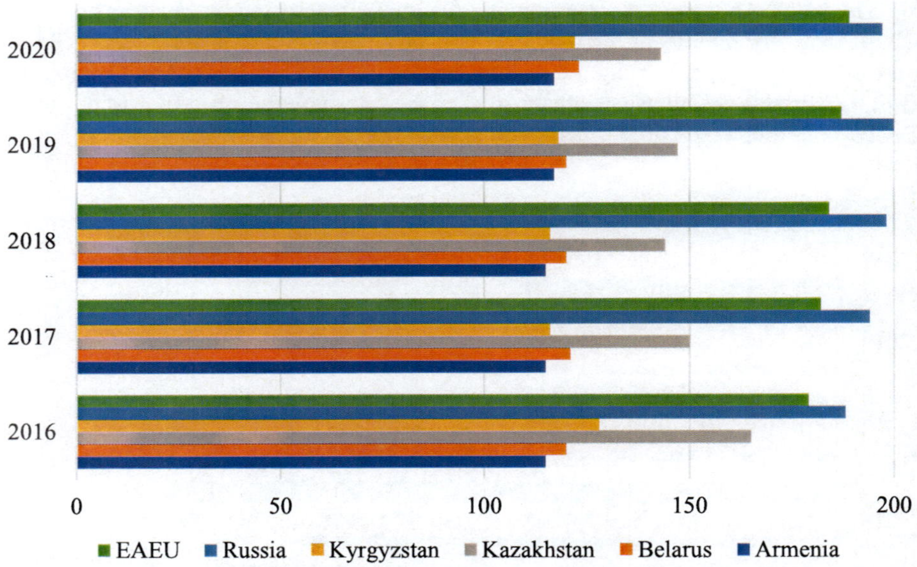

Fig. 4. The number of subscribers of cellular networks per 100 people, subscribers. *Source* Compiled by the authors based on [12].

If we compare the number of Internet users in the EAEU in the international ranking, then by the number of users:

- Armenia is in 87th place (1.75 million, 58.25% of the population);
- Belarus in 61st place (5.9 million, 62.23% of the population);
- Kazakhstan is in 40th place (14 million, 77% of the population);
- Kyrgyzstan is in 101st place (1.1 million, 20% of the population);
- Russia is in 7th place (109.5 million, 76.4% of the population).

Another factor used to assess the current level of digitalization of the EAEU member states is the volume of exports and imports of telecommunications services (Fig. 5).

Analyzing the above data, we can note a significant development of telecommunication, computer, and information services sector in Belarus and Russia, which have a significant volume of exports of these services. At the same time, despite the lag of Kyrgyzstan in the number of users of cellular networks in the international comparison in the EAEU, the country has a sufficient level to implement the digital agenda in the EAEU.

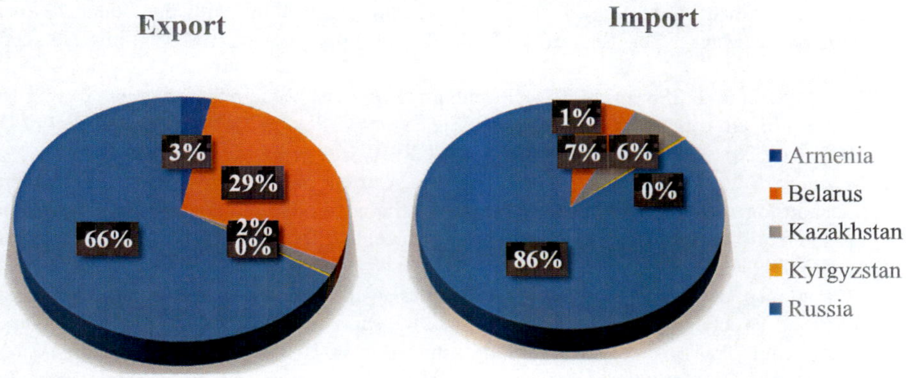

Fig. 5. Volumes of export/import of telecommunications, computer, and information services in the EAEU member states. *Source* Compiled by the authors based on [12].

The target indicators for improving the efficiency of the EDTC in the EAEU space should be a significant reduction in the financial and time costs of international carriers associated with the performance of customs and logistics operations, as well as an increase in the rate of turnover of goods in the EAEU space.

4 Conclusion

Thus, the development of the ecosystem of digital transport corridors based on the introduction of modern customs and logistics technologies will ensure the sustainable operation of the transit flow management system through the EAEU territory using digital services for calculating optimal transports routes, issuing electronic documents, including waybills and international waybills.

Also, the unified digital ecosystem will allow the integration of disparate information systems of participants in foreign economic activity, which will increase the level of coordination in managing the process of goods movement and reduce the level of total logistics costs for customs and logistics operations.

References

1. Dmitriev, A., Plastunyak, I.: Digital platforms for managing transport and logistics systems in the context of sustainable development. E3S Web Conf. **208**, 01007 (2020). https://doi.org/10.1051/e3sconf/202020801007
2. Huang, Q., Yin, W., An, J., Zhou, Y.: A China railway express-based model for designing a cross-border logistics information cloud platform scheme. Appl. Sci. **10**(12), 4110 (2020). https://doi.org/10.3390/app10124110
3. Khakdaman, M., Rezaei, J., Tavasszy, L.A.: Shippers' willingness to delegate modal control in freight transportation. Transp. Res. E Logist. Transp. Rev. **141**, 102027 (2020). https://doi.org/10.1016/j.tre.2020.102027

4. Lee, H.L., Shen, Z.-J.: Supply chain and logistics innovations with the Belt and Road Initiative. J. Manag. Sci. Eng. **5**(2), 77–86 (2020). https://doi.org/10.1016/j.jmse.2020.05.001
5. Los, J., Schulte, F., Spaan, M.T.J., Negenborn, R.R.: The value of information sharing for platform-based collaborative vehicle routing. Transp. Res. E Logist. Transp. Rev. **141**, 102011 (2020). https://doi.org/10.1016/j.tre.2020.102011
6. Malevich, Y.V., Plastunyak, I.A., Aitova, K.A.: Current issues of development of modern transport-logistics and customs technologies in the conditions of economic integration. In: 3rd International Scientific and Practical Conference, DEFIN 2020 (2020). https://doi.org/10.1145/3388984.3390642
7. Myasnikova, O.V.: Development of approaches to the creation of the organizational and functional structure of the ecosystem of digital transport corridors of the Eurasian Economic Union. Digit Trans. **1**(10), 23–35 (2020). https://doi.org/10.38086/2522-9613-2020-1-23-3
8. Nitsche, B.: Decrypting the Belt and road initiative: barriers and development paths for global logistics networks. Sustain **12**(21) (2020). https://doi.org/10.3390/su12219110
9. Pak, E.V.: Transit potentials of Russia and Kazakhstan. World Econ. Int. Relat. **64**(11), 132–138 (2020). https://doi.org/10.20542/0131-2227-2020-64-11-132-138
10. Digital Customs and Logistics Services of the EAEU. https://www.vedomosti.ru/business/articles/2020/02/06/822353-evraziiskii-soyuz. Accessed 20 Feb 2021
11. The Ecosystem of Digital Transport Corridors of the EAEU. http://www.eurasiancommission.org/ru/nae/news/Pages/19-06-2019-2.aspx. Accessed 20 Feb 2021
12. The Volume of Freight Traffic in Russia. Access Mode: https://rosstat.gov.ru/folder/23455. Accessed 20 Feb 2021

Conceptual Model of the Efficiency of Information Security System

Igor V. Salov, Ildar A. Shagapov, and Albina S. Ismagilova(✉)

Bashkir State University, Ufa, Russia

Abstract. The main result of this work is a model of the effectiveness of the information security system. The authors show the feasibility of combining the reconfiguration method and the method of using different types of redundancy elements. To improve the reliability of the information security system, the authors propose to structurally separate two independent systems – the secured information system and the security system of the information system. It is necessary to use subsystems assembled on different components and running different operating systems. The use of the combination of subsystems defined in the developed approach increases the number of independent combinations of reliable subsystems. Thus, it will be possible to get different systems of information security at any given time. The proposed structure of the information security system can operate in three modes (considering impact, not considering impacts, and combined). Thus, switching between subsystems of protection is not subject to external influences. Even though the approach described in this paper will increase the complexity of the information security system, there will be an increase in the efficiency of the information security system with a slight increase in cost.

Keywords: Digital systems · Information systems · Artificial intelligence · Information security system · Threats to information security

JEL Codes: C45 · C55

1 Introduction

The amount of information processed in digital systems is growing worldwide. SEAGATE forecasts that the global data volume will grow to 175 zettabytes by 2025 [7]. This fact necessitates a global migration of information to cloud storage. The authors propose the architecture of modern systems of decentralized data storage and processing. Moreover, the authors argue the applicability of blockchain technology in these systems [12]. A. A. Kogutenko and S. V. Plotnikova reviewed the stages of creating a cross-platform environment with cloud storage [2]. However, there are problems hindering the development of distributed ledger technology [11]. This is not surprising since the importance of digital information is increasing, and it becomes more attractive to hackers. The number of attacks on various information systems, as well as their

E. G. Popkova (Ed.): ISC 2020, LNNS 368, pp. 221–227, 2022.
https://doi.org/10.1007/978-3-030-93244-2_25

complexity and sophistication, is increasing [4]. In this regard, it is necessary to improve the systems for protecting the information placed in the information environment.

The approach to the methods of application of information security systems (ISS) can be divided into two classes: classic ISS and ISS based on new operating principles. Classic ISS include systems based on the so-called "manual" control principle. In their "pure" form, classic ISS are systems where decisions are made based on blacklists, whitelists, regular expressions, and signature methods. The "manual" ISS is based on log-reading, incident reconstruction, etc.

The human decision to respond to an information security incident significantly increases the ISS response time and the likelihood (and extent) of damage. Thus, there emerged the task of automating the decision-making process. One of the solutions to this issue is the emergence of the ISS, where the response to incidents occurs according to new principles. These include the machine-learning-based ISS.

The authors developed a mathematical model for building a defense system using artificial intelligence (AI) [3]. The proposed system is marked with the ability to account for the influence of random factors when choosing a protection option and adapt the protection system to changing environmental conditions. D. Yu. Muromtsev, S. V. Popov, and V. N. Shamkin [6] developed a block to assess the reliability of the ISS on the data of the current operation of the subsystem to ensure the performance of the system of monitoring incidents in information security. Various methods and mechanisms of information protection, as well as mathematical justification, which provide an extreme value of a certain characteristic of a security system under constraints on several other characteristics, are proposed [1].

2 Materials and Methods

AI technologies are used in various subsystems and ISS:

1. Systems based on the analysis of the behavior of subjects in the information system UBA/UEBA [5]. Both systems build models of user behavior and determine deviations from these models in real-time. The difference is that UBA systems are based only on the analysis of the activity of subjects, and UEBA systems process data on the environment of subjects. The result of this analysis is the constructed profiles. If there occur actions other than those described in the profile, an information security incident alarm is generated. There are automatic and manual decision options.
2. Endpoint Detection and Response (EDR) and Network Detection and Response (NDR). Such systems monitor attacks and make quick decisions. The search for Malicious software that does not have signatures on the system is searched using AI. It is noteworthy that these systems detect attacks even at the network level. This is done by analyzing the traffic. This analysis uses statistical data and the threat knowledge base. Decisions to switch to other network configurations are made automatically.

3. Threat detection and assessment systems. Threat Intelligence Platform (TIP) early detection platforms allow one to identify threats and make operational decisions about them. The analysis of a large amount of different data and indicators of compromise based on AI allows effectively detecting unknown threats even at an early stage.
4. Security Orchestration and Automated Response (SOAR) security detection technology apply high-fidelity analysis of system security logs to identify threats. These systems are similar to SIEM systems in their functionality. However. SOAR systems allow automating the decision-making process on the identified threats.
5. Application Security Systems. These systems are designed to provide security for applications. They use AI to automate the process of gathering information about vulnerabilities, infections, and attacks from open sources.
6. Antifraud protection systems. AI allows identifying deviations from established business processes, thereby allowing the necessary decisions to be made automatically.
7. Security Information and Event Management (SIEM) systems monitor information systems. These systems use AI to analyze security event logs in the system. As a result, all sorts of anomalies are detected, and the number of false positives is reduced. It is also possible to automate the decision-making process [9].

The use of AI can significantly reduce the time for making decisions on responding to information security incidents and partially automate the decision-making process. However, it does not sufficiently increase the reliability of the protection system. When analyzing the information security technologies used in Russia and the CIS countries [8], we can conclude that Russia mostly uses manual technologies. The systems using AI, which allow to automate decision-making and reduce response time, are insignificant and fragmented (at the level of individual subsystems). In this aspect, the significant price of such systems plays a significant role. Despite high costs, the reliability of such systems still does not satisfy many users. Practice shows that even the most difficult ISS can be hacked with careful preparation, especially stretched in time, and the study of the protection system.

From our point of view, it is possible to significantly increase the reliability of the information security system by using fault-tolerant approaches. However, a significant increase in reliability will occur only if several theses and conditions are met.

We understand the fault tolerance of technical systems as their ability to perform the intended functions in real operating conditions in the presence of disturbing internal influences – failures, breakdowns of technical components, and software errors [10]. The creation of fault-tolerant systems implies several redundant and backup subsystems in it. The main idea of improving the ISS effectiveness is a combination of two methods described in the theory of reliability of complex systems: the method of reconfiguration and the method of using different types of redundancy elements. At each point in time, the security system of the information system is marked with a set of protection subsystems. At a certain point in time, the ISS represents a set of elements "Subsystem ISS 1," "Subsystem ISS 2," …, "Subsystem ISS N" (Fig. 1).

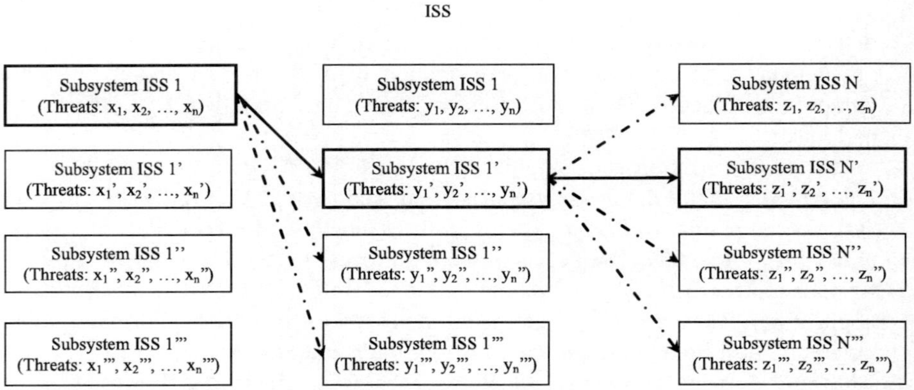

Fig. 1. Architecture of the conceptual information security system. *Source* Compiled by the authors

Reconfiguration is intended to automatically change the architecture of the ISS to ensure its smooth operation.

In contrast to the theory of fault tolerance, we understand the use of heterogeneous elements as the use of subsystems with a different set of threats typical for the information security of a subsystem (different subsets X, Y, Z).

3 Results

1. To improve the reliability of the ISS (Fig. 1), the ISS must be structurally separated from the secured information system. We get two structurally independent systems – the secured system (SS) and the information security system (ISS). Thus, we get the opportunity to back up the subsystems of the security system without interfering with the secured system. Although, it does not cancel the duplication (redundancy) of protected information in the SS. In addition to improving the efficiency of the ISS, this structure allows isolating the failed subsystems of information protection. Additionally, the separation of the protected and protecting systems, when attacked on the subsystems of the ISS, can reduce the negative impact on the protected system. That is, an increase in the number of security subsystems should not increase the perimeter of attack for the user, but it increases the number of lines of defense that the attacker must overcome.
2. Let us assume that there are m redundant security subsystems, all of which are "equal" to each other. Since combinations of ISS, consisting of a set of N subsystems with a level of redundancy m, are constantly changing, then an attacker would have to break into all N^m combinations of information security systems at once to implement an attack successfully.
3. Each of the m subsystems has a different set of vulnerabilities $\{A_k\}$, where $A_k = X_n$ for multiple subsystems of the ISS 1, $A_k = Y_n$ for multiple subsystems of the ISS 2, …, $A_k = Z_n$ for multiple subsystems of the ISS N (Fig. 1). To improve the ISS

efficiency, the intersection of $\cap A_k$ must tend to zero. The effectiveness of the ISS decreases as the intersection $\cap A_k$ increases. When A_k reaches a value equal to n, we get equivalent subsystems, which perform only the function of duplication. Thus, the number of combinations will be uncritically reduced.

4. It was proposed to use the method of different types of redundancy elements not to reduce the number of combinations. Thus, the sets of $\{A_k\}$ vulnerabilities will be different. A similar approach is used to create control systems in aviation. The individual control subsystems (main and standby) are based on different principles – a combination of electric and hydraulic control systems. It is recommended to use subsystems assembled on different components and running different operating systems. It is possible to choose between computing facilities based on Intel, AMD, and Qualcomm processors running different builds of Linux, Windows, etc.
5. Cryptography uses the Kerckhoffs principle, which states that the effectiveness of a cryptographic system should be based on the secrecy of the key, not the secrecy of the principle of the algorithm. Likewise, for the ISS: the effectiveness of the information security system should not be based on the secrecy of the algorithm of the system but on an effective combination of reliable subsystems of the ISS. Using the above subsystem combination principle, we increase the number of possible combinations of reliable subsystems. At any given time, we actually have a different ISS. Thus, even knowing the logic of the ISS, the hacker does not get any advantage in carrying out an attack.
6. The switching time between the protection systems must tend to zero. It is possible to use the transaction principle: the ISS 1 subsystem stops its operation only after the start of the ISS 2 is confirmed.
7. The key point in the functioning of the proposed structure of the ISS is the switching between subsystems. There are two parameters: the method of switching (trigger) (e.g., time or some indicator) and the choice of subsystems. Thus, this structure can operate in three modes: considering impact, not considering impact, and combined.

The "not considering impact" mode involves constant switching between protection subsystems. It is possible to use several cryptographic sequences of pseudorandom numbers. For example, the first sequence is used to select a combination of subsystems, and the second sequence is used to select the time to switch subsystems. Additionally, it is possible to use sequences of random numbers since there are no time limitations, and the sequences can be calculated in advance. A distinctive feature of this option is a certain unpredictability.

The "considering impact" mode assumes that switching occurs under some external action (e.g., a signal from an intrusion detection system or SIEM system). Thus, the switching is done according to some predetermined algorithm (schedule, functional dependence, etc.). In this case, there appears certain predictability.

The third mode is a combination of the first two modes. Basically, the system operates using the algorithm, but it goes to some particular mode depending on the impact when receiving a certain signal (from an intrusion detection system or SIEM system).

This option seems the most relevant since it is possible to implement a switching control subsystem independent of the ISS when using a "not considering impact" mode. In this case, the switching subsystem is not subject to external influences.

The proposed protection is mainly aimed at counteracting the attacker and does not consider unauthorized access to the surveillance subsystem.

4 Conclusion

The application of the proposed approach is possible to achieve two different goals.

First, it will increase the efficiency of the information security system with a slight increase in cost. The protection systems with this architecture can use systems based on classical principles (proven low-cost solutions), i.e., without AI. In this case, the cost of the ISS can be significantly offset. The improved effectiveness of protection is achieved not at the expense of high-cost complex solutions but at the expense of a certain unpredictability, dynamism, variability, and reliability of the ISS.

Second, the proposed approach allows increasing the reliability of the security system of critical facilities. In this case, the cost of the security system is not a matter of principle (e.g., nuclear facilities, railroad control systems, etc.)

It may also be noted that this approach will lead to a complication of the information security system as a whole. However, given that the redundancy of subsystems is still required, this point is not so significant.

References

1. Karganov, V.V.: Methodology for information security in current information systems. In: National Security of Russia: Current Aspects, pp. 21–27. Natsrazvitie Publishing, St. Petersburg (2020)
2. Kogutenko, A.A., Plotnikova, S.V.: Development of cross-platform working environment with the cloudy depository of information. Inf. Cybern. **2**(12), 20–24 (2018)
3. Kosmacheva, I.M., Davidyuk, N.V., Sibikina, I.V., Kuchin, I.Y.: The model for evaluating the effectiveness of an information security system configuration based on genetic algorithms. Model. Opt. Inf. Technol. **8**(3), 40–41 (2020). https://doi.org/10.26102/2310-6018/2020.30.3.022
4. Malyarevsky, A.: Trends 2021: Cyberthreat Relief Becomes More Complex (2020, Dec 23). https://www.crn.ru/news/detail.php?ID=150817. Accessed 13 Mar 2021
5. Matveev, A.: Overview of the market for behavioral analysis systems– User and Entity Behavioral Analytics (UBA/UEBA). (2017, Nov 23). https://www.anti-malware.ru/analytics/Market_Analysis/user-and-entity-behavioral-analytics-ubaueba. Accessed 13 Mar 2021
6. Muromtsev, D., Popov, S.V., Shamkin, V.N.: Improvement of the information security subsystem in the bank information security monitoring system. Trans. Tambov State Tech. Univ. **26**(2), 176–187 (2020). https://doi.org/10.17277/vestnik.2020.02.pp.176-187
7. SEAGATE. Rethink Data: put more of your business data to work—from edge to cloud (2020). https://www.seagate.com/files/www-content/our-story/rethink-data/files/Rethink_Data_Report_2020.pdf. Accessed 13 Mar 2021

8. SERCHINFORM. Study of the level of information security in companies in Russia and the CIS in 2020 (2020). https://searchinform.ru/survey/global-2020/. Accessed 13 Mar 2021
9. Shabanov, A.: Application of artificial intelligence technologies in information security (2020, June 17). https://www.anti-malware.ru/analytics/Technology_Analysis/using-artificial-intelligence-technologies-in-information-security. Accessed 13 Mar 2021
10. Shubinsky, I.B.: Functional reliability of information systems. Methods of analysis. Printing Yard, Ulyanovsk (2012)
11. Yavlyanskaya, E.A., Asadullina, A.V.: Problems of blockchain technology development in Russia. In: Orlova, G.A., Kaplyuk, T.S., Ogneva, Yu.V. (Eds.), New Realities of the Global Economy: Sanctions and Trade Wars, pp. 276–290. Moscow, Russia (2020)
12. Zegzhda, D.P., Moskvin, D.A., Myasnikov, A.V.: Cyber resilience of distributed storage processing systems with application of blockchain technology. Probl. Inf. Secur. Comput. Syst. **2**, 74–79 (2018)

Methodology for Assessing Intellectual Capital of a Business

Angelina V. Tarasova[1(✉)] and Yulia V. Vertakova[2]

[1] Southwest State University, Kursk, Russia
[2] Kursk Branch of the Financial University under the Government of the Russian Federation, Kursk, Russia

Abstract. The formation of modern digital and innovative technologies, the increased importance of knowledge, and modernization of the management of the business process are mandatory components for obtaining competitive advantages in international and national markets. Therefore, these new components pose new challenges to modern organizations in many aspects of their activities. Business, organization, and management are closely coupled since the operation of the organization is impossible without some proper administration and planning of a precise structure. The authors do not separate these concepts but rather combine them into a concept of business organization with the managerial elements necessary to maximize its effectiveness, provide innovative modern development, and control the entire system of the organization. This paper focuses on the concept of "intellectual capital" (IC), its structure, and its impact on the world market for adequate reporting to interested parties, using the example of the Russian agro-industrial organizations. The research subject is the assessment of the quality of (1) using IC in business organizations, (2) improvement of financial performance, (3) increasing business productivity, and (3) optimal management of resources and growth of organizational value through the key components of IC. The paper includes a brief overview of the works studying the position and role of IC (definitions and classification) and its impact on the international stock market and the value of the business organizations themselves. The authors provide a modern definition for the concept of "intellectual capital." Additionally, the authors show the mechanism for the operation of a business organization with the development of IC and distinct, interrelated concepts. The authors proposed considering the VAIC (Value Added Intellectual Coefficient) method for assessing IC in business organizations involved in the production of raw milk. The authors summarize the obtained indicators, come to several conclusions, and provide recommendations for working with IC, which will help external stakeholders to evaluate it in selecting the organizations with a high performance of IC.

Keywords: Intellectual capital · Digital economy · Intellectual property · Tangible assets · Intangible assets · Agro-industrial complex · Methods of intellectual capital

JEL Codes: E22 · G12 · G24 · G31 · G32 · J24 · O13 · O15 · O34

E. G. Popkova (Ed.): ISC 2020, LNNS 368, pp. 228–237, 2022.
https://doi.org/10.1007/978-3-030-93244-2_26

1 Introduction

The role of intangible assets increases in the current "new economy." Intangible assets are aimed at the formation and promotion of knowledge, as well as at the building of competencies, abilities, and skills for achieving the full development of IC, thereby allowing to regulate the very process of creating the value of the business organization [14].

In many large Western and Russian business organizations, the valuation of IC exceeds half the cost of the valuation of the entire organization. This suggests that employees and their professional qualities are the main assets in such business entities. With the retreat of tangible assets to secondary importance, the production of own products requires not only expensive machinery and equipment but also experienced employees and their knowledge. Moreover, it is necessary to use and develop the latest know-how and computer software. That is, we are witnessing a revolutionary shift in decision-making for managers of organizations.

Timely disclosure of information about the existing IC elements is necessary to ensure the sustainable growth of a business entity. The information provided in the standard reporting forms is not enough to assess the importance of intellectual resources and give an adequate assessment of the organization's development. It is necessary to create automated systems of business organization, which would solve various organizational changes affecting the profit and the competitiveness of the organization and its products. This suggests that people and their professional qualities are the main assets in such business entities.

2 Materials and Methods

The development of intellectual capital (IC) serves as a critical element in creating value for business organizations. To gain an advantage in the emerging international and Russian markets, business organizations need to develop information technology, implement innovation, and increase the importance of knowledge, i.e., to raise the level of skills and motivate employees.

The essence and concept of intellectual capital have attracted many researchers over the centuries [5, 7, 9–11, 13, 18]. However, despite this diversity of opinions, it is pointed out that there is no standard definition for the concept of intellectual capital. The authors propose the modern definition for a concept of IC and the scheme for the classification of business operations, taking into account intellectual capital.

Additionally, the authors analyze the current methodological research in evaluating IC in the structures of business entities and identify the advantages and disadvantages of different approaches. The authors also review and evaluate the classifications of methods for evaluating IC. The authors use the VAIC methodology (Value Added Intellectual Coefficient) since the main information for the calculations is contained in the annual and accounting financial statements [15, 16]. It is concluded that the method is applicable to agro-industrial businesses engaged in the production of raw milk.

3 Results

From the 17th century to the present, each researcher has their own perspective on intellectual capital. Researches use different terms such as "human capital," "knowledge capital," "personnel intellectual potential," "structural capital," or "intellectual activity." Thus, there is a lack of a common definition for the concept of IC.

This study offers a modern understanding of the concept of IC. Intellectual capital is the capital of a business entity expressed as a set of interactions between employees, customers, development, and research, the functions of which are aimed at prospects for ensuring competitive advantage in the market.

The authors argue that the main goal of managers is to provide a qualified team motivated by the optimal system of remuneration to achieve all components of the organization's assets (including intangible ones).

For the full operation of a business entity, the authors propose the following mechanism of work through the development of IC (Fig. 1).

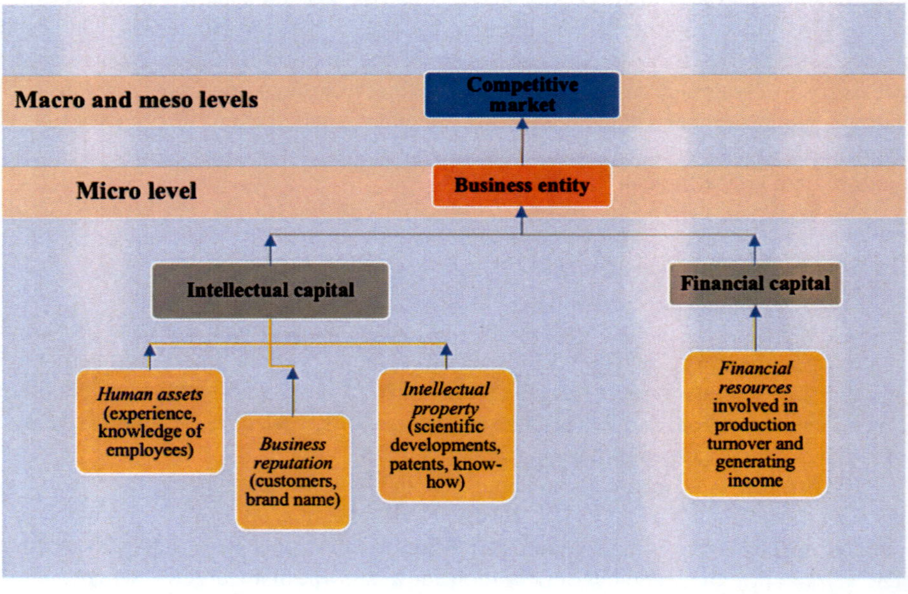

Fig. 1. Author's classification of the operation of business entity, taking into account the objects of intellectual capital. *Source* Compiled by the authors

In the context of a complete understanding of IC in the work of a business entity, this classification considers the development prospects, which assists in studying the intellectual portfolio of resources and allows one to calculate the amount of planned investment for its implementation through internal research developments of the organization or borrowings from outside (acquisition of patents, license agreements, etc.).

Such classification can be combined into a system of intellectual capital management (ICM), which has progressive and science-based principles and approaches (Fig. 2).

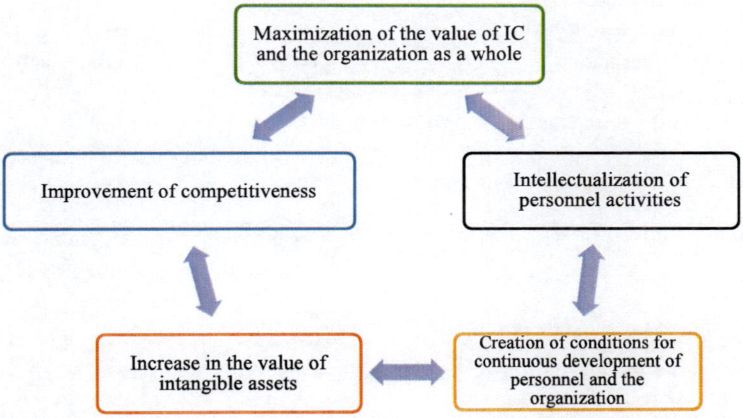

Fig. 2. Author's classification of intellectual capital management (ICM) in business organizations. *Source* Compiled by the authors

In the absence of constituent qualifiers of IC and financial capital (FC), full-fledged activity of the business entity is impossible, and the implementation of achievements and high results to enter the competitive national and international markets is difficult. Meanwhile, unlike financial capital (valued in financial and monetary turnover), IC is subject to valuation in monetary and non-monetary terms (Table 1). Managers use IC in monetary terms for the following purposes:

- Monitoring of a business organization with significant elements of the influence of intellectual activity in the creation of its products;
- Mini-studies of the values of intangible assets and personnel of the business organization;
- Establishing the market sustainability of the business organization to attract interested parties (e.g., stakeholders, investors, and government support organizations).

Table 1. Difference between financial and intellectual capital

Main features	Financial capital	Intellectual capital
Physical form	Yes	No
Availability of cost	Yes	Yes
Possibility of value estimation using the additive method	Yes	No
Ability to use several business entities at the same time	No	Yes
Possibility to be contributed to the share capital, to be the object of purchase, sale, or lease	Yes	Only some elements of intellectual capital
Subject to obsolescence	Yes	Yes
Subject to physical deterioration	Yes	No
Possibility of depreciation	Yes	Yes, in terms of intangible assets
Participation in the formation of the cost of the final product	Yes	Yes
Participation in the formation of the value of assets	Yes	Yes
Ownership	Organization owns all the capital	Organization owns only part of the capital

Source Compiled by the authors [12]

It is necessary to note that not every existing organization currently understands the importance of IC and its presence in the resource portfolio. Accounting standards do not reflect the value of intellectual property rights (IPR) and intangible assets in the organization's accounts. Therefore, IPR are often not fully valued, used, and managed at the proper level. The main task of any organization is to obtain a stable profit in the absence of material investment in the development of its employees, the modernization of technology, and the creation of their own (or acquired) intellectual resources.

We can assume that this misunderstanding is caused by the vast number of viewpoints on the concept of IC. The lack of a unified concept negatively impacts practical and research interests in the field of IC. Many people associate the concept of IC with the concepts of "intellectual property" or "intangible asset." It should be noted that these concepts are related in sound but imply entirely different characteristics and refer to different areas.

Intellectual property (IP) is a legal category, which plays an essential role in developing the global economy and the emergence of a knowledge economy. Intellectual property is "protected means of individualization and the results of intellectual activity in accordance with legislative documents such as Part 4 Section 7 of the Civil Code of the Russian Federation [1], Convention Establishing the World Intellectual Property Organization (WIPO Convention) [4], Paris Convention for the Protection of Industrial Property [2], and Berne Convention for the Protection of Literary and Artistic

Works [3]. As a criterion in an economic category, IP is not essential since the ability of an intellectual object to bring profit is not directly related to the registration of ownership of this object.

Intangible assets are displayed in the form of inventions, results of creative activity, various databases, software, trademarks, goodwill; they are an important functional element, penetrating the global value creation structure of the business entity. These assets are “exclusive rights to the results of intellectual activity and means of individualization (inventions, utility models, industrial designs, trademarks, service marks, appellations of origin, computer programs, databases, and other objects of copyright and related rights). Intangible assets are located within a business organization because the decision to locate production processes in a particular location and establish long-term partnerships is largely determined by how the business entity manages its intangible capital [8].

The notion of intellectual capital in the legislative system does not exist in any international convention, code, law, or bylaw. Nevertheless, it is freely applied in describing the studied processes and the purposeful impact on business, economy, politics, and society.

Based on our analysis of the concepts, we propose a scheme (Fig. 3), showing that intellectual capital, intangible assets, and intellectual property must work together to ensure the successful functioning of the business entity, since economic and legal support are inseparable from each other.

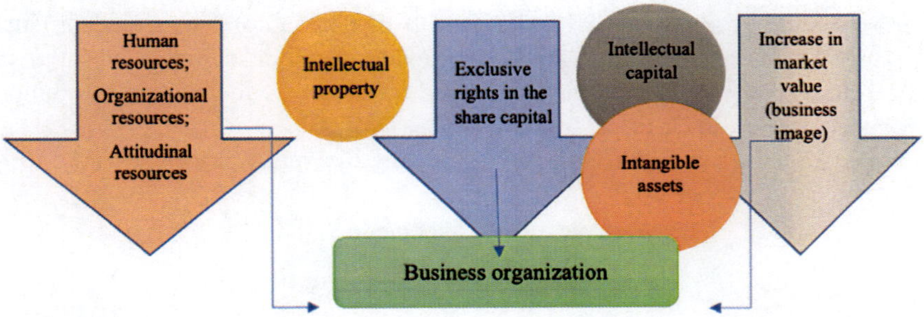

Fig. 3. Classification of the interaction process of intellectual capital, intangible assets, and intellectual property in the structure of a business entity. *Source* Compiled by the authors

Many researchers dealing with the essence, content, and evaluation of the concept of IC in the theoretical direction have a significant interest in the analytical review of methods for assessing IC in different groups (from the chosen industry, the market size of the organization, the territory of the study, etc.).

The reporting of intellectual capital is one of the critical problems. Reporting to the tax authorities reflects only the section of intangible assets. The indicators of IC are reflected only in the internal documents of organizations. To form a positive opinion, some organizations (e.g., PJSC Gazprom, PJSC Sberbank, LLC Rusagro, and others) openly publish their annual reports in the form of booklets reflecting every subject considered for each category).

When compiling ratings on the use and development of IC in business organizations, only large market entities are taken into account. Nevertheless, it is also necessary to conduct analytical studies in smaller or other industries to understand IC development in Russia. Therefore, we decided to use the VAIC model, proposed by A. Pulic, to analyze the business entities involved in raw milk production [15, 16]. This method is a common way of assessing IC in enterprises of all industries. The advantages of this method include the simplicity of measuring values. The calculation is based on corporate financial report information without the use of specific market indicators. Therefore, it is more objective than other valuation methods since the data used and the results obtained are easy to verify. Since Russia has a poorly developed stock market (especially for market valuation), it is recommended to use the methods based on financial statements to assess Russian organizations.

Considering the high importance of cognitive components for business organizations, the authors propose to conduct a detailed study of the IC of business entities operating in the agro-industrial sector for the production of raw milk in Russia. We take the 2019 ranking "Top 20 largest producers of raw milk in Russia" as the basis for our further research and make an overview of their development [17].

Improving the dairy industry is of strategic importance for every country due to the demand for milk and dairy products by population. This industry is steadily increasing worldwide. In the large-scale global market, Russia is one of the most promising producers in the industry. For example, in 2019, Russia accounted for about 8% of the world's share of dairy production, which is 2% higher than in 2018.

Thus, we can conclude that the high performance of the Russian dairy sector in the reporting period is associated with the growth of demand and increased production turnover due to modern, large-scale, and modernized industries. Currently, most farmers have highly productive and advanced technologies for housing, feeding, and reproduction of livestock [6].

There are three main indicators for calculating the valuation of the VAIC method (Fig. 4).

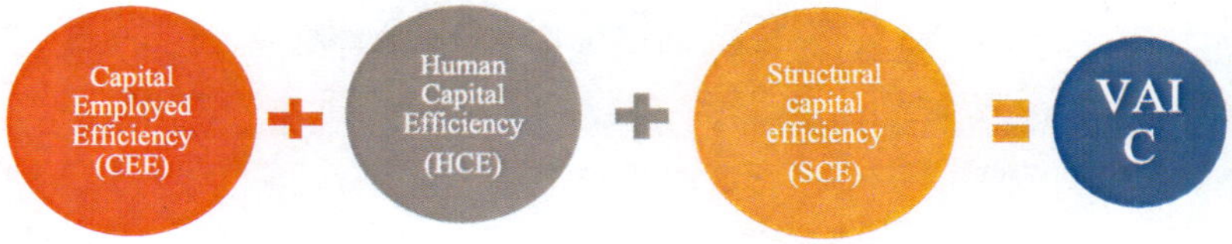

Fig. 4. Formula for calculating the VAIC. *Source* Compiled by the authors

Table 2. Calculation of IC of raw milk production organizations by the VAIC method

No	Organizations	Regions	The VAIC = CEE + HCE + SCE
1	EkoNiva-APK	Central Federal District, Volga Federal District, Siberian Federal District, Northwestern Federal District	5.11 (0.12 + 4.23 + 0.76)
2	Tkachev Agrocomplex	Krasnodar Territory	4.02 (0.27 + 3.08 + 0.68)
3	Piskarevsky Dairy Plant LLC	Leningrad Region	4.91 (1.29 + 2.96 + 0.66)
4	Krasny Vostok AGRO	Republic of Tatarstan	1.13 ((−0.01) + (−0.93) + 2.07)
5	Ak Bars Holding	Republic of Tatarstan	3.01 (0.37 + 2.12 + 0.53)
6	RUSMOLOKO (Damate)	Penza Region, Tyumen Region, Republic of Bashkortostan	−217.10 (−219.62 + 2.02 + 0.50)
7	Concern Pokrovsky	Krasnodar Territory	3.40 (0.41 + 2.40 + 0.58)
8	Kirov Dairy Plant	Kirov Region	3.25 (0.35 + 2.33 + 0.57)
9	Agrofirma Doronici JSC	Kirov Region	4.69 (0.41 + 3.56 + 0.72)
10	Green Valley Group (EFCO Group)	Belgorod Region	4.44 (0.26 + 3.47 + 0.71)
11	Concern Detskoselsky	Leningrad Region, Voronezh Region, The Republic of Belarus	No HCE indicator
12	Avida Group of Companies	Belgorod Region	2.02 (0.52 + 1.28 + 0.22)
13	Slaktis (Kabosh Group of Companies)	Pskov Region	10.01 (0.18 + 8.94 + 0.89)
14	Molvest Company	Voronezh Region	22.18 (17.48 + 3.95 + 0.75)
15	Rodina CJSC (Dominant Group)	Voronezh Region, Lipetsk Region, Krasnodar Territory	2.80 (0.78 + 1.63 + 0.39)
16	Agrosila Milk LLC	Republic of Tatarstan	No HCE indicator
17	Kuban Agroholding (Basic Element Group)	Krasnodar Territory	5.05 (2.28 + 2.22 + 0.55)
18	Agropromkomplektatsiya Group of Companies	Kursk Region, Tver Region	7.07 (0.13 + 6.10 + 0.84)
19	Dairy Product LLC (Russian Agrarian Company)	Ryazan Region	5.17 (0.36 + 4.06 + 0.75)
20	INTERCROSS CENTER LLC	Tula Region	2.57 (0.03 + 2.03 + 0.51)

Source Compiled by the authors based on indicators taken from the accounting and annual reports of these organizations for 2018

Thus, Table 2 shows that the evaluation of IC using the VAIC method is specific. The VAIC indicator ranges from 2 to 15. Therefore, we can conclude about the low and medium levels of human capital efficiency (HCE). The advantage of this methodology is the availability of calculations and the openness of indicators. In terms of the use of accounting data, this method does not allow to consider such important values as client capital (business relationships) and other properties of the IC, as well as the future value of the organization.

4 Conclusion

In the global economy of the 21st century, the influence of IC can be attributed to the most effective competitive advantage. The main task of IC lies in improving the efficiency of economic activity and influencing the value structure of the business organization. Unfortunately, in practice, only the largest business entities in Russia, listed on the world stock market, disclose information about the IC.

When analyzing IC in dairy businesses, it is necessary to consider several factors influencing and providing the development of new industries. In the agro-industrial complex of Russia (AIC), there are some problems holding businesses back from obtaining a record level in the global world market for food production. This is since the significant potential of the industry (not involved agricultural land) and the accumulated development of Russian agricultural science often remain unused. For the AIC business entities to move up to the global level, it is necessary to improve the system of reports using IC indicators for internal and external stakeholders.

There are currently no factors influencing the disclosure of IC that can evaluate IC disclosures among organizations.

Our research used the content analysis of the annual reports of the top 20 largest raw milk producers in Russia for 2018. IC was assessed using the VAIC method. The most disclosed type of capital was human capital (HCE). The least disclosed is structural capital (SCE).

This study provides a generalized view of the disclosure of IC in business entities involved in the production of raw milk in Russia. The authors suggest using not only accounting and financial reports but also other sources of information about business entities: e.g., press releases and final annual figures published on the official website of the organization.

References

1. Russian Federation. Civil Code of Russia (Part 4): Chapter 69, Article 1225 "Protected results of intellectual activity and means of individualization" (December 18, 2006 No. 230-FZ), Moscow, Russia (2006). https://legalacts.ru/kodeks/GK-RF-chast-4/razdel-vii/glava-69/statja-1225/. Accessed 29 Jan 2021
2. WIPO. Paris Convention for the Protection of Industrial Property (as revised at Stockholm on July 14, 1967; as amended on September 28, 1979). World Intellectual Property Organization, Paris (1883, Mar 20). https://wipolex.wipo.int/en/text/288514. Accessed 29 Jan 2021
3. WIPO. Berne Convention for the Protection of Literary and Artistic Works (revised at Paris on July 24, 1971; as amended on September 28, 1979). World Intellectual Property Organization, Berne (1886, Sept 9). https://wipolex.wipo.int/en/text/283698. Accessed 29 Jan 2021
4. WIPO. WIPO Convention (as amended on September 28, 1979). World Intellectual Property Organization, Stockholm (1967, July 14). https://wipolex.wipo.int/en/text/283854. Accessed 29 Jan 2021

5. Abdikeev, N.M., Bogachev, Y.S.: Calculation of the cost of the intangible assets based on intellectual property. Account. Anal. Audit. **4**, 16–26 (2017). https://doi.org/10.26794/2408-9303-2017-4-16-26
6. AgroVestnik. Overview of the Dairy Products Market in Russia in 2019. MilkNews Analytical Center (2020, June 14). https://agrovesti.net/lib/industries/dairy-farming/obzor-rynka-molochnykh-produktov-v-rossii-za-2019-god.html. Accessed 29 Jan 2021
7. Alexashina, T.V.: Evolution of scientific concepts in development of the theory of the intellectual capital. Transp. Bus. Russia **3**, 38–41 (2015). https://cyberleninka.ru/article/n/evolyutsiya-nauchnyh-kontseptsii-v-razvitii-teorii-intellektualnogo-kapitala/viewer. Accessed 29 Jan 2021
8. Bespalova, N.A.: Professional competencies of personnel as a factor in the formation of the intellectual capital of the organization. Dissertation of Candidate of Economics, All-Russian Research Institute of Work, Moscow (2018)
9. Gribov, A.Yu.: How to Make a Billion or the Intellectual Capital of an Entrepreneur. RIPOL Classic, Moscow (2016). http://www.gribov.ru/Gribov_Milliard2.pdf. Accessed 29 Jan 2021
10. Klimov, S.M.: Intellectual Resources of Society. St. Petersburg Institute of Foreign Economic Relations, Economics, and Law; "Obshchestvo Znaniye", St. Petersburg (2002)
11. Leontiev, B.B.: The Price of Intellect: Intellectual Capital in Russian Business. Publishing Center "Aktsioner", Moscow (2002)
12. Lukicheva, L.I., Egorychev, D.N., Egorycheva, E.V.: Fundamentals of intellectual capital management: textbook. In: Aniskin, Yu.P. (Ed.). (2012). http://emirs.miet.ru/oroks-miet/upload/normal1/00ssc8r30q3rcs/LUKIZEVA2.pdf. Accessed 29 Jan 2021
13. Luzin, N.A.: Evolution of intellectual capital theory. Chelyabinsk Humanitarian **2**(23), 21–29 (2013). https://cyberleninka.ru/article/n/evolyutsiya-teorii-intellektualnogo-kapitala/viewer. Accessed 29 Jan 2021
14. Oskolkova, M.A.: Intellectual capital in assessing the investment attractiveness of companies. Corp. Finance Manag. **6**(54), 348–358 (2012). https://perm.hse.ru/data/2013/01/30/1304756069/UKF_6_2012_Осколкова.pdf. Accessed 29 Jan 2021
15. Pulic, A.: VAIC^{TM} – an accounting tool for IC management. Int. J. Technol. Manage. **20**(5–8), 702–714 (2000). https://doi.org/10.1504/IJTM.2000.002891
16. Pulic, A. MVA and VAIC™ Analysis of Randomly Selected Companies from FTSE 250. Austrian Intellectual Capital Research Center, Graz; London (2000)
17. The DairyNews. Top 20 largest raw milk producers in Russia, according to the analytical center MilkNews (2018, Aug 15). https://www.dairynews.ru/news/top-20-krupneyshikh-proizvoditeley-syrogo-moloka. Accessed 29 Jan 2021
18. Yaburova, D.V.: Management of intellectual capital to improve the competitiveness of the company. Dissertation of Candidate of Economics, St. Petersburg University, St. Petersburg (2017)

Organizational Culture of Automotive Enterprise as a Condition of Its Efficiency in Conditions of Market Integration

Tatiana E. Lebedeva(✉), Evgeny E. Egorov, Maria P. Prokhorova, Angelika A. Shkunova, and Tatiana V. Krylova

Minin Nizhny Novgorod State Pedagogical University, Nizhny Novgorod, Russia

Abstract. The main purpose of this research is to study the relevance and satisfaction of the organizational culture at the machine-building enterprise in modern conditions.

The authors, based on the research of leading scientists, consider the organizational culture as a fundamental component of the company's efficiency in the market, analyze the possibilities of its formation and the importance for increasing the efficiency of the enterprise.

It has been established that the organizational culture and the factors of its formation provide equal opportunities for all groups of personnel both for building individual development and for solving organizational issues and managing the production process.

The authors' research is based on a competent approach, which determined the choice of methods: extrapolation and modeling.

It is shown that the staff of PJSC «GAZ» company, despite the difficulties, highly appreciate the existing programs of support of the company's employees, but would like to participate more in management and decision-making, dive more deeply into solving issues and develop/maintain an organizational culture in a company with a rich history; The high level of digitalization and use of the information protocol for the purpose of transparency of personnel information and personnel information for the management of the enterprise has been identified; constraints to the development of organizational culture have also been identified and overcome.

The research has significant prospects for further research related to the study of the impact of digital services and the information protocol on the organizational culture of the enterprise.

Keywords: Organizational culture · Automotive enterprise · Management style · Electronic services · Personnel

JEL Codes: D22 · D23

E. G. Popkova (Ed.): ISC 2020, LNNS 368, pp. 238–245, 2022.
https://doi.org/10.1007/978-3-030-93244-2_27

1 Introduction

Organizational culture does not lose its relevance and is the driver of the development of any company in the market [3]. The ongoing changes in all areas of the economy related to the COVID-19 pandemic have forced managers of all large companies to reconsider views on the organizational culture and its importance in the effectiveness of the company in the market. Since it was no longer there, and real changes will cover all corners of the economy and business [1].

However, it should not be believed, as it was at the dawn of the formation and «engraftment» of organizational culture in Russia, that old concepts and stamps can be used to solve problems in new limited/lockdown conditions. The efficiency of enterprises in the new conditions determines the flexibility of organizational culture, the provision of opportunities for development, self-approval, the realization of abilities [15].

Organizational culture has a direct impact on employee motivation, being perhaps the strongest catalyst (and in cases of failures and the biggest drag) in the implementation of strategic plans developed by management [17].

The culture of the organization is the embodiment of the attitude of the company's employees to the key values, beliefs and traditions that manifest themselves in their behavior.

The relevance of the research is determined by the need to conduct research, test applied methods, in order to identify weaknesses and errors in the created organizational culture.

A special place is currently occupied by large industrial enterprises, for which the organizational culture should be built 100%, only this can keep the staff and survive in the modern market [2, 8].

All this indicates the relevance of studying the formation of an organizational culture. Since the topic of the study is quite broad, in this work we will focus only on the assessment of factors in the formation of organizational culture for PJSC «GAZ».

2 Methodology

The issue of the factors of organizational culture is currently relevant and debatable [5, 10, 11]. The research of organizational culture was carried out by such scientists as: E. S. Markarian, E. Shane, O. Rodin, T. Peters and R. Watermen, B. Clark, E. Martin, M. Schulz, W. Ouchi.

E.S. Markaryan considers an important condition for the effective functioning of the company is the presence of an organizational culture and value orientations. E. Shane claims organizational culture reflects the historical formation of the company [4].

B. Clark, E. Martin argue the formation of an organizational culture is extremely important for production enterprises, although in practice they ignore the mandatory element of the company's activities.

The purpose of the research is to study the level of satisfaction and formation of organizational culture using the example of PJSC «GAZ».

Research objectives:

- to study the main factors of organizational culture formation;
- to research digital technologies, assessments of organizational culture.

At the first stage of the research, a questionnaire was conducted using standard Google forms of management, core and support personnel. The questionnaire was conducted in November 2020–January 2021.

At the second stage, conclusions were drawn on the use of the information protocol of computer programs for assessing the formation of an organizational culture.

3 Results

To diagnose an organizational culture, it is needed to determine which key figures to use. Among the most commonly used, such as the degree of job satisfaction, employee attitude to work, age composition, etc.

Let's consider these indicators at PJSC «GAZ» enterprise. To begin, we will find out their degree of satisfaction with the work of employees. To analyze this indicator, the organization conducted a survey of respondents. Among the respondents to the survey were 75% of workers.

Based on the results obtained, it can be judged that a larger number of satisfied staff between the ages of 31 and 47, that is, a rather adult contingent. Mostly these are people with families who are aimed at stable work and wages. There are also quite a few young employees. This means that many young people are now more interested in stable earnings and working conditions.

Having studied statistics on the www.gderabotaem.ru website [9], it can be assumed many people working there are generally satisfied with work, but still there are both advantages and disadvantages in the management system.

An analysis of the results of reviews about the company made it possible to formulate a number of positive aspects of work at the enterprise: work in this company brings results. The company allocated corporate apartments to employees if they had housing problems, there was a developed network of kindergartens, hospitals and even universities were sponsored. Company management and organization managers are paid for various seminars and courses, sometimes held in other cities, to improve professional level and subsequently introduce new methods of personnel management in the sphere.

The following negative moments are distinguished:

- despite the fact the organization is trying to introduce foreign personnel management systems, to use innovative methods of organizing production, these methods are not perceived by everyone and not in the proper amount.

Thus, it can be judged the methods of organization at the enterprise have long been outdated, and require development without the participation of Western companies. There are also shortcomings associated with the complexity of the interaction of functional departments. Some employees note excessive bureaucratization.

From the record from the website www.orgpage.ru [13], GAZ plant was given an average score of 3,5 points out of 5 possible. This means despite some different improvements, both in production and management, there is still space to strive. Also, according to employees: - what was made of the new? This was a significant modernization of production sites, plus the introduction of a new quality system. The reliability and comfort of manufactured cars were also increased. Employees have also become much more qualified.

Organizational values are the determining factors of organizational culture. They define the standards of staff behavior in the organization and unite them. Figure 1 shows the value orientations of PJSC «GAZ» employees.

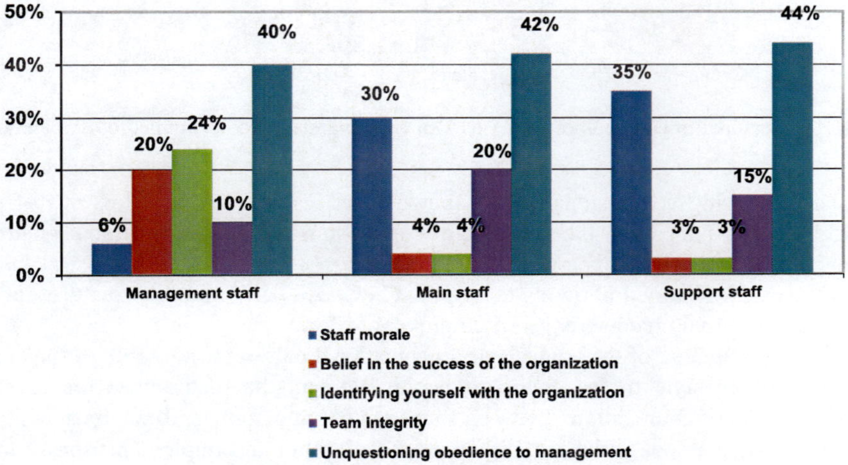

Fig. 1. Value orientations of PJSC «GAZ» employees. Source Compiled by the authors

Almost an equal number of people from management (40%), main (42%) and 44% of support staff consider subordination to management to be a key value. A quarter of management personnel identify with the organization and 20% believe in its success.

Representatives of the main staff attach importance to the moral and psychological climate (30%), the integrity of the team (20%), but they are little interested in the fate of the enterprise, that is, there is no revenge as a management staff. An analysis of the positions of the auxiliary staff revealed a similar distribution of relations moral and psychological climate (35%), team integrity (15%), and also the fate of the enterprise is not a priority. This attitude will negatively affect the relationship between the employees of PJSC «GAZ».

Further, the personal orientation of the workers was analyzed, the results are presented in Fig. 2.

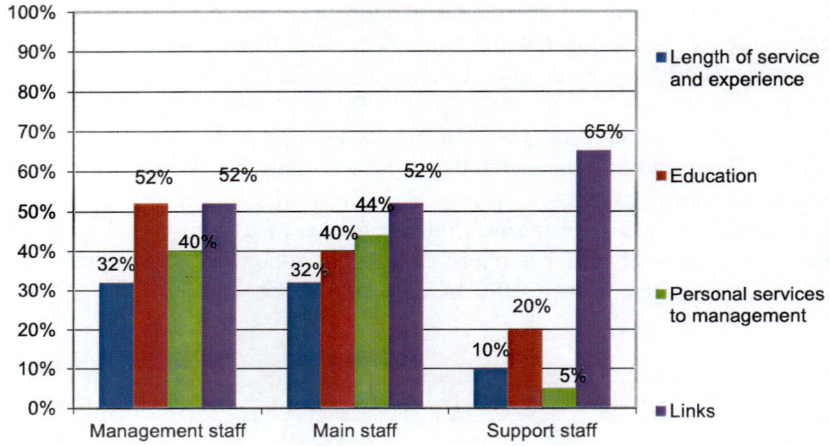

Fig. 2. Personal orientations of PJSC «GAZ» employees. Source Compiled by the authors

The following conclusions can be drawn from Fig. 2. For the categories of personnel studied, education, the presence of connections is important; personal merit to management, as well as experience and experience.

The results obtained allow us to say PJSC «GAZ» has formed a «picture» of the necessary personal characteristics of employees.

The next element of the analysis of organizational culture, the authors of the article highlighted the style of leadership. It is this element that determines the level of relations and communications between management and staff; and manifests itself in how the manager speaks, listens, organizes the adoption and implementation of management decisions.

To determine the style of management, a survey of employees was conducted at PJSC «GAZ». Employee responses are shown in Fig. 3.

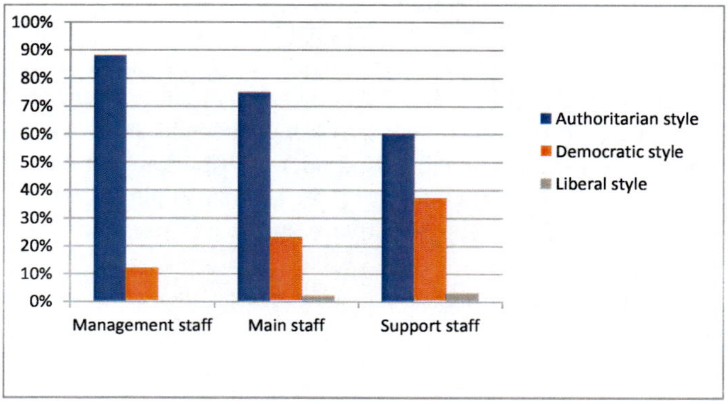

Fig. 3. Opinion of employees on the style of management of PJSC «GAZ». Source Compiled by the authors

The analysis of the results presented in Fig. 3 makes it possible to reach the following conclusions.

1. The authoritarian characteristic of management is given by 88% of management personnel (sole personality in decision-making, distance of management from the company's personnel, criticism of initiative); 78% of core and 60% of support staff.
2. Leadership's propensity for a democratic style is determined by 12% of management; 23% of the main and 37% of the support staff.

Thus, the leader of the automotive industry PJSC «GAZ» is dominated by an authoritarian style. This is due to the fact that, under the current conditions, a strong center with sole management decision-making is needed, quickly implementing the necessary steps to achieve the goals set.

Thus, an analysis of the strengths and weaknesses of the organizational culture of PJSC «GAZ» was carried out, which revealed the absence of individual elements of the culture and thereby confirmed the need for serious changes in the company. The authors of the article see an exit through the introduction of an information protocol in the company.

Of course, free access to information determines a powerful organizational culture, employees feel like an integral part of a strong holistic group with unified aspirations and values, well aware of each other's business [16].

At the PJSC «GAZ» enterprise, management is currently trying to introduce computer tools for diagnosing organizational culture [12, 14]. To do this, use the «Enterprise culture» program. Obviously, the state of corporate culture needs to be constantly monitored, risks associated with inconsistency of employee views on a matter that is significant for organizational processes should be identified in time, and, identifying such «conflict zones», warn them in time [7].

So, this implemented computer program in the organization of PJSC «GAZ» will allow an audit of corporate culture.

Among one of the latest computer utilities, I want to highlight such as Staffcop Enterprise. Using Staffcop Enterprise, you can analyze the full picture of the workforce, strengthen discipline, optimize employee performance, and effectively manage your company with minimal effort [6].

Thus, due to this product, it is possible to form an idea of each of the individual employees and create an overall picture.

4 Conclusion

Taking into account the research results, it is possible to determine a number of areas for improving the organizational culture of PJSC «GAZ»: the introduction of components of a democratic management style, the use of expert computer systems, an inventory of personal results of mental work, a new orientation of the professional sphere, the formation of internal structural divisions, coordination and educational work with personnel, the use of professional audit, the formation of the concept of motivation for the work of employees, etc.

It must be borne in mind without exception all companies are dependent on their own customers and, for this reason, must be aware of their current and upcoming needs, implement their conditions and try to surpass their hopes. For this reason, PJSC «GAZ» can be advised, in addition, to carry out systematic questionnaires among previously available, as well as possible customers of their own services.

Democratic style components are to be implemented. All this will allow employees to better understand the goals of organization, make them more motivated. It should also make employees more interested in research. This will allow managers to quickly find a common language with subordinates and better understand their opinion and needs.

As a result, a significant formation of organizational culture.

The prospects of this research are to clarify the possibilities of total digitalization of organizational culture assessment using the information protocol of activity analysis.

References

1. Abramova, N.S., Vaganova, O.I., Smirnova, Z.V., Chaikina, Z.V., Kaznacheeva, S.N.: Formation of socio-economic culture of future engineers. In: Popkova, E.G. (ed.) Growth Poles of the Global Economy: Emergence, Changes and Future Perspectives. LNNS, vol. 73, pp. 1041–1047. Springer, Cham (2020). https://doi.org/10.1007/978-3-030-15160-7_104
2. Blinov, A.O.: Improving the corporate culture of the automotive industry based on human resource loyalty management. Person. Manag. **12**, 29–31 (2012)
3. Bodrova, M.I., Kizyan, N.G.: Development of organizational culture at enterprises in the Russian Federation - a driver of economic growth in the country. Russ. J. Entrepr. **20**(1), 341–356 (2019). https://doi.org/10.18334/rp.20.1.39765
4. Business Portal About the Foundations Of Organizational Culture (2020). https://delovoymir.biz/ru/biznes_statyi/tag/organizacionnaya-kultura/. Accessed 30 Nov 2020
5. Cameron, K.S., Quinn, R.E.: Diagnosing and Changing Organizational Culture Based on the Competing Values Framework. Jossey-Bass, San Francisco (2011)
6. Description of the Computer Diagnostics Program. StaffCop Enterprise (2020). http://www.anti-malware.ru. Accessed 30 Nov 2020
7. Description of the Computer Diagnostics Program «Corporate Culture» (2020). http://www.orgpsiholog.ru/culture.htm. Accessed 30 Nov 2020
8. Guskova, N.D., Malyasova, M.M.: Formation of a management ethics system aimed at developing the organizational culture of an industrial enterprise. Creat. Econ. **11**(83), 75–82 (2013)
9. Information Site about Jobs and Vacancies (2020). http://gderabotaem.ru. Accessed 30 Nov 2020
10. Kozlova, E.P., Garina, E.P., Romanovskaya, E.V., Andryashina, N.S., Egorova, A.O.: Transformation of economic systems under the conditions of technical and technological complexity of transformed processes: organizational and management decisions. In: Popkova, E.G., Sergi, B.S. (eds.) Scientific and Technical Revolution: Yesterday, Today and Tomorrow, pp. 140–149. Springer, Cham (2020). https://doi.org/10.1007/978-3-030-47945-9_16
11. Kozlova, E.P., Morozova, G.A., Kuznetsov, V.P., Romanovskaya, E.V., Andryashina, N.S.: Interaction of human and smart technologies in modern conditions. In: Popkova, E.G., Sergi, B.S. (eds.) ISC 2020. LNNS, vol. 155, pp. 609–618. Springer, Cham (2020). https://doi.org/10.1007/978-3-030-59126-7_68

12. Listikova, A.V., Egorov, E.E., Lebedeva, T.E., Bulganina, S.V., Prokhorova, M.P.: Research of the best practices of artificial intelligence in the formation and development of personnel. In: Popkova, E.G. (ed.) Growth Poles of the Global Economy: Emergence, Changes and Future Perspectives. LNNS, vol. 73, pp. 1345–1352. Springer, Cham (2020). https://doi.org/10.1007/978-3-030-15160-7_137
13. Official Site of OJSC «GAZ» (2020). http://gazgroup.ru. Accessed 30 Nov 2020
14. Romanovskaya, E.V., Andryashina, N.S., Kuznetsova, S.N., Smirnova, Z.V., Ivonina, O.G.: Digital technologies in Russia: trends, place and role in economy. In: Popkova, E.G., Sergi, B.S. (eds.) ISC 2020. LNNS, vol. 155, pp. 344–351. Springer, Cham (2020). https://doi.org/10.1007/978-3-030-59126-7_38
15. Sedykh, E.P., Zhitkova, V.A.: model of an automated system for assessing the effectiveness of the activity of heads of educational institutions of the city of Nizhny Novgorod. Vestnik Minin Univ. **7**(4), 1 (2019). https://doi.org/10.26795/2307-1281-2019-7-4-1
16. Smirnova, Zh.V., Kochnova, K.A.: Training employees of service enterprises using information technology. Vestnik Minin Univ. **7**(1), 5 (2019)
17. Yukhneva, E.A.: Assessment of organizational culture and models of collective behavior in the strategic planning system (2014). https://creativeconomy.ru/articles/35654/. Accessed 30 Nov 2020

Digital Technologies for the Development of Marketing Communications

Kirill D. Karpunin[1](✉), Tamirlan R. Magomaev[2], Natalya Y. Veselova[3], Natalia N. Zubareva[4], and Pavel P. Kapustin[4]

[1] G.R. Derzhavin Tambov State University, Tambov, Russia
[2] Grozny State Oil Technical University named after Academician M.D. Millionshchikov, Grozny, Russia
[3] Kuban State Technological University, Krasnodar, Russia
[4] Belgorod National Research University, Belgorod, Russia

Abstract. The purpose of the research is to study modern technologies for the development of marketing communications in the digital economy, to analyze their application in the activities of Uber Russia, as well as to justify the vector of improvement of the methods used. In the chapter, the authors reveal the specifics of the digital technologies used at various stages of the development of marketing communications (digital technologies for market analysis, digital technologies for segmentation of the target audience, digital technologies for promoting goods and services). The authors analyze the strategy of development of marketing communications of Uber Russia, the digital technologies used by it, and the ways of promotion. The authors concluded that Uber Russia uses the full range of digital technologies for the development of marketing communications in the direction of market research, segmentation of the target audience, and promotion of services. However, many ways to present information in advertising messages, as well as some shareware channels can also be used by the company to attract the attention of customers.

Keywords: Marketing communications · Marketing · Digital technologies · Uber Russia · Promotion · Luxury concept · Storytelling · Social networks

JEL Code: M31 · M37 · O33

1 Introduction

The logic of digital development is manifested in a qualitative change in approaches to achieving competitive advantages of companies, in a change in business methods, in a transformation of the ways consumers and manufacturers interact. The development of mass production and mass marketing, the use of flexible production technologies and niche marketing technologies, the formation of network communities, and the emergence of new ways to study consumer preferences and improve the mechanisms of influence on the consumer have already become a manifestation of this changeover

E. G. Popkova (Ed.): ISC 2020, LNNS 368, pp. 246–256, 2022.
https://doi.org/10.1007/978-3-030-93244-2_28

[37]. At the stage of global digital transformations, companies are constantly searching for more effective ways to promote products and services, develop marketing communications, attracting the attention of the target audience, which becomes a scarce resource [5, 23]. New ways and technologies of marketing communications should reflect the changing approach to the emergence of the need itself [26]. In the modern world, the need arises under the influence of the network community of consumers, their independent assessments and views, as well as the inclusion of consumers in the process of creating goods and services through interaction with the manufacturer is carried out through digital technologies and is of crucial importance today in terms of creating value [18]. Therefore, the main condition for maintaining the competitiveness of companies is the choice of effective tools for communication with the consumer, taking into account personalization and partnership, focusing on long-term relationships that increase response, conversion, and revenue [2, 24].

2 Literature Review

Burnet et al. [7] understand marketing communications as the distribution of messages transmitted in different forms and various ways: controlled communications and uncontrolled communications. Schultz et al. [27] attributed the reasons for the intensive development of marketing communications in modern society to the loss of the effectiveness of traditional advertising messages, the emergence of new information channels, the increase in the number of producers, the change in consumer motives, the need to achieve consumer loyalty, and the globalization of the market. Trofimenko and Sergeeva [33] point to the need for a transition to two-way communication, that is, to the exchange of information between consumers and producers and the establishment of long-term positive relations between them. Arkhipova and Gurieva [3] reveal the reasons why companies focus on digital marketing technologies: the "flow" of consumers into digital channels, the ability to target, the ability to collect and process huge amounts of information about consumers and on this basis make them an offer that is fully adapted to their tastes and needs [16, 19].

3 Methodology

The purpose of the research is to study modern technologies for the development of marketing communications in the digital economy, as well as to justify the vector of improving the methods and digital tools used on the example of Uber Russia.

Research Stages: (1) Reveal the specifics of the digital technologies used at various stages of the development of marketing communications; (2) analyze the marketing promotion strategy of Rubber Russia and systematize the digital technologies and promotion methods used by it; (3) form recommendations on the use of methods and technologies for the development of marketing communications. Research methods: literature analysis method, systematization method, comparative analysis, economic analysis, system approach, induction, and deduction.

4 Results

Marketing communication is one of the most important marketing tools that allow businesses to identify the needs and expectations of the market regularly and create long-term partnerships [21]. The authors understand marketing communication in this study as a set of conscious actions of a company to promote information about itself and its products in the market environment to generate customer interest and improve its image characteristics [7, 20, 25].Digitalization makes traditional forms of marketing communications ineffective due to their ubiquity and repeatability in terms of form and content, in particular, reducing the sensitivity of consumers to advertising in the media [9]. At the same time, companies have expanded opportunities to use digital technologies for the development of marketing communications. According to the authors, the entire complex of digital technologies for the development of marketing communications can be conditionally divided into the following components: 1) digital technologies for market research; 2) digital technologies for segmentation of the target audience; 3) digital tools for promoting goods and services.

Digital market research technologies allow the company to virtually structure the socio-economic reality, interpreting and positioning any processes, events, phenomena with the help of appropriate labels, study competitors and the specifics of their activities, as well as create a portrait of an ordinary Internet user for further promotion. For this purpose, social networks augmented reality technologies, and Internet technologies are used to assess the company's competitive positions on the Internet "own and acquired channels" [30, 41]. Various digital technologies can be used to obtain information about the target audience and its further segmentation [31]. Digital technologies for promoting products and services include Internet technologies for delivering advertising messages to consumers [12, 15, 40]. A systematic view of the available digital technologies for the development of marketing communications is presented in Table 1.

Table 1. Digital technologies for the development of marketing communications

Digital technology	Content	Example
1) Digital technologies for market research		
Social network	The achievement of marketing goals of market research using social networks is based on direct interaction with both competitors and consumers through communication on discussion forums, participation in specialized communities to collect interesting information [30]	Research of user profiles of social networks Facebook, LinkedIn, and Vkontakte for informal document analysis and content analysis
Augmented reality technologies	Augmented reality technology uses a camera and special modules to recognize the location of mobile devices. The technology allows you to use market signals for subsequent analysis of competitors and determine the target audience	Implementation of projects using AR/VR technologies for large customers, for example, Sberbank, Sibur, Rosatom, Gazprom [14]

(*continued*)

Table 1. (*continued*)

Digital technology	Content	Example
Evaluation of the company's competitive position on the Internet using "owned and earned channels" [6]	The technology is based on evaluating the net traffic on competitors' websites and the balanced number of mentions of the brand or product name by users on the network [13]	Use of technology for research of the Russian market of perfumery and cosmetic products of the luxury segment [13]
2) Digital technologies for segmentation of the target audience		
SMM technologies	SMM allows the company to survey to obtain information about target conversions, age, preferences, devices used by consumers for market segmentation based on specified criteria through online surveys in electronic form, studying the profiles and behavior of buyers on the Internet	Facebook Instagram or Facebook survey on any topic of interest
Thematic Internet forums	Specialized platforms for discussing issues of concern to consumers. Forums help to identify the "pains" and needs of the client for each of the issues raised in the conversation	For example, on the forum about electric scooters, the company can collect information about popular models, accessories of electric scooters [10]
Yandex. Wordstat, Google Adwords	The technology allows the company to identify which queries people are looking for the company's product. This helps to collect information about the interests of the consumer before buying, about the popularity of products from the range	For example, when you enter the word "coffee" in the search bar, the system shows that 3,951,138 consumers searched for it in a month [38]
"Yandex. Audiences", Google Analytics [4]	Yandex.Audience and Google Analytics machine learning technologies help you collect segments for targeting based on the list of contacts of regular customers, as well as to segment the target audience by age, gender, region, and device type	For example, the Look-alike function searches for a "segment of similar users" for a company
Digital Internet VALS [34]	The technology of creating and implementing marketing strategies based on the study of consumer preferences (based on the study of public opinion) and market research	It is used by BORK, Mercedes-Benz, Sport-Master, Miratorg, Ascona, Sloboda to identify and segment the target audience through SERM tools, monitoring, and timely response
3) Digital tools for promoting goods and services		
Advertisements and messages delivered by e-mail (direct-e-mail)	A special feature of the technology is its cost-effectiveness, instant delivery, the ability to quickly receive feedback, the ability of the company to form its unique database of addresses of potential customers for a specifically planned marketing strategy, as well as to deliver messages to them	The company National Geography (USA) with the help of the service @loha @ctive e-mail received from 23,000 new users a CTR of 32%. And this is even though the CTR of the company's banners was only 3.1% [22]

(*continued*)

Table 1. *(continued)*

Digital technology	Content	Example
Websites	The technology allows you to analyze statistics and monitor changes in user behavior, has a low development cost and high conversion rate (from 5 to 10% on average, in some cases 30%), and has a high download speed	For example, We are Social hosts an annual industry survey of data and information on social media, e-commerce, the Internet [36]
Community marketing	The technology allows the company to collect information about the market through participation in niche communities interested in a particular brand or product, joining clubs, and discussion forums	Community Marketing uses Web 2.0 interactivity, such as Online forums, Wikis, Social Networks, Blogs, and related distribution (RSS) to gather information about the state of the market. Webcasts, Podcasts, and e-newsletters are used to encourage community participation by Nike Inc., Starbucks, and Google
Online advertising in search engines (SEO)	Search engine optimization or SEO - a set of measures to raise the site's position in the search engine results for certain user queries to promote the site (Yandex. Metrica, Google Analytics) [1]	If social buttons (Google +, VKontakte, Facebook) with a large number of approving votes are placed at the bottom of the content page
Banners on mobile or websites	An advertising banner is a specially designed picture, animated photo, or video insert. They include hyperlinks that send the target audience to a specific web resource address	Google and Yandex have their tools for creating banners – Google Web Designer
Digital promotions	A technology that encourages advertising activity and has an informational impact on the audience to promote a particular product	Invite a friend, a photo on the background of an advertising product, the best comment on a post/note/news
Video, photo, and contextual advertising	Text, photo, and video ads are displayed to Internet users based on their search queries, interests, or online behavior on various websites, mobile apps, and other resources	Campaigns that are conducted in the Yandex advertising network or the Google Adsense display network
Native integrations	The technology of placing publications on the Internet, which organically combines useful information for subscribers with the mention of the advertiser. Users are more loyal to this advertising format due to its informative content	In the football community, a table with the results of matches in the colors of the mobile operator MTS and with a small logo appears on the cover
Viral marketing	Create and present messages on the Internet (emails, videos, interactive games, photos, animations, etc.), intended for convenient transmission to other Internet users [30]	Adidas asked participants in the Boston Marathon to share their photos and videos of the experience, which garnered more than 100,000 views in the first two days after its release [35]

Source Compiled by the authors

Uber Russia is part of the international public company Uber Technologies Inc. (Uber), which uses digital technologies to search, call, and pay for taxis or private drivers and food delivery, as well as tracking the geolocation of potential customers, their behavioral characteristics. The company positions itself as a service aimed at improving urban transport systems by reducing the need for parking spaces, reducing traffic, and air pollution. The conceptual basis of the company is an online platform that unites many small sellers and many small buyers, reducing transaction costs. Uber Russia standardizes services and prices, issuing them as recommendations, and also achieves a network effect by collecting and organizing the various elements that make up the service, making this service available to customers under a single brand. Such an aggregator business model is also called the on-demand delivery model or Uber for the X model. The company's growth is driven by the following factors: branding, continuous service improvement, localization, and expansion of the customer base. Uber Russia uses digital technologies for the development of marketing communications:

1) Digital Technologies of Market Research. The Uber Russia company has accounts in various social networks (Facebook, Instagram, Vkontakte, Twitter, LinkedIn) (Fig. 1).

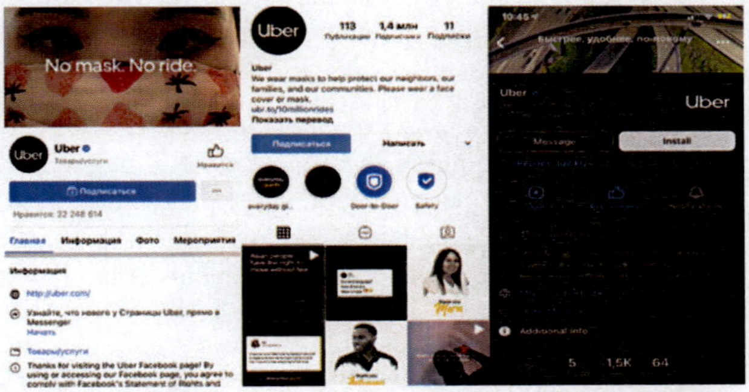

Fig. 1. Facebook Instagram, Vkontakte Uber Russia social media accounts *Source* Developed and compiled by the authors

Since 2017, there has been a tendency to reduce subscribers to Uber Russia social networks [17]. The company uses social networks in the direction of research of potential drivers. Social networks allow tracking through the analysis of complaints concerns about the fact that lower prices affect the income of drivers. On the other hand, social networks identify those factors that better meet the needs of the customer and reduce the cost, for example, the impact of an overabundance of taxi drivers on overcrowding and waiting time. The basis for analyzing the Uber Russia market and identifying growth areas is Big data. At the same time, in 2018, Uber patented a driver search system through augmented reality technology, which allows the company to assess the potential of the market, navigate faster in space and find drivers in crowded places. Many data sets can be analyzed to help the company's analysts identify and succeed in growing markets. An important stage of the analysis is the comparison of

Uber's direct competitors in Russia - Yandex. Taxi and Gett, as well as their expansion strategies. Digital technologies allow the company's analysts to analyze the main resources and key competencies of Uber. One of Uber's competitive advantages is patents. Uber's operating activities are almost identical to those of its competitors, but access to additional capital allows Uber to spend more on attracting drivers, continuing geographical expansion, and building a better infrastructure [8].

2) Digital Technologies for Segmentation of the Target Audience. Uber Russia uses the full range of digital technologies to segment its target audience. The segmentation of the target audience is based on age, gender, geographical, and income characteristics. This allowed the company to develop the following tariffs: UberStart (a popular economy class fare), UberX (a tariff rate among users who order a car through the Uber app), UberSelect (comfort-level rides on higher-class cars with a minimum order time), UberBlack (high-level comfort fare at a high price with a luxury car), UberLux (luxury cars at a high price), UberVAN (a fare for ordering a taxi for a group of 7–9 passengers), UberKids (for customers with children). The list of tariffs demonstrates the use of the luxury concept in the implementation of the company's marketing communications development strategy. Currently, the Uber taxi service covers more than 250 cities in Russia. The fare range also varies depending on the size of the city and the target audience. The analysis of the target audience allowed the company to diversify its activities, in particular, to highlight the new direction of the UberEats food courier delivery.

3) Digital Technologies for the Promotion of Goods and Services. At the initial stage, Uber's marketing was focused on quickly and effectively building a user base [29]. The company's marketing communications development strategy is focused on conversion to app downloads and making the first trip. The strategy implementation tools are referral programs, performance marketing, and local events (Fig. 2).

Fig. 2. Tools for implementing the strategy of marketing communications of Uber Russia. *Source* Compiled by the authors based on [32].

The meaning of the Uber referral program is to provide a free ride to both the referrer and the new client who came on the recommendation. This mechanism encourages consumers to recommend Uber services to their friends and acquaintances, as well as creates favorable conditions for new program participants to get acquainted with the company's service and removes all associated risks. So, the average monthly fee of an active Uber user is $95 for moving around the city, of which 25% are the services of the service [11]. Uber successfully uses performance marketing tools to achieve specific financially measurable business results in the shortest possible time: search engine promotion, contextual advertising, lead generators, targeted advertising in social networks, RTB, display advertising [39]. The company is known for its local events. For example, Uber Russia together with bloggers made an Instagram series about traveling in Russia. The Instagram series shows the most unexpected places with the help of local influencers [28]. Uber has its website that allows the company to analyze statistics and monitor changes in user behavior with adjustments. The company actively uses digital banners on mobile or websites (Fig. 3).

Fig. 3. Example of a digital banner on a website

Uber Russia actively uses the potential of community marketing, pays great attention to the opinions and wishes of consumers, and collects them using feedback forms after using the app. The company uses the potential of digital technologies to organize advertising campaigns, for example, Programmatic Out-of-home [32]. In addition, the company uses video hosting technologies to promote its services. In particular, the company places advertising videos on the YouTube channel. Uber interacts with well-known bloggers to strengthen its presence in the information field of cities with millions of people, creating role models for using the service. Thus, the company uses a native advertising resource [28]. Another example of native advertising is the participation of Uber Russia in Google Maps, where the services of this company are offered when plotting a route. Viral marketing technologies are also used in the work of Uber.

5 Conclusion

The analysis showed that Uber Russia uses the full range of digital technologies for the development of marketing communications in the direction of market research, segmentation of the target audience, and promotion of services. Nevertheless, there are techniques and methods, the introduction of which in the process of marketing communications will expand the target audience of the company. For example, the use of storytelling in the formation of advertising texts on the Internet. Its main advantages are the challenge of trust (consumers trust the comment, personal story), interest in reading the story, and keeping the attention of consumers (an interesting story attracts the reader and leads). Author's example of storytelling for Uber: "Heavy rain". A night city, an area of private houses, a dark street with many lights. One of them starts blinking restlessly. A young girl in a yellow hoodie is running along the road as fast as she can. Her face shows that she is very tired, but her legs continue to carry her forward. Chaotic route changes, wet sneakers, and 5% on the smartphone screen. At this moment, on the other side of the city, a man gets into a car and starts driving fast. After a couple of minutes, the girl realizes that she is at a dead end. Sobbing moans mixed with horror fill all her thoughts. Behind her, she hears the sharp screech of brakes and sees the shadow of headlights on the wall to the side. Sigh, she closes her eyes. *Vibrate your phone with a barking notification sound*. The girl automatically pulls her phone out of her pocket and sees a reassuring message: "The driver is waiting for you". Dark screen. Male voice: "Uber. It's safe with us". In addition, the company should use shareware channels like Telegram Channel and Tik-Tok to attract the attention of customers. Although Uber Russia has an account in Tik-Tok, the promotion of the company in this resource is not conducted.

References

1. Afanasyeva, E.: Peculiarities of internet advertising through search systems, websites, social media, blogs and video hosting. Bull. Inst. Econ. Manag. NOVSU **4**(29), 17–27 (2018)
2. Anokhov, I.: Game aspect of economics. Bull. Irkutsk State Econ. Acad. **2**(88), 5–9 (2013)
3. Arkhipova, N., Gurieva, M.: Modern trends in the development of digital marketing. RSUH/RGGU Bull. Ser. Econ. Manag. Law **1**(11), 9–21 (2018)
4. Audience.yandex. About the Service (2021). https://audience.yandex.ru/. Accessed 01 Apr 2021
5. Ba, S., Stallaert, J., Whinston, A.B.: Optimal investment in knowledge within a firm using a market mechanism. Manage. Sci. **47**(9), 1203–1219 (2001)
6. Benway, J., Lane, D.: Banner Blindness: Web Searchers Often Miss "Obvious" Links (1999). http://www.ruf.rice.edu/~lane/papers/banner_blindness.pdf. Accessed 01 Apr 2021
7. Burnet, D., Moriarty, S., Wells, W.: Advertising: principles and practice. Transl from Eng by Artyukh, M., Bugaeva, M., Bugaeva, E., Huseynova, I.P., St. Petersburg (2011)
8. Chershembeeva, N.: Uber's Expansion Strategy in the Russian Market by Developing Third-Party Services (2018). https://dspace.spbu.ru/bitstream/11701/11226/1/VKR_CHershembeeva_Natalya.pdf. Accessed 01 Apr 2021
9. Danilina, M., Deinekin, T.: Complex approach to marketing communications on the Internet. Competence **1**(122), 40–45 (2015)

10. Electro.club. Electric Scooters (2021). https://electro.club/forum/elektrosamokaty. Accessed 01 Apr 2021
11. Gafarova, D.: How did Uber's Referral System Provide them with Significant Growth? (2019). https://epsi-rating.ru/uber-retention-strategy. Accessed 01 Apr 2021
12. Godin, V., Terekhova, A.: Digital advertising as a tool to promote goods or services. Project implementation experience. E-Management **2**(3), 13–21 (2019)
13. Golovanova, E.: Evolution of the theory of competitive analysis. Evaluation of competitors' positions on the Internet. Mod. Compet. **3**(63), 79–94 (2017)
14. Golovin, S.: Analysis of the augmented and virtual reality market in Russia. Bus. Educ. Knowl. Econ. **3**, 26–29 (2019)
15. Grigorieva, V., Antonov, S., Vorobyov, P.: Digital Marketing Technologies: Creating an Advertising Campaign. Levsha St. Petersburg, St. Petersburg (2017)
16. Isaeva, E., Voronov, A., Magomaev, T., Veselova, N., Maksaev, A.: Digital marketing: opportunities for innovation development and commercialization of innovations, in business 4.0 as a subject of the digital economy. Advances in Science, Technology & Innovation. Springer, Switzerland (2022)
17. Jagajam. Brand Community statistics. Uber Russia(2021). https://jagajam.com/ru/statistics/twitter/detail/uber-russia-511125644?items=&days=180. Accessed 01 Apr 2021
18. Karpunina, E.K., Yashin, S.N., Molchan, A.S., Lisova, E.V., Derkacheva, E.A.: The price of pleasure: the challenges and threats of the digital economy. In: Kolmykova, T., Kharchenko, E.V. (eds.) Digital Future Economic Growth, Social Adaptation, and Technological Perspectives. LNNS, vol. 111, pp. 363–373. Springer, Cham (2020). https://doi.org/10.1007/978-3-030-39797-5_37
19. Karpunina, E.K., Salikov, Y.A., Suslyakova, O.N., Gorbunova, O.N., Avdeeva, R.A.: The way to perfection and safety: about motives of business development in the digital environment. In: Kolmykova, T., Kharchenko, E.V. (eds.) Digital Future Economic Growth, Social Adaptation, and Technological Perspectives. LNNS, vol. 111, pp. 95–104. Springer, Cham (2020). https://doi.org/10.1007/978-3-030-39797-5_10
20. Khitrova, T., Chernikov, D.: Modification of the marketing information system based on intellectual components. Baikal Sci. J. **6**(4) (2015). https://doi.org/10.17150/2411-6262.2015.6(4).19
21. Khlebovich, D., Tokareva, I.: Marketing communications for the promotion of services of theater organizations: features of choice. Educ. Right **2**(43), 195–201 (2018)
22. Kotsarev, O.: E-mail as a channel for distributing advertising information (2013). http://polykur.com.ua/clause/upravlenie_predpriyatiem/elektronnaya_pochta_kak_kanal_rasprostraneniya_reklamnoj_informacii_-.html. Accessed 01 Apr 2021
23. Kozyrev, A.: Mathematical models and forecasting in the digital economy (2018). https://medium.com/cemi-ras. Accessed 01 Apr 2021
24. Okoli, A.: Why brands must start listening in the conversation economy (2018). https://dma.org.uk/article/why-brands-must-start-listening-in-the-conversation-economy. Accessed 01 Apr 2021
25. Polyakova, N., Obukhova, A.: Theoretical and Practical Problems of Production of Information and Technological Services. BSUEP Publishing House, Irkutsk (2014)
26. Ramaswamy, V.: It's about human experiences…and beyond, to co-creation. Indust. Mark. Manag. **40**(2), 195–196 (2011)
27. Schultz, D., Tannenbaum, S., Lauterborn, R.: A new marketing paradigm Integrated marketing communications. Trans. from Eng, INFRA-M, Moscow (2004)
28. Sostav. Uber together with bloggers made an Instagram series about traveling in Russia (2019). https://www.sostav.ru/publication/uber-vmeste-s-blogerami-pokazal-neobychnye-mesta-v-rossijskikh-gorodakh-41102.html. Accessed 01 Apr 2021

29. Spark. The effectiveness of offline advertising: the Uber case (2017). https://spark.ru/startup/sociate-ru/blog/34283/effektivnost-offlajn-reklami-kejs-uber. Accessed 01 Apr 2021
30. Tatarinov, K.: Modern aspects of marketing communications in the digital society. Azimut. Sci. Res. Econ. Manag. **8**(1), 307–312 (2019)
31. Teryaeva, E.: Segmentation of the target audience (2017). https://blog.calltouch.ru/segmentatsiya-tselevoj-auditorii/. Accessed 01 Apr 2021
32. Tikhonova, E.: The effectiveness of offline advertising: the UBER case using the example of an Uber advertising campaign, we understand whether offline marketing tools have effectiveness and potential (2019). https://blog.sociate.ru/effektivnost-offlajn-reklamy-kejs-uber. Accessed 01 Apr 2021
33. Trofimenko, E., Sergeeva, A.: IMK or transition from unilateral communications to bilateral relations. Bull. SUSU Ser. Econ. Manag. **7**(4), 144–149 (2013)
34. VALSDigital. Reputation Management. Big Data Analytics, Creating an Ideal Image (2021). https://vals.digital/. Accessed 01 Apr 2021
35. Vayongroup. 25 best examples and methods of viral marketing for 2019 (2019). https://ru.vayongroup.com/3094-top-25-viral-marketing-examples-and-techniques-for-20.html. Accessed 01 Apr 2021
36. Wearesocial. We are a socially-led creative agency (2021). https://wearesocial.com/. Accessed 01 Apr 2021
37. Webber, A.: What's So New About the New Economy? (1993). https://hbr.org/1993/01/whats-so-new-about-the-new-economy. Accessed 01 Apr 2021
38. Wordstat. Cofee (2020). https://wordstat.yandex.ru/#!/?words=кофе. Accessed 01 Apr 2021
39. Zagrebelny, G.: Performance-Marketing: Make the Internet Work for You. Alpina Publisher, Moscow (2016)
40. Zarrella, D.: Internet Marketing by Science. What, Where, and When to do to Get the Maximum Effect. Mann, Ivanov and Ferber, Moscow (2014)
41. Zemskova, E.: Analysis of consumer behavior in the digital economy from the perspective of generation theory. Euras. Sci. J. **5**(11) (2019)

Use of Augmented Reality (AR) Technologies in Retail

Andrey A. Dashkov[1](✉), Dmitry V. Buzin[2], and Dmitriy A. Slobodyanyuk[2]

[1] Mytischi Branch (MB) of Bauman Moscow State Technical University, Moscow, Russia
[2] "KRIT" Ltd, Saint Petersburg, Russia
{dmitriy.buzin, dmitriy.slobodyanyuk}@krit.pro

Abstract. The article provides an overview of existing systems that use augmented reality to automate business processes, and considers the possibility of their application to address current needs in the retail sector. The practice of using AR applications for domestic and industrial use by various domestic companies is investigated for this. The use of AR-technologies in retail to control the state of the outlet and the quality of goods to improve the quality of the work performed is substantiated. The process of interaction of a salesroom employee with AR equipment is developed. The results of comparative studies of equipment for augmented reality and systems that use augmented reality in the creation of "smart assistants" are presented.

Keywords: Augmented reality · Smart assistant · Retail · Quality control · Stockpile management · Innovation in retail · Business process automation

JEL Codes: C88 · L81 · L86 · O14 · O32

1 Introduction

One of the technologies, which included in the list of the main end-to-end digital technologies listed in the program "Digital Economy of the Russian Federation", is augmented reality technology [1]. Augmented reality (AR) is interpreted as the ability to digitally expand human senses in the physical world using sensors, devices for saturating the user space with relevant, interesting and useful information, using display devices (for example, a mobile device screen, helmet, glasses). Augmented reality integrates the digital world into the real one, but doesn't replace this, as it happens in virtual reality, but expands with the display of digital objects, which has a positive effect on workflows.

An augmented reality app can improve employee training, speed up maintenance and make collaboration easier. Such applications are gaining popularity in the service sector, warehouse logistics, healthcare and retail.

In 2020, the global market for augmented and virtual reality increased by 78.5% compared to 2019 and reached $18.8 billion [8]. According to IDC forecasts, one of the areas, that will attract the largest investments by 2024, will be retail ($2.7 billion), and

E. G. Popkova (Ed.): ISC 2020, LNNS 368, pp. 257–266, 2022.
https://doi.org/10.1007/978-3-030-93244-2_29

by 2022, 5% of the entire augmented and virtual reality technology market will be concentrated in the retail sector [8].

The volume of the augmented and virtual reality market in Russia, according to forecasts of the "Statista" global data platform, will be more than quadruple by 2025: from 1.4 in 2020 to 7 billion rubles [3].

The growth in the use of augmented reality technologies in retail is conditioned by the following factors:

- The growth in the number of stores. The number of supermarkets and hypermarkets in the Russian Federation has grown from 15,762 facilities in 2016 to 28,124 in 2020 [5]. As retail chains grow, they are demanding automation of areas that, with the previous size of the company that didn't need to be automated.
- Increasing the competitive environment and the need to increase customer loyalty. 71% of respondents surveyed by Rosstat for the 4th quarter of 2020 note that high competition is a factor limiting the main activity in retail trade [5].

According to the research of the market of industrial VR/AR solutions in Russia presented by the TAdviser center, most of all companies from the oil and gas, mining, energy and other industries of Russia are interested in and experiment with VR/AR technologies without disseminating the results of their developments to the general public [20].

The main Russian companies that have already had and/or are developing applications using augmented reality are: Severstal, Gazprom Neft, Sibur, Magnitogorsk Iron and Steel Works (MMK). These companies are interested in creating systems for industrial use.

2 Methodology

The information base of the research was the available open sources of information; the methodological basis was the general scientific research methods: analysis, synthesis, comparisons, analogies, generalizations.

It's necessary to take into account the user experience for the successful application of augmented reality technologies in order to optimize internal processes. According to the research results of the executive director of "The Augmented Reality for Enterprise Alliance" Mark Sage [17], while designing an AR system, it's necessary to:

1. Identify a problem in existing business processes;
2. Determine the main use cases and user restrictions;
3. Select a complete set/model of AR equipment;
4. Develop the concept of user interaction with the device.

3 Results

According to the Federal Law of the Russian Federation "On the Quality and Safety of Food Products" [11] and the Federal Law of the Russian Federation "On Protection of Consumer Rights" [12], the seller of food products is obliged to monitor the quality of the products that he sells. In the area of retail, the procedure of a daily round of the sales area is used to control the state of the outlet and the quality of the goods. For example, in the hypermarkets of the X5 Retail Group, the "Hour of Quality" program has been introduced, within the framework of which employees daily monitor the quality and freshness of all products on the shelves in the sales area [24].

The following employees are involved in the daily monitoring of the state of a store and the quality of the goods:

1. The employee who is responsible for the product – an executor and makes the rounds. He submits to the manager of the sales area.
2. Sales area manager - an executive, who organizes and controls the correctness, order and timeliness of displaying goods in the sales area. He regulates the work of the employees responsible for the goods.

According to the results of research carried out by the authors at "KRIT" company, employees, who are responsible for the product, face the following problems:

1. Being on the sales area, an employee doesn't have the opportunity to quickly determine where the expired goods are located, which goods ran out of stock and other problems related to the goods;
2. They don't have a digital tool for making rounds and recording the results of inspecting racks with products, the report is filled in on paper;
3. While working with a product, they have to use paper carriers and data collection terminals, so their hands are constantly busy and it's inconvenient for employees to work with the products.

Shop floor managers are constrained by many factors such as:

1. Lack of tracking the quality of the walk through the sales area by employees responsible for the goods: in the absence of close control, employees can fill out a walk-through report without leaving the sales area.
2. While being on the sales area, it's not possible to determine the specific location of expired goods and/or with insufficient stock on the sales area.

Thus, for the business process of bypassing the sales area, it's possible to offer a software and hardware tool that will be optimal for solving the above problems of retail employees.

For the accounting of goods and trade automation at retail enterprises, ERP systems are used. The leading players in the world market for ERP systems are SAP and Oracle [19]. In the Russian market, the top three also include 1C [18].

A single interface of the ERP system combines all the processes of the company, which provides high-quality analytics and transparency of business processes (minimization of errors related to the product). Integration of mobile applications with such a system allows you to optimize the work of an employee who controls the goods in terms of conveniently displaying relevant information on the goods on the device screen and obtaining analytics of deviations.

However, existing applications on PDA and mobile phones don't solve the problem of freeing the employee's hands, because during the operation he is forced to hold a mobile device in his hands. Augmented reality glasses allow you removing these restrictions from the process.

According to the latest forecasts, CCS Insight [9] will have sold about 14 million AR devices worldwide by 2024. In general, the VR and AR equipment market will grow more than 6 times by 2024. Let's consider the major device manufacturers providing AR technology for corporate use:

1. Google Glass augmented reality glasses were created mainly for large economic segments of the modern market, offering a device, which can adapts to various software [2]. The glasses are lightweight and compact with an adjustable lens.
2. Zebra Company classifies its AR series of displays as industrial grade, which means built-in moisture and dust protection [7]. The equipment runs on its own operating system, which causes a limitation in the development of an application adaptable to various devices.
3. AR-glasses Vuzix Blade Commercial Edge fit snugly to the head, suitable for industrial use [23].
4. Epson has a line of AR-devices Moverio BT [4]. The glasses come with a control device, which has a built-in processor, and the glasses are used as a display for displaying information.
5. Smart glasses Rokid Glass 2 come with an external battery and a thermal scanner [16]. The glasses get very hot during use.

The following characteristics were used to compare AR equipment:

1. OS - operating system. The criterion is important for consideration from the point of view of scalability and convenience of further development of additional components for an AR application.
2. Battery (number of hours of active use/power). For long-term autonomous use, the size of the battery must be taken into account.
3. Weight of the device. The lighter the device, the fewer loads on the body and the user's head.
4. Cost category. There are three categories for the average price of AR-glasses models on the market (relevance for March 2021):
   - 1: up to $1500; 2: $1500–$2000; 3: more than $2000
5. Touch panel. This specification discusses the ability to control the device using the touch panel built into the glasses.
6. The presence of a microphone. This characteristic is required for voice control.

For the analysis, only AR equipment models that allow displaying information in high resolution, have alternative manual control modes, are convenient for long-term active use, are available for purchase in Russia and are intended for use in industry were selected (Table 1).

Table 1. Comparative analysis of AR equipment

Glasses model	OS	Battery (hours of active use/power)	Weight (g)	Cost category	Touchpad	Microphone presence
Google Glass Enterprise Edition 2	Android Oreo	Built-in battery (820 мAh/3 h)	46	2	+	+
Zebra HD4000	Android	Powered by a host device	30 (without cable)	2	+	+
Vuzix Blade Upgraded Smart Glasses	Android OS	Built-in lithium battery (2 h)	50	1	+	+
Epson Moverio BT-350	Moverio OS	Built-in battery (6 h)	119	1	Additional control device	Additional headset required
Rokid Glass 2	Android	External battery (10000mAh/8 h)	96	3	+	+

Source compiled by the authors based on [2, 4, 7, 16, 23]

According to the results of a comparative analysis of AR equipment and a review of trends in the use of AR technologies in the retail market, it can be assumed that the use of functions such as displaying navigation and information on goods is also applicable to the use of sales floor employees in order to optimize the process of walking around the premises. Also, the possibilities of alternative interaction with AR equipment expand the list of available control commands, which allows you to design a highly targeted algorithm for using AR equipment by an employee while walking around the sales area:

1. The user puts on augmented reality glasses with the installed software. The user can use alternative sources of control depending on the device, thus freeing his hands.
2. He launches a specialized application;
3. He goes through authorization;
4. The user selects a task from the list of available ones;
5. He begins to carry out tasks for roaming around the sales area;
   - The system helps the user to navigate the room;
6. The user scans the product;
7. He looks through the information through AR equipment. With the help of augmented reality technologies, the employee is displayed information on product's shelves there are goods that need to be paid attention to (low level of stock, expiry date, the price has been changed and price tags need to be replaced, etc.):
   - If deviations are found, the system displays recommendations to the user for further work with the product.
   - Printing price tags. The user can print the labels using a mobile printer.
8. The employee continues to process the goods

9. He performs a tour of all shelves with products that are indicated in the assignment;
10. He fixes the results of the inspection;
11. The employee completes the scheduled walk.
12. The data is saved in the ERP system of goods accounting. Data will be automatically saved, excluding the use of paper media.

Thus, an employee reduces the time that is usually spent on performing a walking around the sales area using augmented reality glasses. At the beginning of a walk, the system can immediately display the information necessary for work on the augmented reality display and support the user in performing of the sequence of actions, optimizing his labor costs.

4 Discussion

In the Russian market, AR technologies are used in retail only in test modes or are at the development stage. For example, a large Russian retailer X5 Retail Group is testing the training of sales assistants through Oculus Go VR glasses to immerse employees in the workflow [13, 15]. The training takes place in three stages. At the first stage, the employee exactly repeats the script for communicating with the customer, and the system gives hints. On the second, he says the same thing using keywords. On the third, he helps to choose a product on your own, without any prompts from the system. At the end of each module, the algorithm outputs a learning grade. If the result is unsatisfactory, the employee repeats the topic again.

In addition to using AR technologies in production processes, “Perekrestok” is testing the navigation application “Perekrestok.Edalov” with augmented reality for customers [22]. Users will quickly navigate in sales areas and find the products they need with the help of it. The system works due to special tags hung around the supermarket. When a customer walks into a store, they scan their location and pass the data to the app. A cursor appears in the middle of the screen that points to the desired section or product. You can customize navigation by product category and even by recipe. In April 2019, a similar system was launched by the M.Video network [14]. The “Krit” company is developing a similar product combining augmented reality technology, indoor navigation and interactive recommendations to improve the efficiency of routine inspections in the sales area [10].

In other countries, similar solutions are in the stage of the testing phase. DHL is testing an AR application integrated with Knapp, SAP, and Ubimax [6]. The solution consists of an AR display, a camera, a laptop, and an additional battery. The software recognizes objects in real-time, reads product barcodes, supports user navigation on the premises, and communicates with a warehouse management system (WMS). UPSKILL Company has released an AR application for quick picking of goods [21]. Information about the selected product, such as its preview - image, location, and quantity, is displayed on the display of smart glasses. By highlighting the areas, the application prompts the next required action, for example, scanning a barcode. Next, the user scans the items using the built-in camera, Bluetooth ring, or Bluetooth scanner. In post-processing, the employee uses voice commands to enter picking variances and quantities.

A comparative analysis of the software was carried out according to the following criteria and is presented in Table 2:

1. Functions. Available AR solution functions.
2. Information is displayed on the AR display. The data can be used to improve the user experience of the application, make assumptions about the usability of the information, and also determine what data should be displayed on the screens.
3. Useful effect. The values of the expected or obtained as a result of testing the effect of the use of AR equipment compared with the indicators before using.

Table 2. Comparison table of systems using augmented reality

List of systems	Functions	Information displayed on the AR display	Useful effect
"AR – assistant" company "Krit"	Automation of business processes associated with making rounds and recording the results of inspection of shelves with products	1. Interactive allocation of sites in the sales area 2. Information data about the product	1. Reducing the processing time of goods by 20% 2. Reducing the number of write-offs by 10%
JSC "Trading house "PEREKRESTOK""	Navigation of visitors in the retail space	1. Orientation of visitors in the retail space; 2. Building a route to the desired product	Optimization of the shopping process of visitors
PJSC "M.Video"	Navigation of visitors in the retail space	1. Building a route for a visitor to the desired category of goods	Optimization of the orientation of visitors in the retail space
DHL Smart glasses worker assistance	Finding the optimal path and navigating along with it Scanning goods for collection	2. Package list	1. Reduction of picking errors by 40%; 2. Reducing the search time for the next item in the picking list
UPSKILL	Scanning barcode and displaying product information	Scanned item information	1. The time spent on training an employee is reduced and the company can hire temporary workers 2. The picking accuracy increases by 46%

Source compiled by the authors based on [6, 10, 14, 21, 22]

Considered in the Table 2 solutions of the Russian market in the area of retail are focused primarily on the tasks of end customers. However, based on foreign solutions, it can be assumed that in order to automate the processes performed by the employees of the sales area, it's necessary to additionally provide for the following components:

1. Integration with the goods accounting system (for example, SAP, 1C).
2. The indoor navigation system in the room, including the construction of the shortest routes for the user.
3. Adaptation of the logic of the application to the business processes of retail, namely, the execution of rounds of the sales area by internal employees.
4. Fast scanning/recognition of product barcodes.
5. Display on the display of AR equipment of the necessary information for navigation in the room and viewing data on goods.

5 Conclusion

Today, the employees of the sales areas, who are responsible for the goods, as a rule, can't quickly determine where the goods with expired expiration dates are located or the goods for which the stock on the rack has run out. While working with goods, they have to use paper carriers and data collection terminals, so, the employees' hands are constantly busy. It's inconvenient for employees to work with the product, which leads to a decrease in the speed and quality of the process. With a wide range of use cases, the use of augmented reality technologies to optimize internal retail processes such as bypassing the sales area will eliminate existing restrictions. Among the advantages of optimizing the traversal of the sales area by retail employees using AR equipment, let's single out the next ones:

1. Continuous monitoring of racks.
2. Control of the shelf life of the goods.
3. Simplification of personnel logistics.
4. Reducing the cost of training new employees.
5. Fast adaptation of employees to changes in exposed positions.
6. Availability of constant information support in the form of instructions and schemes of actions.
7. Freeing the hands of sales personnel when making rounds of the hall, for other actions.

Optimization of internal business processes is an important factor in the success of an enterprise. AR technologies will show where the problem product is, as well as support user navigation to it. In addition, you can simply scan the product and get all the information you need using an AR device. All that remains for the employee is to read the information and follow the further recommendations of the system. All this will undoubtedly improve the quality of work and ultimately increase the satisfaction of end customers.

References

1. About the Approval of the Digital Economy of the Russian Federation Program. Order of 28 July 2017, No. 1632-p. http://government.ru/docs/28653/. Accessed 24 Mar 2021
2. CHIRONIX: Glass Enterprise Edition 2 Specialty Bundle. https://www.chironix.com/product/specialty-bundle/. Accessed 17 Mar 2021
3. Elagina, D.: AR/VR Market Size in Russia 2020–2025. Statista. https://www.statista.com/statistics/1209495/ar-vr-market-size-russia/. Accessed 16 Mar 2021
4. Epson: Epson Moverio BT-300. https://epson.ru/catalog/moverio/epson-moverio-bt-300. Accessed 22 Mar 2021
5. Federal State Statistics Service: Retail trade, Public Services, Tourism. https://rosstat.gov.ru/folder/23457. Accessed 29 Mar 2021
6. Glockner, H., Jannek, K., Mahn, J., Theis, B.: Augmented reality in logistics. Vol. 28. DHL: Customer Solutions & Innovation. https://www.dhl.com/con-tent/dam/dhl/global/core/documents/pdf/glo-core-augmented-reality-trend-report.pdf. Accessed 30 Mar 2021
7. HD4000. https://www.zebra.com/content/dam/zebra_new_ia/en-us/solutions-verticals/product/Mobile_Computers/Wearable%20Computers/HD4000/fact-sheet/hd4000-fact-sheet-a4-ru-ru.pdf. Accessed 22 Mar 2021
8. IDC: Worldwide Spending on Augmented and Virtual Reality Forecast to Deliver Strong Growth Through 2024, According to a New IDC Spending Guide. https://www.idc.com/getdoc.jsp?containerId=prUS47012020. Accessed 29 Mar 2021
9. Koytcheva, M.: Pandemic Makes Extended Reality a Hot Ticket. CCS Insight. https://www.ccsinsight.com/blog/pandemic-makes-extended-reality-a-hot-ticket/. Accessed 24 Mar 2021
10. KRIT: Innovative Products. https://krit.pro/products/ar-assistent-dlya-riteyla/. Accessed 24 Mar 2021
11. Federal Law of the Russian Federation "On the Quality and Safety of Food Products". https://docs.cntd.ru/document/901751351. Accessed 24 Mar 2021
12. Federal Law of the Russian Federation "On Consumer Rights Protection". https://docs.cntd.ru/document/9005388. Accessed 24 Mar 2021
13. Oborot.Ru: Prodavcov X5 Retail Group Will be Taught in Virtual Reality. https://oborot.ru/news/x5-retail-group-nadenet-na-prodavcov-ochki-virtualnoj-realnosti-i101690.html. Accessed 22 Mar 2021
14. Polyakova, A.: M. Video Launched Navigation with Augmented Reality in Stores. Rusbase. https://rb.ru/news/mvideo-ar/. Accessed 24 Mar 2021
15. Retail & Loyalty: Perekrestok Introduced Navigation with Augmented Reality. https://retail-loyalty.org/news/perekrestok-predstavil-navigatsiyu-s-dopolnennoy-realnostyu/. Accessed 24 Mar 2021
16. Rokid Glass 2. www.rokid.com/en/rokid-glass-2/. Accessed 13 Apr 2021
17. Sage, M.: Creating augmented reality experiences for enterprise: good practices, lessons learned, and technological insights. IEEE Consum. Electron. J. **6**, 42–44 (2017). https://doi.org/10.1109/MCE.2016.2614651
18. TAdviser: ERP-System in Russia: Drivers, Barriers and Major Market Players. https://www.tadviser.ru/index.php/. Accessed 25 Mar 2021
19. TAdviser: The ERP System Market has Added 10% Over the Year. One Change in the Top Five. https://tadviser.com/index.php/Article:Enterprise_Management_Systems_(Russian_ERP_Market). Accessed 25 Mar 2021
20. TAdviser: The market of industrial VR/AR solutions in Russia is expected to grow 11 times. https://tadviser.com/index.php/Article:The_market_industrial_solutions_VR/AR_in_Russia_(a_research_of_TAdviser). Accessed 22 Mar 2021

21. UPSKILL: Using Augment Reality to Manage Peak and Seasonal Surge (2017)
22. Vprok.Ru: Augmented reality will help "Perekrestok" customers save time. https://www.vprok.ru/news/edalov. Accessed 22 Mar 2021
23. VUZIX: Vuzix Blade Smart Glasses for Work & Play. https://vuzix-website.s3.amazonaws.com/files/Content/pdfs/Vuzix-Blade-Upgraded-Smart-Glasses-d08.pdf. Accessed 22 Mar 2021
24. X5 RETAIL GROUP: Annual Report 2018. https://www.x5.ru/ru/PublishingImages/Pages/Investors/ResultsCentre/X5_Annual_Report_2018_RUS.pdf. Accessed 24 Mar 2021

Technological Convergence and the Future of Copyright

Yuliya S. Kharitonova[1](✉) and Rimma Sh. Rahmatulina[2]

[1] Lomonosov Moscow State University, Moscow, Russia
[2] Financial University under the Government of the Russian Federation, Moscow, Russia
RSHRahmatulina@fa.ru

Abstract. New technologies make it possible to create and introduce new objects of digital art and put them into circulation to use new solutions of technological nature. Based on the analysis of the main approaches to the definition of digital art, the chapter formed legal approaches to determining the nature and characteristics of the new objects of copyright. The convergence of digital technologies and their impact on creating digital objects in the cultural industries were investigated. Special attention is paid to the tokenization of art as a way of introduction and transfer of copyrights. In turn, this trend implies a different kind of intellectual property rights action for digital cultural objects [7]. More generally, in the art world, blockchain technology may be "the key to overcoming one of the biggest problems" - lack of transparency.

Keywords: Digitalization · Copyright · Digital art · Non-fungible token

JEL Codes: K01 · K15 · K19

1 Introduction

In today's context, many copyright objects are created, exist, and are sold in a virtual environment. The development of technology led to the emergence of a new direction in art called "digital art"[14]. There are different approaches to the definition of this direction of art. Often the following types of digital art are distinguished: Computer Graphics; Animation and 3D in Art; Hacktivism; New Art; Interactive Objects and Art in Public Space; Computer Games and Art [9]. Blockchain technology is recognized as the leading technology for securing the circulation of exclusive rights in digital art.

Directions for the development of digital art are becoming more and more ramified. The peculiarity of all these objects is that the form of expression of the author's ideas is inseparable from its content [8]. This distinguishes digital art from other works of art and copyright [2]. In this article, we will try to understand what is the influence of technology on art and what legal consequences can arise in connection with it.

E. G. Popkova (Ed.): ISC 2020, LNNS 368, pp. 267–274, 2022.
https://doi.org/10.1007/978-3-030-93244-2_30

2 Methodology

The methodological basis of the study includes a combination of general scientific (analytical, deductive, dialectical, inductive) and special scientific methods (comparative-legal, dynamic, formal-legal, interpretative).

3 Results

Digital Technologies and Their Convergence in Creative Industries
Digital art is rapidly developing due to technology. Creating work in digital form, the author uses various digital tools - these are updatable software products, technologies, graphic editors. At the same time, digital technology is often the basis for their creation. In this regard, let us highlight the significant cross-cutting digital technologies most changing the modern world and digital art objects created with their help, namely: virtual and augmented reality technologies, 3-d design, distributed ledger technology.

Digital technologies give a creative impulse and a different meaning to works by changing collage, montage, creating images by scanning, which is more detailed and different from photographic images. Digital art objects are the result of creative activity, and the technology used affects the content of the object and its form. Copyright protects the form of the digital object, with the essence of the digital art object being its content - what the author wanted to express, and the form is the system of means and methods in which the content is embodied. The form of digital art objects identifies the object; it is capable of expressing the idea; it is a concrete and definite fixer of it. The object is expressed in a particular digital form and becomes protected. Form in digital art objects is inseparable from content, which identifies and defines form, which is influenced by attributes such as visualization and interactivity, as seen in the types of digital art objects presented. Visualization is a toolkit that allows us to see the result in digital form, and interactivity refers to the perception and assimilation of the diverse forms of digital art objects using technology.

For example, *Virtual Reality (VR) technology* refers to the reality created by a computer program that immerses users in a world where they interact with virtual objects. Virtuality is a state of potential interaction, where reality is experienced through specific images. At the same time, virtual reality is a set of ideas, images, symbols, which serve as a cognition tool. Following paragraph 3.9 of State Standard R 57721–2017 [18], virtual reality is a highly developed form of the virtual environment, which has a high degree of reliability of visualization, imitating both the impact on the studied object and reactions to this impact. Digital objects help us immerse ourselves in the virtual world through various digital and physical interaction types. Examples of such digital interactions are various games, pieces of training, webinars, etc., which have a physical impact and influence perception through virtual communications. Virtual space provides new opportunities for creativity using various digital tools through user interaction, which opens new perspectives for business, education, medicine, etc.

One of the first objects of virtual reality technology is a computer game, a complex integrated digital object of copyright (Complex Object in the meaning of Art. 1240 of the Civil Code of the Russian Federation).

Improvements in digital technology in games, affecting several participants simultaneously, served as the basis for developing *Augmented Reality (AR) technology*. Augmented reality does not change the world around a person. It only complements it by adding artificial graphical elements, combining the real world with the artificial one. These technologies were first used in PokemonGo, Minecraft, although later began to be used in other areas. Digital devices have improved different possibilities for artistic realization, complementing physical objects through audio-visual and graphic images.

As an object of the virtual game industry, a computer game includes various digital virtual objects using virtual and augmented reality technologies: virtual characters, digital design objects, video clones (holograms) - these are both characters and natural phenomena, objects based on optical illusion technology (which are made from photographs, analog drawings), protected as works. Holograms are three-dimensional images using light beams, which are also referred to as graphic or audiovisual works [11]. Holograms can imitate real people (artists, politicians, singers, etc.), literary, cartoon characters, can be created using digital devices that provide different opportunities for artistic creativity.

The use of virtual and augmented reality technologies has its perspectives on creating digital art objects and interacts perfectly with *Artificial Intelligence (AI) technologies*. The Deep Art app creates paintings from photos by transforming them with artificial intelligence. New media and the Internet continue to provide new opportunities for creativity. Artificial intelligence argues with traditional notions of authorship [1]. The creation of computer games is a creative process represented by advanced digital technologies and, since it is not standardized, it is of great interest to the creators of artificial intelligence (AI) games. This technology is also developing a learning AI based on neural networks, creating characters that memorize data and systematize them in the game. The various algorithms used in AI bring complex problems to this digital technology.

So in the creative industries, a wide variety of technologies are actively used. They include artificial intelligence, machine learning, 3D design, and others. So we believe that in creative industries, we can see more than anywhere else the convergence of technologies between each other to create a new product. It means that there are several layers of applying technology to the creation of a single digital art object on the market.

What is the convergence of digital technologies in the creative industries? Digital convergence is understood very broadly. Convergence can be represented as the interaction of different digital technologies [13]. In some cases, convergence is understood as the interaction of technology and infrastructure. An examination of historical developments leading up to the Internet era reveals mutual dependence between technical infrastructures and diverse social arrangements including industry, regulatory, and market structures [16]. In our opinion, digital convergence in the creative industries is possible in both directions: through technological interaction as well as through the interaction of technology and infrastructure. In the first step, we are talking about the fact that several technologies are used to create digital art (AI,

Machine Learning), in the second - the application of technology (AI) and infrastructure solutions (3-d design).

The convergence of technologies brings to the fore the question of who is the creator of digital art and who owns the rights to digital objects. By Russian law, an author is a physical person (article 1257 of the Civil Code of the Russian Federation). At the same time, the object of copyright is created by the creative work of the author. The content and form of fixation of objects created with the help of technologies raise the question of the importance of the developer's contribution to the technology and the end-user to the created innovative product.

NFT as a New Generation of Digital Art

The State Hermitage Museum in 2021 will hold an exhibition of works of art made in the format of so-called Non-Fungible Tokens (NFT). The Hermitage project will be the first exhibition research in Russia in the field of NFTs and their use in the artistic sphere, which is today the most topical agenda in contemporary art. NFTs have become a new form of securing the right to a work of art using blockchain technology, allowing precise control of copyright and ownership rights.[1] In our opinion, the application of blockchain technology can be called the next level of technology convergence to create unique digital art objects.

Digital collectibles, such as CryptoKitties[2], are examples of asset-backed tokens that "generate value and scarcity" [5]. They are assets, "vanity items earned in the online game", [3], whose turnover is managed through central databases. These kinds of assets have traditionally been difficult to prove ownership of, as they are often simply represented as contracts with a software vendor [7]. Blockchain can offer specialized trading platforms for digital collectibles created using tokens and secured by assets.

At the same time, мany digital artists today are thinking about the possibility of facilitating the circulation of their work in a digital environment. So they create their own NFTs. They are most interested in creating a digital art token, which allows the interaction of multiple technologies to bring creative industries objects to the market. Media files corresponding to digital art are closed into NFT on the blockchain. To create NFT, you need Ethereum Wallet (Meta Mask or other); Opensea.io website; The content itself, which you want to write into the token (media file). The media file can be a picture, a video, a 3D model, a sound. Acceptable formats are JPG, PNG, GIF, SVG, MP4, WEBM, MP3, WAV, OGG, GLB, GLTF. No more than 100 MB. Also, add the token name.

Next, it is possible to add various characteristics to the token. Alternatively, it is also possible to add locked content, which the buyer can only see (the owner of this token). Next, the number of instances in which the token is issued is specified (1 by default). Usually, on Opensea.io, sales are conducted in an auction format.

The digital file associated with the unique token was converted into a digital file by the artist Banksy Morons (White), which the blockchain company Injective Protocol bought from the Taglialatella Gallery in New York for $95,000 and then burned, organizing the broadcast.

[1] https://hi-tech.mail.ru/news/53455-pochemu-ves-mir-pomeshalsya-na-nft-tokenah/.

[2] https://www.cryptokitties.co/.

In this way, blockchain technology is merging with media file creation technology. Besides, machine learning and artificial intelligence can also be used to generate content. In general, all these technologies generate a single object of rights - a token with media content.

NFTs allow users to use digital content, whether posting it on social networks, using it as an avatar, etc. They are also crucial for confirming the ownership and right to use digital assets. These are unique digital items such as computer art, digital collectibles, and online games. Ultimately digital art (only in the form of a digital file with no other physical medium) was one of the first uses of unique tokens, as they provide proof of the digital object's authenticity and ownership. NFT objects can be sold and purchased on blockchain platforms: OpenSea, Niftygateway. Without NFTs, it is challenging to verify the legitimacy of distributing such art over the Internet.

It should be noted that Russian law does not regulate the creation and circulation of NFT. One would assume that they fall under the concept of digital rights in the sense of Article 141.1 of the Civil Code of the Russian Federation. However, a particular act, the law on digital financial assets, do not talk about the tokenization of intellectual property.

Simultaneously, analysis of foreign legislation allows us to conclude that one of the most convenient rules is contained in the Token and TT Service Provider Act of Liechtenstein (TVTG). According to the TVTG, a token is a piece of information in a Trusted Technology (TT) system (can represent participants' claims or rights to another person, property rights, other absolute or relative rights) which are transferred to one or more TT identifiers. The technology-neutral definition of tokens is entirely consistent with the Token Container Model (TCM) concept. So, any asset can be "put" into a token: shares, real estate, rights, liabilities, currency, metal, or you can put nothing and leave it "empty" as a digital code. This approach makes the legal certainty of a token dependent on its "content" and contributes to determining the place of tokens in the legal sense.

This approach is convenient for determining the legal regime of the NFT as an object that contains the copyright and the possible implementation of the licenses.

Tokenization is a Way to Put Digital Art into Circulation and Protect Its Rights
We believe that in the digitalization context, the creation of protected digital art tokens allows us to solve several problems at once, which are often difficult to solve through law.

Tokenization allows fixing the ownership of the rights by the exclusive right holder to sell the copyright further. Also, tokenization allows to accelerate, simplify the transfer of access to a digital art object, and guarantee payment. Besides, tokenization gives the possibility to protect the holder's rights in case of its infringement using technological solutions, up to blocking the token. It was argued, the combination of blockchain-based protocols with established ambitions of intellectual property policy yields hybrid conceptual-computational financial technologies (such as self-enforcing smart contracts attached to digital artifacts) that are unlikely to empower artists but which serve to financialise digital creative practices as a whole, curtailing the critical potential of the digital as an inherently dynamic and potentially uncommodifiable mode of production and artistic expression [17].

Also, tokenization of intellectual property is a "way of financing art" [4]. Attracting investment in the arts is possible through the direct granting of rights to use intellectual property by multiple token holders and for the results of intellectual activity, which exist as unique assets that can become a reason to attract investment by splitting the rights of the holders. We are talking about the split of the right between several right holders - token holders, which seems to be the most promising for the tokenization of the sphere in question. The economic effect of intellectual rights tokenization is the possibility to provide financing of creative industries in different ways.

Intellectual property assets, such as licenses for works, trademarks, patents, and royalties from music and media rights, typically have low liquidity and currently have almost no secondary market for trading. In this sense, the concept of tokenization of intellectual property involves splitting the right to the result of intellectual activity as such or fractionating the right to the result of intellectual activity itself, rather than the intellectual right. Attention should be paid to the popularity of tokenization due to intellectual property's inherent gravitation towards the registration system of rights confirmation [15].

The peculiarities of intellectual property tokenization in the case under consideration are connected with its initially intangible nature. The intangible nature of intellectual property as an object of civil turnover is very well combined with the speed and extraterritoriality of digital assets' turnover based on blockchain.

In contrast, asset-backed tokens can represent "exclusive and irreplaceable assets, such as collectibles" [6], and are therefore distinguishable from one another by design. Accordingly, each token will be unique, creating a digital scarcity because blockchain network participants know how many are in circulation and their distinctive features (unique characteristics).

We believe that the creation of the NFT makes it possible to see what the convergence of technologies for the creative industry is all about.

4 Conclusion

Copyright Perspectives in the Protection of Digital Art Rights: Applicable Mechanisms and Challenges

The object created with the help of technology, by applicable law, has the characteristics of the object of copyright. The objects of digital art have the attributes of protectability. Therefore, given the application of new technologies and their convergence in the creative industries, unique objects will be created that will still be subject to protection against unapproved copying. Also, we believe that the developer of AI technology software should be given the subjective exclusive right - to prohibit copying the network-generated object.

The convergence of digital technologies for the arts can go in several directions. In any case, a unique object with the signs of copyright protection is created. Its circulation can be most effectively established by tokenization. The two functions of the blockchain network are most evident in the tokenization of intellectual property: the ability to authenticate the origin of an art object or intellectual right and verify it in transactions.

That confirms our belief that the development of blockchain technology and smart contracts enables artists and other creators to protect their works from misuse and expropriation. A robust system for the management of digital art objects is being established. In doing so, the goal of limited art token releases is usually not to impose stricter restrictions on use or a new form of digital rights management but to create new types of tradable digital assets.

Acknowledgments. This research was supported by the Russian Foundation for Basic Research via grant No. 18-29-16145 MK "The Mechanism of legal regulation of relations using the Distributed Ledger Technology".

References

1. Buccafusco, C.J.: A theory of copyright authorship. Virginia Law Rev. **102**, 1229–1295 (2016)
2. Colson, R.: The Fundamentals of Digital Art. Bloomsbury Publishing (2007)
3. David, G., Huemer, C., Oosterlinck, K.: Art dealers' strategy: The case of Goupil, Boussod & Valadon from 1860 to 1914 (2009). https://cepr.org/active/publications/discussion_papers/dp.php?dpno=13941
4. De Filippi, P.: Blockchain-based crowdfunding: what impact on artistic production and art consumption? Observatório Itaú Cult. (2015). https://ssrn.com/abstract=2725373
5. Evans, T.M.: Cryptokitties, cryptography, and copyright. AIPLA Q. J. **47**(2), 219 (2019)
6. Hobbs, D.: Crypto collectibles: The non-fungible token craze (2018). https://www.twotalltotems.com/blog/index.php/2018/05/22/crypto-collectibles-the-nonfungible-token-craze/
7. Kharif, O.: CryptoKitties Mania Overwhelms Ethereum Network's Processing. Bloomberg (2017)
8. Kwastek, K.: Aesthetics of Interaction in Digital Art. Mit Press (2013)
9. Lieser, W., Baumgärtel, T., Dehlinger, H., Herzogenrath, W.: Digital art. Tandem (2009)
10. Michalska, J.: Blockchain: how the revolutionary technology could change the art world. The Art Newspaper 282 (2016). https://www.theartnewspaper.com/analysis/blockchain-how-the-revolutionary-technology-behind-bitcoin-could-change-the-art-market
11. Moreno, V.: Retos legales a los que se enfrentan las nuevas tecnologías (2019)
12. O'Dwyer, R.: Limited edition: producing artificial scarcity for digital art on the blockchain and its implications for the cultural industries. Convergence **26**(4), 874–894 (2020).https://doi.org/10.1177/1354856518795097
13. Pagani, M. (ed.): Multimedia and Interactive Digital TV: Managing the Opportunities Created by Digital Convergence: Managing the Opportunities Created by Digital Convergence. IGI Global (2000)
14. Paul, C.: Digital Art, vol. 224. Thames & Hudson, London (2008)
15. Polatay, V., Ovcharenko, A., Bondarenko, I., Gavkalova, N.: Crypto currencies in the system of international private monetary law. Some issues of virtual assets taxation and accounting in Ukraine. Rev. ESPACIOS **41**(8) (2020)
16. Tilson, D., Lyytinen, K., Sorensen, C.: Desperately seeking the infrastructure in IS research: conceptualization of "digital convergence" as co-evolution of social and technical infrastructures. In: 2010 43rd Hawaii International Conference on System Sciences, pp. 1–10, IEEE, January 2010

17. Zeilinger, M.: Digital art as 'monetised graphics': enforcing intellectual property on the blockchain. Philos. Technol. **31**(1), 15–41 (2018)
18. State Standard R 57721–2017. http://docs.cntd.ru/

Integration Mechanism for Improving Planning for Innovative Engineering Enterprises

Svetlana N. Kuznetsova(✉), Elena V. Romanovskaya, Natalia S. Andryashina, Elena P. Kozlova, and Maksim M. Kutepov

Minin Nizhny Novgorod State Pedagogical University, Nizhny Novgorod, Russia
kutepov_mm@mininuniver.ru

Abstract. In the article, the authors analyze and develop an integration mechanism for improving planning on the example of the innovative Nizhny Novgorod enterprise «Metintegro», the main activity of which is the production of silencers, resonators and exhaust pipes for exhaust systems of cars and trucks (GAS, ZIL, UAZ, PAZ).

Purpose: The authors of the article indicate that the main direction of improving the company's sales planning is to create a continuous operating system. LLC «Metintegro» is recommended to take advantage of sales planning model covering all sales management levels, expanding methods and planning methods, interconnections of forecasts for different periods.

Methodology: The authors determine that extrapolation of historical sales data is the main method that is currently used by an analyzed enterprise. However, this method does not take into account the features of current market conditions and the possibilities of the development of the enterprise. Thus, the analysis of the sales planning process of the enterprise LLC «Metintegro» must be improved.

Findings: An innovative approach to improving the system of internal plans of the studied engineering enterprise is to create a specialized division responsible for making and monitoring the implementation of plans.

Originality: This necessary measure determines the success of the planning system; however, the managers often still significantly underestimate it. It is also necessary to improve the planning system by implementing an improved monitoring system.

Keywords: Integration · Mechanism · Innovation · Planning · Sales · Forecast · Extrapolation · Economic efficiency

JEL Codes: L5 · O1 · N6

E. G. Popkova (Ed.): ISC 2020, LNNS 368, pp. 275–282, 2022.
https://doi.org/10.1007/978-3-030-93244-2_31

1 Introduction

Planning is a design of alternative options for future condition of enterprise in terms of rationality of achieving the set goals. It is planning that is characterized by optimal use of production capabilities. The purpose of the study is to consider planning on the example of enterprise LLC «Metintegro».

2 Methodology

An important stage in the formation of sales policy of an enterprise is planning of sales system. The company «Metintegro» conducts economic activity through custom system, that is, parts for trucks are produced to a specific order with specified functions and assigned amount [1].

Most of the products sold by the enterprise (95–97%) are silencers for trucks «Gazelle», let's assume that conditionally averaged unit of production is the silencer MAK 2705. There is the following data on the dynamics of sales volume of LLC «Metintegro» (Table 1):

Table 1. Dynamics of sales volume in 2017–2018 quarterly, pcs.

Period	1st quarter, 2018	2nd quarter, 2018	3rd quarter, 2018	4th quarter, 2018	1st quarter, 2019	2nd quarter, 2019	3rd quarter, 2019	4th quarter, 2019
Sales volume, pcs	573	621	578	565	579	641	603	581

Source Developed and compiled by the authors

From the given data it can be seen that sales peaks come for April-June, i.e. 2 blocks, and at the end of the year the sharp decreases demand and sales volume. The length of one cycle is a year, while in 2019 recorded a sharp increase in the total sales volume, and in 2020 its slowdown was observed.

The company LLC «Metintegro» predicts the volume of sales of the company for the coming year, based on an analysis of the trends in the values of the implementation of the implementation, by means of analytical alignment of the original series [2]. Planning is built according to the following algorithm:

The selection of the trend, the closest approximate empirical data. By deducting from the empirical values of the implementation of the value calculated on the trend formula, the values of the seasonal component are found. To calculate the parameters of the trend model, linear regression is used. We will make an auxiliary table to calculate the parameters (Table 2):

Table 2. Calculation of the parameters of the trend linear model

№	t	y	t^2	t*y
1	1	573	1	573
2	2	621	4	1242
3	3	578	9	1734
4	4	565	16	2260
5	5	579	25	2895
6	6	641	36	3846
7	7	603	49	4221
8	8	581	64	4648
Sum	36	4741	204	21419

Source Developed and compiled by the authors

Based on the obtained equation, we calculate the trend values (Table 3):

Table 3. Calculation of theoretical values of sales volumes, units

№	t	$Y_{factual}$	$Y_{theoretical}$
1	1	573	586
2	2	621	588
3	3	578	590
4	4	565	592
5	5	579	594
6	6	641	596
7	7	603	598
8	8	581	600

Source Developed and compiled by the authors

To determine the dimensions of the components, we will determine the differences between actual sales volumes and trend values (Table 4):

Table 4. Determination of difference between actual sales volumes and trend values

Period	Trend value	Sales volume in fact, pcs	Deviation
1st quarter, 2018	586	573	−13
2nd quarter, 2018	588	621	33
3rd quarter, 2018	590	578	−12
4th quarter, 2018	592	565	−27
1st quarter, 2018	594	579	−15
2nd quarter, 2019	596	641	45
3rd quarter, 2019	598	603	5
4th quarter, 2019	600	581	−19

Source Developed and compiled by the authors

To obtain the values of the components, we group the deviations by quarters and determine the average values (Table 5):

Table 5. Definition of scope components

Period	Deviation in 2017	Deviation in 2018	Average seasonal variation
1st quarter	−13	−15	−14
2nd quarter	33	45	39
3rd quarter	−12	5	−3
4th quarter	−27	−19	−23

Source Developed and compiled by the authors

Using this model, we calculate the final forecast sales figures for 2019 (Table 6):

Table 6. Calculation of the total projected indicators of sales volumes for 2019

Period	Forecast trend value	Average seasonal variation	Final forecast value
1st quarter, 2020	602	−14	588
2nd quarter, 2020	604	39	643
3rd quarter, 2020	606	−3	603
4th quarter, 2020	608	−23	585

Source Developed and compiled by the authors

As a result, the next sales and price plan for 2020 was formed quarterly (Table 7).

Table 7. Sales and price forecast

Sales and price forecasts	1st quarter	2nd quarter	3rd quarter	4th quarter
Expected volume (pcs.)	588	643	603	585
Expected unit price, rub	1020	1025	1030	1040

Source Developed and compiled by the authors

The sales forecast is adjusted by both production capacity and limited demand [3]. Based on the received sales volume forecast, the final sales plan and cash receipt schedule are generated (Table 8).

Table 8. Sales plan and money receipt schedule

Values	1st quarter	2nd quarter	3rd quarter	4th quarter	Per year
Expected volume, pcs	660	710	745	765	2880
Expected price, rub	1020	1025	1030	1040	
Expected sales revenue, thous. rub	599,8	659,1	621,1	608,4	2488,3
Debtors' accounts, beginning of year	76				76
Cash flow from sales, 1st quarter	527,8	72,0			599,8
Cash flow from sales, 2nd quarter		580,0	79,1		659,1
Cash flow from sales, 3rd quarter			546,6	74,5	621,1
Cash flow from sales, 4th quarter				535,4	535,4
Total cash flow	603,8	652,0	625,6	609,9	

Source Developed and compiled by the authors

At the enterprise, LLC «Metintegro» simultaneously with the preparation of a detailed sales plan, it is necessary to make a schedule for receiving funds using the sales forecast, prices for finished products and the coefficient of collection [4].

The balance of receivables from the initial balance (at the end of 2019) enters the 1st quarter.

The share of payment for money in a given quarter (coefficient of collection) responsible for the conclusion of contracts is 88% of the volume. Thus, in the first quarter it is planned to receive receivables at the beginning of the planned year (76 thousand rubles) and 88% of the 1 block revenue. In the 2nd quarter, it is planned to receive 12% of the 1 block revenue and 88% of the 2nd quarter revenue. In 3 and 4 quarters, calculations are made similarly. Based on research on demand for the production planned for the production, competition in the market, quality and prices, alleged sales volumes, a decision is made to issue products. The production budget is being developed, on the basis of which the budgets of the procurement and use of materials, labor and other expenses are developed [5].

Based on the policy of enterprise reserves management policies, the planned balance of finished products at the end of the period is defined as 12% of the volume planned for sale in the next period.

At the end of the year, stocks are planned in the amount of 74 pcs. The planned reserves at the end of the first period are planned reserves of the start of the second period. The need for materials is made taking into account the residues of warehouse stocks.

Estimate calculation is prepared by sales department (Table 9).

Table 9. Administrative and marketing costs budget

Values/Period	1st quarter	2nd quarter	3rd quarter	4th quarter	Per year
Expected sales volume, (pcs.)	588	643	603	585	2419
Variable costs per unit, thous. rub	0,01	0,01	0,01	0,01	
Estimated variable costs, thous. rub	5,88	6,43	6,03	5,85	24,19
Advertising, thous. rub	2	2	2	2	8
Marketing managers' salary, thous. rub	25	25	25	25	100
Insurance payments, thous. rub	2	2	2	2	8
Total fixed costs	29	29	29	29	116
Total planned costs	34,88	35,43	35,03	34,85	140,19

Source Developed and compiled by the authors

A profit and loss plan is calculated on the basis of the prepared estimates. Results of calculations are given in Table 10:

Table 10. Planned profit report for 2020

Values/Period	1st quarter	2nd quarter	3rd quarter	4th quarter	Per year
Expected sales revenue, thous. rub	599,8	659,1	621,1	608,4	2488,3
Production cost of sold products	538,9	589,4	552,7	536,2	2217,2
Administrative and marketing costs	34,88	35,43	35,03	34,85	140,2
Gross profit	25,9	34,3	33,4	37,4	130,9
Property tax	14	14	14	14	56,0
Taxable profit	11,9	20,3	19,4	23,4	74,9
Income tax	2,4	4,1	3,9	4,7	15,0
Net profit	9,5	16,2	15,5	18,7	60,0

Source Developed and compiled by the authors

3 Results

The analysis carried out in this work showed that the sales planning process at the LLC «Metintegro» under study has a number of significant shortcomings.

The main method currently used by the company under investigation is extrapolation on the basis of sales data for previous periods. However, this method does not take into account the peculiarities of the current market conditions, the opportunities for the development of the enterprise. Thus, the analysis of the sales planning process at LLC «Metintegro» needs to be improved.

The main direction of improving the sales planning of the enterprise is the creation of a continuous system. It is recommended LLC «Metintegro» use a sales planning model that will cover all parts of sales management, expand planning methods and tools, interconnect forecasts of different periods, etc.

Currently, the product sales planning system at LLC «Metintegro» is «second» and is only an additional function of several departments of the company - marketing, sales and accounting.

The main proposed innovation for the enterprise, which can improve the system of internal plans, is the creation of a specialized unit responsible for the preparation and monitoring of the implementation of the plans. This necessary measure determines the success of the planning system, but is often still greatly underestimated by managers [6–10].

The planning system should also be improved through the introduction of an improved monitoring system.

We will analyze the cost-effectiveness of the proposed measures. The application of all proposed measures to improve sales planning (creation of a continuous planning system, coverage of all links of sales management, improvement of methods and models used to calculate forecast indicators, introduction of a control system, etc.) will contribute to the growth of sales profitability by 0.5%, which will be: 1,4 + 0,5 = 1,9%.

All other things being equal, net profit will increase by: 2673 * 1,9% - 38 = 12,8 thousand rubles.

4 Conclusion

In this study, the sales planning system in LLC «MetIntegro» was analysed. Based on the enterprise activity data, a comprehensive enterprise planning system was analysed, which includes forecasting based on data from previous periods and market conditions, a sales plan, a production plan, a cost budget, a planned profit report, a cash budget.

The analysis of the sales forecasting process at LLC «Metintegro» also needs to be improved. The main method currently used by the company under investigation is extrapolation based on sales data from previous periods. It is recommended the company supplement the sales forecasting system with structural and intuitive methods [7].

References

1. Popkova, E.G.: Economic and Legal Foundations of Modern Russian Society. A volume in the series. In: Popkova, E.G. (ed.) Advances in Research on Russian Business and Management, Information Age Publishing, Charlotte, NC, USA (2017)
2. Vaganova, O.I., Smirnova, Z., Kozlova, O.P., Vishnevetskaya, N.A., Prostov, A.V.: Formation of professional competencies of future builders in technical school. IOP Conf. Ser. Mater. Sci. Eng. **828**(1), 012029 (2020)
3. Vaganova, O.I., Smirnova, Z.V., Vovk, E.V., Kapina, A.A., Chelnokova, E.A.: Comparative analysis of cloud technologies. Int. J. Adv. Trends Comput. Sci. Eng. **9**(3), 2563–2566 (2020)
4. Popkova, E.G., Sergi, B.S.: Will industry 4.0 and other innovations impact Russia's development? In: Sergi, B.S. (ed.) Exploring the Future of Russia's Economy and Markets: Towards Sustainable Economic Development, pp. 51–68. Emerald Publishing, Bingley, UK (2018)

5. Kuznetsov, V.P., Garina, E.P., Romanovskaya, E.V., Kuznetsova, S.N., Andryashina, N.S.: Organizational design and rationalization of production systems of a machine-building enterprise (by the example of the contract assembly workshop). Espacios. **39**(1), 25 (2018)
6. Romanovskaya, E.V., Garina, E.P., Andryashina, N.S., Kuznetsova, S.N., Garin, A.P.: The process of technological re-equipment planning of an enterprise in a complex industrial production. Lect. Notes Netw. Syst. **129**, 280–288 (2020)
7. Sergi, B.S., Scanlon, C.C. (eds.): Entrepreneurship and Development in the 21st Century. Emerald Publishing, Bingley, UK (2019)
8. Garina E.P., Kuznetsov V.P., Egorova A.O., Romanovskaya E.V., Garin A.P.: Practice in the application of the production system tools at the enterprise during mastering of new products. In: Contributions to Economics, pp. 105–112 (2017). № 9783319606958
9. Kuznetsova, S.N., Romanovskaya, E.V., Artemyeva, M.V., Andryashina, N.S., Egorova, A. O.: Advantages of residents of industrial parks (by the example of AVTOVAZ). Adv. Intell. Syst. Comput. **622**, 502–509 (2018)
10. Garina, E.P., Romanovskaya, E.V., Andryashina, N.S., Kuznetsov, V.P., Shpilevskaya, E. V.: Organizational and economic foundations of the management of the investment programs at the stage of their implementation. Lect. Notes Netw. Syst. **91**, 163–169 (2020)

Project Management Based on PMBOK 7.0

Valery V. Maslennikov, Elena V. Popova(✉), and Yury V. Lyandau

Plekhanov Russian University of Economics, Moscow, Russia
{Maslennikov.VV, Popova.EV, Lyandau.YUV}@rea.ru

Abstract. Project management is one of the topical issues in the world because issues of effective management arise in small and large companies. Some world market leaders create internal management systems according to the experience of the most promising projects, which are described in an unspoken uniform set of rules, but there can be projects that build an "ideal" approach based on their mistakes.

The article discusses the key features of project management based on the PMBOK 7.0 standard. PMBOK 7.0's comprehensive approach to project management is a kind of business value because these projects can be beneficial and remain the main means of increasing business value by achieving corporate goals, which are set by the company. It's important to note that PMBOK focused on large projects before version 6, while version 7 attempts to cover everything, including small projects. Project management is improving and moving to a new level, the 7th edition of PMBOK turns the usual management towards principles, but not processes.

Effective implementation of the project approach is possible through the use of the principles defined in the PMBOK 7 standard (responsible planning and management, team, stakeholders, value, holistic thinking, leadership, adaptation, quality, etc.). These principles meet the modern requirements of project management in commercial and government organizations.

Keywords: Project · System · Management · Team · Change · Standard

JEL Codes: O22 · O32

1 Introduction

Quite a few methods and technologies have been created for project management, such as Scrum, Agile, PMBOK and others. Every one of them differs from each other in its concept and direction, focus on certain aspects in the implementation of the project. One of the most popular project management techniques is the PMBOK-based technique.

PMBOK is a knowledge base on project management, which describes the mean of project management processes in terms of integration between processes and interactions between them, and also the purpose that they serve [3].

PMBOK is a publicly available published document that belongs to the category of reference literature and can be used in the practical implementation of the project. One of

E. G. Popkova (Ed.): ISC 2020, LNNS 368, pp. 283–289, 2022.
https://doi.org/10.1007/978-3-030-93244-2_32

the most important qualities of this method is that its concept and rules are applicable in a fairly large number of situations and categories of projects. The fact that generations come out with a certain frequency confirms that this document can be attributed to reference books of production and practical use, which are characterized by a systematized structure. At the moment, 7 editions have been published and the last 7th generation shows a completely non-classical project management system. The first edition was released in 1996 and had 176 pages, subsequent editions increased their content and were issued every 4 years and in 2017, the 6th edition had 726 pages, and has already released in the digital version of the document. The 7th edition is very different from the previous ones in its concept, as well as in the method of its publication and studying.

This scientific article will focus on the conceptually new directions adopted for implementation in the last edition and how the PMBOK system has changed in comparison with previous editions. The main task is to explore a new methodology for the project management system and identify conceptual differences in project management in comparison with previous editions.

The fundamental foundations of the scientific issues studied in this article are laid in the works of Burdukin [1], Cybulski [2], Polkovnikov and Dubovik [7], Temchina [9], and Verzukh [11].

The authors also used the materials from Guide to the Project Management Knowledge Base [4], PMI is responding to practitioners' needs for reliable professional tools [6], and The Standard for Earned Value Management [10] in this article.

2 Materials and Methods

The main difference between the 7th edition of the PMBOK and the previous 6th is that the concept with a focus on processes changes to the principles of working with the project. Now there are 12 principles of work and building projects at the head, which must be adhered to. Three main pairs of terms that aren't opposed to each other, but are often mentioned side by side, can also be noted, they cover those key differences and the path of transition from processes to principles (Fig. 1).

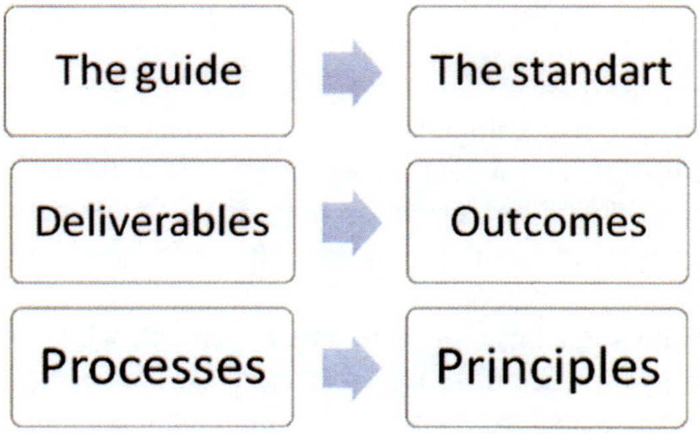

Fig. 1. Key changes. *Source* Review of the 6th edition of the Guide to the Project Management Knowledge Base [8]

The first pair of terms is the transition from the knowledge base (the guide) to the American national project management standard (the standard). It can be said that the project management standard was in the previous versions of PMBOK.

The second pair is two concepts that aren't opposed to each other but have significantly different meanings. There was a transition from a focus on the result to a focus on a valuable output (value creation) for all stakeholders of the project. This carries some of the value that the project creates. This isn't just a completed project, but responsibility for the result.

And, finally, the third pair is a change in the approach to project implementation. The process-oriented approach is being replaced by a principles-based approach. Before that, some processes rigidly described the predictive life cycle of a project. The new edition offers 12 principles that cover all project life cycles and can be applied to a wide variety of types of projects using any tool and framework. The shift from processes to principles is the creation of a wider and more universal information area for the implementation of projects.

After performing a semantic analysis of PMBOK 7, one of the most common words in the document is the word "command" can be seen. So, the new edition of the project management system departs from the concept about one leader project, which is also the key factor for success, and brings to the point that now the key factor is the team and collective responsibility for the result.

This standard consists of the following parts:

- Introduction - why such conceptual shifts have occurred;
- Value delivery system - a chain of value creation for all stakeholders and the parent organization.

The value chain is a cycle that begins with a strategy, moves on to the goals of the project, then to initiatives, implemented initiatives, then to the competencies of the company (project), results, project benefits, and, finally, to the value obtained at the output (Fig. 2).

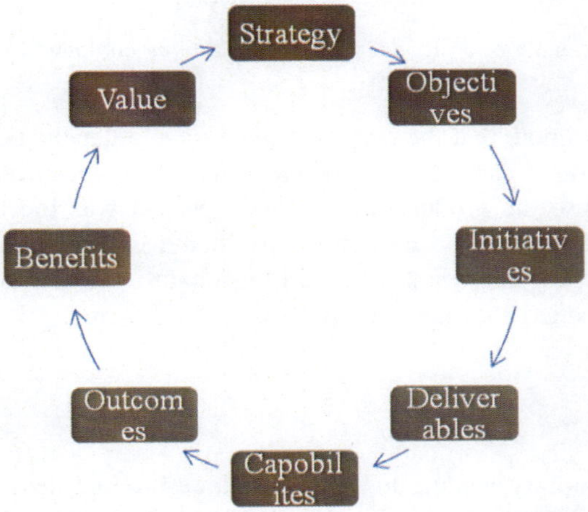

Fig. 2. Value delivery system. *Source* Review of the 6th edition of the Guide to the Project Management Knowledge Base [8]

The main tension in this chain is in the area of deliverables and outcomes, where dreams and costs turn into costs and profits. It's suggested that the result of the project isn't the final result in this standard, but what it was all created for. Project managers should understand how much the project life cycle is increasing, as well as the fact that the boundaries of responsibility are expanding to the consequences of the result created, and the profit that the company will receive. It also needs to be understood how the criteria for project success are changing.

If, before the present moment, the criteria for the success of the project fit into the successful observance of the constraints (deadlines, negotiated aspects of implementation, etc.), but the boundaries of the project's success are Outcomes now.

Let's consider a company focused on innovation for a clearer example (Fig. 3).

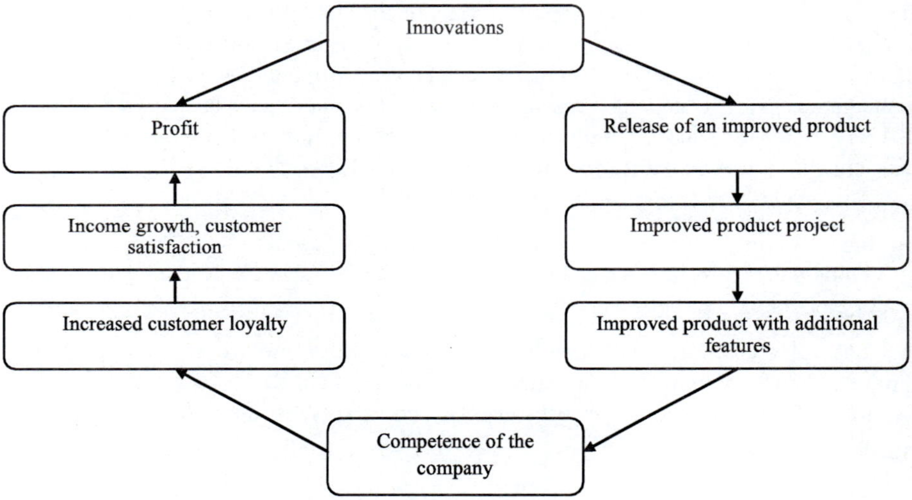

Fig. 3. Example of a product value chain. *Source* compiled by the authors

It's important to note that the next step in this cycle depends on the previous one. For example, a project can't be able to create an improved product with additional functionality because the product improvement project was too expensive to implement. Or this success may be one-time, due to the company doesn't have enough competencies to repeat this result. Value in this chain is the net tangible or intangible result of realized benefits minus the cost of obtaining them.

3 Results

A project is a temporary venture to create a unique product or result [5]. In turn, the results of the project support the achievement of benefits that may not be fully realized upon completion of the project. Therefore, project teams work with business operations or program managers to ensure that the project results in the desired outputs that will lead to the realization of benefits and value.

The term management can be incompatible with the environment, goals and results of some types of projects because this word means "routine by algorithm". Therefore, this standard uses the term project delivery for including all types of project environments.

In the new edition of the PMBOK, there is no such thing as a project leader, as it is believed that this role may be occupied by the Project lead. This decision was made because a project leader isn't necessary, and, according to the new concept, the team is still put in the first place. In addition, the project lead can combine its role with the Facilitator/coach, who is a kind of "leader" at conventions, meetings, and is also engaged in the development of competencies. The rest of the roles in the team - team member, customer, business representative, domain subject matter expert, sponsor - remained unchanged, as well as the definition and types of environments.

The main change in the new edition is precisely the emergence of principles. Their important difference from the processes, which were described earlier, is that the processes are clear guidelines written for certain situations, and it tells the employee what exactly needs to be done. Principles tell what to do, describe behaviour and thinking, and answer the question "why?".

Let's consider them in more detail.

1. Responsible planning and management (honest relationship with the Customer, the interest of the project team primarily as a result, and not in fulfilling the terms of the contract, focus on long-term cooperation with the Customer).
2. Team (the human factor is a key factor in the success of the project and at the same time the key risk of obtaining a negative result. The motivation of team members, aimed at creating the team's interest in achieving the result, and not a constant requirement according to job descriptions).
3. Interested parties (taking into account the personal characteristics of people, the characteristics of all project participants, the ability to negotiate, and not refer to the clauses of the contract).
4. Value (understanding the wishes of the Customer. Superficially, the Customer's requirements may not be what he wants. The Customer needs help to understand what he wants to get from the project).
5. Holistic thinking (solving the Customer's problem, but not focusing on individual tasks or performing certain functions, clarifying the wishes of the Customer, and not just fulfilling the points of the technical task).
6. Leadership (leadership should be shown not only by the project manager but also by all members of the project team, all members of the project team should show initiative, motivate other participants, act as mentors).
7. Adaptation (it's fundamentally wrong decision to drive any projects into the rigid framework of a specific methodology, each project is unique, so it's important to implement it, taking into account the characteristics, requirements and wishes of the Customer, and not try to replicate the previous project results).
8. Quality (the result of the project must be of high quality, i.e., the goal of the Customer must be achieved, all project processes must be performed efficiently, the

project must use high-quality resources from which the desired result can be obtained, but not the projects that can be bought at the lowest price).

9. Complexity (it's necessary to be able to effectively manage the complexity of the project, analyze intermediate results, collect all information on the project, and take into account the characteristics of all project participants).
10. Opportunities and threats (identification of opportunities in the implementation of the project, identification of threats that may negatively affect the achievement of the required result, to determine opportunities and threats, the correlation SWOT analysis, as well as risk management tools - identification of risks and their gradation ("red", "yellow", "green" zones), development of measures aimed at minimizing threats can be used).
11. Adaptability and viability (the ability to promptly and adequately respond to changes in the external environment, the foresight of the project's subject area, as well as factors affecting the implementation of the project, and the result of the project, and also making the right decisions as a result of the impact of environmental factors).
12. Change management (involving all stakeholders in the implementation of the project, agreeing with the Customer the intermediate results of the project, but not just presenting the final result at the final stage of the project, adapting the project based on the intermediate results of it, and also environmental factors affecting the project).

It's important not only to perceive these principles as some theoretical foundations of modern project management, but to apply them in the implementation of projects of any orientation, complexity, regardless of the status, legal form, and sphere of activity of the organization or the team implementing the project. These principles should be "embedded" in the design activities of any project team.

4 Conclusion

In the new edition, major conceptual changes that have been analyzed in this article can be seen. They also can be divided into three conditional groups:

1. The primacy of the standard over the set of values and tools.
2. The supremacy of 12 principles over 40 + processes.
3. The primacy of the value of the project outputs for stakeholders over the delivered results.

As a result, the new edition of the project management system made a breakthrough, while catching the current trends. Transitioning project management from standardized processes to principles will take time and will be challenging due to major changes in the management concept, but it will provide an opportunity to improve production efficiency through the use of a more flexible system in the new edition than in previous editions.

PMBOK 7 is a new flexible control system that has been redesigned for modern concepts, and its rapidly changing world. Due to the introduction of various digital technologies and artificial intelligence in the company, this concept of the control system will fit very well. The main difference, as was written above, is the replacement of standardized processes with 12 principles, which just give the very same flexibility and some freedom in making managerial decisions.

Among other things, a whole platform with various courses is being prepared for learning new functions and features of the new edition, which will help to quickly teach employees and company managers to new principles. This function greatly simplifies the implementation of a new management system in the company.

PMBOK 7 can be called a completely new project management system. Although it retained its old appearance but filling, concept and philosophy have changed. And all of this suggests that it's necessary to move away from the standardized control systems of the "old world" and move to more flexible ones for the introduction of modern technologies.

References

1. Burdukin, A.: The Scrum Guide via PMBoK: A PMP View. (2020). https://vc.ru/hr/125686-scrum-guide-cherez-pmbok-vzglyad-pmp. Accessed 26 Feb 2021
2. Cybulski, R.: PMP PMBOK Study Guide! PMP Exam Prep! Practice Questions Edition! Crash Course & Master Test Prep to Help You Pass the Exam Publishing House: House of Lords LLC (2020)
3. Grinn, D., Stellman, A.; Professional project management in examples, tasks, illustrations. SPb: Professional Literature, 3rd ed. (2020)
4. Guide to the Project Management Knowledge Base: PMBOKR + AGILE manual. A practical guide. Olymp-Business, Moscow (2019)
5. Newton, R.: Project Management from A to Z. Alpina Publisher, Moscow (2019)
6. PMI is responding to practitioners' needs for reliable professional tools. Here's what you'll see in our evolving standards (2021). Project Management Institute PMI. https://www.pmi.org/pmbok-guide-standards. Accessed 20 Feb 2021
7. Polkovnikov, A.V., Dubovik, M.F.: Project Management. Complete MBA Course. Olymp-Business, Moscow (2018)
8. Review of the 6th edition of the Guide to the Project Management Knowledge Base. Project Services (2017). https://www.pmservices.ru/project-management-news/obzor-shestogo-izdaniya-rukovodstva-k-svodu-znanij-po-upravleniyu-proektami/. Accessed 20 Feb 2021
9. Temchina, M.: 7th PMBoK – is the end of classic project management? (2020). https://infostart.ru/1c/articles/1184273/. Accessed 26 Feb 2021
10. The Standard for Earned Value Management: Project Management Institute PMI (2019). https://www.pmi.org/pmbok-guide-standards/foundational/earned-value-management. Accessed 26 Feb 2021
11. Verzukh, E.: Project Management. A Crash Course in the MBA Program. Dialectic-Williams, Saint Petersburg (2019)

Determining the Success Potential of Innovative Product Solutions Using the Example of Network Product Development

Marat R. Usmanov(✉)

LINK LLC, Nizhny Novgorod, Russia

Abstract. The purpose of this article is to systematize and summarize the experience of introducing and developing innovative product solutions, R&D systems using the example of network product development. The methodology proposed by the author is determined by the following provisions. A critical flaw in the project is identified and eliminated at an early stage of development, or the development of product solutions is terminated. The process of creating innovative products is seen in the delegation of R&D to specialized companies, vendor stakeholders, contributing to the development of network forms of product development, in particular, the Russian engineering market. Interpretation of the results of the implementation of innovative product solutions using the example of the research Object showed that the technology for achieving the indicators is realizable, the calculation results are comparable. The methodology is based on the principles of cost minimization, optimal number of functions, economic interchangeability, least variety. The author proposes to use indicators to assess the potential: technical feasibility; cost index; execution index; economic indices: economic aspects of costs, complexity of production, assembly, investment costs, etc., comparability index. The application of the proposed methodology is illustrated by the example of the development strategy of a business unit of Russian enterprise in the oil and gas processing and petrochemical industry. The application of the methodology makes it possible to assess the potential for success of innovative product solutions in the aspect of the development of network forms of business.

Keywords: Product solutions · Engineering · Network forms · Product development

JEL Code: O32

1 Introduction

The competitiveness of manufacturers in the markets for high-tech products is ensured primarily not through cost-cutting and savings programs, but through product innovations, marketing of new products. Closely related terms are often used in product development: «inventions» (research, R&D), «innovation», «new product development» (NPD process), etc.

E. G. Popkova (Ed.): ISC 2020, LNNS 368, pp. 290–298, 2022.
https://doi.org/10.1007/978-3-030-93244-2_33

Inventions are ideas for a new product, the result of research. According to the Frascati manual [1], the main criterion for determining R&D is the presence of a noticeable element of novelty and the resolution of scientific and / or technological uncertainty in the production of new materials, products, devices, in the formation of new PD-processes, systems, or a significant improvement in existing ones. Certain activities, such as equipment setup, process development, design and prototyping, may also contain a significant R&D element without starting a new production or redesigning an existing manufacturing process. This makes it difficult to accurately determine what should or should not be defined as R&D, how the concepts of «R» (research) and «D» (development) are integrated and delineated, where R&D ends and NPD begins. The early stages of product conceptualization often involve a significant amount of R&D in theoretical models of the NPD process [2]. There is also empirical overlap between applied research and NPD [3]. Separation of R&NPD categories is suggested by Soderquist, Godener [4]:

- «research» refers to the creation of new knowledge and technologies applied in the creation of new products, integrating the results of product development (research activities) with operational development work.
- «new product development» refers to the process of creating products: the stages of conceptualization, product planning, product design, process design, and production ramp-up. This process includes an intense iterative stream of progressively evolving specifications, prototypes and layouts, accompanied by intense communication and coordination of the players involved.

The key element of the term «innovation» is novelty associated with better performance of tasks and relative measure of superiority over competitive products. This is in terms of: obtained competitive advantage, achieved market share, the relative advantage of the product from the point of view of the consumer, development costs and development time. The definitions of product innovation also imply that the product must not only be new, but must be successfully introduced to the market.

Typologization and systematization of innovative product solutions are solved in different ways. Thus, product innovations can be grouped by types of innovations according to their content and related functional and design areas (Table 1).

Table 1. Typology of innovations

Innovation type	Content of innovation
1. Basic innovation	• new products in a new market
2. Improving of innovations	• change of use
3. Obvious innovation	• creating obvious novelty, without real improvement
4. Novelty for consumers	• product is new to buyers
5. Manufacturer innovation	• product is new to manufacturers
6. Innovation that transforms needs	• product design has been radically changed to meet new needs
7. Innovative products	• Product change, • Price change, packaging change, • Product differentiation, • Main product modification, • Diversification

Source [4]

The term «technical innovation» means both product innovation and process or technological innovation.

The purpose of this article is to study the existing scientific groundwork, systematize and summarize the experience of introducing and developing innovative product solutions, R&D systems by the engineering center of enterprise in the oil and gas processing and petrochemical industry. The task of the research is to form a vision of the current state of the issue.

2 Methodology

The analysis methodology is based on diagnosing the current success potential of innovative product solutions formed in the context of distributed product development, using the example of network interaction.

Object of research: LLC LUKOIL-Nizhegorodniinefteproekt, R&D Design Institute (abbreviated as L-ESC), whose task is to consolidate corporate experience in R&D, in the formation of best practices on topical issues of development and improvement of technologies, etc.

The PEST analysis of the factors influencing the development of technical innovations in the petrochemical industry in the 2010s is shown in Table 2.

Table 2. PEST analysis of factors influencing the development of technical innovations in the petrochemical industry in the 2010s*

№	Factor name	Influence of factor	Expert assessment of influencing factors	Weight-corrected score	Industry change
1	Level of innovation and technological development of the industry	3	5	0,18	Industry digitalization is expected to rise sharply
2	Research and development costs	3	5	0,18	R&D is a growth driver in the industry
3	Technological equipment of the industry	3	5	0,18	High
4	Development and penetration of the Internet, development of mobile devices	3	5	0,18	High
5	Access to the latest technology	3	3	0,11	Engineering is the leader of the economy
6	Degree of use, implementation and transfer of technologies	3	3	0,11	High

* *The degree of influence of each factor on the sales and profit of the company was assessed on a 3-point scale, where: 1 point - the influence of the factor is small, any change in the factor practically does not affect the activities of the company, 2 points - only a significant change in the factor affects the sales and profit of the company, 3 points - the influence of the factor is high, any fluctuations cause significant changes in sales and profits of the company*

Source Interview with NIINP employees, experts; BCG analysis. Systematized by the author

Another classification feature of product solutions is the degree of novelty (innovativeness). The novelty of innovation fluctuates between the extremes: easy (gradual) or fundamental (radical). The criteria for including innovation between the extreme poles are based on how many elements of the product have been recycled and how large the impact of new scientific and technological discoveries is. Newly created product variants in which innovations are offered through a variation effect are also considered innovations. However, its innovative value can be quite minimal. In this regard, it is often a statement that is often found in the literature: changes in design are perceived by buyers as innovative.

«Innovation process» is a collective term for an innovative event that occurs in stages. The prerequisites for the success of the innovation process are: in research and development: knowledge, in implementation: technical know-how, applied approaches, resource availability, [5], network organization of R&D processes [6].

The stages of the innovation process are presented in Table 3 using the example of the research Object.

The process of creating innovative products must be verifiable at every stage and therefore manageable. Several models of the innovation process are contrasted in the literature, assessment procedures are carried out at separate stages and in separate areas. In particular, the procedures for assessing the degree of innovativeness of product solutions are presented [2, 7–9]:

1) The technical and economic assessment of the solution in accordance with the VDI 2225 directive, proceeding «from the opposite», assesses the proportion of standardized components in the product (degree of implementation).
2) A usefulness analysis (NWA) of systems engineering is a formalized procedure for evaluating the utility of a proposal based on the perceived value and purpose defined by decision makers. The analysis is based on the specific goals of the company and the project and consists of the following stages: 1. Creating a system of goals, 2. Weighing the evaluation criteria, 3. Creating cost functions and cost tables, 4. Identifying and evaluating alternatives, 5. Calculating utility and determining the rank, 6. Sensitivity analysis.
3) Pairwise comparison and ranking procedures - in the sense of criteria (... is better than ...) or by weighting in which the criteria should initially be ranked according to their importance.
4) Failure modes and effects analysis, FMEA. The main goal of the FMEA is to reduce and prevent risks through the systematic detection of possible errors, as well as the assessment and assessment of the associated risks. This should potentially identify potential defects during the product creation phase, analyze their causes and consequences (consequences of failure) and thus prevent the actual occurrence through countermeasures.
5) QFD method «Quality Function Deployment» - planning and development of product quality functions in accordance with the quality characteristics required by consumers. This aims at early, comprehensive and structured clarification of the problem in preparation for finding a solution.

Table 3. Stages of organizing work on R&D projects and development of innovative product solutions *

Stages	Key variables
Organization of work on R&D projects	A project-based approach to work is applied: planning, provision of resources, control of progress R&D projects are prioritized and resources are allocated accordingly Qualified performers are allocated for complex tasks Records of employees' working time and assignment of it to projects are kept
Evaluation of developed solutions in terms of efficiency	The main development parameters affecting efficiency are discussed with the customer A critical flaw in the project is identified and eliminated at an early stage of development. If it is impossible to eliminate a critical flaw, development is terminated Several options for technical solutions are proposed, efficiency is assessed Regular (at least once every 3 months) meetings with the customer are held to assess the development status and discuss options
Provision of work with laboratory and analytical equipment, IT	Laboratory and pilot equipment is available to conduct research on a wide range of refining processes Analytical equipment for a wide range of works has been installed Modern IT tools are used to support research
Availability of KPI system and bonus system associated with it	Balanced KPI system, including qualitative and quantitative KPIs, approved for employees Target KPI values, rules for recalculating KPI levels into premiums are set The consequences of non-fulfillment of KPIs are regulated
Organization of interdisciplinary interaction	Interdisciplinary teams are created to work on projects Mutual information about plans and their adjustments is carried out External experts are involved (if necessary) Organization of interaction aimed at decision-making
Development and advanced training of employees engaged in R&D *	Each employee has an individual development plan There is access to patent and other databases, to search engines in databases Employees participate in conferences and seminars on relevant topics

* *Business development is mainly determined by the willingness and ability of its participants to innovate and the willingness to apply innovative methods. The high level of innovations in the company is mainly due to its employees, employee-centered leadership style and the climate in social relations between individual members of the company*

Source Compiled by the author

The determination of the advantage of a product innovation is performed in comparison with other product innovations or the predecessor model in terms of indicators [10–12]:

1) technical significance (technical feasibility);
2) market comparison index: target requirements for a product must not be lower than competing products in order to be attractive to the market;
3) cost index;
4) fulfillment index: there is a risk of potential non-fulfillment of customer wishes;
5) economic index: economic aspects of costs, complexity of production, assembly, investment costs, etc. - must be realistic, taken into account during the product development phase.
6) ideality index: takes into account how necessary additional functions are to achieve the goal. The ideality of the F&E solution is defined as the quotient of the sum of all useful functions by the sum of all harmful functions. Useful functions are understood as all functions that support a common function, harmful functions that affect the performance of a common function. Ideally, this is the greater, the larger the coefficient, because harmful functions consume unnecessary resources and entail additional costs.

The paths to the ideal solution can be described through:

1. Focused overview of opportunities:
2. Applying of optimal principles of creating a new product: principle of minimizing costs, principle of the optimal number of functions, principle of economic interchangeability, principle of least diversity (principle of standardization).

It is necessary to identify key variables (indicators) to determine the potential for success and the degree of innovation.

3 Results

Interpretation of the results of the implementation of innovative product solutions using the example of the Object of study showed the following:

1. Evaluation results, success potential and degree of innovation are relatively high compared to other pilot results. For comparison, using the example of the Object of research: 1) R&D projects are implemented with an economic effect, with an innovative component. The value of an innovative project is defined as the EMV of the project in the specified macroeconomic indicators. Where EMV is the expected value: NPV of the project, taking into account the probabilities of both development success and development termination at different stages; 2) Production optimization projects (increasing productivity, product quality, reducing costs, optimizing repairs, etc.) are implemented with an economic effect, but without an innovative component; 3) Projects for technical support of production are implemented without an obvious economic effect, however, they are necessary for the normal operation of production and compliance with legal requirements. There is no accepted quantitative performance indicator for projects. For prioritization, the «importance – urgency» matrix is used based on expert judgment.
2. The technology for achieving indicators is realizable. The value of an R&D project is the main quantitative indicator that characterizes the value it creates. The

mechanism for determining the value of project in innovation portfolio based on probabilities of passing milestones and estimated NPV of final scenarios (million rubles for NPV, % for the probabilities of passing the milestones) is shown in Fig. 1.

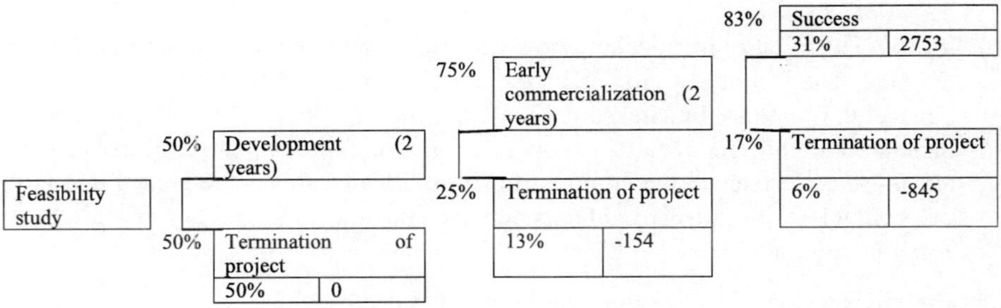

Fig. 1. Mechanism for determining the value of project in innovation portfolio based on probabilities of passing milestones and estimated NPV of final scenarios (million rubles for NPV, % for the probabilities of passing the milestones). *Source* Compiled by the author

where:

1. The aggregate of the values of the projects in the portfolio (adjusted for project connectivity) constitutes the total value of the company's R&D portfolio.
2. Quantitative indicators for creating value have been introduced into the KPIs of the process participants (Table 4):

Table 4. KPIs of process participants*

Science and technology center KPIs	Weight, %	Purpose
Increasing the value of the R&D portfolio	45	Contribution to the increase in the value of the company
Increasing the coefficient of project promotion	15	Quality and timing of projects and milestones
Efficiency of solving problems on the current operation of production	20	Providing prompt and high-quality technical support to production Deadlines
Enforcement of regulatory requirements	20	No violations

* *The weight of indicators characterizing success in the field of R&D is 60%. Core KPI: increasing the value of the R&D portfolio*

Source Compiled by the author

3. Customers' requirements for the product are taken into account. Customer relationships are assessed by sub-parameters:
   - the presence of formalized relations. The practice includes: a) a clear definition of the expected results (composition of documents, costs, terms, quality, penalties...), b) formulated obligations of the parties (for example, terms of approval of documents, composition of initial data, relations with local authorities ...), c) formalization of procedures for organizing an interface with a client/contractor within a project, d) availability of a convenient system for the exchange of project documentation that allows tracking the status of consideration/development;
   - maintaining customer loyalty. The practice includes: a) allocating office space for customer representatives, b) organizing regular meetings (at least once a month) to review the progress of the project based on performance indicators, c) approving a team of executors at the customer;
   - organization of control over the progress of work. The practice includes: a) clearly defining KPIs for assessing the performance of the unit, b) organizing regular meetings (at least once a month) to review the progress of the project based on performance indicators, c) having a system for collecting feedback from customers, d) having a reporting system by the status of development of a separate document (specification, drawing, etc.).
4. The calculation results are comparable.

4 Conclusion

Determining the potential for success and the degree of innovativeness of product ideas from the standpoint of the value created is the result of research. The work touches on three main points. First, the research area of product innovation was studied in accordance with the capabilities and assessment procedures. It was determined that there is a problem of taking full account of product innovations, such as novelty and extended benefits for customers and manufacturers. Second, since innovation is associated in its definition with «new» and «successful», «measurability of innovation» means quantifying the size of these innovative elements. Several methods for analyzing customer requirements for a product and their inclusion in a product are suitable as a basis for this procedure. Third, the results allow us to conclude that the widespread use of innovations contributes to an increase in the efficiency and effectiveness of the development of companies, in particular, from a scientific and technical point of view. However, it is necessary to define, weigh individual projects and include them in additional criteria, such as company strategy, core competencies, market trends, etc., for a reasoned overall entrepreneurial decision.

References

1. OECD: The measurement of scientific and technological activities, proposed standard practice for surveys of research and experimental development, Frascati Manual, (1994). https://www.oecd.org/sti/inno/Frascati-1963.pdf. Accessed 20 Apr 2021
2. Cooper, R.G., Kleinschmidt, E.J.: Benchmarking the firm's critical success factors in new product development. J. Prod. Innov. Manag. **12**(6), 374–391 (1995)
3. Hauser, J.R., Zettelmeyer, F.: Metrics to evaluate R. D&E, Res. Technol. Manag. **40**(4), 32–38 (1997)
4. Soderquist, K.E., Godener, A.: Performance measurement in R&D and new product development: setting the scene. Int. J. Bus. Perf. Manag. **6**(2), 107–132 (2004)
5. Batkovsky, A.M., Batkovsky, M.A., Kravchuk, P.V.: Economic efficiency of the implementation of R&D results in an innovative-active industry sector. Act. Prob. Mod. Econ. **1**, 375–385 (2020)
6. Fomin, E.P., Fomina, N.E., Akchurina, A.R.: Network organization of industrial R&D processes: innovative entrepreneurship. Econ. Sci. No. **150**, 63–66 (2017)
7. Meyersdorf, D., Dori, D.: System modelling of the R&D domain through the object– process methodology: a practical tool to help R&D satisfy its customers' needs. R&D Manag. **27**(4), 333–344 (1997)
8. Soderquist, K.E., Nellore, R.: Information systems in fast cycle development: identifying user needs in integrated automotive component development. R&D Manag. **30**(3), 199–211 (2000)
9. Tokareva, S.A., Remishevskaya, K.V., Zakharov, D.Y.: Influence of the R&D management system on the value chain of oil and gas companies. Vestnik USPTU. Sci. Educ. Econ. Ser. Econ. **1** (27), 68–73 (2019)
10. Tolstykh, T.O., Shkarupeta, E.V., Malkova, T.B., Alpeeva, E.A., Garina, E.P.: Algorithm for assessing the efficiency of innovational technologies of industrial enterprises. In: Popkova, E.G. (ed.) Growth Poles of the Global Economy: Emergence, Changes and Future Perspectives. LNNS, vol. 73, pp. 463–471. Springer, Cham (2020). https://doi.org/10.1007/978-3-030-15160-7_47
11. Potashnik, Y.S., et al.: Determining the value of own investment capital of industrial enterprises. Adv. Intell. Syst. Comput. **622**, 170–178 (2018)
12. Garin, A.P., Garina, E.P., Batsyna, Y.V.: Technical re-equipment as a condition for ensuring the competitiveness of the enterprise. Moscow Econ. J. **12**, 57 (2020)

Regional Differentiation of the Relationship Between E-commerce and the Ability of Transport Infrastructure Facilities to Meet the Needs of Buyers

Lyudmila V. Oveshnikova[1(✉)], Elena V. Sibirskaya[1], Galina D. Slepneva[2], Svetlana N. Bukhantseva[3], and Olga V. Ledneva[4]

[1] Plekhanov Russian University of Economics, Moscow, Russia
{Oveshnikova.LV, Sibirskaya.EV}@rea.ru
[2] Nestle Rossiya LLC, Moscow, Russia
[3] Russian Presidential Academy of National Economy and Public Administration, Moscow, Russia
[4] Moscow Financial and Industrial University «Synergy», Moscow, Russia
oledneva@synergy.ru

Abstract. The target of the study is to determine the relationship between the availability of information technology, the status of the transport infrastructure and the e-commerce development in the Russian regions. The regional differentiation of the dependence of e-commerce on the capabilities of the transport infrastructure and the level of digitalization in the regions of Russia were revealed by authors. The following quantitative indicators were selected for the analysis: dynamics of the total retail trade turnover, freight turnover of rail and road transport, dynamics of the e-commerce volume, the digitalization level of the Russian Federation population. Noteworthy that all the studied indicators, with the exception of road freight turnover, showed a positive growth trend, which indicates an economic growth in the research field. It has been established that the development of e-commerce is facilitated by high level of digitalization and development of transport infrastructure. In this regard, it is advisable to monitor and adjust indicators reflecting the prospects for the compliance of electronic and transport services with regional needs in order to smooth out regional inequality in the e-commerce development in the constituent entities of the Russian Federation. A statistically direct significant relationship has been proven between cargo turnover and the volume of offline and online retail trade in the Russian Federation, which indicates the need for constant updating of transport systems, vehicles and modernization of existing logistics processes.

Keywords: Regions · E-commerce · Transport · Freight turnover · Infrastructure · Differentiation

JEL Codes: R01 · F22 · F63 · J15 · J61 · O15

E. G. Popkova (Ed.): ISC 2020, LNNS 368, pp. 299–307, 2022.
https://doi.org/10.1007/978-3-030-93244-2_34

1 Introduction

Active dynamics of freight turnover is one of the principal factors of economic development between territories. In addition, the processes of globalization and international integration in the world become possible taking into account the use of various objects of transport infrastructure [1]. The population of the whole world is increasingly forced to adapt to the need to use the Internet resource not only with the help of a computer, but also with a smartphone on a daily basis, which is the major factor in the e-commerce growth.

The development of e-commerce determines the increase in the need for information technology. The complexity of determining the mutual e-commerce influence and the ability of transport system opportunities to overcome the needs of buyers is due to the multifaceted nature of these concepts, as well as the presence of various factors that must be mentioned in the analysis.

2 Methodology

The methodological basis of the research consists of scientific and statistical methods: the scientific generalizations method, the grouping analysis, the procedure of correlation analysis, the method of linear regression modeling, the method of rating. The information base of the study included official statistical reports used with the Federal State Statistics Service (Rosstat, 2014 - 2020), the Ministry of Economic Development of the Russian Federation, as well as calculated indicators based on these data and information obtained from periodicals and resources of the global Internet. The analysis result of these methods was carried out using such programs as MS Excel, Statistika.

3 Results

E-commerce is one of the most significantly developing market segments with a 10.9% share in the total retail turnover in Russia [2]. The rise in the cost of shipping goods from China, as well as the willingness of Russians to spend more money after a period of savings in 2020, could be a driver for the recovery of retail trade in Russia in 2021 [6]. The historical maximum change in the total retail trade turnover was recorded in 2019 compared to 2018 – the goods turnover volume is increased by 6, 5% and amounted to 33 643,3 million rubles in 2019 in Russia [10]. The lowest increase in the retail trade turnover volume is recorded at the level of 28 240,9 million rubles in 2016, which is 2, 6% points lower than in 2015, and 5, 1% points lower than in 2017 (Fig. 1).

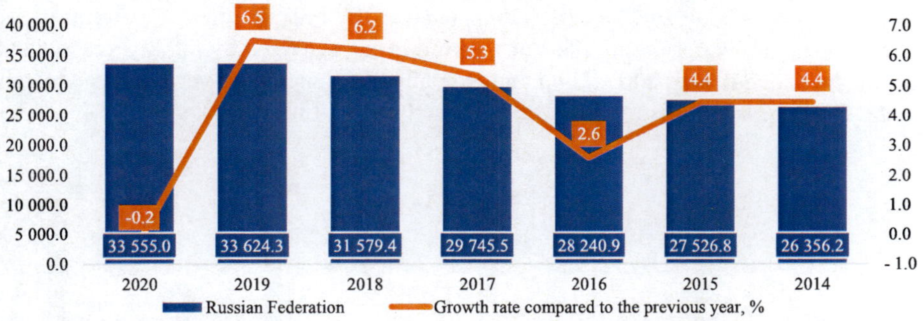

Fig. 1. Dynamics of the total retail trade turnover in the period from 2014 to 2020 in the Russian Federation, million rubles. *Source* Developed and compiled by the authors

The indicators named above indicate a positive impact of the provision of online services to the population on the company's turnover.

The Russian Federation online delivery market is demonstrated by the indicator of freight turnover. Freight turnover statistics are presented only by two modes of transport: rail and road, but it is also relevant to consider the transportation of goods by air. The need to monitor freight turnover by air is due, for example, to the statistically approved reality that the bulk of online store orders are delivered by air. According to Wildberries, the largest marketplace in Russia with a turnover of 100 billion rubles only among shoes and clothing, the audience of which is over 20 million people, about 1.5 million orders are placed on average per day during peak periods - this is almost 1,5% of the total population of the country.

Thus, the company receives a number of orders by volume equal to the total Russian Federation population on average every three months. This is the turnover of only one company, and today there are TOP-10 marketplaces in Russia. We get a huge turnover, in which almost every Russian is involved, through simple mathematical calculations. And the bulk of Wildberries' orders are delivered by air. Thus, statistics on the delivery of goods by air is necessary for a full understanding of the picture of Russian trade [3].

The statistics of freight turnover by rail and road transport are presented by the histograms below. The total delivery of goods by road and rail in 2019 amounted to 3 130,0 million tons of cargo, which is 17, 6% points less than in 2014 (3 778,2 million tons), the decrease was due to a decrease in the growth rate of freight turnover by road transport by 27, 9% points from 2014 to 2019, and notable rise in the growth rate of freight turnover by rail transport by 1, 7% point [5]. More than 50% of cargo was delivered by road, which is the reason for the overall decrease in freight turnover by the two types of transport studied.

Delivery of goods by rail was strongly prevalent in the Siberian Federal district in comparison with other federal districts for the entire study period, 457,9 million tons of goods were delivered by trains in 2019, which is 11, 08% more than in 2014 [5]. The smallest amount of goods was delivered to the North Caucasian Federal district by rail in 2019 - this is 11,6 million tons, the transportation of goods by rail to this Federal district has the lowest turnover for the entire study period. Moreover, freight turnover decreased by 2,9 million tons (by 20%) in 2019 compared to 2014. Freight turnover by trains in the Central Federal district during the study period is characterized by a fairly stable level within the boundaries of 200 million tons annually, the freight turnover generated by rail in the Central Federal district in 2019 amounted to 198,4 million tons (Fig. 2).

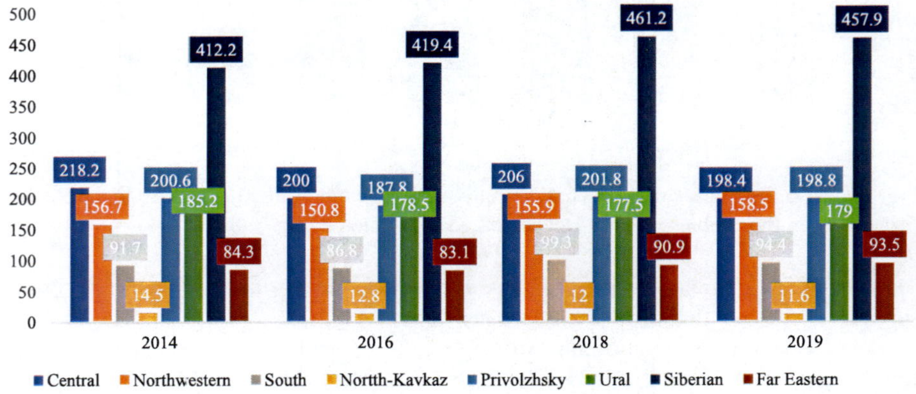

Fig. 2. Dynamics of goods shipped by rail in the Russian Federation for the period from 2014 to 2019, million tons. *Source* Developed and compiled by the authors

In general, the number of goods shipped by road to the Russian Federation decreased by 27, 9%, from 2 402,8 million tons to 1 731,4 million tons in 2019 compared to 2014. The negative dynamics of the delivery of goods by road to the federal districts of the Russian Federation is shown in Fig. 3. The decrease in freight turnover by road transport occurred to a greater extent in the Ural, Siberian and North Caucasian Federal Districts (a decrease by 43, 8%, 38, 4% and 41, 4%, respectively) in 2019 compared to 2014. Delivery of goods by road has changed to a lesser extent in the Central Federal District - the decrease was 10, 2%.

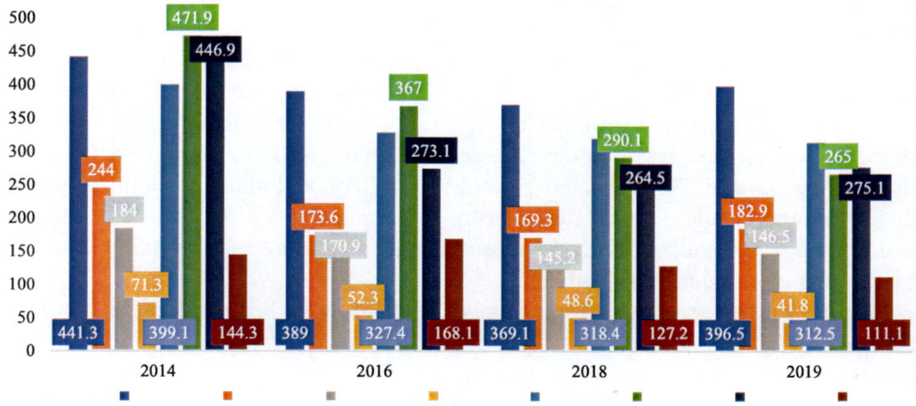

Fig. 3. Dynamics of goods shipped by road in the Russian Federation during the period from 2014 to 2019, million tons. *Source* Developed and compiled by the authors

Freight turnover research allows to estimate indirectly the turnover of e-commerce. In this regard, the indicator of the volume of e-commerce in the country was investigated. This indicator reflects the retail turnover generated by the Internet active population. The volume of e-commerce increased by 8, 82% points in 2019 compared to 2018 and by 55% points compared to 2014, and amounted to 28 879,06 million rubles. The graph shows a progressive growth in the e-commerce volume in the whole country (Fig. 4).

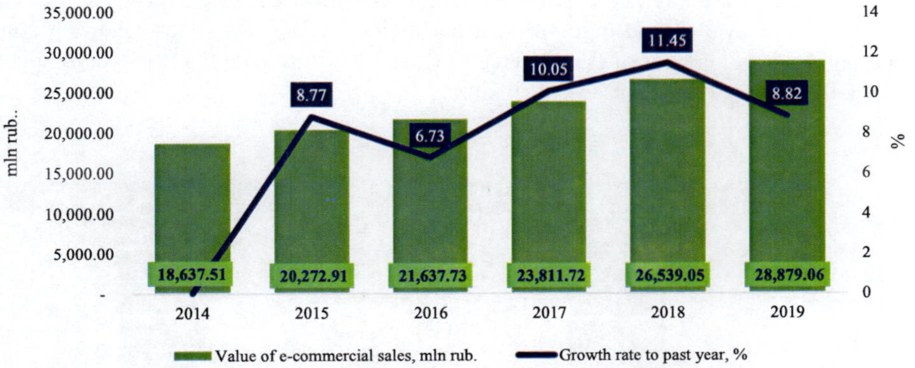

Fig. 4. Dynamics of the e-commerce volume during the period from 2014 to 2019. *Source* Developed and compiled by the authors

Noteworthy, the volume of e-commerce has increased 1,5 times over the analyzed 6 years, this can be characterized by a rather sharp «jump», and, undoubtedly, such a development of trade, «created» by the Internet active population, had a positive effect on the retail trade volume in the Russian Federation [7].

The analysis shows that the «displacement» of regions took place in 2015 and 2019. Two regions - the Central and Volga Federal districts - are the absolute winners in terms of the e-commerce volume for the studied period from 2014 to 2019. The volume of trade generated by the Internet active population in these federal districts increased by 59,09% and 51,53%, respectively, and amounted to 10 242,15 million rubles and 4 998,77 million rubles in 2019 compared to 2014. The Far Eastern and North Caucasian Federal Districts are absolute «outsiders» in terms of their contribution to the total e-commerce turnover in the Russian Federation. The Far Eastern Federal district occupied the penultimate line among the list of participants in the regional rating of Russia in terms of e-commerce turnover during the entire study period, and in 2015 it «went down» to the last place, acquiescented the North Caucasian Federal District. Also, the Urals Federal District ranked third in terms of Internet sales in the beginning of the analyzed period, but then the temp of e-commerce development in this region was characterized by a weak speed in comparison with other regions, but then the rate of e-commerce development in this region was characterized by a weak rate in comparison with other regions and the district dropped to sixth place with the value of the volume of 1,875.48 million rubles by 2019 (Fig. 5).

In general, the e-commerce volume in the country raised by 54,95% for the studied 6 years. The Central Federal district and the Volga Federal district make the greatest importance in the share of the total e-commerce turnover volume, and the regions of the Far East and the North Caucasian District are zones of active growth of the indicator under consideration.

The share of the Internet active population in Russia is growing every year, but the Internet is not available at high speed in all regions, this fact is decisive when considering the development of e-commerce in Russia, because people choose high-speed Internet for more efficient use of their time.

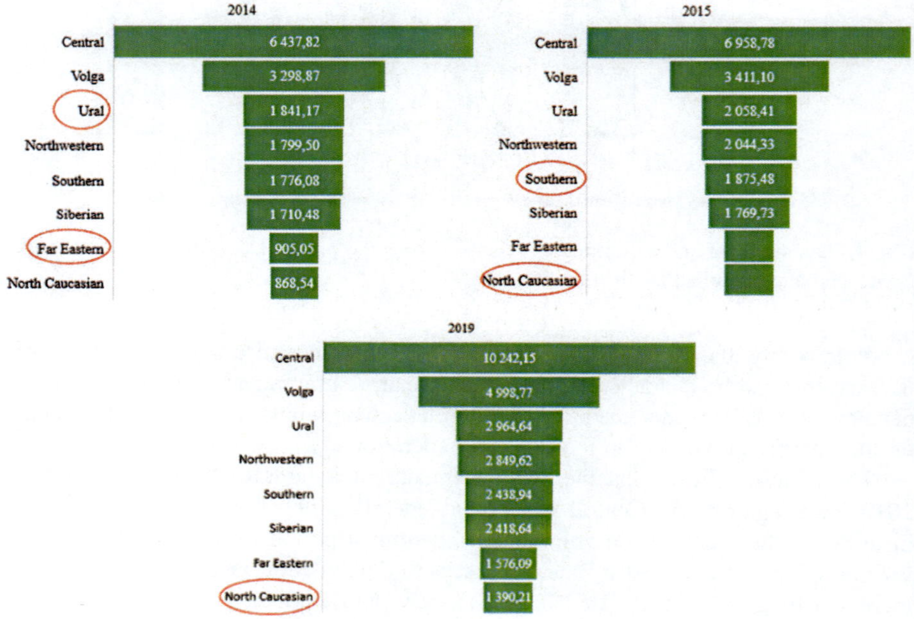

Fig. 5. Rating of federal districts by e-commerce sales during the period from 2014 to 2019, million rubles. *Source* Developed and compiled by the authors

If we study the structure of the Russian Federation population considering Internet availability, a significant increase in people who used the Internet (by 15, 1% points) is noted in 2019 in comparison with 2014 (Fig. 6).

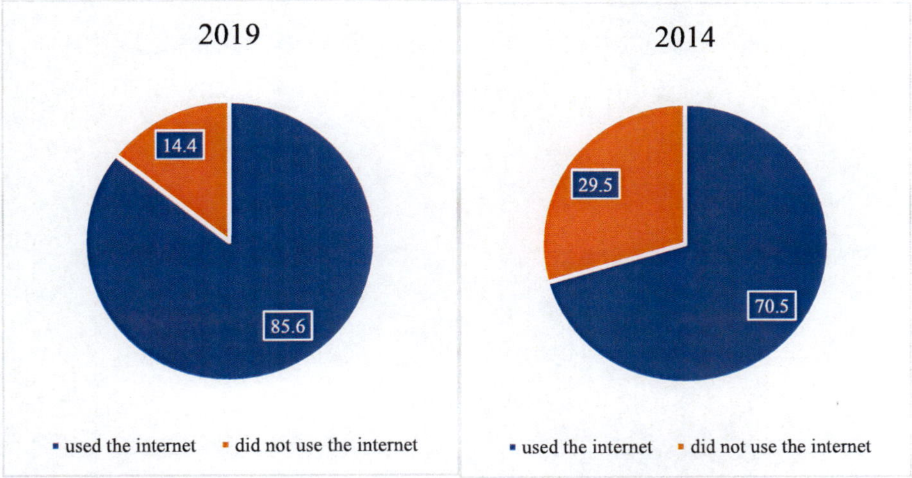

Fig. 6. The structure of digitalization population in the Russian Federation during the period from 2014 to 2019, %. *Source* Developed and compiled by the authors

The increase of the Internet active population confirms the growth of the e-commerce volume over the regions [9].

Consequently, the digitalization of Russian population by 2020 shows a growth trend. In this regard, the relationship between the raise in the digitalization level of the population and the freight turnover was investigated by the scatter diagram construction for the constituent entities of the Russian Federation, where the X-axis is the population who used the Internet and the Y-axis is the transportation of goods by cars (because the transportation of goods by cars is large share in the entire volume of freight turnover) (Fig. 7).

In addition, the Central and Volga Federal districts subjects demonstrate their leadership in terms of the volume of e-commerce. The Krasnodar Territory and the Tyumen Region were also included in the number of «prominent» subjects, which demonstrates a high level of carriage of goods by cars and a small number of people who used the Internet.

Factor models of the dependence of the freight turnover of the Russian Federation on the e-commerce volume and offline sales were built in order to identify the dependence of the volume of freight turnover on the volume of retail turnover created by the Internet active population and the population without access to the Internet [4].

The model has the following form for the whole of Russia: Y = −1309,81 + 0,08 * x1 + 0,45 * x2 (where x1 – e-commerce volume, x2 – the volume of retail turnover created by the Internet passive population) R^2 = 0,8475, which means 84,75% is described by factors x1 and x2. The model is statistically significant because the p-value (F-criterion) is 0,06, which is bigger than the critical value of 0,05. All features

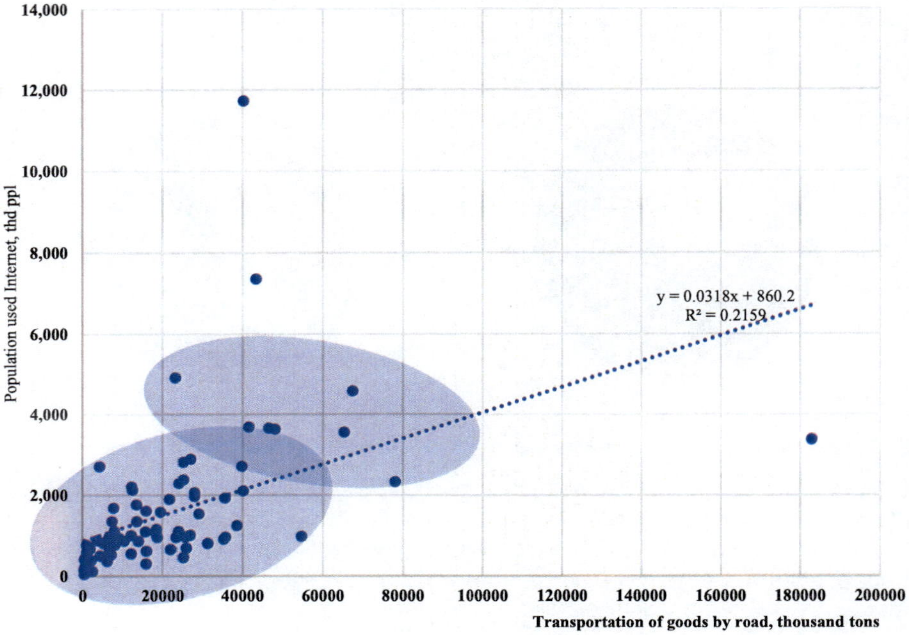

Fig. 7. Scatter diagram of the Russian Federation subjects in 2019. *Source* Developed and compiled by the authors

describing the model are significant, as evidenced by the p-value for each feature (0,49 - for x1, 0,27 - for x2) [8]. The constructed model explains that freight turnover will increase by 0,08 million tons with an increase in the e-commerce volume by 1 million rubles. If the volume of offline trade decreases by 1 million rubles, the freight turnover volume in the Russian Federation will decline by 0,45 million tons [10].

Factorial models of the dependence of freight turnover were built. Those models that are adequate, describe well the changes in features and are suitable for forecasting are presented in Table1.

Table 1. Factor models of the dependence of the freight turnover of the Russian Federation on the e-commerce volume sales and offline sales

Region	Constant term	Coefficient X1	Coefficient X2	R^2	p-value
Russian Federation	−1309,81	0,08	0,045	0,8475	0,06
Central	633,75	−0,008	0,03	0,3893	0,48
North-Western	878,89	−0,11	−0,52	0,5010	0,35
North Caucasian	211,69	−0,10	−0,087	0,9601	0,008
Volga	−489,49	0,133	0,41	0,8412	0,063
Siberian	−1449,61	0,45	2,11	0,6811	0,18
Far East	−167,97	0,09	0,78	0,7548	0,12

Source Developed and compiled by the authors

It was revealed during the analysis that all the constructed models are suitable for forecasting, except for two - for the Southern and Ural Federal Districts. Thus, we can note the Central and Volga Federal districts, analyzing different indicators of federal districts in the context of considering the leaders of e-commerce. Moreover, the freight turnover of the Volga Federal district is largely described by the change in the e-commerce turnover due to the factor analysis. Moreover, that the North Caucasian and Southern Federal Districts are among the regions that require significant development in terms of online trading.

4 Conclusion

To conclude study, it was identified that some regions do not have opportunities for the creation and development of e-commerce. The provision of information technology and wireless forms of communication is the driving force in the development of e-commerce, while it is important to properly maintain of the transport infrastructure state in the regions of Russia. Digitalization and the level of transport infrastructure can be considered as challenges for the e-commerce development in the regions, which is an important way for increasing competitiveness and stimulating the economic development of territories.

References

1. Berezhnaya L.Y.: Interrelation of transport infrastructure and regional development. Bull. Eurasian Sci. 3 (2019). https://esj.today/PDF/23ECVN319.pdf
2. Grigorash, I.: The future belongs to niche marketplaces (2021). https://marketing.rbc.ru/articles/12271/. Accessed 12 Mar 2021
3. Electronic resource: Logistic trends 2020–2021. Life after the pandemic (2021). https://www.globaltrademag.com/2021-logistics-transportation-forecast-heres-what-to-prepare-for-in-the-new-year/. Accessed 18 Mar 2021
4. Tikhomirov, N.P.: Methods of econometrics and multivariate statistical analysis: textbook. Tikhomirov, N.P., Tikhomirova, T.M., Ushmaev, O.S. (eds.) Economics, Moscow p. 647 (2014)
5. Electronic resource: Federal State Statistics Service. https://rosstat.gov.ru/ Accessed 10 Apr 2021
6. Electronic resource: Non-Food retail trade turnover in Russia in 2020 decreased for the first time since 2016 both in monetary and commodity terms. https://infoline.spb.ru/news/?news=207190. Acessed 14 Mar 2021
7. Electronic resource: Freight market growth expected to slow in 2021. https://www.supplychainquarterly.com/articles/4238-freight-market-growth-expected-to-slow-in-2021. Accessed 19 Mar 2021
8. Kartaev F.S.: Introduction to Econometrics: Textbook. Prospect, Moscow, p. 472 (2021)
9. Electronic resource: Internet 2020 in Russia and the world: statistics and trends. https://vc.ru/future/109699-internet-2020-v-rossii-i-mire-statistika-i-trendy. Accessed 12 Mar 2021
10. Sadovnikova N.A., Shmoilova R.A.: Time series analysis and forecasting: textbook. Moscow Financial and Industrial University «Synergy», p. 152 (2016)

Features of Technology Transfer Systems of Asian Region's Technology Leaders

Yuliana V. Solovieva[1] and Mingjun He[2]

[1] RUDN University, Moscow, Russia
[2] Kunming Polytechnic University, Kunming, China

Abstract. The article discusses regional features of the formation and development of national innovation systems (NIS), technology transfer systems of the Asian region's technology leaders. The authors analyze the innovation and technological systems of countries that occupy leading positions according to the Global Innovation Index (GII). In the work, the authors use a systematic approach, comparative and statistical analysis. The article identifies the conditions for interaction of institutional structures (state, fields of science and education, industry) in the process of technological transfer. Based on the analysis, the authors conclude that regional leaders are looking for ways to increase the effectiveness of the innovation component, transfer technologies, and the formation of an organizational system aimed at ensuring the process of obtaining and implementing the results of research and development.

Keywords: National innovation system · Economic development · Asian countries · Technology transfer · Scientific and technological development

JEL Codes: O32 · O5

1 Introduction

The economic growth of any state depends on some factors. As economies develop, the list of these factors is adjusted, assessed, and reassessed. The leading countries of the world, as a priority, determine for themselves the innovative and technological sphere, the development of the knowledge economy, and technological transfer. The production and transfer of high-tech products, new knowledge, and technologies reach key positions of industrialized countries. In the context of modern socio-economic transformations, it is the level of innovation and technological development that creates the world economic superiority of the country.

The purpose of the article is to analyze the national technology transfer systems of technology leaders in Asia, because each region has its leader, whose innovation system is influenced by regional characteristics (climatic, cultural, socio-economic, etc.).

E. G. Popkova (Ed.): ISC 2020, LNNS 368, pp. 308–314, 2022.
https://doi.org/10.1007/978-3-030-93244-2_35

2 Methodology

The main methods of research in the work were statistical, systematic approach, comparative analysis of indicative values of innovation and technological development of leading countries of technological development. So, the system approach found reflection in the application of all elements of the innovative process (the states, spheres of science, the industry, etc.) that allows revealing prerequisites of technological development of the economies of the leading countries.

Investigating problems of technological transfer in the countries, the author analyses both statistical data and the legislative base, which is the cornerstone of the formation of national innovative systems of the leading technological development countries. Methods of the comparative and statistical analysis of data are for this purpose applied.

The Global Innovation Index (GII) is calculated annually by the Economist Intelligence Unit to assess the innovative development potential of the world. So, in 2020, the study covered 131 countries, which account for 99.4% of world GDP (Table 1). Russia, which was in 62nd place in 2013, reached 47th place in 2020.

Table 1. GII (technological leaders), 2020

Country/Economy	Score (0–100)	Rank	Region	Region rank
Switzerland	66.08	1	EUR	1
Sweden	62.47	2	EUR	2
USA	60.56	3	NAC	1
United Kingdom	59.78	4	EUR	3
Netherlands	58.76	5	EUR	4
Denmark	57.53	6	EUR	5
Finland	57.02	7	EUR	6
Singapore	56.61	8	SEAO	1
Germany	56.55	9	EUR	7
Republic of Korea	56.11	10	SEAO	2

Source Table compiled by authors based on data [9]

In the calculation of the Index, countries are divided into regions, in each of which their leader is distinguished (Table 2).

Table 2. The global innovation index (countries - regional technology leaders), 2020

Region	Country/Economy	Rank	Score (0–100)
Europe	Switzerland	1	66.08
Northern America	USA	3	60.56
South-East Asia. East Asia and Oceania	Republic of Singapore	8	56.61
Northern Africa and Western Asia	Israel	13	53.55
Latin America and the Caribbean	Chile	54	33.86
Central and Southern Asia	India	48	35.59
Sub-Saharan Africa	Mauritius	52	34.35

Source Table compiled by authors based on data [9]

3 Results

Given the dynamics of economic and scientific and technological development of Asian countries, let us consider the features of the formation of national innovation systems (NIS) of countries - technological leaders representing this region. According to GII, these include the countries shown in Table 3. Each of the countries has gone its way of forming an NIS, a technological transfer. Consider the key features of key leaders.

Table 3. Leading countries of the Asian region according to GII, 2020 г.

Region	Country	Score	Rank	Region rank
South-East Asia, East Asia and Oceania	Republic of Singapore	56.61	8	1
	Republic of Korea	56.11	10	2
	Hong Kong	54.24	11	3
	China	53.28	14	4
Northern Africa and Western Asia	Israel	53.55	13	1
	Cyprus	45.67	29	2
	United Arab Emirates	41.79	34	3
Central and Southern Asia	India	35.59	48	1
	Iran	30.89	67	2
	Kazakhstan	28.56	77	3

Source Table compiled by authors based on data [9]

The economic and technological development of leading countries forms regional models of development, taking into account natural-climatic, cultural and historical characteristics, the level of socio-economic and technological development. We will consider the main features of innovative development and a transfer of technologies in the leading technological development countries. As note by Edler & Fagerberg, "There is a narrow perspective, considering invention only, and there is a broader, more holistic perspective, which emphasizes the importance of looking at the entire innovation cycle from the creation of novel ideas to their implementation and diffusion" [2].

Consider in more detail the countries that hold the leading positions in their region, namely, Singapore, Israel, and India.

3.1 Singapore

The formation of Singapore's national innovation system began in the late 1970s when the Government identified the development of knowledge-intensive industries as a national policy priority. As a result of the decision on the need to develop NIS, the first scientific and production park of the country (1981) is being organized based on the University of Singapore, which later became Singapore's leading innovation centre and the largest industrial technology development centre. Afterwards, the Government of Singapore identified areas of R&D such as medicine, genetics, biochemistry, biotechnology, etc., as priorities.

The scientific and industrial parks of the country had a significant impact on the implementation of the country's development programs. The basis "for the formation of high-tech companies in the country and the centre of some national programs and R&D aimed at developing industries that are a priority for the country (biomedicine, information technology, software development, telecommunications, electronics, food technologies, chemistry, and materials)" [8] was the Singapore Science Park.

In 2000, the Government is implementing a new technology park project, One-North Science Habitat, aimed at creating a multilateral research community. To this end, two "technopark structures are being formed: Biopolis - specializing in biotechnology, and Fusionopolis - focused on information and fundamental sciences" [1].

Organizations participating in the development of technopark structures are granted some benefits by the State (halving the income tax when investing it in research activities and other tax incentives).

The Government of the Republic of Singapore plans to "increase the number of technoparks with a focus on the creation and development of modern agricultural production technologies, its export to Asia-Pacific region, as well as new developments and technologies are in promising plans" [6].

3.2 Israel

The formation of the NIS took place in the country in several stages. We will identify the main ones (Table 4).

Table 4. Stages of the formation of the NIS and the system of technological transfer of Israel

Period, years	Characteristic/Description
1952	Creation of the Foundation for R&D by the Polytechnic Institute of Haifa (Technion)
1959	Organization of Yeda by the H. Weizmann Institute to commercialize technologies developed by its employees
1969	Creation of the Office of the Chief Scientist to support industrial R&D
1970–1984	Development of the business sector capable of R&D Launch of the Israeli-American program (Binational Industrial Research and Development Fund) as a result of the restructuring of the defense-oriented electronic industry
1985	The Industrial R&D Promotion Act came into force. The role of private companies is growing, and there is a formation of a competitive approach to the commercialization of scientific developments

(*continued*)

Table 4. (*continued*)

Period, years	Characteristic/Description
1986–1992	The model of start-ups is developing and venture capital activity is beginning. A significant number of startups are being sold (about 300), which entails an increase in demand for venture capital and contributes to the development of the industry not only "from above" (through a state initiative), but also "from below" (on the initiative of projects, business, individuals) During this period there is an increase in the volume of grant financing of R & D and start-ups (by 66%), stimulation of international ties, creation of technological incubators
1993–2000	There is a rapid increase in high-tech products, development of the venture capital sector. The state program of Yozma is being implemented (1993–1998). The activity of foreign venture funds is increasing. Increased expenditure on grant programmes and incubator support. Horizontal sowing programs aimed at the emergence of a wide range of innovations are beginning to work interlinked in targeted programs aimed at the development of certain projects
2001–present	During this period, in Israel there is a crisis in the high-tech sector, the venture capital market is recovering. The State creates special funds to finance sowing projects. Privatization of process incubators begins. The high-tech cluster and venture sector are being diversified

Source Compiled by the authors

Considering the features of the formation of Israel's national system, we note its central core, which became small innovative companies. In terms of per capita population, the state has the largest "number of start-ups - about 1 per 1970 people. The activities and interests of companies include almost the entire high-tech spectrum: software, information technologies, electronics, semiconductor products, bio- and Internet technologies, alternative energy sources, etc." [5, p.49].

The scientific and research sphere in Israel has considerable state support carried out by OCS and its divisions in key ministries of the country: agriculture and village development, communications and rear protection, defense, education, environmental protection; health care; immigration and absorption; economies, etc.

BATAT takes an active part in the development of the scientific and innovative technological sphere of Israel - Planning and Budget Commission of the Council for Higher Education. It's "activities are mainly aimed at the interaction between state bodies and universities, including coordination of financial issues between them. In addition, the Commission participates in the TELEM Forum, which includes some ministries and agencies coordinating their actions to develop innovative infrastructure in the country" [5]. Thus, it significantly affects the formation of stable integration links between the fields of science, education, applied industrial R & D and institutions of state power.

The most promising sources of development, creation and transfer of high-tech products in Israel are considered research centers of universities, formed the basis of the scientific and technological complex of Israel.

A characteristic feature of technological transfer companies established at Israeli universities is the presence of not only scientists, but also the active participation of specialists from various fields of production and business. This allows solving several important tasks simultaneously. For example, developers and creators of new technology do not need to search for finance and opportunities to sell their products, while investors reduce costs associated with the complex process of finding and selecting a project for investment. Thus, both research and entrepreneurship participants of the innovation process in the high-tech market of Israel are stimulated.

There are currently 17 technology transfer companies in Israel. Since 2004, at the initiative of the United States-Israel Science and Technology Foundation (USISTF), "most of them have been merged into the umbrella structure Israel Technology Transfer Network, which lobbies for their interests in various state structures of the country, as well as promotes their portfolio technologies outside it" [3].

The Rotem Technology Transfer Center, established in the early 1980s, is distinguished from the structure described. Its characteristic is the initial independence from any structural units involved in the innovation process (universities, research institutes, etc.), and the independent search for research developments for their subsequent commercialization.

Israel's innovation system is developing quite rapidly and has all the prerequisites for maintaining the status of one of the technology leaders. According to the Ministry of Economy and Industry, the percentage of Israeli citizens participating in R & D, as well as the number of expenditures in this area relative to GDP, are among the highest in the world [4].

3.3 India

India's NIS has a significant infrastructure with both modern facilities for R&D in various areas (biotechnology, medicine, electronics, communication technologies, etc.) and interacting with enterprises to implement research results.

Choosing the model of the Massachusetts Institute of Technology (USA) in the 1950s, India begins the creation of technological institutes. The purpose of forming these institutes was to organize centers of advanced knowledge and technologies ready for the rapid introduction of research results into production.

In 1990, the Indian government created the Software Technology Park of India (STPI) as the first technopark whose activities were aimed at developing the software industry. Priority "problems of its functioning were: assistance and development of potential in the export of software products; innovative activity; rendering full support to the companies releasing software products and the equipment for information technologies; rendering telecommunication and infrastructure services to the companies occupied with development and export of software products and which are a part of science and technology park" [8].

Currently, STPI has 36 branches located throughout the country. More than 7,500 companies specializing in export-only software development are members of STPI.

Technology parks in India have some benefits. So, they are completely exempt from import tax, and for five years - from paying domestic taxes and fees.

As noted by Shkvarya et al., “emerging economies should launch large-scale structural reforms, but the only major economy in Asia where such reforms are being implemented, at least to some extent in India. China and India are two emerging economies; the second and ninth-largest economies in the world, respectively” [7].

4 Conclusions

Despite the extensive theoretical basis of scientific research on innovative processes, there are still many debatable and little-studied issues. Such areas include the formation of effective mechanisms for interaction between participants in the innovation process, technology transfer, dynamics, and prospects for the development of the scientific, innovative, and technological potential of various states, etc.

Recently, we have seen governments of countries that are technological leaders in the Asian region realize the need to increase the effectiveness of the innovation component, technology transfer. The creation of an organizational system based on the integration of all participants in the innovation and technological process will contribute to the introduction and dissemination of R&D results, the creation of high-tech competitive products.

References

1. Analysis and assessment of the state and trends of development of legislation on innovation activity and science abroad. Information and analytical materials of the State Duma (2012). http://iam.duma.gov.ru/node/10/4960/20221. Accessed 14 Apr 2021
2. Edler, J., Fagerberg, J.: Innovation policy: what, why, and how. Oxf. Rev. Econ. Policy **33**(1), 2–23 (2017). https://doi.org/10.1093/oxrep/grx001
3. Israel Technology Transfer Organization. www.ittn.org.il/index.php. Accessed 12 Apr 2021
4. Israeli Ministry of Economy and Industry. www.economy.gov.il/English/InternationalAffairs/ForeignTradeAdministration/Pages/AboutFTA.aspx. Accessed 12 Apr 2021
5. Maryasis, D.: Israel’s technology commercialization system. Sci. Innovat. **10**(152), 49–52 (2015)
6. Report of the Economic Strategies Committee: Ministry for Trade and Industry Singapore (2011). www.mti.gov.sg/ResearchRoom/Pages/Report-of-the-Economic-Strategies-Committee.aspx. Accessed 14 Apr 2021
7. Shkvarya, L., Grigorenko, O., Strygin, A., Rusakovich, V., Shilina, S.: Impact of the global economic crisis on Asian technology markets (India, China). Central Asia Caucasus [Centralnaya Aziya i Kavkaz] **19**(2), 119–130 (2016)
8. Solovieva, Yu.V.: National technology transfer systems of leading technological development countries. In: Analysis of International Relations 2020. Methods and Models of Regional Development, Katowice, pp. 210–224 (2020)
9. The Global Innovation Index (2020). www.wipo.int/edocs/pubdocs/en/wipo_pub_gii_2020.pdf. Accessed 10 Apr 2021

Legal Regulation of Artificial Intelligence in the Area of Investment in the Economy

Olga M. Kunitskaya(✉)

Belarusian State University, Minsk, Belarus

Abstract. The article is devoted to the hot issues of legal regulation of artificial intelligence and robotics (AI and RT) of various countries, including the Russian Federation and the Republic of Belarus. The article shows the principal results of the analysis of existing approaches to the legal regulation of AI and RT in investment in the economy. From the said positions, the author researches the basic ideas in the formation of legislation on AI and RT systems. Simultaneously, the existing and potential legal problems arising in the Digital Age and competition in attracting investment are investigated and identified.

The author also has developed proposals for further discussion of such issues as legally relevant problems for the growth of the AI and RT; creation of a concept of legal regulation in AI and RT, necessary for the development (revision) of legislation concerning the formation, exploitation, distribution of robots and systems with AI, and the fundamental principles and limits of legal regulation of these relations. The foundation for identifying the directions of legislative development should be the principle of "stimulation before regulation", with the balance of interests of individuals, society, and the state, concerning the proper provision of security and protection of rights and interests associated with the development of innovations.

Keywords: Investments · Investment climate · Digital economy · Digital technologies · Artificial intelligence and robotics systems · Approaches to legal regulation · Legal personality · Responsibility for damage caused by AI and RT systems · Safety · Intellectual property law · Harmonization of law

JEL Codes: O33 · E22 · E65

1 Introduction

The conditions in a country that affect capital inflows are made up of a wide range of factors, combined under the general wording of the "investment climate." The scale and efficiency of attracting investments are directly related to the creation of a favorable investment climate in the country, which, in turn, depends on the elaboration and timeliness of the adoption or adaptation of legislative acts.

In the digital economy or "data economy", "Internet economy" investment is expected to intensify in the presence of attractive conditions for investors. Since investments are made in the digital economy, the legal regulation of relations arising in such an economy should become attractive to investors [47].

E. G. Popkova (Ed.): ISC 2020, LNNS 368, pp. 315–332, 2022.
https://doi.org/10.1007/978-3-030-93244-2_36

The investment area is a crucial element of the economic system. Traditionally, the investment sphere includes the spheres of 1) capital construction; 2) innovation; 3) circulation of financial capital; 4) realization of property rights of subjects of investment activity [51].

In the information society, the law is faced with a choice: either to prohibit or to integrate into the new increasingly complex mechanisms of legal regulation of social relations, conditioned by the functioning of "smart systems," including AI and RT systems. The decision of this issue can significantly change the position of the investor when deciding on an investment.

Since its introduction in the 1950s, the term "artificial intelligence" about 340 thousand inventions were recorded. Half of them have been registered since 2013, which allows speaking of a "wave of AI systems". Shumilin [74] underlines the active role of machine learning in the use of AI systems.

The scientific research "Artificial Intelligence and Life in 2030" indicated fields in which artificial intelligence will be actively used soon: transportation, services, healthcare, education, public safety, work and employment, home economics, entertainment, and others [70].

The exploitation of artificial intelligence, for example, in the real estate sector can enhance the efficiency of operational tasks and change decision-making processes. By knowing relationships and patterns in big data, one can explore the implications of possible future scenarios in more detail. Algorithms can check the completeness of virtual data warehouses or if the database is sufficiently developed, identify suitable investment targets according to ance the portfolio matrix [50].

The convergence, mixing of people, things, and technologies form new types of relations, in which, on the one hand, traditional objects and subjects are involved; on the other hand, technologies that influence relations, their nature, and directions [62]. A rather significant problem at the present stage is the lack of the required legislative framework, which can accelerate the adaptation of society's needs to develop new technologies.

As in many countries, in the post-Soviet space, including the Republic of Belarus, the legal basis for this phenomenon has not yet been developed enough, despite the serious steps taken in recent years in the formation of legal regulation of the digital economy [37].

To create enabling environment for attracting investments, fundamental studies of the basic principles are necessary, including the content of the components of both general and special legal regimes of AI and RT technology, which first requires the creation of the Concept of legal regulation of the use of these technologies, which serves as a benchmark for the legislator. It is extremely essential to turn the digital economy into a reliable and safe space for business development [43].

2 Materials

In world practice, legal support for the exploitation of A has already become the subject of research. Pagallo [55] and several other scholars in law, engineering, computer science, and philosophy have considered legal problems in the utilization of AI and robotics in their works.

Tripathi and Ghatak [76] have also considered AI's problem in terms of intellectual property law. The prospects of endowing AI with a certain degree of legal personality were considered by [63]. Atabekova and Yastrebov [21] investigate the perspective of considering AI from the perspective of autonomous legal personality as an independent subject of law and control. Gasson and Koops [33] and Johnson [39] studied the issues of improving approaches to determining responsibility concerning the development of innovations.

Russian legal scholars have investigated some issues of using AI systems based on law. We also want to mention the works by [57] dealing with the problems of the legal status of objects with AI, in relation to the functionality, implementation features, and autonomy measures. Arkhipov [18], Neznamov [53] investigated the issue of differentiating the status of robots and robotic agents, and Laptev [44] did it for the issues of legal responsibility. Ponomareva proposed criteria for distinguishing between a subject and a quasi-subject of law. She established those criteria that are absent in robots as subjects of law [59].

The specifics of AI and some legal problems of its application in certain areas and the overall legal personality of AI in intellectual property law (civil law problems) were analyzed by [49].

The results of a detailed analysis of existing approaches to the legal regulation of AI are given in the work of [41].

Russian scientists pay considerable attention to one of the problem areas of using AI - the sphere of copyright. Various approaches to the definition of creativity are analyzed, and the pros and cons of different points of view on determining the ownership of intellectual rights to a work created by AI are considered [35], the potential problem of mixing unique works of authorship with works created by the program is investigated [17], as well as other problems of authorship and the legal nature of the results of intellectual activity created by artificial intelligence. The assumption is substantiated that the results of AI activities cannot be qualified as an object of copyright, and therefore is not subject to protection [79].

Among the Russian legal scholars who are studying the issue of the need for the legal regulation of artificial intelligence, whose works should be paid special attention, one can also name [24, 32, 42, 52, 53, 60, 68, 77, 80].

In the Republic of Belarus, at the current stage, scientists define legal problems associated with AI technologies, such as 1) issues of legal control of the procedure of creating new AI technologies; 2) issues of the legal protection of artificial intelligence technologies as objects of intellectual property, dividing them into two different areas: legal protection of objects and technologies of artificial intelligence and the right to new objects created using artificial intelligence; 3) the problem of liability for decisions and actions taken by such systems [75]. Bochkov [26] raises the methodological issues of using AI that gives rise to problems and risks of moral and legal regulation. Ablameiko and Ablameiko [15] consider the growth and integration of legislation in AI, suggest further steps for the legislative support of this segment.

Savenok [64] considers it necessary to come to a common understanding of the approaches to the subjects of crimes that have developed in the theory of criminal law and the corresponding provisions of technical normative legal acts on the use of robots with elements of artificial intelligence.

Belsky et al. [26] believe that further improvement and use of AI and RT is possible only if a system-integrated approach is implemented, which, on the one hand, will take into account the need to develop legislation that is at the junction of several branches of law, on the other hand, will provide an opportunity to ensure the accounting and harmonization of approaches formed in scientific doctrine and legislation, including within the framework of integration formations.

The global experience of legal regulation of relations appearing due to the development of AI has been studied, starting from adopting strategic documents and principles of using AI in specific areas and ending with the norms of special legislation. Strategic development documents, as a rule, present the current level of development of AI technologies globally, the key sectors of their implementation; stages, tasks, and goals of the development of AI technologies in a particular country; a plan of significant activities aimed at the elaboration of technologies in general; major ethical challenges and issues; expectations about technology expansion in the short-term, medium-term and long-term perspective; the arrangement of funding of the industry, and some others.

The ethical documents on the rules for using AI (guidelines, codes, and others) provide the fundamental principles of regulation: security, confidentiality, non-discrimination, accountability, and alike. Some of the most famous are the Azilomar Principles of AI [20].

Attention is paid to the technical regulation of AI and RT. Within the scope of the International Organization for Standardization (ISO https://www.iso.org/store.html), a special technical committee on artificial intelligence has been established.

In some countries, there are legal acts dedicated to specific types of AI systems used in automated vehicles, healthcare, the implementation of smart city concepts, the financial sector, and public administration. The experience of legal regulation of the relationship between humans and artificial intelligence in South Korea, which has become a pioneer in the normative consolidation of the main provisions (https://law.go.kr/eng/engMain.do), is also considered.

An example of an up-to-date conceptual approach to the normative regulation of AI and RT technologies is the Russian Concept for the development of regulation of relations in the artificial intelligence and robotics technologies [73].

3 Methods

During the survey, the basic research (dialectical) method of cognition, common techniques of formal logic (analysis and synthesis, deduction and induction, analogy, and comparison, and alike), and associated distinctive scientific methods were used: historical, technical-legal, comparative-legal, and alike.

4 Results

Researchers have not yet agreed on what kind of legal regulation of artificial intelligence should be: 1) proactive universal-total or 2) to manage specific AI systems that are being developed, since this field is constantly progressing and it is hardly possible to know for sure where this development will lead to after a while [57].

Besides, to date, the world has developed several basic approaches to measuring the measure (boundaries) of legal regulation. The essence of the first one, sometimes referred to as legal or legalistic (Legalistic Approach), is to rely on the existing legal system and define lists of problems resulting from the exploitation of robots and issues concerning liability, confidentiality, cybersecurity. The second approach (conditionally, it can be called technological) describes "movement" that became the result of technological changes: the appearance of drones, robots - personal assistants, exoskeletons, etc. Proponents of the technological approach insist on the secondary nature of law. Artificial intelligence does not require a particular legal framework. It will penetrate many areas, so it is not worth adopting a general law since many technologies differ in the transport sector, medicine, and further. It will be enough to ensure the responsibility of machines for their actions since interest from the economic effect while exploiting should be subtracted from a particular fund, from which the damage caused by the robot will be compensated [40]. From the perspective of those who support the legalistic approach, it would be more reasonable to impose responsibility not on the robot itself, but on the person who controls the mechanism. However, adherents of the technological approach do not support this idea since such measures will significantly slow down the development of robotics in countries where the legalistic approach is used and there will be a severe economic lag in countries where the technological approach is leading [38].

There is also a third approach, which implies a combined type that separates the ethics of exploiting robots and robotics. Legal regulation will consider only the first one. In any case, one of the main tasks of forming legislation in the field of AI and RT is to ensure a balance of interests [48].

Depending on the scope of regulation, legal regulation approaches can be divided into the following levels1) private initiative and research institutes; 2) self-regulation in the industry; 3) national regulation; 4) supranational regulation [53]. It is proposed to include such elements in the system of legal regulation of artificial intelligence [67].

At the same time, the need to apply self-regulation mechanisms for the growth of digital technologies is based on several reasons and factors: ineffectiveness of traditional mechanisms of legal regulation; diverse affiliation of sources (constitutional, administrative, civil, financial, labor, procedural, etc.) of legal regulation of relations on the use of digital technologies; the need for flexibility in the digital technology regulation system, due to the active development of the information society [45].

Concerning AI, it is rightly noted that the need for legal regulation is necessary first of all to solve urgent problems in the current legal field since they are a serious obstacle to the development of technologies. The need for legal regulation is also due to the desire to accelerate the process of introducing technologies into the digital economy, which requires the creation of mechanisms for legal regulation [72].

The authors believe it is necessary to form framework legal documents containing the methodology and standardization of the development and implementation of AI and RT since a situational response to emerging digitalization requests will significantly reduce the social role and purpose of the law [47].

We consider it necessary to generalize and define the most controversial issues from the point of view of the law, which the exploitation and expansion of AI systems can cause: 1) data confidentiality (personal data can be disclosed through decisions made by artificial intelligence); 2) the functioning of "big data" technology; 3) the problems of determining the legal personality of the AI system and intellectual property (the potential "non-protection" of the results of activities obtained using AI systems can discourage the development and deployment of such systems); 4) security and problems of civil and criminal liability.

1. But all the issues described above require a clear definition of the concept of "artificial intelligence."

The term "artificial intelligence" was introduced in 1956 by the famous American expert on the theory and practice of computers, John McCarthy, who defined AI as "the science and technology of creating intelligent machines, brilliant computer programs; the property of intelligent systems to perform creative functions that are traditionally considered the prerogative of a person." Since then, the understanding of AI has been greatly transformed. The difficulty in defining the term lies in the fact that there is still no precise understanding of natural intelligence [47].

AI technologies form a set of technological solutions that make it possible to imitate human cognitive functions (such as self-study and search for solutions in the absence of a predetermined algorithm) and, when performing tasks, achieve results equal to the results of human intellectual activity, or significantly surpassing them [31]. Artificial intelligence is understood as a fully or partially autonomous self-organizing computer-hardware-software virtual or cyber-physical, including a bio-cybernetic system with the capability and ability to think, self-organize, learn, independently make decisions, and alike.

Artificial intelligence trains independently, taking into account the already accumulated experience, which allows it to choose an algorithm of action depending on previously completed actions. This is the reason why the effectiveness and potential of AI are often unclear [29].

Analysis of existing approaches to the term"artificial intelligence" allows us to highlight the following features: 1) the presence of a technical device (cyber-physical system) capable of perceiving information and transmitting it; 2) a certain degree of autonomous work without human participation (subjectivity) in the absence of life of such a system; 3) the ability to analyze, and generalize information, develop intelligent decisions based on the studied data (thinking), self-awareness; 4) the ability to learn, independently search for information and make decisions depending on this information [77].

Simultaneously, these characteristics of artificial intelligence can be considered problematic: autonomy is the unpredictability for a person of decisions made by AI, and it is difficult for a person to exercise control over a machine programmed to act with a significant degree of autonomy; discreteness - the ability for individual components of the AI system to be developed in different places and at different times without any conscious coordination. Opacity is that the inner workings of an AI system can be kept secret and not subject to reverse engineering [65].

V. Laptev showed the interconnection of the categories of artificial intelligence, robot, and cyber-physical system, as well as their interpenetration, assuming that shortly artificial intelligence will exist in isolation from specific robots or computers, for example, in a virtual "cloud," moving through the Internet [44].

2. Concerning civil law relations today, AI systems are the *object* of such relations. Nevertheless, the object is quite specific. It has always been believed that creativity is available only to humans. Nevertheless, the ability to perform creative functions is also considered by some researchers as a property of AI [68]. There is also an opinion that in the current civil law regulation, Speaking about AI as technologies and smart robots, it should be mentioned that they can act in several forms - as an object of commodity circulation (thing), an item of copyright (program), a source of increased danger (a potentially dangerous thing), and the question of vesting it with offenses within the framework of the existing legal system is impossible [41].

The approaches to establishing the legal standing of artificial intelligence can be reduced to the following: 1) independent legal personality (it is a very remote prospect, but feasible); 2) an item or a certain object of civil rights (there is also the case with subordinate intelligence); 3) an analogy with the legal standing of animals (for more complex systems of subordinate artificial intelligence, but still lacking signs of autonomy). In connection with the foregoing, the rights for the results of AI activities as well as responsibility for the actions reproduced by them are determined [66].

Currently, scientific works propose criteria for distinguishing between the subject and the quasi-subject of law and determine those that are absent in robots as subjects of law: 1) the ability (potential or natural) to independently exercise subjective rights and legal obligations, commit lawful and illegal actions; 2) the ability to bear legal responsibility; 3) the ability to make legal decisions, to be the decisive authority in law; 4) the presence of their legal interests, needs, legal aspirations, and claims; 5) property isolation [59].

The problem of legislative regulation of the protection of intellectual rights to works created by artificial intelligence is urgent since the creators and owners of AI systems are interested in commercializing the results of their use.

The questions that arise in this case cannot be ignored by legal scholars: who will be considered the author if an artificial intelligence system creates any work? Who will be entitled to the results of the program? Who will own the right to a part of the program code written by the program itself as part of the self-learning process? [17].

3. The main problem of legal regulation in artificial intelligence is the distribution of responsibility for harm caused by the actions of a robot. There are two potential avenues for reimbursing victims of robots: claims against the owner or the manufacturer. It causes objections from some stakeholders - development companies, making it less attractive to invest in AI projects [47].

Balkin [22] reckons that the legal regulation of AI activities will cause problems in various areas of law, since some people may use robotics to violate human rights, taking into account the "substitution effect", when living people are replaced by robots.

As one can see, there are plenty of challenging issues for legal scholars. And so far, there is no ready-made, and most importantly, shared solutions for them. The unspoken ethical standards for robotics developers are not sufficient for modern society. Moreover, the more advanced AI and RT systems become, the higher the need to develop unique legal principles and norms governing their use.

In this regard, it is logical to formulate approaches to the concept of legal regulation of artificial intelligence and identify ways to solve the problems considered. Simultaneously, as [36] that the legal regulation of the creation of AI is necessary to achieve two main goals: 1) to protect a person, namely to his rights and freedom, the protection to society, the state, and humanity as a whole; 2) to create a favorable environment for the development of the latest technologies to the extent that it corresponds to the first goal.

Consequently, one should study the experience of their reflection and implementation in the legislation of countries and regulatory documents.

The most profound and detailed national strategic development documents are the Plan for developing artificial intelligence technologies of a new generation [10, 16, 55, 57]. For example, according to the provisions of the French National Strategy for the Development of AI, legislative regulation is supposed to focus on the transparency of algorithms used in socially significant areas and the responsibility of persons who create relevant products and technologies.

There are no comprehensive regulations with explicit norms dedicated only to AI in the world. South Korean legislators who are pioneers in the detailed regulation of the activities of the creators of programs for the functioning of robots and those involved in their development and production, use, and destruction came relatively close to their creation [68]. In other countries, attempts have also been made to resolve this issue [55].

In February 2017, the EU Parliament adopted Resolution 2015/2103 (INL) Civil Law Rules on Robotics [1]. The document deals primarily with robotics, but it is evident from the content and logic that it also refers to AI technologies. The resolution outlines an approach to regulating liability for harm caused. It is proposed to create a European system for registering "smart" robots. The resolution gained fame thanks to the idea of endowing robots with AI with the status of electronic personalities. It is also devoted to the issues of liability for the actions of artificial intelligence. According to para. 59 of the EU Commission, it is necessary to consider any possible consequences from a legal point of view a) creation of an obligatory system of insurance; b) guaranteed use of the resources for compensation, if the insurance company has not covered the damage caused in connection with the use of AI technologies; c) the possibility of partial shifting of responsibility for the actions of AI from the developer, manufacturer, user or owner of the robotics, in case if they deposit funds or joint liability insurance; d) a special compensation fund can be formed for all autonomous robots taken together or for each type separately. The fund can be replenished either once when the robot is just put into operation or permanently during the exploitation of the robot. e) assignment of an individual registration number to each robot, entered in an independent register; f) the prospect of giving the most advanced autonomous robots a special legal status (electronic persons responsible for the damage caused by them).

Most of the ethics documents in the AI cite the critical principles of regulation: security, confidentiality, non-discrimination, accountability, and alike. One of the best known is the Asilomar AI Principles [20], which emphasizes the importance of the confidentiality of personal information and ensures the basic rights of society and the individual. In the future, the creation and development of "superintelligence" will be possible only under strict control which will allow them to conform to established ethical ideals and serve for the benefit of humanity. This approach will also help to minimize possible risks in working with AI at the stage of forecasting.

For the most part, the Asilomar Principles became the basis for the ethical aspects of the Montreal Declaration for Responsible Artificial Intelligence [4] – Montréal Declaration: Responsible AI; Ethics guidelines for trustworthy AI.

An important aspect that must be considered for approaches to regulating AI is the cross-border nature of services for the organization of which AI is used, which requires interaction and coordination of states and all stakeholders at the international level. Thus, the Report on the Ethics of AI of the World Commission on the Ethics of Scientific Knowledge and Technology of UNESCO is devoted to recommendations on the structure and content of such a possible international instrument [61].

To date, the European Union has elaborated some policy documents for the advancement of artificial intelligence, including Policy and Investment Recommendations for Robust Artificial Intelligence [57]; Artificial Intelligence White Paper: A European Approach to Excellence and Trust [80].

In our opinion, when forming the legal basis for the use of AI and RT technologies, it is vital to work out the goals, objectives, and results of legal regulation. Indicative in this sense is the national program "Digital Economy of the Russian Federation" [5]. One of its tasks is to create a system of legal regulation of the digital economy based on a flexible approach in each area and the introduction of civil circulation based on digital technologies.

The foundations for regulating artificial intelligence in Russia were established by Decree of the President of the Russian Federation dated 10.10.2019 No. 490 "On the development of artificial intelligence in the Russian Federation", which approved the "National strategy for the development of artificial intelligence for the period up to 2030" [8]. It provides for the phased creation of a flexible regulatory system that guarantees the population's safety and aims to stimulate the development of AI technologies.

The Government of the Russian Federation approved the Concept for the development of regulation of relations in the AI and RT technologies until 2024 [73]. The major vector of regulation development in the designated period is "incentives as a basis for regulation": regulatory measures should be applied only in cases where there are high risks of harm to the life and health of citizens. The Concept defines the industry-wide tasks of regulating the use of AI and RT technologies:

1. Creation of mechanisms for simplified implementation of products using AI and RT technologies (experimental legal regimes (“regulatory sandboxes”, “regulatory testbeds”). Federal Law “On Experimental Legal Regimes in the Sphere of Digital Innovation in the Russian Federation” [31] allows the introduction of “regulatory sandboxes”, within which it is possible to develop technologies that are not yet regulated by law.
2. *Liability* in using AI and RT systems (for harm caused by AI and RT systems, which have a high degree of autonomy when they make decisions).
3. *Improvement of the data circulation regime* (revision of information confidentiality and personal data protection regimes to ensure favorable legal conditions for safe and responsible access of developers of AI and RT systems to data and secure exchange of various types of data).
4. Improving *the export regime for AI and RT systems* (it is necessary to avoid the formation of dual regulation - separately for the national market and separately for international markets. Measures restricting exports should be applied only in areas affecting national security interests).
5. Development of *insurance institutions* (it is necessary to determine the causes and conditions of compulsory insurance of liability for damage caused by the use of robots or AI and RT systems, and the criteria for referring to AI and RT systems).
6. Development and clarification of *terms and definitions* in the field of AI and RT systems technologies to determine the subject of regulation of regulatory legal acts.
7. *Ensuring security*, including information security (general requirements for the security of goods, works, and services in some cases should be clarified, and depending on the degree of risk, *specific security requirements* can be established).
8. Advancement of regulatory legal acts in the field of AI and RT systems internationally (active *involvement in the development of international documents* in the AI and RT systems, primarily in the Organization for Economic Cooperation and Development, the Council of Europe, the United Nations and its specialized agencies.)
9. Creation of *regulatory conditions* for the use of AI systems when making *legally significant decisions* (point “delegation” of specific decisions to AI systems, where it is objectively expedient and does not pose a threat to fundamental human rights and freedoms, national defense, and state security).
10. Proposal of AI and RT systems as a *product* (work, service) that corresponds to personal and family needs of consumers (introduction of functional features into AI and RT systems that provide, for example, the consumer’s ability to stop the AI system in a critical situation (“red button”) and/or to ensure the fixation of facts allowing to establish the circumstances of harm (“flight recorder”), to stimulate the development and consolidation of appropriate systems of ethical standards by developers and manufacturers).
11. Improvement of t*echnical regulation* and conformity assessment (for certain areas of application of AI and RT systems - updating and maintenance throughout the life cycle).
12. Improvement of the process for the protection of rights to the results of intellectual activity.

In addition to the above, sectoral directions are also required to improve the regulation of AI and RT technologies (in the area of health care, public administration, industry, transport, urban planning, and the financial sector, or others).

If we compare the Concept with the White Paper of Artificial Intelligence [80], intended for public discussion, then the Russian Concept is a more multifaceted document that considers some issues related to the legal regulation of AI and RT systems [25].

Also in Russia, some standards have been adopted that regulate a particular area in which AI systems are used (Robots and robotic devices. Classification [9]; Information support of technology and operator activities, Artificially intellectualized human-information interaction, General provisions [18]; Means for monitoring behavior and predicting people's intentions. Terms and definitions [7], and some others.

In the Republic of Belarus, the development and application of methods and means of artificial intelligence have been going on for a long time. The Institute of Technical Cybernetics (now - the United Institute of Informatics Problems of the Academy of Sciences of Belarus) was the pioneer in image recognition, digital mapping, speech recognition, and synthesis back in the 70s and 80s of the last century [15].

In the Republic of Belarus, one of the most severe legal problems associated with the regulation of AI systems is the breadth of possibilities and spheres for its application and the associated need to adapt many regulations. While there is currently no single normative document regarding AI development, projects in this area are already actively implemented.

In the Strategy for the Development of Informatization in the Republic of Belarus for 2016–2022 [71] AI and RT are considered to be one of the main directions in information and communication technologies. However, the State Program for the Development of the Digital Economic and the Information Society for 2016–2020 [13] did not contain special and clear instructions on developing AI and RT and improving legal regulation in this area.

In the National Strategy for Sustainable Socio-Economic Development of the Republic of Belarus for the term up to 2030 [10], developments to improve artificial intelligence come to the fore: computer eye contact, smart speech processing, recognition of other biological characteristics.

The Program of Activities of the Government of the Republic of Belarus for the term up to 2025 [7] describes the task of creating and implementing "projects of the future", including robotics.

Within the limits of the State Program "Digital Development of Belarus" for 2021–2025 [67], it is supposed to deal with such problems as creating favorable conditions for ensuring and supporting digital development processes on an integrated basis. During the realization of the State Program, the duration of the period of formation (adoption) of the relevant regulatory framework becomes a serious obstacle that will complicate the achievement of not only its consolidated goal but also the target indicators.

The Strategy "Science and Technology: 2018–2040" [70], which defines the long-term conceptual vision of the country's development based on a predominantly intellectual factor, includes the full-scale implementation of digital technologies that form the technological core of the intellectual economy, a component of which is, among other things, software based on systems of artificial intelligence and supportive machine learning.

The conditions for the advancement of the IT industry are created by the Decree of the President of the Republic of Belarus "On the Development of the Digital Economy" [2]. Within the Hi-Tech Park (HTP) framework, it is planned to conduct an experiment to test new legal institutions for the possibility of their implementation into the country's legislation. The new HTP legal mechanism greatly increases the number of possible business models for its residents and also meets the requirements and needs of international organizations that will be able to establish offices with a wide range of functions: unmanned vehicle management systems; business solutions based on artificial intelligence, and neural networks [43]. In this case, the Hi-Tech Park acts as a training ground where various experimental legal regimes can be applied, which can also be extremely in demand in other AI systems, for example, in healthcare, transport, and logistics, public administration, for construction and fuel-energy complex. As part of a legal experiment, it is possible to form a legal field with more flexible state regulation, contributing to the formation of a stimulating environment with the simultaneous creation of mechanisms to protect consumers of digital services and products that are in the zone of increased legal risk [34].

We will note that all the above-described documents set themselves the main goal in the form of stimulating the development of AI technologies due to their efficiency from an economic point of view. Nevertheless, the legal regulation remains an unresolved issue and all the proposed regulation options are a scientific discussion. This situation is largely due to the lack of established clear approaches to legal regulation - the Concept of legal regulation of the use of AI and RT systems technologies.

5 Conclusions

Currently, the implementation of investments is taking place in the digital transformation of the economy, which cannot become less favorable than in other economic relations. In order not to reduce the degree of investment activity and investment protection, public relations associated with the use of AI and RT should be regulated according to specific approaches and when resolving the following issues.

1. What is essential in predicting the directions of legal regulation of the implementation and use of AI and RT systems concerning the tasks of the investment sphere of the economy, and what should be understood in this case?

1.1 It is essential to understand and set a specific goal of legal regulation. The goal of regulating relations in AI and RT should be to stimulate the introduction and use of these technologies, which will contribute to the achievement of high rates of economic growth. Hence, it is necessary to remove legal barriers that prevent the development and application of AI systems in different fields of the economy and social sphere. Simultaneously, government regulation in its quest for complete control should not hinder the development of innovative technologies.

1.2 It is essential to understand that the regime of legal regulation of AI and RT systems should consider the need to guarantee the protection of human rights and freedoms and the introductory provisions that are universal: security, liability, and risk insurance issues.

1.3 When developing legislation on AI and RT, it is necessary to consider the general directions of legal regulation of the digital economy as a whole concerning the conditions of a particular country. At the same time, it is necessary to integrate and diversify existing conceptual approaches, theories, and methods of legal regulation, including self-regulation of public relations in using digital technologies. The introduction of foreign experience should consider the national system and the specifics of the law-making process.

2. What is the optimal solution in the formation of legal regulation? The optimal solution for countries that are at the start of the path towards forming the necessary legal framework will be creating a state concept for the development of legal regulation of the use of AI and RT systems, containing problems, general approaches, and directions of regulation.

Among the *conceptual problem areas* of regulation, the following can be distinguished: maintaining a balance between the demand for the protection of personal information and the need to use them for training AI and RT systems; determination of the subject and boundaries of regulation of the scope of use of AI and RT systems; identification of the AI and RT systems when interacting with a person, including informing him about interaction with such a system; legal "delegation" of solutions to AI and RT systems; liability for causing harm using AI and RT systems; the legal regime of the results of intellectual activity created using AI and RT systems; the use of probabilistic estimates for decision-making by AI and RT systems and the impossibility, in some cases, of a complete explanation of their decision (the problem of algorithmic transparency of systems).

The result of the approval and implementation of the concept will be the achievement of legal certainty in the AI and RT, the development of a regulatory framework that provides the necessary degree of protection of human and civil rights and freedoms that corresponds to the interests of society and the state, and promotion of the development of AI and RT technologies.

3. What are the tasks facing legal science?

Legal science should first develop a general circle of the essential rules that will create the ground for the advancement of legal regulation or individual norms of those modifications that will impact industry legislation and the conceptual apparatus.

3.1 The major principles that must be described should involve: the primacy of human rights and their protection in the use of AI and RT systems; the responsibility of the developers and owners of the technologies of the AI and RT systems for what they do (the higher the autonomy of the AI and RT, the less responsibility the proprietor of the robot should bear); AI systems and a robot are only a thing, they cannot be humanized or considered a human (if in the ethical sphere it is possible to use vague wording, then in the sphere of the law the concept should be clearly defined); introduction of insurance compensation in case of harm caused by AI and RT technologies; strict control of the creation and distribution of robotics by the state, since the creation of robots that can independently decide to kill an enemy is quite likely.

We believe that over time, a set of principles will be supplemented that will form the basis of legal regulation of the area of artificial intelligence, cyber-physical systems, neural networks based on it, and robots and robotics objects.

3.2 Currently, the terms and definitions in the field of technologies of AI and RT systems (to define the subject of regulation) are trying to be modeled in such a way as to cover the broadest possible range of activities for the future development of artificial intelligence, and not to give a clear and concise explanation of the basic concept from a technical point of view. The best way in the state would be to experiment with artificial intelligence to determine how the formulated concepts should be changed.

4. Legal regulation of the use of AI and RT systems technologies can begin and be carried out within the legal systems of different or several states. However, to become effective and large-scale, it must be internationally recognized and applied. Accordingly, its development, distribution, and action are possible through international law and integration law, ultimately having a global focus. Simultaneously, integration mechanisms should be launched first on the most critical issues related to the prospects for human survival in competition with artificial intelligence in the digital age.

References

1. Civil Law Rules on Robotics European Parliament resolution of February 16 2017 with recommendations to the Commission on Civil Law Rules on Robotics (2017). https://www.europarl.europa.eu/doceo/document/TA-8-2017-0051_EN.html
2. Decree of the President of the Republic of Belarus “On the Development of the Digital Economy” No. 8 (2017). https://pravo.by/document/?guid=12551&p0=Pd1700008&p1=1&p5=0
3. Federal Law of July 31, 2020 “On Experimental Legal Regimes in the Field of Digital Innovation in the Russian Federation (2020). http://www.consultant.ru/document/cons_doc_LAW_358738/
4. Montréal Declaration: Responsible AI (2017). https://www.montrealdeclaration-responsibleai.com/the-declaration
5. National program “Digital Economy of the Russian Federation”: minutes of the meeting of the Presidium of the Council under the President of the Russian Federation for Strategic Development and National Projects (2019). https://digital.gov.ru/ru/activity/directions/862/
6. National standard of the Russian Federation: Information support for equipment and operator activities. Artificially intellectualized human-informational interaction. General Provisions (2017). http://docs.cntd.ru/document/1200146327
7. National standard of the Russian Federation: Means for monitoring behavior and predicting people’s intentions. Terms and definitions (2019). http://docs.cntd.ru/document/1200170425
8. National Standard of the Russian Federation: Robots and Robotic Devices. Classification (2016). http://docs.cntd.ru/document/1200142401
9. National Strategy for Artificial Intelligence.Denmark. Ministry of Finance and Ministry of Industry, Business, and Financial Affairs (2019). https://eng.em.dk/media/13081/305755-gb-version_4k.pdf
10. National Strategy for Sustainable Socio-Economic Development of the Republic of Belarus for the Period up to 2030: Minutes of the Meeting of the Presidium of the Council of Ministers of the Republic of Belarus (2017). https://www.economy.gov.by/ru/dejst_prognoz_dok-ru/
11. National strategy for the development of artificial intelligence for the period up to 2030: Decree of the President of the Russian Federation No. 490 (2019). http://www.consultant.ru/document/cons_doc_LAW_335184/

12. State Program "Digital Development of Belarus" for 2021–2025: Resolution of the Council of Ministers of the Republic of Belarus No. 66 (2021). https://pravo.by/document/?guid=3871&p0=C22100066
13. State Program for the Development of the Digital Economy and the Information Society for 2016–2020: Resolution of the Council of Ministers of the Republic of Belarus No. 235 (2016). https://www.pravo.by/document/?guid=12551&p0=C21600235&p1=1
14. The program of activities of the Government of the Republic of Belarus for the period up to 2025: Resolution of the Council of Ministers of the Republic of Belarus No. 758 (2020). https://pravo.by/document/?guid=12551&p0=C22000758&p1=1&p5=0
15. Ablameiko, M., Ablameiko, S.: Legal regulation of the interaction of artificial intelligence systems and humans. Sci. Innovat. **1**(203), 40–44 (2020)
16. AI for Humanity. French strategy for artificial intelligence. France (2018). https://www.aiforhumanity.fr/en/
17. Alekseev, N.V.: Artificial intelligence and robotization: legal problems of intellectual property. Intell. Proper. Law **3**, 42–46 (2019)
18. Arkhipov, V.V., et al.: Regulation of robotics: an introduction to "Robo-law". Legal aspects of the development of robotics and artificial intelligence technologies. Neznamova, A.V. (ed.) Infotropic Media, Moscow, p. 232 (2018)
19. Artificial Intelligence Index Report 2019: In: R. Perrault, Y. Shoham, E. Brynjolfsson et al. (eds). Stanford University, Stanford p. 290 (2019)
20. Asilomar AI Principles: Future of Life (2017). https://futureoflife.org/ai-principles/?cn-reloaded=1
21. Atabekova, A., Yastrebov, O.: Legal status of artificial intelligence across countries: legislation on the move. European Res. Stud. J. XXI **4**, 773–782 (2018)
22. Balkin, J.M.: The path of robotics law. Calif. Law Rev. **6**, 45–60 (2015)
23. Balykina, E.N., Popova, E.E., Buzun, D.N.: Fundamentals of computer science and information technology: teaching methodological complex. Part 2. BSU, Minsk, p. 96 (2008)
24. Baranov, P.P.: Legal regulation of robotics and artificial intelligence in Russia: some approaches to solving the problem. North Caucasian Law Bul. **1**, 39–45 (2018)
25. Begishev, I.R.: The concept of development of regulation of relations in artificial intelligence technologies and robotics: analysis of the main provisions. Baikal Res. J. 11 (40) (2020). http://brj-bguep.ru/reader/article.aspx?id=24234
26. Belsky, V., Markevich, D., Satolina, M.: Conceptual directions of legal support for implementing artificial intelligence. Sci. Innovat. **1**(201), 58–63 (2019)
27. Bochkov, A.A.: Methodological and legal foundations of artificial intelligence and robotics in foreign countries and the Republic of Belarus. Law. Econ. Psychol. **3**, 3–13 (2019)
28. Calo, R., Froomkin, A.M., Kerr, I.: Robot Law. Edward Elgar Publishing, p. 424 (2016)
29. Čerka, P., Grigienė, J., Sirbikytė, G.: Liability for damages caused by artificial intelligence. Comput. Law Secur. Rev. **31**(3), 376–389 (2015)
30. Ethics guidelines for trustworthy AI (2019). https://ec.europa.eu/digital-singlemarket/en/news/ethics-guidelines-trustworthy-ai
31. Filipova, I.A.: Legal regulation of artificial intelligence: textbook. Nizhny Novgorod, Nizhny Novgorod State University, p. 90 (2020)
32. Gadzhiev, G.A.: Is the robot agent an entity? (search for legal forms to regulate the digital economy). J. Russian Law **1**(253), 15–30 (2018)
33. Gasson, M.N., Koops, B.-J.: Attacking human implants: a new generation of cybercrime. Law Innov. Technol. **5**(2), 248–277 (2013)
34. Gavrilchenko, Y.P., Khvatik, Y.A.: Experimental legal regimes: economic and legal view. Bank. Bull. **12**, 3–8 (2020)

35. Grivsky, K.K.: Creativity and artificial intelligence. Legal aspects of creating a work. Intellect. Prop. Copy. Relat. Rights **1**, 55–60 (2020)
36. Ibragimov, R., Suragina, E.: The law of machines. How to bring a robot to justice. Corporate Lawyer **11**, 10–17 (2017)
37. Inshakova, A.O., Goncharov, A.I., Ershova, I.V.: Digital Blockchain Registration of legally significant stages of complex good's export-import supplies by business entities of the EAEU and BRICS jurisdictions. In: Competitive Russia: Foresight Model of Economic and Legal Development In The Digital Age: Proceedings of the International Scientific Conference in Memory of Oleg Inshakov (1952–2018). In: Inshakova, A.O., Inshakova, E.I. (eds.). Cham: Springer Nature, pp. 328–336 (2020). https://doi.org/10.1007/978-3-030-45913-0
38. Inshakova, A.O., Goncharov, A.I., Mineev, O.A., Sevostyanov, M.V.: Amendments to the Civil Code of the Russian Federation: Contradictions of Theory and Practice. Contributions to Economics. Russia and the European Union: Development and Perspectives Number XIII. 147–153 (2017)
39. Johnson, D.G.: Technology with no human responsibility? J. Bus. Ethics **127**(4), 707–715 (2014). https://doi.org/10.1007/s10551-014-2180-1
40. Kalinina, A.E., Inshakova, A.O., Goncharov, A.I.: Polysubject jurisdictional blockchain: electronic registration of facts to reduce economic conflicts. Stud. Computat. Intell. **826**, 205–213 (2019)
41. Khisamova, Z.I., Begishev, I.R.: Legal regulation of artificial intelligence. Baikal Res. J. 10 (2) (2019). http://brj-bguep.ru/reader/article.aspx?id=23011
42. Kholodnaya, E.V.: On perceptive directions of legal regulation in the field of AI technology. MGYuA Bull. **12**, 89–96 (2019)
43. Kunitskaya, O., Ezhova, A.V.: New tools of the digital economy: legal problems of application in business. In: Digitalization of the Economy and Society: Problems, Prospects, Security: Conference proceedings, March 29, 2019. Donetsk. 1, pp. 130–135 (2019)
44. Laptev, V.A.: The concept of artificial intelligence and legal responsibility for its work. Law J. High. School Econ. **2**, 79–102 (2019)
45. Legal regulation of digital technologies in Russia and abroad. The role and place of legal regulation and self-regulation in the development of digital technologies: monograph. In: Minbaleeva, A.V. (ed.). Saratov, Amirit, p. 207
46. Mamychev, A., Miroshnichenko, O.I.: Modeling the future of law: problems and contradictions of legal policy in the field of normative regulation of artificial intelligence systems and robotic technologies. Legal Policy Legal Life **2**(125–13), 3 (2019)
47. Matytsin, D.E.: Internet-investing as a remote algorithm of the retail investment financing. In: Modern Global Economic System: Evolutional Development vs. Revolutionary Leap. Springer book series "Lecture Notes in Networks and Systems". Cham: Springer Nature (2021). https://doi.org/10.1007/978-3-030-69415-9
48. McCarty, L.T.: Finding the Right Balance in Artificial Intelligence and Law. In book: Research Handbook on the Law of Artificial Intelligence. Chapter: 3. Edward Elgar Publishing (2017)
49. Morhat, P.M.: Artificial Intelligence: A Legal Perspective. BukiVedi, Moscow. p. 257 (2017)
50. Mustafina, A.F.: Artificial intelligence technology in the context of the business environment. Bus. Strateg. Anal. Forecast, Manag. **7**, 8–12 (2019)
51. Neshitoi, A.S.: Investments: Textbook. 5th edn. Rev. and add. Moscow p. 372 (2007)
52. Neznamov, A.: Laws of Robotics. How to Regulate Artificial Intelligence. Forbes (2018). https://www.forbes.ru/tehnologii/355757-zakony-robototehniki-kak-regulirovat-iskusstvennyy-intellekt

53. Neznamov, A.: Regulation of AI in world practice (2020). https://ethics.cdto.center/3_8#link198
54. Notice of the State Council Issuing the New Generation of Artificial Intelligence Development Plan. China. Foundation for Law and International Affairs (2017). https://flia.org/wp-content/uploads/2017/07/A-New-Generation-of-Artificial-Intelligence-Development-Plan-1.pdf
55. Pagallo, U.: The Laws of Robots: Crimes, Contracts, and Torts (Law, Governance and Technology Series). Springer Science & Business Media, p. 181 (2013). https://doi.org/10.1007/978-94-007-6564-1
56. Pan-Canadian Artificial Intelligence Strategy. Canada. CIFAR (2018). http://www.jaist.ac.jp/~bao/AI/OtherAIstrategies/Pan-Canadian%20Artificial%20Intelligence%20Strategy.pdf
57. Policy and investment recommendations for trustworthy Artificial Intelligence (2019). https://ec.europa.eu/digital-single-market/en/news/policy-and-investment-recommendationstrustworthy-artificial-intelligence
58. Ponkin, I.V., Redkina, A.I.: Artificial intelligence from the point of view of the law. RUDN Bull. Ser. Legal Sci **22**(1), 91–109 (2018)
59. Ponomareva, E.V.: Subjects and Quasi-Subjects of Law: Theoretical and Legal Problems of Delimitation. Ph.D. thesis. Yekaterinburg, p. 208 (2019)
60. Popova, A.V., Gorokhova, S.S., Aznagulova, G.M.: On the issue of the need to determine the goals and objectives of legal regulation in AI, cyber-physical systems, robots, and robotics objects. Legal State Theory Pract. **3**(57), 31–36 (2019)
61. Report of COMEST on robotics ethic: UNESDOC. (2017). https://unesdoc.unesco.org/ark:/48223/pf0000253952
62. Rusakova, E.P., Frolova, E.E., Gorbacheva, A.I.: Digital rights as a new object of civil rights: issues of substantive and procedural law. Adv. Intell. Syst. Comput. **1100**, 665–673 (2020)
63. Saripan, X.: Are robots human? Rev. Legal Person. Mod.World Appl. Sci. J. **34**(6), 824–831 (2016)
64. Savenok, A.L.: Self-driving injury: skill issues. Bull. Acad. Minist. Internal Affairs Repub. Belarus **2**, 153–156 (2018)
65. Scherer, M.U.: Regulating artificial intelligence systems: risks, challenges, competencies, and strategies (May 30, 2015). Harvard J. Law Technol. 29(2) (2016)
66. Shapsugova, M.D.: Artificial intelligence in business law. Mod. Sci. **8–2**, 167–173 (2019)
67. Shestak, V.A.: Modern Needs of Legal Support of Artificial Intelligence: a view from Russia. All-Russian Criminol. J. **13**(2), 197–200 (2019)
68. Somenkov, S.A.: Artificial intelligence: from object to subject? MSLA Bull. **2**, 75–85 (2019)
69. Stone, P. et al.: Artificial Intelligence and Life in 2030. One Hundred Year Study on Artificial Intelligence: Report of 2015–2016. Stanford, Stanford University (2016). http://ai100.stanford.edu/2016-report
70. Strategy "Science and Technology: 2018–2040": Resolution of the Presidium of the National Academy of Sciences of Belarus (2018). http://www.gknt.gov.by/news/aktualno/strategiya_nauka_i_tekhnologii_2018_2040/
71. Strategy for the development of informatization in the Republic of Belarus for 2016–2022: minutes of the meeting of the Presidium of the Council of Ministers of the Republic of Belarus (2015). http://nmo.basnet.by/concept/strategia2022.php
72. The AI Index 2018: Annual Report. Shoham, Y., Perrault, R., Brynjolfsson, E. et al. Stanford University, Stanford. p. 94 (2018)

73. The concept for the development of regulation of relations in the field of artificial intelligence and robotics technologies until 2024: Order of the Government of the Russian Federation of August 19, 2020. https://www.garant.ru/products/ipo/prime/doc/74460628/ (2020)
74. The results of a comprehensive forecast of the scientific and technological progress of the Republic of Belarus for 2021–2025 and for the period up to 2040. (2020). In: Shumilin, A.G. (ed.). Minsk, p. 92
75. Tolochko, O.N.: Trends in the legal regulation of objects and technologies associated with artificial intelligence. Just. Belarus **3**, 35–39 (2019)
76. Tripathi, S., Ghatak, C.: Artificial intelligence and intellectual property law. Christ Univ. Law J. **7**(1), 83–97 (2018)
77. Vasiliev, A.A., Shpoper, D.: Artificial intelligence: legal aspects. IzvestiaAltSU. Legal Sci. **6** (104), 23–26 (2018)
78. Vitko, V.S.: Analysis of scientific ideas about the author and the rights to the results of artificial intelligence. Intell. Prop. Copyright Relat. Rights **2**, 5–20 (2019)
79. White Paper on Artificial Intelligence: a European approach to excellence and trust (2020). https://ec.europa.eu/digital-single-market/en/news/white-paper-artificial-intelligence-european-approach-excellence-and-trust
80. Yastrebov, O.A.: Legal personality of an electronic person: theoretical and methodological approaches. Proc. Inst. State Law Russian Acad. Sci. **13**(2), 36–55 (2018)

New Management Technologies in the Context of Digitalization

Olga B. Digilina[1(✉)], Irina B. Teslenko[2], and Inna V. Andronova[1]

[1] People' Friendship University (RUDN University), Moscow, Russia
[2] Vladimir State University named after Alexander and Nikolay Stoletovs, Vladimir, Russia

Abstract. The purpose of the chapter is to determine the prospects for the development of new management technologies in the context of digitalization. These technologies include simulation modelling.

The authors attribute to the advantages of simulation modelling the ability to describe a complex system at a high level of detail; high adequacy between the physical essence of the process and its model; broader scope of research than analytical modelling; the ability to assess the behaviour of the system, both in stationary and transient dynamic modes; getting more data about the object; no restrictions on the relationship between the parameters of the model and the most attractive ratio of the result and costs in comparison with analytical and physical modelling.

The authors argue that the development of digitalization in Russia has made it possible to start actively using IM. The construction industry has made a serious step forward in this sense: a new urban planning approach is being introduced in Russia - the Building Information Model.

Keywords: Simulation · Digitalization · BIM technologies · Digital platforms

JEL Codes: L05 · L74 · L96

1 Introduction

The rapid development of digitalization of all aspects of society's life has led to a change in approaches to management. High adaptability, the use of new methodological and instrumental methods and decision-making techniques are becoming more and more characteristic of the modern management paradigm.

This is because the opened possibilities of modern information technologies make it possible to assess the influence of different ideas on the results of an organization's activities, taking into account many environmental factors [10].

Uncertainty and the multifactorial nature of the environment explain the demand for models instead of real objects for assessing possible alternatives for their development. And simulation technology (IM) is best suited for this.

They are used when: an experiment on a real object is very expensive; it is impossible to build an analytical model or there are no methods for solving the

E. G. Popkova (Ed.): ISC 2020, LNNS 368, pp. 333–339, 2022.
https://doi.org/10.1007/978-3-030-93244-2_37

resulting model; you need to understand the behaviour of the system in time. In these cases, they resort to simulators (simulation modelling) to research the design, analysis, and assessment of the functioning of the object [9]. Simulation allows predicting scenarios with the introduction of additional parameters ("what will happen if…") and see the real consequences of the identified alternatives.

Unlike the usual modelling methods (for example, Excel), the simulation model uses the methods of agent modelling (in the model, instead of numbers, agents or objects are represented that can move in space, interact with each other, exchange messages), discrete-event modelling (in the model, as in life, there is a time in which events occur that are related to each other) and system dynamics (a method of describing the dynamics of complex business systems using streams and storage devices). This makes the IM technology attractive for both state enterprises and the commercial sector [2].

2 Materials and Method

Modelling is the informational perception of a particular physical object, and the resulting models are its informational images. Robert Shannon defines imitation modelling as follows: it is the process of constructing a model of a real system and setting up experiments on this model to either understand the behaviour of the system or evaluate (within the constraints imposed by some criterion or a set of criteria) various strategies that ensure the functioning of this system [11]…

The purpose of this article is to determine the place of simulation in the system of modern project management in various industries. To achieve this goal, the authors carried out a content analysis of publications on the INTERNET, articles by Russian and foreign authors, websites of large enterprises in Russia.

3 Results

IM is used in various areas of government and business. The objects of simulation are production, infrastructure, factor and industry markets, competition, road, air, sea traffic, ecology, information security, etc.

Subways, train stations, airports use simulation models to increase throughput: they help organize the operation of services in such a way that people do not have to stand in crowds and queues for a long time.

In modern conditions, a model of office work is in demand to minimize the risk of infection of employees.

For the operation of railways, models for scheduling the scheduling of locomotives are relevant, for industrial enterprises - models for the rational placement of shop equipment. Aircraft cabin models help optimize infrastructure to serve passengers faster and not block aisles [2].

Many Russian companies, such as "Magnit" or "Eldorado", use IM in solving problems in the sphere of logistics. In the area of customer service, the technology is used primarily by banks.

IM is used for the examination of projects to identify bottlenecks, optimize costs and calculate effective operating modes [12].

IM is used not only in production. An interesting project was the IM Polytechnic Museum in Moscow. The goal of the project was to simulate the flow of visitors in the museum; understand where and in what quantity to place services: ticket offices, ticket terminals, wardrobes, and even toilets, to avoid the "crush" and queues, to increase the level of comfort for museum visitors [2].

The use of digital platforms generally makes it possible to design large-scale projects. By integrating information models of objects, engineering systems and the external environment, it is possible to create a single "digital twin" of the city.

Proactive analytics and a set of models embedded in digital platforms make it possible, even before the start of construction work, to determine electromagnetic permeability (to organize high-quality communication and ensure its availability for industrial users and citizens), wind load parameters (to determine the peak values of dynamic changes in load on carriers) building structures to determine the design solutions and the choice of materials) and temperature and humidity conditions (for the design of water supply systems, air conditioning and, in part, power grids).

Without a geological study, it is possible to simulate the characteristics of the soil of a construction site to immediately determine the use of certain construction technologies.

The "digital twin" of the city helps to calculate and equip the storm sewer systems, volumetric models of the location of communication stations, electrical networks, water pipelines in advance. Thus, IM changes the level of urban planning, making it possible to ensure a high quality of life for the population [5].

In general, the benefits of MI are obvious. These include the ability to describe a complex system at a high level of detail; high adequacy between the physical essence of the process and its model; broader scope of research than analytical modelling; the ability to assess the behaviour of the system, both in stationary and transient dynamic modes; getting more data about the object; no restrictions on the relationship between the parameters of the model and the most attractive ratio of the result and costs in comparison with analytical and physical modelling.

At the same time, MI has certain limitations.

From the point of view of MI as a technology, it is based on assumptions that can lead to deliberate manipulation or common mistakes.

It is impossible to simulate reality as it is due to a large number of parameters. It is required to simplify it and focus on precisely those parameters that have or may have a significant impact on the system under certain circumstances.

For simulation, accurate input parameters are important (the accuracy of the model depends on this) and the choice of the correct level of abstraction (model detail), corresponding to the problem being solved [10].

In general, the solution obtained on the simulation model is always private, since it corresponds to fixed elements of the structure, algorithms of behaviour and values of the parameters of the system [15].

If we talk about the practice of using IM, then this technology was widely used in our country back in the 1960s–1980s. Since the late 1980s and into the 1990s, simulation technology has barely evolved.

In the countries of Europe and the USA, the situation is different. Numerous consulting companies have been working there for a long time, providing services for the development of simulation models for businesses, societies and information portals are functioning to disseminate the world experience of IM in various fields of activity [3].

Examples include such societies as the Europe Council for Modeling and Simulation (Europe), The Society for Modeling & Simulation International (USA).

Over the past decade, the situation with MI in Russia has improved. Researchers began to use the GPSS World simulation system, foreign software.

Back in February 2011, the National Society for Simulation was created in the Russian Federation. Its goal is to establish communication between specialists in the field of modelling in Russia, cooperate with foreign colleagues. In 2015, the National Society for Simulation of Russia was admitted to the permanent members of the European Association of Societies for Simulation EUROSIM.

The unfolding digitalization in Russia made it possible to start active use of IM. The construction industry has made a serious step forward in this sense: a new urban planning approach is being introduced in Russia - the Building Information Model (BIM) [8].

The use of this technology allows you to monitor the state of an object throughout the entire life cycle, improves the quality of construction, helps to reduce the risk of serious errors and losses in the implementation of large-scale projects.

The concept of BIM appeared in the Urban Planning Code of our country in 2019. The model is a collection of information, documents and materials that are collected in electronic format at all stages of the "life" of an object - from construction to reconstruction or demolition.

In September 2020, Russia adopted Resolution No. 1431, which approves the rules for the formation and maintenance of the information model, as well as the composition of the information included in it [1].

So far, the practice of using BIM technologies in Russia lags behind foreign ones.

A PwC PropTech study found that the cumulative global BIM market in 2019 approached $5.2 billion. It is expected that by 2027 it will reach the level of 15.5 billion. The volume of the Russian BIM market at the end of 2019 ranges from 67–77 million dollars [14].

In Russia, not so long ago, it was decided that now the construction of social facilities at the expense of the budget will be carried out with the obligatory use of the BIM information model. According to expert estimates, this will lead to a reduction in costs by 2%, construction time and errors in design documentation - by 10% [14].

According to the Ministry of Construction, more than 50% of all new buildings in the country are constructed using information modelling. Despite the slow implementation, there are examples of the successful use of BIM in Russia. These are some of the Olympic facilities in Sochi; stadiums built for FIFA-2018, airports in Moscow and St. Petersburg, coronavirus centre in the Moscow region [4].

As for developers in the field of housing construction, according to the ministry, there are about five thousand of them in the Russian Federation. Of these, only about a dozen companies professionally use information modelling (in the volume of construction capacity, they occupy about 20%, while, for example, in the UK in 2019, the level of use of BIM technologies reached 70%).

Experts believe that construction management processes need to be technologically advanced as much as possible, to switch to a smart construction management system to ensure the efficiency of project implementation and their safety.

Although the introduction of BIM increases the cost and time of design, in the next stages due to its use, cost savings are provided and the degree of validity of decisions increases.

The implementation of BIM takes about a year, the training course for software skills is 20–35 thousand rubles, the price of a single-user version is from 65 thousand rubles per year. It's not cheap, but if the project is large-scale and atypical, then a potential mistake can ultimately lead to more significant losses than the cost of BIM.

According to Russian officials, the formation and introduction of the information model are planned everywhere no later than 2021, and it becomes mandatory for construction projects under the state order. Moscow will become the first region to use information models [7].

In some foreign countries, BIM technologies are mandatory for projects under government orders and have been used, for example, in the UK since 2011, in Spain - since 2018, in Australia - since 2016. In Singapore, they are mandatory for the construction of buildings with an area of more than 5,000 square meters [4].

To summarize, it should be said that a simulation is a convenient tool for analyzing complex systems. In different areas of business, simulation gives a clear idea of the object and helps to choose the best solutions.

In addition to the effect of reducing costs, the model presents different scenarios of possible risk situations and makes it possible to choose a set of measures to minimize the consequences of risk.

Using the digital twin enables businesses to predict and create a competitive advantage. Applying the technology of simulation in practice, it is possible to calculate the behaviour of an object at different stages of life.

The modern development of digital technologies makes it possible to receive input data with the help of such new technologies as the "Internet of Things".

BIM technologies in Russia are beginning to be actively promoted. While the use of this tool is not mandatory, participants in the construction industry can use it based on economic feasibility.

But these technologies have undeniable advantages. It is no coincidence that the introduction of digital construction technologies, such as building information modelling, is one of the tasks of the Digital Economy national project [13] in the Digital Construction part. It is assumed that the use of BIM technologies in the implementation of the Housing and Urban Environment national project will lead to a reduction in construction costs by 20%, and the construction time by 30%.

4 Conclusion

The development of IM, no doubt, will take place in the future. Researchers are already talking about such main directions of development of MI technology as object-oriented MI systems, embedded MI, applications that support optimization, modules of resource planning tasks in an enterprise, emulation (imitation of the operation of one system through another without loss of functionality and distortion of results), simulators and web applications.

Information modelling technologies can be used even more efficiently in conjunction with such innovative technologies as virtual and augmented reality to visualize an object in a future location; Internet of Things (to replace traditional SCADA systems in the future); blockchain (for posting payments to contractors in real-time, to ensure supply chain management).

The researchers believe that BIM technologies can be used to solve the problems of overcoming the "office syndrome" (deterioration in the health of citizens after a long stay or residence in a building due to poor lighting, insufficient ventilation, etc.); to improve the energy efficiency of buildings and create smart homes from scratch [6].

It seems that the advantages of using information modelling technologies will be appreciated by Russian businesses and will begin to be actively applied in the practice of economic activity, and the results of its application will improve the quality of life of Russians.

Acknowledgement. This paper has been supported by the RUDN University Strategic Academic Leadership Program and Grant of RUDN University No. 060119-0-000.

References

1. Decree of the Government of the Russian Federation No. 1431 of September 15, 2020 "On approval of the Rules for the formation and maintenance of an information model of a capital construction object, the composition of information, documents and materials included in the information model of a capital construction object and presented in the form of electronic documents, and requirements for the formats of these electronic documents, as well as amending clause 6 of the Regulation on the implementation of engineering surveys for the preparation of design documentation, construction, reconstruction of capital construction facilities". https://www.garant.ru/products/ipo/prime/doc/74544278/
2. A digital predictor for business (2019). https://www.it-weekly.ru/it-news/tech/156712.html
3. Babina, O.I.: Analysis of the current state and prospects for the development of simulation. Stat. Econ. **6**, 205–210 (2014). https://doi.org/10.21686/2500-3925-2014-6-205-210
4. BIM technologies in 2020. Construction trends - Geoline. https://geoline-tech.com
5. City simulation is real today (2020). https://www.securitymedia.ru/news_one_10927.html. Accessed 29 June 2020
6. Digitalization in construction. Market in Russia and the world in 2020 (2019). https://geoline-tech.com/bim-construction-trends-2020/
7. Implementation of information modelling in Russia. For the construction of which objects it is obligatory to use TIM (2020). https://vc.ru
8. Mikhail Mishustin signed a decree on information modelling in construction (2020). http://government.ru/news/40424/
9. Simulation modelling (2019). https://mybiblioteka.su/2-42632.html
10. Simulation modelling for solving production and technological problems: the main advantages and prospects (2020). https://integral-russia.ru/2020/01/17/imitatsionnoe-modelirovanie-dlya-resheniya-proizvodstvenno-tehnologicheskih-zadach-osnovnye-preimushhetsva-i-perspektivy/

11. Simulation modelling. Robert Shannon (2020). https://present5.com/imitacionnoe-modelirovanie-robert-shennon-imitacionnoe-modelirovanie-est/
12. Tapes of OmSTU (2019). https://www.omgtu.ru/l/?eid=4890324.10.2019
13. The Cabinet of Ministers will clarify the cases of using information modelling technologies in construction (2019). https://tass.ru/ekonomika/10307553
14. The Ministry of Construction of the Russian Federation intends to apply BIM models in urban planning (2019). https://rg.ru
15. The process of simulation (2019). https://poisk-ru.ru/s32613t9.html

Digital Transformation of Organizations as a New Prerequisite for Effective Operations

Elina A. Pakhomova[1], Elena V. Popova[1(✉)], Andrey R. Tikhonov[1], Ivan A. Antonyuk[2], and Alexander A. Lisanov[3]

[1] Plekhanov Russian University of Economics, Moscow, Russia
Popova.EV@rea.ru
[2] "AS ITR" LLC, Moscow, Russia
[3] "GRTPRODUCTION" LLC, Moscow, Russia

Abstract. The paper examines the digital transformation of organizations and the problem of low efficiency of organizations with a low level of digital development. The authors study the dependence of the growth rate of the company's digitalization on demand for consumer interaction between the company and the consumer. Additionally, the authors note the relationship between the type of company and the rate of its digitalization, as well as the potential risks of digitalization and its benefits in the interaction between individuals and the government. The paper aims to review and identify the relationship between organizational performance and digital development. The data reviewed confirm this relationship. Moreover, the data indicate that digitalization provides new opportunities for organizations to interact with customers and users and increase the customer inquiry on the digital level of organizations, which creates the need for digitalization of organizations. The authors identify the risks of digitalization, such as the reduction of low-skilled jobs, which, combined with economic shocks, can lead to a significant social stratification of citizens without government intervention. It is also noted that the COVID-19 pandemic has been a significant impetus for accelerated digital development.

Keywords: Management · Digital economy · Modernization · Digitalization · Competitive environment

JEL Codes: P0 · O10 · O30

1 Introduction

Nowadays, technology is rapidly developing, and consumer requests are becoming more demanding. Therefore, organizations are forced to offer better products and services and become more customer-oriented at every stage. They have to offer the most optimal final products and make every stage of interaction with the client more elaborated (i.e., convenient and comfortable for clients) and effective. These conditions create the problem, which lies in the lack of digitalization of organizations and companies and can cause a competitive loss in the market despite the quality of products and services [5].

E. G. Popkova (Ed.): ISC 2020, LNNS 368, pp. 340–348, 2022.
https://doi.org/10.1007/978-3-030-93244-2_38

Digitalization encompasses the activities of various organizations directly or indirectly related to digital technology – companies directly involved in the development of digital technologies and solutions and organizations that require these solutions to carry out their activities.

2 Materials and Methods

An apparent indicator of the need to develop digital technology is the practice of transferring employees to remote work. Companies that managed to transform their business processes to work remotely or companies that were initially focused on the digitalization of labor and reducing the direct location of personnel at the workplace saw only a slight reduction in their productivity. In turn, companies that found it impossible to build remote work due to several factors (e.g., the specifics of the business), such as confectionery or restaurant businesses, as well as internal organizational and technical problems, experienced great difficulties.

Nevertheless, when developing digital technologies at the level of organizations leading in the regions in terms of employment, it is necessary to consider the factors noted in this research, such as the elimination of low-skilled jobs from 20% to 30% [1], which, combined with the increased unemployment rate associated with the COVID-19 pandemic [7], can lead to social stratification and tension in economically prosperous countries.

3 Results

Changes in the digital component are becoming more common for each stage of interaction between organizations and customers (Fig. 1).

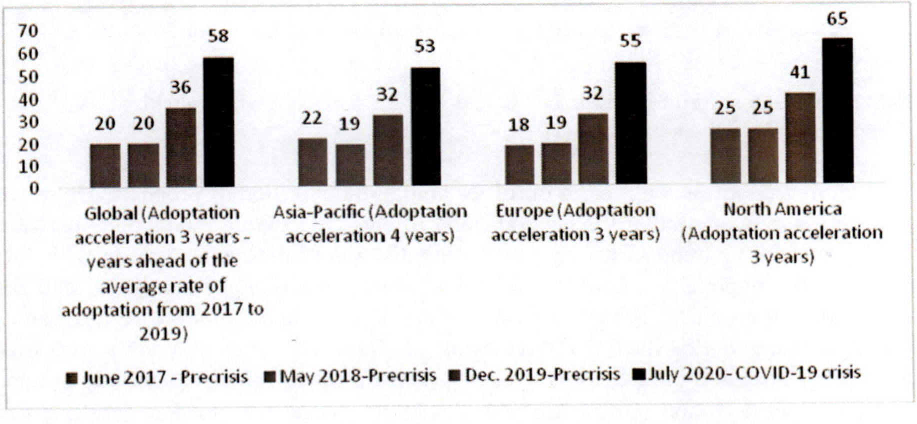

Fig. 1. Average share of digital interactions with customers, %. *Source* Compiled by the authors based on [6]

Each of the regions has seen a dramatic increase in digital consumer interactions in recent years. COVID-19 has considerably influenced digitalization. Nevertheless, even the year before the pandemic had already seen a significant increase in digital interactions. Thus, on the one hand, the COVID-19 pandemic has accelerated the growth of digital interactions – customers and companies have been forced to adapt to events and change their behavior as customers and producers, which has resulted in increased digital interactions. On the other hand, we can assume that this growth has been due to the existing digital base for interactions in the digital environment, which has already had a certain potential, and COVID-19 has served as the impetus for its active development.

As noted above, customer inquiries affect changes in the products. Simultaneously, the habituation of customers to the updated products forms a new environment for them and their interactions. In turn, competition between companies and products forces companies to constantly improve the products, which also increases customer inquiries [8]. The growth of digital interactions causes the creation of digital products. And vice versa, the digitization of products implies digitization of interactions – the creation of a single image of company behavior and its digital image (Fig. 2).

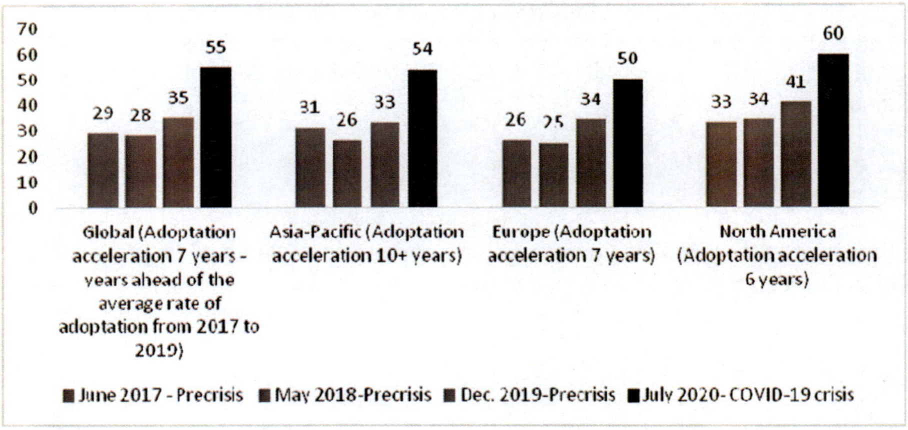

Fig. 2. Average share of products or services that are partially or fully digitized, %. *Source* Compiled by the authors based on [2]

Figure 2 confirms the growth of full or partial digitalization of products. However, it is necessary to note that the general trend of changing digital interactions between customers and the trend of full or partial digitalization of products coincide. This fact confirms the hypothesis about the relationship between digital interactions and the digitization of products. There is also an upward trend in digital products in 2019 and a significant jump during the COVID-19 pandemic. This fact indicates a faster process of technology adoption once there is a market demand for it. Moreover, we can assume that rapid digitalization will continue in the future due to new market demands and customer habituation to the new digital level, which will create the need for organizations to implement further digitalization to operate in a competitive market (Fig. 3).

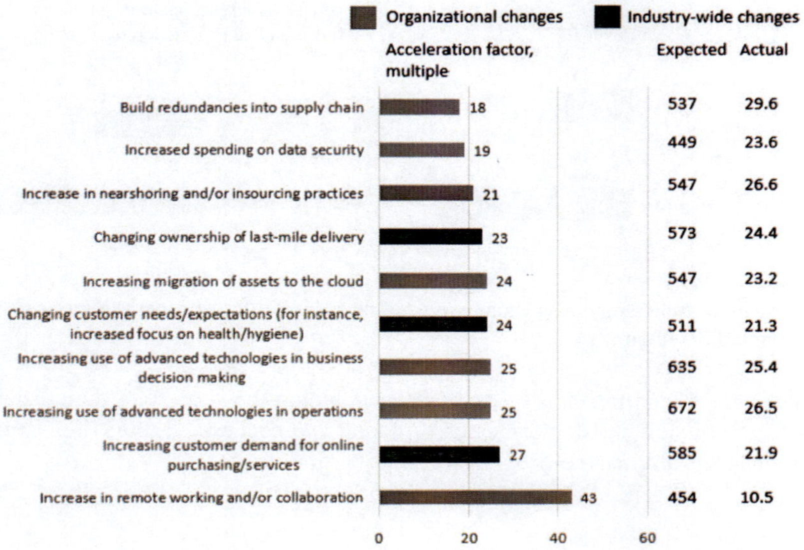

Fig. 3. Time needed to respond or implement changes, expected or actual number of days. *Source* Compiled by the authors based on [6]

It is worth noting that emerging market conditions become a significant impetus for development and accelerated response. The expected time to respond to or implement changes is several times greater than the actual time for all the changes cited, which indicates that the response depends on its necessity. That is, if there is no need for instantaneous changes, they are delayed many times over, and the estimated time can also be assumed based on the conditions normal for the moment. However, with a significant change in the market, the market starts to dictate new conditions; the normality and the reaction change. Planned changes become an abrupt need, and the process of implementing these changes is adjusted for their most rapid and effective implementation. Therefore, based on the figures discussed above, it is worth noting that the COVID-19 pandemic was precisely the condition for a dramatic change in the level of digitalization with the existing base.

In this case, the adaptation of technology also depends on the initial orientation of the company, its division, and the organization (Fig. 4).

The percentage of technology adoption by digital incumbents is higher than that of traditional incumbents. That is, the initial orientation to the digital processes, the already existing level, and the specifics of activity also affect the implementation of changes. Digital incumbents implement changes faster. Thus, they are more likely to implement changes for digitalization faster because they are focused on it initially. This fact affects the future market situation, determining who will be more successful in the market.

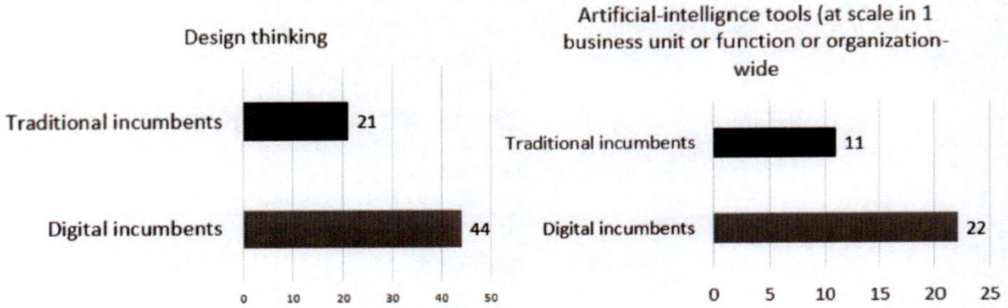

Fig. 4. Scale of technology adaptation by type of business, % of respondents. *Source* Compiled by the authors based on [2]

The success of companies in the market, including the success of digitalization, is also affected by how the companies use different approaches and tools (i.e., apply and implement them in practice) (Fig. 5).

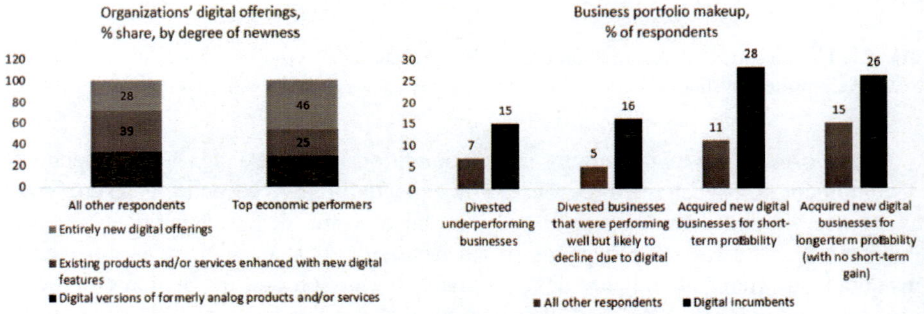

Fig. 5. Organizations that will use digital offerings, such as design thinking, in times of crisis. *Source* Compiled by the authors based on [2]

Figure 5 shows that new digital offerings for top economic performers make up a considerable share. That is, such companies can create fully digital solutions more efficiently. Thus, they occupy leading positions in the digital market, can set further digital trends, and influence the level of digitalization. That is, the initial level of organizations also determines their further digital leadership.

Using the information in the World Bank databank, we made a selection from the database "World Development Indicators," which includes the data on the following indicators for all countries:

- "Unemployment with higher education (% of the total labor force with higher education) (SL.UEM.ADVN.ZS)";
- "Unemployment with basic education (% of the total labor force with basic education) (SL.UEM.BASC.ZS)";
- "Unemployment with secondary education (% of the total labor force with secondary education) (SL.UEM.INTM.ZS)."

The selection period is 2016–2020. From the selected information, we selected Canada, South Korea, the USA, and Vietnam because they have information for 2020 (Table 1).

Table 1. Percentage of unemployment by category

No.	Country	Education	Year				
			2016	2017	2018	2019	2020
1	2	3	4	5	6	7	8
1.	Canada	Basic	14.05	12.95	11.85	11.59	16.74
2.		Secondary	8.29	7.44	6.96	7.16	12.83
3.		Higher	5.46	5.02	4.62	4.41	7.48
4.	South Korea	Basic	2.44	2.61	2.99	3.44	3.93
5.		Secondary	3.36	3.12	3.73	3.62	3.95
6.		Higher	4.21	4.32	4.12	3.89	3.93
7.	USA	Basic	7.49	6.55	5.87	5.4	11.72
8.		Secondary	6.47	5.77	5.17	4.84	10.46
9.		Higher	2.91	2.69	2.45	2.38	5.59
10.	Vietnam	Basic	1.17	1.27	0.86	1.78	1.71
11.		Secondary	2.66	2.78	1.76	2.63	2.66
12.		Higher	4.63	4	2.08	2.48	4.49
13.	Russia	Basic	14.02	13.51	12.69	10.85	5.70
14.		Secondary	5.00	4.68	4.22	3.97	
15.		Higher	5.43	5.04	4.79	4.43	

Source Developed and compiled by the authors

Based on the data obtained, we can conclude that the most significant increase in the number of unemployed is observed in the USA and Canada. Due to insufficient data, we can assume a significant increase in unemployment in Russia among the category of workers with a basic level of income. Nevertheless, given the absence of official statistics on the number of officially unemployed workers, we do not provide any quantitative data on the growth of unemployment in Russia in 2020. The official statistics for the USA and Canada are shown in Table 2.

Table 2. Growth of unemployment by category

No.	Country	Education	Increase in the number of unemployed by 2019, %
1	2	3	4
1.	USA	Basic	117
2.		Secondary	116
3.		Higher	135
4.	Canada	Basic	44
5.		Secondary	79
6.		Higher	70

Source Developed and compiled by the authors

South Korea has not seen a significant increase in the number of unemployed people. In Vietnam, there has been a significant relative increase in the number of unemployed people with higher education (81%).

Thus, despite the clear economic benefits of digital transformation and the transition to "Industry 4.0," it is necessary to carefully calculate the social component of the transformation and consider the impact of digital trends on other areas.

Global trends and new digital demands are forcing a restructuring of the training system. Currently, higher education institutions are rapidly restructuring educational programs to meet the needs of the labor market [9]. The COVID-19 pandemic changes not only educational programs but also the universities themselves. The management system of higher education institutions, the building of organizational processes, and the ways of conducting educational activities are undergoing dramatic and rapid changes [11]. Digitalization of the educational process makes learning more accessible.

During the COVID-19 pandemic, adults had an opportunity to improve their soft skills. According to the Center for Skills Development and Vocational Education Inobra of the Higher School of Economics, the number of students taking online courses increased two and a half times [10].

COVID-19 has had a negative impact on the global economic situation. Nevertheless, it has also stimulated a rapid spurt in the development of digitalization. The increase in the number of students taking online courses indicates that universities must develop the market for online education and increase the partnership with small, medium, and large businesses to build a system of training specialists who are ready to adapt to the current realities quickly [6].

It is worth noting that education is the most progressive sector among the social spheres in the field of digitalization. These data are provided by the Institute of Statistical Research and Knowledge Economy of the National Research University Higher School of Economics in the statistical collection "Indicators of the Digital Economy" [2].

The government plays a significant role in the digitalization of public educational institutions. The extent to which the government sticks to the development of digital solutions also determines conditions that organizations will have for the development in the digital environment. It is necessary to note the relationship between educational institutions and businesses regarding staff readiness to work in the digital space.

The indexes of digitalization of government, social sphere, and business demonstrate the readiness of higher education institutions to meet the needs of business in terms of training and providing the capacity of digital infrastructure.

During the COVID-19 pandemic, not every company managed to go online quickly. The industries most at risk are those in which the level of automation is less than 50% of all processes. Organizations engaged in manufacturing offer their employees risk maps, which are used to assess the possibility of office visits – those involved in manufacturing will remain employed offline, while office workers may be offered to work online. So far, the lack of communication is also solved by digitizing business processes to ensure the smooth operation of the business through the point-to-point inclusion of employees in the technological cycle [12]. Digital twins are also used to solve problems where management and office are separate from production [3]. Digitalization of organizational management structures sets a new trend, the so-called

Digital HR. The effective implementation of this concept requires a set of methods, tools, and techniques that must be introduced to organize a company's Digital HR [4].

Regional companies that export their goods had to face restrictive measures and, consequently, a drop in revenues. The problems of digital transformation of these companies are being solved at the national level, providing point support to economic sectors and certain large enterprises.

Government support is now reduced because strong companies try to find solutions as quickly as possible to help them in the competitive space. At some point, those playing by the old rules will have to quit the market. This issue is solved by transforming processes, digitizing all decision-making models, and considering risks. The ecosystem approach is also becoming popular. Following this approach, more digitized companies integrate with smaller companies to expand, which also helps stabilize the economy.

4 Conclusion

In the near future, there will be a strong business transformation towards digitalization. Digitalization will stabilize the economic situation and solve global problems (i.e., the problem of product quality, unemployment, and global economic development) because it responds to the issues regarding the acceleration of current development processes after a protracted crisis.

The paper examines how digitalization affects the development of various organizations and what conditions have influenced digitalization. The COVID-19 pandemic was a serious condition and impetus for the development of digital technology and the digitalization of organizations. However, this would not have been possible without the digital base that has already been formed and developed in recent years. Because of the dramatic shift to remote working and education and the use of a large number of digital services, companies must develop these services. Organizations with an insufficient level of digitalization are less competitive in the market.

Organizations will strive to digitize their businesses even more rapidly since the transition to new technology is currently especially relevant. There will be new competitive rivalries around digital approaches. The earlier a company invests its resources in such approaches, the earlier it will gain a competitive advantage in the market. This will not come immediately, the optimal cycle of the first results will be in a year, and the effect will only grow in the future.

References

1. Abdrakhmanova, G., et al.: Digital economy indicators in the Russian Federation: 2020: Data Book. In: Gokhberg, L., Kislyakov, E., Kuzminov, Ya., Parshin, M., Shapoval, I. (eds.) HSE University, Moscow, Russia (2020). https://doi.org/10.17323/978-5-7598-2194-6

2. Blackburn, S., LaBerge, L., O'Toole, C., Schneider, J.: Digital strategy in a time of crisis. In: The Next Normal: The Recovery will be Digital, pp. 11–19. McKinsey & Company (2020). https://www.mckinsey.com/~/media/mckinsey/business%20functions/mckinsey%20digital/our%20insights/how%20six%20companies%20are%20using%20technology%20and%20data%20to%20transform%20themselves/the-next-normal-the-recovery-will-be-digital.pdf. Accessed 17 May 2021
3. Harris, K., Kimson, A., Schwedel, A.: Labor 2030: The collision of demographics, automation and inequality, 7 February 2018. https://www.bain.com/insights/labor-2030-the-collision-of-demographics-automation-and-inequality. Accessed 17 May 2021
4. International Labour Organization (ILO). ILO Monitor: COVID-19 and the world of work. Seventh edition: Updated estimates and analysis, 25 January 2021. https://www.ilo.org/wcmsp5/groups/public/@dgreports/@dcomm/documents/briefingnote/wcms_767028.pdf. Accessed 17 May 2021
5. Klochko, E.N., Fomenko, N.M., Nekrasova, V.V.: Modeling of network mechanisms of management in the conditions of organizational development. Int. Rev. Manag. Mark. **6**(S1), 101–106 (2016)
6. LaBerge, L., O'Toole, C., Schneider, J., Smaje, K.: How COVID-19 has pushed companies over the technology tipping point – and transformed business forever. McKinsey & Company, 5 October 2020. https://www.mckinsey.com/business-functions/strategy-and-corporate-finance/our-insights/how-covid-19-has-pushed-companies-over-the-technology-tipping-point-and-transformed-business-forever. Accessed 17 May 2021
7. Maslennikov, V.V., Lyandau, Y., Kalinina, I.A.: Digital HR. Vestnik Plekhanov Russian Univ. Econ. **17**(1), 87–92 (2020). https://doi.org/10.21686/2413-2829-2020-1-87-92
8. Popkova, E.G., Zhuravleva, I.A., Abramov, S.A., Fetisova, O.V., Popova, E.V.: Digitization of taxes as a top-priority direction of optimizing the taxation system in modern Russia. In: Gashenko, I.V., Zima, Y.S., Davidyan, A.V. (eds.) Optimization of the Taxation System: Preconditions, Tendencies, and Perspectives. SSDC, vol. 182, pp. 169–175. Springer, Cham (2019). https://doi.org/10.1007/978-3-030-01514-5_20
9. Popova, E.V., Nikiforova, D.I.: Manufacturing digitalization process. Use of digital twin technology in industry. Econ. Constr. **4**(64), 52–58 (2020)
10. Tikhonov, A.R., Karaya, M.Z., Bolotova, E.S.: Management of digital projects in education. Meridian **2**(36), 99–101 (2020)
11. Tikhonov, A.R., Samsonov, A.N., Bolotova, E.S.: Educational organization management. Curr. Stud. **12**(15), 72–74 (2020)
12. Timofeeva, A.A., Bulganina, S.V., Fomenko, N.M., Khachaturova, M.S., Nekrasova, M.L., Zakharova, E.N.: Features of the implementation of integrated marketing communications: Resistance to change. Int. Rev. Manag. Mark. **6**(S1), 27–32 (2016)

Technologies for Promoting a Company on the Internet Using Telegram Messenger

Andrei G. Voronov, Dmitry G. Voronov, Gennady B. Voronov(✉), and Ivan Yu. Nefedov

Russian Technological University (MIREA), Moscow, Russia
voronov_g@mirea.ru

Abstract. The authors focus on the issues of promoting companies using telecommunication technologies to find tools for the effective promotion of the brand, products, and services. Particularly, the authors study the issue of increasing brand awareness using marketing technologies in Telegram messenger. The research methods include collecting, processing, analyzing, and systematizing publicly available information on the discussed issue. The authors analyze the most popular messengers in Russia in 2020 and state that Telegram is the only suitable application for the integrated promotion of a company. The study summarizes the most promising approaches for the company to work with the audience in Telegram messenger. The best combinations of these approaches are offered to achieve maximum effect. The authors consider the technologies of promotion in this messenger, including the creation of a channel (group, bot), the purchase of advertising in other channels and groups, and promotion on external platforms. Moreover, the authors describe the main problems encountered by organizations during the promotion of their brand (products, services) in the Telegram messenger. The recommendations for the solution to these problems are provided. The study shows that the creation of a company's Telegram channel, group, or chatbot is the most effective and financially optimal algorithm for promoting a company. The authors systematize the independent experience of three years of work with the technologies provided by Telegram messenger and the experience of marketing specialists working in this messenger.

Keywords: Telegram channel · Group · Chatbot · Company promotion · Internet marketing · Community management · Advertising market in telegram channels · Effective tools for business

JEL Codes: M30 · M31 · M37 · M39

E. G. Popkova (Ed.): ISC 2020, LNNS 368, pp. 349–362, 2022.
https://doi.org/10.1007/978-3-030-93244-2_39

1 Introduction

In the current environment of active transfer of trade and services on a remote basis, the methods of developing the company by using the telecommunication technologies of the Internet are of great importance. The widespread development of social media has made traditional marketing less relevant. Managers and entrepreneurs look for new, efficient, and effective ways for promoting technology products and services. There is an urgent need to study the impact of marketing methods that use the existing social networks to raise awareness about the brand, good, or service (viral marketing) on the purchasing demands of users of mobile applications (*Facebook*, *Instagram*, *Line*, *LinkedIn*, *Pinterest*, *Telegram*, *Twitter*, *Viber*, *WeChat*, and *WhatsApp*), as well as the impact of viral marketing on the intention to make a purchase [7].

The switch from social networks to messengers (*WhatsApp*, *Telegram*, and *WeChat*) is logical since messengers are much easier and faster to use. Moreover, their use is safe, which allows users to save their confidential information [22].

The authors of this research discussed some aspects related to the organization of remote work and the problems of developing Internet services for small businesses in their previous works [5, 28]. Distance learning, which requires modern means of information support of educational activities (information and communication tools to support the learning process, software products used in the educational work of students), saw rapid development. G. Yu. Yalamov and G. B. Voronov substantiate the expediency of using distance learning in training specialists of secondary vocational education, demanded in the IT industry [30].

The development of technology necessitates the organization of remote work and internet marketing, which aims to maximize the effective promotion of the company brand, its products, or services in the current conditions. The traditional approach to organizing promotion is the use of contextual advertising in search engines and social networks and the method of targeted advertising (the ads are displayed to particular users according to several parameters, including gender, age, geography, interests, income, etc.). The role of Internet marketing in the current reality and the attitudes of end-users of products and services are discussed in [9, 16].

Among the abundance of Internet marketing technologies, marketing using Internet messengers has recently become particularly important due to the rapid development of information technology, allowing the instant exchange of text, audio, and video messages [2, 15].

2 Materials and Methods

According to the analytical data provided by the *Telega.in* service, in 2020, *WhatsApp*, *Viber*, and *Telegram* occupied the largest share of all Internet messengers in Russia. Table 1 compares the number of people using messengers, the number of people using social networks, and the total number of Internet users in Russia [1, 18].

Table 1. Active Russian audience of messengers compared to the total Internet audience of the Russian Federation, 2020

Indicator	Absolute value	Relative value (in % of the population of the Russian Federation)	Relative value (in % of the number of Internet users in the Russian Federation)
Total number of Internet users in Russia	118 million people	80.6	100.0
Number of users of social networks	70 million people	47.8	59.3
Active audience of *WhatsApp* messenger	70.6 million people	48.0	59.8
Active audience of *Viber* messenger	36.2 million people	48.3	30.7
Active audience of *Telegram* messenger	26.6 million people	18.2	22.5

Source Compiled by the authors

According to Table 1, the audience reach of each messenger does not exceed 60% of the total number of Russian Internet users and 50% of the Russian population. Therefore, there exists the potential for further growth of the audience of messengers. It is important to clarify that the same user may have several different applications installed at once.

At first glance, the audience of some messengers may not seem very significant, but modern companies cannot ignore this audience and trends in the development of these applications. Accordingly, companies need to gradually move from the use of email marketing and marketing in social networks to promotion in online messengers [20].

It is important to note that of all messengers listed above, only *Telegram* is suitable for promoting a company since the other applications lack the necessary tools or these tools are of insufficient quality. The widespread use of *Telegram* is due to the platform's usefulness, security, and privacy. These technical qualities make *Telegram* especially convenient as a source of personalized information and the channel of simple and original news [8].

Table 2 provides detailed information about the functionality of these applications in terms of company promotion.

Table 2. The functionality of messengers to promote the company

Messenger	Functionality and other aspects
Telegram	• Creating a channel by a user (registration to a private entity) • Creating a group (chat) to communicate with the company's customers and integrating this group with the company's channel • Developing a chatbot to expand the messenger's functionality and solve specific problems • Developing and distributing a set of stickers developed by the company (to increase the awareness of a brand, product, or service) • Fine-tuning of administrator rights in channels and groups • Possibility to edit and delete messages • Possibility to purchase advertising in other channels (groups, bots) • Fast messaging technology (reposts) • Messages are stored on cloud servers and are accessible from any device (computer, phone, tablet, etc.) • Possibility to set up comments on posts published on the company's channel • Deep customization of the channel (group) types and the ability to create short links to these resources that are easy to remember • Push notifications to users about posts published on the channel (group or bot) • Built-in channel and group statistics • Effective third-party statistical tools for channels and groups (*Telemetr.me* and *Tgstat.ru* services), etc • Audio and video calls in mobile and PC versions • Maximum number of devices connected to one account is unlimited • Maximum number of participants of the group chat is 200 thousand people • Possibility to conduct surveys on channels and in groups • Search for public channels and groups • Possibility to view posts published on the channel not only from the app but also through direct links with the *t.me* prefix • Audience of the messenger is accustomed to the above functionality and actively uses it
WhatsApp	• Creating a channel by a user (unlike *Telegram*, it is impossible to set a short link to the channel, there are no flexible settings of admin rights, no possibility to add chatbots to the channel, and no other important functions) • Creating a group (chat) to communicate with the company's customers (the problems are similar to those described for the channels) • Creating a bot to expand the functionality of the messenger and solve specific problems • Possibility to develop and distribute a set of stickers • No possibility of editing messages and insufficiently high quality of deleting messages (it leaves "Message deleted" inscription, which clogs the message feed) • Fast messaging technology (reposts) (no statistics on views of forwarded messages) • Messages are stored on users' devices rather than on the servers of the application, which results in the need to copy the message history from the previous device

(*continued*)

Table 2. (*continued*)

Messenger	Functionality and other aspects
	• No opportunity to comment on posts posted on the channel • No settings of channel type (public, private, etc.) • Push notifications to users about posts published on the channel (group or bot) • No built-in statistics for channels and groups, no effective third-party tools for its analysis • Audio and video calls are available only in mobile versions of the app • A maximum of 10 devices can be connected to one account • The maximum number of participants of the group chat is 256 • No possibility to conduct surveys on channels and in groups • No possibility to search for channels and groups • No public accounts • Constant changes in corporate policies (e.g., the recent requirement to transfer all user data to *Facebook*) lead to an active outflow of the audience to other messengers
Viber	• Messenger's functionality is similar to *WhatsApp*, except for some aspects [14]: • Audio and video calls are available in all versions of the app • No web version of the application • Maximum number of devices connected to one account is unlimited • Maximum number of group chat participants is 250 • Possibility to conduct surveys on channels and in groups • Possibility to search for public channels and groups • Presence of public accounts • Messenger has free and paid stickers

Source Compiled by the authors

According to Table 2, it is recommended to focus primarily on working with the audience of Telegram when planning the company's promotion in 2021–2022.

New features introduced in *Telegram* (e.g., channels, bots, supergroups, and advanced sharing mechanisms) have taken the messenger to a higher level. *Telegram* is a new paradigm between social networks and messengers. The instruments of user behavior analysis allow for measuring the quality of a group in *Telegram* [10].

Telegram allows for selecting a sorted list of channels related to a user's query and published messages in the channels, using a semantic approach to extend the query, which can improve channel search performance [12]. *Telegram* can be used for developing the concept of the social communications network to consider the special structural properties of communications in instant messaging systems [23].

Let us indicate several facts about this application. *Telegram* was created by Pavel and Nikolai Durov (the developers of the social network *VKontakte*). The first English version of the app appeared in 2013, and the Russian-language version was released in early 2017 [25].

The main difference between *Telegram* and previously released messengers was the appearance of channels (a separate space where only administrators can write messages and users only view the published content; it is similar to the media) [17]. This feature has become so popular that by 2021 the messenger has registered more than 200,000 Russian channels with an audience of more than 100 subscribers.

Nowadays, companies and individuals (bloggers, journalists, executives, etc.) have *Telegram* channels. In some of these channels, the audience exceeds 100 thousand people. One of the examples of an effective application of this tool in marketing is the tourism industry. There are many *Telegram* channels on this topic and, accordingly, a large advertising market [3].

Another important aspect of *Telegram* is the so-called Instant-communication (rapid exchange of information between the administrator of the channel or group and its subscribers).

This type of communication is made possible by two main components:

1. Push-notification system (users are instantly notified of posts published on the channel if they have this feature enabled in the app settings);
2. No recommendation system (the so-called smart feed). Therefore, the algorithms of the application do not choose the content which is shown and hidden from the user. The reader will receive information about the publication of new content on all subscribed channels (unlike *YouTube*, *Instagram*, and *VKontakte*).

Finally, *Telegram* has an application programming interface (API), allowing third-party developers to create services that significantly expand the range of tasks handled by the messenger (chatbots). Community size and the use of user-created chatbots are interrelated [13]. Chatbots are entities controlled by algorithms that act like humans in conversations via Twitter, Facebook, chat rooms, or Q&A sites. They can affect online communication and cause harm [4].

Let us give a few examples of such services: bots for publishing content on the channel, moderating groups, selling airline tickets, buying goods and services without leaving the messenger, showing stock market index quotations, weather forecasts, etc. For example, an automated news chatbot extracts content from various news sources and creates conversations with the user. Key components of this bot include the automatic organization of news articles in topic-based chats, integration of automatically generated questions into the conversation, and a new method for selecting information to display, which allows avoiding repetitive messages [21]. The study conducted by A. Ivanovskaya et al. [11] presents the development of a software chatbot for analyzing texts for the library based on the question-answer system. Moreover, this study provides an overview of modern platforms for creating chatbots. Additionally, the authors describe the results of experiments with a trained software agent for text analysis. The trained agent fully provided the correct information during the experiments.

The importance of chatbot technology for business and its specifics in the digital economy are discussed by E. A. Kushkov in his research “Messengers for Business” [19]. In order to improve the level of service of transport infrastructure, customers, etc., it is suggested that messengers or chatbots use the results of the analysis of a relatively short time series of tone scores to determine the negative or positive trend of customer feedback [24].

Over the past few years, *Telegram* saw an active growth in its active audience and the market volume for advertising in *Telegram* channels. For comparison, the advertising market did not exceed 1 billion rubles per year in 2017 and, in 2020, this figure increased to approximately 12.8 billion rubles. This assessment is presented in the study by Dr. D. A. Kornilov in conjunction with the advertising exchange *Telega.in* [18]. Moreover, the volume of this market may even be underestimated, which is detailed in the methodology of our research [29].

The active development of *Telegram* in Russia has not stopped. Over the past year, several well-known brands have registered in the messenger (or are planning to register in the future); for example, Russia's largest bank *Sber* has opened its own *Telegram* channel (@sberbank).

The developers of *Telegram* are constantly working to improve functionality and expand business opportunities. Plans for 2021 include the launch of monetization procedures, including the creation of a built-in advertising exchange, special paid opportunities for corporate teams (using *Telegram* in their work), easy transfer of channels and groups from other messengers, the introduction of paid stickers (which will allow artists and designers to make profits.), etc. [6].

The aspects mentioned above make working with the audience of *Telegram* one of the most essential tasks when promoting a company.

Further, we list the main approaches of working with the audience of the messenger:

1. Purchasing the advertisement of products (works, services) in third-party channels and groups;
2. Creating the company's *Telegram* channel and its promotion;
3. Creating a group (chat) of the company for direct communication with customers;
4. Creating a chatbot and its promotion.

In our opinion, the company can achieve the maximum effect by applying the second, third, and fourth approaches.

Let us take a closer look at the features of the audience in *Telegram*.

The activity of users is very high, so the audience views new information published on the channels within 1–3 days.

At the same time, subscribers prefer to click internal links (for example, to other *Telegram* channels) rather than open external resources (including websites, social networks, *Twitch* channels, etc.). The cost of advertising via external links is significantly higher (by 15%–200% compared with advertising containing internal links).

There is also another aspect – if the advertisement promotes *Telegram* resources, it would be possible to inform the attracted audience about the company's products (works, services) free of charge for a long time. In turn, the advertisement of an external project allows for providing only the necessary information once – as long as the post is available.

Summarizing the above, we conclude that the company can significantly save on promotion by creating its own channel, group, or bot in *Telegram*. This is most relevant for small organizations with advertising budgets smaller than the budgets of large companies.

There are successful examples of promotion using any of the four approaches mentioned above. Moreover, these approaches can be combined to achieve the set objectives.

Table 3 presents real-life examples of the implementation of basic approaches to promoting companies in *Telegram*.

Table 3. Examples of promoting companies on *Telegram*

Approach	Examples of companies that have implemented this approach
Advertising on external links: The first approach – purchasing the advertisement of products (works, services) in third-party channels and groups	• Advertisement for the bank card of *Tinkoff Bank*; • Advertising of educational programs of online schools *Skillbox*, *GeekBrains,* and *XYZ School*; • Advertising for *Yandex.Music*, *Yandex.Taxi,* and *Yandex.Plus* services; • Advertisement for *Teremok* restaurant; • Advertisement for *Perekrestok* store; • Advertising for the streaming service *more.tv*
Promotion with the creation and promotion of internal *Telegram* resources: The second approach – creating the company's *Telegram* channel and its promotion; The third approach – creating a group (chat) of the company for direct communication with customers; The fourth approach – creating a chatbot and its promotion	• The official channel of the online game *World of Tanks* (@worldoftanks_official); • Internet media *Mash* (@breakingmash), *RIA Novosti* news agency (@rian_ru), *RT* in Russian (@rt_russian); • *Xbox Wire in Russian* channel (@xbox_wire_russia); • *Promotions of Moscow restaurants – RestoRaids* channel (@restoraids); • *Aviasales* Air Ticket Service: channel, group, and bot (@aviasales); • The official channel of *Telega.in* advertising exchange (@telegain); • The official channel of *Buka* video game publisher (@buka_ent); • The official channel of *Tinkoff Bank* (@tinkoffbank); • The official channels of educational schools *Skillbox* (@skillboxru) and *GeekBrains* (@geekbrains_ru); • Radio *Govorit Moskva*: channel and bot (@radiogovoritmsk); • The official channel of *METRO Cash & Carry* stores in Russia (@metroccrus); • The official channel of the online action game *Caliber* (@caliber_game), etc.

Source Compiled by the authors

3 Results

Based on the results of this research, let us summarize the technology for promoting the company using each of the four approaches.

The first approach is – the promotion of the company using links to external resources.

When using this approach, it is necessary to keep in mind that the advertising post is placed once and, in most cases, is deleted within 24–72 h after its placement by the administrator of the channel (group, bot). If the advertising campaign has no effect, no one will make the necessary adjustments to the published material.

Accordingly, before planning the placement of advertising, it is necessary to adhere to the following algorithm:

1. Studying (1) the topic of the channel in which advertising is purchased; (2) the quality of content in this channel; (3) the approximate effectiveness (audience gain) that similar advertising companies have had on this channel; and (4) the style of content writing by the author to prepare the most appropriate advertising post;
2. Analysis of the cost of advertising on this channel compared with its competitors to determine the feasibility of advertising placement;
3. Development of the concept of advertising material (e.g., a picture, a text, and a button; a picture and a text; text integration into the author's material, etc.);
4. Writing a promotional post and creating (if necessary) a cover (picture or video) for it;
5. Analysis of the activity of the channel audience to select the best date and time of publication of advertising material, its time at the top (i.e., the period during which the channel will not publish new posts, e.g., 1 h), and the total time of the post on the channel (e.g., 24 h);
6. Agreeing on the parameters of advertising material with the channel's administrator;
7. Checking the correctness of advertising;
8. Analysis of the effectiveness of the advertising campaign to identify groups that have shown the best results for further advertising.

The second approach is the creation of the company's *Telegram* channel and its promotion.

In this case, the algorithm for the company's promotion will consist of the following steps:

1. Choosing an interesting and memorable channel name and a link to it;
2. Creating a high-quality channel avatar (cover image) and a competent description of the project (channel);
3. Analysis of the competitive advantages of the company, which should be focused on when writing content on the channel;
4. Formation of the concept of published materials, based on the following parameters:

   - Post format (a picture in combination with a text, a text in combination with a picture, a video in combination with a text, a text in combination with a video, text material, etc.);

- Total number of publications per day;
- Estimated time interval between published posts;
- Time of publishing content on the channel (e.g., from 10 a.m. to 10 p.m.), etc.

5. Connecting a special bot (e.g., @controllerbot or @telepostbot) facilitating the publication of messages;
6. Placing a link to the company's *Telegram* channel after publishing at least ten posts in other sources available to the company (including the official website, social media, handouts, and other sources);
7. Launching an advertising campaign in channels of identical or similar topics when the channel reaches the 100 subscribers mark (the algorithm is described in the first approach);
8. Purchasing channel advertising from external sources (websites, thematic groups in social networks, *YouTube* channels, etc.);
9. Regular promotions to maintain audience interest (discounts on company products or services, contests, giveaways, etc.).

Telegram has no natural audience growth (the so-called organic traffic). Therefore, regular advertising campaigns on the channel are necessary to stop the decrease in the number of subscribers and their activities, which will eventually affect the number of views of published materials. This is a fundamental difference between this messenger and such projects as *VKontakte*, *YouTube*, and *Yandex Zen*.

Another important feature is the problem associated with the fact that *Telegram* channels are currently registered predominantly to individuals, and most legal entities are unlikely to register them in their own names.

The third approach is the creation of a group (chat) to communicate with customers. In this case, the algorithm is as follows:

1. The creation of chat rooms is reasonable only when the company's *Telegram* channel has a certain number of regular subscribers interested in the products or services of the promoted company;
2. Working out the concept of the chat before its registration in the messenger, based on the following parameters:
   - Topics on which users will communicate;
   - Tasks of administrators and the total number of administrators;
   - Whether or not administrators are anonymous;
   - Type of chat – private (chat is visible only to those who have a special hard-to-remember link) or public (chat is visible to all, accessible by a short memorable link, and can be found in the search).
3. Developing chat rules and informing all administrators about them;
4. Ensuring regular communication with customers and timely response to their questions;
5. Connecting a special bot that simplifies the process of administering the group (e.g., free @groupbutler_bot or paid @combot);
6. Integrating the group with the channel (if it is necessary to comment on the materials published on the channel);

7. Providing a link to the chat in the description of the company's channel and in external sources (preferred);
8. Buying group ads in other *Telegram* channels and groups (chats) in accordance with the algorithm from the first approach (if necessary);
9. Regular promotions to keep customers interested in the company's products or services.

The third approach is most effective only when used in combination with the second approach.

The fourth approach is the development of a *Telegram* bot.

In this case, the algorithm of actions consists of the following steps:

1. Working out the range of tasks to be solved by the chatbot;
2. Analysis of existing solutions for compliance with the tasks to be solved (there may exist free solutions, or it will be more reasonable to buy access to a ready-made paid bot, rather than to develop new bots from scratch);
3. Drawing up a detailed technical inquiry with a description of all necessary functions that the bot will be endowed with (if a new bot is created);
4. Development of a chatbot and its testing (it is recommended to provide access to the bot to regular customers of the company);
5. Correcting errors in the bot's operation and fine-tuning its functionality (if necessary);
6. Publication of the link to the bot in all free-access sources (including channels, chat rooms, directories, websites, social networking groups, promotional handouts, etc.);
7. Advertising the company's chatbot in other *Telegram* channels and groups (according to the algorithm described in the first approach);
8. Regular bot promotions to maintain customer interest in the company's products and services.

4 Discussion

If a company has the required budget, the most effective promotion is achieved by combining all four approaches. However, if the budget is limited, it is recommended to use a combination of the second and third or the second and fourth approaches.

We should also add that having its own resources in Telegram, a company can promote its own goods and services, conduct joint promotions (so-called special projects) with other organizations, or sell advertising.

Based on the above, we can draw the following conclusions.

First, *Telegram* is actively developing (e.g., over the past three years, the Russian audience of the app has doubled, and there is significant potential for further growth).

Second, the messenger is now an effective way to promote the company on the Internet and expand the client base.

Third, *Telegram* is used primarily by an adult and solvent audience (including CEOs of large corporations, investors, popular journalists, bloggers, government officials, programmers, designers, web developers, etc.), which increases sales.

Fourth, the emergence of messengers and their active development created new specializations (e.g., community marketing), replacing the telephone channel of communication with customers in the future. More details about this promising format of working with clients are described in the speeches of M. Tokovinin and F. Skuratov at the conference TALKS 2021, held on March 4, 2021 [26, 27].

5 Conclusion

After reviewing and summarizing the main approaches to working with the audience of the *Telegram* messenger, the authors propose the best combination of these approaches to maximize the effect and achieve an effective promotion of the brand, products, and services of the company. It is suggested to create a company's own *Telegram* channel, group, or bot. The prospect of further research lies in exploring the possibility of replacing the telephone channel of communication with customers with a new specialization – community marketing.

References

1. Advertising market in Russian-language Telegram channels (2020). https://www.docs.google.com/presentation/d/1L4kK0aWLvdhAFLC1xf6hq2lRlxiLGL8dLQqSNfXoE-I/present?ueb=true&slide=id.gb62f2a5f1a_0_0
2. Andrianova, E.G., Golovin, S.A., Zykov, S.V., Lesko, S.A., Chukalina, E.R.: Review of modern models and methods of analysis of time series of dynamics of processes in social, economic and socio-technical systems. Russian Technol. J. **8**(4), 7–45 (2020). https://doi.org/10.32362/2500-316X-2020-8-4-7-45
3. Belyakova, M.Yu., Kozenko, I.A.: Analysis of the market of travel Telegram-channels, defining the trajectory of the development of this topic and creating evidence-based recommendations for regional travel telegram-channel implementation. Sci. Notes Russian Acad. Entrep. **19**(4), 82–95 (2020). https://doi.org/10.24182/2073-6258-2020-19-4-82-95
4. Daniel, F., Cappiello, C., Benatallah, B.: Bots acting like humans: Understanding and preventing harm. IEEE Internet Comput. **23**(2), 40–49 (2019). https://doi.org/10.1109/MIC.2019.2893137
5. Danilkina, Yu.V., Voronov, A.G.: Development of Internet services in small business. In: International Scientific and Practical Conference "Organization Management: Problems and Prospects of Development, pp. 81–84. MIREA, Moscow (2011)
6. Durov, P.: As Telegram approaches 500 million active users, many of you are asking the question – who is going to pay to support this growth? [comment of Telegram founder], 23 December 2020. https://t.me/durov/142
7. Fard, M.H., Marvi, R.: Viral marketing and purchase intentions of mobile applications users. Int. J. Emerg. Mark. **15**(2), 287–301 (2020). https://doi.org/10.1108/IJOEM-06-2018-0291
8. Gonzales Sanchez, H.M., Moreno, J.M.: Telegram as a tool for journalists: perception and use. Revista de Comunicacion-Peru **19**(2), 245–261 (2020). https://doi.org/10.26441/RC19.2-2020-A14
9. Grishkova, N.S., Chicherin, Y., Agaeva, A.N.: Marketing research on determining the relationship of consumers to online advertising as an instrument of integrated marketing communications. Herald Belgorod Univ. Coop. Econ. Law **2**(81), 235–246 (2020). https://doi.org/10.21295/2223-5639-2020-2-235-246

10. Hashemi, A., Zare Chahooki, M.A.: Telegram group quality measurement by user behavior analysis. Soc. Netw. Anal. Min. **9**(1), 1–12 (2019). https://doi.org/10.1007/s13278-019-0575-9
11. Ivanovskaya, A., Aksyonov, K., Kalinin, I., Chiryshev, Yu., Aksyonova, O.: Development of the text analysis software agent (chat bot) for the library based on the question-and-answer system TWIN. In: ITM Web of Conferences, vol. 30, p. 04006 (2019). https://doi.org/10.1051/itmconf/20193004006
12. Jalilvand, A., Neshati, M.: Channel retrieval: finding relevant broadcasters on telegram. Soc. Netw. Anal. Min. **10**(1), 1–16 (2020). https://doi.org/10.1007/s13278-020-0629-z
13. Kiene, Ch., Hill, B.M.: Who uses bots? A statistical analysis of bot usage in moderation teams. In: Proceedings of the CHI EA'20: Extended abstracts of the 2020 CHI Conference on Human Factors in Computing Systems. Honolulu, HI (2020). https://doi.org/10.1145/3334480.3382960
14. Klimov, A.: Viber and WhatsApp: Which messenger is better. Messenge.ru, 7 May 2019. https://www.messenge.ru/viber-ili-whatsapp-chto-luchshe/
15. Kolesnikova, E.V., Degtyar, O.N., Trishchenko, D.A.: Internet messengers as a way of brand promotion in the digital environment. Herald Belgorod Univ. Coop. Econ. Law **1**(86), 217–227 (2021). https://doi.org/10.21295/2223-5639-2021-1-217-227
16. Konev, D.M.: Internet marketing and its possibilities in modern world. Mod. Sci. **12**(4), 100–105 (2020)
17. Kontzarenko, F.: Instructions for Telegram channels: Why we need them, how to use them, what to read. Vc.ru, 10 July 2016. https://vc.ru/social/16807-telegram-chanels
18. Kornilov, D., Raitsin, M., Zaitsev, D., Sviridenko, A.: Study "The volume of advertising in Russian-language Telegram Channels in 2020." Telega.in, 10 January 2021. https://www.telega.in/blog/ob-em-rynka-reklamy-v-russkoyazychnyh-telegram-kanalah-v-2020-godu
19. Kushkov, E.A.: Instant messengers for business. Role and impact on the Digital Economy. Universum: Econ. Jurisprudence **4**(61), 4–6 (2019a). https://7universum.com/pdf/economy/4(61)/Kushkov.pdf
20. Kushkov, E.A.: The evolution of customer communication channels: Replacing e-mail marketing with instant messengers and social media. Sci. Forum: Innov. Sci. **3**(21), 35–40 (2019)
21. Laban, P., Canny, J., Hearst, M.A.: What's the latest? A question-driven news chatbot. In: Proceedings of the ACL 2020: 58th Annual meeting of the Association for Computational Linguistics: System Demonstrations, pp. 380–387. Association for Computational Linguistics, Stroudsburg (2020). https://doi.org/10.18653/v1/2020.acl-demos.43
22. Perkov, T.: The application of messengers as an additional blended learning tool for teaching practitioners. In: Proceedings of the INTED-2020: 14th International Technology, Education and Development Conference, pp. 9065–9068. IATED, Valencia (2020). https://doi.org/10.21125/inted.2020.2479
23. Sahafizadeh, E., Ladani, B.T.: A model for social communication network in mobile instant messaging systems. IEEE Trans. Comput. Soc. Syst. **1**, 68–83 (2020). https://doi.org/10.1109/TCSS.2019.2958968
24. Sydorenko, V., Kravchenko, S., Rychok, Y., Zeman, K.: Method of classification of tonal estimations time series in problems of intellectual analysis of text content. Transp. Res. Procedia **44**, 102–109 (2020). https://doi.org/10.1016/j.trpro.2020.02.015
25. Telegram (software). In Wikipedia, 27 April 2021. https://en.wikipedia.org/wiki/Telegram_(software)
26. TextBack sales in messengers. Mikhail Tokovinin from amoCRM at the TALKS 2021 conference [Video file], 11 March 2021a. https://www.youtube.com/watch?app=desktop&v=NifiUVTq6YA

27. TextBack sales in messengers. Fedor Skuratov from Telegram at the TALKS 2021 conference [Video file], 11 March 2021b. https://www.youtube.com/watch?v=YCfsZzLTU2k
28. Voronov, A.G., Voronov, D.G.: Features of the management process of remote work. In: International Scientific and Practical Conference "Organization Management: Problems and Prospects of Development", pp. 92–95. MIREA, Moscow (2010)
29. Voronov, G.B., Voronov, A.G., Flegontov, A.A.: Aspects of the advertising market assessment method Telegam-channels. Digital Econ. **1**(13), 41–44 (2021). https://doi.org/10.34706/DE-2021-01-04
30. Yalamov, G., Voronov, G.B.: Means of information support of specialists training by professions of secondary vocational education that are in demand in the IT industry. Pedagogical Inform. **1**, 96–103 (2018)

E-Commerce Market: Intensification of Development During the Pandemic

Oksana N. Podorova-Anikina[1], Evgeniya K. Karpunina[2(✉)], Zoya O. Gukasyan[3], Natalia P. Nazarchuk[4], and Tatiana A. Perekatieva[5]

[1] A.I. Herzen Russian State Pedagogical University, St. Petersburg, Russia
[2] G.R. Derzhavin Tambov State University, Tambov, Russia
[3] Kuban State Technological University, Krasnodar, Russia
[4] Tambov State Technical University, Tambov, Russia
[5] Ryazan State University named after S.A. Yesenin, Ryazan, Russia

Abstract. The study provides a comparative analysis of the impact of the pandemic on the situation of the e-commerce market in various countries and also identifies areas for improving the structure of the e-commerce market in Russia in the long term. The authors have established the prerequisites for the intensification of the development of the e-commerce market, including the expansion of access to the Internet, the growth of users of mobile devices and social networks; the gradually weakening offline retail sector; the implemented state policy of regulating the e-commerce market. The advantages of e-commerce over other forms of offline retail are established. It is shown that China and the United States are leaders in terms of the volume of the e-commerce market and its share in the retail segment. The authors analyzed the dynamics of the development of the e-commerce market in some countries during the COVID-2019 pandemic. It is proved that during the pandemic, the increase in Internet connections, quarantine measures and institutional restrictions of states provoked the growth of the e-commerce market in most countries. It is emphasized that the largest growth of the e-commerce market during the pandemic occurred in the United States, Great Britain, Poland, and Russia. The authors justified the directions of the state policy on improving the e-commerce market in Russia.

Keywords: E-commerce · Pandemic · Retail · Digital technologies · Online sales · Government support

JEL Codes: O11 · O33

1 Introduction

The COVID-19 pandemic is a new phenomenon for researchers. Its impact on the socio-economic development of states and regions of the world has not yet been fully studied.

However, it is now proven that each new wave of the pandemic has had a specific, unique impact on the economy and society. In particular, the first wave of the pandemic

E. G. Popkova (Ed.): ISC 2020, LNNS 368, pp. 363–373, 2022.
https://doi.org/10.1007/978-3-030-93244-2_40

(March-May 2020), associated with the introduction of a widespread lockdown, the suspension of international traffic, the introduction of strict quarantine measures and social distancing, had a destabilizing character, put health at risk and disrupted economic activity [9, 21]. Normal economic relations in some OECD countries collapsed during the self-isolation regime by 20–30%, and the global economy declined by 6% in annual terms in 2020 [3, 13]. The first wave had a prolonged effect, aggravated by the introduction of new economic restrictions during the second wave of the pandemic. This was manifested in the form of a rapid increase in unemployment around the world, a jump in consumer prices, a decline in the standard of living of the population and an increase in social instability [24]. According to expert estimates, the second wave has already led to a drop in global GDP by 7.6% due to the high rate of spread of the disease.

However, the impact of the pandemic cannot be considered unambiguously negative. There are sectors of the national economy that received an impulse to a new stage of development and demonstrated an increase in indicators already in the first wave of the pandemic [20, 32]. This applies to the sectors of the digital economy, in particular, to the e-commerce market. It was the e-commerce market that reflected the changes in the consumption system during the pandemic and responded as quickly as possible to new challenges. It has formed modern communication channels between the manufacturer and the consumer in the form of specialized open Internet platforms with certain tools and rules. In the pandemic year 2020, the volume of the global e-commerce market reached 4.2 trillion dollars [32], and the number of buyers on the market was 2.14 billion people [29]. Such consumer activity and the intensification of the use of e-commerce platforms create conditions for increasing competition in the market, further expanding the influence on the audience and grow the market share. The development of the e-commerce market in various countries before the pandemic was uneven. There were world leaders, such as China and the United States, and outsiders-developing countries, which included Russia. The pandemic will likely be able to significantly change the situation, reformat the structure of the market and give the necessary impulse to turn the e-commerce market into a tool for ensuring economic growth in Russia [12].

2 Literature Review

Digital platforms for the promotion of goods and services have long been a tool for increasing business competitiveness and a source of additional profit [33]. Digital platforms reduce the cost of development, production, storage and distribution of goods and services, as well as reduce the costs associated with document management, optimize indirect costs by using an organized supply chain In addition, e-commerce tools are very adaptive and allow subjects to instantly track changes in market conditions and the dynamics of external factors. This makes them more attractive for both the consumer and the producer [17].

The pandemic that broke out in 2019 brought a lot of uncertainty to the market. In particular, the quarantine measures introduced by the states first transformed the habitual behaviour of the consumer, creating prerequisites for making online purchases, and then changed the content of consumer preferences themselves [14]. Manufacturers had to adapt to the new conditions. The most adaptive of them have built new digital channels of interaction, used e-commerce tools, and also reformatted traditional business schemes. This made it possible to protect business structures from bankruptcy and support their financial well-being [9].

Thus, the growth of the e-commerce market during the pandemic was a reflection not only of the increased activity of consumers, the multiple expansion of their opportunities to use the Internet to meet their needs but also evidence of the adaptability of manufacturers to the changed economic conditions [11, 13, 15].

3 Methodology

The purpose of the study is to conduct a comparative analysis of the impact of the pandemic on the situation of the e-commerce market in various countries, as well as to determine the directions for improving the structure of the e-commerce market in Russia in the long term.

Research objectives:

1) identification of prerequisites for the intensification of the development of the e-commerce market in various countries;
2) analysis of the dynamics of the development of the e-commerce market in the leading countries and Russia during the COVID-2019 pandemic;
3) justification of the directions of the state policy on improving the e-commerce market in Russia.

Research methods: theoretical analysis, economic analysis, graphical method, comparative analysis, system approach.

4 Results

There are at least three prerequisites for the intensive development of the e-commerce market [19]:

1) expanding access to the Internet, increasing the use of mobile devices and social networks;
2) the gradually weakening offline retail sector;
3) implemented state policy.

The continuous growth in the number of Internet users in the world (4.54 billion people in 2020), easier access to computers, the development of digital infrastructure, as well as the widespread use of smartphones have created conditions for expanding online shopping opportunities [28].

The timeline below shows the dynamics of the growth in the number of digital buyers around the world from 2014 to 2021. It is obvious that in 2016 there were 1.66 billion global digital buyers in the world, and by 2021 their number has increased to 2.14 billion people (Fig. 1).

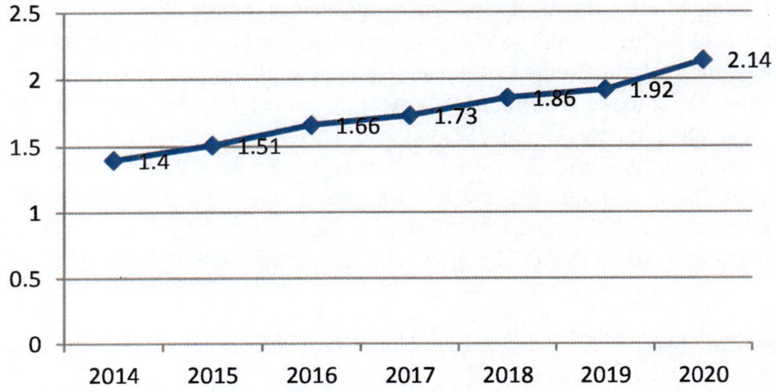

Fig. 1. The number of e-commerce users (billion people), 2014–2021. *Source*: Compiled by the authors based on [29, 30]

The purchase of goods and services using e-commerce tools is becoming a consumer habit [30]. Its formation is due to a wide range of goods and services on online platforms, the convenience of making online purchases, the efficiency of operations, competitive prices on online platforms, the ability to collect information about the consumer characteristics of goods and services.

E-commerce is based on the close relationship between producers and consumers. Producers promptly and efficiently serve customers, get the most complete picture of their preferences, decision-making factors, consumer requests from digital data about their usual purchases, personal information, online behaviour, etc. [8]. This is facilitated by big data processing technologies. For example, the D2C technology, successfully used by Nike and Tesla Motors for sales development, has such potential.

Affordable digital online payment services for goods and services, for example, PayPal (more than 40% of online buyers in the world use this method) add to the advantages of e-commerce [29].

E-commerce forms an ecosystem, that is, it ensures the optimal interaction of business partners and creates an integrated value chain [37]. A feature of ecosystems is their ability to combine products/services, creating additional value when using more than one product/service in terms of convenience and/or financial attractiveness for their suppliers and consumers, based on technological platforms [4]. This ensures a strong penetration of e-commerce [6, 27, 33].

These absolute advantages of e-commerce tools make it more attractive to consumers and weaken the position of the offline retail sector. In particular, in 2020, the volume of retail sales in the world decreased by 3% and amounted to $ 23.839 trillion, while the dynamics of e-commerce shows stable growth (Fig. 2).

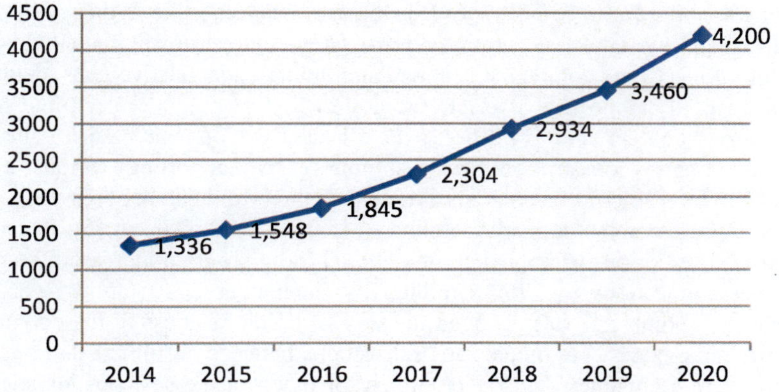

Fig. 2. The volume of the global e-commerce market, million dollars, 2014–2020. *Source*: Compiled by the authors based on [32]

Figure 3 shows the ranking of countries by the share of e-commerce in total retail sales (Fig. 3).

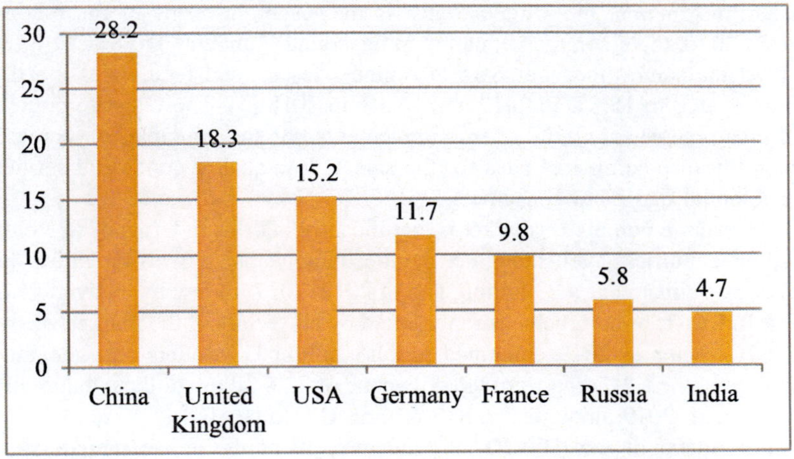

Fig. 3. The share of e-commerce in total retail sales by country, 2019. *Source*: Compiled by the authors based on [10]

Figure 3 shows that in China, e-commerce accounts for almost a third of the total retail segment – 28.2%. The UK follows the leader (18.3%). In the US, the volume of e-commerce accounts for 15.2% of total retail sales.

The development of e-commerce in the regions of the world is differentiated. The Asia-Pacific region and North America are leading in regional e-commerce indicators. For e-commerce, China's dominance means that 62.6% of all digital sales will occur in the Asia-Pacific region. North America and Western Europe are far from each other-19.1% and 12.7%, respectively [5]. In many respects, this is due to the implemented state policy. Thus, the current policy of China's Government in the field of legal regulation of electronic commerce is aimed at eliminating legislative gaps identified in the process of dynamic use of electronic platforms that mediate electronic trade in various goods and services [2]. The 13 Five-Year Development Plan for the period 2016–2020 focuses on cross-border e-commerce by supporting various business models [2]. The result of this policy is a more than three-fold gap between the countries in terms of the e-commerce market compared to the US following it [5]. In India, the efforts of the authorities to implement the Digital India program increased the number of Internet connections to 776.45 million by September 2020 (about 40% of the population). E-commerce has changed the way of doing business in the country. The most common e-commerce models are transactions between customers and businesses B2B (software companies develop programs for other businesses, and logistics companies help businesses deliver their products), B2C (a traditional retail paradigm where companies sell goods or services directly to consumers), C2C (a marketplace between consumers for trading, buying and selling things in exchange for a small commission paid to the site) and C2B (trading platforms like Fiverr, freelancer and UpWork, professionals with a wide range of skills in various fields and industries put their services up for auction). India's e-commerce sector is projected to grow from US $ 38.5 billion in 2017 to the US $ 200 billion by 2026 [35].

In North America, the United States is the leader in terms of the e-commerce market. In 2020, the e-commerce market in the country amounted to 861.12 billion US dollars. At the same time, the share of online sales in total retail sales in 2020 was 21.3% compared to 15.8% in 2019 and 14.3% in 2018 [2]. The US state policy in the field of e-commerce is currently being implemented on the principles of increasing the responsibility of e-commerce sites for the sale of low-quality goods and services that pose a potential danger to consumers [25].

The Russian e-commerce market is significantly inferior in terms of its volumes to the leading countries, even though it demonstrates stable growth dynamics [22]. In 2019, it was estimated at $ 23 billion, and in 2012–2017 it grew by an average of 21% per year. In Russia, e-commerce is represented by all sectors of the market: the revenue of the B2C sector in 2019 amounted to 21.8 billion US dollars and generated 425 million orders; the C2C sector provided revenue of 7.76 billion dollars; the retail online export sector in 2019 amounted to 817 million US dollars [12].

In the context of the COVID-19 pandemic, all of the above prerequisites have intensified their effect.

Firstly, the introduction of self-isolation regimes and social distancing forced consumers to stay at home and increased their time using the Internet. There was an increase in the use of mobile devices and the number of registered users of social networks. Thus, by January 2021, 4.66 billion people around the world began to use the Internet - 316 million (7.3%) more than in 2020. In 2020, the total number of mobile connections grew by 72 million (0.9%) and reached 8.02 billion by the beginning of 2021. In 2020, the number of social network users in the world rose by 490 million to 4.20 billion [34].

Secondly, the reduction of working hours and the temporary closure of shopping centres, the restriction of visits to shops and public places prompted the population to use

Internet platforms to order food and essential goods. Thus, the retail shopping segment lost 30–60% of visitors, depending on the region, and the majority of buyers during this period changed the perception of the shopping process and got used to making them online [16, 36]. In addition, trade enterprises faced the problem of postponing and extending delivery times, dependence on courier delivery, control of social distance, as well as the inability to meet the needs of customers in personal protective equipment.

Thirdly, the measures introduced by the states in the framework of protecting the life and health of the population directly pushed citizens to use digital tools for purchasing goods and services, stimulating the growth of the e-commerce market in all countries.

The world's largest e-commerce companies showed an increase in indicators in 2020. In particular, sales on the world largest trading platforms, such as Alibaba, Amazon, eBay, etc., accounted for 62% of total Internet sales in 2020. The three leading trading platforms — Taobao, Tmall and Amazon-account for almost two-thirds of the 2.67 trillion US dollars received by hundreds of the largest online retailers. One hundred percent sales growth in 2020 was noted by Etsy in the USA and Ozon in Russia [7]. Revenue growth will continue for all the largest e-commerce companies in various global markets. The forecast of experts concerns, first of all, the world giants: Chinese Alibaba, Japanese Rakuten, German Zalando, and English ASOS [1].

Thus, the pandemic led to an increase in the global volume of online sales in 2020 by 27.6% and exceeded the mark of 4.2 trillion US dollars [7]. E-commerce developed at the fastest pace in 2020 in Latin America, where the volume of e-commerce increased by 36.7%. In some countries, for example, in Argentina, online sales increased by a record 79%. Singapore is slightly behind in this indicator with an annual growth of 71.1%.

The dynamics of the growth of e-commerce markets in the leading countries in 2020 is shown in Fig. 4.

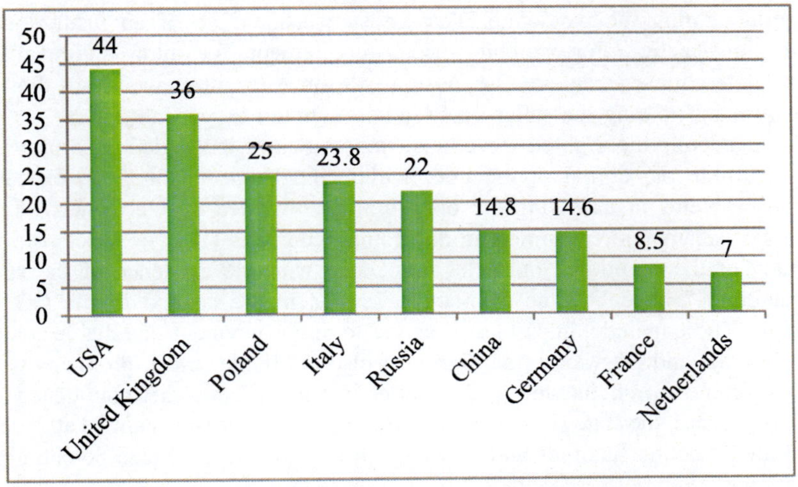

Fig. 4. Growth of the e-commerce market by country, 2020, in % compared to 2019. *Source*: Compiled by the authors based on [2]

The largest growth in the e-commerce market was demonstrated by the United States (44% compared to 2019), the United Kingdom (36%), and Poland (25%). In China, growth was recorded at the level of 14.8%. In Russia, the growth of the e-commerce market under the influence of the pandemic was 22%. According to the forecast of Statista, in 2021, the country is expected to further grow the segment to 30% [31].

In 2020, there were significant changes in the development of the e-commerce market in Russia [23]. There was a reformatting of consumer behaviour. During the initial period of the pandemic, the Russian consumer made online purchases, primarily of health products and personal protective equipment against the virus. Fears of uncertainty have prompted Russians to make online purchases of products with a long shelf life. The easing of restrictive measures led to surge in-store visits, an increase in the volume of the consumer basket. However, during the new wave of coronavirus (autumn 2020), there was a new stage of growth in online sales due to strict restrictions on shopping trips, as well as price increases due to a shortage of goods [12]. The return of Russians to their usual life was accompanied by a shift towards the regular use of e-commerce tools.

Changes in the structure of the e-commerce market during the pandemic are country-specific. Thus, in Russia, the increase in online sales relates primarily to the group of goods for everyday consumption and household maintenance (food, electronic and household appliances, and accessories for working from home, goods for children) [26]. In China, there was an increase in online sales of long-term storage products, in the United States, health products and products from related categories were in demand.

The pandemic and the intensification of e-commerce provoked by it have created a reason for further improvement of this segment of the economy in Russia. In particular, the experience of countries with a developed e-commerce market shows that Russia needs to implement a set of measures aimed at optimizing the "rules of the game" in the market in the conditions of a post-pandemic shock. This refers to the need to introduce measures to ensure the security of online shoppers in connection with the growth of cybercrime during the pandemic. The measures should be of an institutional and organizational nature. For example, stricter requirements for online trading platforms that promote counterfeit goods that pose a real threat to public health. In this case, a mechanism for holding the seller liable in court should be put into effect.

The pandemic highlighted the crucial importance of the digital economy and the need to bridge the digital divide both within and between countries, exacerbating existing problems in accessing the opportunities provided by e-commerce by small businesses, sellers and consumers in developing countries [18].

Many of the traditional obstacles associated with the existence of the so-called "digital divide" have become even more serious in the context of the COVID-19 pandemic. These include, in particular, access to online payments, a reliable connection to the Internet and electricity, the high cost of cross-border trade, problems with visibility in online search, as well as difficulties in gaining consumer confidence.

In this regard, the state should continue to implement measures aimed at developing digital infrastructure facilities and investing in logistics to maximize potential growth opportunities.

5 Conclusions

Firstly, the authors systematized the prerequisites for the intensification of the development of the e-commerce market in various countries: the expansion of access to the Internet, the growth of the use of mobile devices and social networks; the gradually weakening offline retail sector; the implemented state policy of regulating the e-commerce market.

Secondly, the authors investigated the dynamics of the development of the e-commerce market in various countries during the COVID-2019 pandemic. It was revealed that during the pandemic, the audience of e-commerce users expanded. There was a reduction in the volume of the offline retail market, and restrictive measures introduced by the governments of most countries of the world to the strengthened growth of the e-commerce market. It is proved that the largest growth of the e-commerce market has been achieved in the USA, Great Britain, and Poland. In Russia, the growth of the e-commerce market under the influence of the pandemic amounted to 22%, and further growth of the segment is expected in the country in 2021.

Thirdly, the directions of the state policy on improving the e-commerce market in Russia are proposed. The first direction is the policy of developing digital infrastructure facilities and investing in logistics to maximize potential growth opportunities. The second direction focuses on optimizing the "rules of the game" in the market in the conditions of a post-pandemic shock and involves taking measures to ensure the safety of online buyers in connection with the growth of cybercrime during the pandemic.

References

1. Abdelrhim, M., Elsayed, A.: The Effect of COVID-19 Spread on the e-commerce market: the case of the 5 largest e-commerce companies in the world (2020). https://ssrn.com/abstract=3621166. Accessed 08 Sept 2021
2. Belretail: How did the e-commerce markets grow in different countries in 2020? (2021). https://belretail.by/news/kak-v-godu-rosli-ryinki-e-commerce-v-raznyih-stranah. Accessed 08 Sept 2021
3. Chaika, I.: Novelties of legal regulation in the field of cross-border electronic commerce in the People's Republic of China. Actual Probl. Russian Law **9**(70), 176–185 (2016)
4. Drobot, E., Makarov, I., Nazarenko, V., Manasyan, S.: Impact of the COVID-19 pandemic on the real economy. J. Econ. Entrep. Law **8**, 2135–2150 (2020)
5. Economy.gov.ru: The concept of general regulation of the activities of groups of companies developing various digital services based on one "ecosystem" (2021). https://www.economy.gov.ru/material/file/cb29a7d08290120645a871be41599850/koncepciya_21052021.pdf. Accessed 08 Sept 2021
6. Emarketer: Global e-commerce 2020 e-commerce decelerates amid global retail contraction but remains a bright spot (2020). https://www.emarketer.com/content/global-ecommerce-2020. Accessed 08 Sept 2021
7. Fraumeni, B.: E-commerce: measurement and measurement issues. Am. Econ. Rev. **91**(2), 318–322 (2001)

8. Gklike: From offline to online: the volume of global e-commerce increased by more than 27% (2021). https://gklike.ru/news/iz-oflajna-v-onlajn-obem-mirovoj-internet-torgovli-uvelichilsya-bolee-chem-na-27/. Accessed 08 Sept 2021
9. Golovenchik, G.: The essence, classification and features of e-commerce. Sci. Innov. **4**(206), 39–45 (2020)
10. Gukasyan, Z., Tavbulatova, Z., Aksenova, Zh., Gasanova, N., Karpunina, E.: Strategies for adapting companies to the Turbulence Caused by the Covid19 Pandemic. In: Business 4.0 as a Subject of the Digital Economy. Springer, Cham (2021)
11. HSE: The Russian e-commerce market: results of 2019, trends of 2020 (2020). https://www.hse.ru/mirror/pubs/share/373094071.pdf. Accessed 08 Sept 2021
12. Karpunina, E., Butova, L., Sobolevskaya, T., Badokina, E., Pliusnina, O.: The impact of the Covid-19 pandemic on the development of Russian national economy sectors: analysis of dynamics and search for stabilization measures. In: Proceeding of the 37th IBIMA Conference, Cordoba, Spain, 30–31 May 2021, pp. 1213–1226 (2021)
13. Karpunina, E.K., Isaeva, E.A., Galieva, G.F., Sobolevskaya, T.G., Rodin, A.Y.: E-commerce as a driver of economic growth in Russia. In: Popkova, E.G., Sergi, B.S. (eds.) ISC 2019. LNNS, vol. 198, pp. 1622–1633. Springer, Cham (2021). https://doi.org/10.1007/978-3-030-69415-9_179
14. Karpunina, E., Moskovtceva, L., Zabelina, O., Zubareva, N., Tsykora, A.: Socio-economic impact of the Covid-19 pandemic on OECD countries. In: Current Problems of the World Economy and International Trade, Research in Economic Anthropology. Emerald Publishing Limited (2021)
15. Karpunina, E., Ruzhanskaya, N., Podorova-Anikina, O., Zubareva, N., Luchaninov, R.: Transformation of consumer behavior during the COVID-19 pandemic. In: Geo-Economy of the Future: Sustainable Agriculture and Alternative Energy, GeoPlanet: Earth and Planetary Sciences. Springer, Cham (2022)
16. Korolyuk, E., Rustamova, I., Kuzmenko, N., Khashir, B., Karpunina, E.: Diagnostics of regional economic security problems during the 2020 Crisis. In: Proceeding of the 37th IBIMA Conference, Cordoba, Spain, 30–31 May 2021, pp. 5248–5257 (2021)
17. KPMG: The realities of retailing in a COVID-19 world (2020). https://home.kpmg/xx/en/home/insights/2020/03/realities-of-retailing-in-covid-19-world.html. Accessed 08 Sept 2021
18. Laudon, K., Traver, C.: E-commerce, 9th edn. Prentice-Hall, New York (2012)
19. Lola, I., Bakeev, M.: The pulse of the global e-commerce market in the context of the COVID pandemic-19. HSE, Moscow (2020)
20. Lpgenerator: An excursion into the Chinese e-commerce market (2015). https://lpgenerator.ru/blog/2015/02/28/ekskurs-v-kitajskij-rynok-elektronnoj-kommercii/. Accessed 08 Sept 2021
21. Malysheva, G.: Socio-political aspects of the pandemic in the digital network society: the Russian experience. Bull. Moscow Region State Univ. (e-journal) **3**, 60–74 (2020)
22. McKibbin, W., Fernando, R.: The global macroeconomic impacts of COVID-19: seven scenarios. CAMA Working Paper, 19 (2020). https://doi.org/10.2139/ssrn.3547729
23. Morgan Stenley Research: Russia eCommerce: Last but not least (2018). https://www.shopolog.ru/u/6890-morgan-stanley-ru-ecom-internet.pdf. Accessed 08 Sept 2021
24. Nielson: Coronavirus and FMCG-trends in Russia: a new reality and new opportunities (2020). https://www.nielsen.com/wp-content/uploads/sites/3/2020/03/Nielsen-Covid-19-webinar-27.03.2020.pdf. Accessed 08 Sept 2021
25. NVO: The first wave of the pandemic threw the economy back 5 years (2020). https://nvo.ng.ru/economics/2020-06-10/4_7883_prognosis.html. Accessed 08 Sept 2021
26. Proright: Experience in regulating the e-commerce market in the United States (2021). https://proright.ru/2020/09/11/the-rule-ecommerce/. Accessed 08 Sept 2021

27. RAEC: The economy of the Runet in the era of COVID-19 — growing and transforming (2020). https://www.shopolog.ru/metodichka/analytics/issledovanie-raek-ekonomika-runeta-v-epohu-covid-19-rastem-i-transformiruemsya/. Accessed 08 Sept 2021
28. Schneider, G.: Electronic Commerce, 10th edn. Cengage Learning, Ohio (2012)
29. Statista: A global number of internet users 2005–2019 (2020). https://www.statista.com/statistics/273018/number-of-internet-users-worldwide/#statisticContainer. Accessed 08 Sept 2021
30. Statista: The number of digital buyers worldwide from 2014 to 2021 (in billions) (2020). https://www.statista.com/statistics/251666/number-of-digital-buyers-worldwide/. Accessed 08 Sept 2021
31. Statista: Number of internet users worldwide from 2009 to 2019, by region (in millions) (2020). https://www.statista.com/statistics/265147/number-of-worldwide-internet-users-by-region/. Accessed 08 Sept 2021
32. Statista: Forecast e-commerce market growth in Russia influenced by the coronavirus (COVID-19) pandemic from 2020 to 2024 (2021). https://www.statista.com/statistics/1140733/russia-forecasted-e-commerce-market-growth-under-covid-19-scenario/. Accessed 08 Sept 2021
33. Statista: Retail e-commerce sales worldwide from 2014 to 2023 (2021). https://www.statista.com/statistics/379046/worldwide-retail-e-commerce-sales/. Accessed 08 Sept 2021
34. Subramani, M., Walden, E.: The impact of e-commerce announcements on the market value of firms. Inf. Syst. Res. **12**(2), 135–154 (2001)
35. Web-canape: All statistics of the Internet and social networks for 2021 — figures and trends in the world and in Russia (2021). https://www.web-canape.ru/business/vsya-statistika-interneta-i-socsetej-na-2021-god-cifry-i-trendy-v-mire-i-v-rossii/. Accessed 08 Sept 2021
36. Worldscholarshipforum: The 30 Best E-Commerce Businesses in India for 2021 (2021). https://worldscholarshipforum.com/ru/богатство/бизнес-электронной-коммерции-в-Индии/. Accessed 08 Sept 2021
37. Zakharov, A., Startovaya, Y.: Overview of the global e-commerce market in the world economy: challenges and consequences of the pandemic. Russian Foreign Econ. Bull. **12**, 18–32 (2020)

Check for updates

Prospects for the Development of Mobile Technology in the Global Market in the Digital Age

Shakizada U. Niyazbekova[1,2](✉), Svetlana P. Anzorova[3], Anton Zh. Zubets[1], Ainur Zh. Issayeva[4], and Tana A. Abylaikhanova[5]

[1] Financial University under the Government of the Russian Federation, Moscow, Russia
[2] Moscow Witte University, Moscow, Russia
[3] Moscow University for Industry and Finance «Synergy», Moscow, Russia
[4] Zhetysu State University named after I. Zhansugurov, Taldykurgan, Kazakhstan
[5] Sarsen Amanzholov East Kazakhstan University, Ust-Kamenogorsk, Kazakhstan

Abstract. The paper considers and analyzes the sales of the largest manufacturers of smartphones (e.g., Samsung, Xiaomi, Apple, and Oppo) in the global smartphone market. In the global market, Samsung, for the first time, managed to sell the largest number of smartphones in Q3 2011, replacing the former market leader – Nokia. By the beginning of 2021, Apple outperformed Samsung in the global smartphone market for five consecutive quarters. The South Korean company managed to break the 30% mark several times. This dominance was observed several years ago. Samsung's lead is shrinking and is especially overtaken by Chinese manufacturers. As of 2021, there is an increase in the number of smartphone users. The conclusion highlights the importance of smartphones in a digitized environment.

Keywords: Users · Smartphone · Global supply · Technology companies · Digital population

JEL Codes: M11 · M15 · O32 · R59 · Z33

1 Introduction

The financial sector and technology have progressed very quickly. Smartphones, the leaders of the digital age, have taken over the global market.

Currently, the number of smartphone users worldwide exceeds six billion; this number is projected to increase by several hundred million over the next few years. China, India, and the USA are the countries with the largest number of users [23].

The smartphone market still has high growth potential since the penetration of smartphones is still below 70% in many populous countries (e.g., China and India).

E. G. Popkova (Ed.): ISC 2020, LNNS 368, pp. 374–380, 2022.
https://doi.org/10.1007/978-3-030-93244-2_41

Despite stagnant device sales, revenues in the global smartphone market have continued to grow over the past few years due to the rising average selling price of smartphones.

Nowadays, the leading smartphone manufacturers are Samsung and Apple. These companies account for about half of the world's smartphone supplies.

In 2020, annual sales of smartphones in Russia reached 570 billion rubles, which equals about $7.5 billion. Revenues grew steadily over the monitored period. In 2020, the number of smartphones purchased in Russia increased by 31.5 million.

Even though 5G technology was not yet widely available in Russia, smartphones with its support were actively sold on the market. In 2020, revenues from 5G-compatible smartphones in Russia amounted to 42 billion rubles.

Sales of such devices considerably increased after the release of the Apple iPhone 12 in Q4 2020.

In 2020, revenues of the global smartphone market exceeded $409 billion, declining due to the crisis caused by the COVID-19 pandemic.

However, the industry was projected to rebound in 2021 since smartphone sales to end-users worldwide were expected to grow significantly compared to 2020, even though projected revenue was slightly lower than in 2019.

According to Statista [18], as of January 2021, there were 4.66 billion active Internet users worldwide – 59.5% of the world's population. Of that total, 92.6% (4.32 billion) accessed the Internet via mobile devices.

2 Literature Review

A basic literature review provides an introduction to the topic of smartphone sales research.

H. Verkasalo [27] thoroughly analyzed the introduction of mobile apps, user satisfaction, trends, and customer segmentation. Moreover, he analyzed the dynamics of global smartphone sales [1–14, 27].

D. Cho conducted an empirical analysis of the spread of smartphones in a global context. He concluded that smartphones are one of the most widely and rapidly introduced digital devices, where global sales have reached enormous revenues [3].

Researchers from different countries Baburin [2], Pluzhnik [12], Kuksin [6], Moldashbayeva, Zhumataeva, Mezentseva, and Shirshova [10], Kurmankulova, Anzorova, Goigova, and Yessymkhanova [7], Ivanova [5], Nurpeisova, Smailova, Akimova, and Borisova [11], and Maisigova, Isayeva, and Dzholdosheva [8] considered the development of the digital economy as an essential factor for economic growth of the country, digital transformation of public procurement, innovative development of the country's economy, and the specifics of the relationship between public authorities, business, and society in the digital economy. Additionally, these authors paid considerable attention to other issues of digitalization.

3 Methodology

This research is based on the methodology of the basic principles of the systems approach, which creates a rational policy in the field of smartphone sales and management in the digital economy.

4 Results

Mobile devices have become an integral part of everyday life for millions of people. Globally, Internet-enabled devices (e.g., smartphones and tablets) have become indispensable tools for communication, information exchange, and entertainment.

In 2020, the number of active mobile internet users 4.28 billion, which was more than 90% of the world's Internet population using mobile devices to access the Internet [18].

The number of mobile device owners [24] and Internet usage is predicted to increase in the future because mobile technology has become more accessible than ever [16]. This upward trend in mobile web penetration is especially evident in emerging digital markets, where mobile networks are the primary means of accessing the Internet. Currently, the traffic of mobile Internet accounts for more than 55% of total web traffic [25]. In mobile-oriented markets (e.g., Asia and Africa), mobile connections account for a larger share of web browsing.

The mobile revolution continues to transform the behavior and frequency of using the Internet around the world. One of the most notable trends in mobile usage is in-app preference. In 2020, consumer spending on mobile apps on Google Play and Apple App Store totaled $102 billion worldwide [1–14, 27].

The leaders of the app market reported 142 million app downloads in 2020 [21], with social networking, gaming, and messaging apps accounting for the largest share of downloads. Overall, applications of almost all categories experienced a surge in demand in 2020 following the outbreak of the COVID-19 pandemic [1, 4, 14].

Social networking is one of the most popular online activities among mobile users. For over a decade, social networks, messengers, and photo and video sharing platforms have been among the leading drivers of mobile interaction worldwide (Fig. 1).

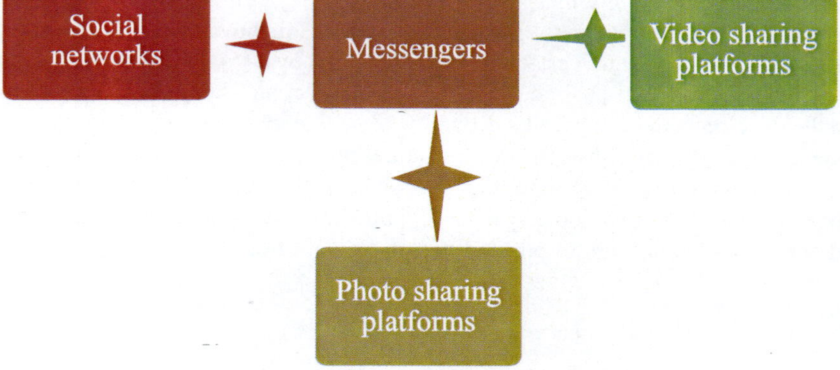

Fig. 1. Drivers of mobile interaction. *Source* Compiled by the authors

In 2019, the penetration of mobile social networks worldwide was 42% [15]. This figure is not surprising given that free mobile services facilitate communication in regions lacking (digital) infrastructure [1–14, 27].

Facebook remains the world's leading social network, with more than 2.7 billion active users, and WhatsApp ranks first among the most popular messengers [22].

Global mobile data traffic is growing in double digits every year. This trend is supported by the rapid growth of the usage of mobile media. In 2020, video apps accounted for more than 66% of the world's data [17]. By then, smartphones had long since overtaken desktops and smart TVs as the primary devices for watching online video. In line with this development, mobile SVoD subscriptions are expected to account for 40% of the world's total subscribers by 2025. This fact highlights the significant impact of mobile technology on global multimedia consumption. In addition to visual media, smartphones have also become the most widely used device for audio entertainment. Spotify is the most downloaded music and audio app worldwide [20].

The number of smartphone users in the Asia-Pacific region has increased over the past six years. Analysts estimate that the number of smartphone users in the Asia-Pacific region will reach more than 120 million by 2022.

Smartphones, mobile phones with more advanced computing capabilities and connectivity, came onto the consumer market in the late 1990s. Nevertheless, they became mainstream with the introduction of the Apple iPhone in 2007.

The iPhone has revolutionized the industry by offering customer-friendly features such as a touchscreen interface and a virtual keyboard. The first Android smartphone was introduced to the consumer market in late 2008. By 2021, 45% of the world's population is expected to own a smartphone.

Satisfaction and brand continuity (Fig. 2) allow smartphone users to make the most of the wide range of features offered by these devices, particularly their ability to run software applications.

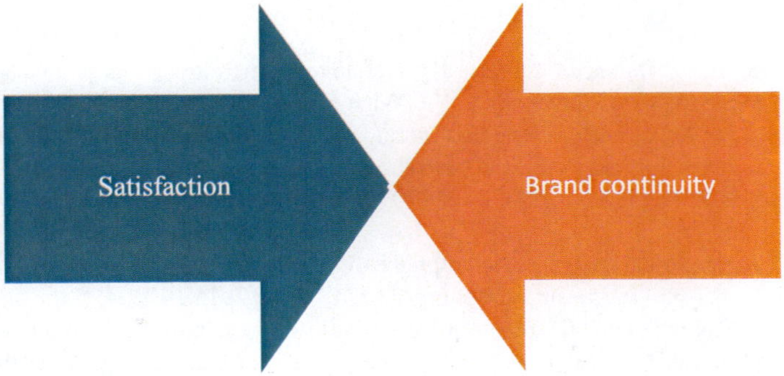

Fig. 2. The range of functions of smartphone devices. *Source* Developed by the authors

The global smartphone industry is steadily evolving and growing in terms of market size, models, and vendors. It is projected that the supply of smartphones worldwide in 2022 will reach about 1.6 billion units, and the number of smartphone users will exceed three billion worldwide. In Germany, the number of smartphone users in recent years has continued to grow, reaching almost 60 million in the country, with revenues from the sale of devices exceeding 12 billion euros [1–14, 27].

The German market is dominated by Samsung and Apple smartphones, each with about 40% and 30% of the market, respectively. Huawei invariably holds third place, while other manufacturers account for the rest of the market. As for operating systems, the clear winner is Android, which accounts for about 70% of the market.

In 2012, Android devices overtook Apple's iOS as the leading mobile operating system [1–4, 6, 8, 9, 11–14, 23, 27]. However, when looking at specific models of smartphones, the latest Apple products are often well-received and popular with consumers.

The analysis of smartphone consumption in Germany shows that more than 90% of smartphone users are satisfied with the chosen brand. A third of users are unlikely to change the brand next time [19].

5 Conclusion

Thus, smartphones are highly demanded in all countries because this device provides many different purposes [26], including access to social networking platforms, tracking of health and well-being, mobile payments. Moreover, the ability to purchase industrial and food products and various transport tickets is highly valued.

We can see the expansion of the capabilities of smartphones. It has led to increased mobile Internet traffic due to the growing number of users connected to faster networks and consuming large amounts of data. This progress has posed new challenges for network and service providers and the entire telecommunications industry because they seek to meet the ever-growing demand.

References

1. Agapitova, E.A.: Balanced scorecard as an investment attractiveness management tool. Bulletin Moscow State University S. Yu. Witte. Series 1: Econ. Manag. **2**(8), 57–66 (2014)
2. Baburin, S.N.: Legal and spiritual and moral foundations of the digital future of mankind in the era of increasing environmental catastrophe. Bulletin Moscow State University S. Yu. Witte. Series 2: Legal Sci. **4**(22), 6–12 (2019). https://doi.org/10.21777/2587-9472-2019-4-6-12
3. Cho, D.: An empirical analysis of smartphone diffusions in a global context. J. Contemp. East. Asia **14**(1), 45–55 (2015). https://doi.org/10.17477/jcea.2015.14.1.045
4. Gavrilova, E.N.: "Green" financing in Russia: specifics, main tools, development problems. Bulletin Moscow State University S. Yu. Witte. Series 1: Econ. Manag. **2**(33), 48–54 (2020). https://doi.org/10.21777/2587-554X-2020-2-48-54
5. Ivanova, O.S., Suleimenova, B., Yerzhanova, S.K., Berstembayeva, R.K.: Oil and gas investment opportunities for companies in modern conditions. In: Popkova, E.G., Ostrovskaya, V.N., Bogoviz, A.V. (eds.) Socio-Economic Systems: Paradigms for the Future, pp. 669–676. Springer, Cham (2021). https://doi.org/10.1007/978-3-030-56433-9_70

6. Kuksin, I.N.: Digitalization – a new reality in law. Bull. S. Yu. Witte Moscow University. Series 2: Legal Sci. **1**(23), 6–12 (2020). https://doi.org/10.21777/2587-9472-2020-1-6-12
7. Niyazbekova, S.U., Kurmankulova, R.Z., Anzorova, S.P., Goigova, M.G., Yessymkhanova, Z.K.: Digital transformation of government procurement on the level of state governance. In: Popkova, E.G., Ostrovskaya, V.N., Bogoviz, A.V. (eds.) Socio-economic Systems: Paradigms for the Future. SSDC, vol. 314, pp. 663–667. Springer, Cham (2021). https://doi.org/10.1007/978-3-030-56433-9_69
8. Maisigova, L.A., Niyazbekova, S.U., Isayeva, B.K., Dzholdosheva, T.Y.: Features of relations between government authorities, business, and civil society in the digital economy. In: Popkova, E.G., Ostrovskaya, V.N., Bogoviz, A.V. (eds.) Socio-economic Systems: Paradigms for the Future. SSDC, vol. 314, pp. 1385–1391. Springer, Cham (2021). https://doi.org/10.1007/978-3-030-56433-9_144
9. Mamitova, N.V.: Digital state: problems of construction in the Russian Federation. Bulletin S. Yu. Witte Moscow University. Series 2: Legal Sci. **4**(22), 13–20 (2019). https://doi.org/10.21777/2587-9472-2019-4-13-20
10. Niyazbekova, S.U., Moldashbayeva, L.P., Zhumatayeva, B.A., Mezentseva, T.M., Shirshova, L.V.: Digital economy development as an important factor for the country's economic growth. In: Popkova, E.G., Ostrovskaya, V.N., Bogoviz, A.V. (eds.) Socio-economic Systems: Paradigms for the Future. SSDC, vol. 314, pp. 361–366. Springer, Cham (2021). https://doi.org/10.1007/978-3-030-56433-9_38
11. Nurpeisova, A.A., Smailova, L.K., Akimova, B.Z., Borisova, E.V., Niyazbekova, S.U.: Condition and prospects of innovative development of the economy in Kazakhstan. In: Popkova, E.G., Ostrovskaya, V.N., Bogoviz, A.V. (eds.) Socio-economic Systems: Paradigms for the Future. SSDC, vol. 314, pp. 1773–1779. Springer, Cham (2021). https://doi.org/10.1007/978-3-030-56433-9_184
12. Pluzhnik, G.N.: Marketing approach to management – the key to the economic well–being of enterprises. Bulletin S. Yu. Witte Moscow University. Series 1: Econ. Manag. **4**(15), 64–68 (2015)
13. Schmidt, L.: Das Erfolgsgeheimnis des Apple App Stores: Eine diffusions diffusionstheoretische Analyse. GRIN Verlag, München (2011). https://www.grin.com/document/200472. Accessed 24 July 2021
14. Simonin, P.: The influence of institutions on the standard of living and human resources. Bulletin Moscow State University S. Yu. Witte. Series 1: Econ. Manag. **1**(28), 75–81 (2019)
15. Statista. Global mobile social network penetration rate as of January 2019, by region (2019). https://www.statista.com/statistics/412257/mobile-social-penetration-rate-region/
16. Statista. Number of internet users worldwide from 2005 to 2019 (2020). https://www.statista.com/statistics/273018/number-of-internet-users-worldwide/
17. Statista. Distribution of global monthly mobile data volume as of January 2021, by category (2021). https://www.statista.com/statistics/383715/global-mobile-data-traffic-share/
18. Statista. Global digital population as of January 2021 (2021). https://www.statista.com/statistics/617136/digital-population-worldwide/
19. Statista. How likely are you going to change your (primarily used) smartphone brand on the next possible occasion? (2021). https://www.statista.com/forecasts/998786/willingness-to-change-smartphone-brand-in-germany
20. Statista. Leading music and audio apps in the Google Play Store worldwide in July 2021, by number of downloads (2021). https://www.statista.com/statistics/691256/leading-google-play-music-worldwide-downloads/
21. Statista. Mobile app downloads worldwide from 2018 to 2025, by store (2021). https://www.statista.com/statistics/1010716/apple-app-store-google-play-app-downloads-forecast/

22. Statista. Most popular social networks worldwide as of July 2021, ranked by number of active users (2021). Retrieved from https://www.statista.com/statistics/272014/global-social-networks-ranked-by-number-of-users/
23. Statista. Number of smartphone users by leading countries as of May 2021 (2021). https://www.statista.com/statistics/748053/worldwide-top-countries-smartphone-users/
24. Statista. Number of smartphone users from 2016 to 2021 (2021). https://www.statista.com/statistics/330695/number-of-smartphone-users-worldwide/
25. Statista. Percentage of mobile device website traffic worldwide from 1st quarter 2015 to 1st quarter 2021 (2021). https://www.statista.com/statistics/277125/share-of-website-traffic-coming-from-mobile-devices/
26. Statista. Which of these services have you used in the past 12 months to pay in stores, restaurants or other points of sale with your smartphone? (2021). https://www.statista.com/forecasts/998738/mobile-payments-by-brand-in-germany
27. Verkasalo, H.: Analysis of smartphone user behavior. In: Proceedings of the ICMB-GMR 2010: Ninth International Conference on Mobile Business and Ninth Global Mobility Roundtable, pp. 258–263. IEEE, Athens June 2010. https://doi.org/10.1109/ICMB-GMR.2010.74

Evaluation of the Effectiveness of Innovative Services at Service Enterprises

Zhanna V. Smirnova(✉), Svetlana N. Kaznacheeva, Natalia V. Bystrova, Elena V. Romanovskaya, and Natalia S. Andryashina

Minin Nizhny Novgorod State Pedagogical University, Nizhny Novgorod, Russia
z.v.smirnova@mininuniver.ru

Abstract. Innovation processes are considered as processes that permeate all scientific, technical, production, marketing activities of manufacturers and, ultimately, are focused on meeting market needs. The purpose of this article is to study the process of designing a service and developing innovative services at a service enterprise. The author gives an example of designing the provision of services in the field of international transportation by the airline «International airport Nizhny Novgorod» JSC. Nizhny Novgorod airport is one of the oldest airports in the country. The need to introduce innovations has been identified. In the process of researching the use of innovative services, a survey was conducted to assess the effectiveness of the proposed innovative activities. The author analyzed the proposed innovative services that should increase the efficiency of the company. The author describes the advantages and disadvantages of the applied innovative technologies in airport services as a result of the research.

It was found that the need for innovation is stimulated by competition and a number of other market requirements. The types and stages of design were also reviewed and studied. All proposed innovations will be effective from both the economic and the social perspective.

Keywords: Innovative services · Service activities · Design · Efficiency

JEL Code: R11 · R12 · R58 · Q13 · Q18

1 Introduction

Statistics on the quantitative growth of social services that are consumed by a modern person quite confidently proves the assertion that the modern world primarily creates services. The service process is typical for all service enterprises.

The need to introduce innovations is stimulated by competition and a number of other market requirements.

E. G. Popkova (Ed.): ISC 2020, LNNS 368, pp. 381–388, 2022.
https://doi.org/10.1007/978-3-030-93244-2_42

Currently, innovative technologies for designing the content of the process of organizing the activities of service enterprises need to be introduced in the service sector.

In the process of researching the design of the process of providing services, we considered a social service for the population – international transportation. One of the most common types of transportation is air transport, a network of international airports.

Nizhny Novgorod airport is one of the oldest airports in the country. At this time, Strigino International Airport creates comfortable conditions for airlines and passengers. New international-class passenger terminals are being commissioned, high quality of passenger service is being maintained, and a flexible customer-oriented policy is being implemented [1, 3].

Airline design process research requires evaluating innovative services to improve company performance.

2 Methodology

Innovation processes are considered as processes that permeate all scientific, technical, production, marketing activities of manufacturers and, ultimately, are focused on meeting market needs.

The most important condition for the success of innovation is the presence of an innovator-enthusiast who is captured by a new idea and is ready to make every effort to bring it to life, and an entrepreneurial leader who has found investments, organized production, promoted a new product to the market, took the main risk and realized his commercial interest [2].

The scheme of parameters for assessing the quality of airport services in cargo and passenger transportation (Fig. 1) was developed in the research process.

Fig. 1. Parameters for assessing the quality of airport services. *Source* Developed and compiled by the authors

Airport service standards have always existed. However, modern requirements due to the transition to a market economy have provided the creation of additional value for such a concept as «quality». The high quality of service of the airline, passengers, meeting and seeing off passengers means meeting their expectations and, at the same time, creating an advantage over competitors, which is the key to a long-term partnership.

The modern Russian airport «IANN» JSC is an element of the country's transport infrastructure, which faced the need to take care of its efficiency, independently applying the regulatory mechanisms available to it [7, 8].

The methodological basis of the study is the methods of expert assessments, comparative analysis, statistical comparison and the theory of operations research. Legislative and normative acts of the Russian Federation, including sectoral normative documents and instructions, instructive, statistical and analytical materials in the field of quality of services, were used in the course of the study.

Scientific works of domestic and foreign authors in the field of quality management of provided services were studied and analyzed in the process of research.

Evaluation of the effectiveness of the proposed innovative services was carried out based on the application of innovative technologies in the management of transport services of the airline airport «IANN» JSC.

A survey on the application of innovative services in this area was carried out among the population, the calculation of the profitability of the use of these services (both economic and social efficiency) was carried out to achieve the set goal of assessing the effectiveness of innovative services of «IANN» JSC.

The expected result of the study is the profitability of one of the offered services (parking) of «IANN» JSC, which will be 68%.

The project will pay off in 3 years with such indicators. And this is a good indicator.

3 Results

It is necessary to focus not only on the accumulated experience of other airports, but also on the expectations and needs of consumers (passengers in this case) depending on the specifics of Nizhny Novgorod and the Nizhny Novgorod region in order to provide a high level of service quality. This will mean creating an advantage over competing airports that fosters long-term partnerships [13].

A questionnaire to assess the effectiveness of the proposed innovative measures was carried out in the process of researching the use of innovative services.

The following data (Fig. 2) were obtained after processing the survey results.

The diagram in Fig. 2 shows that not all customers are satisfied with the current level of services at Strigino airport. In general, passengers are satisfied with the speed of service delivery. But not everyone is satisfied with the comfort in the waiting room and the level of security at the airport. It can be concluded that these are the service sectors that need to be improved.

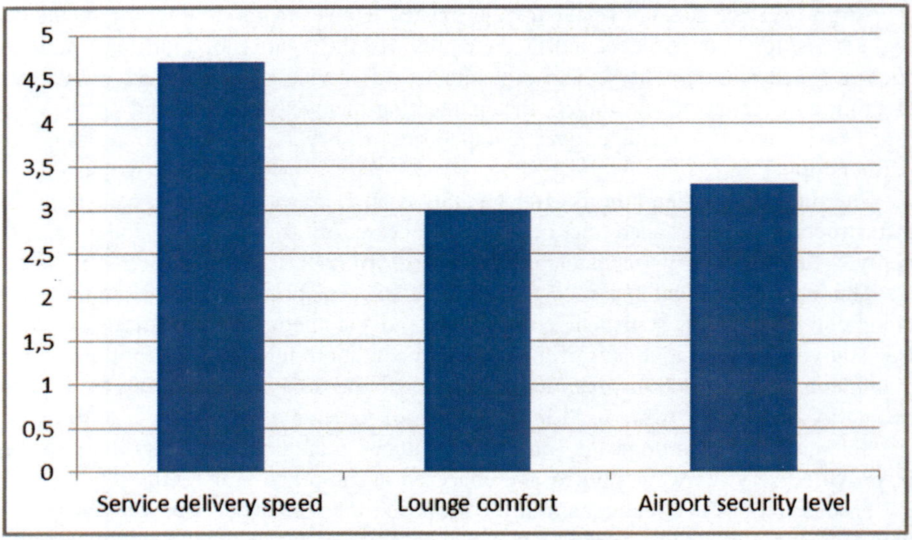

Fig. 2. The result of answering the question «Assessment of the quality of airport services». *Source* Developed and compiled by the authors

Let's move on to the analysis of the main question of our interest: «How do you feel about the need to introduce additional services?». All the options for the innovative development of airport services were outlined precisely in this issue (Fig. 3).

It can be seen from the results of answers to the question that passengers respond positively to the proposed innovations (Fig. 3). Passengers are a little confused by paid services, but nevertheless, their use is unavoidable because the airport cannot operate at a loss [4–6]. But in general, all innovations have high marks, not less than 4 points.

In fact, according to Russian law, there should be no competition between airports, because airports are subjects of natural monopolies (their activities are regulated by the federal law «On Natural Monopolies»). In practice, everything happens differently. Airports in all cities are competing for customers.

Currently, some of the passengers who could be direct customers of Strigino airport often prefer to fly from Moscow airports. «IANN» JSC increases its competitiveness and attracts passengers by implementing the recommendations proposed in the previous section.

The proposed options for innovation are primarily social, not commercial, because most of the innovations are free services. But on the other hand, the more free services the Strigino airport provides, the more customers it receives.

One of the most important innovations, in my opinion, is the introduction of a new hand luggage scanning system using the new Bottle Scanner technology [11].

Airport security is still a separate area. After all, the modern airport «IANN» JSC should not only provide passengers with high-quality services, but also provide them with a high level of protection. Improvement of the security system at «IANN» JSC took place in 2018 as part of a comprehensive public transport safety program.

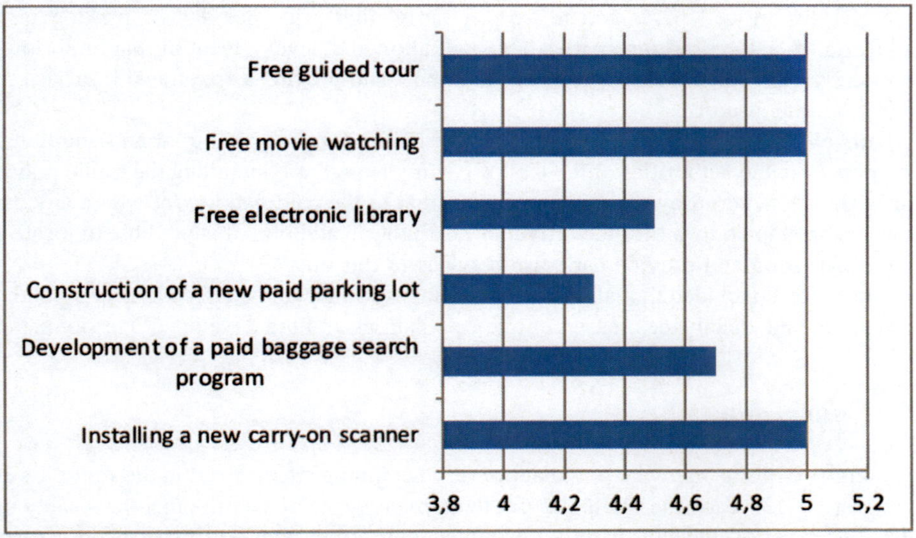

Fig. 3. The result of answering the question «How do you feel about the need to introduce the proposed additional services?». Source Developed and compiled by the authors

Thus, the introduction of a new system for scanning carry-on baggage is very important.

Using the scanner will help speed up the screening process and increase security at the airport and on airplanes. The presence of such a scanner is necessary because the danger of terrorism is increased during such events. In addition, increasing competition among airports, growing globalization and technological progress are leading to a rapid increase in traffic and passenger numbers, that ultimately increases the requirements for modern information systems for airports. The use of the latest scanner will help «IANN» JSC to increase its competitiveness in relation to the nearest airports in Moscow and Kazan. There will probably not be a direct economic effect from buying a hand luggage scanner [9]. The cost of such a scanner varies from 3.5 to 4.0 million rubles. Of course, the costs are quite high, but the cost of the scanner will directly affect the cost of air tickets. The planned payback will be about 2 years if the passenger traffic is high.

Another equally important innovation is the introduction of a baggage tracking system. This is the most important thing for the passenger during the flight, after the safety of the flight, because the suitcases of most passengers contain valuable and important things.

People have to spend a long time at the airport, because the procedure for preparing for a flight is not simple enough and takes a lot of time [10]. According to statistics, 40% of passengers are transit. Flight waiting times can take several hours.

Such offered services such as free library and movie screenings are great options for wasting time.

Let's take a closer look at the efficiency of creation of paid parking.

The airport will be forced to keep a fairly high price per day of parking in a situation of lack of parking spaces. This will allow the development of market of low-cost car parks which are located near the airport and provide a free transfer service to and from the airport.

«IANN» JSC will receive additional advertising for the parking lot and additional support from the authorities of the city of Nizhny Novgorod, including the traffic police of Nizhny Novgorod by organizing an area next to the paid parking lot where any car owner can stop and wait for the arrival of his flight. In addition, it is possible to locate a tire repair shop and provide car wash services in this area.

It can be concluded that all the proposed innovations will be effective from both the economic and social sides, based on the calculations and the developed questionnaire.

4 Conclusion

A number of tasks that were posed at the very beginning were solved in the study. Over the past 10–15 years, the world has mastered many types of services that the society of the Soviet period did not directly encounter (consulting, engineering, leasing, games and show business, etc.).

It was found that the need for innovation is stimulated by competition and a number of other market requirements [12].

Innovations in Russian conditions are becoming the main means of survival in the market for service companies.

The service organization exists to serve the customer, and systems and service personnel exist to support the service delivery process.

Designing a service delivery process begins with analyzing the demand for services that a service company can offer to its customers, i.e. with marketing research.

The object of the research is «IANN» JSC. Nizhny Novgorod airport is one of the oldest airports in the country. The airport has international status since 1993 and is certified for all types of aviation activities. It is an alternate aerodrome for the airports of the Moscow aviation hub.

Strigino International Airport creates comfortable conditions for airlines and passengers.

The standard that regulates the system of high-quality passenger service was developed at «IANN» JSC.

All services provided by the company were studied in the research process. The list of services provided by the airport is very large. The process of providing services is also analyzed in detail.

The main services for passengers are, of course, the purchase of tickets and a number of pre-flight actions.

The proposed possible innovative activities for «IANN» JSC can contribute to the development of the process of providing airport services.

This includes the following services:

- Free library;
- Free city tours;

- Free screening of films;
- Baggage tracking by smartphone;
- Organization of paid parking.

The proposed options for innovation are primarily social, not commercial, because most of the innovations are free services. But on the other hand, the more free services the Strigino airport provides, the more customers it receives.

One of the important innovations, in my opinion, is the introduction of a baggage tracking system. And also the opening of a paid parking lot will bring not only customer satisfaction, but also a great profit to the airport.

The effectiveness of the proposed measures is described based on the results of the study.

Net profit of the parking lot after annual operation is assumed to be 9 480 375 rubles.

All proposed innovations will be effective from both economic and social sides.

References

1. Basovsky, L.E.: Quality management. In: Basovsky, L.E., Protasiev, V.B. (eds.) INFRA-M, 212 p. (2012)
2. Batalova, Y.S.: Service Quality Scorecard. SGEA, Samara (2014)
3. Belobzhetsky, V.N.: Methods for assessing the quality of services. In: Belobzhetsky, V.N. (ed.) RISK: Resources, Information, Procurement, Competition, no. 1, pp. 374–379 (2012)
4. Garina, E.P., Kuznetsov, V.P., Egorova, A.O., Romanovskaya, E.V., Garin, A.P.: Practice in the application of the production system tools at the enterprise during mastering of new products. In: Popkova, E.G. (ed.) Overcoming Uncertainty of Institutional Environment as a Tool of Global Crisis Management. CE, pp. 105–112. Springer, Cham (2017). https://doi.org/10.1007/978-3-319-60696-5_13
5. Garina, E.P., Kuznetsov, V.P., Romanovskaya, E.V., Andryashina, N.S., Efremova, A.D.: Research and generalization of design practice of industrial product development (by the example of domestic automotive industry). Qual. - Access Success **19**(S2), 135–140 (2018)
6. Garina, E.P., Romanovskaya, E.V., Andryashina, N.S., Kuznetsov, V.P., Tsymbalov, S.D.: Methodology for creating a product development system based on complexity management strategy. In: Popkova, E.G., Ostrovskaya, V.N., Bogoviz, A.V. (eds.) Socio-economic Systems: Paradigms for the Future. SSDC, vol. 314, pp. 1661–1669. Springer, Cham (2021). https://doi.org/10.1007/978-3-030-56433-9_172
7. Kozlova, E.P., Morozova, G.A., Kuznetsov, V.P., Romanovskaya, E.V., Andryashina, N.S.: Interaction of human and smart technologies in modern conditions. In: Popkova, E.G., Sergi, B.S. (eds.) ISC 2020. LNNS, vol. 155, pp. 609–618. Springer, Cham (2021). https://doi.org/10.1007/978-3-030-59126-7_68
8. Kuznetsova, S.N., Romanovskaya, E.V., Artemyeva, M.V., Andryashina, N.S., Egorova, A. O.: Advantages of residents of industrial parks (by the example of AVTOVAZ). In: Popkova, E.G. (ed.) HOSMC 2017. AISC, vol. 622, pp. 502–509. Springer, Cham (2018). https://doi.org/10.1007/978-3-319-75383-6_64
9. Novatorov, E.V.: International models of marketing services. In: Novatorov, E.V. (ed.) Marketing in Russia and Abroad, no. 3, pp. 91–96 (2015)

10. Semakhin, E.A., et al.: Improvement of management efficiency of material resources in service sphere. Int. J. Innov. Technol. Explor. Eng. **8**(12), 1740–1745 (2019)
11. Smirnova, Z.V., Bystrova, N.V., Golubeva, O.V., Lebedeva, T.E., Kaznacheeva, S.N.: Analysis of the socio-psychological climate in the work collective. Int. J. Manag. **11**(5), 465–475 (2020)
12. Smirnova, Z.V., Vaganova, O.I., Cherney, O.T., Andryashina, N.S., Bystrova, N.V.: Calculation of automatic submerged arc welding. Int. J. Emerg. Trends Eng. Res. **8**(7), 3697–3699 (2020)
13. Volkova, L.P.: Airport Operations Management. Part 1. Legal Framework for Airport Management: Textbook, 2096 p. Moscow State Technical University of Civil Aviation (2016)

Business Management and Business Integration in the Digital Economy

Prospects for Using Model Lawmaking for Legal Protection of Scientific Discoveries in the Commonwealth of Independent States

Tatyana V. Shatkovskaya[1,2(✉)], Tatyana V. Epifanova[2], Galina B. Vlasova[3], and Svetlana V. Denisenko[3]

[1] South-Russian Institute of Management – Branch of the Russian Presidential Academy of National Economy and Public Administration, Rostov-on-Don, Russia
[2] Rostov State University of Economic (RINH), Rostov-on-Don, Russia
[3] Rostov Branch of the Russian State University of Justice, Rostov-on-Don, Russia

Abstract. In this study, we aim to explore the effectiveness of model lawmaking in ensuring legal protection of scientific discoveries at a regional level on the example of the Commonwealth of Independent States (CIS). We believe that the increasing openness of science and cross-border commerce requires promoting universalization for the legal protection of scientific discoveries. By analyzing, synthesizing, and modeling several Soviet legal provisions, we identified the features that discovery must possess to be protected. Discovery should be internationally novel, theoretically grounded, experimentally proven, and economically and technically viable. Moreover, it should possess potential or real value for science.

The systemic approach and the methods of generalization and reconstruction allowed us to organize the existing scientific approaches to the legal protection of scientific discoveries. We identified the place of scientific discoveries in the system of civil law. Moreover, we proved that the feasibility of the main provisions of the model act "On scientific discoveries." The act was adopted in 2010 by the CIS Interparliamentary Assembly and several CIS member states. However, we propose making several amendments to the model act. The social significance of scientific discoveries and their potential use in commerce makes them an independent type of intangible objects of civil law and not intellectual property. We believe that national academies of science should conduct registration procedures, provide scientific expertise, and issue discovery certificates.

Keywords: Author · Intellectual property · Model act · Scientific discovery · Commonwealth of Independent States · Law

JEL code: K10

1 Introduction

Recent regulation of science dialectically "intertwines" two main trends. On the one hand, there is a clear need for universalized legislation and international forms of legal communication, driven by active scientific-technological and economic cooperation

E. G. Popkova (Ed.): ISC 2020, LNNS 368, pp. 391–398, 2022.
https://doi.org/10.1007/978-3-030-93244-2_43

among states. On the other hand, national regulations of science are still diverse due to the differences in the historical development of socio-economic, political, and cultural features [16].

Expanding the regulation of science past national borders and the creation of a common digital space raises the question of universalizing regulatory practices. Moreover, these factors force the regulators to reconsider the established approaches to the legal protection of scientific discoveries, including both the methods and forms of law applicability in transnational unions. Science "without borders" should not lead to the disenfranchisement of scientists.

The issue of the legal protection of scientific discoveries (LPSD) is especially relevant since scholarly knowledge only relatively recently found its commercial applications. The merging of science and technology and the use of scientific achievements in everyday transactions create a dilemma. On the one hand, a copyrighted result of research may only include "discovery" as an unprotected element. Therefore, all rights belong to a person that found commercial application for the discovery – this approach encourages entrepreneurial innovation but discourages scholars. This means that the personal rights of a scholar lie outside the law and are forced to rely on the ethics and moral qualities of a person that commercialized the discovery.

On the other hand, scientific works cannot be covered by the patenting regime. First of all, applying copyright to scientific discoveries will promote the emergence of exclusive rights, which violates the principle of academic freedom. Secondly, the existing patenting systems are outdated and do not correspond to the scope and pace of scientific-technological development. Thirdly, patent law protects only specific types of technical solutions that are of practical rather than theoretical importance.

These factors shaped the aim of this study: identify the potential of model law-making in LPSD at the regional level on the example of member states of the Commonwealth of Independent States (CIS).

The study proves that open science and cross-border commerce require developing universalized standards for LPSD. The main ways of producing such standards are:

- Increasing the priority of legal acts of transnational unions (such as CIS);
- Adopting main provisions of transnational acts into national legislations.

2 Materials and Methods

Both Russian and foreign scholars have been trying to find a model for LPSD since the 19th century. Scholarly opinions on the issue differ drastically, with some authors claiming that LPSD contradicts scientific freedom and that ethics and morals provide adequate protection of scholars' rights [11]. Other authors [10, 14, 15] argue for the development of LPSD and propose models for international legal protection. However, no such model has received international approval. For example, the World Intellectual Property Organization (WIPO) does enshrine the term "scientific discovery" but fails to define it [7]. The concept was to be detailed in the Geneva Treaty on the International Registration of Scientific Discoveries [8] but failed to receive needed support from member states.

The scientific substantiation of the term "scientific discovery" appeared in the Soviet jurisprudence in the 1930s. The Soviet leadership declared scientific-technological progress as a priority of state development. For example, N. A. Raigorodsky asserted, "Any discovery, taken as a result of learning the patterns and regularities of the external world phenomena, expands the possibilities of our mastery of the world." To ensure the priority of Soviet authors' discoveries, he proposed establishing a checkbox registration system that does not scrutinize the scientific novelty and practical applications of discoveries [12].

During the second codification of Soviet civil law, a special commission consisting of prominent lawyers (S. N. Bratus, A. I. Omelchenko, etc.), inventors, and scholars was created by the Technical Terminology Committee of the Academy of Sciences of the USSR to develop several terms, including "scientific discovery." The commission suggested that scientific discovery should be defined as "the establishment of previously unknown objectively existing phenomena, properties, and patterns of the material world" that introduce fundamental changes to scientific knowledge (i.e., various inventions and proposals). This excludes scientific discoveries in the social sciences, archaeology, mineral deposits, and paleontology [1]. This definition is to this day used as a normative-legal definition of "scientific discovery."

By analyzing, synthesizing, and modeling several Soviet legal provisions, we identified the features that discovery must possess to be protected [4]:

- International novelty of discovery and its ability to introduce fundamental changes to the knowledge about objective patterns, features, and phenomena of the material world;
- Theoretical grounding, experimental proofs, and reliability of the discovery;
- Real or potential practical value and technical-economic efficiency.

In this study, we used the systematic approach, methods of generalization and reconstruction, as well as the structural-functional and institutional methods. This allowed us to organize the existing scientific approaches to LPSD and express them in Fig. 1.

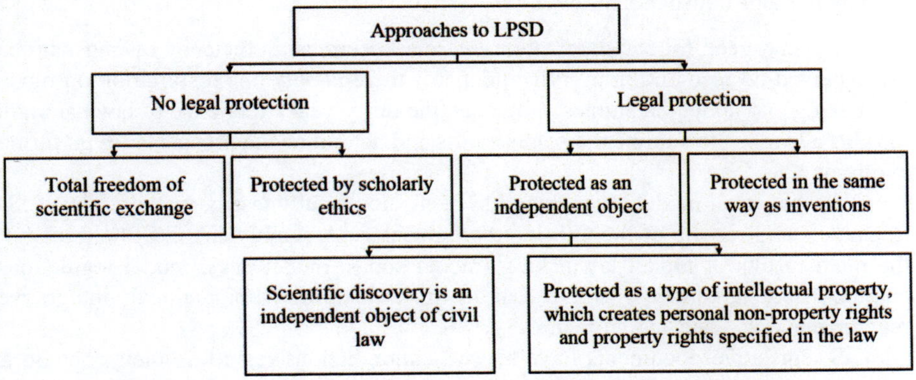

Fig. 1. Scientific approaches to LPSD. *Source* Compiled by the authors

Registration of scientific discoveries in the USSR involved academia and university scholars. On average, only 14 out of 900–1300 annually submitted applications were approved [9]. Despite all the shortcomings of this system (labor-intensive scientific expertise, lack of clear legislative criteria of protection, state-wide scale of legal protection, insufficient incentives for scholars), the USSR was the pioneer and leader of LPSD globally. After the fall of the Soviet Union, the system of state protection collapsed. Scholars were no longer granted exclusive rights to discoveries out of fear that access to scientific knowledge would be limited.

Still, there is a clear need for an LPSD mechanism. Some national legislations cover LPSD: in the USA, scientific discoveries are traded as inventions (US Patent Law) [6], while the civil law of Ukraine maintains the Soviet model of LPSD (Articles 457–458 Chapter 38 of the Civil Code of Ukraine) [5].

In Russia, state registration of scientific discoveries was abolished in 1992. However, the Russian Academy of Natural Sciences, jointly with the International Academy of Authors of Scientific Discoveries and Inventions, maintains a system of public expertise and registration for scientific discoveries, ideas, and hypotheses. Over the years, the organization has issued 161 certificates for scientific discoveries in natural sciences [13]. The applications are submitted by scholars from different countries (Belarus, Israel, Kazakhstan, the United States, Chile, and Ukraine). Approximately 500 discoveries were registered, bearing in mind that the applicants pay for the expertise and that the certificates are not enforceable.

3 Results

Creating a system of LPSD in the CIS entails addressing several issues of model lawmaking efficiency. These include:

- Understanding the limits of transnational law universalization;
- Developing methods for combining national and universalized legislation to prevent the law from being declarative;
- Creating a mechanism for developing and enforcing universalized acts on the national and transnational level.

The rising need for standardization, harmonization, and efficiency of transnational law received no recognition in jurisprudence – there is only one dissertation on model lawmaking and about ten studies. However, the thirty-year experience of lawmaking in the CIS allows us to summarize some results and determine the prospects for its further evolution.

Over the years, model lawmaking has been successfully tested, including in LPSD. Legal acts and reports of the CIS Interparliamentary Assembly (IA CIS) help identify the main results of model lawmaking (model codes, model laws, model regulations, charters, and agreements) and reveal its main features: supranational, integrative, recommendatory, and model nature.

CIS legislative documents have several contradictions regarding their content, as well as inconsistencies between doctrinal approaches to lawmaking and its results. Moreover, there is a lack of distinction between model laws and recommendations.

Both these types are of recommendatory nature; they aimed to standardize the regulation of specific social relation types. They propose member states harmonize their national legislation on common issues with the CIS legislation and other international treaties and agreements. The difference lies in the concretization of recommendation types, which makes model laws a type of recommendation.

Despite the diversity of scientific approaches in jurisprudence, the essence of law does not change. In a formal sense, a law is a concentrated expression of the will of a public authority via mandatory provisions within its jurisdiction. The law establishes a legal rule that creates the duties and obligations of subjects. Two interpretations arise from these facts:

1. CIS model laws are not enforceable and can be regarded not as a low, but as a model or an example of a law;
2. The recommendatory nature of model laws allows (but not forces) member states to adopt model laws, taking into account the specifics of the national legislature.

Therefore, we need to identify the legal nature of model laws and determine the conditions that make it an effective legislative means for aligning socio-economic, political, and cultural activities of CIS member states.

We believe that model law should be regarded as a means of universalization of law at the transnational level. One should also distinguish between universalization and unification. Unification aims to simplify and standardize laws, while universalization tries to create to cover many legal provisions and standardize them. Universalization serves to accommodate the joint modeling of common transnational law. Therefore, the model law must have an adequate structure and serve as a normative legal model for the laws of member states.

The model law is more variable than national laws; it includes norms-principles, norms-definitions, norms-incentives of permissive nature. These norms lay down unified approaches to state regulation of specific social relations and allow national parliaments to choose an interpretation of the model law.

The development of a model law has some technical-legal peculiarities. Using model laws in CIS documents is needed to their relevance and ability to ameliorate any gaps and inconsistencies in national regulation of some social relations. Doing so would bring legislation of CIS member states in tune and clarify some of the transnational agreements.

The efficiency of using model laws at the CIS level is explained by several factors:

1. Joint agreement of national parliaments on the adoption of such a law;
2. Voluntary self-declaration of states to take the model law as the basis of national law and to ensure that national legislation does not contradict its basic principles;
3. The fact that model laws can be transformed during their adoption into national legislatures;
4. Mandatory nature of adopting model law. If a model law is not mandatory, there is no way to distinguish between it, a bill, and a recommendation as the two latter acts lack enforceability.

The only real guarantee of model law implementation is its effective use in real social relations – the achievement of social effect "as a result of the creation of a tangible or intangible product demanded by a social group or society as a whole that does not contradict the moral, ethical, and political foundations of a given society" [17].

The model law should be regarded as a means of harmonizing the actions, prescriptions, and measures of states united by voluntary commitments to socio-economic, political, and cultural cooperation. The ultimate aim of a model law is to create favorable conditions for achieving the common goals of transnational unions. The role that model laws play in normative consolidation is obvious, especially during digital transformations and the transition to digital document circulation. The recommendatory nature of model laws does not mean that a member state may refuse to implement its provisions. Model law does not lose its binding effect on people covered by it.

In the decentralized approach to the regulation of relations inside transnational unions, the recommendatory law should be perceived in the context of the term "permissive law." Permissive laws imply transformation and account for the national characteristics of the entity that implements the law. Legal means such as multilateral and bilateral consultations, interstate agreements, parliamentary hearings, etc., are used to ensure this approach [17].

Throughout the CIS existence, IA CIS has adopted more than 350 model laws. These include the model law "On Protection of Rights to Scientific Discoveries" No 34-9 dated April 7, 2010. [2]. Article 1 defines scientific discovery as "the establishment of previously unknown, but objectively existing regularities, properties, and phenomena of the material world, which introduce fundamental changes to scientific knowledge" with the exception of the objects specified in Paragraphs 3 and 4 of Article 10. The law basically reproduces the Soviet model of LPSD. The authors are granted personal non-property rights (the right of authorship, the right to assign a name to the scientific discovery, the right to be credited or remain anonymous), as well as property rights to remuneration for scientific discovery, and the right to participate in its practical implementation (Article 27).

However, even though the model law was adopted with the active participation of the Parliament of the Russian Federation, it has not become a subject for discussion in the authorities. The law is not included among the sources of modern research on LPSD. As a result, the proposed LPSD models disregard this and other model laws proposed at the IA CIS. Thus, the author's priority for scientific discoveries, proposed as an innovation [14], proposed as an innovation, is already regulated in the IA CIS model law "On rationalization activity" No 37-5 dated May 17, 2012 [3]. The law establishes not only the right of authorship and the right to remuneration (Paragraph 2 of Article 3) but also the right of priority.

4 Conclusion

The given reasons prove the need for the return of the registration model for LPSD. This can be done via improving CIS model laws and implementing them in the national legislations of member states. First and foremost, we believe that it is necessary to redefine the USSR experience in LPSD so that it may be renewed and adapted to the new digital environment, as it not only provides new means for scientific exchange but also pose threats to the rights of scholars.

Taking into account the effectiveness of regional unions (e.g., the European Union, the Eurasian Economic Union); the common experience in socio-economic, political, cultural, and scientific cooperation; and the legal traditions of the CIS member states, we advise implementing the model law "On scientific discoveries" into national legislations of member states, including Russia.

The adoption of LPSD laws will provide a legal basis for scientific-technological integration and cooperation of the CIS member states. Moreover, it will serve to harmonize national legislations and provide wider protection of scientific discoveries. However, several provisions of the said law need to be edited. Among the fundamental points, we will highlight the following:

1. Scientific discoveries should not be protected by the intellectual property law, as it does not result from intellectual activity, which has commercial value and practical relevance. However, the social significance of scientific discoveries and their potential use in everyday transactions calls for protecting the rights of scholars. These facts highlight that scientific discoveries are an independent type of intangible objects of civil law;
2. Registration procedures, scientific expertise, and issuing discovery certificates should be delegated to national academies of sciences, as they would involve the professional expert community and the public in these activities. This approach will not only de-bureaucratize the registration procedures but also revitalize science and stimulate cooperation, exchange, and mutual assistance;
3. We believe that an author of scientific discovery should not be guaranteed payments from the state budget, as stipulated by Paragraph 3 of Article 27 of the model law. Such payments are a burden on the budget. In addition, the volume of state payment could not possibly be calculated by taking into account the significance of the scientific discovery. The author should be given the right to demand remuneration for the commercial use of the discovery by business entities and other interested parties on a contractual basis.

References

1. Council of Ministers of the Soviet Union: Decree "On Approval of the regulations on discoveries, inventions, and rationalization proposals" (1973). https://docs.cntd.ru/document/9004990. Accessed 29 May 2021. (August 21, 1973 No. 584, as amended September 8, 1990)
2. Interparliamentary Assembly of Member Nations of the Commonwealth of Independent States (IPA CIS): Model law on the protection of rights on scientific discoveries, St. Petersburg, Russia (2010). https://iacis.ru/public/upload/files/1/265.pdf. Accessed 29 May 2021. (April 7, 2010 No. 34-9)
3. IPA CIS: Model law "On Rationalization Activity", St. Petersburg, Russia (2012). https://iacis.ru/public/upload/files/1/281.pdf. Accessed 29 May 2021. (May 17, 2012. No. 37-5)
4. Numerical modeling of dynamic workflows in socio-technical systems (2019). Standartinform, Moscow. https://docs.cntd.ru/document/1200169686. Accessed 29 May 2021. (November 22, 2019 GOST R 57700.20-2019)

5. Ukraine: Civil Code of Ukraine, Kyiv, Ukraine (2003). http://www.wipo.int/wipolex/en/text.jsp?file_id=181865. Accessed 29 May 2021. (January 16, 2003 No. 435-IV)
6. United States of America: United States Code Title 35 – Patents (Consolidated Patent Laws as of May 2014) (1952). http://www.wipo.int/wipolex/en/text.jsp?file_id=335758. Accessed 29 May 2021
7. WIPO: WIPO Convention (as amended September 28, 1979). World Intellectual Property Organization, Stockholm, July 1967. https://wipolex.wipo.int/en/text/283854. Accessed 29 May 2021
8. WIPO: Geneva treaty on the international recording of scientific discoveries. World Intellectual Property Organization, Geneva, March 1978. https://www.wipo.int/edocs/pubdocs/en/wipo_pub_279.pdf. Accessed 29 May 2021
9. Androshchuk, G.A.: Scientific discoveries: role and importance in the innovation economy. Intellectual property. Ind. Prop. **2**, 32–41 (2014)
10. Boguslavsky, M.M.: Copyright Issues in International Relations. Nauka, Moscow (1973)
11. Ionas, V.Y.: Works of Creation in Civil Law. Legal Literature, Moscow (1972)
12. Raigorodsky, N.A.: Discovery and its protection in Soviet civil law (on the issue of copyright of researchers). In: Cherpakhin, B.B. (ed.) Issues of the Soviet State and Law, no. 187, issue 6, pp. 120–158. Leningrad University Press, Leningrad (1955)
13. Russian Academy of Natural Sciences: Discovery list (n.d.). https://raen.info/activities/registracija-nauchnyh-otkrytii/spisok-otkrytii. Accessed 29 May 2021
14. Salitskaya, E.A.: Legal protection of intellectual rights to research results (Dissertation of Candidate of Laws), Moscow, Russia (2018)
15. Serebrovsky, V.I.: Legal Protection of Scientific Discoveries in the USSR. Publishing House of the USSR Academy of Sciences, Moscow (1960)
16. Shatkovskaya, T.V., Epifanova, T.V., Pozdnishov, A.N., Fedorenko, N.V.: Legal standards of free utilization of intellectual property in the sphere of digital economy. Eur. Res. Stud. J. **XII**(1), 25–33 (2018). https://doi.org/10.35808/ersj/1155
17. Shatkovskaya, T.V., Epifanova, T.V., Vovchenko, N.G., Romanenko, N.G.: A legal mechanism for regulating digital economy. In: CBU International Conference on Innovation in Science and Education, vol. 6, pp. 763–767 (2018). https://doi.org/10.12955/cbup.v6.1246

Value Creation in the Process of Product Development Within the Framework of the Network Interaction of System Participants

Ekaterina P. Garina(✉), Alexander P. Garin, Victor P. Kuznetsov, Natalia S. Andryashina, and Elena P. Kozlova

Minin Nizhny Novgorod State Pedagogical University, Nizhny Novgorod, Russia
direktor@reklamapovolzhya.ru

Abstract. The purpose of this article is to systematize and summarize the experience of creating value in the process of product development within the framework of the network interaction of system participants.

The methodology for assessing the cost of a product proposed by the authors is determined by the following provisions. Firstly, it is the choice of approach to the technical implementation of product development system based on two possibilities: function-oriented and component-oriented approaches. Secondly, it is the choice of a cost management system - administration, accounting, cost estimation methods based on the implementation of alternatives: methods based on the functions implemented by the product; methods based on the characteristics of the components of the product. Thirdly, the calculation of the cost of PD- by: summation and decomposition of the functional structure; calculating the cost of the production process, calculating the cost of materials, etc. Interpretation of the results of the implementation of product solutions on the example of a network of stakeholders of the research Object showed that the method is realizable, the calculation results are comparable.

The application of the proposed methodology will make it possible to assess the potential for success of product solutions in the development of networked forms of business, in particular, by highlighting the hierarchical structure in the value chain, as well as the shares of costs attributable to the participants in the system.

Keywords: Product development · Value creation · Cost management · Networking

JEL Code: O32

E. G. Popkova (Ed.): ISC 2020, LNNS 368, pp. 399–407, 2022.
https://doi.org/10.1007/978-3-030-93244-2_44

1 Introduction

The condition for the implementation of an approach to modeling a product development system is the need to include cost as an active parameter of the project, but not a factor arising from this process. Hence the importance of its assessment at the stage of conceptualization.

«Overlapping» value chains is common in the formation of a product development system because: a) each participant in the system forms its own value profile based on the developed relationship between the internal structure of the product and the process; b) each participant in the system has its own stages of cost occurrence in the product value chain (Fig. 1).

In addition, modern value chains are presented in the form of a hierarchical structure, and the shares of costs attributable to the participants in the system have been transformed. At the top of this pyramid, the large ODM members of the R@D company control the design process and most of the manufacturing operations, accumulating a large share of the value (Complex systematic product development processes (PDP) create more value during the design phase) [1].

Product development	Planning	Supply
→	→	→
– models/layouts/system drawings; – resource provision; – prototyping/system testing; – product, systems, tools specifications; – primary commercialization of the product	– system maintenance; – performing planned calculations, tasks; – increase in the cost of samples	– rise in prices for the purchase of materials; – resource provision; – expenses for additional processing of deliveries, their admission to production; – packaging
Production	Logistics	Marketing & Sales
→	→	→
– increase in processes; – increase in production waste; – increase in the share of work in progress; – additional resource support.; – additional work schedules.	– increase in storage space; – increased handling of inventory flows; – increase in transportation costs; – increase in stocks; – increase in the cost of transportation, etc.	– additional resource support; – increase in the cost of promoting goods; – increase in errors in order processing / consumer complaints; – strengthening of documentation support of processes

Fig. 1. Points of occurrence of costs in the product value chain. *Source* Developed and compiled by the authors

According to open sources, the cost of the product development process, which was initially 10–15%, has actually fluctuated in the range of 80% over the past two decades [2]. The gradation between the cost of the manufacturing process and the cost of the product development process is evidence of the growing role of the product development stage in the overall value chain (Fig. 2), proving that the cost of product development increases as estimated and then as actual as the product develops [3].

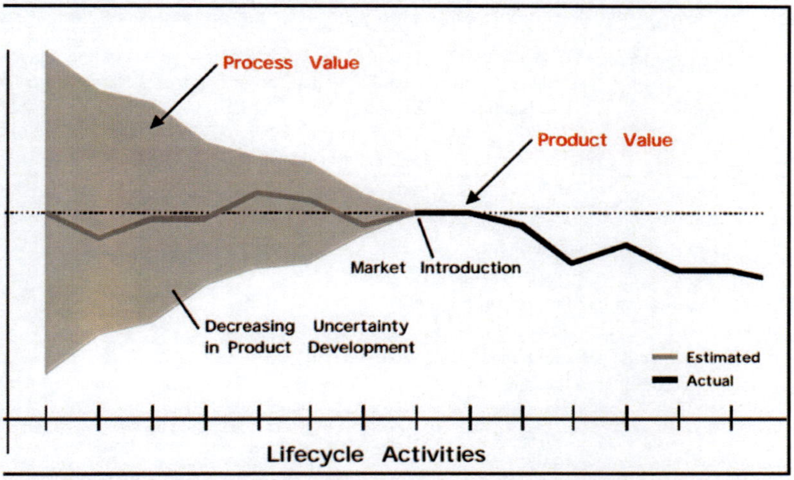

Fig. 2. Product cost depending on the process in product development. *Source* [4]

Moreover, the category of «value» in this aspect is multidimensional and includes quantitative and qualitative characteristics, acting as: 1) complex system value is associated with the satisfaction of all stakeholders, which implies a flawless product/mission delivered with minimal cost, in the shortest possible time [5]; 2) as the sum of the financial values of the interim results of the project. This is used as an indicator of the project's effectiveness by comparing the cost of planned work with the delivered cost [6]; 3) A product delivered at the right time to downstream processes/consumers where it is quantified in form, performance, functionality and timeliness [7]; 4) Perceived quality, corrected for the relative price of the product, etc. [8].

Common concepts of determining the cost in product development are [9]:

1. Acquisition cost is the unit cost of a product, expressed as the sum of research, development, manufacturing and construction costs.
2. Cost of use is the unit cost of a product, as the sum of the costs of operation, maintenance, decommissioning and disposal.
3. The life cycle cost of a product includes the unit cost of purchasing and using the product.
4. Target cost is the value of the life cycle cost of a product, determined at the beginning of the design process by target values. Actions are aimed at finding design solutions to prevent the actual cost from exceeding the target.

The variety of approaches and methodologies for designing a PD system makes it difficult to form a clear cost paradigm. In addition, the described approaches are limited in use, because they allow standard deviations of the calculated values (Table 1), do not take into account the costs for maintaining the product in working order (consumables, spare parts, inspections, processing, etc.) [10, 11], the value of which is often higher than acquisition cost, determining costs from the perspective of their relationship to «designer functions», determining costs from the perspective of lean thinking, except for relatively simple products.

Table 1. Error in the design cost estimation process at the product development stage

Stage of the design process		Granularity and specification	Error (%) according to the American Association of Engineers	Error (%) according to AACE International	According to Bubel (Bubel 2004)	According to Sabery (Sabery 2004)
Product planning	Initial stages of planning	Value ratio	±40	±50	±50	±30…50
	Development order	Value ratio	±30	±30	±50	±30…50
Product development	Clarification of the task	Performance criteria	±30	±20	±20%	
	Understanding of the design	Preliminary sketches of assemblies, process and equipment	±20	±20…30	±20%	±20…30
	Preliminary draft	Technological process, equipment and parts details	±10	±20	±10%	±20…30
	Detailed project	Exact data	±5	±15	±5%	±5…15

Source Systematized by the authors based on open source data

The purpose of this study is to systematize and summarize the experience of creating value in the process of product development within the framework of the network interaction of system participants. The task of the research is to form a vision of the current state of the issue.

2 Methodology

The analysis methodology is based on the gradation of function-oriented and component-oriented approaches in the development of products formed in the context of distributed product development, using the example of network interaction.

According to the authors, the methodology for assessing the cost of a product involves the following stages:

1 Stage: The choice of approach to the technical implementation of a product development system from two: 1. functionally oriented (aspects and factors related to production, assembly and product cost are evaluated based on the cost of use) and 2. component-oriented (the cost of system components is estimated based on the acquisition cost). The choice should be seen as the starting point for calculating the cost.

2 Stage: The choice of a cost management system for the administration, accounting, cost estimation methods based on the implementation of alternatives: 1. Methods based on the functions implemented by the product; 2. Methods based on the characteristics of product components (Table 2).

Table 2. Cost management adaptive to PD activities

Cost control methods	Characteristic
Methods based on the functions implemented by the product	
«Target cost» method (TC-Target costing)	«Allowable» production costs for the product and its components, target profit margins, non-production costs are determined. Suitable for products whose price is a key competitive factor in industries with low profit margins
«Value Engineering» method (VE-Value engineering)	The main functions of the product and its target cost are determined; a prototype is being developed, product costs are structured by elements in several options. If the cost of the final prototype is acceptable, it is subjected to reliability tests and provided to the customer. If not, then the product design changes until an acceptable target value is reached
«Quality function deployment» method (QFD-Quality function deployment)	The customer's request is translated into technical attributes or product features, and the «allowable» costs are allocated among the product elements
«Functional cost analysis» method (FCA-Functional cost analysis)	Product cost structures are assessed in the context of improving product design, manufacturing process with subsequent cost reduction while maintaining functionality and performance. The goal is to provide the necessary functions at a low cost
Methods based on characteristics of product components	
Cost breakdown structure (CBS)	Displays the cost elements of a product or system throughout its life cycle in a hierarchy (using numbering, indices). Costs are broken down to provide cost transparency for management at discrete stages such as system design, development, manufacturing, etc.
«Component commonality» method (CC – Component commonality)	Commonality includes trade-offs in «strategies for assembling different products, their components», providing cost savings through economies of scale. At the same time, developing components that are adaptable to several different products can be relatively expensive per component, leading to higher unit costs. The total profit may decrease due to the decrease in differentiation, although the production costs are reduced using the generality
Product modular design method (MD - Modular design)	A wide variety of end products is obtained using a limited number of modules in various combinations. Modular designs are larger subsystems than individual components, with similarities to the «Component commonality» approach, and therefore the resulting trade-offs become more meaningful
Product platforms (PP - Product platforms)	The approach is another step in shaping the architecture of products, thus influencing the cost of the product during its development. The idea implies that the architecture of the product is developed from scratch, simplifying the creation of a range of final products, their generations

Source [12–14]

3 Results

Applied to the research object:

- the implementation of a product development system based on a function-oriented approach is presented in the following form (Table 3):

Table 3. Cost of production for the project by years, taking into account inflation, thousand rubles without VAT

Name of the event	2016	2017	2018	2019	2020	Total
Cost price	19 218	91 921	99 388	268 859	324 719	804 105
Variable:	16 464	87 995	95 462	262 749	318 610	781 280
Materials	5 088	27 145	29 333	85 286	103 824	250 676
Transport and procurement costs (6,158% of materials)	313	1 672	1 806	5 252	6 393	15 437
Wage fund	5 730	30 801	33 667	97 418	117 475	285 091
Electricity	711	3 728	4 019	11 629	14 023	34 110
Other services	1 874	9 999	10 805	24 276	29 553	76 508
Auxiliary materials and consumables for technology and transport	872	4 651	5 026	14 613	17 789	42 951
Equipment, tools (OREH)	1 874	9 999	10 805	24 276	29 553	76 508
Constant:	1 527	1 474	1 474	1 474	1 474	7 423
Wages fund	1 474	1 474	1 474	1 474	1 474	7 370
ATD and AFA	53					53
Depreciation (7 years)	1 226	2 452	2 452	4 635	4 635	15 402

Source Interviews with experts; BCG analysis. Systematized by the authors

- component-oriented (Table 4).

Table 4. Cost of production for the project by years, taking into account inflation in 2020, thousand rubles without VAT

Components	Price for 1 piece, thousand roubles (without VAT)	Quantity, pcs.	Amount, thousand rubles (without VAT)	VAT	Amount, thousand rubles (VAT included)
Test bench	2 310	1	2 310	416	2 726
Piston compressor	1 860	1	1 860	335	2 194
Tape machine	761	1	761	137	898
Electric pipe bender	351	1	351	63	414
Dead end powder paint polymerization oven	347	1	347	62	409

(*continued*)

Table 4. (*continued*)

Components	Price for 1 piece, thousand roubles (without VAT)	Quantity, pcs.	Amount, thousand rubles (without VAT)	VAT	Amount, thousand rubles (VAT included)
Milling machine with gear	337	1	337	61	397
Video surveillance system	289	1	289	52	341
Welding inverter	285	2	571	103	674
Dead end powder paint spraying booth	248	1	248	45	293
Powder atomizer Electron	139	1	139	25	164
Licensing	693	1	693	125	818
Site modernization	1 155	1	1 155	208	1 363
Crane with construction and installation work	14 553	1	14 553	2 620	17 173
ATD and AFA sections	4 043	1	4 043	728	4 770
……					
Cost price					324 719

Source Developed and compiled by the authors

The choice should be seen as the starting point for calculating the cost.

3 Stage. Calculating the cost of PD- by: 1. Summation and decomposition of the functional structure; 2. Calculation of the cost of the production process, calculation of the cost of materials, etc.

Improving accuracy is possible through the use of the following methods in the PD-construct: a) catalog method; b) scaling method; c) analog method. Calculation errors are proposed to be reduced by decomposition of work and structural breakdown of costs obtained using regression procedures, which links the final cost of product development with a set of its physical and technological characteristics. It is recommended to use the ratio of costs for similar implemented projects in the absence of a number of indicators (Table 5).

The specified list of life cycle costs from the point of view of the percentage contribution of each item to the product is proposed to be used when assessing the costs of alternative value creation solutions from the point of view of the analog method in the conditions. Vendor stakeholders have no experience of practical application with the resulting output data when individual cost items for product conceptualization are not known, or when they are individual participants in the product development system.

Table 5. Relative contribution of costs determined by product creation processes

Product life cycle costs	Relative costs (%)
1- Product planning	10,3
a - Research phase of product inquiries, product conceptualization	1,5
b - System conceptualization	0,7
c - Design	5,4
d - Project documentation	0,8
e - Product software	0,7
f - Product testing and evaluation	1,2
2 - Product development/production	45,3
a - Development management	0,8
b - Understanding the design	1,1
c - Preliminary draft	35,4
d - Detailed project	5,3
e - Quality control	0,9
f - Initial response	1,8
3 - Operation and maintenance cost	39,9
a - Product lifecycle management	1,5
b - Product operations	3,2
c - Lean thinking practices	8,8
d - Product maintenance	15,2
e - Documentary support	4,7
f - Operation and maintenance training	5,6
g - Technical data	0,9
h - Product modification	...
4 - Decommissioning and disposal costs	4,5
Total percent	100

Source [9]

4 Conclusion

Observations from case studies show that: Operations in the early stages of the value chain, such as idea generation and design, tend to be prioritized for innovation mechanisms. Operations in the middle stages of the value chain, such as development, manufacturing and supply, often prioritize efficiency-oriented mechanisms. Operations later in the value chain, such as service and support, tend to prioritize flexibility-oriented mechanisms.

The product development process can be broken down into a number of tasks, then into a number of value attributes with their subsequent distribution to project participants – vendor stakeholders. Tasks are often far removed from the consumer's demand in product development, so it is more useful to focus on process cost in process improvement efforts, counting on «delayed» utility (where process value is the enterprise's approach to delivering the desired product for the customer, constant shareholder return, and satisfaction. life-long for an employee). The cost can be divided into the cost of the product and the cost of the process as part of the chain formation initiated by manufacturers.

A concept of value can be expressed in more than just quantitative metrics. The value is embodied in a structured approach to the system, which is a combination of qualitative and quantitative tools to maximize the value of the PD process.

References

1. Kuznetsov, V.P., et al.: Systems for Creating a Product in Industry and Their Development: Monograph, 176 p. Minin University, Nizhny Novgorod (2014)
2. Shipshova, O.A., et al.: The innovative component of the formation of competitive advantages in production systems in the context of changing technological structures: monograph, 142 p. Moscow (2017)
3. Budde, L.: Integrated complexity management in manufacturing companies. A model for assessing complexity/dissertation, 321 p. Difo-Druck GmbH, Bamberg (2016)
4. Vasholts, M.: A systematic approach to economically-oriented conceptual design of intelligent technical systems/dissertation, 247 p. Heinz Nixdorf Institute, University of Paderborn, Paderborn (2015)
5. Vdovin, S.M., Guskova, N.D., Neretina, E.A.: Conceptual approaches to sustainable development of socio-economic systems. Regionology **3**(76), 32–39 (2011)
6. Maksimova, A.A., Rogozhkina, E.M., Garina, E.P., Garin, A.P.: Theoretical foundations for diagnosing the economic state of development of an enterprise in the industry: category content, element saturation. Moscow Econ. J. (12), 59 (2020)
7. Romanovskaya, E.V., Kuznetsov, V.P., Andryashina, N.S., Garina, E.P., Garin, A.P.: Development of the system of operational and production planning in the conditions of complex industrial production. In: Popkova, E.G., Sergi, B.S. (eds.) ISC 2019. LNNS, vol. 87, pp. 572–583. Springer, Cham (2020). https://doi.org/10.1007/978-3-030-29586-8_66
8. Garin, A.P., Artashina, I.A., Zaitseva, S.A., Smirnova, Z.V., Potashnik, Y.S.: Research of the tools of influence on the behavior of market subjects. In: Popkova, E.G., Sergi, B.S. (eds.) ISC 2019. LNNS, vol. 91, pp. 156–162. Springer, Cham (2020). https://doi.org/10.1007/978-3-030-32015-7_19
9. Ferreira, K.M.: Evaluation of the cost of a product at the stage of conceptual design: methodology for choosing a functional structure and an alternative solution/dissertation, 163 p. Federal University of Santa Catarina, Florianópolis (1997)
10. Garina, E.P., Andryashina, N.S.: Complex development of a product in terms of its parallel design and production. Econ. Ind. **10**(2), 98–106 (2017)
11. Garina, E.P., Kozlova, E.P., Sevryukova, A.A.: Study of alternative strategies and methodological tools for the development of complex systems in the context of the product being created. Azimuth Sci. Res. Econ. Adm. **5**(2), 58–62 (2016)
12. Siyam, G.I., Wynn, D.C., Clarkson, P.J.: Review of value and lean in complex product development. Syst. Eng. **18**, 192–220 (2015)
13. Kirner, K.G.M.: Relationship between product development efficiency and option management - influence model and analysis method/dissertation, 368 p. Munich Technical University (2014)
14. Tolstykh, T.O., Shkarupeta, E.V., Kostuhin, Y.Y., Zhaglovskaya, A.V., Garin, A.P.: Scenarios for the development of industrial complexes in the digital economy. In: Popkova, E.G. (ed.) Growth Poles of the Global Economy: Emergence, Changes and Future Perspectives. LNNS, vol. 73, pp. 1255–1261. Springer, Cham (2020). https://doi.org/10.1007/978-3-030-15160-7_128

Method of Complex Analysis of Time Series to Predict the Dynamics of Changes in Product Groups in Foreign Economic Activity

Aleksey N. Kislyakov[1(✉)], Natalya E. Tikhonyuk[1], Natalia M. Filimonova[2], Dmitry V. Kochanov[1], and Alyona A. Susina[1]

[1] RANEPA (Vladimir Branch), Vladimir, Russia
[2] All-Russian Scientific Research Institute of Labor, Moscow, Russia

Abstract. The purpose of this article is to investigate the possibility of using special time series analysis methods for clustering product groups and its testing based on foreign trade statistics.

The article examined various tools for hierarchical clustering of time series, factor analysis, and identified the features of using some of them. To evaluate the possibilities of clustering, the author uses a three-stage clustering model based on the Gini index and the DTW algorithm.

The authors are exploring the possibility of identifying patterns in the behavior of product groups during export and import by compiling a matrix of pairwise comparisons based on the Gini index, which allows assessing the tight relationship between significant parameters for describing the behavior of product groups.

The problem of describing the behavior of the time series of foreign economic activity product groups is necessary to build predictive models for the development of export activities at the regional level.

Keywords: Clustering methods · Time series · Foreign economic activity

JEL Codes: C22 · F14 · O47

1 Introduction

One of the stages of strategic planning of the socio-economic development of the region and the country is the analysis of the external environment, during which the main economic indicators, the volume of GDP, the state of the investment climate, etc., are estimated. In this regard, federal and regional state structures are interested in monitoring the foreign trade market and forecasting the economic situation in general and in production sectors for short, medium and long-term strategic planning. Timely, prompt monitoring and accurate forecasting of the indicators of the foreign trade market demonstrate the industries that should be chosen as priority for the development of the economy of the region and the country as a whole.

E. G. Popkova (Ed.): ISC 2020, LNNS 368, pp. 408–415, 2022.
https://doi.org/10.1007/978-3-030-93244-2_45

To date, there are many approaches for assessing the state of socioeconomic systems, ranging from descriptive statistics and factor analysis methods to more advanced tools for analyzing complex multifactorial situations based on hybrid cognitive maps [10] and data mining methods [4]. At the same time, the key goal of such an analysis is to create better predictive models of socio-economic indicators.

The dynamic development and application of machine learning methods allows the design of prognostic models on a large number of features represented by small time sequences, which ultimately requires not only qualitative selection of features to build a regression model, but also interpretation of the results obtained. An example of the object of analysis is the foreign economic activity of the Russian Federation, namely import and export indicators for individual product groups in accordance with the Commodity Nomenclature of Foreign Economic Activity. An urgent problem at the same time is the creation of predictive models for changing import and export volumes for individual product groups, both in monetary terms and in absolute units of quantity.

Traditionally, the task of analysis is to classify individual factors influencing the situation, evaluating the target characteristic based on a prognostic model, as well as identify the causes of the achieved state that has developed under the influence of many external random and controlled factors [6]. Most often, it is more important to investigate the change in factors in dynamics, so usually the data set for each factor is presented in the form of a time series, which can be disclosed in the form of a summary table, both for individual product groups and for individual regions of the Russian Federation. At the same time, the main problem is the choice of a measure to assess the similarity/difference in the time series of import and export indicators in order to combine product groups/territories into clusters with close values of the integral assessment of this measure, which will ultimately allow determining individual classes of product groups.

The main task is that the same values of the selected indicator should be tied not only to the actual assessment of factors in a separate period of time, but also to evaluate the dynamics of their change over time, that is, to allow comparing the time series of factors and giving a generalized assessment of their change relative to the general picture of events.

2 Methodology

The main problems that arise when describing socio-economic processes in the form of time series are the high degree of redundancy of information, and, as a result, the low degree of interpretability of the results [8]. In this regard, the most promising approach is to perform hierarchical clustering of time series as objects.

Hierarchical clustering methods allow you to select from the two merge options shown in Fig. 1.

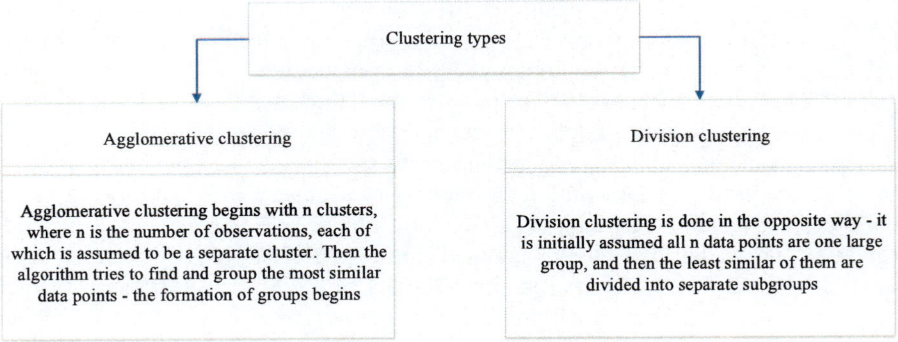

Fig. 1. Clustering options. Source Developed by authors

In the literature, you can find an extensive selection of metrics of distances between objects when performing cluster analysis. Standard measures include Euclidean distance, Manhattan distance, Minkowski distance, Chebyshev distance and Pearson correlation coefficient.

However, the problem of matching time series requires a metric that would allow to qualitatively comparing the similarity of time changes of several features on a different scale. In this case, the task of parallel comparison of these indicators based on the use of DTW distance as a measure is of particular importance. The name of this measure comes from the dynamic time transforming algorithm [7].

The DTW algorithm allows you to transform and scale the timeline to achieve optimal matching and/or optimal alignment of the two dynamic series so that the distance between them is minimal.

The basic idea of this approach is dynamically exploiting the existence of similarities and/or correlations among dynamic series. The greater the similarity among the rows, the less space is required to calculate the DTW among them.

However, this technique takes into account only the similarity of the morphological characteristics of the time series of individual indicators and despite its viability for time series studies, this technique cannot be adequately applied for a generalized comparison of the dynamics of time series presented in different units of measurement.

In this regard, it is necessary to use other similarity indicators, which depend not only on the shape of the time series, but also on the task of assessing the dynamics of a series and variability relative to several units of measure. It is on this that the possibility of a qualitative interpretation of the results depends.

Therefore, for assessment of dynamic characteristics of temporary ranks as the indicator characterizing degree of an imbalance of manifestation of one sign concerning other sign Jeanie's coefficient can be used [3]. The figure shows an example - the dependence in the fraction of points in the ith cluster (on the total number of points in the sample) on the fraction of the cluster number (on the total number of clusters).

For example, for 4 clusters, the fraction of the first cluster will be 0,25 (25%) and this cluster will contain 25 points from 100, so the graph will display a point (0,25; 0,25). If all clusters have the same number of points, then absolute symmetry is observed in the partition groups and the Gini index is zero, respectively, the imbalance is described by the area of the bounded broken Lorentz curve and the absolute symmetry curve and is calculated by the formula:

$$G = 1 - \sum_{k=1}^{n} (X_k - X_{k-1}) \times (Y_k + Y_{k-1}), G \in [0; 1] \tag{1}$$

where n is the number of clusters; is the cumulative fraction of the clusters' number; is the cumulative fraction of the points' number in the cluster (Fig. 2).

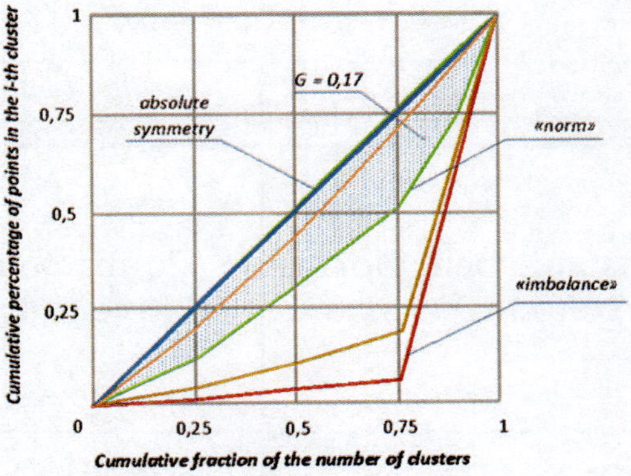

Fig. 2. The interpretation of the Gini index in clustering problems. Source Developed by the authors

The greater the value of the Gini index deviates from zero, the more asymmetry appears in the characteristics of the clusters [2]. The calculation of the Gini index makes it possible to estimate the degree of imbalance in the characteristics of pairs of time series represented in different units.

3 Results

Consider a technique allowing authors to perform a comprehensive analysis of time series and identify the features of their behavior according to two factors: differences within groups of features and estimates of the points' diversity within a series.

A comprehensive time series analysis involves the following steps:

1. Preparing data for analysis synchronizes with time samples.
2. Time series clustering and pair matching based on distance DWT [1].
3. Constructing a map of paired comparisons of a set of time series having different units based on the Gini index.

The first two stages were described in the works [9]. The third step requires a special explanation. To construct a map of paired comparisons of multiple time series based on the Gini index, the area under the Lorentz curve should be estimated for each time series, including the time series of the target indicator represented in two units [5]. The resulting pair of Gini index values is applied to a map of paired comparisons, which as a result allows you to map the target point to factor points.

The approach to the map analysis of paired comparisons of Gini index can be described as follows: the target point forming a kind of cross («sight») is a generalized Gini index for the indicator represented in two dimensions, and divides the resulting map of objects into quadrants, points in which can be characterized relative to the target characteristic in the way shown in the next figure (Fig. 3).

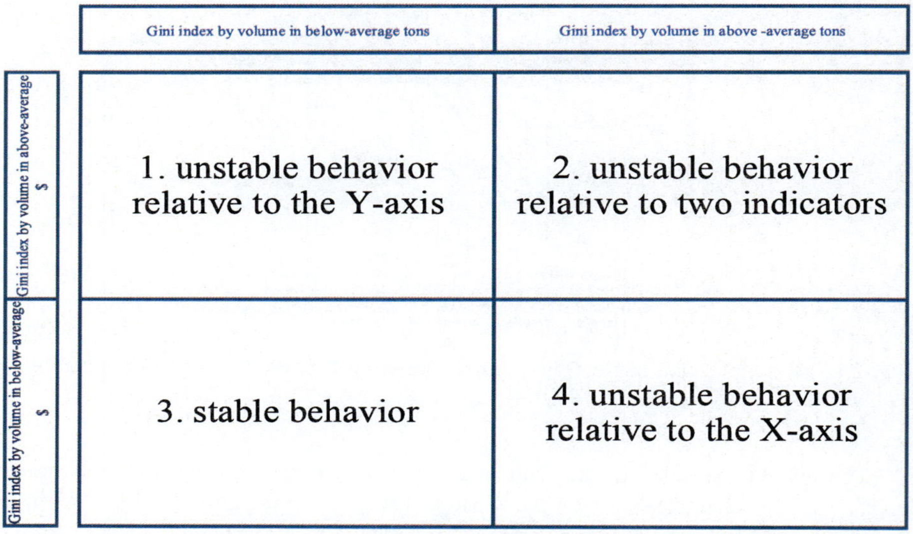

Fig. 3. Pairwise comparison matrix of Gini indexes. Source Developed by the authors

4 Discussion

This technique can be useful for making a model that allows predicting changes in the behavior of product groups during export/import to the Russian Federation.

This task is most relevant for developing the export potential of regional development in order to implement a national project aimed at international cooperation and export activities. Conducting an expert interview with employees of regional export development centers showed that there is an urgent need for government officials to assess product areas, the intensity of their development, see the forecast of the country's import and export results taking into account external factors and make sound management decisions to target support regional enterprises.

The analysis of data on the volume of imports/exports in thousands of dollars starting from 01.01.1998 to 01.01.2020 by product groups, which is structured by aggregated groups of product nomenclature, indicating in a compressed form categories and types of goods. The analysis was carried out in a time series step per month.

The first stage of the study was a cluster analysis, which allows you to identify groups of series with similar behavior in dynamics, but having differences in structure. The results of the analysis are shown in Fig. 4.

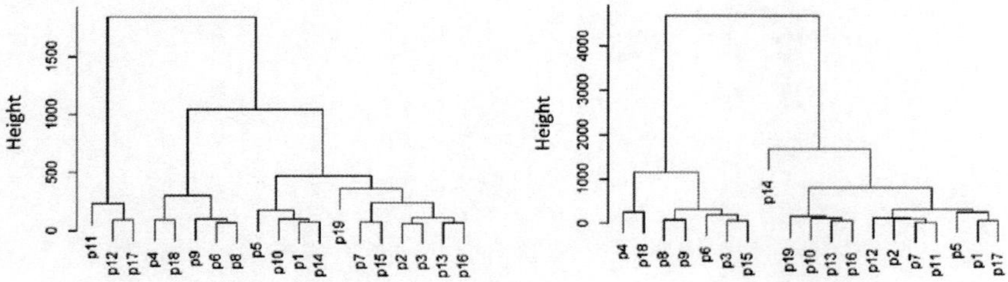

a
Dollars

b
Tons

Fig. 4. Hierarchical clustering diagrams based on export data. Source Developed by the authors

Analysis of hierarchical clustering data showed the presence of three main clusters, within which 19 product groups are heterogeneously distributed.

In the next stage, time series analysis based on DTW was carried out, which made it possible to compare the similarity of the nature of the behavior of time series with each other.

Based on the analysis, a matrix of paired comparison of Gini indexes is constructed (Fig. 5).

Using the Gini index to analyze the dynamics of the time series of the foreign economic activity indicators allows comparing the balance and asymmetry of the change in the indicator. The greater the Gini index value, the greater the variety of points within the series, and accordingly higher the variability of the export/import indicators of the commodity group and, as a result, less accurate forecasting.

The Gini index pairwise comparison matrix allows you to make an initial forecast of the change in the dynamics of the product group. Movement along the vertical axis of the «sight» shows a decrease in prices in the world market, the horizontal axis - an increase in consumption. Most often, dependence is linear (decrease in price - increase in consumption), but uncharacteristic behavior can be observed (1 and 4 matrix quadrates).

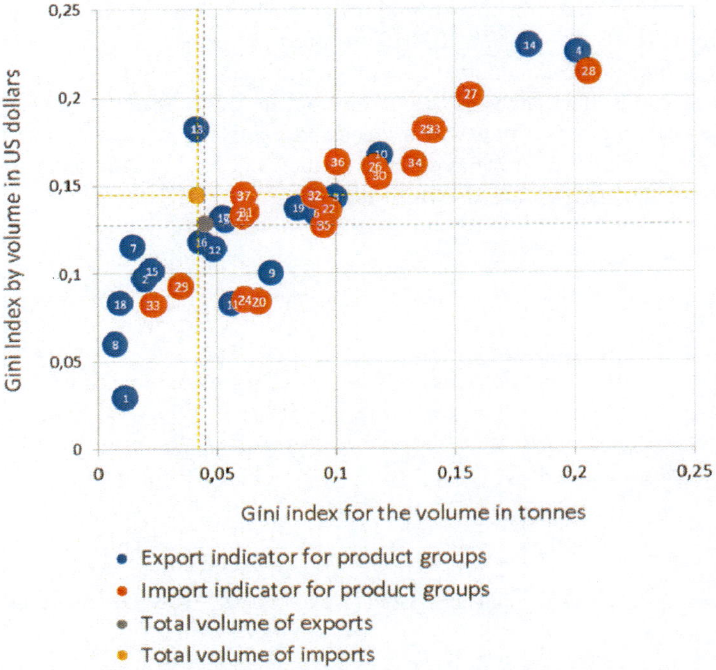

Fig. 5. Matrix of paired comparison of Gini indexes according to baseline data. Source Developed by the authors

5 Conclusion

The compliance of the «sights» on import and export speaks of the balance of their volumes in the Russian Federation, which indicates the possibility of revising the directions of foreign economic activity of enterprises producing products, and possibly directing demand to the domestic, Russian market. You can also see an export trend in those industries where the cost of a product in the world market is higher than in the domestic market. This situation requires the introduction of regulatory measures aimed at reducing the volume of exports if the goods are of high social importance. All of this can be used to build predictive models of the development of foreign economic activity in the region.

Further development of the topic is associated with the use of various prognostic algorithms for selected product groups.

References

1. Biró, T.S., Néda, Z.: Gintropy: Gini index based generalization of entropy. Entropy **22**(8), 879 (2020). https://doi.org/10.3390/e22080879
2. Zheng, C.: A novel classification tree based on local minimum Gini index and attribute partial order structure diagram. Int. J. Comput. Appl. Technol. **64**(1), 33–45 (2020). https://doi.org/10.1504/IJCAT.2020.111083
3. Zamotailova, D.A., Savinskaya, D.N., Nedogonova, T.A.: Application of trend-seasonal models for time series research and forecasting. Mod. Econ. Probl. Solutions **12**, 40–48 (2020). https://doi.org/10.17308/meps.2019.12/2197
4. Zinoviev, V.I., Nesterov, A.V.: Integration of multiple regression into data mining algorithms, to build predictive analytics subsystems, and to make management decisions based on them. Nat. Tech. Sci. (10), 61–73 (2020). https://doi.org/10.37882/2223-2966.2020.10.12
5. Ivanchenko, A.: Graph isomorphism criterion based on resistance distance. University News. North-Caucasian Region. Tech. Sci. Ser. **2**, 13–18 (2020). https://doi.org/10.17213/1560-3644-2020-2-13-18
6. Kislyakov, A.N.: Intelligent analysis of consumer demand in conditions of information asymmetry. Mod. Econ. Probl. Solutions **10**, 8–17 (2019). https://doi.org/10.17308/meps.2019.10/2221
7. Kislyakov, A.N.: Graph clustering of behavioral activity of product users taking into account information asymmetry. News of southwestern state university. Ser. Econ. Sociol. Manag. (3), 152–163 (2020)
8. Magomadov, E.M., Murtazalieva, A.H.: Using economic and mathematical models in the conditions of digital transformation of economic relations. Econ. Legal Issues **141**, 43–47 (2020). https://doi.org/10.14451/2.141.43
9. Malkina, M., Balakin, R.V.: Study of the concentration and uniformity of the distribution of tax revenues in the regions of the Russian Federation based on the Herfindal-Hirschmann, Gini and Teil indices. Taxes Tax. **11**(11), 1010–1023 (2014). https://doi.org/10.7256/1812-8688.2014.11.12546
10. Povidalo, I.S., Yarushev, S.A., Averkin, A.N., Fedotova, A.V.: Cognitive hybrid decision support and prediction systems. Softw. Prod. Syst. Algorithms **4**, 16–24 (2017)

Principles of Selecting Objects for Cost Accounting

Anatolii V. Kholkin(✉) and Irina N. Bachurinskaya

Vyatka State University, Kirov, Russia

Abstract. This article deals with finding a solution to the problem of the absence of clearly developed principles for choosing objects of calculation in scientific works. The principles for selecting objects of calculation, based on a preliminary determination of the factors which influence the choice of the object of calculation, are developed in the article. The solution to this problem is carried out using the following methods: analysis, synthesis, generalization, table method, graphical presentation. The article consists of the following parts: introduction, research methods, results, discussion, and conclusions. The result of the study was the developed selection principles, formulated in the form of a system of principles, adhering to them is necessary for the implementation of a scientifically grounded choice of objects of calculation and the formation of their list is necessary for organizing the cost accounting process and calculating the cost of production. For this, based on the preliminary study of the selection process, the factors influencing the choice of the object of calculation were determined; the principles of selecting the object of calculation were formulated. Then the nature of the selection principle action on the very process of selecting objects of calculation was determined. These principles should be used when selecting objects of calculation. The obtained results can be used in the practical activities of the accounting service of any enterprise, especially these developments are relevant for the production of technically complex products and (or) industries with a large production datasheet.

Keywords: Cost accounting · Cost management · Object of calculation

JEL Code: M41

1 Introduction

The correct choice of the object of calculation is one of the key problems in the practice of organizing cost accounting. The object of calculation is the final or intermediate product in relation to which the cost of the finished product is calculated, and, accordingly, goods sold. That is why the selection of objects of calculation and the formation of their list is one of the first steps when organizing cost accounting since it determines both the process of documenting costs and their reflection in accounting registers and the presentation of information about costs in internal accounts.

E. G. Popkova (Ed.): ISC 2020, LNNS 368, pp. 416–426, 2022.
https://doi.org/10.1007/978-3-030-93244-2_46

Thus, the choice of objects of calculation and the formation of their closed list is the initial task of organizing the cost accounting process and its solution is important both for the entire accounting process and cost management.

The solution to this problem is especially relevant for production branches of production of technically complex products, consisting of various parts and assemblies, as well as for industries with a large production datasheet since for these industries the problem of choosing objects of calculation is important for the distribution of indirect costs.

To solve this problem an appropriate methodology should be developed, which should be based on scientifically grounded principles for choosing objects of calculation, as well as approaches for their competent application.

In the scientific literature, this problem is not fully described. It is indicated that when choosing objects of calculation, one should proceed from technology, product characteristics, and organization structure and calculation goals. That is, specific understandable principles of choice are not formulated. Therefore, each enterprise independently chooses objects of calculation, focusing on the experience and professional judgments of people who manage the accounting service and accounting employees.

It makes it necessary and relevant to develop principles for the selection of objects of calculation and their justification.

Therefore, the study aims to develop scientifically grounded principles for choosing objects of calculation.

The aim was achieved by solving the following tasks:

1. Identifying the factors which influence the composition of objects of calculation and describing their influence
2. Formulating principles of the object selection based on the description of the factors
3. Substantiating the formulated principles.

The solution of the set tasks will allow achieving the goal of the study.

The idea that there are certain rules or principles, the adherence to which will make it possible to scientifically form a list of objects of calculation, in the context of which the collection, grouping and generalization of costs and cost accounting of finished products will be carried out was determined as the hypothesis of the study.

The scientific novelty of the study lies in the clarification of the principles for choosing objects of calculation, as well as a description of the principles and the nature of their action in the process of choosing objects of calculation.

The scientific significance of the study lies in the further development of the methodology for organizing cost accounting.

The practical significance of the obtained results lies in the possibility of a scientifically grounded selection of objects of calculation, which will prevent errors in cost accounting.

2 Methodology

During the research, the following methods were used: analysis, synthesis, generalization, table method, graphical presentation.

The analysis was used to identify factors influencing the choice of objects of calculation and to describe their action.

The synthesis made it possible to formulate the selection principles based on the identified factors.

The use of the generalization method made it possible to generalize the obtained results.

The table method was used to present and summarize the principles of choosing objects of calculation.

The graphical presentation made it possible to demonstrate the obtained result in the form of a graphical image.

The following sources of information were used in the study. Educational publications of the following authors which are devoted to management accounting were studied as part of the literature sources, namely: [1–5, 7–9, 13, 14, 16–19, 22]. The scientific works of the authors were also analyzed: [11, 12, 15, 20, 21, 24].

The scientific articles of the authors were examined: [6, 10, 23].

3 Results

In the course of solving the tasks, the authors obtained the following results.

Firstly, the following factors that affect the choice of objects of calculation are identified:

1. Structure of manufactured and sold products
2. Specificity (type) of production
3. Product manufacturing technology
4. Product design
5. Applied accounting methods and techniques
6. Rationality of accounting
7. Information needs of users.

Secondly, based on identifying and describing the factors, the following principles for choosing the object of calculation were formulated.

1. Taking into account the structure of manufactured and sold products.
2. Taking into account the specifics (type) of production.
3. Taking into account manufacturing technology, design features.
4. Availability of technical capabilities to identify objects of calculation.
5. Compliance with the requirement of the rationality of accounting.
6. Taking into account the information needs of users.

The description of each principle is presented in Table 1.

Table 1. Principles of selecting objects of calculation

The principle	The principle content
1. Taking into account the structure of products	When forming objects of calculation, it is necessary to take into account the composition and components of manufactured and sold products
2. Taking into account the specifics and type of production	When choosing objects of calculation, it is necessary to take into account the specificity and type of production
3. Taking into account manufacturing technology, design features	When selecting objects of calculation, it is necessary to take into account the technological relationship, the complete set of products and the features of the manufacturing technology
4. Availability of technical capabilities to identify objects of calculation	When choosing objects of calculation, it is necessary to take into account the methods and techniques used for cost accounting, which determine the very possibility and rationality of selecting objects
5. Compliance with the requirement of the rationality of accounting	When identifying objects, it is necessary to compare the additional costs arising for accounting and the effect obtained from this
6. Taking into account the information needs of users	When selecting objects, the needs of users should be satisfied to the maximum extent (taking into account the principle of rationality)

Source Compiled by the authors

The relationship between factors and principles can be traced based on Fig. 1.

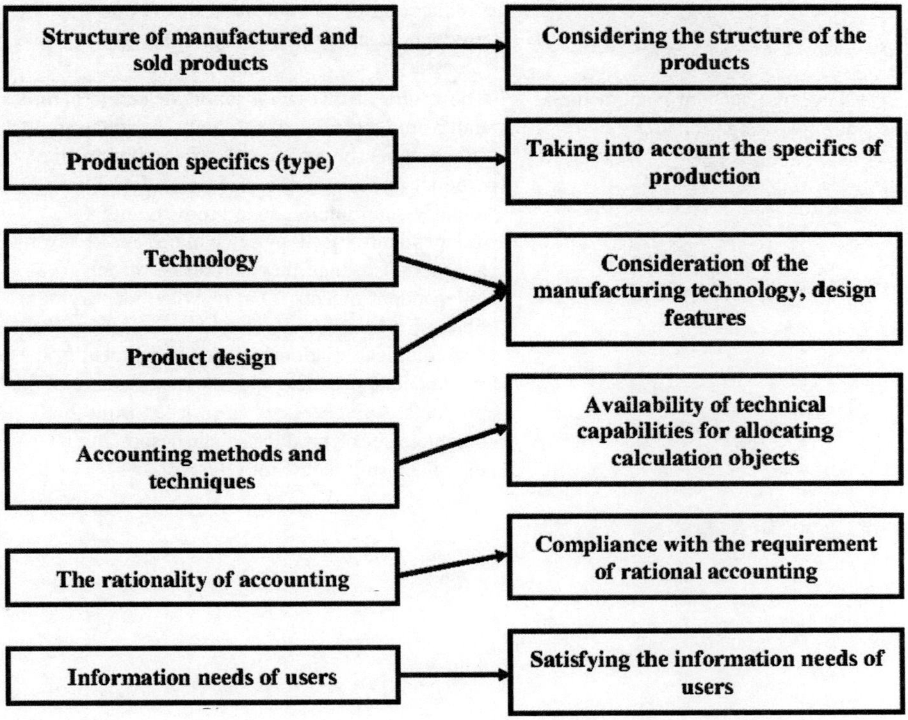

Fig. 1. The interrelation between factors and principles. *Source* Compiled by the authors

Thirdly, the justification through detailed descriptions of the nature of their impact when selecting objects of calculation was carried out. The nature of the action for each factor is shown in Table 2.

Table 2. The action nature of each principle

The principle	The action nature
1. Taking into account the structure of products	The nature of the action is manifested in the presence of intermediate products in the composition of the product. If it is planned to sell intermediate products, then they should be separated into independent objects
2. Taking into account the specifics of production	The nature of the action is manifested in the presence of intermediate products. When making intermediate products independent objects, one should take into account the specifics of production. So, in the conditions of single and small-scale production, the identification of intermediate products is not advisable, since it is possible to identify the production of the intermediate product with the production of the specific final product. In mass multi-product production, the identification of intermediate products as independent objects is advisable and necessary
3. Taking into account manufacturing technology, design features	The nature of this factor is manifested in complex multi-product industries. If only one final product is subsequently produced from the intermediate product (products), then it is not advisable to identify such intermediate products into independent objects of calculation, and if several are produced, then they should be made independent objects. That is, when identifying objects, it is necessary to take into account the technological relationship, that is, to identify these intermediate products with the manufacture of final products. It is necessary to proceed from the complete set of products, their design and manufacturing technology features

(*continued*)

Table 2. (*continued*)

The principle	The action nature
4. Availability of technical capabilities to identify objects of calculation	When selecting objects of calculation, it is necessary to take into account the methods and techniques used for cost accounting, which determine the possibility and rationality of identifying objects. For example, using the method of excluding the cost of by-products would not require sufficient detail. Or, when using the job-order cost accounting method, the object of calculation will be the order being executed. When using the process method of cost accounting, either semi-finished products or finished products will act as objects
5. Compliance with the requirement of the rationality of accounting	When selecting objects of calculation, it is necessary to take into account the need to comply with the requirement of rationality. So, excessive selection of objects of calculations will only contribute to an increase in the labour intensity of accounting without obtaining any effect from the use of data. Insufficiently complete selection of objects will not allow to correctly form the cost of both intermediate products and finished products
6. Taking into account the information needs of users	When selecting objects of calculation, the needs of users must be taken into account, due to the need to obtain information for making decisions in the field of cost management. This also includes ensuring the unity of planning, accounting and cost control. That is, the use of a single list of objects of calculation in planning and accounting

Source Compiled by the authors

4 Discussion

The study of the sources revealed that in foreign literary sources and scientific articles the issues of selection objects of calculation were not fully disclosed over the past 5 years.

In Russian sources, the principles of selecting objects of calculation are disclosed in the works of [14, p. 80]. She defined the following principles for selecting objects of calculation:

1. Features of the applied production process.
2. Product Description.
3. Features of the organizational structure of the enterprise.
4. Purposes of calculation.

However, this author does not specify the content of these principles and the order of their influence on the choice of objects of calculation. This, in turn, makes the application of these principles in practice impossible, since it is not clear how to follow them, to ensure their implementation in practice.

As a result of the study of the articles, it was found that they do not clarify the formation of the principles for selecting objects of calculation.

The following factors that affect the choice of the product, intermediate product, order, or other object selected as an object of calculation were identified.

1. Structure of manufactured and sold products
2. Specificity (type) of production
3. Product manufacturing technology
4. Product design
5. Applied accounting methods and techniques
6. Rationality of accounting
7. Information needs of users.

The structure of manufactured and sold products affects the selection of objects as follows. When producing technically complex products, or if the production process involves sequencing of products through several interconnected stages, then situations may arise when intermediate products, or individual units or parts, as spare parts, can be sold to the buyer. Therefore, it becomes necessary to separate intermediate products that are sold to customers as independent objects of calculation.

The specificity of production, its type affects the choice of objects of calculation as follows, if there are intermediate products. In the conditions of single or small-scale production, the separation of intermediate products as independent objects is seen as inexpedient, since it is almost always possible to identify the release of the final product with the release of intermediate products, parts and assemblies necessary for its manufacture. In conditions of mass production, especially technically complex products, in which a different combination of assemblies and parts will lead to the production of different products, when the same part or assembly can be used in the production of different products and in the manufacture of a part or assembly it may not be known, or it is not obvious in the production of which product the unit or part will be used, then it is necessary to single out intermediate products as independent objects of calculation.

The product manufacturing technology influences the choice of objects of calculation in the following way. In the conditions of mixed production, when it is possible to manufacture several types of products using one intermediate product, or intermediate products are used to manufacture various final products, when it is impossible or difficult to directly identify the manufacture of an intermediate product with the release of a certain final product, then the intermediate product should be chosen as an independent object of calculation.

The design of products affects the choice of an object because in the case of the presence of the same units in the designs of various products, provided that at the time of production it is not obvious that a part or unit is used to complete a certain final product, then such parts or units must be taken into account as an independent object of calculation.

At the same time, the nature of the action of the technology and product design is almost general, which means that when identifying the principles, it is necessary to combine these factors.

The applied methods and techniques influence the selection of objects of calculation as follows. When using methods of excluding the cost of by-products, it is impractical to single out such by-products as independent objects of calculation, except in cases of control and verification of the correctness of determining the estimated cost of by-products. When using the job-order cost accounting method, one must select a separate object of calculation - an open order. When using the process method, intermediate products and semi-finished products must be selected as objects of calculation.

The rationality of accounting affects the choice of objects of calculation by means of the fact that each selected independent object of calculation leads to an increase in the costs of accounting labour. The insufficient number of objects of calculation leads to a distortion of the cost and the impossibility of adequate cost accounting of individual products.

The information needs of users determine the selection of objects of calculation as follows. Cost management and other decision making lead to the need to obtain information about the cost of individual types of products, as well as its parts and intermediate products. Therefore, when selecting objects of calculation, in the context of which costs are accounted for, it is necessary to take into account the needs of users. Also, when selecting objects, it is necessary to take into account the unity of planning, accounting and cost control. That is, the composition of objects of calculation in accounting must correspond to the composition of objects in the context of which cost planning is carried out, which is necessary for analyzing and controlling costs.

The principles for selecting objects of calculation were determined based on the identified cost factors.

So, based on the factor of the structure of manufactured and sold products and the study of its action, the principle of taking into account the structure of manufactured and sold products was formulated.

The factor of the specificity and type of production, research and description of its action led to the determination of the principle of taking into account the specificity and type of production.

The factors of technology and design of products made it possible to formulate the principle of taking into account manufacturing technology and design features.

Based on the factor of applied methods and techniques, the principle of the availability of technical capabilities for identifying objects of calculation was determined.

The factor of taking into account rationality led to the formulation of the principle of compliance with the requirement of accounting rationality.

The factor of information needs of users made it possible to determine the principle of taking into account the information needs of users.

Based on the study of the action of the factors, the nature of the action and the use of each principle of selecting the object of calculation was determined.

So, the principle of taking into account the structure of products is valid only if there are intermediate products in the sales volume. In this case, it is necessary to select intermediate products as part of the objects of calculation.

The principle of taking into account the specificity of production influences the inclusion of intermediate products in the composition of objects of calculation, depending on the type of production, when it is possible or impossible to identify the production of an intermediate product with the production of a specific final product.

The principle of taking into account the manufacturing technology and design features of products affects the choice of objects of calculation by highlighting technological links between intermediate products and final products, that is if the technology provides for further transformation of an intermediate product into a final product or several final products and if this predetermined connection between an intermediate product and final is obvious. The design of products also makes it possible to specify the obviousness of the connection and the predetermination of the manufacture of the final products of intermediate products, or the absence of such a connection.

The principle of the availability of technical capabilities for the selection of objects of calculation affects the selection of objects because the applied accounting methods and techniques either make it impossible or impractical to select objects or impose methodological requirements for obligatory selecting objects of calculation.

The principle of compliance with the requirements of the rationality of accounting introduces a limitation on the number of selected objects by the amount of accounting labour costs and the effect obtained from the use of additionally provided information.

The principle of taking into account the information needs of users affects the choice of objects of calculation by the fact that the selection of objects should be based on the information needs of users. The minimum necessary way to meet these needs is to maintain the unity of the list of objects of calculation win planning, cost accounting and cost control.

As it can be seen from the above principles and the nature of their action, decision making on the choice of objects of calculation is possible not only and not so much based on accounting information, but also based on information on the technology of production and design of products, as well as based on a preliminary study of the information needs of users. That is, such decisions should be made not only by the employees of the accounting service and its management but also by technologists, engineers, designers and management of the organization. This may require the creation of a special commission or a working group.

In general, adherence to these principles will make it possible to compile a scientifically grounded list of objects of calculation, which is necessary, in turn, for the optimal organization of cost accounting and cost management.

5 Conclusion

This article proposes a possible solution to the problem arising when organizing cost accounting begins, when it is necessary to correctly and scientifically select objects of calculation, due to the lack of scientifically grounded principles for choosing objects of calculation.

Therefore, in this article, the following principles for the selection of calculation objects were determined:

1. Taking into account the structure of manufactured and sold products.
2. Taking into account the specificity (type) of production.
3. Taking into account manufacturing technology, design features.
4. Availability of technical capabilities to identify objects of calculation.
5. Compliance with the requirement of the rationality of accounting.
6. Taking into account the information needs of users.

Each principle as well as the nature of its use when selecting objects of calculation are described.

The composition of factors influencing the choice of objects of calculation was also found.

The research results can be used by any organization, regardless of the industry sector and place of business, especially these developments will be useful for organizations with technologically complex and (or) mixed production.

References

1. Agrawal, R.K.: Principle of Management Accounting, 229 p. Education Publishing, Bilaspur (2018)
2. Arora, M.N.: A Textbook of Cost and Management Accounting, 10th edn, 1132 p. Vikas Publishing House Pvt Ltd, New Delhi (2012)
3. Banerjee, B.: Cost Accounting. Theory and Practice, 1044 p. PHI Learning Private Limited, Delhi (2014)
4. Boyd, K.: Cost Accounting for Dummies, 408 p. Wiley, Hoboken (2013)
5. Collis, J.: Management Accounting, 176 p. Palgrave, London (2015)
6. Cohen, S., Kaimenaki, E.: Cost accounting systems structure and information quality properties: an empirical analysis. J. Appl. Acc. Res. **12**(1), 5–25 (2009)
7. Drury, C.: Management and Cost Accounting, 7th edn., 775 p. Cengage Learning Business Press, London (2008)
8. Gill, S.: Cost and Management Accounting, 773 p. Vikas Publishing House Ltd., New Delhi (2015)
9. Horngren, C.T., Datar, S.M., Foster, G.: Cost Accounting, 450 p. Pearson Education Limited, New York (2006)
10. Kalonov, M.B.: Organization of accounting and cost analysis of the provision of cars. Eur. J. Econ. Manag. Sci. 10–12 (2017). https://doi.org/10.20534/EJEMS-17-1-10-12
11. Malina, M.A.: Advances in Management Accounting, 224 p. Emerald Publishing Limited, Bingley (2017)
12. McWatters, C.S., Zimmerman, J.L.: Management Accounting in a Dynamic Environmental, 712 p. Routledge, New York (2015)
13. Mehta, B.K.: Cost and Management Accounting by Dr, B. K. Mehta, 295 p. SBPD Publications, Agra (2020)
14. Nikolaeva, S.A.: Management Accounting. IPB-BINFA, Moscow (2002)
15. Norreklit, H.: A Philosophy of Management Accounting, 286 p. Routledge, New York (2017)
16. Pandey, K.: Cost Accounting. Laxmi Book Publication, Solapur (2016)

17. Rainborn, C.A., Kinney, M.R.: Cost Accounting Principles, 832 p. Cengage Learning Business Press, London (2013)
18. Seal, W.B., Rohde, C.: Management Accounting, 865 p. McGraw-Hill Education, London (2014)
19. Singh, S.: Management Accounting, 572 p. PHI Learning Private Limited, Delhi (2016)
20. Shibaeva, N.A., Zarudneva, A.I., Sozinova, A.A., Shuvaev, A.V., Alekseev, A.N.: Restructuring of tax liabilities as an upcoming trend of economic diversification in modern Russia. In: Gashenko, I.V., Zima, Y.S., Davidyan, A.V. (eds.) Optimization of the Taxation System: Preconditions, Tendencies, and Perspectives. SSDC, vol. 182, pp. 83–89. Springer, Cham (2019). https://doi.org/10.1007/978-3-030-01514-5_10
21. Sozinova, A.A.: Investigating the prospects of new institutional theory in the smart economy. In: Popkova, E.G., Sergi, B.S. (eds.) ISC 2019. LNNS, vol. 198, pp. 2075–2081. Springer, Cham (2021). https://doi.org/10.1007/978-3-030-69415-9_227
22. Tascher, A., Charifzadeh, M.: Management Accounting and Control: Tools and Concepts in a Central European Context, 350 p. Wiley, Weinheim (2016)
23. Urkayev, M.H., Karimov, F.J., Gurbanova, K.V., Karimova, M.N., Karimova, I.J.: Methods of management accounting and economic analysis. Eur. J. Econ. Manag. Sci., 13–24 (2019)
24. VanDerbeck, E.J., Mitchel, M.R.: Principles of Cost Accounting, 624 p. Cengage Learning Business Press, London (2015)

Game-Theoretic Modeling of Competitive Interaction Between Mushroom Farming Companies

Aleksei A. Nabokikh(✉) and Alexandra V. Ryattel

Vyatka State University, Kirov, Russia

Abstract. The paper presents the results of modeling the competitive antagonistic interaction between mushroom farming companies operating in the same regional market of cultivated mushrooms. A model of competition between two companies operating in the market of cultivated mushrooms under the assumption of continuity of time and a model reflecting the interaction of competitive companies in the regional market of mushroom farming in terms of their pricing strategies are proposed.

Keywords: Model · Game theory · Regional market of cultivated mushrooms · Competitive interaction of companies

JEL Codes: C02 · C70

1 Introduction

Cultivated mushrooms became generally available in Russia in the late 90s of the XX century. Due to the expanded supermarket chains and increased purchasing power of the population, the consumption of fresh cultivated mushrooms reached its maximum in 2013 and amounted to 1.26 kg per capita per year. The annual consumption of fresh cultivated mushrooms in developed countries is 2–4.5 kg per capita, on average. According to the Food and Agriculture Organization of the United Nations, the Russian Federation occupies the 25–26th place in the world in terms of the volume of mushrooms produced and is ranked only 46th in terms of consumption of mushrooms per capita [5].

Since 2014, the price of cultivated mushrooms has been rising. For example, the retail price for fresh champignons has increased by 2.5 times (from 103 to 253 rubles/kg). As a result, the consumption of fresh cultivated mushrooms decreased to 0.79 kg per capita in 2016 [11]. In 2015–2020, there was an increase in mushroom production due to the increase in yields on existing mushroom farms and the launch of new ones. For the first time since the early 2000s, in 2017, there were fewer imported fresh cultivated mushrooms in the market than the domestic ones.

In 2020, the domestic production of champignons increased by 67%, while the average annual selling price of champignon producers fell by 13% (to 132 rubles/kg) when the annual food inflation rate was 4.6%. The production volume of 99,700 tons allowed Russian mushroom producers to become the third-largest champignon

E. G. Popkova (Ed.): ISC 2020, LNNS 368, pp. 427–436, 2022.
https://doi.org/10.1007/978-3-030-93244-2_47

producers in Europe, following the Netherlands and Poland. The production of oyster mushrooms compared to last year's figures fell by 15%. The volume of exotic mushrooms grown has increased by 45%, but, in absolute terms, it is still very small – only 323 tons per year [12] (Fig. 1).

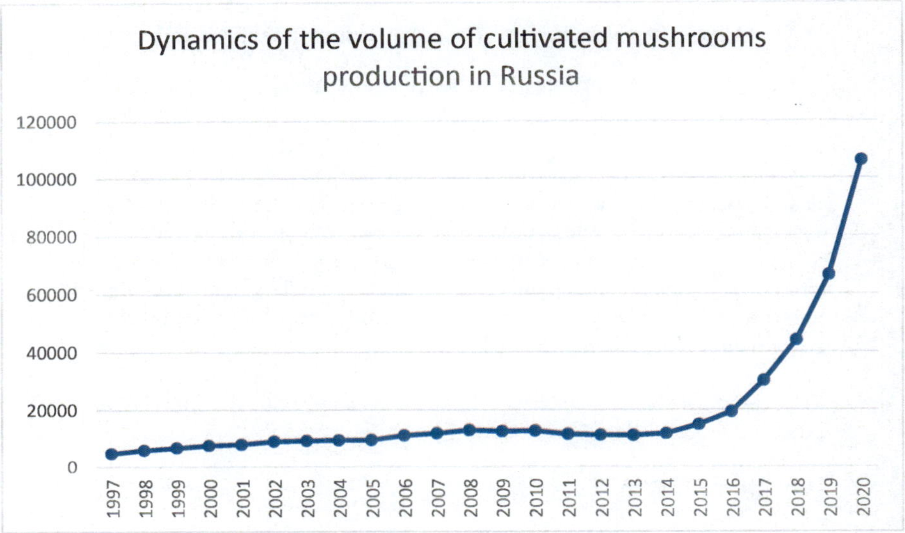

Fig. 1. Dynamics of the volume of cultivated mushrooms production. *Source* Developed and compiled by the authors

In 2021, mushroom production will continue growing, and the producers will have to make serious efforts to improve the price situation. Moreover, according to experts, in 2021, the real income of the population will significantly decrease under the influence of the economic crisis, aggravated by the coronavirus pandemic. In this connection, sales of fresh mushrooms are expected to decrease by 5.5% compared to the previous year. But despite this, Russian economists characterize the production of cultivated mushrooms in the country as an attractive investment process since the total consumption market today is 10–15% higher than the volume of imported products. At the same time, the annual volume of imported mushrooms in the Russian Federation is about 1 billion dollars [13]. The sanctions introduced caused the accelerated implemented measures to increase the scale of domestic mushrooms production – since 2014, cultivated mushrooms, like many other food products, have been under the embargo. Polish mushrooms fell under the restrictions; the latter occupied more than 90% of the Russian market. As a result, domestic mushroom growers got a chance to increase the production volume and fill the resulting niche.

2 Methods

The study of the interaction between the subjects of the cultivated mushrooms market plays a crucial role in planning the optimal use of the limited resources of the mushroom sector to meet population needs.

The main purpose of the paper is to build mathematical models of competitive interaction between companies in the mushroom industry, which allow identifying the patterns of this interaction.

Some aspects of economic and mathematical modeling of the cultivated mushrooms market are described in [3, 16–18, 20–23]. The economic aspects of the COVID-2019 pandemic are explored in [2, 6, 25].

Over the past few years, in the market of cultivated mushrooms, the quantitative and qualitative structure of market participants, methods, principles, and technologies of their interaction has been significantly reformed. Therefore, at the present stage, it is necessary to search for new opportunities to optimize the effectiveness of relations between market participants. One of the ways to solve this problem is to model the interaction of market participants in mushroom farming, taking into account its features.

A market mechanism in which a small number of firms operate, as in the case of the market for cultivated mushrooms, is defined as the competition among a few. At the same time, companies operating in the market have the opportunity to influence the market using their strategies (e.g., to launch various advertising campaigns, implement their pricing policy, etc.). When choosing the optimal strategy, the company should take into account the influence of competitors, in addition to its direct impact on the market. The above assumption makes it possible to consider the interaction of competitors in the market of fresh cultivated mushrooms in the framework of game theory [1, 7, 8, 10, 27]. The number of players, in this case, is the same as the number of competing companies.

The purpose of the methodology for modeling interaction in the market of cultivated mushrooms is to create models that describe the interaction of the subjects of the specified market in a certain situation.

The subjects of the cultivated mushrooms market interact in the situation of limited resources used. The composition, nature, and volume of resources consumed are determined by various functions the subjects perform. In case various subjects need the same limited resource, and the method to achieve the satisfaction of the need for this resource is similar, there is an antagonistic interaction between the subjects.

Various ways of modeling the interactions of these categories following the available strategies are dealt with in-game theory [4, 9, 14, 15, 19, 24, 26]. This section of applied mathematics also describes various types of equilibria in the market under study.

Let us describe a three-level system of competition in the regional market of cultivated mushrooms (Fig. 2), which contains instrumental, object, and target levels.

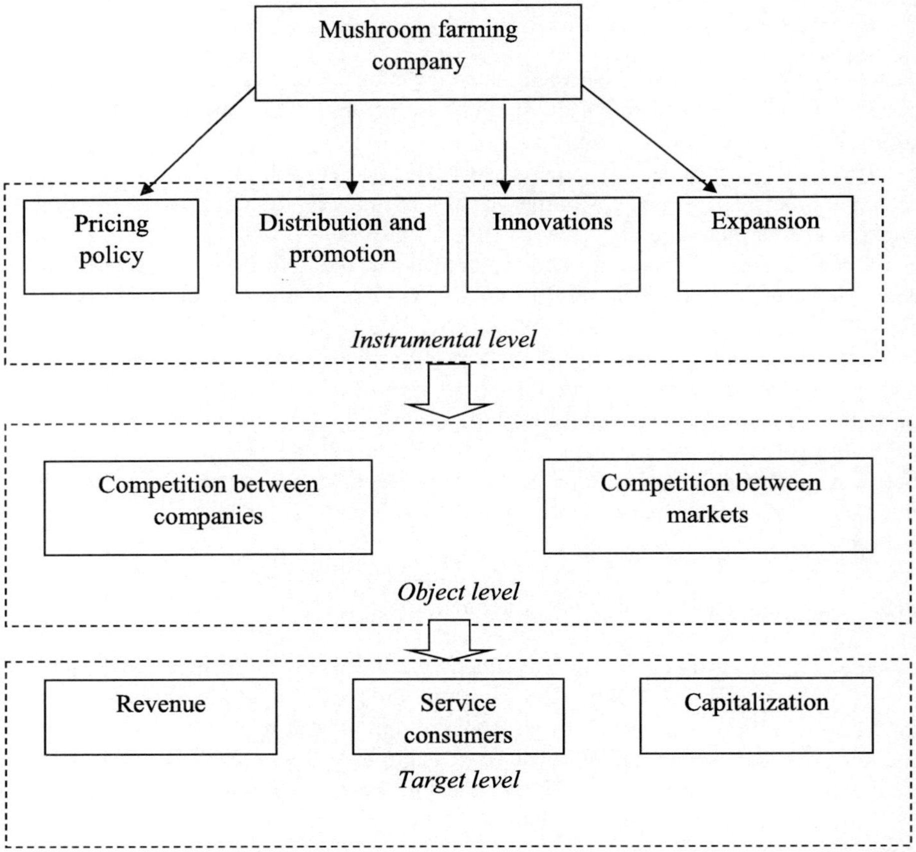

Fig. 2. Scheme of competitive interaction in the mushroom farming market. *Source* Developed and compiled by the authors.

The upper level of the above-mentioned scheme stands for the use of the corresponding model apparatus. At this stage, it is natural to use statistical methods mostly, which allow establishing possible correlations between the regressors and the explained variable. The components of the second level are defined as antagonistic games, where the market capacity acts as a compensating factor. At the third level of the hierarchy, the communication between the firm and the consumer of products is associated with statistical games, where the market performance acts as a statistician.

3 Results

When constructing models of interaction between participants in the market of cultivated mushrooms, the researcher faces a significant problem – the lack of complete, and most importantly, objective information concerning the key performance indicators of the companies' performance and the market of cultivated mushrooms as a whole.

Model 1

Let us build the model of competitive antagonistic interaction between mushroom farming companies operating in the same regional market of cultivated mushrooms – a large corporation planning to enter the market of the specified region (player A) and a local company already operating in this market (player B). We assume that each of the presented players has a few strategies: player A – entering the market under study with little (A_1) or large (A_2) efforts, and player B has responses to a new market subject emerged – producing the product on the same scale (B_1) or reducing activity (B_2).

We introduce the necessary notation. Let x be the percentage of the market occupied by the large company. The payment matrix of the described situation is as follows (Table 1).

If the large player enters the regional market with a small effort, provided that the local producer reduces the volume of the output produced, both players win nothing (situation (A_1; B_2)). In case of Player A enters the market with more effort and Player B's production volume decreases, the player's winnings will be as follows: player A – $0.67x$, Player B – $0.33x$ (situation (A_2; B_2)). The situation is similar to the combination of strategies (A_1; B_1): the winning of player A will be $0.33x$, player B – $0.67x$. And, finally, if the large company enters the market with a great effort, and the local producer does not reduce the production volumes, both players receive the same winnings ($0.5x$ each).

Table 1. Game payment matrix

Large company	Local producer	
	Product produced on the former scale (*)	Decreased activity
Small efforts	$0.67x$	0
	$0.33x$	0
Great efforts (*)	$0.5x$	$0.33x$
	$0.5x$	$0.67x$

Source Developed and compiled by the authors

To analyze the situation, we use the method of majoring (dominating) strategies. We compare the players' winnings when using different combinations of strategies. If the local producer does not change the volume of production, the large player benefits from the strategy of entering the market with a great effort ($0.5x > 0.33x$), and they will do the same if the local producer reduces the volume of the output ($0.67x > 0$). Thus, player A, regardless of the opponent's actions, benefits from the A_2 strategy (marked with*). The local producer will leave the production volumes unchanged ($0.67x > 0$)

with a small effort of player B and will choose the same strategy with a large effort ($0.5x > 0.33x$). Therefore, the B_1 strategy (marked with*) is advantageous for Player B.

Thus, we have obtained an equilibrium situation that provides the maximum utility of players, depending on the opponent's actions. This means that in the described conflict situation there is a Nash equilibrium determined by the situation (A_2; B_1). It should be noted that this outcome of the game is also Pareto-optimal.

Model 2

Let us consider the model of the competition between two companies operating in the market of cultivated mushrooms under the assumption of continuity of time. Any of these companies can begin to supply products to the market of cultivated mushrooms at any time in the interval [0; 1]. We denote the moment of product delivery to the market by the first firm by x, by y – the moment of product delivery by the second firm, by c – the income from the sale of products per unit of time. The later the company's products appear in the market, the higher their quality is – the products are fresher – and, therefore, all other factors being equal, they will be sold.

The winning function of the first firm will be as follows:

$$H(x,y) = \begin{cases} c(y-x) \text{ when } x<y \\ \dfrac{c(1-x)}{2} \text{ when } x=y \\ (1-x) \text{ when } x>y \end{cases}.$$

The optimal price of the game [19] $v = \frac{1}{e}$, the optimal strategy of the first player on the segment $\left[0;\ 1-\frac{1}{e}\right]$ is given by the density $g(y) = \frac{1}{1-y}$. The corresponding optimal strategy of the first player – the distribution function – has a jump of $\frac{1}{e}$ at the point $x = 0$, and the distribution density on the interval $\left(0;\ 1-\frac{1}{e}\right)$ is as follows $f(x) = \frac{1}{e(1-x)^2}$.

Model 3

Let us describe the model of interaction between competitive mushroom farming companies in the regional market in terms of the strategies they use to form the product price. To simplify the model, we will take into account that the price strategy of any company is limited to setting the average price for products.

For the task of the game, we define the sets:

a set of players-producers of mushroom products;

a set of the player's strategies – numbers $x_1, x_2, \ldots, x_n \in R^+$, equal to the average prices for the company's products;

a set of functions of the players' winning $H_1, H_2, \ldots, H_n$.

We introduce some restrictions:

the measure of the player's winning is the company's revenue;

the volume of the output is a decreasing function of the price factor;

the difference in prices for the products of competing firms is an incentive to change the supplier of products for some part of the buyers.

We determine the players' winnings:

$$H_1 = (a_1(x_1) + k_1(x_2 - x_1) + l_1(x_3 - x_1) + m_1(x_4 - x_1)) \cdot x_1,$$
$$H_2 = (a_2(x_2) + k_2(x_1 - x_2) + l_2(x_3 - x_2) + m_2(x_4 - x_2)) \cdot x_2,$$

$H_3 = (a_3(x_3) + k_3(x_1 - x_3) + l_3(x_2 - x_3) + m_3(x_4 - x_3)) \cdot x_3,$
$H_4 = (a_4(x_3) + k_4(x_1 - x_4) + l_4(x_2 - x_4) + m_4(x_3 - x_4)) \cdot x_4,$

where $a_1(x_1)$, $a_2(x_2)$, $a_3(x_3)$, $a_4(x_4)$ are continuously differentiable decreasing functions, the coefficients of the equalities $k_1, k_2, ..., k_n, l_1, l_2, ..., l_n \in R^+$ show the intensity of the transition of buyers from a company with a more expensive product to a company with a cheaper product. In addition, we will also assume that the goal of the players in the game is to get the most winnings.

The necessary condition of the extremum is as follows:

$$\frac{\partial H_1}{\partial x_1} = a_1(x_1) + x_1\frac{da_1}{dx_1} + (k_1\frac{dx_2}{dx_1} + l_1\frac{dx_3}{dx_1} + m_1\frac{dx_4}{dx_1} - 2\cdot(k_1 + l_1 + m_1)) \cdot x_1 + k_1x_2 + l_1x_3 + m_1x_4,$$

where $\frac{dx_2}{dx_1}, \frac{dx_3}{dx_1}, \frac{dx_4}{dx_1}$ is the response of competing firms to the choice and implementation of their strategy by the first firm. Similarly defined are $\frac{\partial H_2}{\partial x_2}, \frac{\partial H_3}{\partial x_3}, \frac{\partial H_4}{\partial x_4}$.

The system of non-linear equations

$$\begin{cases} \frac{\partial H_1}{\partial x_1} = 0 \\ \frac{\partial H_2}{\partial x_2} = 0 \\ \frac{\partial H_3}{\partial x_3} = 0 \\ \frac{\partial H_4}{\partial x_4} = 0 \end{cases} \quad (*)$$

characterizes the state of equilibrium in the regional market of mushrooms. The solution of the latter system depends more on the type of demand functions.

$a_1(x_1)$, $a_2(x_2)$, $a_3(x_3)$, $a_4(x_4)$ and the response of competing firms to the strategy used by the player.

Let us consider the simplest interpretation of this model, when the demand functions are linear, i.e.

$$a_i(x_i) = \alpha_i - \beta_i x_i (i = 1,2,3,4),$$

and the players-firms believe that the change in the product price does not affect the change in the price of competitors (Cournot's assumption), i.e.

$$\frac{dx_i}{dx_j} = 0\,(i \neq j).$$

In this case, the system of equations (*) is as follows:

$$\begin{cases} (2\beta_1 - 2k_1 - 2l_1 - 2m_1) \cdot x_1 + k_1x_2 + l_1x_3 + m_1x_4 + \alpha_1 = 0, \\ (2\beta_2 - 2k_2 - 2l_2 - 2m_2) \cdot x_2 + k_2x_1 + l_2x_3 + m_2x_4 + \alpha_2 = 0, \\ (2\beta_3 - 2k_3 - 2l_3 - 2m_3) \cdot x_3 + k_3x_1 + l_3x_2 + m_3x_4 + \alpha_3 = 0, \\ (2\beta_4 - 2k_4 - 2l_4 - 2m_4) \cdot x_4 + k_4x_1 + l_4x_2 + m_4x_4 + \alpha_4 = 0. \end{cases}$$

This system is a system of linear equations. Such systems are known to have a unique solution if the determinant of the system matrix is different from zero, and this, taking into account the arbitrariness of the model parameters, is an almost reliable event. We emphasize, however, that in the model under consideration, the question of whether there is not just a solution to the system (*), but a positive solution to the system is important. Due to the inequality $\alpha_i > 0$ (i = 1, 2, 3, 4), the negativity of the roots of the characteristic polynomial of the system matrix can serve as a criterion for the existing price equilibrium in the market of cultivated mushrooms. It is almost impossible to obtain a general formula for solving the system described above. But despite this, it is theoretically possible to consider individual special cases and solve the written system, when determining specific numerical estimates of its parameters.

4 Discussion

The pandemic has accelerated and modified many technological processes. As a result, the producers of cultivated mushrooms have faced the tasks of developing new mushroom products, growing new varieties of mushrooms, creating and developing primary and deep processing of products.

The increased competition, changes in consumer preferences due to economic factors (including various crises), changes in the lifestyle of modern people, as well as other factors have caused increasingly severe competition of cultivated mushrooms producers for consumers. However, the level of mushroom consumption in Russia is still 2–3 times lower than in Europe. At the same time, federal networks control only 30% of food retail. Through this channel, consumers get approximately the same proportion of all mushrooms. Electronic commerce, delivery of goods directly to the consumer have opened up new opportunities for the promotion of mushroom products.

5 Conclusion

The paper describes the game-theoretic model of competitive antagonistic interaction between the companies operating in the regional market of cultivated mushrooms. The paper also proposes the model of competition between two firms operating in the market of cultivated mushrooms under the assumption of continuity of time and the model that reflects the interaction of competitive firms in the regional mushroom market in terms of their pricing strategies.

References

1. Algazina, D.G.: Modeling of network interaction in competitive markets with a heterogeneous composition of participants. Ph.D. in Technical Sciences thesis. In: Dubovtseva, D.G. (ed.) Altai State University, Novosibirsk, 22 p. (2014)
2. Ataguba, O.A., Ataguba, J.E.: Social determinants of health: the role of effective communication in the COVID-19 pandemic in developing countries. Glob. Health Action (2020). https://doi.org/10.1080/16549716.2020.1788263
3. Bogoviz, A.V., Kurilova, A.A., Kozhanova, T.E., Sozinova, A.A.: Artificial intelligence as the core of the production of the future: machine learning and intelligent decision support. In: Ram, M. (ed.) Advances in Mathematics for Industry 4.0, pp. 235–256. Academic Press (2021). https://doi.org/10.1016/B978-0-12-818906-1.00010-3
4. Borodin, A.I., Shash, N., Streltsova, E., Kiseleva, N., Bilchak, V.: Game-theoretic interpretation of problems of adoption of administrative decisions at the level of the region. Int. J. Civil Eng. Technol. **10**(2), 1487–1499 (2019)
5. Concept of the development of Russian mushroom farming for the period 2015–2020. https://agrotip.ru/wp-content/uploads/2015/10/Concepcia.pdf Accessed 15 Mar 2020
6. Dente, S.M.R., Hashimoto, S.: COVID-19: a pandemic with positive and negative outcomes on resource and waste flows and stocks. Resour. Conserv. Recycl. **161**, 1049799 (2020). https://doi.org/10.1016/j.resconrec.2020.104979
7. Drokina, K.V.: Game-theoretic modeling of the interaction of regional labor market subjects. Bull. Adygey State Univ. Ser. 5 Econ. **3**, 196–204 (2012)
8. Dubovtsev, A.V.: Modeling of competition in the regional cellular market. Ph.D. in Economics thesis. In: Dubovtsev, A.V. (ed.) Ivanovo State University of Chemistry and Technology, Ivanovo, 18 p. (2011)
9. Dyubin, G.N.: Introduction to Applied Game Theory. In: Dyubin, G.N., Suzdal, V.G. (eds.) Science, Moscow, 336 p. (1981)
10. Gladkova, M.A.: A game-theoretic model of quality management in a competitive environment. In: Gladkova, M.A., Zenkevich, N.A. (eds.) UBS 31-1, pp. 239–262 (2010)
11. Khrenov, A.V.: The capacity of the Russian mushroom market. School Mushroom Farming **2**(104), 20–22 (2017)
12. Khrenov, A.V.: There are more champignons and exotics. Oyster mushrooms are fewer. School Mushroom Farming **6**(126), 3 (2020)
13. Lazareva, T.G., Aleksandrova, E.G.: Analysis of mushroom production and market in Russia. Bull. Eurasian Sci. **11**(1) (2019)
14. Motta, M.: Endogenous quality choice: price vs. quantity competition. J. Ind. Econ. **41**(2), 113–131 (1993)
15. Myerson, R.B.: Game Theory: Analysis of Conflict, 568 p. Harvard University Press, London (1997)
16. Nabokikh, A.A., Barmina, E.A., Ryattel, A.V., Fadeeva, N.Y., Cheraneva, E.K.: Use of the framework of production functions to optimize the operation of the production process. Espacios **38**(49), 6 (2017)
17. Nabokikh, A.A.: Modeling of the market of cultivated mushrooms by methods of discriminant and factor analysis. Eng. J. Don **2**(29), 53 (2014)
18. Nabokikh, A.A.: Model of integration of infrastructure elements as the basis of cluster technology for managing the development of the market of cultivated mushrooms. Econ. Entrepreneurship **6–2**(59), 1040–1043 (2015)
19. Aoki, R., Prusa, T.J.: Sequential versus simultaneous choice with endogenous quality. Int. J. Ind. Organ. **15**, 103–121 (1996)

20. Popkova, E.G., Saveleva, N.K., Sozinova, A.A.: A new quality of economic growth in "smart" economy: advantages for developing countries. In: Popkova, E.G., Sergi, B.S. (eds.) ISC 2020. LNNS, vol. 155, pp. 426–433. Springer, Cham (2021). https://doi.org/10.1007/978-3-030-59126-7_48
21. Popkova, E.G., Savelyeva, N.K., Sozinova, A.A.: Smart technologies in entrepreneurship: launching a new business cycle or a countercyclical instrument for regulating the economic situation. In: Popkova, E.G., Sergi, B.S. (eds.) ISC 2020. LNNS, vol. 155, pp. 1722–1730. Springer, Cham (2021). https://doi.org/10.1007/978-3-030-59126-7_188
22. Saveleva, N.K.: Transnational capital in banking market: making investment decisions with the help of AI in the conditions of competition. In: Popkova, E.G., Sergi, B.S. (eds.) ISC 2020. LNNS, vol. 155, pp. 1714–1721. Springer, Cham (2021). https://doi.org/10.1007/978-3-030-59126-7_187
23. Shibaeva, N.A., Zarudneva, A.I., Sozinova, A.A., Shuvaev, A.V., Alekseev, A.N.: Restructuring of tax liabilities as an upcoming trend of economic diversification in modern Russia. In: Gashenko, I.V., Zima, Y.S., Davidyan, A.V. (eds.) Optimization of the Taxation System: Preconditions, Tendencies, and Perspectives. SSDC, vol. 182, pp. 83–89. Springer, Cham (2019). https://doi.org/10.1007/978-3-030-01514-5_10
24. Silkina, G.Y.: Game-theoretic modeling of the interaction between subjects in the innovative sphere. Saint-Petersburg State Polytechnical Univ. J. Econ. Sci. **2–1**(144), 99–104 (2012)
25. Sozinova, A.A., Sofiina, E.V., Safargaliyev, M.F., Varlamov, A.V.: Pandemic as a new factor in sustainable economic development in 2020: scientific analytics and management prospects. In: Popkova, E.G., Sergi, B.S. (eds.) ISC 2019. LNNS, vol. 198, pp. 756–763. Springer, Cham (2021). https://doi.org/10.1007/978-3-030-69415-9_86
26. Stepanov, L.V.: Modeling Competition in Market Conditions: Monograph, 114 p. Academy of Natural Sciences, Moscow (2009)
27. Vlasov, D.A.: A game-theoretic model of competition for product sales markets. Econ. Manag. Issues **5**(7), 27–29 (2016)

Organizational and Economic Mechanisms of Integration of Water Container Communications of European Countries and the Possibility of Their Use in Russia

Valery I. Mineev[1(✉)], Igor K. Kuzmichev[1], Mikhail V. Ivanov[1], Olga V. Pochekaeva[1], and Evgeny A. Semakhin[2]

[1] Volga State University of Water Transport, Nizhny Novgorod, Russia
vi@vsuwt.ru, rector@vgavt-nn.ru, ivanov@vsawt.com
[2] Minin Nizhny Novgorod State Pedagogical University, Nizhny Novgorod, Russia

Abstract. Currently, the countries of Europe, which are based on the principles of the "green economy", are switching to environmentally friendly modes of transport, including water transport. The purpose of the article is to analyze the organizational and economic mechanisms for integrating inland waterways into the system of container routes operating in European countries and the prospects for their use in Russia.

During the research, a systematic approach was used, which made it possible, based on the analysis of statistical information and transport development programs adopted by the European Union, to identify the organizational and economic mechanisms for integrating water transport into the container transport system.

The authors of the article analyzed the modern organizational and economic mechanisms for the effective functioning of water container lines in European countries, identified the feasibility of creating sustainable freight flows, rigid schedules for the movement of the fleet, and determined the directions for integrating Russian inland water transport into the system of container communications.

The use of foreign experience will promote a more active inclusion of Russian inland waterway transport into the international container shipments system.

Keywords: Waterway transport · European countries · Container lines · Integration efficiency · Foreign experience

JEL Codes: R41 · R42 · O52 · O18 · L95 · L98 · P11 · F55 · R48

1 Introduction

Historically, inland waterways (IW) have been the key transport communications for goods and passenger's transportation and largely determined the economy efficiency of individual regions. Alternative transport, primarily motor one, modes development has

E. G. Popkova (Ed.): ISC 2020, LNNS 368, pp. 437–444, 2022.
https://doi.org/10.1007/978-3-030-93244-2_48

greatly reduced the volume of transportation work performed by inland waterway transport (IWT), but at the same time aggravated the environmental pollution problem. At present, IWT works mainly to carry bulk cargoes. Container lines on the territory of Russia operate without involving IWT, using mainly motor and railroad transport modes.

In its turn, container shipment in the world cargo turnover accounts for a significant share (more than 90% of piece cargo is transported in containers), while their percentage ratio to the total traffic volume is growing from year to year. The cargoes containerization level in the European Union (EU) countries is characterized by the turnover of 140 TEU [Eng. TEU (twenty-foot equivalent unit) - twenty-foot equivalent] (conventional account unit for containers' fleet) per one thousand people a year. In Russia, this figure does not exceed 30 TEU. The difference in these parameters is containerization potential for Russia.

An analysis of foreign experience in organizing container traffic on IW will allow revealing organizational and economic problems and their solving ways for efficient use in Russia. This, in turn, will facilitate container flows transfer to Russian IWT and will have a positive impact on the environmental situation in the country.

2 Methodology

Transport as a branch of economy is its significant part. For example, in the European Union (EU), transport provides employment for about 10 million people and about 5% of IW [1]. At the same time, the current state of the transport complex in Europe does not meet the requirements of society, since this complex refers to industries that negatively affect the environment. This primarily concerns motor transport, which accounts for the main portion of greenhouse gases [18]. Therefore, the EU has developed and is implementing programs to reduce emissions by 2050 by 80–95% below the 1990 level or by 70% below the 2008 level. One of the ways to improve the transport sector environmental performance in Europe is to transfer cargo flows to water transport in general and to IWT in particular, since CO_2 emissions per 1 ton-kilometer on IWT are five times less than on motor transport [8].

The need to intensify the inclusion of IWT in regional transport systems is emphasized in the papers [6, 8, 14, 15, 17–20].

Russia has also given the green light to green economy development. In this regard, a number of decisions were made to develop the transport sector, including the expansion of IWT use sphere. However today, like in most countries of the world, IWT is mainly used for bulk cargo transportation, while container shipment is performed by other transport modes, primarily by motor one. Container flows transfer from motor transport to IWT will contribute to environmental issues solution and load decrease on highways. However, for the efficient inclusion of IWT in the container shipment system a number of organizational and methodological issues needs to be solved.

The method used by the authors is based on a review and analysis of the available scientific and statistical materials with an assessment of the possibility to use advanced technologies regarding the transport system of Russia.

3 Results

Today, goods containerization level in the world is more than 70%; in recent years, the growth rates have reached 2–4% per year [4]. In international traffic sea fleet performs great volumes of container shipment. At the same time, the growth in demand for container shipment and the need to reduce transportation costs in a competitive environment led to an increase of sea container ships' cargo capacity. Alongside this fact, the increase of container ships' cargo capacity created certain difficulties in entering ports due to the shallow depth of straits or their water area; as a result, ocean-going container ships deliver containers to ports that are hubs, and then the containers are reloaded onto other transport modes, including ships of smaller capacity, working on feeder lines.

Abroad, on such feeder lines river transport is used. On European waterways, this refers to the Danube, Rhine and other rivers.

The Danube - the second longest river in Europe, originates in Austria, flows through ten European countries and flows into the Black Sea [6]. In 1992, the Danube was connected with the Main (the Rhine tributary) by a canal, as a result the Trans-European waterway 3503 km long was created. In total, there are about 75 ports with about 200 terminals on the Danube. The cargo turnover distribution in the Danube River ports by different countries in 2019 is shown in Fig. 1.

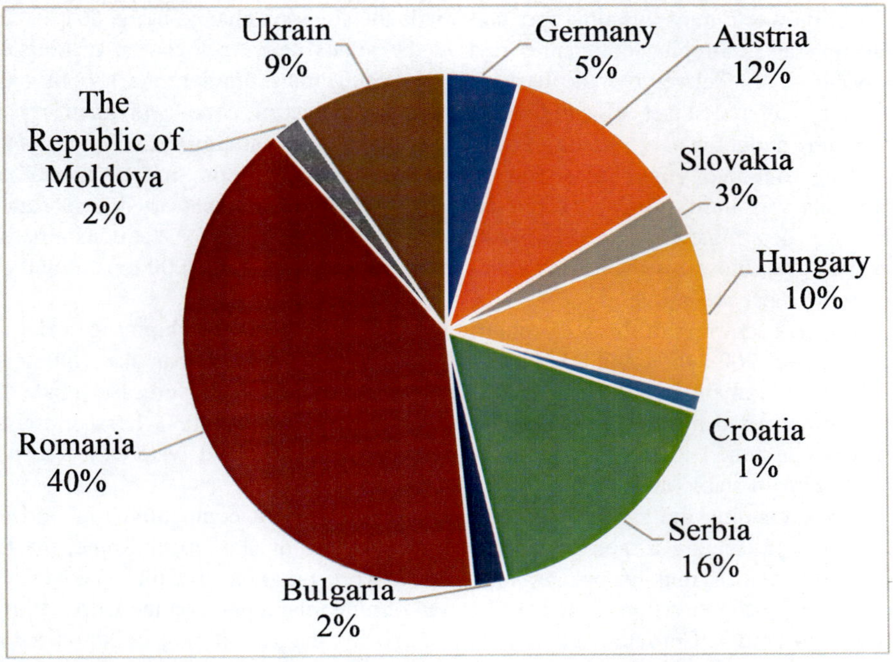

Fig. 1. Cargo turnover distribution in the Danube countries ports. Source Compiled by the authors based on [5]

Containers shipment along the Danube was started in 2005 between Constanta and Giurgiu by two barges with a deadweight of 2000 tons, 54 TEU on each barge. At present, the Krems - Constanta - Krems container line has been created, on which cargo ships with attachments of 222 TEU (90 + 132) total capacity and with a 16 days round trip operate.

Figure 1 shows that Romanian ports are leading in terms of cargo turnover on the Danube. The main port among them is the Constanta Port, from where the main container routes diverge to Germany, Hungary (river transport), Ukraine and Russia (rail transport), Georgia and Turkey (sea transport).

Constanta is the hub for the feeder container line «Tavria Line» of the «Aquarelle» shipping company operating since 2010. Transportation is carried out by river-sea vessels of 2 STK class with a nominal capacity of 112 TEU.

The Rhine originates in the Swiss Alps and flows northward through the territories of 6 countries (Switzerland, Liechtenstein, Austria, Germany, France, the Netherlands). The length of the waterways in the Rhine basin is 3000 km [3].

The first regular container shipments on the Rhine date back to 1968 and is associated with the creation of the Rotterdam - Amsterdam - Basel return container line. A year later, the Rhine Container Line (RCL) company was created, which daily sent container ships from Rotterdam, Amsterdam and Antwerp to Basel, Strasbourg and Manheim.

The company's experience showed the low efficiency of container shipment, the reason for which was unsatisfactory transport process organization. Having done the correction work, transportation operators made the situation change in the 80s. By the mid-seventies, shipping companies revised the goods delivery logistics, making the «door-to-door» delivery as the basis for organizing transportation. As a bonus, the company offered a set of additional services (storage of containers, their repair, informing about the cargo location, customs documents preparation etc.). To master the growing volume of container shipment, a network including more than 20 specialized terminals was built. Vessels' cargo capacity providing the necessary transportation limit was determined: it was 1350–1700 tons. From the point of view of transportation profitability, shipments by vessels with a cargo capacity of 1700–2000 tons turned out to be the most preferable.

By the beginning of the XXI century, the volume of container shipment in Europe has reached 3000 thousand TEU, which is three times higher than the volume of container shipment in 1990. However, a number of long-term forecasts made for Europe promised a 50% growth of IWT shipment volume by 2030. Less optimistic forecasts assumed that a rather modest market share of inland waterway transport would remain stable at best.

Understanding the significance of multimodal shipment competitiveness and the relevance of creating a modern transport infrastructure on the Upper Rhine, the EU approved of the Trans-European Transport Network program (TEN-T). The TEN-T project originally envisioned the coordination of nine inland ports on the Upper Rhine, located in France, Germany and Switzerland. Today, the TEN-T project is focused on transportation corridors: East - West (North Sea - Baltic, Mediterranean, Rhine - Danube), diagonal corridors (Baltics - Adriatic, North Sea - Mediterranean Sea, Middle East, Atlantic), North - South corridors (Scandinavia - Mediterranean Sea, Rhine - Alps) [2].

Within the framework of the project, shipping companies have achieved a significant reduction in container shipment time for the Upper Rhine ports by improving cargo work organization, modernizing the container transportation fleet and, which is extremely important, by adhering to strict schedules.

Special attention should be paid to strict schedules, since such an approach, on the one hand, ensures container line stability, and on the other hand, it is impossible without a certain fleet reserve. A strict vessel operation schedule on a container line implies lower risks, since such an organization of the line's operation allows customers to rely on their activities on a strict schedule of goods' departure, but at the same time requires from them to focus on a higher service rate.

In 2006 the EU Transport Policy Committee approved of the inland navigation support program «NAIADES», and in 2013 – «NAIADES II» [16]. The main directions for the program implementation until 2020 are: improving the quality characteristics (infrastructure, market functioning, environment due to low emissions), innovation, skilled labor force, integration of IWT into multimodal logistics systems.

The reason for drawing attention to IWT is the use of Green Logistics principles (GL) in the EU countries. The current logistics trends in the EU countries define new green chains' models of goods delivery that reduce the negative impact on the environment and are energy-saving [13].

Taking into account the importance of GL, in recent years the interest in IWT has increased also because the cost of inland waterways logistics is 17% of motor transport cost and 50% of rail transport cost. In addition, river transport is seven times more stable than other transport modes [12].

Inland waterway transport as a transport mode [9]:

- economical and environmentally friendly;
- convenient for organizing both container and piggyback transportation due to horizontal transshipment (movement of vehicles over the ship's ramp);
- indispensable in oversized cargo transportation;
- provides communication between seaports and inland ports.

In the authors' opinion, the above-mentioned advantages of IWT also include such characteristics as a high degree of work adaptation under synchromodal logistics conditions [7.10].

Currently in the EU Inland Navigation Europe (INE) is engaged in the IWT development and maintaining its competitiveness on a permanent basis [2]. Logistics measures taken by the EU have brought some results. Thus, since 2015 the traffic volume, and since 2016 the freight turnover of IWT received positive trends (Fig. 2).

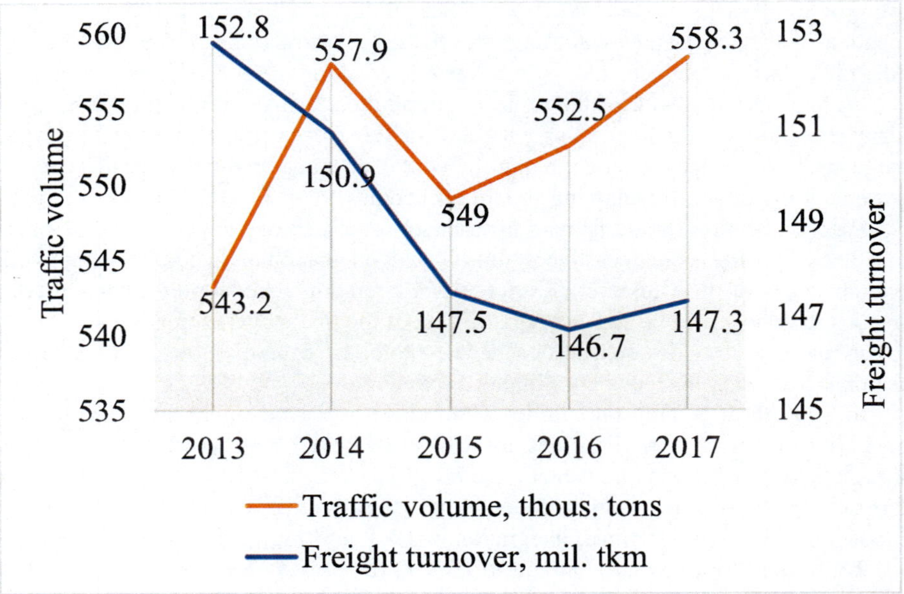

Fig. 2. Traffic volume and freight turnover of the EU IWT. Source Compiled by the authors based on [2]

4 Conclusion

In Russia IWT began to master goods' container shipment in 1949. First, it was mixed rail-water transportation, and since 1952, it became direct water transportation (Moscow - Gorky). The development of river container shipment was due to the organization of a container line between the ports of Gorky and Volgograd (1959). Subsequently, up to 1990, six container lines operated in the European part of Russia. As a result, they covered about 70 ports, wharves and container berths. Specially for these container lines' operation, dedicated catamarans of the R-19 project (the «Brothers Ignatovy2 type) with a cargo capacity of 1000 tons were designed, which together with vessels having 2000 tons cargo capacity, worked on the lines according to a strict schedule [11]. After 1991, the regular river container shipment gradually «came to naught», and at present there are no container lines on the Volga.

The above given critical analysis of the IWT activities in the EU indicates that despite a number of existing problems IW container shipment tends to grow, with the support of state structures can successfully compete with motor and rail transport. However, in order to ensure container lines' competitiveness on IW, the following conditions must be met:

1. Transferring a part of container cargo flows on IWT for the purpose of efficient functioning of multimodal logistics chains, consisting of motor, rail and sea transport.

2. Carrying out door-to-door delivery on a synchromodal basis during a container line organization.
3. Providing regular voyages based on strict schedules.
4. Availability of the necessary fleet reserve in order to execute the traffic schedule.
5. Freight rates on these lines should be lower than road and rail rates, which requires the appropriate government support like in the EU countries.

In recent years, IWT has received a certain renaissance due to the more attentive attitude of society to environmental problems. At the same time, water transport is considered to have some disadvantages, the main of which is a low movement speed. However, if we consider not a ship's movement speed, but the commercial speed, that is the speed of goods' movement, then it is not much less on IWT in than on motor and rail transport. And considering the restrictions on heavy cargo vehicles passage adopted in European countries, it is quite obvious that in this point IWT may have certain advantages, since the fleet unlike cars and trains, does not experience difficulties with the tracks' capacity.

It is obvious that Russia shouldn't not find itself at the back of the process of IWT use intensification, when organizing both domestic and international container lines. The transport market is highly competitive, any delay in decision-making (the experience of such decisions is available) will lead the country to the crisis of the Single European Water System, which is already experiencing serious problems.

References

1. European Commission: White Paper on transport - Roadmap to a single European transport area - Towards a competitive and resource-efficient transport system. Luxembourg: Publications Office of the European Union, 28 p. (2011). ISBN 978-92-79-18270-9. https://doi.org/10.2832/30955
2. An official website of the European Union / MOBILITY AND TRANSPORT/Inland waterway. https://ec.europa.eu/transport/modes/inland_en. Accessed 20 Dec 2020
3. An official website of the Inland Navigation Europe. http://www.inlandnavigation.eu/about-ine/. Accessed 20 Mar 2020
4. Container Trade Statistics: Global container traffic increased by 1.4% this year (2019). https://www.korabel.ru/news/comments/mirovye_konteynernye_perevozki_uvelichilis_na_1_4_za_etot_god.html. Accessed 12 Jan 2021
5. Danube Shipping Market Observation: 2019 Results. Budapest (2020). https://danubecommission.orq/extranet/e-librarv/index.html. Accessed 12 Jan 2021
6. David, A., Madudova, E.: The Danube river and its importance on the Danube countries in cargo transport. Transp. Res. Procedia **40**, 1010–1016 (2019)
7. Giusti, R., Manerba, D., Bruno, G., Tadei, R.: Synchromodal logistics: an overview of critical success factors, enabling technologies, and open research issues. Transp. Res. Part E Logistics Transp. Rev. **129**, 92–110 (2019)
8. Gołebiowski, C.: inland water transport in Poland. Transp. Res. Procedia **14**, 223–232 (2016)
9. Hansut, L., Gasparik, J., David, A.: The critical path method as the method for evaluation and identification of the optimal container trade route between Asia and Slovakia. In: Book Series: Proceedings of International Scientific Conference Business Logistics in Modern Management, pp. 29–42 (2017)

10. He, Z., Navneet, K., van Dam, W., Mieghem, P.V.: Robustness assessment of multimodal freight transport networks. Reliab. Eng. Syst. Saf. **207**, 107315 (2021). https://www.sciencedirect.com/science/article/pii/S0951832020308097. Accessed 1 Oct 2021
11. Kalachev, V.K.: Optimization of the use of river cargo vessels in the transport of goods in containers. Dissertation for the degree of Candidate of Technical Sciences. N. Novgorod, 158 p. (1988)
12. Krcum, M., Plazibat, V., Mrcellic, G.J.: Integration sea and river ports - the challenge of the Croatian transport system for the 21st century. Int. J. Marit. Sci. Technol. **62**(4), 247–255 (2015)
13. Mihic, S., Golusin, M., Mihajlovic, M.: Policy and promotion of sustainable inland waterway transport in Europe - Danube River. Renew. Sustain. Energy Rev. **15**(4), 1801–1809 (2011)
14. Mineev, V.I., Ivanov, V.M., Ivanov, M.V.: Evaluation of the condition and prospects for the development of container transportation by inland water transport. Volga State University of Water Transport, vol. 63, pp. 123–137 (2020). https://doi.org/10.37890/jwt.vi63
15. Montwił, A.: Inland ports in the urban logistics system. Case studies. Transp. Res. Procedia **39**, 333–340 (2019)
16. NAIADES II: An official website of the European Union / Mobility and transport/Inland waterway/NAIADES II. https://ec.europa.eu/transport/modes/inland/promotion/naiades2_en. Accessed 20 Dec 2020
17. Nyrkov, A., Shnurenko, A., Sokolov, S., Chernyi, S., Korotkov, V.: Some methods of increasing the efficiency of river transport system. Procedia Eng. **178**, 543–550 (2017)
18. Vilarinho, A., Bartocci Liboni, L., Siegler, J.: Challenges and opportunities for the development of river logistics as a sustainable alternative: a systematic review. Transp. Res. Procedia **39**, 576–586 (2019)
19. Wiegmans, B., Witte, P., Spit, T.: Inland port performance: a statistical analysis of Dutch inland ports. Transp. Res. Procedia **8**, 145–154 (2015)
20. Wisnicki, B.: Determinants of river ports development into logistics Trimodal nodes, illustrated by the ports of the lower Vistula river. Transp. Res. Procedia. **16**, 576–586 (2016)

Locating an Information and Logistics Center for Innovation and Industrial Clusters in a Federal District

Sergei N. Yashin[1(✉)], Nadezhda I. Yashina[1], Egor V. Koshelev[1], Dmitry A. Sukhanov[1], and Victor P. Kuznetsov[2]

[1] Lobachevsky State University of Nizhny Novgorod, Nizhny Novgorod, Russia
[2] Minin Nizhny Novgorod State Pedagogical University, Nizhny Novgorod, Russia
kuzneczov-vp@mail.ru

Abstract. The subject of this study is to create technologies to identify a center of innovation and industrial clusters, which would provide their effective economic and financial, information and logistics interaction. For this purpose, the paper proposes a method for solving the problem of locating an information and logistics center of clusters in a federal district. This method is based on the use of simulation modeling algorithms. An innovation and industrial cluster center is determined on the basis of minimizing the sum of distances from the planned center to clusters. The maximum remoteness of the center from district nuclear power plants and atom cities is also taken into account. With this aim in view, a genetic algorithm, a simulated annealing method and a pattern search are used.

This approach is tested for the Volga Federal District. As a result, it has been obtained that the information and logistics center of the Volga Federal District should be the city of Kazan.

If the information and logistics center of the Volga Federal District is to be actually located in Kazan, this will considerably reduce transaction costs associated both with regulation of information flows and with traffic streams within the confines of the federal district under investigation. And this, in turn, will lead to a reduction in financial costs in the Volga Federal District and, most importantly, to an increase in the synergy effect of a large innovation system that brings together innovation and industrial clusters on a large territory of the entire federal district.

Keywords: Genetic algorithm · Location problem · Pattern search · Simulated annealing

JEL Codes: C63 · E17 · O21 · O36

E. G. Popkova (Ed.): ISC 2020, LNNS 368, pp. 445–454, 2022.
https://doi.org/10.1007/978-3-030-93244-2_49

1 Introduction

For the sustainable development of the federal district's economy, it is necessary to have new operating conditions that would lead to an additional synergy effect. One of such important conditions includes optimal information and logistic interaction of innovation and industrial clusters within one district. Global optimization of this interaction will allow solving national strategic objectives of the district evolution.

With that in mind, this paper solves the problem of locating the information and logistics center of the federal district clusters. In order to do that, appropriate simulation modeling technologies are used.

The location problem is a classic global optimization problem. There are many scientific articles and books confirming the effectiveness of the methods used to solve it.

Murray and Tong have found that facility siting and related maintenance are the main problems in urban and regional planning [10]. They describe an approach to the problem of covering the spatial demand for services, where potential facilities are located in a continuous plane. The authors show that the weighted demand represented as points, lines or polygons can be optimally served by a finite number of potential facility locations called a polygon intersection point set (PIPS). The developed approach is an extension of the demand point abstraction to more general representations (for example, points, lines or polygons) [10].

Ageyev, Gimadi and Kurochkin have handled a routing graph location problem for identical production capacities of enterprises. Previously, Ageyev had built an exact algorithm solving the problem during the time of $O(m^5n^2 + m^3n^3)$, where m and n are the number of enterprises and the number of demand points respectively. In this paper, we propose to modify this algorithm using the time complexity of $O(m^4n^2)$ that is an order of magnitude less for the both parameters [1].

Vos et al. have obtained an effective way to assign services to each of the potential light bulbs in the city so that all the services are distributed adequately (i.e., meet the requirements of coverage) at minimum cost [11]. A two-stage approach to finding a smart and economical distribution is proposed. In the first stage, the services are distributed among the existing lamp posts taking into account the required coverage and the cost of equipping the lamp post with services. A statement of this problem has been presented, which is called a multiple service set coverage problem. Several heuristic methods are proposed to solve this problem. For best results, a heuristic method is shown based on a consistent solution to the set coverage problem for each of the services. The quality of this heuristics can be partly attributed to the effective method that is implemented to solve the "set coverage problem". The selected locations from the first stage are then accepted as input for the second stage, which can be considered as a cost reduction stage. At this stage, light bulbs are selected that can serve as a hub for at least a certain number of bulbs that are within the required range. For this problem, which is called a wireless network problem, there is a statement. Several approaches are available to solve the wireless network problem. It is shown that the iterated local search method is capable of finding the best solutions. The best solution methods are used in a two-stage approach to provide cities with multiple services [11].

Gimadi has handled a problem of location of a network facility having unlimited production levels [4]. This problem is NP-complicated in general and is known to have an optimal quadratic complexity solution in a tree network. Gimadi has studied an instance of a network represented by an external planar graph, i.e. a graph whose vertices belong to a single (external) face. It is known that this problem has an optimal algorithm with the time complexity of $O(nm^3)$, where n is the number of vertices, and m is the number of possible locations of the facility. Using some properties of external graphs (binary 2-trees) and the availability of an optimal solution with a family of centrally connected service domains, repetition ratios have been obtained to build an optimal algorithm with the time complexity, which is less by the coefficient $\sqrt{m}$ than the time complexity of the previous algorithm [4].

Thus, Huang, Bergman and Gopal have studied an analytical approach to the choice of expansion sites for retailers selling add-on products, the demand for which is determined by the demand for another core product [5]. The demand for an add-on product is only implemented as a supplement to the demand for a core product. In this context, any of these two products can be exposed to spatial autocorrelation when the demand in a given location is affected by the demand elsewhere. Using data from an industrial partner selling add-on products, the authors have created predictive models to understand the derivative demand for the add-on products and an optimization system to automate expansion solutions to maximize expected sales. In this case, the spatial autocorrelation and the complexity of the predictive model affect the complexity and structure of the prescriptive optimization model. The results show that the developed models are very effective in predicting sales of add-on products and that using the structure of optimization based on the predicted model can lead to a significant increase in expected sales compared to the base-case scenarios [5].

Thus, the solution of a location problem is applied to many scenarios of global optimization in the economy, finance and other industries.

To solve such location problems, simulation modeling algorithms are also used, among which we will analyze: 1) genetic algorithm; 2) simulated annealing; 3) pattern search.

Let us describe their advantages in more detail.

1. Evolutionary algorithms (genetic algorithm) are relatively new but very powerful methods used to find solutions to many real search and optimization problems. Many of these problems have multiple objectives resulting in the need for a set of optimal solutions known as effective solutions. It has been found that the use of evolutionary algorithms is a highly effective way to find many effective solutions in a single simulation run [7].
2. The annealing method allows avoiding the "trap" in local extrema of the function being optimized and continuing the search for a global extremum [9]. On top of that, even in case of any lack of computational power for finding a global extremum, the annealing method usually provides a good solution, i.e. one of the local extrema [9]. Finally, when comparing the Adaptive Simulated Annealing (ASA) method and genetic algorithms, it has been obtained that in most problems, the annealing method is not behind genetic algorithms, but is ahead in many problems [6].

3. Direct search (pattern search) is an optimization problem solving method requiring no target function gradient information. Unlike more traditional optimization methods that use gradient information or higher derivatives to search for an optimal point, the direct search algorithm searches for multiple points around the current point looking for a point where the target function value is lower than the value at the current point. Direct search can be used to solve problems for which the target function is not differentiable or even not continuous [2, 8].

As practice shows, among the selected simulation modeling algorithms, the pattern search allows finding a global extremum more correctly, and as a result of fewer iterations.

2 Methodology

In order to create a technology for determining the center of innovation and industrial clusters, we will use a model based on the stages shown in Fig. 1.

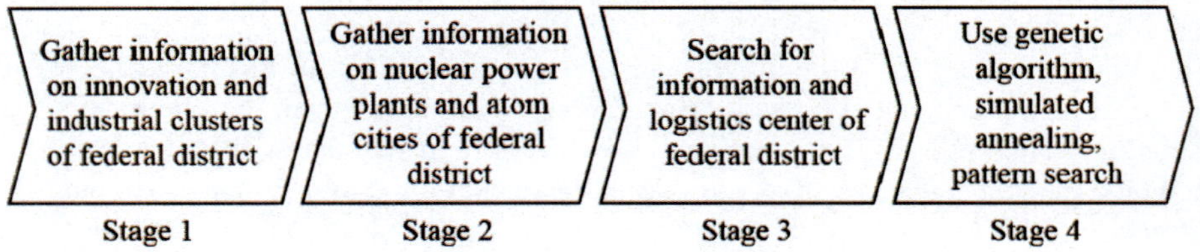

Fig. 1. Stages of solving a location problem for an information and logistics center in a federal district. *Source* Developed and compiled by the authors

Let us analyze the stages presented in more detail.

Stage 1 – Gathering information on federal district clusters. To understand what data will be needed to solve the location problem, we will provide a formula for a function that should be minimized:

$$Z(x, y) = \sum_{i=1}^{n} w_i \sqrt{(x - x_{0,i})^2 + (y - y_{0,i})^2},$$

where w_i - weights, i.e. the population of i cluster centers and cluster centers located relatively not far from nuclear power plants (NPPs) and atom cities (persons);

$(x_{0,i}; y_{0,i})$ - coordinates of i cluster centers as well as of NPPs and atom cities.

In this way, we want to minimize the sum of distances from the information and logistics center of the federal district to cluster centers, i.e. cities. At the same time, the weights w_i are included in this function with a plus for cluster centers and with a minus for NPPs and atom cities. In a similar way, we want to find the coordinates of the district center, which would be as close to the centers of clusters as possible and at the same time as far away from NPPs and atom cities as possible.

Therefore, at the 1st stage it is necessary to collect data on the regions of the federal district where innovation and industrial clusters are located, on the clusters and geographic coordinates of their centers, as well as on the population of these centers (cities).

Stage 2 – Gathering information on federal district NPPs and atom cities. Next, we will need the coordinates of NPPs and atom cities as well as data on the population in the cluster centers located relatively close to them. If more than one of such centers (cities) are obtained for a particular NPP or atom city, the sum of the population of these cities is taken as weights w_i.

Stage 3 – Searching for a federal district information and logistics center. The function $Z(x, y)$, which is minimized, can have both global and local minima. To find exactly a global minimum, we will use appropriate simulation modeling methods. They will be chosen at the next stage.

Stage 4 – Using a genetic algorithm, simulated annealing and a pattern search. Most functions $Z(x, y)$ in the location problem are quite easily minimized in terms of global optimization by all the three algorithms. However, for a noisy stochastic function, for example, the optimization results may turn out to be different depending on the algorithm used. This can also happen with some continuously differentiable functions. For this reason, we use several methods, relying primarily on the results of the pattern search as the most accurate method for reaching the global function extremum [2, 3, 8].

3 Results

Let us illustrate the effect of the presented model as exemplified by the Volga Federal District. According to the list approved by the Government of the Russian Federation, there are 25 pilot innovative territorial clusters operating in the regions of the Russian Federation. We will then consider only those regions (territories or republics) within the Volga Federal District where clusters from this list are located. Then for the purpose of economic and financial foresight of the Volga Federal District innovation system development, let us pass through the stages consistently presented in the model.

Stage 1 – Gathering information on federal district clusters. Information on the Volga Federal District regions where innovation and industrial clusters are located, on clusters and geographic coordinates of their centers, and on the population of these centers (cities) is shown in Table 1.

Table 1. Innovative territorial clusters by regions of the Volga Federal District

Region	Innovative cluster	Cluster center	Center coordinates	Population (persons)
1. Nizhny Novgorod Region	Nizhny Novgorod industrial innovative cluster in automotive and petrochemical engineering	Nizhny Novgorod	56°19′37″N 44°00′27″E $x = 44.1$ $y = 56.33$	1,253,511
	Sarov innovative cluster	Sarov	54°56′N 43°19′E $x = 43.32$ $y = 54.93$	95,388
2. Republic of Mordovia	Energy-efficient lighting engineering and smart lights	Saransk	54°11′N 45°11′E $x = 45.18$ $y = 54.18$	318,578
3. Ulyanovsk Region	ULYANOVSK-AVIA Scientific and Educational Production Cluster consortium	Ulyanovsk	54°19′N 48°22′E $x = 48.37$ $y = 54.32$	627,870
	Dimitrovgrad nuclear innovation cluster	Dimitrovgrad	54°14′N 49°35′E $x = 49.58$ $y = 54.23$	115,253
4. Samara Region	Innovative territorial aerospace cluster	Samara	53°11′N 50°07′E $x = 50.12$ $y = 53.18$	1,156,644
5. Perm Territory	NOVY ZVEZDNY Technopolis innovative territorial cluster of rocket engine building	Perm	58°00′50″N 56°14′56″E $x = 56.25$ $y = 58.01$	1,053,938
	Photonics	Perm		
6. Udmurt Republic	Udmurt machine building cluster	Izhevsk	56°51′11″N 53°12′44″E $x = 53.21$ $y = 56.85$	648,944
7. Republic of Tatarstan	Kama innovative territorial production cluster	Kazan	55°47′27″N 49°06′52″E $x = 49.12$ $y = 55.79$	1,251,969
8. Republic of Bashkortostan	Petrochemical territorial cluster	Ufa	54°44′N 55°58′E $x = 55.97$ $y = 54.73$	1,124,226

Source Developed and compiled by the authors

Stage 2 – Gathering information on federal district NPPs and atom cities. The coordinates of NPPs and atom cities in the Volga Federal District as well as data on the total population in the cluster centers located relatively close to them are shown in Table 2.

Table 2. Nuclear power plants, atom cities by regions of the Volga Federal District

Region	NPPs and atom cities	Near cities	NPP coordinates	Population (persons)
1. Nizhny Novgorod Region	Sarov (Arzamas-16)	Sarov, Nizhny Novgorod, Saransk	54°56′N 43°19′E $x = 43.32$ $y = 54.93$	95,388 1,253,511 318,578 (1,667,477)
2. Republic of Mordovia	Zarechny (Penza-19)	Saransk	53°12′N 45°10′E $x = 45.17$ $y = 53.2$	318,578
3. Ulyanovsk Region	Dimitrovgrad nuclear innovation cluster	Dimitrovgrad, Ulyanovsk	54°14′N 49°35′E $x = 49.58$ $y = 54.23$	115,253 627,870 (743,123)
4. Samara Region	Balakovo NPP (Balakovo)	Samara	52°05′34″N 47°57′14″E $x = 47.95$ $y = 52.09$	1,156,644
5. Perm Territory	Lesnoy (Sverdlovsk-45)	Perm	58°38′N 59°47′E $x = 59.78$ $y = 58.63$	1,053,938
	Novouralsk (Sverdlovsk-44)	Perm	57°15′N 60°05′E $x = 60.08$ $y = 57.25$	1,053,938
8. Republic of Bashkortostan	Tryokhgorny (Zlatoust-36)	Ufa	54°48N 58°27′E $x = 58.45$ $y = 54.8$	1,124,226

Source Developed and compiled by the authors

Stage 3 – Searching for a federal district information and logistics center. Searching for the global minimum of the function $Z(x, y)$, i.e. the sum of distances from the information and logistics center of the Volga Federal District to the centers of clusters (cities). In this respect, we want to find the coordinates of the Volga Federal District center, which would be as close to the centers of clusters as possible and as far away from NPPs and atom cities as possible. The graph $Z(x, y)$ has been obtained in the *Matlab* software (Fig. 2).

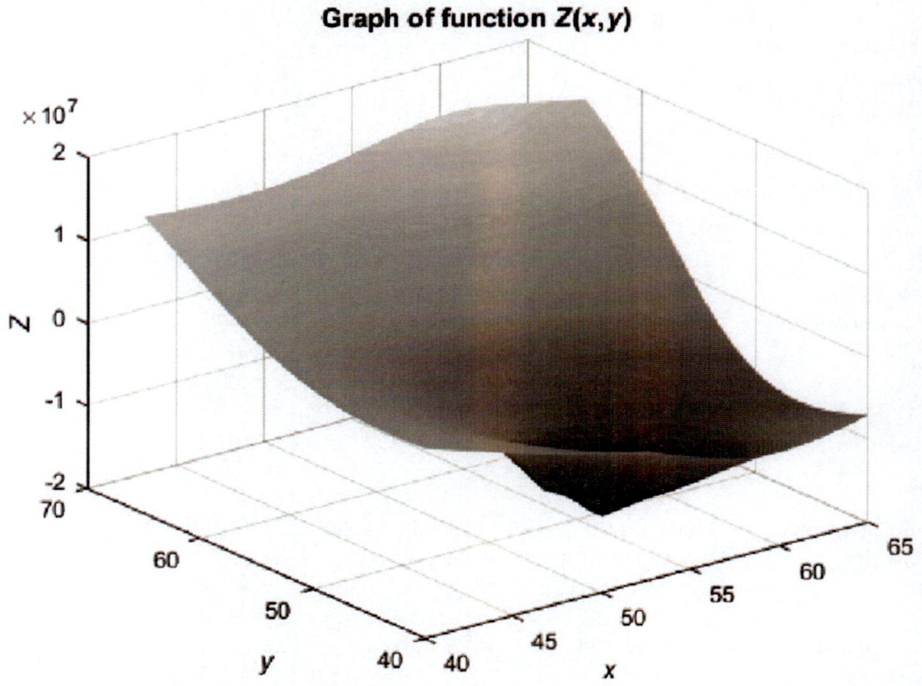

Fig. 2. Graph of optimizable function. *Source* Developed and compiled by the authors

Stage 4 – Using a genetic algorithm, simulated annealing and a pattern search. As noted earlier, we will solve the set location problem in the *Matlab* software using three different simulation modeling methods: a genetic algorithm, simulated annealing and a pattern search.

The results of the simulation modeling are shown in Table 3.

Table 3. Comparing the results of simulation modeling

Algorithm	x	y	$Z(x, y)$	Iterations
Genetic algorithm	49.12	55.79	−1.97	178
Simulated annealing	49.12	55.79	−1.97	1,610
Pattern search	49.12	55.79	−1.97	80

Source Developed and compiled by the authors

Comparison of the results of the three methods (Table 3) leads to the conclusion that all of them have allowed obtaining the same coordinates of the desired information and logistics center of the Volga Federal District, i.e. 55°47′N and 49°07′E. With a small margin of error, this corresponds to Kazan. In this case, the pattern search shows the fastest convergence – across 80 iterations. This is indicative of computational advantage of this method.

If the information and logistics center of the Volga Federal District is to be actually located in Kazan, this will considerably reduce transaction costs associated both with regulation of information flows and with traffic streams within the confines of the federal district under investigation. And this, in turn, will lead to a reduction in financial costs in the Volga Federal District and, most importantly, to an increase in the synergy effect of a large innovation system that brings together innovation and industrial clusters on a large territory of the entire federal district.

4 Conclusion

Finally, let us formulate the most important practical conclusions:

1. For the sustainable economic and financial development of the federal district, it is necessary to have new operating conditions that would lead to an additional synergy effect. One of such important conditions includes optimal information and logistic interaction of innovation and industrial clusters within one federal district.
2. With that in mind, the problem of locating the information and logistics center of the federal district clusters has been solved. In order to do that, appropriate simulation modeling technologies were used.
3. The search for the information and logistics center for innovation and industrial clusters in the federal district was based on minimizing the sum of distances from the planned center to the clusters. The maximum remoteness of the center from district nuclear power plants and atom cities was also taken into account. With this aim in view, a genetic algorithm, a simulated annealing method and a pattern search were used.
4. As a result, it has been obtained that the information and logistics center of the Volga Federal District should be the city of Kazan.
5. This would lead to a reduction in financial costs in the Volga Federal District and, most importantly, to an increase in the synergy effect of a large innovation system that brings together innovation and industrial clusters on a large territory of the entire federal district.

Acknowledgments. The research has been performed with financial support from the Russian Foundation for Basic Research within the confines of scientific project No. 19-010-00932 "Creating an innovation system development model for industrial regions under current conditions of social and economic development".

References

1. Ageev, A.A., Gimadi, E.Kh., Kurochkin, A.A.: Polynomial algorithm for solving a location problem in a chain having identical production capacities of enterprises. Discrete Anal. Oper. Res. **16**(5), 3–18 (2009)
2. Conn, A.R., Gould, N.I.M., Toint, P.: A globally convergent augmented Lagrangian algorithm for optimization with general constraints and simple bounds. SIAM J. Numer. Anal. **28**(2), 545–572 (1991). https://doi.org/10.1137/0728030

3. Conn, A.R., Gould, N.I.M., Toint, P.: A globally convergent augmented Lagrangian barrier algorithm for optimization with general inequality constraints and simple bounds. Math. Comput. **66**(217), 261–288 (1997). https://doi.org/10.1090/S0025-5718-97-00777-1
4. Gimadi, E.: Exact algorithm for solving an outerplanar location problem with improved time complexity. Writings Inst. Math. Mech. Ural Branch Russian Academy Sci. **23**(3), 74–81 (2017). https://doi.org/10.21538/0134-4889-2017-23-3-74-81
5. Huang, T., Bergman, D., Gopal, R.D.: Predictive and prescriptive analytics for location selection of add-on retail products. Prod. Oper. Manag. **28**(7), 1858–1877 (2019). https://doi.org/10.1111/poms.13018
6. Ingber, L., Rosen, B.: Genetic algorithms and very fast simulated reannealing: a comparison. Math. Comput. Model. **16**(11), 87–100 (1992). https://doi.org/10.1016/0895-7177(92)90108-W
7. Kalyanmoy, D.: Multiobjective Optimization Using Evolutionary Algorithms. Wiley, New York (2001)
8. Kolda, T.G., Lewis, R.M., Torczon, V.: A generating set direct search augmented Lagrangian algorithm for optimization with a combination of general and linear constraints (Report No. SAND2006-5315), Technical Report. Retrieved from Oak Ridge, Sandia National Laboratories (2006). https://public.ca.sandia.gov/~tgkolda/pubs/bibtgkfiles/SAND2006-5315.pdf. Accessed 27 Nov 2020
9. Lopatin, A.S.: Annealing method. Stoch. Optim. Inf. **1**, 133–149 (2005)
10. Murray, A.T., Tong, D.: Coverage optimization in continuous space facility siting. Int. J. Geogr. Inf. Sci. **21**(7), 757–776 (2007). https://doi.org/10.1080/13658810601169857
11. Vos, T., Heuvel, W., Phillipson, F., Huisman, D.: Using lampposts to provide urban areas with multiple services, Master's thesis (2016). https://www.google.ru/url?sa=t&rct=j&q=&esrc=s&source=web&cd=&ved=2ahUKEwjIosXV45frAhXOpYsKHYsBT0QFjACegQIAxAB&url=https%3A%2F%2Fthesis.eur.nl%2Fpub%2F34849%2FVos.pdf&usg=AOvVaw0U-81Ezqo1nVr_WwRUftsK. Accessed 27 Nov 2020

Optimization of Logistic, Informational and Economic-Financial Intercluster Interaction Using an Ant Colony Optimization Algorithm

Sergei N. Yashin[1(✉)], Egor V. Koshelev[1], Sergei A. Borisov[1], Natalya A. Yagunova[1], and Elena V. Romanovskaya[2]

[1] Lobachevsky State University of Nizhny Novgorod, Nizhny Novgorod, Russia
[2] Minin Nizhny Novgorod State Pedagogical University, Nizhny Novgorod, Russia

Abstract. Purpose : Propose a model of selecting determination for the optimal center of industrial agglomeration for innovation and industrial clusters located in one federal district.

Design/Methodology/Approach: Intercluster interaction of district regions is optimized in the following sequence: 1) logistic, 2) informational, and 3) economic and financial. At the first two stages, the problem is solved as a quadratic assignment problem using an ant colony optimization algorithm. The third stage is formal, because as a result of the conducted optimization of logistic and informational intercluster interaction, common transportation costs and population loss for the district are significantly reduced. The difference between the previous and new common transportation costs and population loss represents a value of the synergy effect of the performed optimization.

Findings: As a result of using the proposed model of intercluster interaction for the Volga Federal District (VFD), the optimum alternative of the industrial agglomeration Samara-Ulyanovsk-Dimitrovgrad has been obtained. It corresponds to the lowest transportation costs of RUB 24,987.77. Also, the result of information interaction optimization included the same industrial agglomeration, which had the lowest natural loss of population for the VFD amounting to 79,915 people per degree.

Originality/Value: Reducing the total transportation costs and the natural loss of population in the VFD in case of the location of an industrial agglomeration center in the resulting three cities will qualitatively improve the foresight process of evolution of a large innovation system represented by the district under review.

Keywords: Quadratic assignment problem · Intercluster interaction · Ant colony optimization algorithm · Industrial agglomeration

JEL Codes: C63 · E17 · O21 · O36

E. G. Popkova (Ed.): ISC 2020, LNNS 368, pp. 455–464, 2022.
https://doi.org/10.1007/978-3-030-93244-2_50

1 Introduction

Artificial intelligence technologies are already widely used in various management spheres. They allow for optimizing many complex processes using appropriate algorithms, including the ant colony optimization algorithm. It can be used to develop a management model for large territories, such as federal districts.

In this case, it is important to highlight the most important areas of interaction between regions that are part of the federal district. The regional economy becomes successful when innovation and industrial clusters located in the respective regions function effectively. In this regard, it is also appropriate to consider intercluster interaction of regions. Among its priority areas, we include economic and financial, information and logistics interaction.

In order to solve such questions, it is possible to use methods for solving the Quadratic Assignment Problem (QAP), among which the ant colony optimization algorithm is included. Its main advantage is that the artificial intelligence (ants) continues to look for a more optimal path even when the optimum is already found. This eventually allows for an actual significant reduction in the costs of economic-financial, information and logistics intercluster interaction, and, consequently, results in a new synergy effect of functioning of innovation systems of industrial regions in conditions of economic uncertainty. Such effects are extremely important in the foresight process of evolution of the federal district innovation system [1].

The ant colony optimization algorithm is one of the most powerful tools for solving the QAP. During the last but one decade, artificial ants were undergoing rapid development in the research community, mainly for optimization purposes. Monmarché, Guinand and Siarry have presented an overview of the situation in which the ant colony algorithms were found [2]. Artificial ants include methods for solving complex optimization problems and new trends for collective intelligence.

One of the most amazing behavioral patterns exhibited by ants is the ability of certain types of ants to find what programmers call the shortest paths. Biologists have experimentally shown that this is possible by using a bond based on pheromones only, a smelly chemical substance that ants can deposit. It was this behavioral pattern that inspired the programmers to develop algorithms for solving optimization problems. The first attempts in this direction appeared in the early 90's and can be regarded more like "toy" demonstrations, although they are important to indicate the general validity of the approach. Since then, these and similar ideas have attracted a steadily growing number of studies, and ant colony optimization (ACO) is one of the results of these studies. Indeed, ACO algorithms are the most successful and widely accepted algorithmic methods based on the behavior of ants. Their success is evidenced by the wide range of different problems to which they have been applied, and, moreover, by the fact that ACO algorithms are used for many problems among modern high-performance algorithms [3].

In swarm robotics, communication between robots is important. Inspired by biological swarms using pheromones, Ryusuke et al. have proposed the use of chemical compounds to implement the behavior of group foraging in robot swarms [4].

The study by Danielli and Oliveira offers a cellular automata ant memory (CAAM) model that controls a robot swarm while performing a foraging task in a previously known environment with nests [5].

It is noteworthy that the ant colony optimization algorithms are also used in financial studies. Thus, Zhang, Wang and Ji have proposed a hybrid system for predicting corporate bankruptcy [6]. The whole procedure consists of the following four stages: first, a sequential direct selection was used to extract the most important features; second, a rule-based model was selected to match the data set because it could represent a physical meaning; third, a genetic ant colony algorithm (GACA) was introduced; a fitness-scaling strategy and a chaotic statement were added into the GACA forming a new algorithm – fitness-scaling chaotic GACA (FSCGACA), which was used to find the optimal parameters for the rule-based model; and finally, a stratified *K*-fold cross-check method was used to enhance the model generalization. Experiments for modeling data from 1,000 corporations collected from 2006 to 2009 showed that the proposed model was effective. It highlighted 5 most important factors: "net income for broker equation", "quick ratio", "retained earnings for total assets", "shareholders' equity for total assets" and "financial expenses for sales". The total misclassification error of the proposed FSCGACA was 7.9% only, exceeding the results of the genetic algorithm (GA), the ant colony algorithm (ACA) and the GACA. The average time to calculate the model was 2.02 s.

Let us then describe already known ways of applying the ant colony optimization algorithm to solve the QAP. For instance, Tsutsui has suggested several types of parallel ACO algorithms with symmetric multiprocessing (SMP) to solve the QAP [7]. They include the "master-slave" model and the "island" model. The authors have evaluated each parallel algorithm with the condition that the runtime for each parallel algorithm and the basic sequential algorithm is the same. The results show that using the "master-slave" model with increased iteration of ACO algorithms is promising in solving the QAP.

ACO algorithms, more precisely "min-max ant systems" (MMAS) are usually used with two types of stochastic local search (SLS) methods: "2-opt local search" and "tabu local search". Mouhoub and Wang have suggested improving these two methods according to the ACO and QAP properties [8]. As applied to the "MMAS 2opt", they used a new random bypass strategy to avoid a rapid stasis in local optimization. In addition, a forward-looking strategy for a more thorough study of the surrounding area is proposed. As applied to the "MMAG tabu", a random walking strategy is also used to avoid getting stuck in the local optimum. To show the advantages of the proposed methods, the authors conducted experimental tests comparing the "MMAS 2opt" and "MMAS tabu" with and without improvements, respectively. The results show that an improved local method has higher performance in terms of quality of the returned solution than the original ones. Moreover, Mouhoub and Wang have also noted that the improved methods are superior to each other for different classes of problems [8].

The population-based ant colony optimization (P-ACO) algorithm differs from other ACO algorithms because of its implementation of pheromone regeneration. The P-ACO monitors the solution population, which serves as an archive of solutions generated by the ant colony. Pheromone regenerations in the P-ACO are only performed on the basis of solutions that pass in or out of the solution archive. The population-based scheme significantly reduces the computational time required to regenerate the pheromone compared to classical ACO algorithms, such as "Ant System". Oliveira et al. have studied the P-ACO behavior for solving the traveling salesman problem (TSP) and the quadratic assignment problem (QAP) [9, 10]. In particular, they investigated the impact of local search on the P-ACO parameters and performance. The results clearly show that the P-ACO is a very competitive tool, whose parameters and behavior strongly depend on the problem being solved and on whether or not local search is used.

2 Methodology

Considering the requirements of the QAP under review, let us present the stages of the ant colony optimization algorithm implementation as a three-stage model (Fig. 1). Below we will describe these stages in more detail.

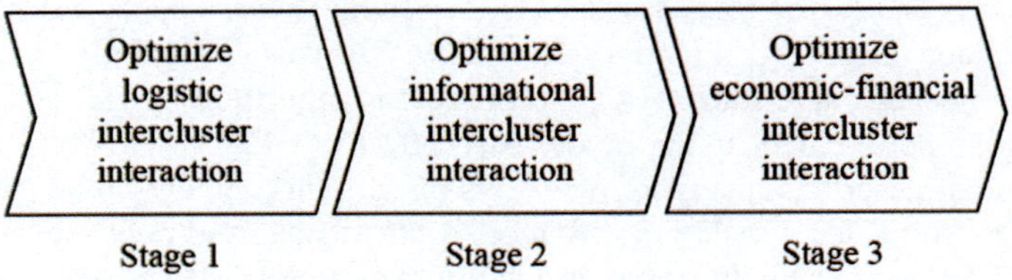

Fig. 1. Stages of implementation of ant colony optimization algorithm to optimize intercluster interaction. *Source* Developed and compiled by the authors

Stage 1 – Optimization of logistic intercluster interaction. The purpose of the presented model implementation is to determine the optimal center of the industrial agglomeration. It should include those innovation and industrial clusters, the location of which will reduce logistics, information and, consequently, economic and financial costs within one federal district. For this purpose, an agglomeration center is first sought to reduce the total logistics costs for the district. These include the cost of transporting cluster products by three modes of transport: 1) by trucks, 2) by rail, and 3) by air. Thus, preliminary clusters are divided into three branches according to the listed types of cargo transportation. The result is a matrix of average transport prices (weight matrix), each item of which shows the average price of transporting products from branch i to branch j.

That done, the ant colony optimization algorithm is used to search for a center of the industrial agglomeration of the federal district, for which it is necessary to find the minimum of the function on the set of clusters P.

$$\sum_{a,b \in P} w_{a,b} d_{f(a),f(b)},$$

where W - weight matrix for clusters a and b;

D - matrix of distances between clusters a and b;

$f(a)$ and $f(b)$ - permutations on the set P.

The result is three cities that are the geographic centers of clusters and correspond to the optimal locations of the centers of the three respective branches.

Stage 2 – Optimization of informational intercluster interaction. The branch centers obtained at the previous stage are taken into account at the stage of optimization of informational intercluster interaction. With that said, the new lists of clusters in the three branches are adjusted.

Then we look for geographic centers for the same three branches to minimize the natural decline in the population in the federal district. It is conditioned by information on the living standards in the regions provided that there is an industrial agglomeration in the district. To do this, the total natural population decline for each branch is first calculated. Then a matrix of the average natural decline in the population (weight matrix) is compiled, each item of which reflects the average amount of decline taking into account possible migrations of the population between clusters from branch i to branch j.

That done, the ant colony optimization algorithm is used to search for a center of the industrial agglomeration of the federal district over again. If it turns out the same set of three cities – the cluster centers, the problem is solved.

Stage 3 – Optimization of economic-financial intercluster interaction. This stage is already formal, because as a result of the conducted optimization of logistic and informational intercluster interaction, common transportation costs and population loss for the district are significantly reduced. The difference between the previous and new common transportation costs and population loss represents a value of the synergy effect of the performed optimization. As a result, the identification and location of the industrial agglomeration center will make it possible to qualitatively improve the foresight process of evolution of a large innovation system represented by the federal district.

3 Results

Let us illustrate the implementation of the presented model as exemplified by the Volga Federal District (VFD). We will only consider those regions of the district where the innovation and industrial clusters from the list of 25 pilot innovation territorial clusters approved by the Government of the Russian Federation are located. Let us describe them in Table 1. With this information, you can proceed to the model implementation stages.

Table 1. Innovative territorial clusters by region of the Volga Federal District

Region	Innovative cluster	Cluster center	Center coordinates
1. Nizhny Novgorod Region	1. Nizhny Novgorod industrial innovative cluster in automotive and petrochemical engineering (automo-tive industry)	Nizhny Novgorod	56°19′37″ N 44°00′27″ E $x = 44.1$ $y = 56.33$
	2. Sarov innovative cluster (nuclear technologies)	Sarov	54°56′ N 43°19′ E $x = 43.32$ $y = 54.93$
2. Republic of Mordovia	3. Energy-efficient lighting engi-neering and smart lights (micro-electronics and instrumentation)	Saransk	54°11′ N 45°11′ E $x = 45.18$ $y = 54.18$
3. Ulyanovsk Region	4. ULYANOVSK-AVIA Scientific and Educational Production Cluster consortium (aircraft engineering)	Ulyanovsk	54°19′ N 48°22′ E $x = 48.37$ $y = 54.32$
	5. Dimitrovgrad nuclear innovation cluster (nuclear technologies)	Dimitrov-grad	54°14′ N 49°35′ E $x = 49.58$ $y = 54.23$
4. Samara Region	6. Innovative territorial aerospace cluster (space industry, aircraft engi-neering)	Samara	53°11′ N 50°07′ E $x = 50.12$ $y = 53.18$
5. Perm Territory	7. NOVY ZVEZDNY Technopolis innovative territorial cluster of rocket engine building (space industry, aircraft engineering)	Perm	58°00′50″ N 56°14′56″ E $x = 56.25$ $y = 58.01$
	8. Photonics (microelectronics and instrumentation)	Perm	
6. Udmurt Republic	9. Udmurt machine building cluster (automotive industry)	Izhevsk	56°51′11″ N 53°12′44″ E $x = 53.21$ $y = 56.85$
7. Republic of Tatarstan	10. Kama innovative territorial pro-duction cluster (automotive industry)	Kazan	55°47′27″ N 49°06′52″ E $x = 49.12$ $y = 55.79$
8. Republic of Bashkor-tostan	11. Petrochemical territorial cluster (chemical industry)	Ufa	54°44′ N 55°58′ E $x = 55.97$ $y = 54.73$

Source Developed and compiled by the authors

Stage 1 – Optimization of logistic intercluster interaction. Taking into account that one degree makes up approx. 111 km for latitude and longitude, let us estimate the cost of transporting the products of the clusters.

1. The price of transportation by truck per 1 km is 33 rubles (3,663 rubles per 1°).
2. The price of transportation by rail per 1 km is 29 rubles (3,219 rubles per 1°).
3. The price of transportation by air per 1 km is 103 rubles (11,433 rubles per 1°).

Then the matrix of average prices for transportation from branch i to branch j will look like this.

$$\begin{pmatrix} 0 & 3,441 & 7,548 \\ 3,441 & 0 & 7,326 \\ 7,548 & 7,326 & 0 \end{pmatrix}.$$

The 1st branch includes clusters 1, 3, 8, 9, 10, the 2nd branch includes clusters 4, 6, 7, 11, and the 3rd branch includes clusters 2, 5 from Table 1.

After that, we solve the quadratic assignment problem (QAP) for the VFD. That is, we look for an optimal industrial agglomeration for it. For solving, we use the ant colony optimization algorithm in the *Matlab* software.

The results are shown in Fig. 2. They mean that the center of the 1st branch should be in Samara, the 2nd branch – in Ulyanovsk, and the 3rd – in Dimitrovgrad. In this case, the total transportation costs for the VFD will be minimum, i.e. 24,987.77 rubles.

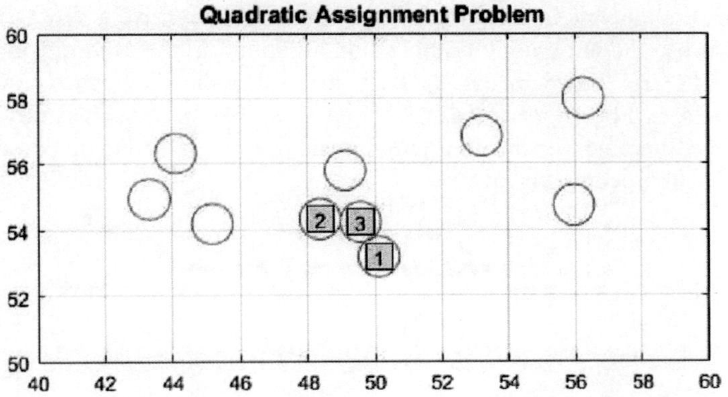

Fig. 2. Optimal industrial agglomeration Samara-Ulyanovsk-Dimitrovgrad. *Source* Developed and compiled by the authors

Stage 2 – Optimization of informational intercluster interaction. Table 2 collects information on the natural decline in the population for the VFD regions in 2019. Then we adjust the new lists of clusters in three branches based on the results of the previous analysis stage. Thus, regions 1, 2, 4, 6, 7 are included into the 1st branch, regions 5, 8 are included into the 2nd branch, and region 3 is included into the 3rd branch from Table 2.

Table 2. Information on the population of regions of the Volga Federal District with clusters

Region	Natural population decline in 2019 (people)
1. Nizhny Novgorod Region	18,107
2. Republic of Mordovia	4,536
3. Ulyanovsk Region	6,197
4. Samara Region	12,454
5. Perm Territory	7,777
6. Udmurt Republic	3,269
7. Republic of Tatarstan	180
8. Republic of Bashkortostan	7,348

Source Developed and compiled by the authors

All in all, the natural loss of population for the branches is 38,546, 15,125 and 6,197 people, respectively.

Then the average natural decline matrix taking into account possible population migrations between the clusters from branch i to branch j will look like this.

$$\begin{pmatrix} 0 & 26,835.5 & 22,371.5 \\ 26,835.5 & 0 & 10,661 \\ 22,371.5 & 10,661 & 0 \end{pmatrix}.$$

Following this, we look for geographical centers for the same 3 branches to minimize the natural decline in the population in the VFD, i.e. solve a new QAP. For that end, we also use the ant colony optimization algorithm in the *Matlab* software.

The results are shown in Fig. 3. They mean that the previous set of cities, the cluster centers, has been preserved, i.e. the problem has been successfully solved. At the same time, there will be the minimum natural decline in the population in the entire VFD, i.e., 79,915 people per degree.

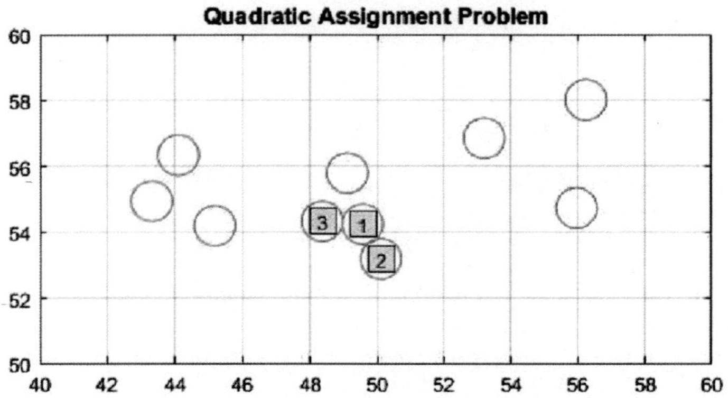

Fig. 3. Optimal industrial agglomeration Dimitrovgrad-Samara-Ulyanovsk. *Source* Developed and compiled by the authors

Stage 3 – Optimization of economic-financial intercluster interaction. Reducing the total transportation costs and the natural loss of population in the VFD in case of the location of an industrial agglomeration center in the resulting three cities will qualitatively improve the foresight process of evolution of a large innovation system represented by the district under review.

4 Conclusion

In conclusion, let us formulate the most important findings.

1. The federal district economy becomes successful when innovation and industrial clusters located in its respective regions function effectively. In this regard, it is appropriate to analyze intercluster interaction of the regions in the following sequence: 1) logistic, 2) informational, and 3) economic and financial.
2. In order to solve such questions, it is possible to use methods for solving the quadratic assignment problem, among which the ant colony optimization algorithm is included. Its main advantage is that the artificial intelligence (ants) continues to look for a more optimal path even when the optimum is already found. This allows for a significant reduction in the costs of intercluster interaction, and, consequently, results in a new synergy effect of functioning of innovation systems of industrial regions in conditions of economic uncertainty. Such effects are extremely important in the foresight process of evolution of the federal district innovation system.
3. The purpose of the presented model implementation is to determine the optimal center of the industrial agglomeration. It should include those innovation and industrial clusters, the location of which will reduce logistics, information and, consequently, economic and financial costs within one federal district. For this purpose, an agglomeration center is first sought to reduce the total logistics costs for the district.
4. Then we look for geographic centers to minimize the natural decline in the population in the federal district. It is conditioned by information on the living standards in the regions provided that there is an industrial agglomeration in the district.
5. As a result of the conducted optimization of logistic and informational intercluster interaction, common transportation costs and population loss for the district are significantly reduced. The difference between the previous and new common transportation costs and population loss represents a value of the synergy effect of the performed optimization.

This approach will enable government agencies and their experts to make better decisions regarding the foresight of development of large territories of the country, such as federal districts.

Acknowledgments. The research has been performed with financial support from the Russian Foundation for Basic Research within the confines of scientific project No. 19–010-00932 “Creating an innovation system development model for industrial regions under current conditions of social and economic development”.

References

1. Yashin, S.N., Koshelev, E.V., Sukhanov, D.A.: Foresight of federal district innovation system development using a multi-objective genetic algorithm. Finan. Credit **26**(6), 1208–1227 (2020)
2. Monmarche, N., Guinand, F., Siarry, P.: Artificial Ants. Wiley-ISTE (2010)
3. Dorigo, M., Stültze, T.: Ant Colony Optimization. The MIT Press, Cambridge (2004)
4. Fujisawa, R., Dobata, S., Sugawara, K., Matsuno, F.: Designing pheromone communication in swarm robotics: group foraging behavior mediated by chemical substance. Swarm Intell. **8**(3), 227–246 (2014). https://doi.org/10.1007/s11721-014-0097-z
5. Danielli, L.A., Oliveira, G.: A cellular automata ant memory model of foraging in a swarm of robots. Appl. Math. Model. **47**, 551–572 (2017). https://doi.org/10.1016/j.apm.2017.03.021
6. Zhang, Y., Wang, S., Ji, G.: A rule-based model for bankruptcy prediction based on an improved genetic ant colony algorithm. Math. Prob. Eng., 1–10 (2013). https://doi.org/10.1155/2013/753251
7. Tsutsui, S.: Parallel ant colony optimization for the quadratic assignment problems with symmetric multi processing. In: Dorigo, M., Birattari, M., Blum, C., Clerc, M., Stützle, T., Winfield, A.F.T. (eds.) ANTS 2008. LNCS, vol. 5217, pp. 363–370. Springer, Heidelberg (2008). https://doi.org/10.1007/978-3-540-87527-7_38
8. Mouhoub, M., Wang, Z.: Improving the ant colony optimization algorithm for the quadratic assignment problem. In: 2008 IEEE Congress on Evolutionary Computation (IEEE World Congress on Computational Intelligence), Hong Kong, pp. 250–257 (2008). https://doi.org/10.1109/CEC.2008.4630807
9. Oliveira, S., Hussin, M.S., Stützle, T., Roli, A., Dorigo, M.: A detailed analysis of the population-based ant colony optimization algorithm for the TSP and the QAP. In: Genetic and Evolutionary Computation Conference, GECCO 2011 - Companion Publication, pp. 13–14 (2011). https://doi.org/10.1145/2001858.2001866
10. Oliveira, S., Hussin, M.S., Roli, A., Dorigo, M., Stützle, T.: Analysis of the population-based ant colony optimization algorithm for the TSP and the QAP. In: 2017 IEEE Congress on Evolutionary Computation (CEC), San Sebastian, pp. 1734–1741 (2017). https://doi.org/10.1109/CEC.2017.7969511

Application of Lean Construction Management Innovation Economy Development Technology in Industrial Parks

Svetlana N. Kuznetsova(✉), Elena V. Romanovskaya, Natalia S. Andryashina, Elena P. Kozlova, and Tatiana E. Lebedeva

Minin Nizhny Novgorod State Pedagogical University, Nizhny Novgorod, Russia

Abstract. This study explores the essence of lean manufacturing in industrial parks. Currently, the development of spaces for the location of production facilities is a topical area, therefore it is recommended to introduce the lean construction of industrial parks approach.

Purpose: Development of implementation and functioning of lean manufacturing is carried out by residents of industrial parks. The work focuses on lean logistics, that is, on the use of a pull system that unites all participants in the value chain.

Methodology: The use of lean manufacturing methods, the development and improvement of production processes provide a positive result for industrial parks with little investment.

Findings: A decision was made to create a production stream in an industrial park on the basis of an integrator, where finished products are manufactured. It is necessary to create a matrix of a family of products, to combine into one family of residents who create important parts, components with similar processing stages.

Originality: The use of lean manufacturing tools will provide more independence for residents and make management more efficient. Lean production in industrial parks will allow implementing import substitution programs and creating new ones. The research problem is the creation of industrial parks, but with a low occupancy rate by residents, as well as a long cycle of construction of industrial parks. The authors see the solution to this problem in the application of Lean Construction Management technologies in the territories of industrial parks.

Keywords: Lean manufacturing · Industrial parks · Savings · Residents · Lean construction

JEL Codes: L5 · O1 · N6

E. G. Popkova (Ed.): ISC 2020, LNNS 368, pp. 465–471, 2022.
https://doi.org/10.1007/978-3-030-93244-2_51

1 Introduction

Lean manufacturing refers to an approach to organizing production systems aimed at minimizing waste in order to create the highest possible value. When implementing the research project, the key goals were selected to achieve - inventory turnover, process time and production. With the help of lean manufacturing tools, diagnostics of the existing production was carried out by residents of industrial parks [1].

Let us consider in the study the experience of the resident of the Novocherkassk Industrial Park LLC «BONUM», which, thanks to lean production, increased the output of products by more than 2.5 times.

As a result, it was possible to achieve an increase in output by more than 2.5 times, stocks of work in progress decreased by 35%, and the process time - by more than 71% [2].

2 Methodology

Today, the lean manufacturing methodology is deeply developed and is applied in completely different areas. A new theoretically grounded methodology of construction management, like lean construction, began to develop in the mid-1990s [3].

Construction projects are managed by dividing the project into separate tasks, assessing the time and financial costs required to complete each task, applying the critical path method to determine the logical sequence of tasks, attracting and monitoring the work of performers [4].

To increase productivity up to 85% in industrial park construction, the authors propose to combine technology, staff motivation, and lean construction methods.

Management is carried out by tracking status and taking corrective action when deviations from the plan, standards or generally accepted practice are detected. More than 90% of the world's infrastructure projects are delivered late or in excess of the budget, the potential for productivity growth in the construction industry due to improved management and labor organization is estimated at 60% [5].

The lean construction approach, on the other hand, is based on management through a systematic process of preparing tasks for the start of execution, combined with an increase in the responsibility of performers for the amount of work undertaken and the final result of the project [6].

Lean Construction Management methods eliminate sources of waste from the construction process. The Lean Construction Management concept is used by many industrial parks, as well as by their residents, for example, GAZ Group, notes savings on equipment operation in the amount of 3 million rubles as a result of the application of Lean-technologies [7].

There are many tools for lean construction, which are more common in foreign practice [8]. The study highlights the following most elaborated tools (Table 1).

The methodology of most of the lean building tools is based on the following basic principles (Table 2): reducing the level of process volatility; preparation of conditions for efficient and uninterrupted work performance; coordinating the work schedule of residents; monitoring the rate of implementation of the plan [9].

Table 1. Developed lean construction management tools

Last Planne System	A comprehensive management system based on the involvement of direct performers (as a rule, line managers responsible for work on the site), which allows to improve the quality of planning, involvement and responsibility for the result
Just-in-time	Products and materials are produced and delivered on time. Eliminate downtime and stockpiling between operations, and reduce inventory and transfer costs
Pull Scheduling/Planning	Joint planning of the construction process. The approach allows you to identify all the necessary preparatory measures, necessary actions and potential problems for each specific task
Visualization tools / management	Visualization of organizational processes or progress status using diagrams for making management decisions. This also includes, for example, the marking of tools or work areas and a schematic representation of the technological processes on the construction site
Concurrent Engineering	Simultaneous development of successive stages, including the study of construction technology. Design time is reduced, including due to intermediate approvals
Teamwork and partnering	Building team spirit on the project and building long-term relationships with contractors. The level of trust between participants and interest in the final result increases
Value Based Management / Value Streaming Mapping	Study and visualization of flows in the course of creating value as materials move through construction processes. Solutions based on detailed analysis of processes can improve their efficiency

Source Developed and compiled by the authors

Table 2. Lean construction management as an example of the practical application of the lean construction tool

Improving risk manageability	64%
Increase in profits / decrease in costs	77%
Increase in productivity	74%
Shortening the project schedule	84%
Improving the quality of construction work	77%
Increased customer satisfaction	80%

Source Developed and compiled by the authors

As a practical application of the lean manufacturing methodology, the company has developed an approach to project management in the real estate sector - Lean Construction Management. For example, within a construction site, the approach includes three stages [10]:

1. General analysis of processes (Overall Process analysis) - various strategies for the implementation of the project are being worked out, a directive schedule is being developed, containing a list of main works and key dates of the project;
2. Planning of processes (Process planning) - the project is divided into zones based on the technical features of the facility, resources and technologies; the indicators of the work performed are determined;
3. Detailed planning.

3 Results

The total number of resident personnel annually increases by 15.4% on average, while 92.7% of residents are small and medium-sized businesses [11]. The number of jobs created by residents exceeded 110 thousand people (Table 3).

Table 3. Activities of residents

Research and development	R&D expenses of a resident	184 600 rubles
	Average cost of a resident on R&D	3040000 rubles
	Number of intellectual property objects	1523 units
Production creation	Total investment of residents	219 700 000 rubles
	Specific investment volume of residents	219 700 000 rubles/ thousand m^2
	Attracted investments of residents for 1 rub. budget investments in infrastructure	9 300 000 rubles
Production and sale of products	The total volume of production of import-substituting products	48,5 billion rubles
	Total exports	21,3 billion rubles
	The amount of tax deductions from residents per 1 rub. total budget investment in infrastructure	1,7 rubles

Source Developed and compiled by the authors

The structure of areas by type (2465.9 thousand m^2): production facilities 51.1%; office premises 22.6%; technological infrastructure premises 4.7%.

The cost of renting 1 m^2 of premises: industrial premises 3711.9 rubles; office space 5007.7 rubles; premises of technological infrastructure 6276.9 rubles.

The study identifies the most effective management companies, infrastructure-equipped sites for the placement and development of high-tech companies [12].

The study presents the average efficiency indicators of Lean technologies implemented in industrial parks (for example, instrumentation): savings of 52 million rubles in 6 months; release of 25% of production space; reduction of the production cycle from 9 to 1 day [13].

Let's consider the results of the rating on the example of the technoparks of the Nizhny Novgorod region: Technopark Ankudinovka. The final score is 4.916 (group A +. The highest level of operational efficiency); Technopark Sarov. The final score is 4.154 (group A. High level of functioning efficiency) [14].

4 Conclusion

When designing industrial parks, it is rational to use building information modeling technologies.

For example, the Ural (Miass) automobile plant, which is part of the GAZ Group, implements Lean technologies, as a result of which it provides savings of 300–400 million rubles annually. Automobile plant KamAZ applies the principles of lean production [15].

The Nizhny Novgorod Region in 2019 took part in the competition of the Ministry of Economic Development of Russia. The selection involved 48 projects from 25 regions of the Russian Federation. Based on the selection results, 17 projects were selected, including three projects from the Nizhny Novgorod region [16].

- Creation of an industrial technopark «Electrotransport» on the territory of AZ GAZ, on a single site of which the production of a line of innovative electric vehicles GAZ and components for it is being realized.
- Creation of an industrial technopark on the territory of the Drobmash plant.
- A project for the development of energy capacities of the Sarov Technopark of existing residents.

The total cost of projects from the Nizhny Novgorod region will amount to 1,733.6 million rubles including: 1,039.78 million rubles - attracted funds from the federal budget [17].

For the implementation of projects it is necessary

- Build value stream maps;
- to build a procedure for assessing resident-participants, as a result of which, with the least investment, residents gain advantages: increased labor productivity, reduced production cycle time, increased quality, equipment operation time in good working order, release of production space, reduced scrap, savings, reduced investment requirements; reduction of the time frame for the implementation of investment projects [18].

Lean implementation of projects is a long transformation process, and cost reduction and an early transition from construction to production is a real prospect for applying the principles of lean project implementation [19].

References

1. Andryashina, N.S., Romanovskaya, E.V., Kuznetsov, V.P., Kuznetsova, S.N., Kozlova, E. P.: Development of the production system of the enterprise on the example of workshop of the PJSC "GAZ." In: Popkova, E.G. (ed.) Growth Poles of the Global Economy: Emergence, Changes and Future Perspectives. LNNS, vol. 73, pp. 747–758. Springer, Cham (2020). https://doi.org/10.1007/978-3-030-15160-7_75
2. Bezrukova T.L., Morkovina S.S., Sui Jiang M.: The development of territorial-industrial clusters: methodological foundations of management, formation factors, support vectors (monograph) (2019)
3. Bodrunov, S.D.: The Russian economic system: the future of high-tech material production. Econ. Rev. Russia **2**(40), 5–16 (2018)
4. Donichev, O.A.: Economic integration as a factor in reducing differentiation in the development of regions. Bull. Vladimir State University named after Alexander Grigoryevich Nikolay Grigoryevich Stoletov Ser. Econ. Sci. **1**, 4–14 (2019)
5. Kharchenko S.V.: Modern information models of accounting support management and profit formation **12**, 377–382 (2018)
6. Knyaginin, V.N., Shchedrovitsky, P.G.: The industrial policy of Russia: who will pay the costs of globalization. Ser. «Ideologies» Europe, 160 (2018)
7. Konishchev, A.S.: Integration of the process of operational logistics planning at the enterprise. Logistic Syst. Glob. Econ. **4**, 142–146 (2019)
8. Kulikov, A.: Lean manufacturing: how LEAN technologies will help your business (2020). https://new-retail.ru/business/berezhlivoe_proizvodstvo_kak_lean_tekhnologii_pomogut_vashemu_biznesu3333/. Accessed 14 Mar 2021
9. Kuznetsova, S.N., Kuznetsov, V.P.: An effective tool for solving conversion tasks. Econ. Mange. Prob. Solutions **1**(10), 113–116 (2018)
10. Kuznetsova, S.N., Kuznetsov, V.P., Egorova, A.O.: Modern logistics systems of industrial sites. Storage Process. Agric. Raw Mater. **1**, 71–74 (2018)
11. Kuznetsova, S.N., Kuznetsov, V.P., Kozlova, E.P., Potashnik, Y.S., Tsymbalov, S.D.: Transformational period of russian development in the digital economy. In: Popkova, E.G., Sergi, B.S. (eds.) ISC 2019. LNNS, vol. 91, pp. 663–669. Springer, Cham (2020). https://doi.org/10.1007/978-3-030-32015-7_74
12. Mansurova, A.F.: Integration as a category of economic science. Econ. Innov. Qual. Control. **4**(9), 165 (2018)
13. Mizikovsky, I.E., Kozlova, E.P., Garina, E.P., Garin, A.P., Kuznetsova, S.N.: Improvement of cost allocation processes for maintenance and operation of industrial enterprise equipment. In: Popkova, E.G. (ed.) Growth Poles of the Global Economy: Emergence, Changes and Future Perspectives. LNNS, vol. 73, pp. 775–784. Springer, Cham (2020). https://doi.org/10.1007/978-3-030-15160-7_78
14. Potashnik, Y.S., Artemyeva, M.V., Kuznetsova, S.N., Garin, A.P., Letyagina, E.N.: The status and trends in innovative activity of industrial enterprises of nizhny novgorod region. In: Popkova, E.G. (ed.) Growth Poles of the Global Economy: Emergence, Changes and Future Perspectives. LNNS, vol. 73, pp. 525–534. Springer, Cham (2020). https://doi.org/10.1007/978-3-030-15160-7_53
15. Romanovskaya, E.V., Garina, E.P., Andryashina, N.S., Kuznetsova, S.N., Artemyeva, M.V.: Studying the experience of reengineering business processes in the practice of domestic enterprises: problems and prospects of application. In: Popkova, E.G. (ed.) Growth Poles of the Global Economy: Emergence, Changes and Future Perspectives. LNNS, vol. 73, pp. 517–524. Springer, Cham (2020). https://doi.org/10.1007/978-3-030-15160-7_52

16. Salikov, Y.A., Barzenkova, A.S., Zenin, A.A.: A conceptual approach to the implementation of multi-purpose regional and sectoral integration based on cluster policy. In: Technologies of the Food and Processing Industry of the Agro-Industrial Complex - Healthy Food Products, vol. 3, pp. 76–81 (2018)
17. Stepchenko, T.S.: Lean-technologies in enterprise management. Modern management technologies **7**(55) (2019). Article number: 5508. https://sovman.ru/article/5508/. Accessed 14 Mar 2021
18. Vikulov, V.A., Butrin, A.G.: Algorithm for the formation and management of an industrial enterprise interactions with suppliers of material resources and consumers of finished products. Basic Res. **8**(part 5), 1141–1145 (2019)
19. Yanovsky, V.V., Proskurnikov, S.V.: Methodological aspects of the impact of investments on business value. Bull. Belgorod University Cooperation Econ. Law **4–2,** pp. 165–169 (2017)

A Lifecycle Approach to Cluster Competitiveness Management

Alla G. Zinchenko[1(✉)], Elena L. Neznahina[1], Sergey B. Sorokin[1], Elena V. Kornilova[2], and Ilya G. Ilyushchenko[3]

[1] Nizhny Novgorod State Technical University named after R.E. Alekseev, Nizhny Novgorod, Russia
alla@zinchenko.com, sorokin@nntu.ru
[2] Lobachevsky State University, Nizhny Novgorod, Russia
[3] Volga State University of Water Transport, Nizhny Novgorod, Russia

Abstract. As a form of organization, clusters act as a means of increasing the competitiveness of the region's economy, ensuring the transition to processes with higher added value. In the Russian realities, the cluster is considered for practical purposes and is represented as an object of support within the framework of regional development strategies. The purpose of the article is to develop a tool for managing the competitiveness of a cluster based on a life cycle model.

Overall, an approach to the cluster is considered as a dynamic system. Analysis of the functioning from a standpoint of the concept of life cycles, as an organizational and economic entity, makes it possible to single out many processes that form the life cycle of the competitiveness.

An approach to the formation of a list of parameters that determine competitiveness has been substantiated. It is shown that at each stage of the life cycle of the competitiveness indicators vary and require different management decisions.

For the analysis of competitiveness, a three-level system of cluster presentation is proposed. Then the competitiveness is presented through the satisfaction of each level of the cluster. Characteristics of the cluster influencing the assessment of satisfaction are determined. Also, formulated the benefits derived by each cluster level. Based on the resulting benefits, described the tools, which provide recommendations for management decisions at various stages of the life cycle of a cluster to improve its competitiveness.

Keywords: Cluster · Cluster effect · The lifecycle of a cluster · Cluster competitiveness · Competitiveness cluster parameters

JEL Codes: R110 · O4 · M1

E. G. Popkova (Ed.): ISC 2020, LNNS 368, pp. 472–481, 2022.
https://doi.org/10.1007/978-3-030-93244-2_52

1 Introduction

In today's economy, organizations are forced to look for ways to ensure competitiveness. The search for ways to improve competitiveness has led to the widespread use of the cluster approach. As a form of organization, the clusters act as a means of improving the competitiveness of the regional economy. They provide a transition to processes with higher added value, through the establishment of cooperation between enterprises of various fields. The aim of this work was to develop an approach to managing the competitiveness of the cluster. However, the challenge lies in the fact that there is still no common approach to the definition of clusters, not only for professionals, but also at the level of normative-legal acts [1].

There are at least two described approaches to studying a cluster [2]:

1. In order to analyze the cluster presented as an alternative object of some plants or industries
2. For practical activities, the cluster is represented as an object of support within the framework of regional development strategies

A. Marshal also noted the connection between the concentration of enterprises and their productivity. He formulated three advantages of localized production of organizations in a particular industry: concentration of qualified specialists and employers with similar specialization, specialization of suppliers, and the rapid dissemination of knowledge. Later this idea was developed by M. Porter, noting that successful organizations tend to be located in a close area and designated the concept of a cluster as "the geographical concentration of competing and cooperating interconnected companies and related institutions." Porter noted the positive effect of geographic concentration, intra-cluster competition, self-regulation, and formulated the factors that form a cluster. The combination of these factors creates the effect of increased competitiveness within the cluster. Practice has shown that the factors and their influence can be different for different clusters. Based on a certain combination of characteristics, six models of territorial associations are distinguished in the world. Each model is named according to the region of formation and has its own characteristics. For example, in the North American model, competition is strongest between firms, and most linkages are market-driven. The Italian industrial cluster is made up of a large number of small firms that come together in various associations to improve their competitiveness. The modern Russian model - market relations and competition are minimized, production is concentrated on large firms. This has a very large role of the state, not only in the area of legal control over its operations, but also in the creation, direction, and development conditions [3].

Also worth noting is that the factors that form the cluster, and their influence can change over time. The changes can have either positive or negative effect until the destruction of the cluster. In order, to ensure the cluster effect, it is necessary to strengthen the factors that have a positive impact on the development of the cluster and reduce possible negative effects, i.e. manage factors according to the stages of the cluster life cycle.

2 Methodology

The choice of control actions on a cluster in order to enhance and prolong "cluster effect" must ensure the equilibrium status of all cluster processes. In addition, it is possible to talk about managing the competitiveness of a cluster by regulating the influence of individual factors, taking into account the stage of the cluster's life cycle.

Today there is no single model describing the development of a cluster. A large number of interrelated elements dynamically changing within a cluster can explain the variety of approaches. This responds to internal and external influences by the fact that a cluster itself is a complex dynamic system. Therefore, a lifecycle model can be applied to analyze the competitiveness of the cluster.

The authors distinguish different stages of cluster development [4–7], but more often are the following:

- Agglomeration - in the region, there are several enterprises of the same industry and other organizations.
- An emerging cluster - a number of agglomeration participants begin to cooperate with each other, forming connections.
- Developing cluster - there are more cluster members, not only industry enterprises are added, but also other companies. Links are beginning to form between enterprises and organizations in various fields of activity.
- Mature cluster - the number of participants has reached a certain level. There are foreign relations, new ventures through startups, joint ventures, etc.
- Transformation - in order to avoid stagnation and decay, the cluster must adapt to external and internal changes. Transformation can lead to the creation of one or more new clusters, centered, perhaps, on other activities, or at least a change in internal or external processes.

In order to analyze the competitiveness, the stage of the Agglomeration can be excluded, since at this stage the connections that will determine the structure of the cluster have not yet been formed.

In the Russian realities of the cluster is seen, first of all, in order to practice and as an object of support in the framework of regional development strategies. Therefore, the mechanism for creating and developing clusters has its own characteristics; in particular, there are methods to stimulate its development [8].

Based on the "classic" definition of the cluster, our research objectives, and taking into account the Russian realities of the formation and management, we formulate the concept of a cluster.

A cluster is an organizational and economic form of interaction between organizations participating in each other's value chain, in which methodological, organizational, expert-analytical, and information support of development is carried out by a specialized organization.

We represent the cluster as a hierarchical system:

Macro level — Government and regional governing bodies.

Meso - Management Organization

Micro level - Organization, in a cluster

Each participant of the system has the characteristic expected benefits from participating. Let us form a summary table (Table 1) of the expected benefits by levels of the cluster system.

Table 1. Expected benefits of cluster participation for each level of the cluster system

Level of economic system	Composition of system levels	Expected benefits
Macro-level	Government and regional governing bodies	Economic development of the territory
Meso	Management organization	Cluster industry development
Micro-level	Organizations that are part of the cluster	Enhancing competitiveness

Source Compiled by the authors

Table 2 shows methods for estimating the expected benefits for each level of the system.

Table 2. Definition of methods for assessing expected benefits

Level of economic system	Method for assessing the expected benefits
Macro-level	Estimated to achieve economic parameters of cluster development, set in the development strategy
Meso	An evaluation of the organizational, administrative, and accompanying measures
Micro-level	Assessment of customer satisfaction of the manufactured product

Source Compiled by the authors

For the management of competitiveness, it is necessary to determine which cluster characteristics will affect the satisfaction scores.

Different researchers identify different underlying characteristics. In studies [9], the internal dynamics of a cluster are based on the change in heterogeneity between cluster residents and how residents use heterogeneity. Heterogeneity refers to technological differences between firms, different management models, and the possession of unique experiences and information.

In their works [10] and later [11], the characteristics of the cluster were identified, which are different at each stage of the life cycle. In their opinion, the fundamental characteristic underlying the functioning of any cluster is the value chain that arises between the organizations of the cluster. The second characteristic is strategic relationships. Developing joint strategies or coordinating existing ones. The third characteristic is the dynamics of the cluster or the change in the number of organizations. Fourth - general areas of activity: education in the region, R&D, marketing, infrastructure construction, standardization of products and technologies. The drivers of success and the role of the government are also indicated by the authors as characteristics that characterize each stage.

Let us define the characteristics of the cluster, which will determine the satisfaction scores for each level of the cluster (Table 3).

Table 3. Characteristics that determine satisfaction from participation in a cluster for different levels

Level of economic system	Characteristics that determine satisfaction levels for different cluster
Macro-level	**External economic effect of the cluster:** The total amount of exports of organizations included in the cluster The total amount of tax deductions of organizations included in the cluster The total amount of revenue of organizations included in the cluster from internal trade Increase in revenue per organization for the selected period (in %) **Cluster dynamics:** The cluster size (the number of organizations participating) The size of the cluster core (the number of organizations included in the core) The total number of cluster organizations industry in the region Dynamics of changes in the number of companies in a cluster Number of employees The number of highly skilled workers Dynamics of changes in the number of highly qualified workers
Meso	The level of strategic relations (%) **Satisfaction of the managing organization:** Quality/satisfaction with cluster benefits Quality/Satisfaction with cluster processes Quality/Cluster staff satisfaction
Micro-level	Consumer satisfaction Financial performance: Profit Profitability Capital structure

Source Compiled by the authors

1. Characteristics showing the degree of satisfaction of the government and regional authorities. They will reflect the economic, social, and institutional areas.

- The external economic effect of the cluster will show the competitiveness of goods produced in the domestic and foreign markets.

For the analysis, we will use the parameter values:

- The total amount of exports of organizations included in the cluster (rubles).
- The total amount of tax deductions of organizations included in the cluster (rubles).
- The total amount of revenue of organizations included in the cluster from domestic trade (rubles).

In order to assess competitiveness, we will use the dynamics of changes in these parameters for the period under study (for example, for a year) to the dynamics of the cluster (change in the number of organizations included in the cluster). We will receive an increase in revenue in rubles per organization.

- Cluster dynamics is the change in the number of organizations and workers in a cluster. Let us consider separately the change in the number of companies and the number of employees. The growth in the number of organizations participating in the cluster shows their interest in the cluster, and therefore shows its effectiveness and competitiveness. A rise in the number of jobs and skilled workers in the cluster will show not only the social orientation but the increase in tax deductions, which ultimately will give the satisfaction of regional governments. It is determined by the ratio of all organizations in the region's industry to the number of participants in the cluster core.

  For the analysis, we will use the parameter values:

– Size (units) - the total number of organizations participating in the cluster.
– Cluster core size (units) - the number of organizations in the "cluster" industry in the region that are part of the cluster
– Total number of cluster industry organizations in the region (units)

The dynamics of the cluster will tend to 1 at the stage of maturity, will show positive dynamics at the stage of growth, and negative at the stage of decline.

– The number of employees is the ratio of the number of employees of organizations in the cluster core to the total number of employees in the industry.
– The number of highly qualified employees is the ratio of the number of highly qualified employees of organizations in the cluster core to the total number of employees in the industry.

- The level of strategic relations is a characteristic of a cluster that reflects the degree of interaction of organizations within the cluster. Determines the degree of development of the cluster, through a chain of value creation. The more mature the cluster, the more cluster organizations are involved in the value chain. To determine the level of interaction will be used for each organization. It is the ratio of the cluster organizations within the value chain to the total number of organizations participating in this chain percentage. For a cluster level strategic relationship will be determined as the average of the level of interaction between all organizations. In dynamics, this indicator will change from a value of 0 or slightly more than 0, at the stage of cluster nucleation, since most of the organizations are fragmented and have few internal connections. In the process of development, the number of connections grows and approaches 100% at the stage of maturity.

2. Parameters of competitiveness that determine the satisfaction with goods/services produced by the organizations included in the cluster. In this case, can be considered the products/services produced by both the external and the domestic market, where consumers are members of the cluster.

There are a large number of methods for assessing the competitiveness of an organization [12]. We propose to consider competitiveness through customer satisfaction and the financial performance of the organization.

Consumer satisfaction can be evaluated by expert evaluations. The overall satisfaction score will be the same for all consumers, but the ways to improve it may be different for consumers in the foreign and domestic markets. In a mature cluster, consumer satisfaction with the products offered by the cluster core organizations will tend to 100%. At the growth and decline stages, satisfaction will have positive and negative dynamics, respectively.

There are many indicators of the organization's financial performance, but in our opinion, you can focus on indicators of profit, profitability, and capital structure.

Changing these indicators at different stages of the cluster will show the benefits for the organization from participation in the cluster.

3. Parameters of competitiveness characterizing the satisfaction of cluster members with the work of the management company. This is a set of assessments related to the activities of the management company. Organizational, managerial and accompanying activities are assessed.

Cluster Member Satisfaction Index (CMSI) can be used as a system for evaluating the activities of the cluster management company. CMSI is a quarterly assessment of the ongoing satisfaction of organizations belonging to the cluster on three components:

- Assessment of satisfaction with the benefits received from the cluster, through activities of Management Company.
- Assessment of satisfaction with the internal business processes of Management Company that cluster members encounter while interacting with it.
- Assessment of satisfaction with the personnel of a management company that cluster members encounter while interacting with it.

An increase in the satisfaction indices will show the effectiveness of the decisions made by the management company at different stages of the life cycle. In the growth stage, there must be a positive trend. At the stage of maturity, satisfaction from each process will approach the maximum. Range of index values is from zero to ten.

3 Results

Based on the formulated criteria, we will give some recommendations to improve the competitiveness of the cluster (Table 4):

Table 4. Recommendations for increasing the competitiveness of the cluster at each stage of its development, depending on the level

Level of economic system	Subject of analysis	Recommendations by stage of the cluster lifecycle			
		Emerging cluster	Growing cluster	Mature cluster	Transformation
Macro-level	Development of measures to support the creation and development of clusters	Development of measures to stimulate the creation of clusters	Development of measures to support cluster members	Development of support measures to preserve the existing cluster	Selection of the most promising areas of newly emerging clusters and their support
Meso	Analysis of activities carried out by a management company with an assessment of their impact on the characteristics of the competitiveness	Conducting activities to form the interest of enterprises from participation in the cluster	Carrying out activities to assist in the development of enterprises included in the cluster	Carrying out activities that contribute to maintaining the achieved level	Carrying out activities aimed at preserving the cluster or reducing losses from its transformation
	Value chain analysis	Search for companies within the cluster to participate in the value chain and	Search for new members the value chain	Optimizing the value chain	Creation of new value chains
Micro-level	Analysis of consumer preferences	Monitor and assess the competitiveness of consumer preferences at each stage of the cluster life cycle			
	Analysis of financial performance	Analysis of the potential benefits of participating in the cluster	Taking measures to reduce costs and increase profits	Formation of a capital structure that increases the stability of the organization	Assessing the organization's ability to create new products or transition to a new technology

Source Compiled by the authors

In order to improve competitiveness at the stage of cluster growth, it is necessary to analyze the value chains of each organization included in the cluster for the use of external resources and their replacement with internal resources. This step will reduce costs at every stage of the value chain. In the first place, it is advisable to make such changes to the organizations included in the core.

To prevent a decrease in the competitiveness of the cluster, it is necessary to monitor the competitiveness of the cluster enterprises. To do this, it is necessary to analyze the satisfaction of consumers of the manufactured products and monitor the indicators of the financial activity of the enterprise.

Active support of the management company is necessary, primarily at the stage of growth and transformation. At the stage of maturity, when enterprises have reached their stability, and the cluster effect is at its maximum. The influence of the management company should be minimized so as not to interfere with the natural cluster effect.

4 Conclusion

Within the framework of the mechanism for managing the competitiveness of the cluster and its members, it is necessary to determine what and at what point in time the management company and the leadership of the organizations need to be undertaken to increase the competitiveness of both individual enterprises and the cluster as a whole.

The goal of managing the competitiveness of a cluster will be to increase competitiveness at the stage of cluster growth, prolong the effect of maturity and minimize losses at the stage of transformation.

In order, to manage the competitiveness of a cluster, it is necessary to determine at which stage of the life cycle the cluster is located [13] and to determine the criteria for which values are unsatisfactory.

The following table provides guidance for managing competition at every level and the cluster for each stage of the life cycle.

References

1. Bakhshian, E.A.: Clusters in a modern economy: essence, characteristics and generated effects. Theor. Appl. Econ. **1**, 64–74 (2019). https://doi.org/10.25136/2409-8647.2019.1.28209
2. Wikipedia. https://ru.wikipedia.org/wiki/. Accessed 8 Feb 2021
3. Zinchenko, A.G., Neznahina, E.L.: Life cycle model as a cluster competitiveness management tool. Vestnik nauki **5**(4), 4–8 (2018)
4. Palt, M.: Features of the stages of the cluster life cycle. Adm. Consult. **3**, 187–197 (2015)
5. Kostenko, O.V.: The stages of the life cycle of economic clusters: state, driving forces and processes. Agric. Sci. Euro-North-East **2**, 74–79 (2016). https://doi.org/10.30766/2072-9081.2016.51.2.74-79
6. Kostenko, O.V., Olenin, O.A.: Life cycle of economic cluster: criteria of development. Fundam. Res. **4**, 92–97 (2018)
7. Andersson, T., Schwaag-Serger, S., Sorvik, J., Hansson, E.W.: The Cluster Policies Whitebook. IKED, Sweden (2004)
8. Order of the Government of the Nizhny Novgorod Region dated 29 December 2016, No. 2169-r «On approval of cluster development strategies» (as amended on January 24, 2019). http://docs.cntd.ru/document/465569531. Accessed 8 Feb 2021
9. Menzel, M.-P., Fornahl, D.: Cluster life cycles – dimensions and rationales of cluster evolution. Ind. Corp. Chang. **19**(1), 205–238 (2009)

10. Klink, V., Langen, P.: Cycles in industrial clusters: the case of the shipbuilding industry in the Northern Netherlands. Tijdschr. Econ. Soc. Geogr. **92**(4), 449–463 (2001)
11. Shin, D., Hassink, R.: Cluster life cycles: the case of the shipbuilding industry cluster in South Korea. Reg. Stud. **45**(10), 1387–1402 (2009)
12. Bolodurina, V.A.: Methods of assessing the competitiveness of enterprises. Int. Res. J. **11** (42), 18–21 (2015). https://doi.org/10.18454/IRJ.2015.42.056
13. Ilyin, K.A.: Influence of the life cycle stage on the result of cluster activity (2013). https://www.hse.ru/edu/vkr/91906373. Accessed 8 Feb 2021

The Influence of Risk Factors on the Dynamics of Gross Regional Product in Conditions of Clustering of the Economy (On the Example of the Nizhny Novgorod Region)

Andrey L. Sochkov(✉), Artem V. Dorozhkin, Olga V. Rokunova, Vyacheslav N. Yasenev, and Oleg V. Yasenev

Lobachevsky Nizhny Novgorod State University, Nizhny Novgorod, Russia
sochkov@iee.unn.ru, yasenev@fnf.unn.ru

Abstract. The purpose of this study is to develop a model of multivariate analysis of regional risk that takes into account both intraregional and global financial and economic determinants.

Regional risk as a specific concept that limits the effectiveness of the development of the regional economic system in conditions of clustering is the object of research in this study. Multivariate modeling of regional risk based on correlation-regression analysis is the subject of research. Systematic assessment of risk factors is carried out using an economic-statistical model, which is an extended version of the Cobb-Douglas function. Its arguments include additional factors that determine the pace and risks of development of any region. The most significant deviations of the actual values of the elasticity coefficients of the model from the reference proportions highlight the most significant factors of regional risk.

A model for assessing the influence of factors of intraregional and external financial and economic risk on the dynamics of the gross regional product is proposed. The proportions of the elasticity coefficients of the model are reasoned. It was tested on the data of the Nizhny Novgorod Region for 2010–2019, recommendations to minimize the level of risk in the context of clustering the economy of this region were formulated.

The model allows highlighting risk factors from the point of view of the dynamics of the gross regional product, ranking them according to the level of significance, and, on this basis, developing recommendations for minimizing the risks of development of the regional economy.

Keywords: Regional economy · Cluster development · Regional risk · Risk assessment · Multivariate model · Multiplicative model · Elasticity coefficients

JEL Codes: R150

E. G. Popkova (Ed.): ISC 2020, LNNS 368, pp. 482–491, 2022.
https://doi.org/10.1007/978-3-030-93244-2_53

1 Introduction

The current stage of development of many regional economies is characterized by the formation of sectoral clusters that contribute to their growth. Its pace is greatly influenced by the quality of regional risk management. The latter is directly conditioned by the effectiveness of the qualitative and quantitative analysis of the level of regional risk, factors, and conditions of its formation, the influence of general macroeconomic processes on it. The study of regional risk is often carried out based on the use of expert assessment tools and sociological surveys of entrepreneurs in the region, which reduces the level of objectivity of this kind of analysis. In addition, when studying trends in regional risk, the dynamics of global financial and economic factors are not sufficiently taken into account.

Regional risk, as a specific economic concept that limits the effectiveness of the development of the regional economic system in modern conditions of clustering, is the object of research in this study. Multivariate modeling of regional risk carried out based on multiplicative correlation-regression analysis is the subject of the research.

The purpose of this study is to develop a model for multivariate analysis of regional risk that takes into account both intraregional and global financial and economic determinants. To achieve it, it is necessary:

- to systematize the main approaches to the analysis of regional risks and identify their disadvantages;
- to develop a model for assessing the influence of factors of intraregional and external financial and economic risk on the dynamics of the gross regional product;
- to test the model on the data of the economy of the Nizhny Novgorod region.

The scientific novelty of the research lies in the substantiation of a multivariate model, in which the main intraregional factors of production and price parameters of all basic segments of the financial market are taken into account. In contrast to the approaches presented in the literature, the most significant regional risk factors are identified based on comparing the elasticity coefficients of appropriate variables with reasoned most preferred (reference) ranges.

There is no unified generally accepted approach to the quantitative analysis of regional risk in the special literature. Its level is often aggregated based on the dynamics of basic financial and economic indicators of the development of the region. So, A.M. King considers the average depreciation rate of fixed assets of the regional economy and the level of capital productivity as such a key indicator in [1]. T.F. Paley proposes to use the proportions of the development of regional transport infrastructure in [2]. In the article [3], the authors consider the degree of accessibility and relevance of economic information in the regional economy as a key indicator. The main disadvantage of this approach is that different indicators can demonstrate different trends often in different ways, contradictingly characterizing the general level of regional risk.

Within the framework of another approach, as a rule, complex surveys or questionnaires of various groups of the regional business community are carried out regarding the level and factors of the formation of various types of risks in the economy of the region [4]. The main disadvantage of this method is the high level of subjectivity.

2 Methodology

The basic methodological principles of multivariate statistical analysis of the risks of the development of economic systems of various levels are argued, in particular, by Jarrow [5]. He substantiates the preference for using a multiplicative model for studying the risks of complex socio-economic systems, which include the modern regional economy during the formation of industry clusters. In general, those variables of the model are the factors of greatest risk, for which the coefficients of elasticity are more significant in absolute value.

The authors in [6] propose a model for assessing regional risk, which is a simple variation of the Cobb-Douglas production function (CDPF). Within the framework of this model, gross output in the regional economy is considered as a function of employment and the initial cost of fixed assets in the regional economy. In our opinion, the use of gross output as the resulting variable of the regional risk research function is less preferable than the more informative indicator of GRP, and the inclusion of the initial (historical) value of fixed assets in the factorial variables does not fully indicate the nature of the use of fixed capital in the regional economy.

T.F. Palei [7] proposes a multivariate model for assessing regional risk based on the interpretation of the VAR method. Within the framework of this model, GRP is considered as a function of the simultaneous influence of labor factors, fixed capital in the regional economy, and infrastructure capital. This approach is of theoretical interest; at the same time, the author's conclusion, made based on his approbation on the materials of the economy of the Republic of Tatarstan about the pronounced elastic influence of all the selected factors on the dynamics of GRP, raises certain doubts. This leads to the fact that they are all determinants of significant risk, which is unlikely.

Within the framework of the approach [8], intraregional cargo turnover is considered as the most significant factor of regional risk; the authors build a one-factor model of the influence of freight turnover on GRP. In our opinion, the dependence, in which GRP acts as an independent variable and regional cargo turnover as a result variable, seems to be more correct.

The authors [9] propose a model for studying the risks of regional development in terms of entrepreneurial activity, insufficient availability of labor resources, excessive concentration of economic agents in the regional economy, etc. At the same time, the model proposed by the authors is additive (unlike the multiplicative model, within its framework, the factor variables are not multiplied, but added), which raises certain doubts about the level of its adequacy for our case.

The authors construct a system of one-factor functions of the mutual influence of regional markets for goods, labor, and capital in the work [10]. In their opinion, the greatest risk factors are those aspects of regional development, in which the actual elasticity of functions differs from the reference type reasoned by them.

In general, all the considered approaches to multivariate statistical analysis of the level of regional risk are characterized by ignoring such factors as the exchange rate, the average rate on bank loans in the national economy, the state stock market index. These factors are significant determinants of risk, especially for regions with a high share of export-oriented industry, regions where a large number of enterprises are significantly dependent on bank loans, regions where the financial condition of large enterprises is significantly dependent on fluctuations in the securities market.

We propose the following economic and statistical model (1) to systematically assess all the noted factors of financial and economic risk:

$$GRP = a0 * I^{a1} * E^{a2} * Ihf^{a3} * P^{a4} * DEK^{a5} * BI^{a6}, \tag{1}$$

where GRP – growth rate of real gross regional product, % to some base year;

I – the growth rate of real investment in fixed assets in the regional economy, % to base year;

E – number of employees in the regional economy, thousand people;

Inf – level of officially registered inflation in the national economy as a whole;

P – the average annual interest rate on loans for organizations in the regional financial and credit market, %;

DEK – the average annual value of the Ruble against the US Dollar and Euro;

BI – the average annual value of the stock index (stock index of the united exchange MICEX - RTS is such an index in the Russian Federation);

a1, a2 … a6 – elasticity coefficients;

a0 – free member.

The model is an extended version of the CDPF extrapolated for the level of the economy of the region. At the same time, we included additional factors, that determine the rates and risks of development of any region in the modern era of clustering, in the arguments.

The reference, most preferred proportions of the elasticity coefficients of the proposed model are presented in Table 1. In this connection, the most significant deviations of the actual values of the elasticity coefficients of the model built on the materials of a particular region from the reference proportions are the most significant factors of regional risk.

Table 1. Preferred values of elasticity coefficients

The most preferred proportions of elasticity coefficients	Argumentation
a1 > 1	Ensuring high efficiency of the investment process in the regional economy
a2 > 1	Ensuring accelerated growth in the level of average labor productivity in the region's economy
a1 > a2	Formation of economic growth in the region mainly due to an increase in the productivity of fixed capital, but not as a result of an inertial increase in employment
a3 → 0	Ensuring relatively insignificant effect of inflationary processes in the national economy on fluctuations in GRP
−1 < a4 < 0	Inverse inelastic effect of interest rates on loans for companies existing in the financial market on fluctuations in GRP
a5 → 0	Ensuring low dependence of the dynamics of real GRP on hard-to-predict fluctuations of the exchange rate
a6 → 0	Ensuring low elasticity of the influence of the dynamics of the national stock market, often formed by speculative market factors, on the formation of the gross regional product

Source Compiled by the authors

3 Results

The proposed assessment model is tested on the materials of the economy of the Nizhny Novgorod region. This territory is characterized by a sufficiently diversified structure of the economy and balanced budget, therefore it can be considered as an average statistical region for model verification. The post-crisis period of 2010–2019 was chosen as the time horizon for approbation. There are two reasons. Firstly, there was an intensive formation of industry clusters in the region according to the electronic fund of legal and regulatory, and technical documentation during this period [11]. Secondly, significant fluctuations in external financial factors of regional risk that took place in 2008–2009 could have a distorting effect on the final parameters of the model (elasticity coefficients). The data on the Nizhny Novgorod region used in the course of model testing are grouped in Table 2.

Table 2. Data on the Nizhny Novgorod region

Time period, year	Real GRP, % to 2010	Real investments in fixed assets in the region, % to 2010	Number of employees, thousand people	The inflation rate, times to the previous year	Average interest on loans to organizations, %	Average exchange rate of the Ruble against the US Dollar and the Euro	MICEX-RTS Index, annual average
2010	100.0	100.0	1,711	1.10	17.7	28.4	1,452
2011	106.9	107.9	1,701	1.09	16.8	32.8	1,658
2012	111.0	112.1	1,703	1.11	16.1	36.2	1,472
2013	113.2	117.8	1,686	1.08	15.7	38.8	1,389
2014	116.5	118.9	1,678	1.07	17.1	54.1	1,412
2015	112.4	117.4	1,654	1.12	18.9	66.2	1,876
2016	115.8	119.3	1,645	1.07	19.1	68.7	1,989
2017	118.5	122.5	1,633	1.05	17.3	63.4	2,203
2018	119.7	125.1	1,629	1.04	15.8	65.1	2,408
2019[a]	121.4	128.3	1,615	1.04	14.9	62.3	2,673

Source Compiled by the authors based on [12]
[a]For 2019, data are preliminary

The following statistically stable model (2) was synthesized using the data from Table 2:

$$GRP = \exp(-15,8) * I^{0,85} * E^{2,13} * Ihf^{-0,41} * P^{0,01} * DEK^{0,05} * BI^{0,06}. \quad (2)$$

Its high representativeness is characterized by close to unity values of the coefficients of multiple correlation, determination, Fisher's Criterion (F = 73.85), as well as a small value of the free member. The econometric characteristics of the model are shown in Table 3.

As follows from function (2) the actual proportions of the elasticity coefficients of the model, which characterize various components of regional risk, significantly differ from the reference ones, namely:

1. The influence of factor I on the GRP of the region under consideration is positive but inelastic. Its increase by 1% entails an increase in the real GRP of the region by only 0.85%. The revealed disproportion indicates the risk of insufficient efficiency of the organization and implementation of the investment process in the regional economy, high risks of misuse of limited financial and investment resources, capital investment not mainly in the direction of the most promising regional projects, which are the main drivers of growth in the gross regional product.
2. The effect of factor E on the GRP variable of the region under consideration corresponds to the reference indicators. At the same time, there is a significant excess of the coefficient a2 over the value of indicator a1 for variable I. This

Table 3. Econometric characteristics of the model

Regression statistics				
Multiple R		0.996632		
R-squared		0.9932754		
Normalized R-squared		0.9798262		
Standard error		0.0082695		
Observations		10		
ANOVA				
	df	*SS*	*MS*	*F*
Regression	6	0.0303025	0.0050504	73.8537431
Balance	3	0.0002052	6.83841E-05	
Total	9	0.0305077		
	Coefficients	Standard error	t- statistics	P-value
Y-intersection	-15.8443423	8.7872705	-1.8031017	0.1691539
Variable value I	0.8584916	0.2618597	3.2784417	0.0464804
Variable value E	2.132282	1.069483	1.9937507	0.1401728
Variable value Inf	-0.416785	0.200486	-2.0788745	0.1291414
Variable value P	0.0109151	0.1302883	0.0837767	0.9385111
Variable value DEK	0.0550438	0.0641025	0.8586843	0.4536503
Variable value BI	0.0611043	0.0503501	1.2135875	0.3117355

Source Compiled by the authors based on the calculation results

suggests that the economy of the Nizhny Novgorod Region is developing according to a scenario in which the growth of GRP is formed mainly not due to the involvement in the regional production process of increasingly large arrays of modern, highly efficient fixed assets, but due to increased employment, including in the segment of personnel with low qualifications.

3. The dependence of the dynamics of the GRP of the Nizhny Novgorod Region on inflation risk is quite high. Thus, an increase in the inflation rate in the national economy of the Russian Federation by 1% entails a decrease in the real gross regional product of the studied region by 0.41%. Such a significant dependence is a consequence of the significant influence of the inflationary factor both on the formation of costs of enterprises in the Nizhny Novgorod region and on the intensity of consumer behavior.
4. The effect of factor P on the GRP variable is weak. This is due to the low use of the potential of bank lending, primarily long-term, by most organizations in the region.
5. The influence of the factors of the exchange rate and the dynamics of the stock market on the dynamics of the GRP of the Nizhny Novgorod Region, in general, corresponds to the reference proportion of the elasticity coefficients and is rather inelastic. This proportion is a consequence of the relatively low level of dependence of most of the region's enterprises on imported raw materials, equipment, and components, as well as the fact that the securities of the region's enterprises are practically not represented on the national stock market.

4 Recommendations

Based on the results of testing the model, it is possible to systematize the following recommendations in the field of improving regional risk management in the economy of the Nizhny Novgorod Region:

- rationalization of the processes of organizing and implementing the regional investment process, primarily in the direction of increasing the level of its transparency, reducing the risks of misuse of budget investment resources, etc.;
- changing the model of regional development from predominantly labor-intensive to capital-intensive, including, for example, through the formation and implementation of a comprehensive regional program for the intensification of labor processes, qualitative improvement of automation technologies and informatization of labor activities, modernization of the equipment fleet in the direction of reducing the labor intensity of regional production as a whole;
- intensification of measures in the field of regional antimonopoly policy, primarily in terms of tightening control over pricing in regional food markets.

Approbation of the model necessitates the activation of the following main strategies for the economic development of the Nizhny Novgorod region in the coming years:

- a) the strategy of innovative modernization of fixed capital in the regional economy with the aim of intensive growth in the level of productivity of the latter and, accordingly, increasing the elasticity of the impact on the dynamics of real GRP (through the implementation of regional programs of preferential leasing of advanced equipment for regional companies in the manufacturing sector, etc.);
- b) formation and implementation of a comprehensive regional anti-inflationary strategy, including through the rationalization of the tariff management system in housing and communal services, activation of antimonopoly control over pricing processes within the framework of large retail chains represented in the Nizhny Novgorod region;
- c) the strategy of orienting commercial banks in the region towards priority long-term lending of investment projects in the real sector of the regional economy (to overcome the revealed abnormally low influence of factor P on the GRP of the Nizhny Novgorod Region).

5 Conclusion

In general, the proposed model allows highlighting the most significant factors of regional risk from the point of view of the formation and dynamics of the gross regional product, ranking intraregional and global factors of financial and economic risk by the level of significance based on comparing the actual values of the elasticity coefficients with reasoned reference proportions and developing of systemic recommendations for minimizing the level of risk in the regional economy on this basis.

The scientific novelty of the proposed model includes the following elements:

- the main regional factors associated with production factors and global factors of financial and economic risk are considered in a unified coordinate system and the latter reflects the price parameters of all the main segments of the national financial market;
- the differentiation of risk factors is carried out based on comparing the coefficients of elasticity of the model with the most preferred (reference) proportions argued by the authors.

The practical implications of the research consist in the fact that such recommendations to reduce the overall level of regional risk as a qualitative reorganization of the regional investment process, primarily in the direction of increasing the level of its transparency, changing the model of regional development from predominantly labor-intensive to capital-intensive, activation of the toolkit of a comprehensive anti-inflationary policy were substantiated for the Nizhny Novgorod Region.

At the same time, the proposed model has some limitations. So, it can be used within the framework of conditions of at least relative financial stability, since the financial crisis, manifested in sharp fluctuations in the parameters of the financial market, worsens the representativeness of the model. In addition, the model in the proposed modification does not take into account such a rather significant factor of potential efficiency and, at the same time, the risk of regional development, as the level of innovative activity in the regional economy.

In the future, the developed model can be supplemented with a toolkit for a comprehensive assessment of the level of socio-economic efficiency of regional development. Such a toolkit can help to carry out a more correct justification of measures to reduce the overall level of regional risk and its components.

References

1. King, A.M.: Internal Control of Fixed Assets, p. 192. Wiley, New York (2011)
2. Paley, T.F.: Methodology of Systemic Infrastructural Incentives for Industrial Growth, p. 186. Artifact, Kazan (2017)
3. Ismagilova, G.N., Danilina, E.I., Gafurov, I.R., Ismagilov, R.I., Safiullin, L.N.: Asymmetric information and consumer demand. Asian Soc. Sci. **11**, 259 (2014)
4. Abrams, R.: Entrepreneurship: A Real-World Approach, p. 412. Planning Shop, Redwood City (2018)
5. Jarrow, R.: Economic Foundations of Risk Management, p. 206. WSPC (2016)
6. Hook, G., Mason, R., O'Shea, P.: Regional Risk and Security in Japan, p. 256. The University of Sheffield, Routledge (2015)
7. Palei, T.F.: Assessing the impact of infrastructure on economic growth and global competitiveness. In: Procedia Economics and Finance, 2nd Global Conference on Business, Economics, Management and Tourism, Rome, Sapienza University, Faculty of Economics, vol. 23, pp. 168 –175 (2015)
8. Sandakova, N., Bulatova, N.N., Korodyuk, I.S.: Model of innovative development of regional transport infrastructure. Int. J. Appl. Eng. Res. **10**, 40834–40840 (2015)
9. Barinova, V.A., Zemtsov, S.P., Tsareva, Y.V.: Entrepreneurship and institutions: does the relationship exist at the regional level in Russia? Econ. Issue **6**, 92–116 (2018)

10. Demyanova, O., Valitov, S.: A multidimensional assessment of the efficiency of the Russian economy, vol. 25, pp. 760–767 (2013). https://doi.org/10.5829/idosi.wasj.2013.25.05.13336
11. Electronic fund of legal and regulatory technical documentation. Order of the Government of the Nizhny Novgorod Region No. 2169-r dated 29 December 2016 on the approval of cluster development strategies (2021). http://docs.cntd.ru/document/465569531 Accessed 8 Feb 2021
12. Regions of Russia. Statistical Yearbook M. Publishing House of the Federal State Statistics Service, p. 812 (2019)

Ecommerce Surge as an Element of a Modern Economy Integration Mechanism Development

Marina V. Artemyeva[1(✉)], Ekaterina P. Garina[1], Svetlana N. Kuznetsova[1], Yaroslav S. Potashnik[1], and Nataliya A. Bezrukova[2]

[1] Minin Nizhny Novgorod State Pedagogical University, Nizhny Novgorod, Russia

[2] Lobachevsky State University of Nizhny Novgorod, Nizhny Novgorod, Russia

Abstract. Purpose: The purpose of the chapter is to investigate the e-commerce development process taking into account its determining factors, including the spread of the new coronavirus infection COVID-19, and possible consequences.

Design/Methodology/Approach: The authors explore statistical data from leading research companies in order to examine not only their results but also the methodological base, new research tools that allow conducting researches taking into account the specifics of e-commerce as a relatively new economic phenomenon.

Findings: It is shown that e-commerce growth is obtained by the large-scale permanent involvement of new consumers and widening the range of product categories they buy. The e-commerce transactions increase is going to surpass the traditional retail growth not only during the crisis period but also after the pandemic is over. E-commerce will stay a key growth opportunity for manufacturers and retailers, underscoring the need to study the features of its development, identify which countries and industries are best positioned to retain the e-commerce surge, plan and implement omnichannel strategies.

Originality/Value: The article combines the results of the latest research and allows to get an understanding of the e-commerce development as part of the new economy and form an opinion about the significance of this phenomenon, the prospects for investment in e-commerce assets, the development of omnichannel strategies.

Keywords: Economic development · Economic integration · E-commerce · Analytics · Economic forecast

JEL Code: F63 · O11 · O33 · R58

1 Introduction

E-Commerce is an area of the economy that includes all financial and trade transactions carried out via the Internet and the business processes associated with them. Primarily it is understood as online trading, but e-commerce also includes any electronic exchange of information, including simple information sites, all marketing communications

E. G. Popkova (Ed.): ISC 2020, LNNS 368, pp. 492–500, 2022.
https://doi.org/10.1007/978-3-030-93244-2_54

provided via the Internet, online banking and other online services, money transfers, e-wallets, and so on.

Manufacturers, supply chains, and distributors all need to work together to capitalize on the growth of e-commerce. Manufacturers have to adapt their product, packaging, pricing, assortment, distribution strategies, and how they partner with traditional retailers to ensure their products can be sold via retailers' online sales channels. They need to purposefully plan and unceasingly optimize their relationships with the leading e-commerce platforms. Choices related to retailer distribution and product search require ongoing efforts to optimize the partnerships manufacturers have with all kinds of retailers.

Omnichannel strategies are increasingly significant as e-commerce will continue to grow as online shopping becomes more appealing to consumers and wider groups of product categories. It will also stay demanded as the very concept of retailing evolves [1].

Omnichannel strategy means mutual integration of dissimilar communication and sale channels into a unified system to provide seamless and continuous relationships with the customer. The full list of omnichannel features is most applicable to businesses that combine online and retail sales.

Every year e-commerce is growing more and more rapidly all over the world. The reasons for this, in turn, are growing exactingness, education, awareness, individualization of the consumers, including older age groups; increasing market differentiation and complexity of competition; technological progress; accumulation of electronic databases and development of the database market; lack of time for consumers; increasing availability and improvement of online payment methods. E-commerce provides a lot of advantages for a consumer and the continuous improvement of all its elements, including trade, payment, and information, helps to eliminate shortcomings. For example, the possibility of free fitting allows you to get rid of the greatest online trading drawback - consumer uncertainty about the product quality.

And today's situation is giving way to what's going to possibly be a permanent e-commerce boost. COVID-19 infection spreading made the digital channel turning into the default for many shoppers. As a result of government-mandated restrictions and non-essential business closures, shopper defrayment underwent unexampled levels of channel shifts [2]. Moreover, these trends differ between the first and second waves of the infection. At first, changes were unexpected for both consumers and businesses and were perceived as something one-off. Stores were in a hurry to implement online ordering services which were far from ideal. During the second wave we found ourselves in a situation where, on the one hand, sellers were forced to develop online commerce and related business processes in order to justify the spring investments and keep up with competitors, and on the other hand, buyers were able to appreciate the advantages of ordering goods via the Internet. Thus, excellent prerequisites were created for the development of e-commerce in almost all sectors of the consumer economy. Nevertheless, the probability of such development is dissimilar in different product categories and across regions. Considering the fact, that for many businesses in a pandemic-driven crisis with reduced consumer spending and possible lockdowns, e-commerce can become a key to survival, there is a need for identifying industries best prepared for sustained e-commerce growth.

2 Methodology

E-commerce represents the direction of improving the economic activity of economic entities, corresponding to the strategy of sustainable development. This is especially true in the current context of the COVID 19 crisis.

The necessity for exploring possible directions of economic development, including sustainable development is widely covered in scientific works [3–8], which form the theoretical base of the research.

The practical base of the research is formed by statistical data from leading research companies, like Euromonitor International - a global marketing research company, creating data and analysis on products and services all over the world, The Economist Intelligence Unit (EIU) - research and analysis division of The Economist Group, the sister company to world's famous The Economist newspaper, GroupM - the world's leading media investment company and Statista - global business data platform.

To examine e-commerce development dynamics, authors appeal to Euromonitor International's Lifestyles Survey, fielded January to February 2020, Digital Consumer Survey, fielded March to April 2020, Euromonitor International's E-Commerce Readiness Model data, etc.

3 Results

Recent researches show that global e-commerce retail sales are going to reach a new high in 2021. E-commerce companies can grow by 265%, from $ 1.3 trillion in 2014 to $ 4.9 trillion in 2021 [9]. This indicates a future steady upward trend with no visible signs of decline.

Statista predicts that by 2023 the retail e-commerce turnover will exceed $6.5 trillion (see Fig. 1).

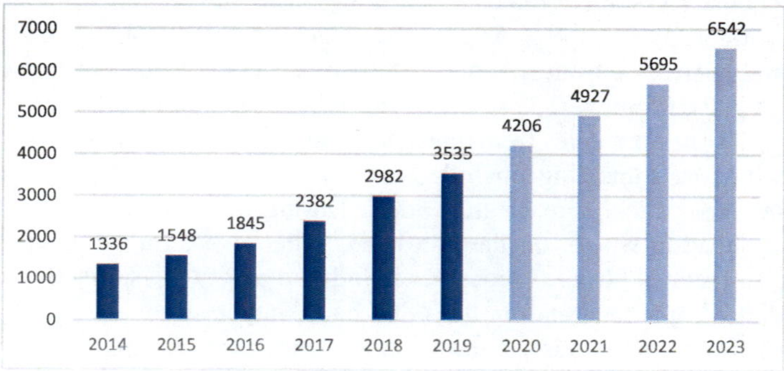

Fig. 1. Retail e-commerce sales worldwide from 2014 to 2023 (in billion U.S. dollars), 2020–2023 – forecast. *Source* Compiled by the authors based on [10]

Group M predicts more modest results and notes that the growth rate of e-commerce will slow down. (see Fig. 2). However, the report also says that the share of e-commerce in the retail trade turnover of the entire planet will reach 25% by 2024, and by 2027 the e-commerce sector will sell goods and services worth $10 trillion. The main beneficiary in the segment has already become China, and in 2021 the share of e-commerce in the country will reach 27.3% [1].

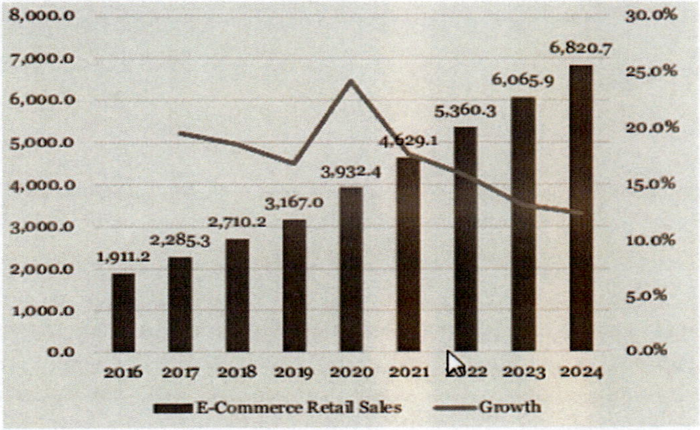

Fig. 2. E-commerce retail sales worldwide from 2014 to 2024 (in billion U.S. dollars), 2020–2024 – forecast. *Source* GroupM [1]

Drastic e-commerce boost through the pandemic was obtained two different ways. Firstly, people started to use already-familiar e-commerce channels more frequently. But besides that, a lot of consumers faced the necessity to explore new (for them) channels, even if they never used any e-commerce tools before. This is especially true for people with relatively low computer literacy. So, e-commerce involved more people and a wider range of products.

This tendency can be observed by a comparison of two consumer surveys conducted by Euromonitor International: Lifestyles Survey, fielded January to February 2020 and Digital Consumer Survey, fielded March to April 2020 (see Fig. 3). During the researches, all connected consumers were divided into 3 categories: minimal online shoppers, who almost didn't use digital or research/purchase across 1 category; moderate online shoppers - research/purchase across 2–3 categories, and heavy online shoppers, researching/purchasing across 4–5 categories.

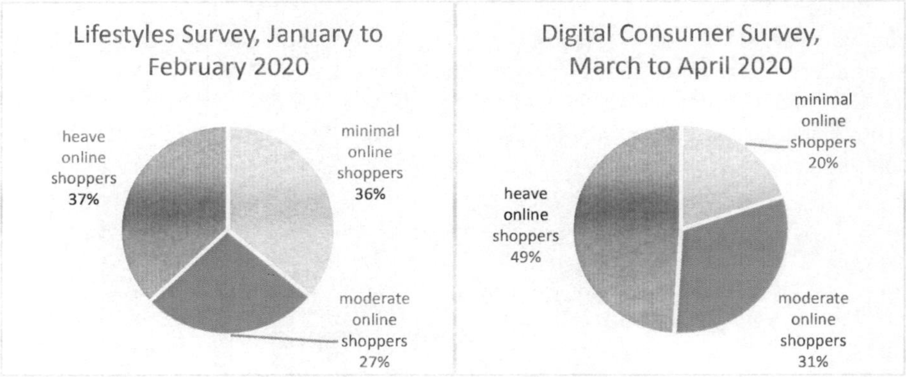

Fig. 3. Consumers researching/purchasing online across a number of industries, comparison of the consumer surveys' results. *Source* Compiled by the authors based on E-Commerce Readiness Model [2]

The results show that a lot of customers have turned to e-commerce during the pandemic. The share of minimal online shoppers decreased significantly, in early 2020 they and heavy online shoppers were evenly split at 36% and 37%, respectively. While at the peak of the pandemic, the minimal online shoppers' group reduced by almost 2 times. By March, there was nearly a 30-percentage point gap between them (20%) and heavy online shoppers (49%) [2].

Euromonitor International consumer surveys regarding the path to purchase habits fielded in 2020 provide insights into these habits' forceful shifts and the resulting surge in e-commerce. Nearly each product category (except beauty, health, and personal care) showed a double-digit increase within the share of shoppers making digital purchases (Fig. 4).

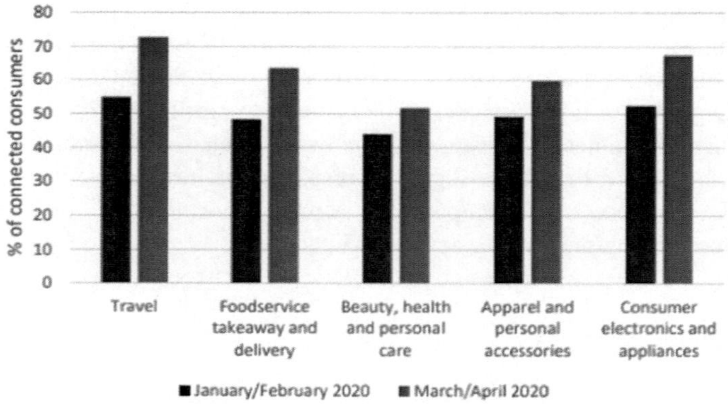

Fig. 4. Digital purchase by industry, 2020. *Source* Euromonitor International's Lifestyles Survey, fielded January to February 2020 and Digital Consumer Survey, fielded March to April 2020. [2]

Growing e-commerce opportunities demand new analytical tools. Further, we'd like to argue the importance of popularizing such tools by the example of some of Euromonitor International's E-Commerce Readiness Model results.

Euromonitor International's E-Commerce Readiness Model is a complicated analytical tool, designed to spot the industries (product categories) and/or regions most suitable for long-term e-commerce growth. It explores 76 countries and 16 industries. The model identifies such market conditions that are most likely to provide a higher share of e-commerce.

It uses seven thousand data points that measure e-commerce sales by category (Fig. 5) and country (Fig. 6) to predict the markets presumably to be ready to support sustained e-commerce growth.

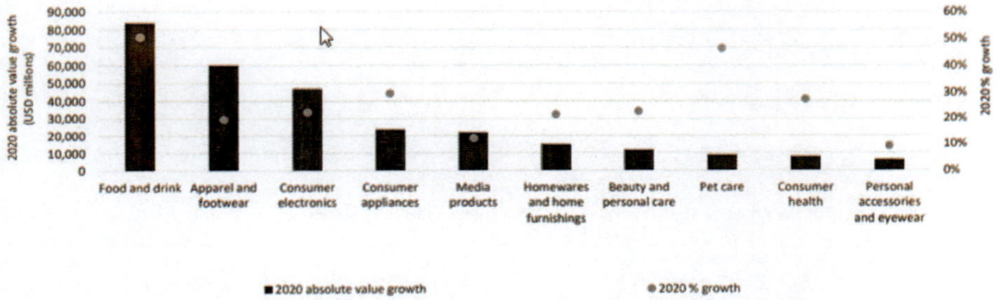

Fig. 5. Growth of e-commerce by product category 2020. *Source* Euromonitor International E-Commerce Readiness Model [11]

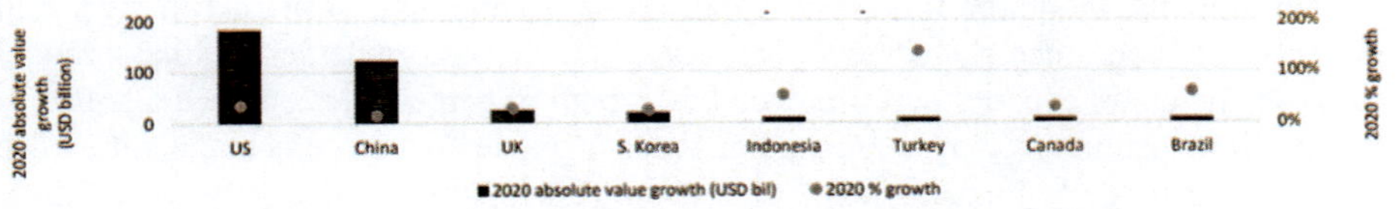

Fig. 6. Growth of e-commerce by country 2020. *Source*: [11]. *Source* Euromonitor International E-Commerce Readiness Model [11]

The primary output of this approach is a foretold share of e-commerce that would have been supported by the market in 2019 before COVID-19 became a world pandemic. Since market conditions would have allowed higher e-commerce activity pre-pandemic, the model helps companies foresee which industries are probably going to meet sustainable e-commerce growth, even after pandemic challenges diminish.

The distinction between foretold share and actual share of e-commerce signals that countries and categories had the foremost unfulfilled e-commerce potential before the pandemic and are most likely presumably to sustained e-commerce growth later on [2].

The following Fig. 7 examine differences between the e-commerce readiness model forecast and the real e-commerce channel share – in percentage and absolute value by the example of the alcoholic drinks industry. Thus, unmet e-commerce channel potential is shown. Percentage indicators seem to be more useful for internal (for a given industry) investors, including company managers, making investment decisions, and absolute – for external. Using this data, companies trying to anticipate where more permanent e-commerce growth may occur can make investment and resource allocations today to be best prepared for the consumer landscape of tomorrow.

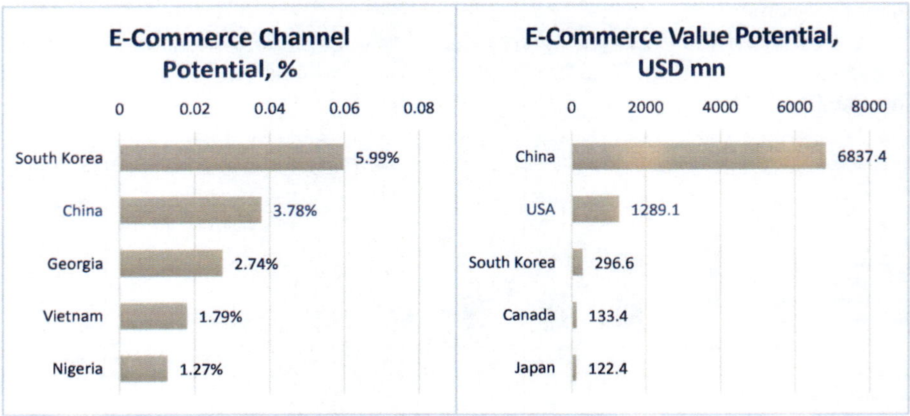

Fig. 7. Unmet e-commerce potential across alcoholic drinks industry. *Source* Compiled by the authors based on E-Commerce Readiness Model [2]

The example of the alcoholic drinks industry shown in Fig. 7 illustrates the principle of how the model performs very well. China has the biggest market in the world in terms of e-commerce channel share. Even so, the e-commerce readiness model suggests the channel could perform even better both in percentages and actuals. It tops the list for absolute value potential, but when it came to the percentage difference between the model's forecast versus the actual share, South Korea surpassed China. Although South Korea is very advanced in terms of consumer use of digital channels in many industries, in the case of alcoholic drinks, only beer can be sold via e-commerce.

4 Conclusion

There will be both winners and losers from the market transformation, depending on retailers' ability to shift online and retain customers. Success will be underpinned by several factors, including competitive product positioning and pricing strategies, user-friendly digital platforms, efficient inventory management, fast and efficient fulfillment processes [12].

As the boundaries between the physical and digital environments blur, the use of multiple channels will become more common among shoppers. This is evidenced by 73% of customers who use several channels to make purchases [9].

The use of omnichannel promotion of goods and services by companies working in the field of e-commerce makes it possible to determine what, where, when, how, and why people buy.

There are many examples of how omnichannel shopping works. For example, people can find all the product information online and then buy from a physical store, or buy products online and pick them up from an offline store. The more channels customers use, the more likely it is that the average order price will increase. For example, customers who used more than 4 shopping channels spent an average of 9% more than those who used only one channel.

Every point of contact with the customer is important because it gathers every piece of information into a whole story about the customer.

Companies should make sure that they integrate their offline and online platforms into a single marketing channel, create convenient offline points for customers who get acquainted with the product on the Internet, and make purchases in offline mode. It is necessary to develop a convenient mechanism through which customers can make purchases online and pick them up at a point near them. It also means that offline and online data should always be synchronized to make business decisions faster and more efficiently.

References

1. E-commerce forecast. GroupM Worldwide, Inc. (2020). https://s3.amazonaws.com/media.mediapost.com/uploads/groupmecommerceforecast.pdf. Accessed 12 Apr 2021
2. Evans, M., Rehorst, C.-P., Curtiss, G.: The digital consumer journey best markets for sustained e-commerce growth (2020). https://go.euromonitor.com/rs/805-KOK-719/images/sbEcommerceGrowthForecast-0.4.pdf?mkt_tok=eyJpIjoiTUdNeU1XRTJNR0l6TnpZMyIsInQiOiJ1aFV5NGFsWmQ1Uzd3a1BGaDNaNFhTS2s2M1RFRnA2NWlSd09UMnlOWEpFaWFRTWUwU1VEb2VlNGlwZjVwaUJYckZNNFc3WFJHTkxObGpxeExNSzRvVVNHam5abTQwNmhMTHNnWTV6ZHQ2cmFvMkxQcldjYTdKWjYzbVJWQ2dnWiJ9. Accessed 14 Apr 2021
3. Efremova, A.D., Kuznetsov, V.P., Artemyeva, M.V., Grigoryev, A.I., Sintsov, E.V.: Disaggregation of industrial enterprises as one of the ways to overcome the economic crisis. In: Popkova, E.G. (ed.) ISC 2018. LNNS, vol. 57, pp. 947–953. Springer, Cham (2019). https://doi.org/10.1007/978-3-030-00102-5_101
4. Kozlova, E.P., Kuznetsov, V.P., Garina, E.P., Romanovskaya, E.V., Andryashina, N.S.: Methodological bases of the assessment of sustainable development of industrial enterprises (technological approach). In: Popkova, E.G., Sergi, B.S. (eds.) ISC 2019. LNNS, vol. 91, pp. 670–679. Springer, Cham (2020). https://doi.org/10.1007/978-3-030-32015-7_75
5. Kozlova, E.P., Kuznetsova, S.N., Garina, E.P., Romanovskaya, E.V., Andryashina, N.S.: The concept interpretation of sustainable development mechanism in the conditions of technological transformation of industrial enterprises. In: Popkova, E.G., Ostrovskaya, V.N., Bogoviz, A.V. (eds.) Socio-economic Systems: Paradigms for the Future. SSDC, vol. 314, pp. 729–736. Springer, Cham (2021). https://doi.org/10.1007/978-3-030-56433-9_77

6. Kozlova, E.P., Potashnik, Y.S., Artemyeva, M.V., Romanovskaya, E.V., Andryashina, N.S.: Formation of an effective mechanism for sustainable development of industrial enterprises. In: Popkova, E.G. (ed.) Growth Poles of the Global Economy: Emergence, Changes and Future Perspectives. LNNS, vol. 73, pp. 545–556. Springer, Cham (2020). https://doi.org/10.1007/978-3-030-15160-7_55
7. Kuznetsova, S.N., Lapaev, D.N., Artemyeva, M.V., Potashnik, Y.S., Kozlova, E.P.: Increase of economic effectiveness of investment projects in industrial parks. In: Popkova, E.G. (ed.) ISC 2018. LNNS, vol. 57, pp. 1138–1144. Springer, Cham (2019). https://doi.org/10.1007/978-3-030-00102-5_120
8. Romanovskaya, E.V., Garina, E.P., Andryashina, N.S., Kuznetsova, S.N., Artemyeva, M.V.: Studying the experience of reengineering business processes in the practice of domestic enterprises: problems and prospects of application. In: Popkova, E.G. (ed.) Growth Poles of the Global Economy: Emergence, Changes and Future Perspectives. LNNS, vol. 73, pp. 517–524. Springer, Cham (2020). https://doi.org/10.1007/978-3-030-15160-7_52
9. The future of e-commerce: 10 current trends in the development of the e-commerce market in the world (2020). Ecom Today - Vc.ru. https://vc.ru/trade/167337-budushchee-ecommerce-10-sovremennyh-tendenciy-razvitiya-rynka-elektronnoy-kommercii-v-mire. Accessed 12 Apr 2021
10. Global retail e-commerce market size 2014–2023 (2020). Statista. https://www.statista.com/statistics/379046/worldwide-retail-e-commerce-sales/. Accessed 17 Mar 2021
11. Evans, M.: Retail in transition: capitalising on future e-commerce opportunities (2021)
12. Online retailing How to navigate the new normal (2021). The Economist Intelligence Unit. https://pages.eiu.com/rs/753-RIQ-438/images/online-retailing-in-2021.pdf?mkt_tok=NzUzLVJJUS00MzgAAAF8S93VsozwEqwEXa02CWAFcr4x7BmvgADkfOM1wn_uZcNWX4EMOPf4vhtXszrwamZ6GbBEtTPboyhenrhB_4iMgQlX14T-ZWGJzGJGgLe4xvv5PA. Accessed 12 Apr 2021

Mechanisms of Integration Interaction as a Determining Factor in the Development of Industrial Enterprises

Elena P. Kozlova(✉), Victor P. Kuznetsov, Sergey D. Tsymbalov, Yaroslav S. Potashnik, and Marina V. Artemyeva

Minin Nizhny Novgorod State Pedagogical University, Nizhny Novgorod, Russia

Abstract. Issues arising in the process of cooperation and integration of industries are becoming a priority. The development of horizontal and vertical integration, the formation of associations for joint production, processing, sales of products, etc., are becoming the key directions for the development of machine-building integration interaction.

The purpose of the study is to analyze the types and elements of integration, as well as to develop an algorithm for implementing the mechanism of integration interaction as one of the key factors in the development of industrial enterprises. The article discusses the benefits arising from the integration of enterprises, describes the types of integration interaction. The research methodology is based on the analysis of information published in the open press and includes empirical, theoretical, cognitive and general logical methods.

An algorithm for the implementation of integration interaction, a fragment of which was tested in the process of merging enterprises of the machine-building industry into one legal entity, was developed as a result of the study. The article notes that the results of such interaction can be expressed through a change in such indicators as: profit growth, an increase in sales income, an increase in profitability, an increase in the number of fixed assets, an increase in authorized capital, etc.

Keywords: Integration · Industrial enterprises · Horizontal integration · Integration structures · Five-factor model

JEL Codes: I 3 · O 1 · O 5

1 Introduction

One of the priority areas in the analysis of the existing conditions for economic development is the study of the development problems of the largest corporations, ways to improve efficiency and find their sources. The key factor in the development of the country's economy, increasing the efficiency of its functioning is the unification of production forces through integration. The emergence of difficulties is mainly a

E. G. Popkova (Ed.): ISC 2020, LNNS 368, pp. 501–507, 2022.
https://doi.org/10.1007/978-3-030-93244-2_55

consequence of the delay at the stage of formation of practical integration mechanism, which would include a plural number of cooperative principles in accordance with the considered sphere of production, the development of legislation and state policy in this area corresponding to these principles.

It can be noted that large organizations are more focused on the formation of integration structures in comparison with small ones, if we are talking about the format of companies for which integration mechanisms are most acceptable.

Improving the mechanism of integration interaction allows opening up promising ways of company development: economy at the production level, cooperation in research and development, separation of emerging threats. Unstable economic conditions, as well as the complexity of the mechanism of interconnection between various production departments in large companies, complicate the process of maintaining a stable position. Organizations can count on the successful implementation of their goals in the process of combining their efforts.

2 Methodology

The issue of integration was considered theoretically by both domestic and foreign authors. Key elements of the development of integration processes have often been the subject of economics. R. Coase, B. Balassa, A.S. Hosni, S. Longy and P. Nykamp., J. Sakamoto, P.J. Wiles, as well as such domestic authors as L.I. Abalkina, O. Yu. Michurina, N.B. Rudyk and others made significant contributions to the study of theoretical issues of integration.

Volkova Yu.A. speaks in her work about a number of benefits arising from enterprise integration. The author identifies several directions [14]:

- benefits for the economy as a whole (structural, financial and investment, scientific and innovative, consumer).
- advantages for the participants of the integrated structure (transactional, transformational, monopoly, synergistic, risky, investment-innovative, financial, independence).

It can be established that there are several basic directions of integration development, which in turn are also suitable for industrial enterprises, based on the works of researchers in the field of integration (Fig. 1).

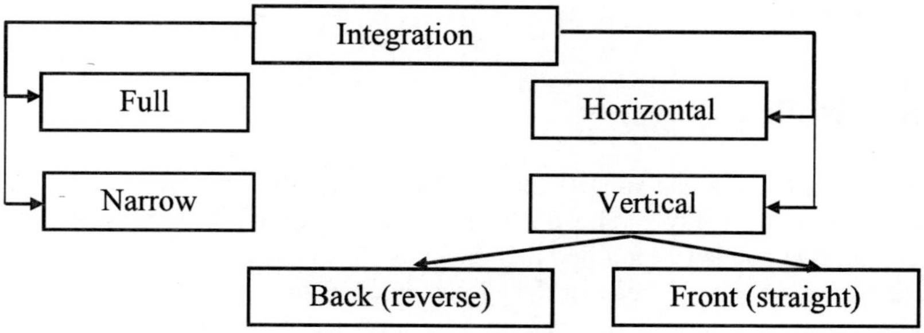

Fig. 1. Possible integration options [12]

For example, the use of vertical integration at the level of the production or technological process of a company allows reducing costs, leads to an increase in the flexibility of the entire system and an increase in the quality of the product. When talking about vertical integration, it is possible to define baselines in the enterprise that are process oriented [3]:

- Strategic planning level: the highest level of management, allows managing the company in various directions (economics, logistics and personnel).
- Company management level: includes planning and analysis of production processes, rationalization of production and product quality control.
- Production process control level: includes production control and diagnostics, production data localization, production element automation, etc.
- Operation level: includes operation control.

Horizontal integration is a combination of enterprises in the same industry, with the same products, through acquisitions or gaining control of one company over another [2]. Moreover, it must be at the same stage of development as the company that it has absorbed.

At the present stage, the possibility of expanding the boundaries between enterprise levels is mentioned more and more often, they will become blurred (vertical integration), and all steps in the process chain will be linked (horizontal integration). It will ultimately lead to increased performance [4].

Within the framework of the study, it was found that the algorithm for developing the mechanism of integration interaction as a determining factor in the development of industrial enterprises currently requires improvement.

3 Results

It can be concluded that a corporation as an organizational form of an enlarged business can become the core of sustainable development of the country's economy as a whole, based on the experience of domestic and foreign researchers. There is an increase in production efficiency, as well as an increase in the competitiveness of enterprises and the country at the international level [1].

Enterprises expect certain results in the process of integration interaction. They can be expressed through a change in such indicators as: profit growth, an increase in sales income, an increase in profitability, an increase in the number of fixed assets, an increase in authorized capital, etc. [9]. The arrangement of priority areas also changes in the subsequent choice of areas of both vertical and diagonal integration based on the existing relations [11]. It is possible to propose an algorithm for the implementation of the mechanism of the integration interaction of Business units on the basis of the studied literature (Fig. 2).

The application of the algorithm for the implementation of the mechanism of integration interaction is presented fragmentarily in this work, because all the stages are quite voluminous and will be considered in the content of subsequent studies.

The merger of enterprises in the engineering industry into one legal entity, let's designate them as Business Unit 1 and Business Unit 2, can be one of the examples of the application of the integration algorithm. The merger took place at the end of December 2020. Both enterprises produce military space defense equipment, have a common management and are located in an adjacent territory [5].

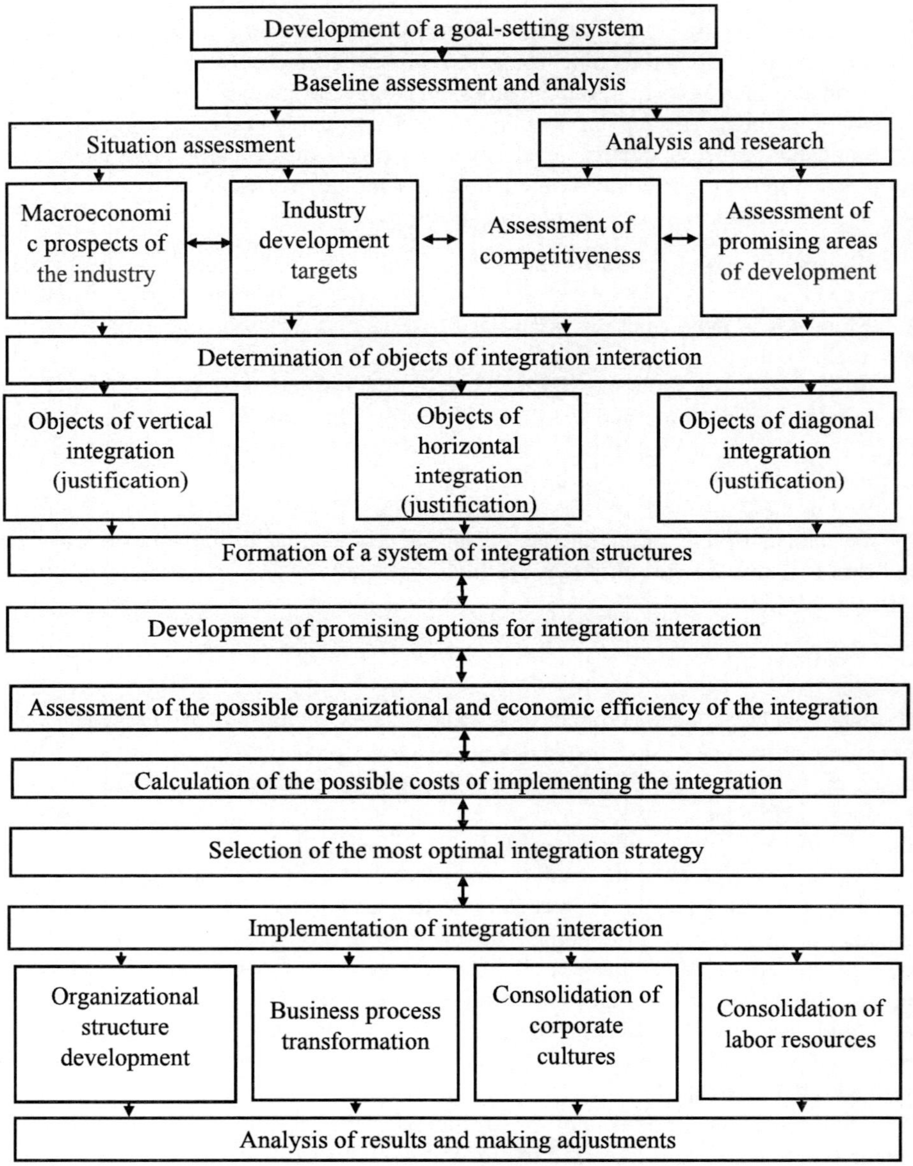

Fig. 2. Algorithm for the implementation of the mechanism of integration interaction. *Source* Developed and compiled by the authors

The prospects of the merger were announced in 2017. At that time, the auditors were already concerned about the further ongoing activities of Business Unit 2, as the enterprise's liabilities exceeded the value of its assets.

The algorithm for implementing the mechanism of integration interaction includes a number of stages, and one of the key stages is «Assessment and analysis of the initial state». It includes an assessment of the situation, as well as an analysis of the study of objects of integration [6]. It is necessary to determine the targets for the development of the industry as a whole and its macroeconomic prospects. And then it is necessary to assess the promising directions of development of the subjects of integration, taking into account their competitiveness.

Competition is one of the most important elements of a modern economy, the state tries to support these processes. The most optimal ways of organizing production, allowing to achieve the most rational exploitation of national resources, arise as a result of competition [8]. Companies that suffer significant losses in the current crisis situation, unable to quickly respond to the impact of external factors, are forced to look for other ways of existence. Bankruptcy can be one of the options, which is an inevitable outcome as a result of the impact of the capital redistribution mechanism. But there are other ways out of this situation, this is horizontal integration, which will allow to preserve part of the production through the merger of two large enterprises into one legal entity [13].

Therefore, it is very important to assess in time the promising directions of development of the subjects and to understand in time whether they have economic difficulties.

One of the tools that can be used in this case is the Altman Model. It allows assessing the probability of bankruptcy according to the balance sheet and income statement [7]. The calculation of the Z-score, on the basis of which it is possible to characterize the state of the enterprise by three boundaries: safe, undefined and dangerous, is carried out. There are several formulas for calculating, it is recommended to use a five-factor model for industrial enterprises. Let's consider her formula:

$$Z = 0,717\,X_1 + 0,847\,X_2 + 3,107\,X_3 + 0,420\,X_4 + 0,998\,X_5$$

where X_1 = working capital/assets;
X_2 = (reserve capital + unappropriated balance)/assets;
X_3 = (profit before tax + percentage to be paid)/assets;
X_4 = equity/obligation;
X_5 = proceeds (net) from sales/assets.

The accuracy of this Altman model is 90.9% in predicting the bankruptcy of an enterprise one year before it occurs.

The interpretation of the results for manufacturing organizations can be as follows: Z less than 1,23 – high degree of bankruptcy; Z between 1,23–2,90 – undefined area; Z more than 2,90 – low threat of bankruptcy.

Let's analyze the reporting of Business unit 1 and Business unit 2 (Table 1).

Table 1. Assessment of bankruptcy according to E. Altman's model

Indicators	Business unit 1			Business unit 2		
	2017	2018	2019	2017	2018	2019
x1	0,69	0,5	0,47	0,49	0,40	0,44
x2	0,21	0,09	0,06	0,01	0,00	0,00
x3	0,12	0,18	0,19	0,02	0,00	0,01
x4	0,23	0,19	0,16	0,10	0,09	0,06
x5	0,39	0,22	0,19	0,40	0,42	0,32
z	1,53	1,29	1,23	0,86	0,74	0,69
Bankruptcy probability	**Undefined area**			**High threat of bankruptcy**		

Source Compiled by the authors

According to the table, we can conclude that the probability of bankruptcy for Business unit 2 was 0,69 in 2019, which is lower than the standard value equal to 1,23, therefore, the probability of bankruptcy of the organization is high. The most preferred option for an enterprise is the integration process with the strongest partner in the industry, which is Business unit 1 (Z in 2019 was 1.23).

As a result of the merger of enterprises, it is planned to actively work on:

- expansion of the range of manufactured products within the framework of the import substitution program;
- development and production of advanced high-tech equipment;
- expansion of joint activities with the largest enterprises of leading industries to establish a joint production process.

Business unit 2 placed an additional share issue, increasing the authorized capital by 26 million rubles, for financial stabilization.

It became possible to create a single team that continues to strengthen the military power of the country, relying on the many years of experience of Business unit 2 and modern technologies underlying the creation of Business unit 1 [10]. At the moment, the enterprises employ more than 5,5 thousand people.

4 Conclusions

Integration of these enterprises is necessary in order to optimize the production process, while the production capacity, product line, contractual and labor relations, rights to the property complex, human resources available today at the reorganized enterprises are preserved. The new integration enterprise cooperates with large Russian firms in the sale of their civilian products.

References

1. Andryashina, N.S., Garin, A.P., Romanovskaya, E.V., Kuznetsova, S.N., Kozlova, E.P.: Analysis of reserves for effective development of production. In: Popkova, E. (eds.) Growth Poles of the Global Economy: Emergence, Changes and Future Perspectives. LNSS, vol. 73, pp. 403–414. Springer, Cham (2020). https://doi.org/10.1007/978-3-030-15160-7_41
2. Balassa, B.: The Theory of Economic Integration. Richard D. Irwin, Homewood (1961)
3. Dedinak, A., Wegerer, K., Haslinger, H., Hadinger, P.: Vertical integration in industrial examples. In: Camarinha-Matos, L.M. (ed.) New Solutions for the Manufacturing Systems of the Future, BASYS 2004. International Federation of Information Processing IFIP, vol. 159. Springer, Boston (2005)
4. Hosni, A.S.: Economic integration theories: a review of economic and political literature. Int. J. Econ. Manage. Soc. Sci. **2**(5), 133–155 (2013)
5. Kozlova, E.P., Kuznetsov, V.P., Garina, E.P., Romanovskaya, E.V., Andryashina, N.S.: Methodological bases of the assessment of sustainable development of industrial enterprises (technological approach). In: Popkova, E.G., Sergi, B.S. (eds.) ISC 2019. LNNS, vol. 91, pp. 670–679. Springer, Cham (2020). https://doi.org/10.1007/978-3-030-32015-7_75
6. Kozlova, E.P., Kuznetsova, S.N., Garina, E.P., Romanovskaya, E.V., Andryashina, N.S.: The concept interpretation of sustainable development mechanism in the conditions of technological transformation of industrial enterprises. In: Popkova, E.G., Ostrovskaya, V.N., Bogoviz, A.V. (eds.) Socio-economic Systems: Paradigms for the Future. Studies in Systems, Decision and Control, vol. 314, pp. 729–736. Springer, Cham (2021). https://doi.org/10.1007/978-3-030-56433-9_77
7. Kuznetsova, S.N., Kozlova, E.P., Kuznetsov, V.P., Shabarov, A.A., Efremova, A.D.: Efficiency improvement of the use of labor resources (on the Example of PAO «GAZ»). In: Popkova, E.G., Ostrovskaya, V.N., Bogoviz, A.V. (eds.) Socio-economic Systems: Paradigms for the Future. Studies in Systems, Decision and Control, vol. 314, pp. 1173–1180. Springer, Cham (2021). https://doi.org/10.1007/978-3-030-56433-9_123
8. Kuznetsova, S.N., Kuznetsov, V.P., Kozlova, E.P., Potashnik, Y.S., Tsymbalov, S.D.: Transformational period of Russian development in the digital economy. In: Popkova, E.G., Sergi, B.S. (eds.) ISC 2019. LNNS, vol. 91, pp. 663–669. Springer, Cham (2020). https://doi.org/10.1007/978-3-030-32015-7_74
9. Kuznetsov, V.P., Kozlova, E.P., Garina, E.P., Romanovskaya, E.V., Andryashina, N.S.: Reporting in the field of sustainable development as a tool of digitalization of the economy. In: Kolmykova, T., Kharchenko, E. (eds.) Digital Future Economic Growth, Social Adaptation, and Technological Perspectives. LNNS, vol. 111, pp. 715–722. Springer, Cham (2020). https://doi.org/10.1007/978-3-030-39797-5_70
10. Longy, S., Nykamp, P.: Economic integration and labor markets: development prospects, pp. 3–24. Regionalization, Growth and Economic Integration. Physica-Verlag, Heidelberg (2007)
11. Pilipenko, A.A.: A reference model for managing the process of enterprise integration development. In: Pilipenko, A.A. (ed.) Scientific works of DonNTU, vol. 30, pp. 118–124 (2006)
12. Perez-Lara, M., Saucedo-Martinez, J.A., Marmolejo-Saucedo, J.A., et al.: Systems of vertical and horizontal integration in the Industry 4.0. Wirel. Netw. **26**, 4767–4775 (2020)
13. Sakamoto, J.: Industrial development and integration of underdeveloped countries. J. Common Mark. Stud. **7**(4), 283–304 (1969)
14. Volkova, Yu.A.: Integration of industrial enterprises in the Republic of Belarus: choice of forms and assessment of efficiency. In: Volkova, Yu.A. (ed.) P.O. Sukhoi State Technical University of Gomel, Gomel, p. 272 (2019)

Check for updates

Formation of Sectoral Clusters as a Mechanism for the Integration Development of the Territorial Economies

Alexander P. Garin[1(✉)], Antonina V. Pavlyukova[2], Arina A. Eroshenko[2], Evgeny A. Semakhin[1], and Anatoly A. Permovsky[1]

[1] Minin Nizhny Novgorod State Pedagogical University, Nizhny Novgorod, Russia
garinsasha@icloud.com
[2] K.G. Razumovsky Moscow State University of Technologies and Management, Moscow, Russia

Abstract. The purpose of the article is to systematize and generalize approaches to the development of territories from the standpoint of regional and local government tools that ensure the transition from «equalizing» to differentiated ways of providing support. Firstly, the authors have systematized the tools for the development of single-industry industrial territories, which made it possible to propose to concentrate on replicating the country's successful experience in the formation of a «spatial structure» based on an even distribution of industry across the territory in order to ensure the most efficient centralized management of production and technological processes, distributed resource base. Secondly, it is proposed to expand the main strategic directions of the development of a part of single-industry towns through the localization of production in accordance with the existing concepts of the development of domestic production and technological potential; formation of mechanisms of cluster economy, creation of industrial (manufacturing) parks. Thirdly, a comprehensive assessment of the level of innovative industrial development of the economy of single-industry entities has been expanded to two groups of indicators - indicators of the first level: the socio-economic development of the territory and the second level: indicators of the growth potential of the economy of a single-industry municipality. The practical application of the proposed solutions in the framework of the implementation of the mechanism of the integration development of the economy of the territories has shown the potential for the growth of the production and technological reserve of the innovative level of the economy.

Keywords: Efficiency · Technological potential · Territory development · Integration interaction

JEL-Code: O32

E. G. Popkova (Ed.): ISC 2020, LNNS 368, pp. 508–515, 2022.
https://doi.org/10.1007/978-3-030-93244-2_56

1 Introduction

Russia is a state characterized by a high degree of distribution of single-industry territories, in which a little more than 10% of the country's population and almost a quarter of the total industrial production are concentrated. Most approaches to the development of monoformations are focused on eliminating the «bottlenecks» of previous periods, do not take into account the specifics of individual territories, its specialization, show inappropriate «spending» on support in the territory with initially high investment attractiveness territory and its long-term survival [1]. Analysis of the sectoral component of the activities of enterprises, concentrated mainly in the field of production, deserves special attention [2]. Support measures and tools for the development of single-industry territories are aimed at their stable development, and providing for the involvement of commercial and non-profit organizations in solving the assigned tasks, «leveling» to differentiated ways of providing support [3].

At the same time, the59 modern development of single-industry towns is associated with a whole range of problems, which include the problems of organizational interaction, largely related to the insufficiently effective mechanism of interaction between all development institutions and management bodies; multidimensionality and multidirectionality of the implemented support measures. The accumulated problems indicate the instability of the current system of state support, arousing an active research interest in this issue. Tasks of the problem of organizational interaction of the participants in the system, associated in many respects with the insufficiently effective mechanism of interaction of all development institutions and management bodies; the multidimensionality and multidirectionality of the implemented support measures have not been resolved, which predetermined the choice of the topic, the purpose of the study and its tasks.

2 Methodology

The analysis of scientific works in the field of methodology for the development of single-profile formations made it possible to formulate and systematize the tools for their development [4–6]:

1. Financial instruments are aimed at reducing budgetary expenditures for financing single-industry territories: 1. Standard lending schemes for an enterprise with sufficient assets as collateral; 2. Leasing tools; 3. Loans and subsidies from the state budget; 4. Co-financing at an interest rate; 5. Co-financing the construction of technology parks and technological innovation zones, etc.
2. Institutional instruments - the development of infrastructure, institutions and diversification of single-industry economies, the components of which are: 1. «Project office», 2. Industrial parks with the status of priority social and economic development area (PSEDA), 3. Technological parks, 4. Intra-regional industrial clusters, 5. Transport, information, innovation, investment, infrastructure, etc.

The study made it possible to identify the features of the use of tools for the development of single-industry territories:

1. A certain consonance of tools, proven mechanisms and organizational forms of their implementation - assessing the effectiveness of the existing system of institutions for managing single-industry education and eliminating "bottlenecks", reducing budgetary spending on financing single-industry territories, diversifying the economy and moving away from the dependence of the territory's economy on the city-forming enterprise, transition to a post-industrial stage;
2. Impact on the socio-economic situation of single-industry cities, strengthening the social component of the issue, such as: creating conditions for a favorable environment and landscaping;
3. Search for diversified sources for the development of single-industry towns in isolation from its city-forming enterprise.

The practice of implementing development tools for single-industry municipalities during the operation of support programs has shown ambiguous results [6]: the unemployment rate in 57,1% of single-industry towns exceeded the average level in Russia, the number of working-age population decreased by 353,6 thousand people, the number of registered individual entrepreneurs decreased by 68,3%, from 2010 to 2019 the number of single-industry municipalities is decreasing extremely slowly - from 335 to 319. An explanation can be an unsystematic consideration of the development issue, the fact of non-fulfillment of certain areas of programs [3, 6]: planned KPI 230 thousand; the actual value is 406.5 (including temporary), created by the PSEDA: planned = 100; actually = 65; the number of single-industry towns: planned - a decrease by 18 units; in fact - an increase in the number of crisis monotowns by 1/4 (from 75 to 93).

3 Results

According to the design of the study, it is important not only to focus on eliminating the «bottlenecks» of previous periods, but also to concentrate on replicating the successful experience of the past of the country in modernizing the approach to managing monoformations, when forming tools for the development of single-industry municipalities. The initial concept of monoformations is the formation of a «spatial structure» of the country based on an even distribution of industry and population over the territory, carried out in order to ensure the most effective centralized management of the chains of production and technological processes, a distributed resource base, which became the basis for the industrialization of the country in the 20th century. It is necessary to expand the main strategic directions for the development of a part of single-industry towns through the localization of production in accordance with the existing concepts for the development of domestic production and technological potential; formation of mechanisms of the cluster economy. In this case, the main drawback of modern approaches to the development of single-profile formations is solved - the disintegration of development programs with strategic national priorities - the creation of highly productive jobs, an increase in labor productivity growth. Instruments and mechanisms of regulation in the implementation of the decision will be within the framework of the support direction «Industrial development and increasing its competitiveness», support services for all stages of investment in the creation or development of an industrial enterprise: technology centers, industrial infrastructure, transformation and modernization of factories into modern urban sites, programs of environmentally friendly technologies, etc.; smart manufacturing technologies; institutionalization of the «full product life cycle» system, etc. [7–9].

An assessment model for predicting the level of innovative industrial development of the economy of single-industry formations in the implementation of the concept of developing the potential of the territory in this case will include [10]:

1. Basic indicators characterizing the level of socio-economic state of a monotown.
2. Indicators characterizing the development potential of single-industry municipalities:

- Demographic processes;
- Growth of the production and technological reserve of the innovative level of the economy of single-industry formations due to the formation of a cluster development territory (a territory intended to accommodate a group of interconnected organizations that complement each other and enhance the competitive advantages of individual companies and the cluster as a whole). In addition, it is possible to implement commercial projects, for example, the creation of industrial (manufacturing) parks;
- Increase in the adaptability of the territory's economy;
- Growth of the innovative level of a single-industry economy;
- Level of social development of a single territory.

The proposed assessment is based on the following indicators [3, 5]:

1. Share of own revenues in the revenue part of the budget of the municipality, %.
2. Average per capita budgetary provision (level of budget expenditures per capita), rubles.
3. Share of unprofitable organizations (excluding small businesses) in all types of activities, %.
4. Balanced financial result profit (+), loss (-) of organizations (excluding small businesses) for all types of activities per capita, thousand rubles.
5. The volume of investments in fixed assets per one employed person (employee) in the city economy, thousand rubles/person.
6. Population growth / decline due to migration per 1000 population, ‰ (ppm);
7. Registered unemployment rate in the municipality, %.
8. Average monthly wages of employees, rubles.

The assessment of the development potential of single-industry municipalities is based on the following indicators:

1. Share in the total number of personnel of enterprises and organizations engaged in research and innovation;
2. How many goods of own production were shipped, how many works and services were performed on their own in current prices, per 1000 inhabitants, thousand rubles.
3. Turnover by enterprises and organizations not related to small business entities, thousand rubles;
4. Indicator of investment activity of enterprises and organizations: by the level of investments attracted by the city's economy in fixed assets per capita,
5. Indicator of innovative activity of enterprises and organizations: by the level of investments attracted by the city's economy in fixed assets per capita, etc.

The proposed comprehensive assessment of the level of innovative industrial development of the economy of single-industry formations in the implementation of the concept of potential development includes two groups of indicators, consisting of indicators of the first level: socio-economic development of the territory and the second level: indicators of the growth potential of the economy of a single-industry municipal formation, determined by the formula:

$$y = a_0 + a_1x_1 + a_2x_2 + a_3x_3 + \ldots + a_nx_n \tag{1}$$

where:

$x_{1,}\ x_{2,}\ x_{3,}\ x_n$ – basic indicators characterizing the level of socio-economic state of single-industry education; or indicators characterizing the potential of their development.

The development of the model is possible through the inclusion in the assessment of indicators characterizing the potential for innovative development of single-industry territories. In this case, indicators can be supplemented.

The proposed changes will create:

- Updated configuration of project management of development of mono-territories;
- Integral organizational system of innovative industrial development of the economy of single-industry entities;
- A system that unites stakeholders in the implementation of projects within the framework of the Integrated «Development of single-industry towns» program;
- Project control and monitoring system.

A set of criteria with reference to the categories of stakeholders is proposed (Table 1) for the purpose of a primary assessment of the effectiveness of the project management system for the development of sectoral clusters.

Table 1. Evaluating the effectiveness of proposals for improving the project approach for stakeholders

Beneficiary category	Performance indicators
Administration of the constituent entity of the Russian Federation	- creation of an integrated project support system in the subject - ensuring organizational and functional unity in the adoption and implementation of decisions in the direction of the development of the territories of single-industry towns Integral assessment based on the results of expert meetings
Investor	- creation of an integral project support system through a single project office Investment size - reducing administrative barriers, increasing the openness of public authorities and local self-government Integral assessment of the number of new industrial entities
Single-industry enterprises	- creation of an integral system of services to support all stages of investment in industrial enterprises of the region Integral assessment - capacity building for technological industrial development Integral assessment - growth and development of the industrial sector of the territory Balanced financial result Investment in fixed assets The number of newly registered items

Source Developed and compiled by the authors

The assessment involves the use of materials obtained in previous projects (implementation of the comparison and analogy method); formation of an expert environment for timely assessment and risk management; maintaining information transparency, assessing compliance with the originally set goals and objectives, regulatory values in terms of:

- mutual influence of the region and the project being implemented,
- identifying threats and bottlenecks,
- assessing the compliance of stakeholders' interests with intermediate results;
- the formation of the final system of indicators and indicators, taking into account the system defined at the planning stage, as closely as possible to the specifics of the project;
- development of methodological recommendations and explanations related to the ongoing monitoring, as well as justification of deviations from the standards identified at the planning stage; data collection; formation of the database and the final report.

4 Conclusion

The result of the study is the determination of the potential for the formation of sectoral clusters as a mechanism for the integration development of the economy of territories belonging to single-industry formations.

The author's classification of tools for the development of single-industry formations is proposed based on the tasks of implementing strategic national priorities. It differs from the others by the introduction of a group of tools for the development of city-forming enterprises of single-industry territories, which predetermine the need to create high-performance jobs: localization of production, formation of a cluster development area, creation of private industrial (manufacturing) parks.

An assessment model has been developed for predicting the level of innovative industrial development of the economy of single-industry formations in the implementation of the concept of developing the potential of the territory. The proposed comprehensive assessment of the level of innovative industrial development of the economy of single-industry municipalities in the implementation of the concept of potential development includes two groups of indicators, consisting of indicators of the first level: socio-economic development of the territory and the second level: indicators of the growth potential of the economy of a single-industry municipal formation in terms of production and technological groundwork, formation of mechanisms of the cluster economy.

References

1. Kozlova, E.P., Kuznetsova, S.N., Garina, E.P., Romanovskaya, E.V., Andryashina, N.S.: The concept interpretation of sustainable development mechanism in the conditions of technological transformation of industrial enterprises. Stud. Syst. Decis. Control **314**, 729–736 (2021)
2. Yashin, S.N., Kuznetsov, V.P., Okhezina, G.M.: Assessment of the Prospects and Feasibility of Process Innovations at an Industrial Enterprise: Monograph, p. 152. Minin University, Nizhny Novgorod (2016)
3. Development of single-industry towns in Russia: Monograph/collection of the author, p. 168. In: Ilyina, I.N. (ed.) Doctor of Economics. Financial University, Moscow (2013)
4. Tolstykh, T.O., Shkarupeta, E.V., Malkova, T.B., Alpeeva, E.A., Garina, E.P.: Algorithm for assessing the efficiency of innovational technologies of industrial enterprises. In: Popkova, E.G. (ed.) Growth Poles of the Global Economy: Emergence, Changes and Future Perspectives. LNNS, vol. 73, pp. 463–471. Springer, Cham (2020). https://doi.org/10.1007/978-3-030-15160-7_47
5. Dmitrieva, E.O.: State support of single-industry towns of the Russian Federation in the conditions of economic growth recovery. Internet Mag. Sci. **9**(1) (2017). http://naukovedenie.ru/PDF/37EVN117.pdf. Accessed 05 May 2021
6. Rubtsov, G., Litvinenko, A.: Development of single-industry towns as a factor of economic and regional growth. St. Petersburg University of the Ministry of Internal Affairs of Russia [Electronic resource]. https://www.e3s-conferences.org/articles/e3sconf/pdf/2020/68/e3sconf_ift2020_08005.pdf. Accessed 01 May 2021

7. Garina, E.P., Kozlova, E.P., Sevryukova, A.A.: Study of alternative strategies and methodological tools for the development of complex systems in the context of the product being created. Azimuth Sci. Res. Econ. Adm. **5**(2(15)), 58–62 (2016)
8. Garin, A.P., Garina, E.P., Batsyna, Y.V.: Technical re-equipment as a condition for ensuring the competitiveness of the enterprise. Moscow Econ. J. **12**, 57 (2020)
9. Potashnik, Y.S., et al.: Determining the value of own investment capital of industrial enterprises. Adv. Intell. Syst. Comput. **622**, 170–178 (2018)
10. Kuznetsov, V.P., Andryashina, N.S.: Innovative activity of industrial enterprises: problems and prospects. Izvestiya V.G Belinsky Penza State Pedagogical Univ **28**, 408–410 (2012)

Modeling of Unbalanced Operating Modes in Low-Voltage Distribution Networks

Igor V. Naumov[1,2](✉), Sergey V. Podyachikh[1], Dmitry A. Ivanov[1], Marina A. Yakupova[1], and Elvira S. Fedorinova[1]

[1] Irkutsk State Agrarian University named after A. A. Ezhevsky, Irkutsk, Russia
[2] Irkutsk National Research Technical University, Irkutsk, Russia

Abstract. The article examines the indicators of the quality of electric energy and additional power losses due to the asymmetry of the three-phase voltage system in rural distribution electric networks 0.38 kV. Violation of the symmetrical mode operation of the electrical network causes a deterioration of the power quality and can lead to fire-hazardous situations. The proposed method of using balancing means modes operation of electric networks is considered as the most effective way to increase energy savings in the power supply in rural areas. The connection of balancing devices (BD) and various types of three-phase symmetric and unbalanced loads, as well as their combinations, are considered on the model of the electric network at various points of power consumption. The parameters of the electrical network and BD are modeled and calculated by the computer complex "Asymmetry-2". The program for the Matlab graphic editor are compiled and diagrams of changes the indicators characterizing the quality and additional losses of electricity from the changing unbalanced load are constructed. A numerical analysis is carried out based on the performed calculation and recommendations are given for determining the most appropriate place for connecting the BD in the electrical network.

Keywords: Power quality · The negative and zero sequences of the voltage unbalance coefficients · Power losses · Voltages and currents unbalance · Power loss coefficient

JEL Codes: L94 · Q49 · P48

1 Introduction

The operation of low-voltage three-phase electrical networks is objectively associated with asymmetric power consumption due to various reasons. Such reasons include the uneven single-phase electric receivers distribution in a three-phase electric network, their probabilistic nature of switching, as well as electric networks emergency modes, leading to the occurrence of incomplete modes of operation in the three-phase distribution of electric energy. A sufficient number of publications are devoted to the study of asymmetric modes [2, 4, 9, 12]. According to published data, it is known that the

E. G. Popkova (Ed.): ISC 2020, LNNS 368, pp. 516–525, 2022.
https://doi.org/10.1007/978-3-030-93244-2_57

three-phase voltage asymmetry system is estimated by the coefficients of negative (K2U) and zero (K0U) sequences voltage asymmetry. The values of these indicators are set by the Electricity Quality Standards of different countries [1, 6, 10]. As established [3, 5, 7, 8, 11, 13, 18, 19, 27], the current asymmetry also leads to an increase in additional power losses, which are estimated by the power loss increase factor (K_P). This coefficient not only leads to overconsumption of electrical energy, but also to the creation of conditions that contribute to the occurrence of fire-hazardous situations [20, 24]. This coefficient leads not only to the electrical energy overconsumption but also to the creation of the conditions that contribute to the occurrence of the fire-hazardous situation [14–17, 21–23, 25, 26]. At the same time, the special means of balance effect [18], aimed at stabilizing an unbalanced power consumption system, has not been studied sufficiently, since the combinations of the changed electric network operating conditions take on numerous variations. As a result, the possibility of modeling such combinations allows us to study the consequences of the emerging asymmetric modes and to develop methods and technical means on their basis that improve the electrical networks functioning. As the most effective symmetry means in numerous literature sources [16], it is proposed to use shunt-balancing devices with minimum zero-sequence resistance. The device power is automatically adjusted depending on the changing of phase current asymmetry level. For the simulation, an electrical network of 0.38 kV is adopted, consisting of a power transformer and a power transmission line with a length of 600 m. Thes line is divided into 4 equal sections, at the end of each of which three-phase unbalanced and symmetrical loads are connected, corresponding to the real electrical energy distribution among consumers.

2 Materials and Method

The electrical network replacement scheme is shown in Fig. 1.

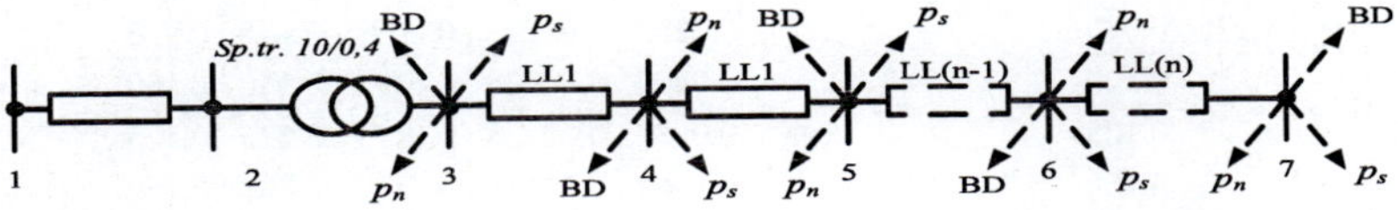

Fig. 1. Electrical network replacement scheme. *Source* compiled by the authors

Figure 1 shows: section 1–2 −10 kV line made by the AC 35 wire, is reduced to a voltage of 0.4 kV and is equal to: $Z_{HV1} = Z_{HV2} = 0.0243 + j0.01 = 0.0263e^{j22,37}$, *Ohms*; section 2–3 – 250-10/0.4 kV power transformer with full complex resistances: $Z_{T1} = Z_{T2} = 0.0094 + j0.0272 = 0.0288e^{j70,9355}$ *Ohms*; $Z_{T0} = 0.0960 + j0.2349 = 0.2538e^{j67,7709}$ *Ohms*; 3 – the buses of 0.4 kV, from which the power line departs, made of self-supporting insulated wire SIP-2: 3 × 70 + 1 × 70 with specific resistances: $Z_{L1} = Z_{L2} = 0.443 + j0.0785 = 0.4499e^{j66.1063}$ *Ohms/km*; $Z_{L0} = 0.363 + j0.06 = 0.3679e^{j9.3861}$ *Ohms/km*; As shown in Fig. 1, in each of the considered load nodes (3, 4, 5, 6, 7) a three-phase unbalanced load (pn) is switched on and a three-phase symmetric load (ps) can be switched on, as well as a balancing device whose parameters are determined by the method described in [16]. The consumers load are connected to the considered electric network varies in accordance with Table 1.

It should be noted that the modules of p_a, p_b, p_c are the ratio of the capacity of single-phase loads in phases to the average full rated power three-phase unbalanced load and module p_s – the ratio of active power symmetric load to its full capacity. The complexes of these values are given by using the arguments: $\varphi_a = \varphi_b = \varphi_c = 25.24^0$; $\varphi_s = 36.87^0$. $p_a + p_b + p_c = p_n$; $p_n + p_s = 1$ - every time when the load are changes. Since the resistance of the negative sequence of a three-phase symmetric load (asynchronous motor) differs from the resistance of the positive sequence, a complex coefficient is used to determine it: $K_{2s} = 0.17 + j0.24 = 0.0735e^{j54,689}$. To each of the nodes (4, 5, 6, 7), and on the transformer busbars also (node 3), a BD can be connected. The BD parameters are defined in accordance with an equal: $Y_{BD1} = Y_{BD2} = 0.0791e^{j90}$; $Y_{BD0} = 0.475e^{j0}$.

Modeling of the incomplete phase mode is carried out by replacing the three-phase unbalanced load with a two-phase one, the change of which is carried out following Table 2. A method for determining the parameters of current and voltage unbalanced understudy was developed for the network model. It was described in detail in [15]. Based on this method, the software package "Asymmetry" was developed. The program allows the calculation of the unbalance voltage coefficients and additional power losses in each load node and on the buses of the 0.4 kV transformer. Let us consider how they change under different studied electrical network operating modes.

Table 1. Changing the modules three-phase unbalanced and symmetric loads

$p_a = p_b$	p_c	p_s	$p_H = p_a + p_b + p_c$
0.0125	0	0.225	0.025
0.0250	0	0.200	0.050
0.0375	0	0.175	0.075
0.0500	0	0.150	0.100
0.0625	0	0.125	0.125
0.0750	0	0.100	0.150
0.0875	0	0.075	0.175
0.1000	0	0.050	0.200
0.1125	0	0.025	0.225
0.1250	0	0.000	0.250

Source compiled by the authors

Table 2. Changing the modules two-phase unbalanced and symmetric loads

p_a	p_b	p_c	p_s	$p_H = p_a + p_b + p_c$
0.0177	0.00425	0.0030	0.225	0.025
0.0355	0.00850	0.0060	0.200	0.050
0.0532	0.0125	0.0091	0.175	0.075
0.0709	0.01700	0.0121	0.150	0.100
0.0887	0.02125	0.0151	0.125	0.125
0.1064	0.02550	0.0181	0.100	0.150
0.1241	0.02980	0.0211	0.075	0.175
0.1419	0.04000	0.0241	0.050	0.200
0.1596	0.03830	0.0272	0.025	0.225
0.1773	0.04250	0.0302	0.000	0.250

The program is written in visual basic 6.0 using specialized libraries (MS Scripting Runtime, MS Excel 10.0 object library) and allows you to evaluate the asymmetry indicators at any point of the model under study with any changes in its parameters and any combination of unbalanced loads and BD. Based on the calculation of the current and voltage asymmetry indicators according to the "Asymmetry" program, the values of the changing coefficients of the negative, zero voltage sequences and coefficient K_P in each of the load nodes are obtained.

3 Results

Figure 2 shows diagrams of the power loss factor change in the absence of BD and its inclusion in various load nodes. The analysis of the presented diagrams showed the following. In all graphs, the last three points of the unbalanced load power change correspond to a significant increase in the power loss factor. This is because at the last points when the unbalanced load increases. There is a sharp decrease in the symmetric load power. Let's consider what BD has an effect when it is turned on in different network nodes. In the BD absence, the average value of the loss factor is 3.64 for all four nodes, that is, the power and energy losses are higher than the similar losses, compared to the symmetric mode almost 4 times. When BD is enabled in the first node, the average ***Kp*** value decreases by 58% in this node, in the second node by 56%, in the third by 33%, and in the fourth by 4%. That is, the greatest simmetrzing BD effect is shown when it is enabled on 1 node. Let us show how the balancing affects the power quality indicators that characterize the asymmetry of a three-phase system. The analysis of Figs. 3 and 4 showed the following. The average value of the ***K2u*** coefficient in all nodes is almost 6% when BD absent in the electrical network. Moreover, its value practically does not change when the BD is turned in 1 node and increases slightly (by 11%) when ones is switched in 2, 3, and 4 nodes.

This is because the voltage's negative sequence does not depend on the BD parameters changing. The zero sequence K0i voltage coefficient is almost 10% when BD is absent, which is almost 5 times higher than the Norms values established.

A significant reduction in this coefficient occurs only when BD in the first node is activated. In this case, the average reduction in the coefficient is 37%. The analysis of Fig. 5 showed that the BD inclusion of any load node does not lead to a decrease in the coefficients of the negative (K2ux) and zero (K0ux) voltage sequences on the transformer buses. Thus, the greatest balancing effect in the three-phase unbalanced load electrical network of the operating mode is possible when BD is turned on in the first load node of the electrical network with a distributed load to reduce additional power losses.

Figures 3 and 4 show the diagrams of negative (K2u – Fig. 3) and zero (K0u – Fig. 4) voltage sequences coefficients changes of the in the corresponding network nodes, and Fig. 5 shows diagrams of the similar coefficients changes on the transformer buses.

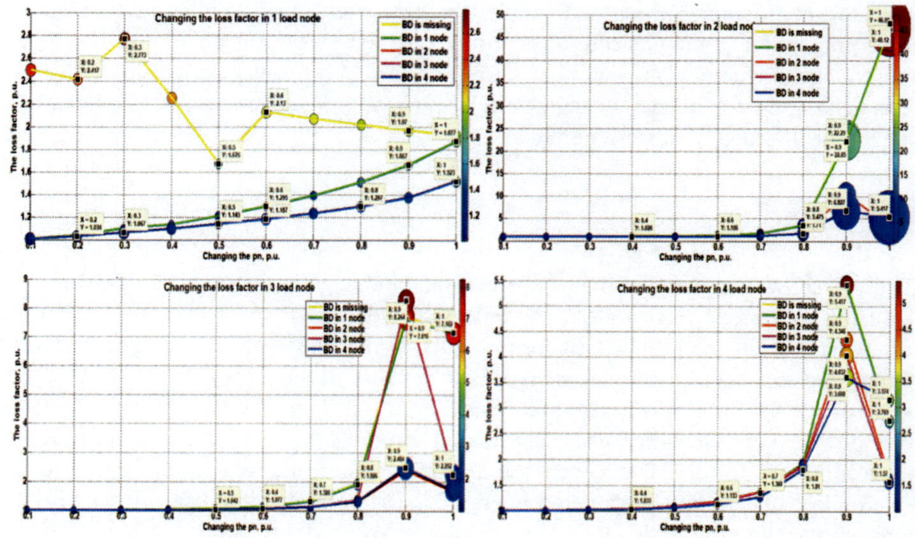

Fig. 2. Change in the power loss factor in the nodes of the considered network with a three-phase unbalanced load and switching on the BD in different nodes. *Source* compiled by the authors

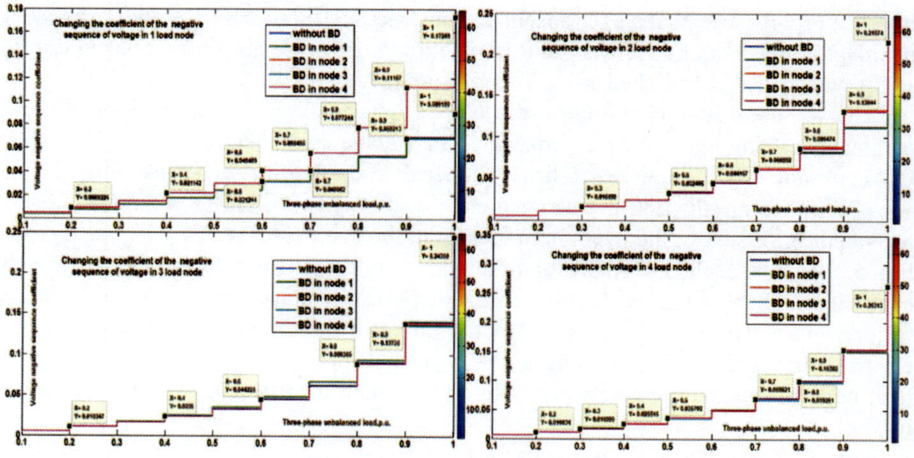

Fig. 3. Change in the reverse sequence coefficient of the voltage in different load nodes when BD is included in them. *Source* compiled by the authors

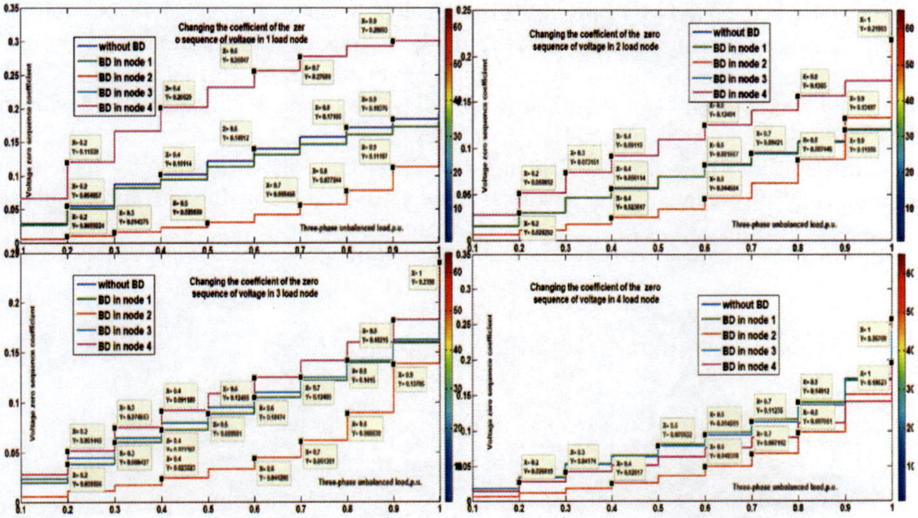

Fig. 4. Change in the voltage zero-sequence coefficients in the network nodes with different connection BD options. *Source* compiled by the authors

Let's consider how the studied parameters change when an incomplete phase mode occurs, for example, as a result of switching off one of the phases (in case of a short circuit or phase break). In this case, the unbalanced load in the two intact phases changes according to Table 2.

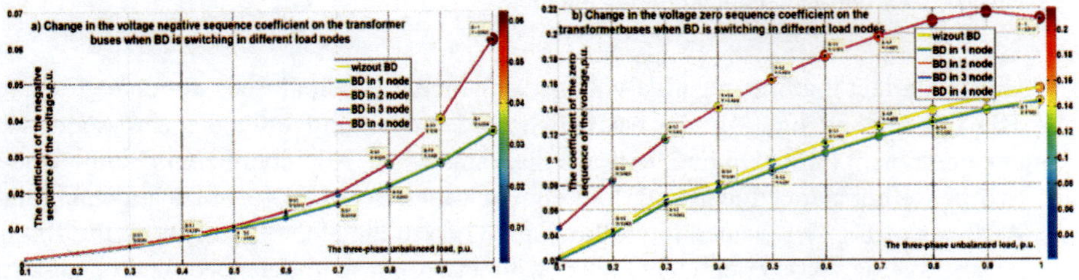

Fig. 5. Change in the voltage negative (a) and zero (b) sequences coefficient on the transformer buses. *Source* compiled by the authors

Diagrams of the power loss factor changes in the load nodes under consideration are shown in Fig. 6. The analysis of Fig. 6 showed the following. When one of the phases of the power line is disconnected (in our experience, this is phase "C") and there is no BD in the network, the average loss coefficient value was 1.123 in the first node, 15.077 in the second, 9.259 in the third and 6.523 in the fourth. As you can see, there is a sharp increase in additional power losses from the last node to the second almost 15 times. When the BD is turned on in the first node, the loss factor changes almost completely, it decreases by only 0.3%. When BD is enabled in other nodes, the Kp coefficient increases. In the second node by 3.3% (regardless of which node the BD was connected to), in the third – by 16.7%, and in the fourth-by 16.1%. Therefore, with the shutdown of one of the phases, it is not advisable to connect a balancing device in any of the load nodes in case of an emergency.

Let's consider how the BD enabling effects on the voltage asymmetry indicators change, that characterizes the electrical network operation in the incomplete-phase mode.

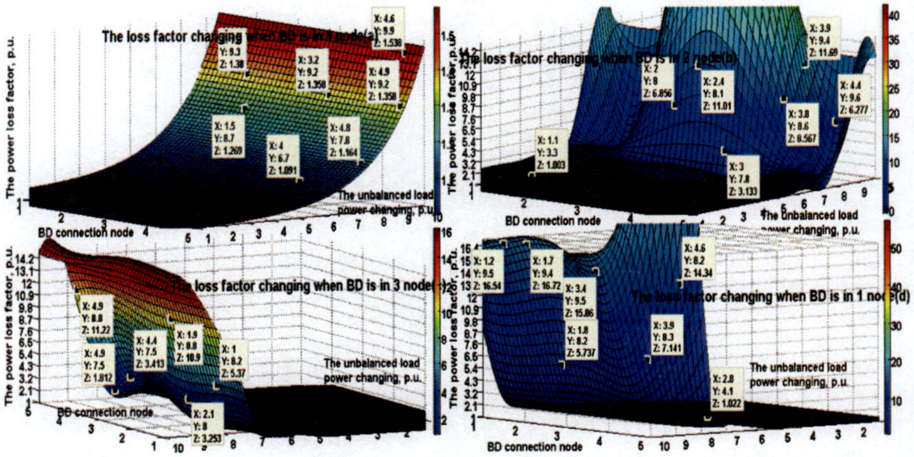

Fig. 6. The power loss factor changing in the electrical network nodes with a two-phase unbalanced load. *Source* compiled by the authors

Under the fact that the negative voltage sequence coefficient does not change when the BD is switched on (Fig. 4), we will consider changing only the zero-sequence voltage coefficient (K0u). Figure 7 shows diagrams of the K0u coefficient change when the two-phase unbalanced load and different combinations of BD enables are changing.

Analysis of Fig. 7 revealed the following. When in the electrical network the BD is absent, the average K0u coefficient value (with a two-phase unbalanced load changing) is the following values: in the first node – 0.895%, in the second – 3.56%, in the third – 5.37% and in the fourth – 6.95%. That is, the value of this indicator increases along the length of the line. In the second node, it increases by 75%, in the third – by 83%, and in the fourth-by 87%. When BD is enabled in any of the nodes, it leads to an increase in K0u in the first node by an average by 5%, in the second – to a decrease by 1%, in the third – to an increase by 1%, in the fourth-to a decrease by 4%. Thus, the greatest balancing device effect in an incomplete phase load electrical network can only be manifested if ones is installed at the end of the power line. But this effect can be considered minimal, given that this will increase the power loss factor.

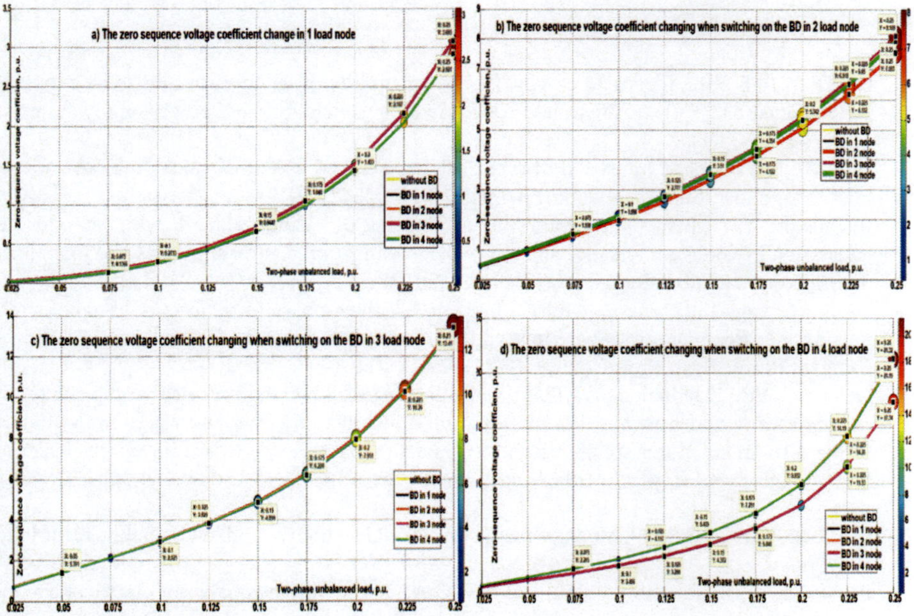

Fig. 7. Diagrams of the zero-sequence voltage coefficients changes in the network nodes when the two-phase unbalanced load changes and various options for connecting the BD. *Source* compiled by the authors

4 Conclusion

Thus, taking into account the above, we can draw the following conclusions.

1. Modeling of asymmetric modes operation electrical networks allows you to assess the level of power quality distortion and its additional losses.
2. The use of balancing devices in the full-phase unbalanced mode of electrical network operation allows improving the quality and reducing electrical energy additional losses.
3. The greatest balancing effect is shown when the BD is turned at the closest to the power source load node.
4. In the event of an incomplete electrical network mode operation (in the event of a break in one of the phases, or its disconnection as a result of an emergency mode), the BD use is not advisable. If BD present in the electrical network, it must be disconnected from the power supply.

References

1. GOST 32144–2013 Electrical energy: The compatibility of technical means is electromagnetic. Standards for the quality of electrical energy in general-purpose power supply systems. in Russia
2. Bao, G., Ke, S.: Load transfer device for solving a three-phase unbalance problem under a low-voltage distribution network. Energies **12**(15), 2842 (2019)
3. Bozalakov, D., Mnati, M.J., Laveyne, J., Desmet, J., Vandevelde, L.: Battery storage integration in voltage unbalance and overvoltage mitigation control strategies and its impact on the power quality. Energies **12**(8), 1501 (2019)
4. Du, F., Cheng, X., Zhou, Q., Xu, F.: Influence of three-phase imbalance and harmonic on line loss of three-phase four-wire low-voltage distribution network. Dianli Xitong Baohu yu Kongzhi/Power Syst. Prot. Control **48**(21), 22–30 (2020)
5. Dulepov, D.E., Kondranenkova, T.E.: Reduction of losses and improvement of the quality of electric energy under asymmetric modes in rural distribution electric networks In: Materials of the VIII International Scientific and Technical Conference "Electric power engineering through the eyes of youth" (Samara: Samara State Technical University), pp. 328–331 (2017)
6. EN 50160:2010/A3:2019: Voltage characteristics of electricity supplied by public electricity networks
7. Fetisov, L.V., Rozhentsova, N.V., Bulatov, O.A.: Improving the quality of electrical energy in low-voltage networks. Izvestiya vysshikh uchebnykh zavedeniy. Energy Prob. **20**(11–12), 99–106 (2018). https://doi.org/10.30724/1998-9903-2018-20-11-12-99-106
8. Fetisov, L.V., Rozhentsova, N.V., Bulatov, O.A.: Improving the quality of electrical energy in low-voltage networks. News High. Educ. Inst. **11–12**, 99–106 (2018)
9. Girshin, A., Goryunov, V., Kuznetsov, E., Safonov, D.: Analysis of asymmetrical modes in medium voltage electrical grids with compensated neutral. In: MATEC Web of Conferences, vol. 70, p. 10008, 1 May 2016
10. IEEE 1159–2019 - IEEE recommended practice for monitoring electric power quality
11. Kholiddinov, I.: Electric power quality analysis 6–10/0.4 kV distribution networks. Energy Power Eng. **8**, 263–269 (2016)
12. Kozlovskaya, V.B., Kalechyts, V.N.: Asymmetrical modes of outdoor lighting lines. Energetika Proc. CIS High. Educ. Inst. Power Eng. Associations **62**(3), 232–246 (2019)
13. Li, R., Wong, P., Wang, K., Li, B., Yuan, F.: Power quality enhancement and engineering application with high permeability distributed photovoltaic access to low-voltage distribution networks in Australia. Prot. Control Mod. Power Syst. **5**(1), 1–7 (2020). https://doi.org/10.1186/s41601-020-00163-x
14. Lukovenko, A.S., Kukartsev, V.V., Tynchenko, V.S., Mikhalev, A.S., Kukartsev, V.A., Bashmur, K.A.: Calculation of balancing and filter compensating devices of the power supply system. In: Journal of Physics: Conference Series. International Scientific Conference "Conference on Applied Physics, Information Technologies and Engineering - APITECH-2019". Krasnoyarsk Science and Technology City Hall of the Russian Union of Scientific and Engineering Associations; Polytechnical Institute of Siberian Federal University, p. 55086 (2019).
15. Naumov, I.: Reducing losses and improving the quality of electric energy in rural distribution networks of 0.38 kV with the help of symmetric devices. Dissertation for the degree of Doctor of Technical Sciences (Saint Petersburg), p. 387 (2002)
16. Naumov, I., Belousova, E.: Choice of the device parameters of balancing in distribution networks 0.38 kV. Bull. Krasnoyarsk State Univ. Krasnoyarsk **1**, 99–107 (2017)

17. Naumov, I.V, Savina, N.V., Shevchenko, M.V.: Modeling of three-phase electric motor operation by the MATLAB system with deteriorated power quality in the 0.38 kV distribution networks. In: E3S Web of Conferences Rudenko International Conference: "Methodological problems in reliability study of large energy systems" Section: "Energy Security, Reliability and Quality of Energy Consumption, Modeling and Information Technology", vol. 58, p. 03016, 1 April (2018)
18. Naumov, I., Karamov, D., Tretyakov, A., Fedorinova, E., Yakupova, M.: Additional electric loss in rural distribution networks 0.38 kV. In: E3S Web of Conferences vol. 209, p. 07007 (2020). EISSN: 2267–1242 ENERGY-21
19. Naumov, I.V.: Additional power losses in low-voltage electrical networks and their influence on people. In: RPTSS 2018. International Conference on Research Paradigms Transformation in Social Sciences. The European Proceedings of Social & Behavioural Sciences EpSBS. Future Academy, pp. 844–853 (2018). ISSN: 2357–1330
20. Naumov, I.V.: An analysis of fires due to violation of the rules for design and operation of electrical plants, and how to prevent them in residential buildings (on the example of the amur region). In: International Science and Technology Conference "EastConf" INSPEC, pp. 1–6 (2019) Accession Number: 18724557
21. Naumov, I.V., et al.: Device for balancing currents and voltages in a three-phase network with a zero wire and self-regulating inductance. Certificate for utility model No. 26699: (51) IPC 7 No. 02 J 3/26. Issued on 03.06.2002 with the formula: (21)2002114471/20 (2002)
22. Naumov, I.V. et al.: Symmetric device for a three-phase four-wire network with adjustable parameters. Patent for utility model No. 61063-Published on 10.02.2007 Byul. no. 4 (2007)
23. Naumov, I.V., Ivanov, D.A., Podyachikh, S.V., Gantulga, D.: Symmetric device for three-phase networks with zero wire. Patent for invention No. 2490768. Registered in the State Register of Inventions of the Russian Federation on 20, August 2013
24. Naumov, I.V., Savina, N.V., Yamshchikova, I.V., Shevchenko, M.V.: Methods for fire risk reduction in residential premises with the help of balancing devices. IPASJ Int. J. Electr. Eng. (IIJEE) **5**(5), 16 (2017). ISSN 2321–600X
25. Oskin, S.V., Makarenko, A.S., Miroshnikov, A.V.: Increased probability of proper operation of filter protection devices against phase failure at voltage asymmetry in rural electrical networks. In: Proceedings - 2020 International Ural Conference on Electrical Power Engineering, UralCon 2020, vol. 9216236 (2020)
26. Xie, Y., Wang, L., Li, P., Li, T., Su, X.: Research on dynamic hybrid compensation technology and device in low voltage distribution area of transformer. Journal of Physics: Conference Series, vol. 1549, no. 5, p. 052053 (2020)
27. Zhilin, M.I., Vorkunov, O.V.: Improving the quality of electric energy the priority directions of development of science a Collection of articles on materials of the international scientific-practical conference in 2 parts. Limited Liability Company "Scientific Partnership "Apex", Moscow, pp. 83–85 (2017)

Digital Communications in Marketing Strategies of Advertisers: Global and Russian Approach

Galina G. Deryabina[1,2](✉) and Nina V. Trubnikova[1]

[1] RUDN University, Moscow, Russia
trubnikova_nv@rudn.ru
[2] Synergy University, Moscow, Russia

Abstract. The article is devoted to changes in the advertising market, which are dictated both by the state of the world economy and the economy of individual countries and by intra-industry factors.

The author systematizes the problems that have been exacerbated in connection with the global crisis caused by the spasmodic course of the COVID-19 pandemic. The dynamics of the advertising market as a whole is a derivative of the dynamics of the consumer market, however, during the crisis, it might not completely coincide with it.

Particular attention is paid to the most dynamic segment of the advertising market: digital advertising. The pace of the transition to digital communications, on the one hand, is determined by the development of the digital economy as a whole; while on the other hand, the technological features and characteristics of communications set it. The authors carried out the analysis of the logic of these processes based on the activities of the main participant and initiator of marketing communications: the advertiser. To remove the attitude of customers to the installation of various digital tools in communication strategies, the authors studied the features of various channels and the historical stages in the development of digital communications in the field of marketing. The authors concluded that in the modern world the digital channels would effectively work on the image and brand building of the advertiser only in combination with classical communication tools, actively influenced by digitalization. The dynamics of advertisers' attitude to digital communications is associated both with the technological aspects of the development of this area, which open up new opportunities, and also with the economic conditions that set the basis for marketing transformations in the modern world, in particular, in the difficult situation of the COVID-19 pandemic.

Keywords: Mediaadvertising market · Digital communications · Advertiser · Advertising budget · Pandemic impact

JEL Codes: F63 · M30 · M31 · M37

E. G. Popkova (Ed.): ISC 2020, LNNS 368, pp. 526–534, 2022.
https://doi.org/10.1007/978-3-030-93244-2_58

1 Introduction

The digital advertising media of today can compete with traditional media in terms of reaching the target audience. Experts in the communications industry must study digital innovations such as Mobile Internet, Cloud technologies, Internet of Things, which are having a revolutionary impact on the economy on a global scale, to anticipate the changes and to develop a new marketing approach [3].

The research questions centered on the digital transformation of marketing approaches and the impact of this process on the advertising market in general and on the activities of advertisers, in particular, are associated with this phenomenon. A crisis is capable of provoking an even stronger fluctuation in the dynamics of the advertising market relative to the consumer market, which can be both anticipatory and expectant. The purpose of the work is to study the dynamics of the attitude of advertisers to digital communications. The theoretical nature of the research question is complemented by a practical component: what tools of digital influence do the participants of the advertising market, represented by top managers of advertisers and the advertising agencies choose during the crisis; and how to develop these tools to find the solution, which might interest the communication customer.

Thus, based on the above goal, a set of tasks can also be distinguished:

1. To study external and intraindustrial factors affecting the advertising market, both globally and in Russia.
2. To analyze the concept of a marketing and communication channel and, in this regard, to clarify the concept of a digital channel.
3. Based on the allocation of digital channels, consider changes in the attitude of advertisers for this type of communication.
4. To focus on the problem of transparency of digital communications and to consider it in the plane of advertising management.
5. To study the possible prospects of a new type of combining online and offline communications and the impact of such a combination on the advertising strategy of the advertiser.

2 Methodology

The information base of this research includes the works of scientists and communication specialists who have made a significant contribution to the development of streamlining various marketing approaches [5, 14, 20].

[17] were among the first at the analytical level, while, without deep dive into the diverse manifestations of digital advertising in practice. [16] emphasized the new role of the user as the initiator of advertising contact and the active subjective relationship of the consumer with the brand. They later continued the research. Starting with classic paid advertising, the modern media landscape has changed significantly. This landscape has a systemic impact on the advertising market and requires in-depth study, and the models of functioning of the media advertising market worked out well abroad are undergoing significant changes in Russian conditions.

The advertising experts Veselov, Kolomiets, Evstafiev, Stolyarova, and other regular authors of the leading industrial media “Advertising Yearbook” about advertising (published for the 20th time this year) have made a significant contribution to the theoretical, practical, and methodological study of the phenomenon of digital advertising in Russia [2].

The work implements an approach focused on obtaining conclusions based on the analysis of the behavior of market entities under the influence of digital drivers, which sets a certain approach to sources. In general, in the world of professional discourse, one can speak of a strictly limited opportunity for an individual scientist to collect “field” material in a designated area both at the macro level and even at the level of individual companies. This limited access to the object of research is explained by the factor of commercial secrecy and high-tech requirements for research tools. Therefore, to understand the processes taking place in the digital advertising market, it is necessary to refer to the comparative analysis of reports and research materials of global companies and authoritative Russian research organizations.

However, the analysis of the implementation of digital technologies at the level of an individual corporation can also enrich the understanding of real processes; therefore, data from surveys of leading advertisers became especially important for the authors’ analysis, based on the results of which the IAB has been initiating a research project “Russia Digital Advertisers Barometer” for several years [7–13].

The authors analyzed the dynamics of advertisers’ attitudes towards channel usage as well as available tools and formats on the digital communications market in the desk research format based on an authoritative study that has been annual, starting in 2014, studying the prospects of interactive advertising in Russia from the point of view of companies that order communication services.

3 Results

External and Internal Industrial Factors in the Development of the Advertising Market

Among the factors influencing the advertising market, the authors distinguish a complex of factors external to the industry and intraindustrial.

The most important external factor for the advertising market is the situation of a pandemic. Its course and consequences for the development of the communication sphere cannot be quickly predicted using economic tools, since this crisis has not economic roots, but a medical and social origin. Another important external factor, negatively and for a long time affecting the economy of our country, is still the economic sanctions of political origin, introduced for Russia since 2014 by some economically most developed countries and which, in the light of recent events, were not removed, but even intensified.

Significant foreign advertisers and advertising agencies have seriously limited their investments in the Russian communications economy and media, which has led to a decrease in the dynamics of the advertising market development. This harmed the dynamics of the development of the Russian market, although it gave a chance for the development of domestic communication structures and brands in a territory free from global structures. Despite the above two factors, the integration of the domestic advertising market with the global advertising market is very high. The digital-advertisers-barometer study analyzed by the authors represents the opinion of big advertisers, including international ones, for which the processes, happening in Russia are part of their marketing and communication strategy.

Intraindustrial factors, in contrast to those external factors that the advertising industry essentially cannot influence and must adapt to them, are compensatory and, with proper management of the communication market, can weaken the influence of external factors. We mean here the activities in the field of advertising of the main subjects of the advertising market: advertisers, advertising agencies, advertising distribution media, and advertising consumers, which showed active subject functions. To a large extent, the volumes and dynamics of the advertising market depend on the actions of advertisers in terms of the volume and areas of advertising investments.

In the context of this study, communication channels are understood as a type of marketing channel. Both types of channels, communication and sales channels [14], have common tasks in the strategic management of the company, but on the tactical front, they have a pronounced specificity. While sales channels facilitate operations, communication channels create comfortable access to content for clearly delineated target audiences. The blurring of the boundaries between advertising and PR ("fourth p", that is, promotion) and other areas of marketing activities taking place in the digital environment makes it difficult for the advertiser to understand what results in he can and wants to achieve using communications in the digital environment.

In earlier periods an advertising campaign and a sales promotion campaign were independent with their budget, nowadays, the budget becomes general, and its distribution is determined by the realities of the omnichannel approach and specific tactical decisions on the situation. In modern conditions of the synergy of online and offline channels, for the target audience to perform the target action faster and more accurately, the advertiser needs strategic thinking and understanding of the prospects of his business.

As already mentioned before, the authors used digital-advertisers-barometer data for analysis to consider these issues. The title of the study emphasizes that it captures the changing "weather" in the "advertising market", trends and prospects for the development of attitudes towards digital communications in the business environment. Its organizer, IAB RUSSIA (The Interactive Advertising Bureau Russia), determines the scope of the research. This Association for the Development of Interactive Advertising is part of the international network of interactive advertising associations and is a unique platform where representatives of leading agencies and advertisers come together to solve pressing industry problems.

This study of the association is not strictly panelized: the respondent companies rotate 15–20% annually. The 2020 study ran in the most challenging conditions to estimate and predict and has a sample bias towards those whose businesses survived the pandemic. But the representation of spheres of activity, the size of the business, the number of respondents (about 50) as a whole remains the same, therefore, a comparative analysis of data for seven years of the study turned out, in the opinion of the authors, to be representative and led to significant results. It is important that the study on Russia was attended by international companies whose business is international in nature and develops in the local conditions of a particular country according to global scenarios.

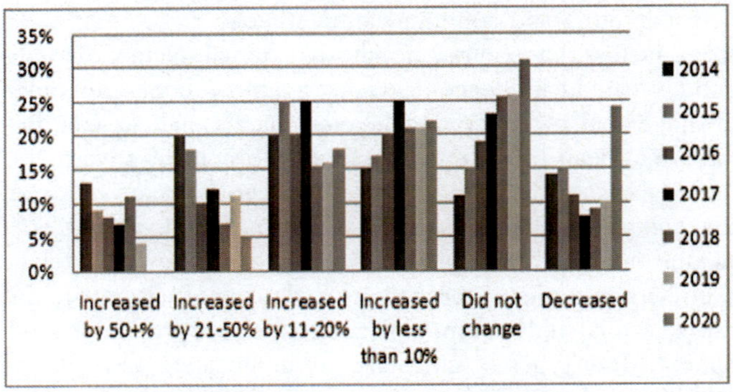

Fig. 1. Dynamics of advertisers' budgets for promotion in digital channels, 2014 to 2020. Source Compiled by the authors based on [7–13].

Figure 1 shows the summary statistics on the dynamics of budgets for promotion in digital channels from 2014 to 2020. The data allows us to assess the activity of advertisers in this area and to notice that share of "extreme" ones, revising their strategy towards an expressive increase in the use of digital tools in the media mix (growth above 50%), is constantly decreasing, no more than 3–4% of such companies in 2019, and no one in 2020. The market can be assessed as having signs of sufficient development due to the "conservative layer", that is those advertisers who keep the ratio between investments in digital and the rest of their budget at a stable level. The proportion of those who maintain proportions is growing steadily in 2020, more than 30% of advertisers belonged to this category.

A marker of the impact of the pandemic was the increase in the share of advertisers, which reduced budgets for digital promotion: more than 20% of advertisers. This trend is observed in the entire world. Moreover, in many ways, because the many modern types of communications do not provide advertisers with tools to control how effectively funds are spent [18].

It is becoming clear that the dynamics of digital communications must be viewed in light of the current reduction in marketing spending in general. Perhaps the investment of advertisers in offline communications has decreased even more. In a crisis, companies do not have enough funds for promotion in general, and for many businesses in the face of changing demand, such promotion is simply irrelevant.

The absolute numbers are shown in Fig. 2. Here we can observe an interesting pattern, a simultaneous increase in the number of those advertisers who reduce the share of digital advertising in the media mix in favor of the classic type and those who increase this share, while the number of "moderate" advertisers who are not inclined to either decrease or increase this part is decreasing. This polarization speaks not only of the uneven development of the digital advertising market, but also of the processes that occur among advertisers, and which are associated with the inability to fully solve the problem of transparency of digital tools and clear criteria for their effectiveness [19].

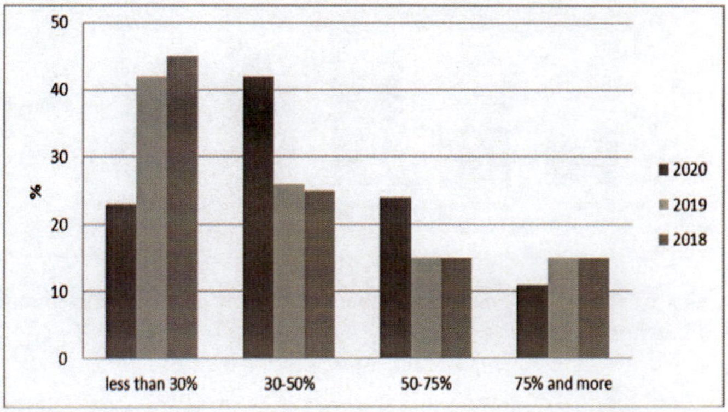

Fig. 2. Promotion budget spent on digital advertising (advertisers estimate). *Source* Compiled by the authors based on [7–13].

The number of digital channels chosen by advertisers for their promotion (Fig. 3) grew from 2018 to 2019. The number of advertisers who invest in only one channel was minimized to an insignificant few percent, and more than 40% of advertisers began using five or more channels by diversifying investments into different instruments. Understandably, the pandemic 2020 has shown a more cautious attitude of advertisers towards multichannel. Comparative analysis showed that the standard "menu" of the advertiser has become somewhat impoverished. It happened due to both the overall reduction in promotion costs caused by the reduction of many business segments, and the desire of advertisers in difficult times to choose proven tools with proven effectiveness.

This idea is supported by the data in Fig. 4, where the percentage of advertisers willing to use non-standard formats and experiment with different types of influence, surprising their target audience, has dropped significantly. The number of those radicals who are ready to invest more than 30% of their budget in experiments has halved, now only every tenth advertiser makes it. More than 40% of companies will spend a modest 10–20% of their digital communications budget on innovative tools. The greatest interest among the respondents is caused by native advertising, the capabilities of which are trusted by about 40% of the respondents.

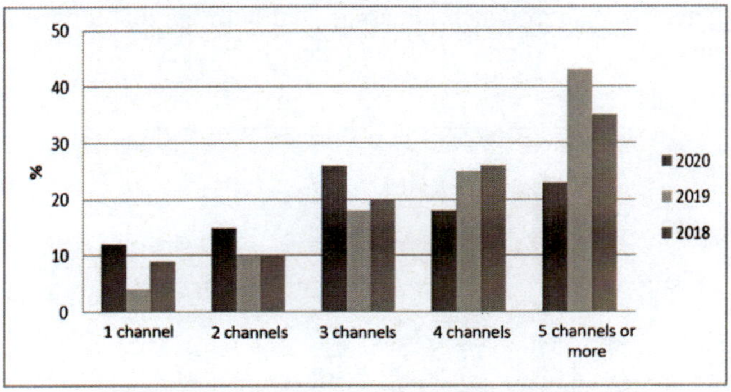

Fig. 3. Dynamics of the use of various numbers of digital channels by advertisers. *Sources* Compiled by the authors based on [7–13].

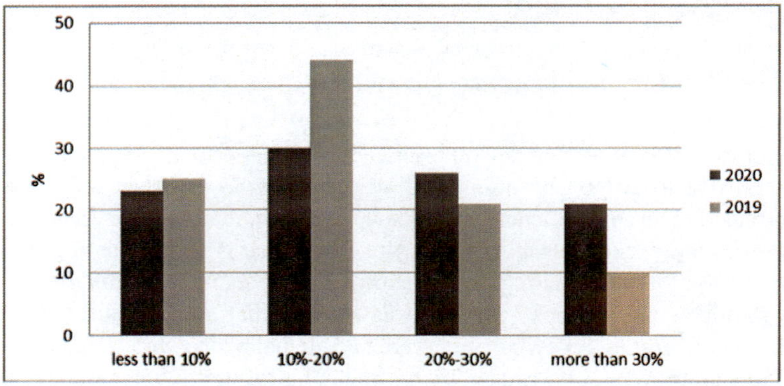

Fig. 4. Dynamics of budget used by advertisers for promotion in non-standard formats (2019–2020). *Sources* Compiled by the authors based on [7–13].

4 Conclusions

A comparative analysis of advertisers' attitude towards digital communications during crisis changes, including those related to the COVID-19 pandemic, carried out by the authors, proves an inextricable link between the processes taking place in the communications industry and general trends in the development of the global economy. Despite a more cautious attitude towards various digital communications than the previous years, the budgets in this area continue to grow. This indicates the progressive development of this segment and the advertisers trust the technological opportunities and the growth drivers providing the new types of marketing communications compared with the classic promotion.

The pandemic affected the rethinking of the role of digital advertising, which is no longer perceived by advertisers as a tactical tool, aimed exclusively at increasing conversions and sales (performance). In previous years, the advertisers believed that digital advertising was not suitable for solving branding tasks and increasing brand awareness, nowadays, with the development of various tools, the advertisers choose this type of advertising not only for sales, but also to form a loyal audience. Moreover, these tools can be used for the correct alignment of the so-called customer journey [6], for example, various popular types of video advertising (instream, outstream, mobile video) are especially suitable for solving such problems today.

Traditional marketing continues to play an important role in raising awareness and engagement with the target audience at the early stages of a brand's engagement with an audience. Following the brand's relationship with the consumer evolving, when a closer relationship with the customer is built, digital marketing begins stimulating the audience to take action and further spread positive information about the brand. At this stage, the omnichannel approach comes into force, combining offline and online promotion [4].

Recording the dynamics of the advertiser's attitude to communications in the digital environment in a study based on the opinions of representatives of the companies makes it possible to optimistically assess the prospects for the development of the Russian media advertising market as an integral part of the world economy. The mutual enrichment of digital and classic tools will create a synergy effect and open up complex tools for advertisers that will allow them to take a fresh approach to solve the entire spectrum of marketing problems.

References

1. Advertising market ratings: Sostav. https://www.sostav.ru/ratings/business (2020). Accessed 10 Apr 2021
2. Advertising Yearbook 2019 Moscow (2020). in Russia
3. Benavides, L., Khanam, R., Lefort, F., Lovera-Perez, O.: Breakthrough technologies fundamentally change the game. McKinsey & Company. https://www.mckinsey.com/industries/consumer-packaged-goods/our-insights/breakthrough-technologies-fundamentally-change-the-game (2017). Accessed 15 Mar 2021
4. Critical Channels of Choice: Meeting the customer expectation for omnichannel relationships. CMO Council. https://www.cmocouncil.org/thought-leadership/reports/critical-channels-of-choice (2021). Accessed 12 Apr 2021
5. Davis, J.: Advertising research theory and practice. Williams (2003)
6. Edelman, D.C., Singer, M.: Competing on Customer Journeys, pp. 88–100. Harvard Business Review, Brighton November 2015
7. IAB Russia Digital Advertisers Barometer – Moscow 2014. in Russia. https://pdfslide.net/documents/iab-russia-digital-advertisers-barometer-2014 (2014). Accessed 10 Mar 2021
8. IAB Russia Digital Advertisers Barometer – Moscow 2015. in Russia. https://present5.com/iab-russia-digital-advertisers-barometer-2015-perspektivy/ (2015). Accessed 10 Mar 2021
9. IAB Russia Digital Advertisers Barometer – Moscow 2016. https://datainsight.ru/sites/default/files/DI-IABBarometer2016.pdf (2016). in Russia
10. IAB Russia Digital Advertisers Barometer – 2017. https://datainsight.ru/sites/default/files/DI-IABBarometer2017.pdf (2017). in Russia
11. IAB Russia Digital Advertisers Barometer – 2018 Moscow (2018). https://www.datainsight.ru/sites/default/files/IAB_Barometer_2018.pdf in Russia

12. IAB Russia Digital Advertisers Barometer – 2019 Moscow (2019). in Russia. https://www.datainsight.ru/sites/default/files/DI-IAB_Barometer_2019.pdf
13. IAB Russia Digital Advertisers Barometer – 2020 Moscow (2020). in Russia. https://datainsight.ru/sites/default/files/IAB_Barometer_2020.pdf
14. Kotler, P., Kartajaya, H., Setiawan, I.: Marketing 4.0. Moving from traditional to digital, Moscow (2019)
15. Rodgers, S., Thorson, E.: Advertising Theory. Rutledge, New York (2012)
16. Rodgers, S., Thorson, E.: The interactive advertising model: how users perceive and process online ads. J. Interact. Advertising **1**(1), 42–61 (2000)
17. Schumann, D.W., Thorson, E.: Advertising and the Internet. Lawrence Erlbaum Associates Inc, Hills Dale (1999)
18. Stewart, R.: FMCG giants still waste $50bn a year on digital despite media clean-up. the drum (2019). https://www.thedrum.com/news/2019/08/07/fmcg-giants-still-waste-50bn-year-digital-despite-media-clean-up
19. Veselov, S.V.: On the issue of transparency of Internet advertising Advertising: theory and practice **4**, 246–256 (2019). in Russia
20. Wells, W., Burnett, J., Moriarty, S.: Advertising: Principles and Practice 3rd edn. St. Petersburg (2001)

Using Scripts and Overcoming Barriers to Effective Communication Between Staff and Guests in Hospitality Organizations: Problem Analysis

Larisa Kh. Gazgireeva[1(✉)], Ludmila A. Burnyasheva[1], Ivan P. Kulgachev[2], Larisa A. Savinkina[3], and Marina G. Gadzhimuradova[4]

[1] Russian State University of Tourism and Service, Moscow, Russia
[2] Plekhanov Russian University of Economics, Moscow, Russia
[3] Moscow State University of Sport and Tourism, Moscow, Russia
[4] Pyatigorsk State University, Pyatigorsk, Russia

Abstract. Hospitality can be considered a powerful industry, developing during the period of the experience economic growth. The Horeca sphere is constantly replenished with millions of professionals, trying to satisfy the desires of customers.

This article analyzes the barriers in intercultural communication, reveals the main reasons for their occurrence, describes the rules and principles of building effective communication operating in the Horeca sphere, analyzes the intercultural barriers of the sanatorium-resort complex of the "Mashuk Aqua-Term" enterprise, develops measures to overcome the barriers to effective communication of this sanatorium-resort enterprise, and also substantiates the necessity of drawing up and using ready-made scripts .

The focus of hospitality industry enterprises on developing the ability of employees to intercultural business communication and the ability to overcome barriers to effective communication between staff and guests allows the authors of this article to concentrate on the specifics of accounting for the functioning of certain types of barriers and the use of some scripts (speech modules). The results obtained and the measures developed to overcome the barriers to effective communication will improve the process of interaction between staff and guests not only in the "Mashuk Aqua-Term" health resort complex, but they also can become a guide to action for personnel of other hospitality organizations.

Keywords: Guest · Staff · Scripts · Speech modules · Hospitality industry · Intercultural communication · Barriers · Effective communication

JEL Codes: Z30 · Z31 · Z32 · Z33

E. G. Popkova (Ed.): ISC 2020, LNNS 368, pp. 535–545, 2022.
https://doi.org/10.1007/978-3-030-93244-2_59

1 Introduction

The actuality of the topic of this article is dictated by some circumstances: firstly, in the community, there is a need for emerging intercultural relations, including government institutions and services, the media, business communities, the population of the host country as a whole; secondly, in the hotel industry there is a growing need to adapt programs and routes for national linguocultural target markets, their support, and information support, training staff to communicate effectively with the guest, conduct training, obtain professional education in the area of hotel service - the ability to build a communicative dialogue between staff and guests, use scripts, constantly improve knowledge, skills, and abilities to overcome barriers to effective communication between staff and guests in hospitality organizations.

So, let's turn to the definition of "communication barriers". Communication barriers are obstacles caused by natural, social, and psychological factors that arise in the process of communication. In the modern world in the context of human interaction, in various areas of activity, difficulties arise in the process of communication [7].

The information above allows confirming the need for the development and application of theoretical provisions describing the algorithm for overcoming communication barriers in hotel and sanatorium enterprises. Consequently, the purpose of this study is to consider the problem of the emergence of communication barriers and identify the causes of their occurrence, as well as develop recommendations for overcoming them, where a special role is given to the observance of scripts and speech modules for communicating between hotel staff and guests. Strengthening the position of using scripts and speech modules will contribute not only to effective communication between hotel staff and guests but also to the development of a smart concept in hospitality enterprises.

2 Materials and Methods

The theoretical and methodological basis of the study was formed by the works of not only many well-known domestic, but also foreign scientists in the area of the theory of the provision of hotel and tourist services in the context of the transformation of socio-economic systems in the modern period of society development.

The methodological basis of the study is the fundamental provisions of the scientific direction - the economy of the sphere of hotel and tourist services. This scientific research is carried out from the standpoint of consistency of views, taking into account the structural, functional, and managerial aspects, as well as the principles of determinism, complexity, and consistency.

In this article, the authors rely on general scientific research methods - expert assessments, analysis, and synthesis, induction and deduction, a systematic approach, analogy, and classification, economic methods for assessing the quality of management in the sphere of hospitality and tourism.

3 Results

Minenkova believes that taking into account the development of the economy, the hospitality industry hasn't lost its relevance, since for many centuries the problem of intercultural interaction will be in the first place among others because the specifics of communication between staff and guests will depend on the knowledge and experience of hotel employees and their possession of certain competencies [9]. Therefore, according to the demands in the Horeca industry, the need to improve the quality of service delivery, in particular in the area of communication, will always arise. The territorial analysis demonstrates, that the tourist potential of the studied regions is very large [5], but such factors as lack of information, undeveloped and not adapted to the needs of tourists infrastructure, the inability of the hotel staff to overcome communication barriers in communicating with the guest reduces their value [3].

Barriers exist at the level of interpersonal and at the level of organizational communication.

The authors characterize the enterprise of the sanatorium-resort type "Mashuk Aqua-Term", which is located in the village of Inozemtsevo on the border of the resorts of Zheleznovodsk and Pyatigorsk, between the mountains Zheleznaya, Zmeika, Mashuk, and Beshtau. Sanatorium "Mashuk Aqua-Term", which is located in the specially protected ecological-resort region of the Russian Federation Caucasian Mineral Waters, is of particular value for guests, who want to relax in one of the buildings of this enterprise [11].

Having analyzed the barriers using the example of the "Mashuk Aqua-Term" sanatorium-resort complex, which is located in the city of Zheleznovodsk, the following conclusions can be drawn. To begin with, many of the terms used in intercultural communication are unfamiliar. The authors conducted an anonymous survey to understand the level of knowledge of certain terminology of guests and employees of "Mashuk Aqua-Therm" – they interviewed 50 guests of this sanatorium and 10 employees of this enterprise. Several well-known terms were proposed, among which the words that are known for their lexical meaning have to be chosen.

The results are presented below the next text.

1 What terms are unfamiliar to you: verbal, non-verbal; sender (generator of ideas, addressee, communicator, and forwarding agent), encoding/decoding (in the technical sense), message (meaningful and encoded information), communication channel, effect, recipient (interpreter of ideas, addressee, and communicator, recipient), feedback, communication barriers, context; rapport, congruence, pacing, affiliation, attraction, "stroking" according to Bern (Fig. 1).

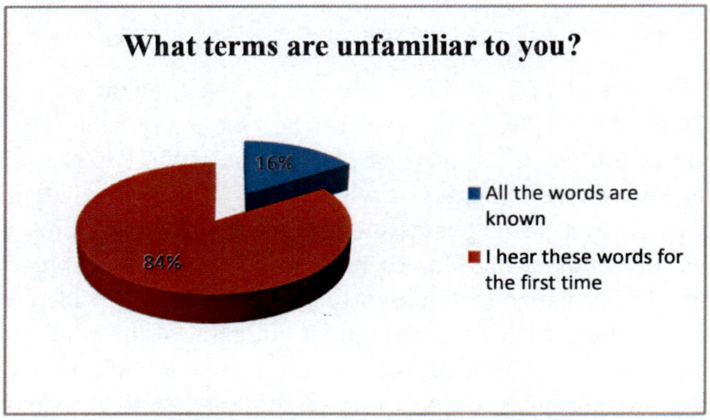

Fig. 1. The structure of the responses of guests and employees about understandable terms encountered in business communication (for example, the sanatorium "Mashuk Aqua-Term") *Source* compiled by the authors

An analysis of the responses of guests and employees of the "Mashuk Aqua-Term" sanatorium indicates that the majority of respondents (84%) don't know the meaning of such words as rapport, congruence, pacing, affiliation, attraction, "stroking" according to Burn, 16% are familiar with these concepts. Therefore, it can be concluded that the lack of knowledge in the area of effective communication terminology may have its consequences - the presence of certain barriers to constructive communication.

Most of the recipients pointed to the following, unfamiliar to them, words that play a significant role in building effective communication between communication opponents: rapport, congruence, pacing, affiliation, attraction, and "stroking" according to Bern.

2 In your opinion, from whom should you not expect excessive polite phrases reflecting the prospect of building effective communication, according to the gender aspect?

Figure 2 is a visualization of responses to a gender question in the use of polite phrases by sanatorium staff.

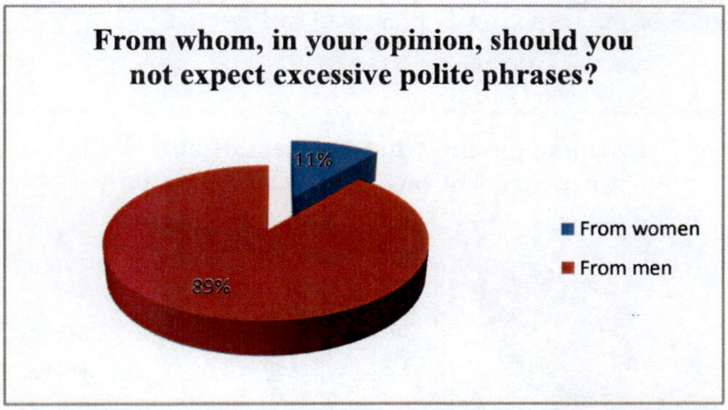

Fig. 2. The structure of the responses of guests and employees about the gender priority of the use of polite forms of speech (on the example of the sanatorium "Mashuk Aqua-Term") *Source* compiled by the authors

About 11% noted that excessive polite words from women shouldn't be expected, 89% believe that men are stingy with excessive use of polite etiquette phrases.

3 In your opinion, what's the level of impact of non-verbal communication?

The structure of the responses is presented in Fig. 3.

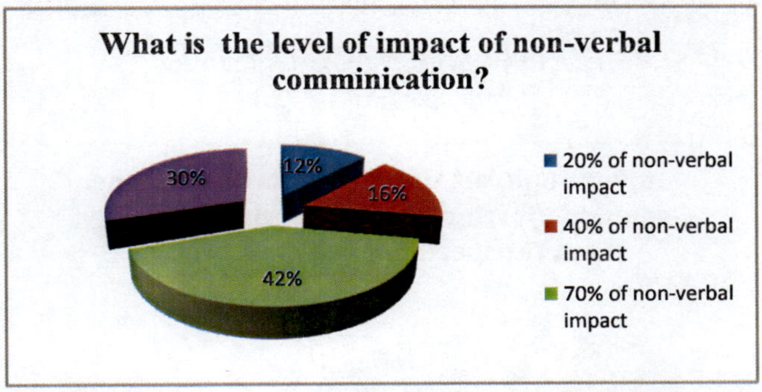

Fig. 3. The structure of the responses of guests and employees about the impact of non-verbal communication (on the example of the sanatorium "Mashuk Aqua-Term") *Source* compiled by the authors

The respondents' answers were surprising: exactly 70% of non-verbal communication is present in our verbal communication - and this answer is correct. 42% of guests and employees of the sanatorium also think so.

4 In your opinion, what barriers and barriers-tactics in business communication often arise: at the level of interpersonal communications or the level of organizational communications?

The structure of the responses is presented in Fig. 4.

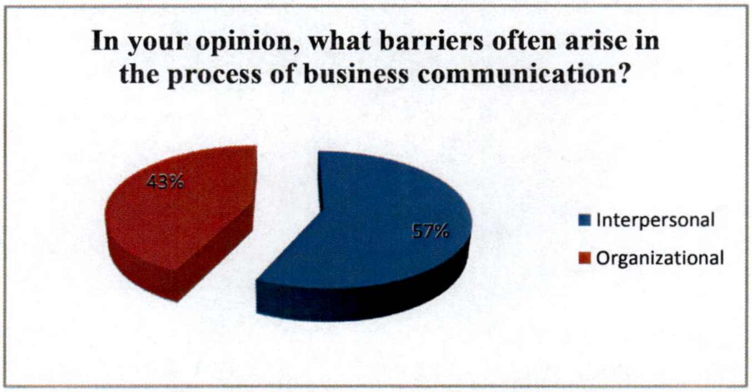

Fig. 4. The structure of the responses of employees of the Hotel Company and guests about the frequency of occurrence of a certain group of barriers (for example, the sanatorium "Mashuk Aqua-Term") *Source* compiled by the authors

About 57% noted that there are interpersonal barriers, 43% - organizational. From the data obtained, an almost equal ratio can be seen.

5 In your opinion, do representatives of other cultures/foreigners need deep knowledge about the specifics of communication?

The structure of the responses is presented in Fig. 5.

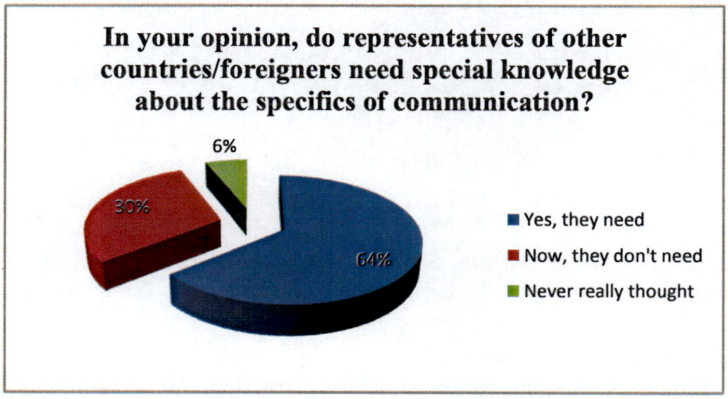

Fig. 5. The structure of the responses of guests and employees about the frequency of occurrence of a certain group of barriers (for example, the sanatorium "Mashuk Aqua-Term") *Source* compiled by the authors

An analysis of the responses of the guests and employees of the "Mashuk Aqua-Term" sanatorium indicates that 64% believe that such knowledge is needed, 30% believe that it isn't needed, 6% didn't think about this question at all. It follows from this that the respondents don't pay special attention to such concepts as "culture shock", "inter-ethnic conflict", "multicultural environment", etc., which causes particular disappointment.

After analyzing guest reviews/cookies/information about the "Mashuk Aqua-Term" sanatorium, taken from website booking.com, the following conclusions can be drawn. As a result of their analysis, it became clear that the guest puts interpersonal barriers in the first place, i.e., first of all, he assesses "congruence", which means harmonious communication between the patient and the staff of the sanatorium, and then the professional qualities of therapists, for example. It's a paradox, but it's a fact. That's why it's so important to know the technology of overcoming various kinds of barriers [10].

Guests and employees of the sanatorium were tested, and this test was called the "Test of knowledge of sign language". The results surprised us. The test consisted of 8 questions, where 70% of the respondents answered correctly. Consequently, recognition of the meaning of non-verbal communication with guests allows the staff of the "Mashuk Aqua-Term" sanatorium to intuitively establish rapport, congruence, pacing, affiliation, attraction, and "stroking" (according to Burn).

And the final stage of the study of the problem of overcoming barriers was the test "Know your barriers in communication". The results were as follows: most of the answers reflect the adherence of respondents and employees of the sanatorium to negative stereotypes (bias, prejudice, etc.). This, of course, has a negative effect on the construction of an interpersonal dialogue with the guest of the sanatorium.

Knowledge of how to overcome barriers to effective communication is certainly necessary. The personnel of the sanatorium have lacked this knowledge. The analysis showed that guests know the words "friendly staff", "polite therapists", "patient teachers", and nothing else. But the staff is polite; this is a pleasing aspect that gives a general positive characteristic to the staff of the "Mashuk Aqua-Term" sanatorium.

Considering the text above, it should be concluded, that it's necessary to take timely measures to eliminate barriers in communication and the use of scripts (speech modules), which inevitably lead to non-constructive interaction between hotel staff and guests, as well as ineffective management of the hotel enterprise. Therefore, this article substantiates the need to compose and use ready-made scripts in order to "correlate person as an employee with the whole company" in which he works [12], due to hotel employee/hotel sales manager will be able to achieve mastery in the art of communication with a guest using a script - a developed scenario of interaction with a client for successful communication. A speech module or speech preparation is a ready-made phrase for a specific case [8].

Moreover, the authors name several terms, the realization of which plays an essential role in establishing contacts between personnel in the tourism and hospitality sector. These concepts are: "rapport", "congruence", "pacing", "affiliation", "attraction", and "stroking", according to Bern.

In order to achieve *rapport*, harmoniously *congruence* should be demonstrated - the ability to be in agreement not only with the timbre of the voice, facial expressions, and gestures (unconscious communication) but also with the selection of appropriate words (conscious communication) while showing absolute sincerity.

So, *pacing* (reflection of feelings) gives the staff reason to show their feelings towards the guest. Especially if this guest is staying at the hotel for the first time.

Affiliation - a sense of the importance of a person, a desire to please - should always be taken into account by the hotel staff in the course of communication with the guest.

Attraction, external and internal - attracting attention - can add complementarity and support to the dialogue.

"Stroking", according to Bern and his theory of transactional analysis, can be accompanied by approval, acceptance of the interlocutor through the use of a benevolent intonation, an attractive look, etc. [1].

Let's analyze the features of the use of scripts. As a rule, in hotels with a small number of rooms, approved scripts don't exist. In hotels with a more solid number of rooms, the use of scripts is regulated.

The scripts are classified as follows: 1) outgoing cold calls; 2) incoming calls; 3) identification of needs; 4) execution of the transaction (booking); 5) presentation of services and others.

Hospitality scripts are necessary because they:

- Help to systematize work and improve control at all stages of sales;
- Allow the hotel employee to achieve the desired results in a shorter time;
- Facilitate the work of the hotel employee [2].

An effective script is distinguished by the following features: 1) it must be clear what task it solves; 2) the algorithm for using the script must be spelt out for several options for the interaction of the hotel staff with the guest; 3) the script must be tested on real clients with adjustments already made to it; 4) the script should be a guide in continuing a constructive conversation and completing a successful transaction.

There are also mistakes in the development of scripts: 1) scripts may not always cover the main "optimization zones"; 2) scripts try to completely "close" the entire process of interaction between a hotel employee and a guest; 3) the need to talk more than the client; 4) there may be insincerity in their performance.

4 Conclusion

Raising the issue of overcoming the barriers to effective communication in the "Mashuk Aqua-Term" sanatorium, some measures should be developed to improve the communication skills of the employees of this enterprise: the ability to establish contact with a newly arrived guest, to predict a communicative dialogue with the patient, to resolve conflict situations, to determine the type guest of the sanatorium, and also to build personalized relationships with work colleagues and guests of the sanatorium. Moreover, despite the productive work of the personnel of individual departments, the cohesion of the team of the hotel enterprise, control over the functioning of the hotel to take into account certain of its indicators, as well as a single mechanism for functional modelling of the processes of providing hotel services in conditions of socio-economic development are needed [4]. In addition, the hotel, which is part of the chain, gets great opportunities to improve service, relying on schemes, which have already been used, client programs, and their support [6].

So, the following directions for overcoming intercultural barriers in the organizations of the hospitality industry need to be proposed, in particular in the "Mashuk Aqua-Term" sanatorium and resort enterprise:

1. Accept as a reminder and indisputable list of postulates necessary in the professional work of an employee of the Horeca sphere:
- *Problem (perceptual barrier).* The indifferent attitude of the staff, disinterest, and "ostentatious" politeness create a sceptical attitude in the guest towards the success of business communication. Guests lose all desire to conduct a conversation with a hotel employee because of this reason. Solution. It's necessary to start the conversation with a positive intonation and try to keep it throughout the entire conversation with the guest. Moreover, effective communication will be facilitated by moderate gestures: a sincere smile and eye contact with the interlocutor.
- *Problem (behavioural barrier).* Excessive self-confidence, a pompous tone can cause hostility from the interlocutor. Conversely, low self-esteem and lack of confidence in actions can also be the reason for a barrier in communication. Solution. It's necessary to use in communication not only the strategy of "take the other as your equal", but also not to forget about the smile, eye contact, and a benevolent tone, expressing, when the opportunity arises, positive emotions concerning the work done in good faith.
- *Problem (language barrier).* The use of professional jargon in the speech by the employees of the hotel enterprise in a conversation with a guest or in a conversation with newcomers to the Horeca industry will make it possible to judge professional unsuitability. Solution. It's necessary to "adjust" to the interlocutor, simplifying speech to the level of a guest, without showing arrogance towards people, who are characterized by high self-esteem.
- *Problem (emotional barrier).* Due to the excitement, uncertainty and irritability of the staff in communicating with the guest, the ability to receive information, its objective interpretation, and analysis can be lost. Solution. The person should tune in to a neutral communication style that doesn't affect the mood of other people. The working dialogue should be smooth
- *Problem (cultural barrier).* Lack of knowledge, skills, and abilities in relation to correct communication with a person of a different confession and nationality can offend his beliefs; due to the phenomenon of hedonism and permissiveness in his life have a special meaning. Solution. Don't be afraid, if such knowledge and skills are absent; ask the guest face to face about some cultural moments.
- *Problem (gender barrier).* The influence of stereotypes can often interfere with the work process. For example, a male leader can be sceptical about the professional skills of women, because of the stereotypes that work in his mind. Solution. Interlocutors shouldn't be separated by gender [13].

2. Create a group of employees with some additional knowledge in the area of psychology, business communication to resolve certain situations that have occurred for a month inside of this organization in order to find a compromise, an optimal solution to a particular problem.

3. Every three years, the management of the “Mashuk Aqua-Term” sanatorium-resort enterprise should identify potential employees to improve their qualifications or professional retraining.
4. Nobody cancelled the self-education of the manager's personality. It’s necessary to determine the list of sources from the area of psychology, management, and business communication. Constantly set a goal - to read one source every three months.
5. Visiting master classes, trainings, for example, “Training by Allan Piz”, “Training by Irina Khakamada”, “Training by Radislav Gandapas”, and trainings by other gurus in the area of effective communication technologies.
6. Constantly analyze cookies/information about the sanatorium (guest reviews) “Mashuk Aqua-Therm”, which are taken from the different sites, such as TripAdvisor, booking.com.
7. Start every day with a smile and the memory of the next dream!
8. Develop and hone the skill in the use of scripts (speech modules) in order to effectively build the interaction of hotel employees and guests, because the script is an effective tool in overcoming communication barriers.

Thus, in the course of personnel training to implement any management process, it’s necessary to form certain competencies, relying on professional standards in the area of hospitality, because various kinds of barriers may arise in the process of communicative interaction among employees of sanatoriums or hotels. Unfortunately, such communication barriers often interfere with constructive dialogue, distorting the meaning of the message received from the communicator.

References

1. Bern, E.: People who play games. Eksmo, p. 608 (2020)
2. Bokareva, V.: Sales scripts and speech modules: specificity in hotel activities. Welcome Times. https://welcometimes.ru/opinions/skripty-prodazh-i-rechevye-moduli-specifika-v-deyatelnosti-oteley (2021). Accessed 13 July 2021
3. Gorbunov, A.P., Gazgireeva, L., Burnyasheva, L.A., Rud, N.: Brand of the region in the system of marketing factors of regional competitiveness. Bull. Kabardino-Balkarian Sci. Center Russ. Acad. Sci. **3**, 73–78 (2016)
4. Gorbunov, A.P., Kolyadin, A.P., Gazgireeva, L.K., Burnyasheva, L.A.: Man, his spiritual and moral potential and the limitless nature of postmodern hedonism in the provision of hotel and tourist services in the context of the transformation of socio-economic systems. In: Popkova, E.G., Sergi, B.S. (eds.) ISC 2019. LNNS, vol. 198, pp. 47–55. Springer, Cham (2021). https://doi.org/10.1007/978-3-030-69415-9_6
5. Gorbunov, A.P., Kolyadin, A.P., Burnyasheva, L.A., Gazgireeva, L.K., Kosenko, O.Y.: Tourist and recreational clusters as an organizational and economic mechanism of control of formation and development of the innovative capacity of the North Caucasus Federal District. Amazonia Inv. Periodicidad Bimestral. Univ. Amazon **7**(17), 60–71 (2018). https://amazoniainvestiga.info/index.php/amazonia/article/view/326. Accessed 13 July 2021
6. Kulgachev, I.P., Babakhanova, I.B., Kosolapov, A.B., Savinkina, L.A., Dracheva, E.L.: Russian hotel chains: formation and development trends. J. Environ. Manage. Tourism **11**(2(42)), 290–298 (2020). https://doi.org/10.14505//jemt.v11.2(42).07

7. Leontovich, O.A.: Introduction to intercultural communication. Gnosis, p. 368 (2007)
8. Mazilkina, E.I.: Organization of sales of a hotel product. Alpha-M: INFRA-M, p. 207 (2016)
9. Minenkova, V.V.: Fundamentals of the hotel business. Kuban State University, p. 160 (2010)
10. Reviews of the sanatorium "Mashuk Aqua-Therm": Booking.com. https://www.booking.com/reviews/ru/hotel/dddegnndo-ddod2ddeg-dc-dund1-4.ru.html (2021). Accessed 13 July 2021
11. Sanatorium "Mashuk Aqua-Therm", Zheleznovodsk. http://www.sanatory-mashuk.ru/ (2021). Accessed 13 July 2021
12. Schmitt, B.: Empirical marketing: how to make a client feel, think, act, and also relate himself to your company, p. 388. FAIR-PRESS, (2001)
13. Six barriers to effective communication and ways to overcome those (2021). https://lifehacker.ru/6-barriers-to-communication/. Accessed 13 July 2021

Integrated Marketing Communications that Determine Consumer Behavior Models: Management Implementation Strategies

Natalia P. Ketova and Irina Yu. Granovskaya(✉)

Southern Federal University, Rostov-on-Don, Russia
nketova@sfedu.ru

Abstract. The prerequisites for the study, the results of which are presented in the article, are that the modern concept of integrated marketing communications (IMC) permeates all stages of market reproduction - from the idea of creating a product to implementation, in order to optimize the formation of communication. The purpose of the study is to conceptually substantiate management strategies for the formation of integrated marketing technologies that predetermine the main models of consumer behaviour. Research methods include analysis of the marketing environment in which the IMC is formed and implemented, factor and strategic analysis, methods of sociological surveys (expert survey), systematization of empirical and factual information in the form of tables, and market segmentation. The results and novelty of the article consist in the author's presentation of the possibilities and prospects for the development of IMC and technologies in the digital environment, building a model of modern communications, substantiating the total positive result of integrating marketing communications, forming a new marketing landscape, including a new direction in marketing, designated as neuromarketing.

Keywords: Integrated Marketing Communications · Neuromarketing · Effective communications · Strategy management · Co-creation marketing · Consumer choice

JEL Codes: D10 · D21 · E71 · M20 · M30 · M31 · M37

1 Introduction

The relevance of the research problem is predetermined by the fact that in the conditions of the rapidly emerging demand of the business community for the development of algorithms for coordinating consumer behaviour of target groups in the digital space, innovative methods of implementing communication systems in marketing are actively developing. The search for a modern format of the marketing strategy of companies is carried out in many areas, including in the area of integrated marketing technologies. This is due to the need to re-subordinate various communication channels to the goals of a single, coordinated approach to solving the problem of positioning a company or a specific brand on the market for goods, services, and innovations. The idea of synthesizing the presented communication environments, the formation of a single

E. G. Popkova (Ed.): ISC 2020, LNNS 368, pp. 546–554, 2022.
https://doi.org/10.1007/978-3-030-93244-2_60

communication area to achieve internal and external integration is implemented in practice in the form of a system of integrated marketing communications while delivering information about the company's products or services to the target audience.

The increasing complexity of search algorithms in the area of modern communications is an additional challenge in the area of marketing, predetermining the relevance and significance of this study. Indeed, the formation of effective communication requires the introduction of innovative developments, but their limitations can become an insurmountable obstacle for modern marketers. The problem lies in a certain shortage of effective methods and tools of the IMC, as well as limited competencies for the use of modern forms and methods of communication interactions by the personnel of companies, incl. Internet communications [16].

The problem of the effective formation of integrated marketing technologies that determine consumer behaviour models consists in a simplified perception of integrated communication as a way to optimize advertising products (services), without its relationship with the positioning of companies in the market and systematic accounting of consumer behaviour models. This approach is determined by the lack of rational management algorithms and competencies for the implementation of the IMC system in the activities of modern companies.

In this regard, the purpose of the study is to generate the idea of using integrated marketing technologies, which are needed to optimize management strategies for the implementation of IMC in innovation-oriented companies, with the prospect of increasing the functionality, flexibility, programmability of their behaviour. The article is focused on the development of an IMC model for an innovation-oriented company.

The methodological base of the research is formed in the context of the imperatives of the system-functional approach and is based on the following theories: mass communication, digitalization, as the leading modern trend in social development, marketing management, co-creation marketing, interactive marketing and consumer behaviour models, as well as neuromarketing. The methodology is concretized through the use of modern analysis tools and the formation of innovative marketing ecosystems. Research methods used to obtain significant results of studying IMCs include factorial and comparative analysis, benchmarking, a method for justifying algorithms, systematizing empirical-factual information in the form of tables, etc.

The scientific novelty of the results, which are presented in the article, lies in the fact, that the study of the accumulated experience of using the tools and private methods of IMC to form effective communication is considered as a stage in the development of methodologically grounded aspects of the marketing policy of companies. Of undoubted scientific interest is the innovative nature of effective communication, the generation of which within the IMC is carried out based on co-creation marketing: companies - manufacturers, consumers, suppliers, and distribution channels. They form a network of innovations, which is focused on creating communication of trust. The increment of innovative scientific knowledge is also determined by the emphasis on the synergy of integrated marketing communications, the variability of the choice of the IMC strategy, as well as the proposal of unified management algorithms for the implementation of the IMC in companies.

2 Discussion

The range of modern marketing technologies and tools is successfully complemented by ideas for the implementation of IMC. Some modern studies indicate the significant effectiveness of integrated communications in comparison with the separate use of each communication. The synergy effect is a fundamental one in IMC. It's ensured by the coordination of various types of communication activities, which creates the preconditions for obtaining the maximum in conditions of macro environment market and economic efficiency, emphasized by [23]. The understanding of IMCs and their growing capabilities have passed some stages. The first definitions of integrated marketing communications are associated with [5]. They belong to the early 90s of the 20th century. (Although the idea of integration in communications was expressed in the 30 s, this didn't receive timely development). The "deployment" of the concept of integrated marketing communications determined their interpretation as a phenomenon related specifically to marketing (for example, integrated communications, used as a more general concept, are used in a fairly large number of social disciplines, while IMC is used only in marketing). IMC is characterized by such approaches as "one vision - one voice", "induction style", and "focus on the consumer". The priorities are media integration, coordination, and agreement of many channels of communication, messages, content [1].

The theoretical aspects of integrated marketing communications were developed by different authors in the framework of several directions: clarification and expansion of the essence and effectiveness of IMC, the typology of messages in them – [6]; characterization of IMC as an opportunity for strategic coordination of relationships with stakeholders – [5]; interpretation of IMC as a process of optimization of communication channels for the delivery of a clear and consistent message – [18].

The study of IMC capabilities allowed the researcher Schultz to reveal the dominant coordination of IMCs, which ensures the orientation of consumers to such communications as the flow of information from non-differentiable sources [22]. The interaction of all communications, which are integrated with the use of other marketing tools and supported by technologies, to achieve maximum economic efficiency, is noted [13].

In search of new methods, principles, and tools of IMC, marketers turn to their innovative forms to increase the effect of communications. Some authors are investigating the practice of applying neuromarketing techniques to stimulate the behavioural motivations of consumers with the possibility of subconscious management of consumer choices [11]. This is also advisable to include digital marketing communications tools as a strategic resource.

Thus, the development of the methodological base of marketing and a positive assessment of the prospects for integrated communication for thirty years has supported the interest of researchers in the IMC. It's also stimulated by the adopted course towards digitalization of relations. As Melent'eva and Tatarenko emphasize, the building of modern marketing communications within the framework of new Internet technologies is considered as the radical transformation of socio-economic systems with giving them completely new qualities [20]. According to the authors, the process of integrating marketing communications on various kinds of digital platforms makes it possible to more actively use their potential.

3 Results

The study of the IMC problematic factors in modern marketing determines one of the significant vectors of its development. It's updated as companies understand the prospects for the systematic combined use of the entire available set of information links in practice [7].

Modern concepts of promoting goods (services) are formed in the context of the main paradigm - the shift of interests from mass marketing to targeted marketing. So, the accelerated development of integrated marketing communication channels and tools for promoting goods, services, and projects contribute to it. Their application implies the formation of a well-thought-out message, which is the result of the compilation of numerous communication channels and the coordination of the work of many departments of the companies.

3.1 Model Representation of a Complex of Integrated Marketing Communications (IMC)

The IMC complex is presented to the authors of the article as a project-oriented complex of actions of modern companies to implement their interactions with the entire set of subjects of the marketing system with a reasonable strategy for using a complex of communicative means (communicative mix). Table 1 shows the author's model of the IMC complex, which includes the main communications: advertising of all types, sales promotion, personal sales, public relations, branding, direct marketing, and sponsoring. Their systematic use in practice provides a stable and directional formation of consumer behaviour options, which is desirable for the company.

In this regard, McDonald's and Coca-Cola companies, which for many years have been using joint advertising and brand campaigns that successfully work on the image of these world-famous companies, are convincing examples [3]. They provide information, recognition, a reminder of their highly demanded (especially among young people) products, as well as the formation of preferred options for consumer behaviour [14].

Representation of IMCs in the form of a model makes it possible to analyze them not only in a single complex but also to take into account the motivation for active interaction of business partners at each communicative level [15].

An example of the effective use of an integrated marketing communications strategy is the Russian company Aeroflot, which occupies a leading position in the Russian air transportation market and operates more than 60% of international flights. Aeroflot is currently actively investing in its image and improving its business reputation. Aeroflot has reoriented its strategy through integration. Work on attracting and retaining passengers, ensuring the growth of the consumer value of the product, as well as strengthening the brand was carried out. Aeroflot successfully implements the principle of "exceeding customer expectations" by providing professional service. This

Table 1. Model of a complex of integrated marketing communications that stimulate the formation of preferred options for consumer behaviour

No.	Components of the complex of integrated marketing communications (IMC)	The essential content of the components of integrated marketing communications (IMC) and the functions, which are performed in the communications complex
1	Advertising of all types, on different media, incl. internet advertising	The appeal of sellers of goods, services, projects, etc. to a potential consumer, formulated and paid for by the advertiser, currently placed mainly on the Internet (up to 47% of all advertising), for a targeted impact on the target audience
2	Sales promotion	Incentives for consumers, resellers, manufacturing companies to encourage sales growth
3	Personal sales carried out for the oriented presentation of the offered goods	A component of marketing communications that provides a personal presentation of goods or services to the buyer (in a shopping centre) or their Internet offer to potential consumers for stimulating their sales
4	Public relation, PR, blogs	Communication as a management system for political, public, cultural, and entertainment spheres, interaction with the public to create a positive image
5	Branding	Brand formation, which means loyalty and consistent preference of customer groups (students, groups of young families, and athletes)
6	Direct Marketing	Active interactions of specific companies, shopping centres, beauty centres, restaurants with consumers through building relationships, with use of email newsletters, chatbots, personal sales

Source Compiled by the authors based on [1, 2, 7, 8, 12, 13, 17, 19]

approach made it possible to significantly increase Aeroflot's rating in the rating lists of the world's airlines [4].

3.2 Basic Principles and Methods of Implementing the IMC Model in Modern Companies

In order to achieve tangible results in stimulating the growth of the efficiency of the company's activities by realizing the potential of effective marketing communications, it's important to have a clear understanding of the factors that need to be taken into account in their formation. Table 2 presents the author's systematization of the principles and methods of accounting for the factors that should be used during the process of building an effective IMC.

Based on the materials presented in the table, it's legitimate to conclude that integrated marketing communications can be considered as an effective operational tool, as well as a tool for strengthening interactions with target audiences and partner companies. In the context of digitalization, the IMC create conditions for the implementation of network relations and connections of any level of complexity, and this is the most important feature. Thus, marketing communications become an effective means of organizing industrial and social connections, as well as emerging relationships within the framework of companies (internal marketing) and, at the same time, with external partners and consumer audiences [21].

It should be also emphasized that in the conditions of the dominance of Internet technologies, IMCs are embedded in the modern information and communication space and provide a combination of effects, strengthening each of them. Consequently, the importance of IMC can be rightfully assessed in the future as an opportunity to build more effective interactions between companies with business partners and target audiences of consumers, taking into account the assessments of opinion leaders in social networks.

Table 2. Priority principles and methods of implementing the IMC model of a modern company

Basic principles	Implementation methods
Credibility	The basis for the development of integrated marketing communications is the real activity of the company. It's should be noted that manipulation and an attempt to artificially, which stimulate a positive reaction, are unacceptable
Openness	It's required to provide the company with communication links with stakeholders, to have representatives responsible for operational contact with partners
Dynamism	Communications must meet the conditions and constantly evolve, it's necessary to master new channels and technologies
Trust and involvement	It's required to promptly record the expectations and preferences of key customer groups and business partners. Prioritize loyalty and create a supportive atmosphere for the company
Consistency	It is advisable to maintain integrated marketing communications between various levels and divisions of the company for building its interactions with the external environment
Consistency	It's necessary to create consistent information messages for target audiences and partners
Initiative	Initiative in all "communication signals" about the company should be taken, the loss of initiative in the information area leads to instability with unpredictable consequences

Source Author's interpretation based on [10, 21]

However, with the expansion of penetration of Internet technologies into marketing, IMCs often give way to other tools. Integrated marketing communications are still seen as a powerful marketing tool with all the changes.

One of the significant obstacles to the creation of an effective IMC system in modern companies is insufficient understanding by specialists of algorithms for the effective implementation of a set of management decisions for their construction and timely modification [21]. In addition, the difficulties in the formation of effective IMCs include functional isolation, lack of creativity, temporary internal conflicts, and the lack of clear management algorithms. This, according to the authors of the article, is overcome by correctly predicting trends in the development of the market macro-environment, creating a clearly expressed corporate style, ensuring the coordinated work of departments, correctly assessing the level of consumer demand and competently identifying target audiences.

It's advisable to identify marketing management with rationally organized work on the formation and maintenance of IMC [9]. It also seems rational to use the following standardized methods: expert assessments, brainstorm, structured interviews, test and simulation data recording, questionnaires, assessment of previous experience, as well as assessments from other projects.

Marketing management in the organization of the IMC system is a set of actions aimed at the systematic implementation of solutions and the implementation of relevant competencies within the framework of innovation-oriented marketing management, the formation of "smart" services, as well as information interaction in the digital environment at any time and anywhere.

4 Conclusion

The results of the research allow summarizing it as follows.

1. Integrated marketing communications in the modern conditions of digitalization, the development of the regulatory and legal Internet space, information and telecommunication networks, the exponential growth of Internet resources, the expansion of the presence in networks as a variety of companies, business structures, associations of business partners, etc., individual consumers and their groups, represent integrated interactions between subjects of marketing systems for coordinating tactical and strategic decisions, using innovations, through the integration of different marketing communications, such as advertising, sales promotion, personal sales, public relations, branding, direct marketing, and sponsoring.
2. The advantages of the IMC system create conditions for increasing competitiveness, dominance in the markets, and the investment attractiveness of companies using them; contribute to the strengthening of business reputation, forming important components of the commercial success of business structures. Effective marketing communications that form a rational macro-environment act as an effective mechanism for influencing target audiences.
3. The organization of the IMC system in a company requires the implementation of a set of management decisions, including the formation of an effective marketing management strategy; re-subordination of the organizational and administrative structure of the company to the goals of the IMC; development of competently supported measures for the implementation of IMC; organization of monitoring and

control over the effectiveness of integrated marketing communications; maintaining the company's business reputation; solving problems of ensuring innovative activity and high creativity of personnel.
4. Among the new tools that are appropriate for inclusion in the IMC, it's legitimate to consider neuromarketing, which represents a set of methods, and with the use of them, marketers manage to study how certain stimuli affect the activity of the brain and cause differentiated consequences. The data obtained in the course of such studies, combined with the results of studying other elements of the IMC, are used to format effective marketing strategies and optimize advertising campaigns.

4.1 Results of the Research and Its Practical Significance

The conducted research confirms that integrated marketing communications remain in high demand for some years. Their implementation involves the search and testing of new management techniques and tools for creating modern effective integrated communications in the context of digitalization and the development of network interactions. The further study of the statement presented in the article about the possibility of classifying the neuromarketing complex as one of the innovative IMC tools is also of scientific interest.

The search and development of management solutions for the implementation and realization of the IMC system, available for practising marketers, using cognitive-behavioural psychological technologies designed for specific elements in the structure of the company's target audience, seems promising.

References

1. Akhmedov, A.E., Smolyaninova, I.V.: The mechanism for the implementation of innovative marketing activities of agro-industrial complex enterprises based on communication policy. Ostrovsky Readings **1**, 165–168 (2017)
2. Best, R.: Marketing from the Consumer. Mann, Ivanov and Ferber, pp. 34–96 (2019)
3. Blagikh, N.V.: The problem of evaluating the effectiveness of specialists in integrated marketing communications. Service in Russia and abroad, pp. 38–40 (2017)
4. Competent application of PR technologies in the formation and restoration of the image of airlines. https://sci-article.ru/stat.php?i=1576782706. Accessed 2 July 2021
5. Duncan, T.: The concept, process, and evolution of integrated marketing communication. In: Caywood, C., Thorson, E., Moore, J. (eds.) Integrated Communication: Synergy of Persuasive Voices, pp. 13–35. Lawrence Erlbaum Associates (1996)
6. Duncan, T.: Client perceptions of integrated marketing communications. J. Advertising Res. **3**, 30–39 (1993). Everett, S. (ed.)
7. "Emotional" packaging design. https://www.worldinpack.com/design/ti-che-takoi-emotcionalniy-a/. Accessed 7 Apr 2021
8. Eser, Z., Bahar, F., Metehan, T.: Perceptions of marketing academics, neurologists, and marketing professionals about neuromarketing. J. Mark. Manag. **27**(7–8), 854–868 (2011)
9. Integrated communications: strategic PR, marketing or something else? http://www.advlab.ru/articles/article201.htm. Accessed 17 Apr 2021

10. Kaftandzhiev, H.: Integrated marketing communications. Humanitarian Center, pp. 18–46 (2021)
11. Ketova, N.P., Granovskaya, Y.: Possibilities of implementing neuromarketing to enhance the promotion of goods and services to consumers. Creative Econ. **14**(1), 73–92 (2020)
12. Ketova, N.P., Natkhina, M.V.: Marketing of co-creation: features of accounting the values and preferences of consumers in the context of the theory of generations. In: Bagieva, G.L., Yuldasheva, O.U. (eds.) Monograph: Marketing of Co-creation and Global Communication of Trust, pp. 29–41. Publishing house of Saint Petersburg University of Economics (2020)
13. Key, T.M., Czaplewski, A.J.: Upstream social marketing strategy: an integrated marketing communications approach. Bus. Horiz. **60**(3), 325–333 (2017)
14. Klimasheva, A.D.: Evolution of marketing concepts. The scientific community of students of the XXI century. In: Collection of Articles Based on the Materials of the VII Scientific Conference, vol. 9(57), p. 37 (2017)
15. Kornyushin, Yu.V.: Marketing communications: curriculum: study-method. complex. Center for Distance Education Technologies of Moscow University named after S.Yu. Witte (2015). http://sa.vscc.ac.ru/article/1704?info=references. Accessed 28 Apr 2021
16. Kotler, F., Amstrong, G., Wong, V., Sondders, J.: Fundamentals of Marketing, 5th edn., p. 45. LLC "I.D. Williams", Moscow (2014)
17. Kotler, F., Kartadzhaya, H., Setiavan, A.: Marketing 4.0. U-turn from traditional to digital: Internet promotion technologies. Bombora, pp. 56–96 (2019)
18. Kotler, P.: Marketing Management. Hall International, London (2000)
19. Larry, D.L.: Will we need a whole new breed of professionals? Commun. World **12**(7), 26 (1995)
20. Melent'eva, N.I., Tatarenko, V.N.: On some approaches to the study of socio-economic aspects of Internet communications. In: Bagieva, G.L., Ponomarevoy, E.V. (eds.) Collective Monograph, pp. 209–217. Publishing house of Saint Petersburg University of Economics (2021)
21. New amendments to 214-Federal Law: what buyers need to know? https://www.l1-stroy.ru/PressDetail?Id=1124. Accessed 15 Apr 2021
22. Schultz, D.E., Tannenbaum, S.I., Lauterborn, R.F.: Integrated Marketing Communications: Pulling it Together and Making it Work, p. 57. NTC Business Books (1999)
23. Smith, R.P.: Marketing in questions and answers. Balance club, pp. 46–102 (2004)

Features of the Implementation of Mechanisms for Reorganization Development of SMEs in Modern Conditions: Strategic Planning of Market Integration, Difficulties and Prospects

Olga V. Konina and Alexander V. Tekin(✉)

Volgograd State Technical University, Volgograd, Russia

Abstract. Within the framework of this article, the authors attempted to consolidate the main directions and features of the implementation of mechanisms of reorganization development (especially, mergers/acquisitions) of small and medium-sized enterprises in modern conditions of transformation of the process of strategic planning of business activities and technologies, which are used for this (budget forecasting (as an integrated technology of strategic business planning)). For this purpose, the article discusses issues related to directions of reorganization of modern business in general and transactions on mergers and/or acquisitions (M&A-transactions) in particular; identification of features, main difficulties and promising general conceptual directions for the development of the domestic market for reorganization (mergers, acquisitions) in the segment (for entities) of small and medium-sized enterprises; strategic planning of the potential market integration of small and medium-sized enterprises (in the M&A market) and its financial, as well as economic determinants through the use of basic and specific budget forecasting tools. According to the text of this article, it's clarified that potential mergers and/or acquisitions can be the subject of strategic planning of entrepreneurial activities of small and medium-size enterprises, the specifics of the determinant-instrumental budget forecasting of which will depend on the stages of such a transaction (from planning to the completion of the last transaction).

Keywords: Merger · Acquisition · M&A – transactions · M&A – market · SME · Reorganization · Mechanism · Planning · Budget forecasting · Integration

JEL Codes: L26 · G34 · O2

1 Introduction

Reorganization processes and transformations of various nature and directions are generally characteristic of the capitalist market model. Since the middle of the 20th century, a strategic focus on integration processes, implemented to incorporate a business (and in a supranational context in the process of globalization), has gained

E. G. Popkova (Ed.): ISC 2020, LNNS 368, pp. 555–565, 2022.
https://doi.org/10.1007/978-3-030-93244-2_61

particular popularity in the corporate and in the business environment. On the other hand, since the beginning of the formation of the knowledge and information economy of the 21st century, against the background of business liberalization and an orientation toward "innovative creativity" in entrepreneurial activity, integration has been replaced by reorganization and disintegration processes, expressed in "de-corporatization" (disintegration of corporations into more mobile, innovative, and creatively oriented business entities with increased market adaptation capabilities). The established practice has shown that the popularity of business enlargement slightly increased only in times of crisis until 2020. But the pandemic of the new coronavirus infection and the newly emerging economic realities have confirmed the cyclical nature of interest in integration transformations, which is growing against the background of the lack of high financial stability (innovation-oriented, but small in terms of financial scale and capabilities) of business entities now, especially in the sector of small and medium-sized entrepreneurship (hereinafter referred to as SME).

Thus, the need to use market integration and reorganization mechanisms is especially high in the difficult (and, to a certain extent, unique) conditions of the recent pandemic years. Given the unpredictable nature of further paths of market changes and transformations, the study of the features, planning opportunities, and application of such mechanisms by SME business structures (given their importance for ensuring the "health" of the national economy) are becoming increasingly practical.

2 Reorganization of Modern Business. M&A Transactions

The scale of some reorganization transactions, their market importance, and the possibility of a significant change in market conditions due to reorganization processes and transformations, necessitated the appearance in the historical retrospective of state regulation of such activities. The ways (forms) of modern business reorganization, within the framework of the domestic economic and legal model of market regulation, are determined by the norms of the current legislation of the Russian Federation [1–3]. They include:

- Transformation (formation of a new legal entity (hereinafter - LE) from some ceasing to exist);
- Allocation (formation (actually - budding) of a new LE from another legal entity that continues its existence);
- Division (formation of new legal entities from one legal entity that has ceased to exist);
- Accession (it's also absorption in the "market" terminology - the termination of the independent activity of one or LE with the corresponding transfer of property, rights, and obligations to another pre-existing LE that continues its activities),
- Merger (termination of independent activities of two or more LE with the corresponding transfer of property, rights, and obligations to another newly created LE).

Under certain circumstances, this list can be expanded by similar forms: transfer of assets by succession [15], unification of independent LE (clustering, collaboration, creation of consortia, B2B or public-private partnerships, unions, etc. [6]) while maintaining legal, managerial, and financial autonomy or with their limited loss; unification of partially or completely dependent legal entities (creation of syndicates, groups of companies, etc. [6]). Such forms aren't reorganizational, but they can become these forms in the future. For example, some peculiarities of LE interaction within the framework of modern consortia allow regarding this as a kind of preliminary "analogue of merger" and a form of "intention to merge". The potential transfer of the practical implementation of such forms of cooperation from an unregulated channel to a planned one provides a basis for studying possible forms of intentions for potential reorganization and peculiar forms of "pre-reorganization" – "possible planned reorganization" (for example, in the case of the aforementioned consortium).

According to the experts of BCS Company LLC (part of the BCS financial group), mergers, and acquisitions are the main types of reorganization transactions on a global scale [4].

Mergers and acquisitions (M&A) are classic reorganizations and integration processes of capital consolidation, business enlargement, and the potential incorporation of individual business units, including those of an entrepreneurial type. Along with the development of related markets for securities, products, technologies, and innovations, the modern M&A market seems to be quite wide and diverse. M&A - transactions are the object of close attention of state authorities, private consulting, rating, marketing agencies, as well as specialists in the area of strategic planning and business forecasting, business analysts, defining the strategic aspect of the national and global market agenda for decades. To understand the scale and specifics of the domestic M&A market, the authors will briefly analyze the basic digital material:

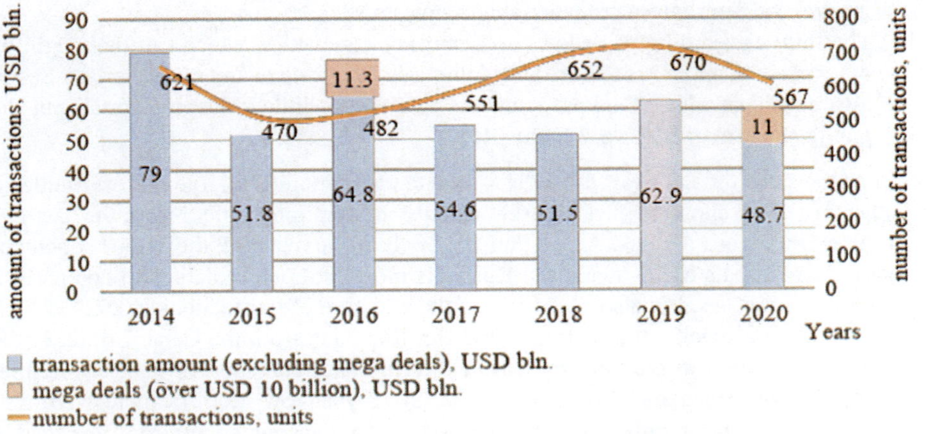

Fig. 1. Russian M&A market from 2014 to 2020 (according to KPMG analysts) *Source* Research "The market of mergers and acquisitions in Russia in 2020" [12]

Analysts of the global auditing and consulting company KMPG, assessing the dynamics of the volume and number of M&A transactions in the Russian market (Fig. 1), note: "…Even in such a difficult year for the entire world economy, the Russian M&A market was able to stay in the number and the total volume of transactions at the level of the 2017 year. Compared to 2019, which was the most successful for Russia in terms of investment activity, due to the imposition of sanctions, in 2020 the number of transactions decreased by 15% (to 567 transactions), and their total value decreased by 5% (to 59.7 billion US dollars)" [12]. They also admit that the situation in the domestic M&A market could be even more critical, but despite the difficulties of the end of 2019 and 2020–2021, there are noted: "…investors are ready to adapt to work in new conditions, change their business strategies, and look for new opportunities… At the same time, they act more carefully…" [12].

The statistics of the domestic M&A market corresponds to the main ways of sectoral "migration" of "large" capitals abroad as a whole (first of all, in countries with developed market economies), but with some time delay (3–5 years), towards "Innovations and technologies", "Communications and media", as well as "Health and pharmaceuticals" [12].

Co-director of the Moscow office of the Goldman Sachs Group financial conglomerate (Sedov) has a different point of view on the specifics of the domestic M&A market; he highlights some problems of its development:

- The first is the lack of private institutional financial capital, which is expressed by a decrease in the number of M&A transactions - in the market, the frequency of ownership changes, a decrease in the number of consolidation transactions, etc.;
- The second - the low number of M&A deals as a result of the lack of direct investment from international strategic investors;
- The third - according to Sedov: "… for many companies, which can be bought by other players, an IPO is a more affordable way to monetize and raise capital". Sedov also considers this a "unique situation": "There is little strategic capital, but the public stock market is very lively" [5].

At the same time, the last aspect is associated with more/less large players, while all subjects and stakeholders of the market under consideration "suffer" from the "action" of the two previous aspects. As a summary of all of the material, the author concludes that the small-scale M&A market in Russia is more "soft and volatile" in comparison with other countries. It's also noted that in the structure of transactions in 2020, $ 33.9 billion fell on a specific transaction to transfer Sberbank from the Central Bank of the Russian Federation to the Government of the Russian Federation (in fact, M&A prevented the Russian market from sinking to an 18-year low (2002)); there were also transactions involving only Rosneft in the top-3 - they accounted a third of the market ($ 17.96 billion) [5].

On the other hand, against the background of such a "specificity" of the domestic M&A market, systematic legislative activities are being carried out to develop and support it, including in conditions of sanctions pressure - the legal system for its support is gradually improving [10]. Against the background of a difficult economic situation, one of the proposals of the President of the Russian Federation Putin at SPIEF 2021 was the cancellation of the approval of the antimonopoly authorities for M&A deals worth less than 800 million rubles. The President of the Russian Federation stressed: "It's necessary to free small and medium-sized enterprises from clearly excessive antimonopoly control. Many thresholds, which are established here, haven't been revised for a long time and they also don't correspond to modern economic realities" [11].

It can be seen that the help and support of the M&A actors - the Russian market extends not only to large business entities. In general, the key players in the M&A market are usually large enterprises and/or business structures with state participation. But, the instruments and mechanisms of general reorganization nature and M&A transactions, in particular, have been actively penetrating in the practice of mutual (sometimes joint) activities, as well as in the processes of inter-subjective interaction of SME structures (in local markets).

3 Features, Difficulties and Prospects for the Development of the Reorganization (Mergers, Acquisitions) Market in the SME Segment

In the author's opinion, one of the most promising reserves for the growth of M&A volumes in the Russian market in the current conditions is the expansion of M&A - the capital of the SME segment most severely affected by the pandemic, and equivalent transactions between similar business entities (which can be objects of planning their mutual development in a strategic context).

M&A market researcher Musatova notes that in the pre-pandemic 5–6 years "…the structure of the Russian M&A market had a steady trend towards the development of private investments in the medium-sized business sector. The growth of M&A deals with the assets of small and medium-sized enterprises was recorded in the public and private sectors" (note by the author - which is confirmed by a statistical source [14]). During this period, there was also active integration in trade and other industries and spheres of activity, because of the acquisitions [10].

As in the case of large business entities, the main goals of M&A transactions in the SME segment are to maintain and/or increase competitiveness, protect their market positions, expand new market segments or reduce (curtail) independent activities, and capitalization (business value growth as in a merger, as well as during absorption), etc. At the same time, the agenda of voluntary takeovers as part of the adoption of strategies for liquidating SMEs will remain relevant for them shortly, given the colossal existential difficulties that arose during the pandemic. Some analysts agree that one of the growth drivers of the M&A market will be a series of bankruptcies of SME structures in 2021–22 years (especially retail players, owners of commercial real estate, and service providers) [9]. An alternative to complete bankruptcy can be "market integration" with a larger business entity.

It should be also be accounted that SMEs in the M&A market have a number of the following features:

- They often act as a "product", "object" of the market (they are sold and have value, are the subject of transactions) as separate (or affiliated) business units;
- They are focused on using the liquidation strategy through takeover in "difficult" times of crisis ("integration" into a larger business entity) as the only alternative to complete de facto bankruptcy; in "peacetime" - they build authentic protective mechanisms in the process of competing with large players, "psychologically" not ready for voluntary integration;
- They are initially liquidated to reorganize individual entrepreneurship if it's necessary (due to legal circumstances);
- They pursue similar strategic goals (as large business entities), use similar strategic approaches, strategies, and market integration tools;
- They are more adaptive to implement merger transactions if it's necessary (as the closest analogue of M&A transactions without losing autonomy) with the subsequent formation of clusters, collaborations, the creation of consortia, B2B or public-private partnerships, unions, associations, cartels, etc.

The mechanisms of reorganization development of SMEs (in fact, types of possible transactions) in the M&A market will be Horizontal mergers and acquisitions; Vertical mergers and acquisitions; Conglomerate (adjacent) mergers, acquisitions, and diversification; Generic mergers and acquisitions.

In this case, the main forms of associations of such entities can be Association of partially dependent enterprises; Consolidation of affiliated enterprises; Partnerships; Joint project activities.

Other forms of reorganization can be Change of organizational and legal form; Change of tax regime; Formation of a startup company; Company spin-off.

As in any other market, there are general and specific difficulties and promising areas for its subjects. The authors systematized the main of them in Table 1:

The system-wide difficulties and prospects presented in Table 1 are typical for the domestic M&A market in general, and specific for the SME segment in particular. The basis of specific difficulties is the economic behavior of some owners of SME, who not only deny the objective financial and economic feasibility of a strategic focus on reorganization and integration development but also deny the need for its strategic planning on this basis. On the other hand, the unification and development of accounting and analytical approaches and technologies for strategic planning in the modern business allow this planning to be carried out more productively and efficiently, which is reflected in the increase in the relevance of its use.

Table 1. Difficulties and prospects for the development of the domestic reorganization market in the SME segment

Difficulties	Prospects
System-wide	
Structural deficit (small share) of SMEs in the economy, underdeveloped communications in the business environment	Unification of accounting and analytical approaches and systems, transition to digital decision-making, expansion of digital ways of doing/developing business and its infrastructure
Lack of information, market statistics (compared to the market of large players)	Liberalization of control over the development of SMEs by reorganization, programs to support SMEs, optimization of the tax burden
Low attractiveness of transactions	Expansion of opportunities for debt financing of transactions (high liquidity in the banking system)
Lack of financial solvency of investors	
Lack of investors, incl. private and foreign investors	
Specific	
Opportunistic, irrational economic behavior of owners under conditions of possible stable development	High probability of bankruptcy due to a pandemic (this factor forces to integrate with big business)
Psychological orientation towards protectionism and protection of business based on attachment to it	Development of new forms of associations of SMEs within the framework of project activities, clustering and zoning, PPP, as well as other circumstances
Limited opportunities for business growth/capitalization, combined with ignorance of investments in fixed assets and low added value	Increased opportunities for restructuring and diversifying its activities into a structurally attractive industry direction for reorganization and integration transactions
High business risks and short life cycle	

Source Compiled by the authors

4 Budgeting as a Technology for Strategic Planning of the Integration of SMEs in the M&A Market

In many cases and subject to the voluntariness of reorganization and integration processes, such transactions, mutually or separately (for each of the parties to a potential transaction - SME structures), can be (and, in the author's opinion, should be) planned - taken into account in the process of developing a strategy, especially against the background of difficulties, as well as the prospects of the considered segment of the M&A-market.

To achieve this goal, the authors recognize budgeting (budget forecasting) as the most versatile technology of strategic planning, which allows producing consolidating subsidiary or solidarity (relative to the parties involved in transactions) strategic planning, including based on the unification of existing [8] and the possibility of developing additional specific [13] forms of budget forecasts. According to this, budgeting of such transactions:

- Should be used as a tool for planning the strategy of an M&A transaction at all its key stages;
- Can be implemented within the existing systems of budgets of business structures of the Ministry of Railways (if they exist), and separately under a transaction (or in two options at the same time), or in aggregate according to a newly created structure;
- Should be based on certain criteria of financial and economic efficiency of such transactions (determinants of reorganization), as a strategic planning tool (M&A transactions).

The full implementation of such requirements is possible only with the adequate use of budget forecasting tools in the strategic planning system of market integration, which is consolidated on large scale (due to the determinants of M&A transactions) in Table 2.

Table 2. Toolkit for budget forecasting in the strategic planning system of market integration of SMEs

Determinants of reorganization (including M&A transactions)	Determinant budget forecasting toolkit in the system of strategic planning of market integration
At the stage of planning a deal and its strategic component:	
Enterprise value	Forecast balance, Budget cash flow forecast
Reorganization costs	Budget forecast of transaction costs, Budget forecast of risk compensation
Financial leverage and financial structure of the reorganization transaction	Forecast balance and budget forecast of the financial condition; business plan in case of debt (credit) financing
Other (including qualitative) parameters of strategic analysis	SWOT - analysis, PESTLE - analysis, Competitive and expert assessments and forecasts
At the stage of completing a transaction and/or while predicting its results:	
Market indicators (market share, competitiveness index of products, enterprises)	Sales Budget Forecast, Financial Results Budget Forecast
Enterprise value	Forecast balance, Budget cash flow forecast
Savings operating costs	Operating budget forecast
Saving transaction costs	Budget forecast of transaction costs
Savings VAT	Fiscal forecast of the total tax burden
Other determinants…	System of budget and management forecasts…

Source compiled by the authors

According to Table 2, it can be seen that the main determinants (characteristic indicators of the financial and economic efficiency of M&A-market transactions) of such processes are the main criteria, as well as indicators of financial and economic performance, business efficiency. They can be constant or can change depending on other stages of the transaction, but in both cases - they can be planned and calculated, planned within the framework of standard and specific budget forecasts of business structures, and also while using standard planning techniques (such as SWOT - analysis or PESTLE - analysis).

For example, to plan savings on transaction costs, the preparation of a common (for all parties to the merger transaction) budget forecast of transaction costs [13]; operating costs (their savings) - a set of budget forecasts of operating activities, and other ones should be used. In general, budget forecasting tools for strategic planning of market integration within the SME M&A market segment can be the following ones:

At the level of one actor-subject of a transaction (for example, a merger):

1. Budget forecast of transaction costs;
2. Development budget forecast;
3. Financial budget forecasts (forecast balance, budget cash flow forecast, budget forecast of income and expenses);
4. Operating budget forecasts;

At the level of all actors (subjects) of the transaction (consolidated): Consolidated forms of the aforementioned budget forecasts, which were drawn up for the period of the transaction completion horizon (and not for a calendar year as in classical budgeting).

So, the use of the budget forecasting system (classical and specific tools for budget forecasting of entrepreneurial activity) as a technology for strategic planning of M&A transactions at the level of one actor - the subject of the transaction (a specific subject of the IPS), and at the level of all its participants, or consolidated for a newly created business entity (for example, as a result of a merger), it seems to be a promising direction for the development of this segment of the national M&A market, including increasing its transparency and investment attractiveness (from not only domestic but also investors of mergers - non-residents of the Russian Federation in the case of construction of budget forecasts on the mechanics and forms of IFRS).

5 Conclusion

Paradoxically, many years of experience in the implementation of the above-mentioned class of integration economic processes (due to evaluating on a global scale) indicates an increase in their relevance (demand by market participants) during periods of strategic orientation of business to use the mechanisms of "protective" development through capitalization and consolidation. The protectionist-expansionary strategic orientation becomes the result of a defensive reaction to external high-risk disturbances and crises (especially which were developing according to a completely unpredictable scenario, complementing the "true uncertainties" of Knight [7]), and it's just advisable to oppose an increase in financial capabilities and strategic growth, business differentiation in the product, as well as the geographic context through mergers and acquisitions for them. This is especially evident in the segment of small and medium-sized enterprises.

On a local scale, individual mechanisms of the M&A market (especially public-private partnership projects, other joint projects and programs for the development/support of business and its infrastructure, zones, and clusters of innovative development, the establishment of the activities of various cooperative entities, consortia and other associations with the primary preservation of legal autonomy etc., as well as classic mergers and acquisitions) have proven their relevance and effectiveness to ensure the progressive development of small and medium-sized enterprises (their entities and regional/national segments of the SME market as a whole). This emphasizes the need to implement and continue the scientific and applied search for ways and mechanisms of strategic planning for the implementation of such transactions once again, including through budget forecasting tools for small and medium-sized enterprises.

References

1. Civil Code of the Russian Federation (current edition). http://www.consultant.ru/document/cons_doc_LAW_5142/. Accessed 03 June 2021
2. Federal Law of the Russian Federation No. 14 Federal Law On Limited Liability Companies (current edition). http://www.consultant.ru/document/cons_doc_LAW_17819/. Accessed 03 June 2021
3. Federal Law of the Russian Federation No. 208 Federal Law On Joint Stock Companies (current edition). http://www.consultant.ru/document/cons_doc_LAW_8743/. Accessed 03 June 2021
4. Dmitriev, A.: M&A transactions. Why do companies carry out mergers and acquisitions? (2019). https://bcs-express.ru/novosti-i-analitika/sdelki-m-a-zachem-kompanii-provodiat-sliianiia-i-pogloshcheniia. Accessed 15 June 2021
5. Goldman Sachs Names Three Problems of the M&A Market in Russia (2021). https://www.rbc.ru/finances/02/04/2021/6065d6479a7947017527f363. Accessed 10 June 2021
6. Isakova, N.: The interactive paradigm of business amalgamation in the context of economic globalization and capital integration. Fundam. Res. **10–1**, 163–167 (2013)
7. Knight, F.H.: Risk, uncertainty and profit. Delo, 360 (2003)
8. Konina, O.V.: Development of economic relations between enterprises of the Russian federation and the EU based on unification of accounting and analytical systems for planning and forecasting joint activities. Financ. Res. **1**(54), 31–38 (2017)
9. Mergers are in quarantine. In 2020, the M&A market showed a record drop (2021). https://www.kommersant.ru/doc/4692046. Accessed 06 July 2021
10. Musatova, M.M.: Structural deformations and metrics of the Russian market of mergers and acquisitions of companies. Econ. Entrepreneurship Law **10**(12), 3075–3092 (2020). http: doi.org/https://doi.org/10.18334/epp.10.12.111451
11. Putin suggested not agreeing with the FAS on mergers and acquisitions in the amount of less than 800 million rubles (2021). https://tass.ru/ekonomika/11567311. Accessed 12 June 2021
12. Research The market of mergers and acquisitions in Russia in 2020 (2021). https://assets.kpmg/content/dam/kpmg/ru/pdf/2021/02/ru-ru-ma-survey-2020-fin.pdf. Accessed 12 June 2021

13. Tekin, A.V.: The role of specific budgets in ensuring the strategic development of modern business structures. Audit Financ Analy. **6**, 95–105 (2018)
14. The market of mergers and acquisitions (with the results of 2019). M&A statistics. http://mergers.akm.ru/stats/29. Accessed 09 June 2021
15. Yanushkevich, E.A.: Subjects of inheritance of an enterprise by will. Bulletin of the Udmurt State University, Series Economics and Law (2007). 2. https://cyberleninka.ru/article/n/subekty-nasledovaniya-predpriyatiya-po-zaveschaniyu. Accessed 10 June 2021

Assessment of the Asymmetry of Development and Support of Small and Medium-Sized Businesses in the Regions of Russia

Sergey A. Tinkov[1](✉), Elena V. Tinkova[2], and Vadim S. Kankhva[3]

[1] Plekhanov Russian University of Economics, Moscow, Russia
[2] Moscow Pedagogical State University, Moscow, Russia
[3] Moscow State University of Civil Engineering, Moscow, Russia

Abstract. The article deals with the problem of asymmetry in the development of small and medium-sized businesses in the regions of Russia. The methodological basis is cluster analysis while identifying asymmetry. The analysis is carried out using the "R" for statistical data processing and graphing. The use of this programming language makes it possible to identify homogeneous groups of regions based on a variety of primary statistical indicators. This makes it possible to differentiate measures of state support and more effectively bring the development of regions to the target indicators of the development of small and medium-sized businesses.

Keywords: Business · Entrepreneurship · Small business · Asymmetry · Cluster analysis · Regions Government support

JEL Codes: R12 · C38 · L26 · R58

1 Introduction

Currently, the development of small and medium-sized enterprises (SMEs) is one of the priority tasks of the Russian economy. In the national project "Small and Medium Enterprises", as well as in the Strategy for the Development of Small and Medium Enterprises in the Russian Federation until 2030, the growth of many indicators of SMEs development is planned: for example, an increase in the number of employees by 10%, an increase in SME turnover by 2.5 times, an increase in the share of SMEs exports in the total volume of Russian exports to 12%. Nowadays, objective conditions have been created for this, which is indirectly reflected in the Doing Business 2020 rating, in which the Russian Federation ranks 28th out of 190 countries participating in the rating.

The assessment of target indicators for the development of SMEs is carried out based on gross statistical indicators throughout the country and also the development of programs to support SMEs is carried out in the same way. However, it should be noted

E. G. Popkova (Ed.): ISC 2020, LNNS 368, pp. 566–572, 2022.
https://doi.org/10.1007/978-3-030-93244-2_62

that there are significant regional differences in the level of development of SMEs and the amount of state support. The territorial asymmetry of Russian regions is a fact that is recognized today in the results of many ratings. The Russian Federation is heterogeneous in terms of economic development. This gives rise to the differentiation of regions. Regional differentiation can become a serious problem. There is a high probability of impoverishment of some regions against the background of the enrichment of others. The development of entrepreneurship is also a factor in the well-being of the regions. Therefore, the imbalance in the development of the entrepreneurial environment is one of the problems that need to be investigated. Measures of state support for SMEs should be differentiated depending on the region or their homogeneous groups. At the same time, homogeneity should be determined only in the context of solving a specific economic problem according to several criteria, ideally, statistically recorded indicators, but not the results of a survey of the population and (or) business entities. Russia includes 85 regions - constituent entities of the Russian Federation, including 22 republics, 9 territories, 46 regions, 3 cities of federal significance, 1 autonomous region, and 4 autonomous districts.

The differentiation of regions is due to some factors, from which natural resources, geographical location and the results of human activity can be distinguished. The regions of the Russian Federation also differ significantly in terms of population.

2 Materials and Methods

Today, statistical compilations reflect a limited number of indicators for assessing the level of development of SMEs. Despite the abundance of studies of different directions of business development, for example, [3, 4, 7], they mostly use only a few primary indicators. But modern technologies make it possible to collect large amounts of data. For their processing and subsequent use, cluster analysis is well suited and is increasingly used. Algorithms for conducting and using cluster analysis are considered in the works of various authors. Rodriguez et al. [11] carried out a comparative analysis of clustering algorithms in sufficient detail. Tomi Kinnunen et al. [9] considers parametric Gaussian mixture model (GMM) and non-parametric vector quantization (VQ) model using the best-known clustering algorithms including iterative. Fraley and Raftery [6] were considering the problem of determining the structure of clustered data, without prior knowledge of the number of clusters or any other information about their composition. Maulik and Bandyopadhyay [10] were evaluating the performance of three clustering algorithms: hard K-Means, single linkage, and simulated annealing. Halkidi et al. [8] conducted a comparative review of well-known clustering algorithms. Davies and Bouldin [5] investigated the use of the measure of data division into clusters.

Cluster analysis is often used to analyze regional differences. So Brauksa [2] used cluster analysis to study the differences in economic performance between regions in Latvia. With the help of cluster analysis, regions are grouped according to a set of indicators into homogeneous groups. Bergman and Feser [1] used cluster theories to identify industrial and regional clusters.

Different functions are used as a measure of distances. The distance measure for cluster analysis of the level of development of small and medium-sized enterprises is the Euclidean distance. Euclidean distance is a geometric distance in multidimensional space. The preliminary standardization of the source data is required for using this metric.

The cluster analysis of the level of support for small and medium-sized businesses in the regions of the Russian Federation was carried out according to nine indicators characterizing the development of small and medium-sized enterprises in the constituent entities of the Russian Federation. The analysis was carried out using the "R" programming language for statistical data processing and graphing.

Indicators: x_1 - number of small enterprises, units; x_2 - number of medium-sized enterprises, units; x_3 - turnover of small enterprises, billion rubles; x_4 - turnover of medium-sized enterprises, billion rubles; x_5 - the number of subsidies to support small and medium-sized businesses, thousand rubles; x_6 - the number of active microloans issued by microfinance organizations, thousand rubles; x_7 - the number of active loans issued against guarantees of guarantee funds, thousand rubles; x_8 - the average number of employees of medium-sized enterprises, thousand people; x_9 - the average number of employees of small enterprises, thousand people.

To conduct a cluster analysis of the level of support for small and medium-sized enterprises in the regions of the Russian Federation, it's necessary to carry out primary data processing for a more correct interpretation of the results.

The scree method, which is also known as the bent knee method, is used for graphically determining the optimal number of clusters. The optimal number of clusters corresponds to the inflexion point of the graph. The abscissa shows the number of clusters k, the ordinate corresponds to the values of the function W (k), which determines the within-group spread depending on the number of clusters. As it can be seen, the regions would be best divided into 4 clusters (Fig. 1).

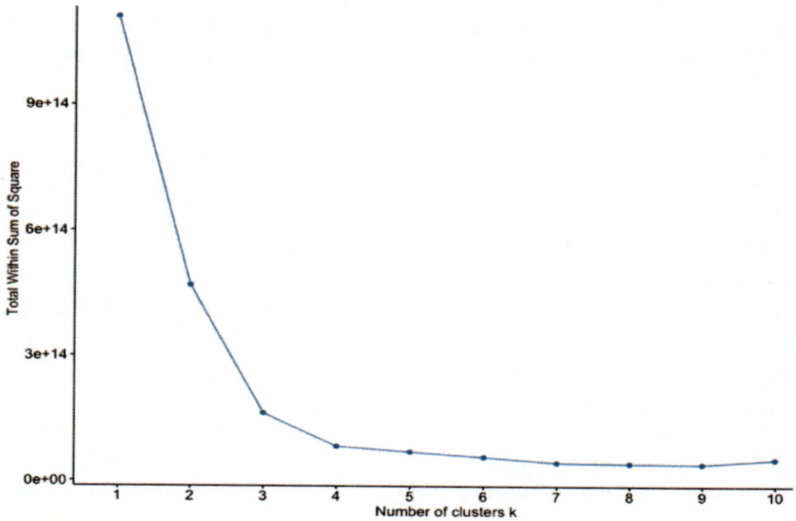

Fig. 1. Determination of the optimal number of clusters based on the "Scree" method of Russian regions by factors in 2018. *Source* Developed and compiled by the authors

The cluster analysis of the level of support for small and medium-sized enterprises in the regions of Russia is carried out based on hierarchical algorithms. In the beginning, all the constituent entities of the Russian Federation are separate clusters; in the subsequent steps of clustering, all similar regions are combined into one cluster.

The unification of regions into clusters was carried out based on the Ward method, in which the objective function is the sum of the squares of the deviations (distances) between the objects under study (in our case, the regions). It's convenient to use this method when clusters can be closely located.

To visualize the merging process, let's imagine the clustering of regions in a dendrogram. The dendrogram allows describing the proximity of individual regions and clusters to each other. A dendrogram based on Ward's method showed that regions could be divided into four groups. The authors estimate the quality of clustering using the correlation coefficient, which was 0.804 for this model (Fig. 2).

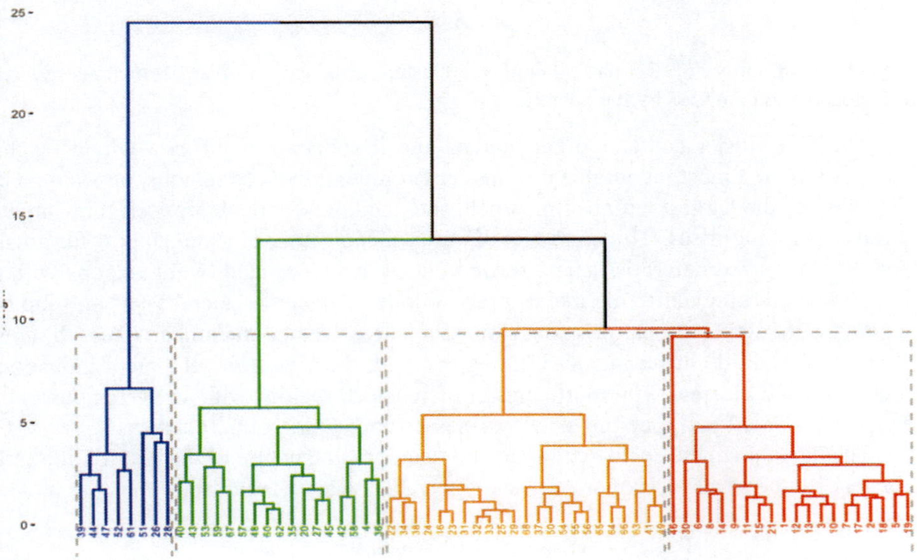

Fig. 2. Dendrogram of uniting Russian regions into clusters. Source Developed and compiled by the authors

3 Results

The result of clustering is the selection of a certain number of groups with similar characteristics. The result of the cluster analysis of the level of support for small and medium-sized businesses was the division of 68 regions (after primary data processing) of the Russian Federation into four clusters. All regions of Russia were numbered for the convenience of visualization and calculations (Fig. 3).

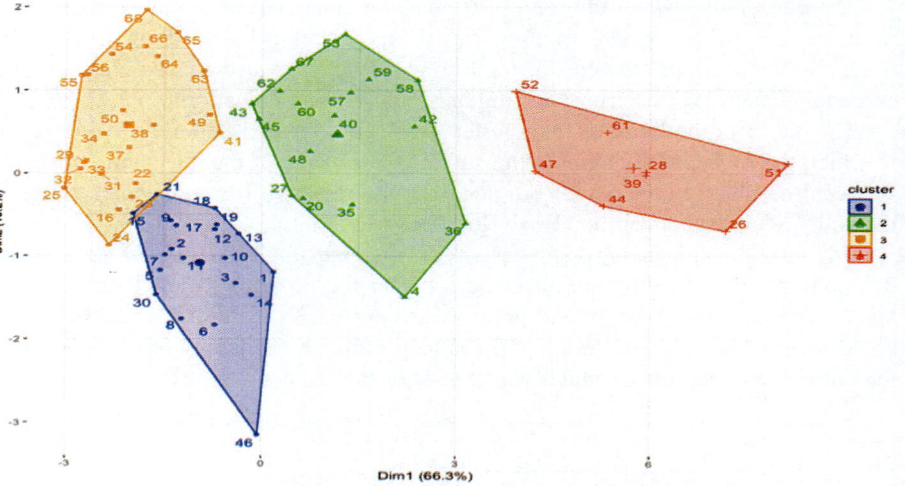

Fig. 3. Distribution of Russian regions by clusters obtained by Ward's method. *Source* Developed and compiled by the authors

The first cluster contains eight regions and it's characterized as a whole by the clusters with the most favourable business environment. In these regions, on average in the cluster, the largest number of small and medium-sized enterprises, the largest turnover of enterprises. There are also significant indicators of state support for small and medium-sized enterprises. The second cluster is represented by 17 regions with a favourable environment. This makes it possible to characterize a cluster as a group of regions with a level of enterprise development above average in comparison with other clusters. The third cluster includes 23 regions with the lowest development indicators relative to 2–3 clusters. The fourth cluster includes 20 regions with an average level of support for small and medium-sized businesses concerning other clusters.

To characterize the socio-economic and demographic nature of the distribution into clusters by the level of support, the average values were used (Table 1).

Table 1. Average values of indicators by clusters

Factors	Average value			
	Cluster 1	Cluster 2	Cluster 3	Cluster 4
x_1	71,099	30,780	7,907	16,762
x_2	408	204	46	92
x_3	1,254	527	114	303
x_4	218	95	21	45
x_5	77,739	55,365	36,662	75,909
x_6	551,740	441,685	225,976	244,582
x_7	11,289,566	6,326,448	2,337,853	4,322,331
x_8	46	24	6	12
x_9	345	163	38	87

Source obtained by the authors based on the results of cluster analysis

It should be noted that the third cluster is the most numerous. For example, this group includes such subjects of Russia as the Chechen Republic, the Republic of Altay, the Republic of Khakassia, the Kamchatka Territory, the Sakhalin Region, the Novgorod Region, the city of Sevastopol, etc. This cluster is characterized by an unfavourable social and entrepreneurial environment. It's worth paying special attention to the analysis of the measures that were taken in these constituent entities of the country.

On the contrary, the first cluster is characterized by the most favourable environment for the development of small and medium-sized enterprises and received support from the state. The cluster is represented by such subjects as the Krasnodar Territory, the Republic of Tatarstan, the Novosibirsk Region, the Rostov Region, etc.

4 Discussion

Thus, as a result of the conducted cluster analysis, the authors identified the differentiation of the regions of the Russian Federation. The social and economic difference is noticeable while comparing the result of the obtained clusters with the grouped regions of Russia in terms of the number of small enterprises.

The study of the level of support for small and medium-sized businesses and the level of its development showed that all interrelated economic processes depend on the social and economic policy of the state. In this regard, the study of economic processes in their close relationship with social and political conditions is of great interest at present.

Analysis of small and medium-sized enterprises in the Russian Federation showed that there is a trend towards an increase in the number of small and medium-sized enterprises for the entire period under consideration from 2008 to 2018. At the same time with the increase, the turnover of these enterprises increases and the average number of employees per enterprise decreases. Human labour is replaced by machine labour. The study of the dynamics showed that the greatest decrease in the number of enterprises occurs after serious external economic processes. This confirms the fact that the number of marriages depends on the state of the country's economy as a whole.

In addition, the number of government subsidies for support significantly decreased over 2008–2018. The reasons for the decline may be socio-economic factors, for example, a change in the government's policy, the realization of the ineffectiveness of the distribution of funds.

Clustering allowed finding regions that are similar in terms of SMEs development indicators and combine them into groups (clusters). The first cluster included eight regions with a favourable business environment; the second cluster included 17 regions with a level of entrepreneurship development above the average in comparison with other clusters; the third cluster - 23 regions with a low level, and the fourth cluster - 20 regions with average indicators.

5 Conclusion

Some social, economic, demographic, and other factors simultaneously influence the sustainability of entrepreneurship development.

To improve the business environment, in particular, for the development of small and medium-sized enterprises in Russia, a certain socio-economic policy should be developed, based on a scientific analysis of trends and characteristics of the environment. Measures to optimize the development of entrepreneurship should be integrated into each region according to the current situation in the constituent entities of the Russian Federation, where the main goals of the regional policy are to provide enterprises with everything necessary for full functioning.

References

1. Bergman, E.M., Feser, E.J.: Industrial and Regional Clusters: Concepts and Comparative Applications. Reprint. Edited by Scott Loveridge and Randall Jackson. WVU Research Repository, 2020 (1999). https://researchrepository.wvu.edu/cgi/viewcontent.cgi?article=1004&context=rri-web-book
2. Brauksa, I.: Use of cluster analysis in exploring economic indicator differences among regions: the case of Latvia. J. Econ. Bus. Manage. **1**(1), 42–45 (2013). https://doi.org/10.7763/JOEBM.2013.V1.10
3. Chaika, N.: Formation of development strategy for industrial enterprise. Qual. J. Manage. Syst. **22**(180), 20–26 (2021)
4. Danilina, M.V., Trifonov, P.V., Surkova, E.V., Tchaika, N.N., Klonitskaya, A.Y., Ermolaeva, E.N.: Managing business in the regions of Russia: Tthreats' analysis. Int. J. Econ. Res. **14**(4), 55–67 (2017)
5. Davies, D.L., Bouldin, D.W.: A cluster separation measure. IEEE Trans. Pattern Analy. Mach. Intell. **2**, 224–227 (1979). https://doi.org/10.1109/TPAMI.1979.4766909
6. Fraley, C., Raftery, A.E.: How many clusters? which clustering method? Answers via Model-Based Cluster Analysis. Comput. J. **41**(8), 578–588 (1998). https://doi.org/10.1093/comjnl/41.8.578
7. Golov, R., Alkhimovich, I., Kazarnovskij, V., Ermolaeva, E.: Development of a mechanism for assessing the economic sustainability of small enterprises. In: E3S Web of Conferences, vol. 91 (2019). https://doi.org/10.1051/e3sconf/20199108052
8. Halkidi, M., Batistakis, Y., Vazirgiannis, M.: On clustering validation techniques. J. Intell. Inf. Syst. **17**, 107–145 (2001). https://doi.org/10.1023/A:1012801612483
9. Kinnunen, T., Sidoroff, I., Tuononen, M., Franti, P.: Comparison of clustering methods: a case study of text-independent speaker modelling. Pattern Recogn. Lett. **32**(13), 1604–1617 (2011). https://doi.org/10.1016/j.patrec.2011.06.023
10. Maulik, U., Bandyopadhyay, S.: Performance evaluation of some clustering algorithms and validity indices. IEEE Trans. Pattern Anal. Mach. Intell. **24**(12), 1650–1654 (2002). https://doi.org/10.1109/TPAMI.2002.1114856
11. Rodriguez, M.Z., Comin, C.H., Casanova, D., Bruno, O.M., Amancio, D.R., Costa, L., et al.: Clustering algorithms: a comparative approach. PLoS ONE **14**(1), e0210236 (2019). https://doi.org/10.1371/journal.pone.0210236

The Role of Small Enterprises of Production and Consumer Services in the Development of the Modern Economy

Evgeniya V. Kazanskaya[1(✉)], Yuliya A. Mazneva[2], Valeriy V. Kotelnikov[1], and Natalya I. Gerasimova[1]

[1] Rostov State University of Economics, Rostov-on-Don, Russia
[2] Russian Customs Academy, Rostov branch, Rostov-on-Don, Russia

Abstract. The work is devoted to the study of the role of small businesses in the sphere of production and consumer services in the economic growth of the country. The research is based on a general scientific methodology, which provides for the application of a systematic approach to solving problems using marketing research tools.

Despite the fact that the Russian market of services is considered one of the most promising business areas, there are urgent problems for this market that need to be solved. These include poor quality of services, outdated technologies, and equipment, low staff qualifications, administrative barriers, low wages, unhealthy competition, etc. It is necessary to regard the scientific and practical features of entrepreneurial and marketing activities arising from the features and specifics of the intangible production of small enterprises in the field of production and consumer services. Such features include sensitivity to market conditions changes, expansion of communications, etc. Only 10–15% of middle-class citizens can use high-quality services of small enterprises in the modern Russian market of services. The trend of economic growth may stop, as the services market will face demand constraints soon.

Despite the fact that Russia is one of the 30 countries participating in international services trade, where there are no customs duties, Russia's part in global services exports and imports is still small. International trade can largely solve the problem of limiting demand for small business services. It is advisable to export and import services by opening branches of service providers in other countries.

Keywords: Economic growth · Service market · Small business · Enterprises

JEL Codes: G00 · R 00

1 Introduction

The problem of trends and economic growth is currently the most important problem of modern macroeconomics. Economic growth is identified with a certain growth and development of the state economy, namely, the real volume of gross domestic product for a certain period [7, p. 1–5]. Currently, we can state the existence of two types of

E. G. Popkova (Ed.): ISC 2020, LNNS 368, pp. 573–579, 2022.
https://doi.org/10.1007/978-3-030-93244-2_63

economic growth. Extensive growth implies an increase in the economic development of the country due to an increase in production resources. Intensive growth is an economic growth in which technologies are used to develop an economy where production reaches perfection, which causes an increase in production volumes for the main production factors [9, p.165–171].

According to Rosstat, the GDP rate in the first half of 2018 slowed by 0.9%, while the unemployment rate continued to rise to 4.2%. There was a slowdown in economic growth due to a decrease in its main principles that create opportunities for increasing wealth and increasing its volume in the economy, and, consequently, a decrease in real monetary incomes of the population by 30% [16]. The estimate of the index of the physical volume of GDP for 2019 was 101.3%, which in absolute terms amounted to 110.046 trillion rubles [14]. According to the first estimate of GDP produced in 2019, its volume was 109.362 trillion rubles [11].

Widespread all over the world small business is a significant economic sector, one of the crucial tax revenue sources. The development of innovative technologies plays an important socio-economic role in the country's economy, and small businesses contribute to this [2, 6, 16]. In comparison to large and medium enterprises, it contains many advantages. Among them are the following: flexibility, high ability to adapt to the external environment, the ability to introduce new technologies, greater freedom of action, and low need for initial capital.

Recently, small business has been actively developing In Russia. We can witness small enterprises of various types of economic growth, even though they face a huge number of problems in their development. However, at the same time, small businesses in Russia also have huge reserves for further development [3, p. 126–134].

The analysis showed that in the sphere of small business, the largest share falls on small enterprises involved in such economic activities as manufacturing, construction, etc. The important role of small business in the economic growth of Russia is well known [7, p. 180–185]. A significant amount of such activity occurs in the market of services - the sphere of services exchange that are the result of the work of non-production enterprises.

2 Methodology

A service is an intangible activity that one party offers to another, possibly related to a tangible product, but does not lead to ownership of anything, as well as to the transfer of property.

The development of commodity-money relations is a historical prerequisite for the emergence of the market of the service. The goods production and services presentation has developed into independent spheres of activity. The service market is completely different from other markets. There are several explanations for this. The service cannot be evaluated until it is provided since such an assessment is possible only by comparing the expected and acquired benefits. On the other hand, the uncertain nature of consumer services makes it difficult for sellers to promote services to the market.

Due to the saturation of the market with goods services production becomes more complex, the demand for services is constantly growing. The continuous appearance of new service types contributes to the service sector's adaptation to market demand. As a result, the manufacturing sector is lagging behind the service sector in terms of growth in most industrialized countries. More than 40% of the foreign companies invested in trade and insurance services, banking. Moreover, in leading countries, the excess of the service segment in the GDP is 70%, as the number of unemployed people in the service sector increases. The service sector accounts for 80–90% of the increase in new jobs [17].

The modern Russian market of services has splendid perspectives in business. More and more people and organizations stop solving many problems independently, trusting qualified specialists. The service sector accounts for a significant part of the gross regional product, while society is provided with thousands of jobs.

The intensive development of service enterprises is facilitated by the growing demand for services in recent years, both from corporate clients and from individuals. The annual increase in the service segment is due to the desire of wealthy people to improve their quality of life and the desire to save their time and money.

Russian market of services differs from the European one. In the sphere of Russian public services, collective, cooperative forms of management, labor collectives, legal entities, and individuals have become common. In Western countries, small private businesses and family businesses are popular. Another specific feature of the Russian services market is the diversification of enterprises associated with the necessity for enterprise survival in the modern competitive environment. Therefore, enterprises often actively develop other non-core activities along with their main activities, which often bring significant additional income.

The typical problems for the small business services sector:

- lack of production space;
- difficulties in obtaining credit resources and attracting investments in the required volume;
- lack of highly qualified personnel;
- shortcomings of legal regulation of the industry on the regional and federal levels;
- the need to improve the level of service and the quality of services provided;
- price barriers for middle-and low-income citizens to receive services;
- insufficient public awareness of small businesses and the services they provide.

Furthermore, we identify the reasons for these problems and consider possible positive trends in the Russian Federation services market development [16]. The world trade in services is dominated by the industrialized countries (Japan, South Korea, countries of North America, EU, etc.) by providing financial, information, telecommunications, medical education, and services. More than 50% of global trade in services is accounted for by these countries. Developing countries provide mainly transport, tourism, and financial (offshore) services; this share is much smaller in the global trade services [5, 10].

In comparison with other countries involved in international trade in services, Russia's share in global exports and imports of services is still small. However, the Russian Federation acts as a promising importer and exporter of services and is of interest to foreign investors providing services in mediation, audit, consulting, tourism, public catering, trade, and hotel business.

The specificity of the export and import of services is that the results of their provision, even material, do not go beyond the customs borders of the country, so there are no customs duties in the world trade in services. Opening of the branches of service providers in other countries helps to export and import the services.

The service market is a kind of commodity market and develops according to the market economy laws. Some specific features determine a special approach to business and marketing activities that meet the demand for services. The service market features consist primarily in the high dynamism of the market processes of providing services and the desire to directly meet the needs of the consumer. The characteristics of the territory covered by a particular market have a constant impact on the conditions of functioning of service enterprises. Thus, the search for the optimal form of service provision that activates the demand of the population, the high speed of capital turnover resulting from a shorter production cycle is the goal of territorial segmentation.

The high sensitivity of small businesses to market conditions changes is explained by the lack of need for storage, warehousing, and transportation of services. Small enterprises that produce services have greater mobility and excellent flexibility to respond to market environment changes, in the case of spatial overlap of production and consumption of services. The specifics of the organization of the production of services and their provision of services is the need to ensure personal contact between the manufacturer and the consumer. on the one hand, all this is a condition for the expansion of the sphere of communication, but at the same time leads to an increase in the requirements for personnel, ethics, and general culture of the manufacturer [8]. Diversification, personification, and individualization of the demand for services, lead to a high level of service diversity and is the most important incentive for the development of innovative activities in the service sector [10]. Since the appearance of non-standard services is due to the complex demand structure, which is explained by the desire of the buyer to search for innovations in the field of interest, and becomes a permanent process that is increasingly developed as market demand becomes saturated. It is impossible to determine the result of the provision of services work with sufficient accuracy because we can estimate the result of the service only after its consumption.

The features and specifics of intangible production determine the most relevant branches of the service sector (scientific and practical features of marketing): audit, consulting, banking, travel services, hotels, and restaurants. Analyzing the sphere of social services we can note with regret that every second or third small business enterprise carries out its activities mainly in the field of trade, public catering, or consumer services.

Currently, companies are developing for which services are a related business that enhances the competitive advantage of the main business. Another effective form of entrepreneurship in the service segment is the network form of business, which is used by enterprises that create their networks through their own or foreign investment. These network companies develop according to the principle of construction of a branch office or using franchising. The network business model allows the participants to create new segments of the service market, thereby increasing their competitiveness.

The secret to the success of a small business often lies in accurately identifying unmet demand that other companies have not noticed, providing a completely new service that provides rapid growth in the number of customers in a short time. This situation makes it possible to build a sustainable business, promising both in terms of creating large network structures.

The modern Russian market of services is formed by network companies and their role will increase even more in the future. Analyzing the current situation of service enterprises we can identify various factors that influence the life of society. The positive factors of economic growth include the relative ease of solving the poverty problem [10]. Negative factors of economic growth: the desire to radically solve the problem of poverty by an equitable distribution of national income; decline in employment and increase in unemployment due to advances in science and technology; change in established values in society.

Among the topical issues for the Russian market include poor quality of services, which is a consequence of the obsolescence of technologies and equipment, insufficient qualification of personnel, low wages; administrative barriers; unhealthy competition, the hiding of profit as a result of incomplete payment of taxes to obtain profits.

For the solution of the Russian services market problems, we consider the following set of measures to be effective:

- improvement of the credit system and lending programs for the service sector;
- active use of concessional lending to small business services, including the small business support fund, along with facilitating access to credit resources;
- provision of tax benefits to small businesses on rent payments;
- improvement of the regulatory regulation of small enterprises.

The creation of systems of voluntary certification of services and quality systems in the field of service provision seems appropriate in order to provide the population with reliable information about the quality of services and improve the level of population service. The insufficient number of small enterprises meets the needs of the people, which does not contribute to the competitive environment development: high prices for services make them inaccessible to the population with middle and lower-middle incomes, as well as practically inaccessible to low-income citizens. Therefore, an important task for the development of services is to ensure their social orientation, for example, preferential services for certain groups of the population and social support in the form of discounts on services provided, etc.

The lack of consumer awareness about the service sector forces the population to turn to private illegal business. This situation reduces the quality of the services provided does not provide consumers with any guarantees, and in some cases creates all kinds of threats to the health and property of citizens. Currently, in different countries, there is a transition from extensive to intensive growth, from the number of products by attracting additional means of production to technologies to increase the productivity of employees.

It is obvious that for the effectiveness of the process of economic growth, it is necessary to strive for the maximum satisfaction of the needs of society. For example, increasing environmental problems in a country may constrain its economic growth. In 2020, the growth of GDP slowed down in comparison to the previous year, the unemployment rate continues to rise, and the real monetary income of the population in the market economy sector of the modern state is declining. Small businesses are inextricably linked to the implementation of independent activities aimed at obtaining a permanent income in the face of the risk of losing their assets. The role that small business plays in the processes of economic growth and development of the state becomes obvious. In the sphere of production and service provision, the small business acts as the most important development factor of the Russian economy [1]. In 2016, 2017 and 2018 the number of small enterprises in Russia was 979,300, 1,032,800, and 1,137,400, respectively. Among them, by type of economic activity related to construction, 109,300, 117,100, 130,700 [12]. Well-known negative social phenomena can be overcome by solving problems developing small and medium-sized businesses reduce unemployment and increase the educational level of small and medium-sized businesses [5].

The economic efficiency of small businesses is determined by the difference between the revenue and the price of conducting events. The increase in tax revenues can be considered as evidence of an increase in income from small businesses. The growth of tax revenues from small businesses has a positive impact on the country's economic development, which is a basic component for the small businesses' development in the service segment.

3 Results

The role of small business in the economy of modern Russia is extremely important, largely because it provides a significant part of the budget (regional and federal) through tax collection. Consequently, the formation of the budget of income and expenses in Russia, which depends on the receipt of a share of tax payments, is directly interrelated with the small businesses' effective development in Russia. In addition, small businesses have an impact on the creation of new jobs, resulting from the number of services growth provided to the population in the market. In light of the above, the state is actively trying to provide small businesses with all possible legal and financial support in various ways.

4 Conclusions

The growing demand for services, both from corporate clients and from private individuals, contributes to the intensive development of small enterprises providing services. The impact on economic trends from small business development programs in Russia is mainly accompanied by an increase in tax revenues from small businesses that can have a positive impact on the country's economic growth. However, currently in Russia, only 10–15% of middle-class citizens are the target audience receiving high-quality services [13]. In this regard, in the next few years, the trend of rapid growth may reduce its growth, since the service market may face demand constraints.

References

1. Federal Law No. 88-FZ of 14.06.1995 (as amended on 02.02.2006) On State Support for Small Business in the Russian Federation. http://www.consultant.ru/document/cons_doc_LAW_6857/. Accessed 17 Dec 12 2020
2. Butenko, A.Y.: Assessment of the competitiveness of a small business company. Working paper. Mark. Rus. Abroad **4**, 126–134 (2017)
3. Buyankina, A.N.: Small Business: State Regulation. INFRA, M, Moscow (2016)
4. Cravino, J., Levchenko, A.: Multinational firms and international business cycle transmission. Q. J. Econ. **132**(2), 921–962 (2017)
5. Dadashev, A., Glovatskaya, N., et al.: Effectiveness of small business support. Working paper. Econ. Issues **7**, 16–18 (2018)
6. Danilov A.V.: The role of small business in the economic growth of Russia at the present stage. Working paper, Young Sci. **21**(259), 180–185. (2019). https://moluch./archive/259/59351/. Accessed 24 January 2021
7. Gamza, V.A.: Small and medium-sized enterprises: financial and credit problems of development. Working paper. Banking **6**, 1–5 (2018)
8. Kazanskaya, E.V., Kotelnikov, V.V.: Marketing communication and interaction systems for small businesses. In: Joint Conference: 20th conference Professional Culture of the Specialist of the Future and 12th conference Communicative Strategies of Information Society, The European Proceedings of Social Behavioral Sciences. Eur Publisher, 2020 (2020)
9. Korchagina, I.V., Korchagin, R.L.: Clusters of small enterprises in the economic space of the region and institutional limitations of their development. Working paper, Econ. Entrepr. **11** (1), 165–171 (2018)
10. Savchuk, S.B.: Research on the formation of the innovative potential of the region working paper. Bull. South Inst. Manag. **1**, 54–58 (2018)
11. Gross domestic product of Russia according to Rosstat. Regional and federal statistics and indicators. https://rosinfostat.ru/vvp/. Accessed 31 January 2021
12. Official website of the State Statistics Committee of Russia. http: www.gks.ru. Accessed 12 Nov 2020
13. Rosstat for the first tie revealed the share of small and medium-sized businesses in the economy. RBC. https://www.rbc.ru/economics/05/02/2019/5c5948c59a794758389cfdf7 Accessed 31 Jan 2021
14. Rosstat presents the second estimate of GDP for 2019. https://rosstat.gov.ru/folder/313/document/81201. Accessed 31 Jan 2021
15. Scenario forecast of macroeconomic dynamics and income of the population until 2024, National Research University Higher School of Economics. https://dcenter.hse.ru/forecast_nep. Accessed 25 Jan 2021
16. Small business in the field of household services. http://bishelp.ru/business/economics/malyy-biznes-v-sfere-bytovyh-uslu. Accessed 15 Dec 2020
17. The first estimate of GDP for 2018. https://rosstat.gov.ru/storage/mediabank/Pervaya-otsenka-VVP-za-2018-god.pdf. Accessed 31 Jan 2021

Economic Security and Sustainable Development of Markets and Companies in the Digital Economy

Identification of Double Combined Knitted Fabrics and Quilted Textiles

Elena G. Ustyugova[1(✉)], Galina Yu. Fedotova[1], Tatyana A. Zakharenko[2], Viktoriya V. Ivanova[2], and Adelina E. Emirova[1]

[1] North-West Institute of Management of the Russian Presidential Academy of National Economy and Public Administration, NWIM RANEPA), Saint Petersburg, Russia
{ustyugova-eg, fedotova-gy, emirova-ae}@ranepa.ru
[2] St. Petersburg Branch of the Russian Customs Academy named after V. B. Bobkov, Saint Petersburg, Russia

Abstract. The identification of goods in international trade is the basis for assigning a product code according to the Commodity Nomenclature for Foreign Economic Activity of the EAEU (CN FEA EAEU). The code provides the application of customs measures of tariff regulation, prohibitions, and restrictions, as well as measures to protect the internal market of the county. Double combined knitted fabrics and quilted textile materials are widely demanded in the international market, but the classification of these products often leads to litigation between customs authorities and participants of foreign economic activity. The identification research of such textile goods is carried out based on a visual study of the structure of these materials, the location of the threads that form the outer surfaces of the web, and the threads that hold the surfaces (layers) of the web together. The analysis of the technology for the production of these materials is applied and includes the sequence and number of the necessary basic operations, the classification of knitting machines on which the fabrics are made, the number of needle cases (fontur) - single-fontured, double-fontured, multi-fontured on these textile machines. The one-thread and two-thread structures of the sewing stitches in quilted materials are investigated. The threads connecting the surfaces or layers of these materials are also investigated, and the direction of twisting of these threads is determined - left S twist or right Z twist. These identification signs made it possible to unambiguously differentiate such textile goods to assign the code number following the CN FEA EAEU and minimize the risks of their unverified classification.

Keywords: International trade · Identification · Double combined knitted fabrics · Quilted textile materials · Classification · CN FEA

JEL Codes: B17 · L67 · L81

E. G. Popkova (Ed.): ISC 2020, LNNS 368, pp. 583–590, 2022.
https://doi.org/10.1007/978-3-030-93244-2_64

1 Introduction

The identification of textile materials allows you to determine their purpose, properties, structure, as well as to establish their price characteristics [18]. Currently, the textile industry is developing in the direction of improving equipment, introducing new technologies and producing textile materials for various purposes [12]. For participants in international trade, the issue of identifying goods is relevant in determining whether it meets the requirements of the contract and if it establishes the conditions for the movement of such goods across the customs border of the EAEU [4]. The commodity code by the CN FEA EAEU is the key legal instrument in customs regulation. The determination of the code to move across the customs border of the EAEU provides the basis for the application of measures of customs and tariff regulation, the application of prohibitions and restrictions, measures to protect the internal market [2]. In turn, this is also an important factor in pricing, organizing logistics operations in the supply chain, business planning [15].

The identification characteristics of the textile materials are reflected in their aesthetic and physical-mechanical properties [11]. And it also depends on the fibrous composition, the structure of the interweaving of textile threads in the fabric, from the method of production, finishing, etc. [9].

Depending on the purpose, different requirements are imposed on textile materials, for example, disposable items or resistance to repeated washing; with increased heat and air permeability or, on the contrary, heat-insulating; resistant to weather factors and the environment or for decoration of premises [10].

According to the manufacturing method textile materials can be knitted, non-woven and woven [13]. The development of modern production technologies is primarily reflected in the emergence of a large assortment of such textile materials as knitted and non-woven [6]. Those are the materials that make up the bulk of the import of textile materials into the territory of the Russian Federation [14]. We are also taking into account the fact that not all customs officials are competent in terms of knowledge of the external distinctive features of the textile goods. It seems important to carry out a comparative analysis in particular for the identification of combined knitted fabrics and quilted textile materials.

Thus, it is a relevant question of practical interest for the suppliers and the end-users to determine the difference between double combined knitted fabrics and quilted textile materials. The quilted textile materials mean that they are obtained by stitch bonding of two or more single knitted fabrics [7].

The purpose of the study is to determine the distinctive features to make it possible unambiguously to identify a double combined knitted fabric and the quilted textile materials and provide the classification of it following the CN FEA of the EAEU.

To achieve the target were set the following tasks to establish the external distinctive features of the materials under the investigation that will provide with the determination of:

- the production technology;
- the equipment for the production of these textile materials;

- the properties of the threads that hold the front and back surfaces of textile materials together;
- the characteristics of the sewing lines (stitches), if any;
- the characteristics of the classification of the double combined knitted fabrics and the quilted textile materials following the CN FEA of the EAEU.

2 Methodology

For the identification of the investigated materials the following methods were used: the visual inspection of the surface, photography, the analysis of the mutual arrangement of the threads in the materials and the direction of the twist of the threads fastening the front and back surfaces of the materials.

The object of the research is the double combined knitted fabrics produced from three systems of threads on the double-fontur knitting machines with two-needle beds. One thread system forms the front surface; the second one forms the back surface of the fabric. And the third thread system is laid across (in the direction of the "weft") between the outer surfaces (front and back) in the process of fabric production. So, it is a complex textured untwisted thread (consisting of elementary threads that have the crimp, which gives it volume and increased extensibility).

It is impossible to disassemble into parts nor the front, the purl and the additional weft thread without violating the integrity of the material. Such features are typical for the double cross-knitted jacquard knitted weft fabric produced by one technological operation. Similar textile materials are produced on modern computerized double-fontur (with two-needle beds) cross-knitting jacquard machines. All the migratory threads, which are fastening the front and back surfaces of the double combined knitted fabric have a left-hand S-twist.

Another object of the research is the quilted textile materials, which are produced from various materials, by the usage of the method of mechanical fastening of stitches by the loops on the knitting-stitching machines or sewing machines.

Quilted textiles have layers that can be similar or different in chemical composition and structure, including those, which can be coated or impregnated with plastic. The sewing stitches of the quilted textile materials can be single-thread, double-thread, or included a larger number of threads.

The one-thread chain stitches and the two-thread stitches (or more) are performed by the knitting-stitching machines, and the two-thread shuttle stitches (or more) are performed by the sewing machines. In this case, the stitch is formed from the additional sewing thread (threads) tucked into a sewing or knitting-stitching machine. These threads must have a right-hand Z-twist.

The manufacturing technology of the objects under the research is different. The double combined fabrics are produced (created) from textile threads on the knitting-stitching machines and the connection (fastening) of the layers by the textile threads into one quilted fabric are carried out on the sewing machines to produce the finished textile product.

The investigated method of the identification of the textile materials allows the customs authorities and the participants of the foreign economic activity to classify the textile products following the CN FEA unambiguously [5].

3 Results

The double combined knitted fabrics are produced on the double-fontur knitting machines with two-needle beds (fonturs). For this, the following groups of machines are used: flat-knitting, knitting, warp knitting [8].

The process of looping by the knitting technique on the hook needles (in the form of an open hook) on the flat knitting machines includes the following operations: conclusion, laying, flatting, carrying out, pressing, applying, connecting, dropping, shaping, pulling. This method of production is provided with the twisting (bending) of the thread after laying it on the needles [8].

The process of looping in a knitting way on tongue needles (with a hook and a swivel tongue) is different in the order of the operations [8].

The looping process on warp knitting machines includes the same operations as in the knitting method, but it differs in that each loop is formed from a separate warp thread [8].

Knitted weaves are classified as combined weaves if it consists of a set of weaves of different classes of the interweaving: main, derivative or patterned [8].

The appearance of double combined knitted fabric under the research is perceived as a material consisting of three layers: two outer layers (knitted fabrics) and one inner layer (parallel chemical threads) (Figs. 1, 2 and 3).

Fig. 1. The appearance of the surface structure of the samples of the double combined knitted fabrics. *Source* compiled by the author

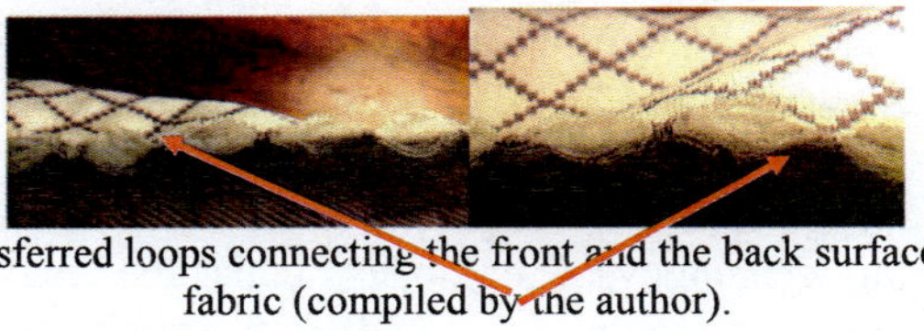

Fig. 2. The appearance of a cut sample of the double combined knitted fabric. *Source* compiled by the author

Fig. 3. The appearance of the inner "weft" textured complex filaments. *Source* compiled by the author

In the production of a double combined knitted fabric, the looping process is carried out simultaneously on two needle beds. The process of creating the fabric is as follows: two threads form the outer surfaces of the fabrics (the front and the back sides), formed from continuously twisting threads that form rows of loops laying in the same direction crosswise the fabric. The loops in adjacent rows are intertwined, forming cells, which is typical for a double cross-knitted knitted fabric.

Both surfaces are containing rows of looped stitches and transferred loops connecting the front and back surfaces of the fabric. The transferred loops are located unevenly, under the rapport, due to which large plant or geometric patterns are formed on the surface of the fabric. The textured weft (having additional volume and extensibility) complex yarns [17] are placed (knitted) into the gaps between the two surfaces of the fabric in the process of the production of the fabric.

It is impossible to disassemble the front and the back and the additional threads into the parts without violating the integrity of the material. These characteristics are typical for the double combined cross-knitted weft jacquard knitted fabric produced within one technological operation.

The embossed pattern on the surface of the fabrics is the appearance of a jacquard knitted weave that has a large rapport (a repeating fragment determined by the number of loops in width and length), which is created during the production (knitting) of the fabric. This pattern is not embroidery, sewing stitches or knitting-stitching stitches.

The rows of the stocking stitch structure in the combination with the rows of the rib structure are provided with high dimensional stability in both longitudinal and the transverse directions of the fabric. The cellular structure of the fabric is obtained by the transferred loops and the threads over (the open loops).

Similar textile materials are produced on modern computerized double-fontur (with two-needle beds) cross-knitting jacquard machines. They are knitting from three systems of threads, with single edges.

The regular distribution of the cells in the structure of the fabric, the volume and elasticity, which are caused by the untwisted textured complex chemical thread, contributes to a decrease in the thermal conductivity. At the same time, the sufficiently large crosscutting porosity of the structure of the fabric increases the vapour permeability. These properties of the fabric are improving the heat and air exchange processes during the usage of such knitwear and, accordingly, increasing the heat-shielding properties, which determines the scope of application of such materials.

All migratory threads that are fastening the front and back surfaces in the double combined knitted fabric have a left-hand S-twist, which is typical for the threads intended for the production of the fabrics. These threads are not sewing threads.

These characteristics are typical for the cross-knitted double weft jacquard (patterned) combined knitted fabrics, obtained within one technological operation [16].

The quilted textile materials are textile fabrics that are produced from various materials by the method of mechanical fastening of stitches with loops on the knitting and stitching machines or the sewing machines.

This kind of material is usually consisting of two outer layers of fabric (knitted or woven or non-woven, or combinations thereof) separated by a layer of the padding material. The quilted textile materials have layers, which can be similar in the chemical composition and the structure, or different ones, including coated or impregnated with plastic [8]. These layers of fabric can be interconnected by stitching (on the sewing or knitting machines), the straight rows of stitches, or by the decorative stitches (not the embroidering patterns). These stitches are mainly used to quilt the garment.

The sewing stitches of the quilted textile materials can be single-thread, double-thread, and also from a larger number of threads.

The names, the graphic image, and the application of the stitches are established both in the international [3] and in the interstate standard.

Double- and single-thread (and more) chain stitches are produced on the knitting-stitching machines with the names of the stitch, like these: single-thread single-line straight chain stitch, two-thread single-line straight chain stitch, two-strand zigzag chain stitch [1].

Two-thread (and more) shuttle stitches are produced on sewing machines with the names of the stitch, like these: two-thread single-line straight shuttle stitch, two-thread single-line zigzag shuttle stitch [1].

In the case of the damage of the textile thread that forms the stitches and connects the outer of the fabrics, the quilted fabrics can be split into two fabrics without damaging the integrity of the surfaces of each fabric.

In addition, a stitch is formed from the additional sewing threads (threads) tucked into a sewing or knitting-stitching machine. These threads, which form the stitches and hold the quilted textiles together, are the sewing threads.

These threads are imposed a special requirement due to the increased mechanical impact during the operation of the quilting of the textile materials.

These threads can be twisted in different directions: left-hand twist S and right-hand Z twist. For any threads of the left-hand twist S - from the bottom on the right up to the left. And for any threads of the right-hand twist Z, the direction of the twist turns is from the bottom on the left up to the right.

It has been experimentally proved that sewing threads exhibit different properties. It depends on the direction of twisting when materials are quilted. The left-hand S twist threads are gradually unwinding and become loose, and the individual components of the thread break. The shortest elementary thread breaks first, and then the rest. In this case, the thread breakage is uneven.

The right-hand Z twist threads during quilting (stitching) gradually twisting and become denser and more durable, withstand heavy loads. And if they break then the thread break is sharply outlined.

Thus, for the sewing and knitting-stitching machines and the quilting (sewing) textile materials, it is advisable to use threads of the right-hand Z twist [3].

4 Conclusion

Due to the research, the method for the identification of the distinguishing features of the double combined knitted fabrics and the quilted textile materials was proposed. These characteristics of the fabrics provided the determination of the code number following the CN FEA of the EAEU.

By visual inspection of the surface of textile materials, by analyzing the mutual arrangement of the threads in the materials and determining the direction of the twist of the threads fastening the front and back surfaces of the studied materials, it was possible to identify the double combined knitted fabrics produced within one technological operation on the double-fontur knitting machines, and the quilted textile materials, obtained by the method of mechanical fastening of various materials with loops of stitches on the knitting and stitching machines or the sewing machines.

It is impossible to disassemble double combined knitted fabrics without violating the integrity of the material; but for quilted textile materials, when the sewing lines (stitches) are removed, the fabrics disintegrate into separate layers without distraction.

All migratory threads that fasten the front and back surfaces in the investigated double combined knitted fabric have a left-hand S-twist.

In quilted textiles, a stitch (s) is formed from additional sewing threads (threads) threaded into a sewing or knitting machine. These threads must have a right Z-twist.

Considering the classification criteria for the researched commodities laid down in the CN FEA EAEU, the key stage is the sequence or simultaneity of operations in the production of such goods, the presence/absence of layers (including soft cushioning) in the structure and the use of established methods of bonding layers.

Thus, the proposed method for identifying researched textile materials allows the customs authorities and the participants of foreign economic activity to unambiguously classify goods following the CN FEA EAEU.

References

1. GOST. GOST 12807–2003 Sewing products. Classification of stitches, stitches, and seams (with Amendment) [Izdeliya shvejnye. Klassifikaciya stezhkov, strochek i shvov (s Popravkoj)], GOST fom 05 July 2005, No. 12807–2003. 7 March 2021 (2003). http://docs.cntd.ru/document/1200040904. Accessed 10 Mar 2021
2. Afonin, D.: Identification of textiles related to goods at risk. Bull. Innov. Technol. **3**(10) (2019). http://bitjournal.ru. Accessed 10 Mar 2021
3. Chervyakov, F., Sumarokov, N. (n.d.).: Sewing machines. Mech. Eng. 7 March 2021. https://www.chipmaker.ru/files/file/7351/. Accessed 10 Mar 2021
4. Denisovich, V.: Legal aspects of the performance of customs operations when moving goods through the customs border of the EAEU. Bull. Pskov State Univ. Manage. **4** (2016). https://cyberleninka.ru/article/n/pravovye-aspekty-soversheniya-tamozhennyh-operatsiy-pri-peremeschenii-tovarov-cherez-tamozhennuyu-granitsu-eaes. Accessed 10 Mar 2021

5. Fedotova, G.: Uniform interpretation of the commodity nomenclature as the basis for regulation of international trade. Admin. Consult. 67–79 (2020). https://doi.org/10.22394/1726-1139-2020-2-67-79. Accessed 10 Mar 2021
6. Kongahage, D., Spinks, G.M., Richards, C.J., Shepherd, D.J., Foroughi, J.: A new approach to develop, characterise and model actuating textiles. Smart Mater. Struct. **30**(2) (2021). https://doi.org/10.1088/1361-665X/abd58d.. Accessed 10 Mar 2021
7. Kosiuk, G., Matusiak, M.: Effect of quilting on selected thermo-insulation properties of textile multilayer packages. Fibres Text. East. Euro. **28**(2), 50–56 (2020). https://doi.org/10.5604/01.3001.0013.7314
8. Kudryavin, L., Shalov, I. (n.d.).: Fundamentals of knitwear production technology. 7 March 2021. https://www.liveinternet.ru/users/4431209/post424408212/. Accessed 10 Mar 2021
9. Nazarov, Y.P.V.: Innovative textiles. The main types and areas of application [Innovacionnyj tekstil'. Osnovnye vidy i oblasti primeneniya]. MNG, 172–174. 7 March 2021. https://cyberleninka.ru/article/n/innovatsionnyy-tekstil-osnovnye-vidy-i-oblasti-primeneniya . Accessed 10 Mar 2021
10. Nikitina, L., Popkova, E.G.: Ensuring the specified properties of materials for light industry products using polymer compositions. Bull. og Kazan Technol. Univ. **16**(18) (2013). https://cyberleninka.ru/article/n/obespechenie-zadannyh-svoystv-materialov-dlya-izdeliy-legkoy-promyshlennosti-s-ispolzovaniem-polimernyh-kompozitsiy. Accessed 10 Mar 2021
11. Petrova, I.N., Fedotova, G.Y.: Problematic aspects of customs identification examinations in the light of innovations in the technology of fibrous materials [Problemnye aspekty identifikacionnyh tamozhennyh ekspertiz v svete innovacij v tekhnologii voloknistyh materialov]. Bulletin of State Treasury Educational Institution of Higher Education "Russian Customs Academy" St. Petersburg branch named after V.B. Bobkov, **1**(41) (2012). https://cyberleninka.ru/article/n/problemnye-aspekty-identifikatsionnyh-tamozhennyh-ekspertiz-v-svete-innovatsiy-v-tehnologii-voloknistyh-materialov . Accessed 10 Mar 2021
12. Provin, A.P., Regina de Aguiar Dutra, A., Machado, M.M., Vieira Cubas, A.L.: New materials for clothing: rethinking possibilities through a sustainability approach - a review. J. Cleaner Product. **282**, 12444 (2021). https://doi.org/10.1016/j.jclepro.2020.124444. Accessed 10 Mar 2021
13. Shagina, N.: New technologies in the textile industry [Novye tehnologii v tekstil'noj promyshlennosti]. Bull. Dagestan State Technol. Univ. Tech. Sci. **13** (2007). https://cyberleninka.ru/article/n/novye-tehnologii-v-tekstilnoy-promyshlennosti. Accessed 10 Mar 2021
14. Statistics, C., (n.d.).: Customs statistics of foreign trade of the Russian Federation. Customs Online - All about customs for foreign trade participants, customs online. 7 March 2021. https://customsonline.ru/search_ts.html. Accessed 10 Mar 2021
15. Tarigan, Z.J.H., Jiputra, J.A., Siagian, H.: The effect of supply chain practices on retailer performance with information technology as moderating variable. Int. J. Data Netw. Sci. **5** (1), 47–54 (2021). https://doi.org/10.5267/j.ijdns.2020.11.003 Accessed 10 Mar 2021
16. Ustyugova, E. (n.d.).: Identified characteristics of light industry goods for classification in accordance with EAEU TN VED: terminological errors and inaccuracies in explanations to positions, sub positions. 7 March 2021. https://www.elibrary.ru/item.asp?id=42916908. Accessed 07 Mar 2021
17. Ustyugova, E., Ivanova, V., (n.d.).: Identification in accordance with ETN FEA EAEU and determination of the amount of customs duty on the example of goods for advertising purposes pillow-toy. https://www.elibrary.ru/item.asp?id=41759329. Accessed 10 Mar 2021
18. Vlasova, J.: Identification of the materials of textile products imported into the Russian federation. Sci. Time **6**(18) (2015). https://cyberleninka.ru/article/n/identifikatsiya-syrievogo-sostava-tekstilnyh-izdeliy-importiruemyh-v-rf. Accessed 10 Mar 2021

Modeling the Demand Function for LNG in the NEA Countries

Margarita M. Kozenyasheva(✉) and Juyoung Ko

Gubkin Russian State University of Oil and Gas, Moscow, Russia
MKozenyasheva@gmail.com, 103100@gubkin.ru

Abstract. The main purpose of this work is to form approaches for developing an effective model of demand for Liquefied Natural Gas (LNG) in the countries of North-East Asia (NEA) - the largest and fastest growing energy market in the world.

The authors analyzed the approaches to assessing the competitiveness of an energy product – LNG and the factors that determine the demand for energy products, set out in the works of Russian and foreign economists.

The concept of a systemic classification of factors is proposed, the main factors, which have the most significant impact on the formation of demand for LNG, are identified. An assessment of the quantitative factors affecting the volumes of LNG supplies to the market of the three NEA countries has been made, and an econometric model of LNG demand has been built.

The methodology that is used in the study contains a set of methods and approaches: systemic, situational, dynamic. In addition to general scientific methods of analysis and synthesis, comparative economic and statistical methods are used, with the help of regression analysis; an assessment of the influence of quantitative factors on the volumes of LNG supplies to the market of the three NEA countries is made. At the same time, the author's proposed approach to building a model of demand for LNG in the NEA countries became the key method. As a result of the research, a several of conclusions were obtained; they made it possible to formulate specific practical recommendations for increasing the competitiveness of the LNG supplier countries to the market of the NEA countries, taking into account the balance of interests of LNG sellers and buyers. Proposals for importing countries on rational diversification of LNG supply sources have been formulated.

Keywords: Competitiveness of liquefied natural gas · Model of demand for energy products · Assessment of the influence of quantitative factors · Prioritization of risks · Assessment of unforeseen factors · Seasonal effect

JEL Codes: F12 · F17 · F40 · O57

1 Introduction

In the context of the energy transition to the era of widespread use of renewable energy sources in the world market of natural energy resources, the role of LNG is increasing. International LNG trade is increasingly transforming into a global energy structure,

E. G. Popkova (Ed.): ISC 2020, LNNS 368, pp. 591–600, 2022.
https://doi.org/10.1007/978-3-030-93244-2_65

which means a restructuring of all the basic foundations of the LNG business. One of the main advantages of LNG is that there is no need to build and maintain a gas pipeline system; LNG can be transported by sea, performing transoceanic supplies to markets that are currently the most competitively attractive. This resource turned out to be extremely important for countries that, not having a sufficient energy base of their own, are pursuing a strategy of strengthening their energy security by diversifying energy supplies. The main major LNG markets are formed at a rapid pace, such as Japan and South Korea, Europe, China and India today. At the same time, the emerging global LNG market is characterized by constant price volatility, increasing uncertainty of the impact of market factors on the demand and supply side. Under these conditions, LNG exporting countries need a scientifically based approach to assessing the dynamics of demand in LNG markets, in particular, in the NEA region, while consumer countries need to rely on a solid toolkit for assessing the competitiveness of LNG suppliers, the range of which is rapidly changing in modern circumstances.

The main hypothesis of the study is the assumption that the dynamics of demand for LNG is currently influenced by a whole complex of interrelated factors that differ not only in quantitative but also in qualitative characteristics, the degree of impact of which is different, ambiguous and changes under the influence of long-term trends in energy transition, and in at the same time, the trends in the formation of the world LNG market, the increasing volatility of conjuncture fluctuations in this market and also the growing role of the impact of unpredictable events on it.

The purpose of the study is to develop a concept for a systemic classification of factors that allows identifying the main ones that dominantly affect the formation of demand, building an econometric model that will identify ways and means of effectively building a system for consistently increasing the competitiveness of LNG suppliers to the NEA market and managing the process of diversifying sources of LNG supply to dynamic market conditions.

The subject of the research is methodological approaches to assessing quantitative factors affecting the dynamics of demand for energy products. The object of the research is the demand for LNG in three countries of North-East Asia.

2 Methodology

The volumes of LNG imports from the purchasing countries are currently influenced by qualitative and quantitative factors. The analysis of quantitative factors affecting the ratio of supply and demand for energy products is an important aspect; a fairly large number of scientific publications have been devoted to its study. The authors grouped and compared them, based on the data from the published works. The identified factors were grouped by authors into macroeconomic, resource, price and others.

Table. 1 Matrix of factors influencing the volume of LNG demand

Factors	Macroeconomic factors								Resource factors				Price factors						Other factors			
Object of study:	GDP	Volume of ind. Production	Index ind. production in Europe	Share of investments in R&D in GDP of countries	Population size	Per capita income	Unemployment rate	Exchange rate ratio	Energy reserves	Oil reserves	Oil production	Import volume of pipeline natural gas	Energy price	Electricity price	World oil price	Coal and natural gas prices	LNG price	LNG transportation costs	Climatic factor	Technology development	Traffic volume	Access to gas transportation infrastructure
Oil consumption in Australia	✓																					
Oil imports in the Republic of Korea	✓																				✓	
Oil imports in China		✓									✓											
Oil imports in South Africa	✓														✓							
Volumes of oil imports from Mexico to the USA						✓	✓			✓												
Volumes of imports of oil and oil products in Europe			✓					✓					✓			✓						
Natural gas consumption in different countries	✓				✓				✓										✓	✓		
Natural gas consumption in China												✓					✓					
LNG trade volumes in different countries																						
LNG imports in the USA																						

Source compiled by the authors

Within the framework of the comparative analysis methodology, the following points were identified:

1 Macroeconomic factors have the greatest effect on the dynamics of demand for energy products, followed by resource factors and others.

2 Most studies are devoted to the analysis of quantitative factors, primarily macroeconomic and price. Assessment of the influence of qualitative factors affecting demand is insufficiently presented.

3 Concerning to the demand for LNG, only a few quantitative factors were analyzed, such as: GDP, the volume of imports of pipeline natural gas, the price of LNG, the cost of transporting LNG and the share of investments in R&D in the GDP of countries.

The next applied method is the systems approach. Within the framework of the research methodology, this means that the demand for an energy product and its change is considered as a system - a complex set of many elements and relationships between them. Thus, the fact that a change in the exchange rate can also affect trade in natural gas in the NEA markets can't be ignoring while analyzing the dynamics of demand. Since the transaction in the international market is carried out not in the national currencies of the three countries, but in dollars, it's also necessary to consider the influence of the dynamics of the exchange rates of the three countries against the dollar. So, this applies equally to natural gas and liquefied gas [1].

It should also be taken into account that the seasonal effect, i.e., the change in the temperature regime, directly affects the volumes of LNG imports, because the demand for "blue fuel" increases in winter due to the need for heating and therefore the demand also depends on the temperature. From the analysis of the study, it follows that in three NEA countries there is seasonal fluctuation, which is especially evident in Japan and the Republic of Korea, in these countries there is a noticeable increase and/or decrease in natural gas consumption depending on the season. However, in China, due to domestic production of this type of energy carrier, the seasonal effect is comparatively weaker. In this analysis, the authors view this factor as a seasonal effect.

Concerning to China, as factors affecting the demand for LNG, the authors, firstly, single out pipeline gas, because the PRC, in contradistinction to Japan and the Republic of Korea, imports not only LNG, but also pipeline natural gas, and also produces "blue fuel" within the country. Therefore, the factor of imports of pipeline natural gas and the dynamics of domestic production of natural gas are also taken into account in our approach.

Further, the effect of the consequences of the accident at the Fukushima-1 nuclear power plant in March 2011 in Japan can't be ignored, because it led to a change in the energy balance of Japan. After this accident, the operation of all nuclear power plants was restored in stages, and only starting in August 2015, the restart of reactors at several nuclear power plants in the country began [2]. In this regard, the effect of the "consequences of the 2011 accident" in this study covers the period from March 2011 to July 2015. Although this accident is not directly a quantitative factor, it can be expressed as a time period; therefore, for the purposes of analysis in this study, this factor was also taken into account.

Having aggregated the results of the study of the above works and the above characteristics in the energy sector of the NEA countries, in this work the authors have identified and evaluated the quantitative factors that are reflected in Fig. 1.

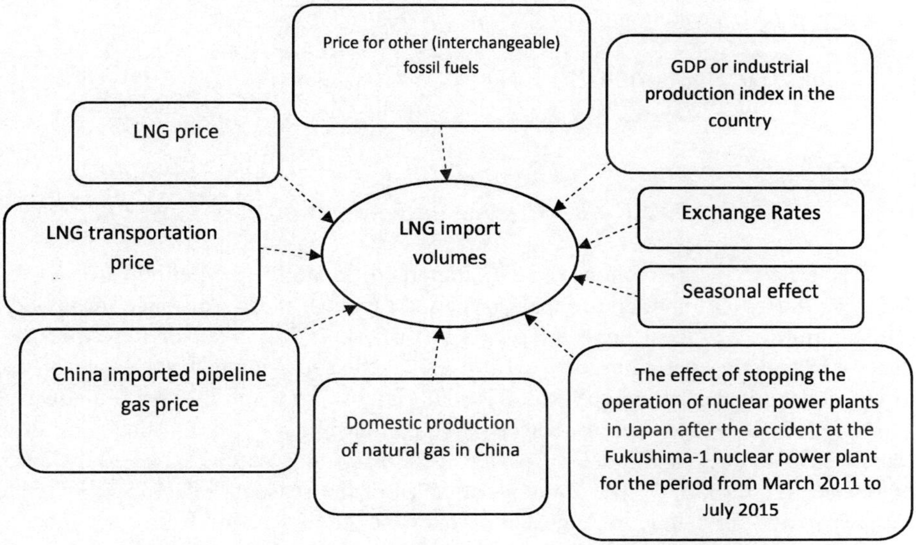

Fig. 1. Quantitative factors affecting the volume of LNG imports in three NEA countries. Source compiled by the authors

Quantitative factors affect the volumes of LNG imports in the markets of the three NEA countries in different ways. Therefore, an econometric method is used and for this, three linear multivariate econometric models of demand for imported LNG in three NEA countries are constructed for empirical analysis in this work. All models include such basic factors as: the price of LNG, the price of fungible resources (oil and coal), industrial production, exchange rates[1]. The seasonal effect is also included in the three country models as a dummy variable.

The other factors affecting LNG demand were also reviewed while analyzing, especially in Japan and China. For example, the Japanese model also includes "the effect of stopping the operation of nuclear power plants in Japan after the accident at the Fukushima-1 nuclear power plant in the period from March 2011 to July 2015" as a dummy variable. As for China, the Chinese model considers the price of pipeline gas imports in the country and the volume of domestic production of natural gas.

The constructed model of the demand for imported LNG from the three NEA countries and their empirical analysis makes it possible to determine which factors among the above have a statistically significant effect on the volumes of LNG imports in these countries.

Considering that Japan, China and the Republic of Korea have their distinctive energy characteristics, accordingly, the LNG demand patterns in each country should look different. In the regression model, we include various quantitative factors that can affect the change in LNG demand in the NEA countries.

[1] It should be noted that the data on LNG prices from Japan, China and the Republic of Korea are official data, they are presented on the websites of the customs services of the three countries, and, accordingly, this price indicator is the CIF price.

Japan's LNG demand models have the following form[2]*:*

$$\begin{aligned} lnLNGJ_t &= \alpha_0 + \alpha_1 lnGPJ_t + \alpha_2 lnOP_t + \alpha_3 lnCP_t + \alpha_4 lnIPJ_t + \alpha_5 lnEXJ_t \\ &\quad + \alpha_6 NU + \alpha_7 S + \alpha_8 F + \alpha_9 W + \varepsilon_t \end{aligned} \tag{1}$$

$$\begin{aligned} lnLNGJ_t &= b_0 + b_1 lnLNGJ_{t-1} + b_2 lnGPJ_t + b_3 lnOP_t + b_4 lnCP_t + b_5 lnIPJ_t \\ &\quad + b_6 lnEXJ_t + b_7 NU + b_8 S + b_9 F + b_{10} W + \epsilon_t \end{aligned} \tag{2}$$

Where $LNGJ_t$ is the volume of LNG imports to Japan for the period t (t.); GPJ_t - CIF price for LNG imported to Japan for period *t* (yen/t.)[3]; OP_t - oil price for period t (USD/barrel); CP_t - coal price for period *t* (USD/t); IPJ_t - index of industrial production in Japan for period *t* (2015 = 100); EXJ_t – the average rate of the Japanese yen to the dollar for the period *t* (yen/dollar); *NU* is a dummy variable for the effect of stopping NPP operation after the accident at the Fukushima-1 NPP for the period from March 2011 to July 2015. (*NU* = 1 if the period is from March 2011 to July 2015, otherwise *NU* = 0); *S*, *F*, *W* - dummy variables of the seasonal effect (*S* = 1 for the summer period from June to August, *F* = 1 for the autumn period from September to November, *W* = 1 for the winter period from December to February, otherwise *S* = 0, *F* = 0, *W* = 0)[4]; α_0 and b_0 are constant variables; ε_t and ϵ_t are random components (or random error) for the period *t*. *ln* - notation for natural logarithm.

Data on LNG imports and its CIF price in Japan ($LNGJ_t$,GPJ_t) are obtained from the Japanese customs website [4]. Data on oil and coal prices (OP_t,CP_t) are obtained from the website of the World Bank [5]. Data on the index of industrial production in Japan (IPJ_t) are obtained from the OECD website [6]. Data on the yen/dollar exchange rate (EXJ_t) were obtained from the IMF website [7].

China's LNG demand models have the following form:

$$\begin{aligned} lnLNGC_t &= d_0 + d_1 lnGPC_t + d_2 lnOP_t + d_3 lnCP_t + d_4 lnIPC_t + d_5 lnEXC_t \\ &\quad + d_6 lnPROD_t + d_7 S + d_8 F + d_9 W + \theta_t \end{aligned} \tag{3}$$

$$\begin{aligned} lnLNGC_t &= h_0 + h_1 lnLNGC_{t-1} + h_2 lnGPC_t + h_3 lnOP_t + h_4 lnCP_t \\ &\quad + h_5 lnIPC_t + h_6 lnEXC_t + h_7 lnPROD_t + h_8 S + h_9 F + h_{10} W + \mu_t \end{aligned} \tag{4}$$

---

[2] Although that in the present work, the analysis used time series data in addition to dummy variables, the unit root test was not carried out. This is due to the fact that the power of a test of common root tests such as the "Dickey-Fuller test" and "Augmented Dickey-Fuller (ADF)" often turns out to be low [3]. Moreover, not too much data from the three countries were collected for analysis, which leads to a decrease in the statistical power of tests for common roots. Since one of the methods of econometric research using time series data is to ignore the nonstationarity of all-time series variables in the models, this work uses time series data without differentiating them.

[3] The price of CIF for LNG includes the cost of transporting this type of energy carrier; therefore the model doesn't include the price for transporting LNG.

[4] A dummy variable is an indicator variable with values {0, 1} to indicate whether a specified condition is satisfied or not.

Where $LNGC_t$ is the volume of LNG imports to China for period t (t.); GPC_t - CIF price for LNG imported to China for period t (USD/t); IPC_t - value-added of Industry[5] in China as a substitute variable for industrial production for period t (YoY); EXC_t is the average exchange rate of the Chinese yuan to the dollar for the period t (yuan/dollar); $PROD_t$ is the volume of domestic production of natural gas in China for period t (100 million cubic meters); d_0 and h_0 are constant variables; θ_t and μ_t are random components for the period t.

It should be emphasized that the Chinese models don't include the price variable for imported pipeline gas to China, because this variable causes the multicollinearity problem due to its high level of correlation with other variables in the models[6].

Data on LNG imports, CIF prices and prices for imported pipeline gas ($LNGC_t$,GPC_t) to China are obtained from the website of the Chinese customs service [8]. Data on industrial value added and domestic production of natural gas in China (IPC_t,$PROD_t$) are obtained from the website of the National Bureau of Statistics of the People's Republic of China (NBS) [9]. Data on the yuan/dollar exchange rate (EXC_t) were obtained from the IMF website [7].

The LNG demand models in the Republic of Korea have the following form:

$$\begin{aligned} lnLNGK_t &= v_0 + v_1 lnGPK_t + v_2 lnOP_t + v_3 lnCP_t + v_4 lnIPK_t + v_5 lnEXK_t \\ &\quad + v_6 S + v_7 F + v_8 W + \rho_t \end{aligned} \tag{5}$$

$$\begin{aligned} lnLNGK_t &= z_0 + z_1 lnLNGK_{t-1} + z_2 lnGPK_t + z_3 lnOP_t + z_4 lnCP_t \\ &\quad + z_5 lnIPK_t + z_6 lnEXK_t + z_7 S + z_8 F + z_9 W + \varphi_t \end{aligned} \tag{6}$$

Where: $LNGK_t$ is the volume of LNG imports to Korea for the period for period t (t.); GPK_t - CIF price for LNG imported to Korea for period t (USD/t); IPK_t - index of industrial production in Korea for period t (2015 = 100); EXK_t is the average exchange rate of the Korean won against the dollar for period t (won/USD); v_0 and z_0 are constant variables; ρ_t and φ_t are random components for the period t[7].

Models (2), (4) and (6) with lagged dependent variables are called "dynamic models". If the absolute values of the coefficients of the lagged dependent variables are below unity, then the autocorrelation problem is asymptotically overcome.

All data used for regression analysis of the models are monthly, except for dummy variables. For the regression analysis of the Japanese and Korean models, data were

[5] Value-added of Industry is the growth rate of value added in the industry compared to the same period of the previous year in China.

[6] Due to the fact that multicollinearity leads to instability of parameter values estimates due to an increase in the standard error, in order to detect multicollinearity in this analysis, the authors calculated the variance inflation factors (VIF). For the variable price of imported pipeline gas to China, VIF = 15.50. In contrast to the Chinese models, the multicollinearity problem wasn't found as a result of the VIF calculations for the variables of the Japanese and Korean models.

[7] The data on LNG imports and its CIF price in the Republic of ($LNGK_t$,GPK_t) are obtained from the website of the Korean Customs Service of the Republic of Korea. The Republic of Korea Industrial Production Index (IPK_t) is from the OECD website. The data for the Korean currency against the dollar (EXK_t) was obtained from the IMF website.

applied from January 2009 to May 2019, and for the analysis of the Chinese model, the data of variables for the period from March 2017 to June 2019 were applied. The short time span for which the data sample was taken from China is due to limited access to Chinese official sources[8].

3 Results

The results of the regression analysis of this work show that the following factors have a statistically significant effect on the volumes of LNG imports in the NEA countries:

- Prices for imported LNG to China and the Republic of Korea have a positive effect on the volume of imports of this energy resource. Such results contradict the law of supply and demand, according to which, all other things being equal, a decrease in the price of a product increases demand for it. This result can be explained by the fact that countries with a high dependence on LNG imports can't promptly respond to price increases, because it's difficult to find an alternative and drastically change the country's fuel and energy balance depending on the level of energy prices in short time.

Table 2. Estimation of coefficients of variables of models (1)–(6)

Explanatory variables	Models					
	Japan (*J*)		China (*C*)		Republic of Korea (*K*)	
	(1)[9]	(2)	(3)	(4)	(5)	(6)[10]
$lnLNG_{t-1}$		0.3641***		–0.1343		0.4553***
$lnGP_t$	0.1312	0.0380	1.3594***	1.2087	0.4204***	0.3838***
$lnOP_t$	–0.2069***	–0.1273**	0.0406	0.0628	–0.2914***	–0.2318**
$lnCP_t$	0.1123	0.0692	–0.3945	–0.2385	0.3467***	0.3085***
$lnIP_t$	0.7698***	0.5291**	–0.5683**	–0.5017*	1.4802***	1.7210***
$lnEX_t$	0.1167	0.0820	-1.2471	-1.0267	0.2711	0.5057
NU	0.1635***	0.1263***				
$lnPROD_t$			–0.2339	0.2434		
S	–0.0004	0.0430*	0.2094**	0.2121	–0.1871***	–0.2607***
F	–0.0491*	0.0035	0.2850***	0.3017*	–0.0180	–0.1423***
W	0.0564**	0.1029***	0.3872***	0.4010	0.0273***	0.1693***
Constant variable	10.5103***	6.9101***	12.8466***	12.1297	3.2009	0.6676
R^2 (Adjusted R^2)	0.9507 (0.9468)	0.6221 (0.5887)	0.8596 (0.7894)	0.8283 (0.7210)	0.6502, 0.6261	0.7662, 0.7477

NOTE: 1) the dependent variables of the six models are $lnLNGJ_t$, $lnLNGC_t$, $lnLNGK_t$, respectively; 2) *, **, *** - the statistical significance of the coefficients at the 10–, 5–, 1% level, respectively; 3) all values are rounded to ten thousandths

Source compiled by the authors.

[8] Official data on China's LNG imports have been available on the country's customs website since January 2017. However, for the analysis of this work, data from March 2017 were used due to the fact that there are no official data for the variables IPC_t and $PROD_t$ for the period January-February 2017, January-February 2018 and January -February 2019. Missing data for the period January-February 2018 and January-February 2019 were filled in using linear interpolation.

[9] Due to the detection of errors autocorrelation problems, the values of the coefficients of this model were obtained by the Prais-Winsten method.

[10] Due to the detection of errors autocorrelation and hetesockedasticity problems, the values of the coefficients of this model were obtained by the Newey-West method.

In addition, it should be noted that according to the results of the regression analysis of the Chinese and Korean models (3), (5) and (6), which include lagged price variables for imported LNG in two countries[11], the prices for imported LNG in China and the Republic of Korea in periods for each of the previous six months (for China) and twelve months (for Korea) don't have a statistically significant effect on LNG imports over period *t*.

4 Conclusion

The conducted research led to the following conclusions:

1. LNG prices don't have a definite effect on the volumes of LNG imports in the three NEA countries in a short time.
2. Rising oil prices have a negative impact on LNG imports in Japan and the Republic of Korea in a short time, and an increase in the price of coal has a positive impact on the Republic of Korea. In other words, a fall in oil prices leads to an increase in LNG imports in the two countries, and a fall in coal prices leads to a decrease in LNG imports in Korea. According to economic theory, oil and LNG in Japan and the Republic of Korea can be considered complementary commodities, while coal and LNG in the Republic of Korea are interchangeable commodities.
3. Influence of the industrial production index. In Japan and the Republic of Korea, industrial development of countries is accompanied by an increase in LNG imports. For example, an increase in industrial production in Japan leads to an increase in LNG imports, because that natural gas is consumed in the Japanese industry in significant volumes. According to the Ministry of Economy, Trade and Industry of Japan, this figure was about 24.7% of the total supply of natural gas in the country. As for the Republic of Korea, this result is due to the fact that natural gas is used in industry in the country in the form of LNG and "city gas", the share of natural gas in electricity production in 2017 was 16.9%. Thus, the growth of industrial production in the country has a positive effect on the volume of LNG imports. In contradistinction to Japan and the Republic of Korea, China's industrial production growth doesn't necessarily translate into an increase in the country's LNG imports. This is due to the following features of the Chinese energy sector. Firstly, China imports natural gas not only by sea, but also through gas pipelines, while the country has its reserves of this type of energy carrier. Secondly, coal still plays a key role in the Chinese industry, and the infrastructure for the use of coal is most developed, the growth of industrial production in the country is accompanied by an increase in the use of coal and even the share of its consumption in the fuel and energy balance of the country.
4. Influence of the factor of unforeseen events in the energy sector. Based on the results of this analysis, during the shutdown of nuclear power plants in Japan in the

[11] For example, $lnGPC_{t-1}, lnGPC_{t-2}, \ldots, lnGPC_{t-6}, lnGPK_{t-1}, lnGPK_{t-2}, \ldots, lnGPK_{t-12}$. In the Chinese model (3), lag variables are included up to t–6 due to the small amount of data for analysis.

period from March 2011 to July 2015, the volume of LNG imports in Japan increased significantly - by almost 12.6–16.3% in comparison with the pre-accident period and the period after the restoration of NPP operation in the country. Thus, unforeseen events in the energy sector can also significantly affect the fuel and energy balance structure and the volume of LNG imports.

5. The influence of the seasonal factor in the three countries is manifested in different ways. In Japan, natural gas is used at peak loads for electricity generation in summer and winter periods, therefore, LNG imports in Japan increase in summer period (model 2) and winter period (models 1 and 2). In the Republic of Korea, LNG imports increase in winter due to the fact that the country uses natural gas at peak loads in electricity generation. In turn, in Japan, the volume of LNG demand in the summer period increases compared to the spring period, which is considered to be the reason due to the use of gas conditioners. In the Republic of Korea, LNG imports decrease during the summer period because heating in residential premises isn't required; in the Republic of Korea, the gas conditioning infrastructure isn't developed. In Japan and the Republic of Korea (models 1 and 6), the volumes of LNG imports in the autumn period decrease compared to the volumes in the spring period, in China they increase, which is caused by the difference in the average air temperature in the autumn period. In the three countries, LNG imports in the winter period significantly increase compared to the spring period due to the heating load, i.e. natural gas is used for heating in the domestic sector of the three NEA countries. In other words, changes in air temperature, i.e. the seasonal effect have a statistically significant effect on the volumes of LNG imports of the three countries.

References

1. Lee, J.H., Kim, Y.K.: The inter-relationship between natural gas imports and foreign reserves: a time series analysis for Korea. J. Soc. Sci. **26**(1), 3–26 (2015). (in Korean)
2. Slade, J.: Japan has restarted five nuclear power reactors in 2018. EIA. 28 Nov 2018. https://www.eia.gov/todayinenergy/detail.php?id=37633. Accessed 13 Dec 2020
3. Damodar, N.G., Porter, D.C.: Basic Econometrics, 5th edition, Korean Language Edition. – Seoul: Jiphil Media, 1120 (2009). (in Korean)
4. Trade Statistics of Japan. Commodity by Country: Conditions' input. http://www.customs.go.jp/toukei/srch/indexe.htm?M=01&P=0. Accessed 13 Dec 2020
5. The World Bank. Commodity Markets. http://www.worldbank.org/en/research/commodity-markets. Accessed 13 Dec 2020
6. OECD Data. Industrial Production. https://data.oecd.org/industry/industrialproduction.htm. Accessed 13 Dec 2020
7. IMF. IMF Exchange Rates. https://www.imf.org/external/np/fin/ert/GUI/Pages/CountryDataBase.aspx. Accessed 15 Dec 2020
8. Customs statistics. Interactive tables. http://43.248.49.97/indexEn. Accessed 15 Dec 2020
9. National Data. National Bureau of Statistics of China. http://data.stats.gov.cn/english/easyquery.htm?cn=A01. Accessed 15 Dec 2020

Provincial Features of Industrial Production Dynamics in the Research of Textile Enterprises' Financial Security in Uzbekistan

Bobir O. Tursunov(✉)

Tashkent State University of Economics, Tashkent, Uzbekistan

Abstract. Regional features of industrial production dynamics in the research of textile enterprises' financial security in Uzbekistan are investigated in this paper. As per the author's thought introductory phase of a quantitative investigation of the monetary strength of material ventures in the Republic of Uzbekistan at the provincial level ought to be the phase of bunching districts to recognize bunches that remember homogeneous areas in terms of mechanical development rates, periods of the financial cyclicality of modern elements. This will permit when looking at the monetary supportability of material undertakings situated in various locales, to consider comparative or various states of modern advancement in the individual domains. This article introduces a hypothetical defense and set of calculations, just as the consequences of computations dependent on information from the State Statistics Committee of the Republic of Uzbekistan, which made it conceivable to arrange locales as per the attributes of development and cyclicality of mechanical creation for the time frame 2010–2018.

Keywords: Financial security · Financial stability · Production · Trend-cyclical component · Clustering regions · Textile enterprises

JEL Codes: L16 · L6 · L32 · L5 · P25

1 Introduction

The economic literature presents works on the analysis of the financial stability of enterprises, taking into account the impact of macroeconomic conditions in specific periods. For example, in the work "Modeling the financial stability of an enterprise taking into account macroeconomic indicators" [1], the authors investigate the financial stability of metallurgical enterprises using mathematical models that include the parameters of dynamic models of macroeconomic indicators characterizing their fluctuation in time. This provided the authors with the opportunity to take into account in the specification of models of financial stability of enterprises the special conditions of their activities in different periods.

The main hypothesis adopted by the authors during this study is as follows: the financial stability of enterprises at the regional level is determined not only by the peculiarities of the manifestation of general macroeconomic factors in time and a given

E. G. Popkova (Ed.): ISC 2020, LNNS 368, pp. 601–610, 2022.
https://doi.org/10.1007/978-3-030-93244-2_66

territory but also by the specifics of mesoeconomic factors that determine the growth and cyclical dynamics of production by industry and sectors of the regional economy. For enterprises of the textile industry on the territory of the regions of the Republic of Uzbekistan, an assessment of the impact of mesofactors on financial stability can be obtained by examining the dynamics of industrial production as a whole. The purpose of the study is to identify and assess the territorial specifics of the economic dynamics of industrial production in the Republic of Uzbekistan. A mathematical and statistical decomposition of the investigated dynamics of industrial production in the regions of the Republic of Uzbekistan is carried out, with the allocation of a trend-cyclical component of dynamics, which is a combination of the main trend (trend) and a cyclical component due to market factors [13]. This allows us to identify regions that have general and specific characteristics of the trend-cyclical components of the dynamics of industrial production as a whole, which will allow taking these features into account in the study of the factors of financial security of textile enterprises located in these regions.

2 Literature Review

In the last twenty years the study of production capacity, the organization of production at industrial enterprises, and the management of production processes were analyzed by [3–5, 7–10] and others [11, 12].

Some of these scientists are investigated, interrelated functions are examined, the implementation of which is the management of production capacity. Methodology for assessing the efficiency of production capacities management at textile enterprises were investigated by Tursunov [11] in other works and this research is logic continue of last research part.

Uzbek professor Burkhanov [2] researched main indicators of textile enterprises' financial security assessment, Kalandarovna et al. [6] studied methodical aspects of establishing a control system over compliance with principles of decent work and social security in textile enterprises. But in upper researches have not been studied regional features of industrial production dynamics in the research of textile enterprises` financial security in Uzbekistan.

3 Methodology

The study is based on a set of mathematical and statistical methods, including:

- assessment of the parameters of the trend-cyclical component of the dynamics of industrial production in the regions of the Republic of Uzbekistan;
- a combination grouping of the regions of the Republic of Uzbekistan by the parameters of the trend-cyclical components of the dynamics of industrial production, combining:
  - grouping of regions according to the type of linear trend (upward or downward);

- grouping of regions according to the parameters of the cyclical components of the time series of industrial production growth rates.

* determination of specific economic types of regions, taking into account the results of the combination grouping.

The initial model for approximating the time series of industrial production growth rates in the *j*-th region of the Republic of Uzbekistan:

$$\overline{y}_J = a_0 + a_1 * \text{t} + a_2 \, \text{Sin}\,(\text{k} * \text{t}) + a_3 \, \text{Cos}\,(\text{k} * \text{t}),$$

here:

$\overline{y_j}$ – calculated values of the trend-cyclical component of the time series of the annual growth rates of industrial production in the *j*-th region of the Republic of Uzbekistan;

t- year number, t = 2000,…., 2018 yy.

a_0, a_1 – model parameters that determine the trend component of the time series;

k– parameter determining the wavelength (period) of cyclic oscillations;

a_2, a_3 – parameters of the contribution of harmonic vibrations to the general model.

To determine the wavelength (period, years), the formula is used:

$$\text{L} = 2\pi/\text{k}.$$

The quality of the model is assessed using the multiple correlation coefficient (R), as well as the multiple determination coefficient (D), which makes it possible to determine the specific weight of the explained variation (the percentage of the actual values of industrial production growth rates in the region corresponding to the model trend-cyclical values).

4 Analysis and Results

Based on the results of building models, it was found that the above formula describes the trend-cyclical component of the studied time series rather reliably: the explained variation across regions was 55–78%.

The graphs of the most reliable models are shown below (Figs. 1–4).

The resulting model for the Republic of Uzbekistan looks like this:

$$\overline{y}_1 = = \mathbf{108.19 + 0.096 * t + Sin\,((-0679) * t) \;\; -1155\,Cos\,((-0.6791) * t)}$$

This model explains 66% of the variation (D) in the actual annual growth rate of industrial production in the Republic of Uzbekistan as a whole. The wavelength is 9.25 years, i.e. its half-life is approximately 4.5 years (Fig. 1).

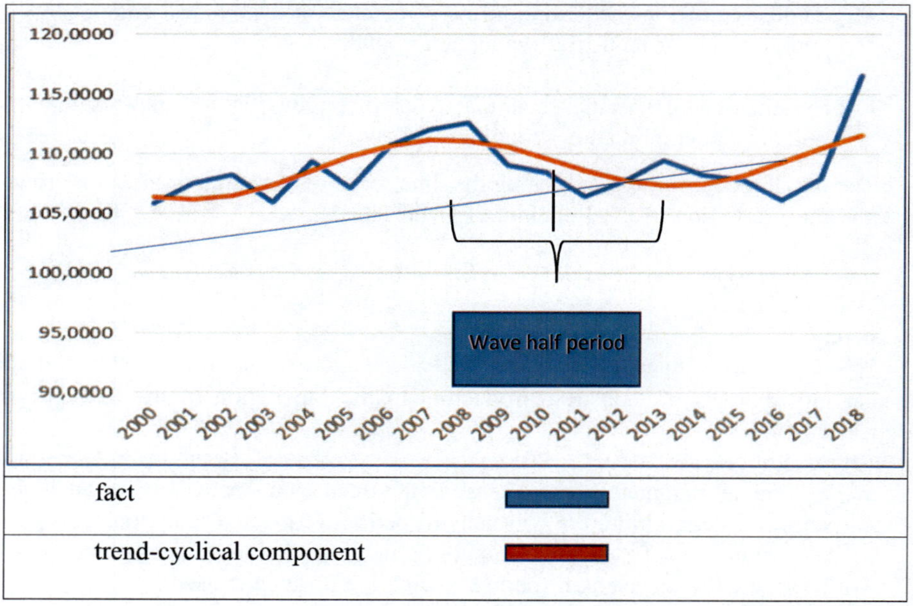

Fig. 1. Graph of the trend-cyclical component of the time series (orange line) of the annual rates of industrial production in the Republic of Uzbekistan, 2000–2018. *Source* Author's calculation based on data for www.stat.uz

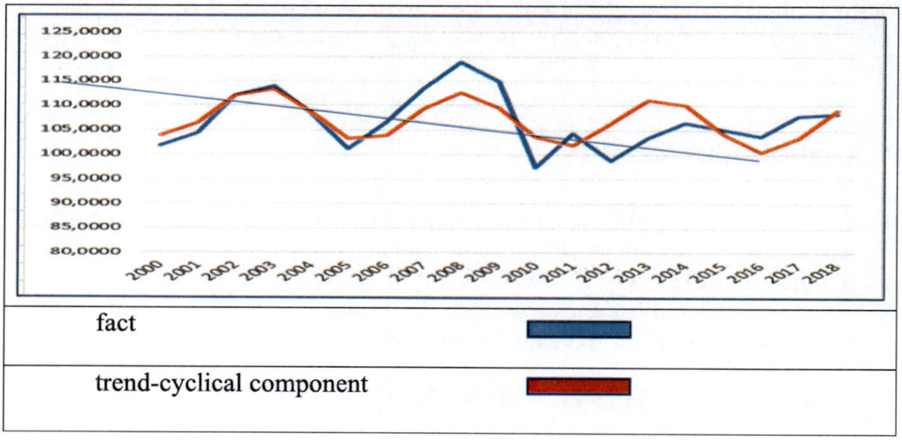

Fig. 2. Graph of the trend-cyclical component of the time series of the annual rates of industrial production in the Kashkadarya region, 2000–2018. (D = 76%). *Source* Author's calculation based on data for www.stat.uz

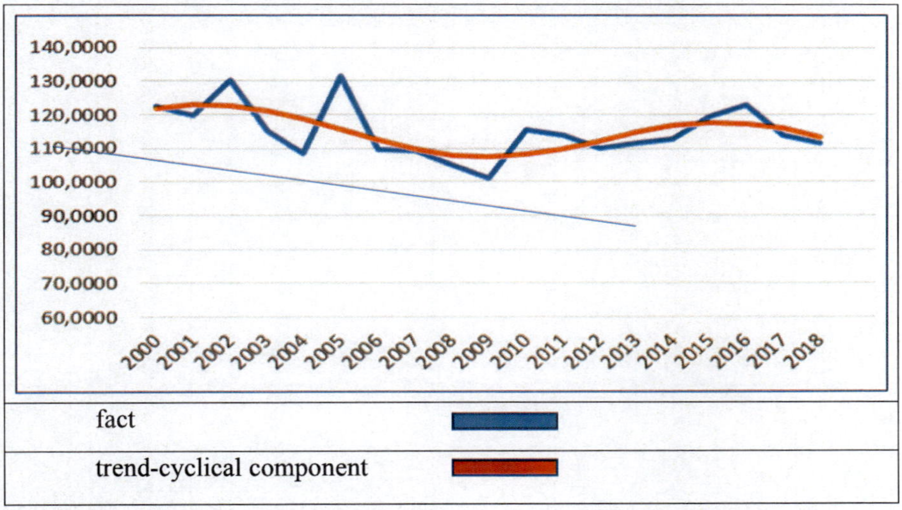

Fig. 3. Graph of the trend-cyclical component of the time series of the annual rates of industrial production in the Jizzakh region, 2000–2018. (D = 71%). *Source* Author's calculation based on data for www.stat.uz

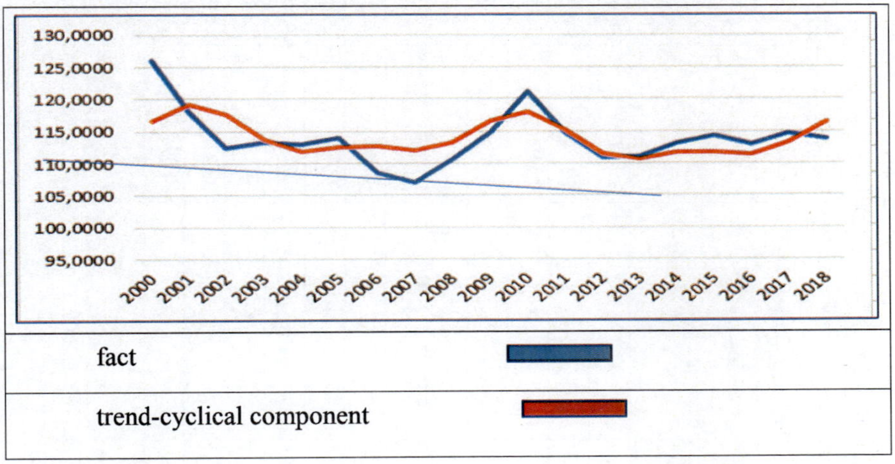

Fig. 4. Graph of the trend-cyclical component of the time series of the annual rates of industrial production in the Namangan region, 2000–2018 (D = 78%). *Source* Author's calculation based on data for www.stat.uz

Table 1. Parameters of the models of trend-cyclical components of the dynamics of industrial production in the regions of the Republic of Uzbekistan, built according to data on annual growth rates for 2000–2018.

Regions		Model parameters				Wave period (years)
	Type of trend: ascending ↗ descending ↘	***a1***	***a2***	***a3***	***k2 (abs.)***	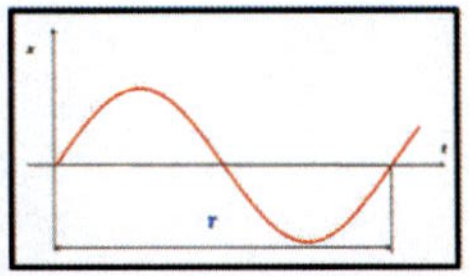
Republic of Karakalpakstan	↘	1.199	-6.687	6.971	2.346	4.94
Andijan	↘	-0.329	9.411	-7.224	5.484	5.72
Bukhara	↗	0.169	1.836	-2.828	2.026	4.46
Jizzakh	↘	-0.294	7.169	-6.88	2.512	8.94
Kashkadarya	↘	-0.287	-5.894	7.649	1.79	4.62
Navoi	↗	0.0009	-1.505	-0.747	3.376	9.37
Namangan	↘	-0.125	2.903	-1.604	0.573	5.47
Samarkand	↗	0.537	2.055	-6.461	5.906	5.31
Surkhandarya	↗	0.515	4.491	2.305	0.471	6.66
Syrdarya	↘	-0.151	2.773	3.885	4.376	7.17
Tashkent	↗	0.621	-1.091	4.057	0.233	7.25
Fergana	↗	-0.094	-1.605	1.001	5.891	5.33
Khorezm	↗	1.02	-4.285	8.115	1.506	6.96
Tashkent city	↗	0.722	7.017	0.982	0.746	4.21

Source Author`s elaboration

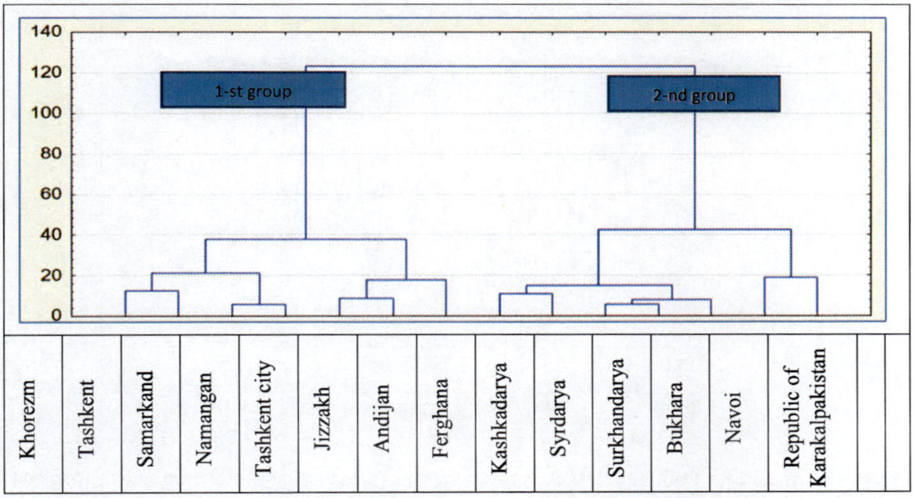

Fig. 5. Dendrogram of the distribution of regions of the Republic of Uzbekistan into homogeneous groups according to the parameters of the trend-cyclical components of the dynamics of industrial production. *Source* Author's calculation based on data for www.stat.uz

Based on the data presented in Table 1, a cluster analysis was carried out, which made it possible to determine the regions of Uzbekistan that are homogeneous in terms of the characteristics of the trend-cyclical components of dynamics (parameters a_0, a_1, a_2, a_3, k).

However, the inclusion of the linear trend parameter in the cluster analysis is impractical, since its positive and negative values will cancel each other out when determining the average values of the cluster-forming variables. Therefore, the regions were subdivided into two groups: with an upward and downward linear trend (in Table 1, they are marked with the corresponding graphic symbols).

The established average values of cluster-forming variables make it possible to determine the calculated values of the rates of industrial production for the selected clusters of regions (Fig. 6).

Note 1: In this figure, the regions of the first cluster are marked in blue (Andijan, Jizzak, Namangan, Samarkand, Tashkent, Khorezm, Tashkent city); in brown - the regions of the second cluster (Republic of Karakalpakstan, Bukhara, Kashkadarya, Navoi, Surkhandarya, Syrdarya, Fergana).

Note 2: The shaded areas indicate regions with an upward type of economic growth: Bukhara, Navoi, Samarkand, Surkhandarya, Tashkent, Fergana, Khorezm, Tashkent city.

It should be noted that in the regions of the 1st cluster and the regions of the 2nd cluster (Fig. 7), the forecast for the trend-cyclical model for 2020 gives a drop in the growth rates of industrial production relative to the average annual level, respectively,

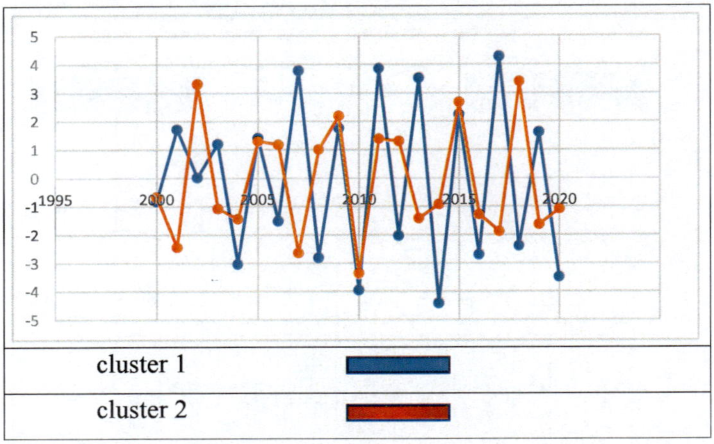

Fig. 6. Estimated for 2000–2020 values of deviations of industrial production rates for selected clusters of regions from the average annual level for 2000–2018,% - points. *Source* Author's calculation based on data for www.stat.uz

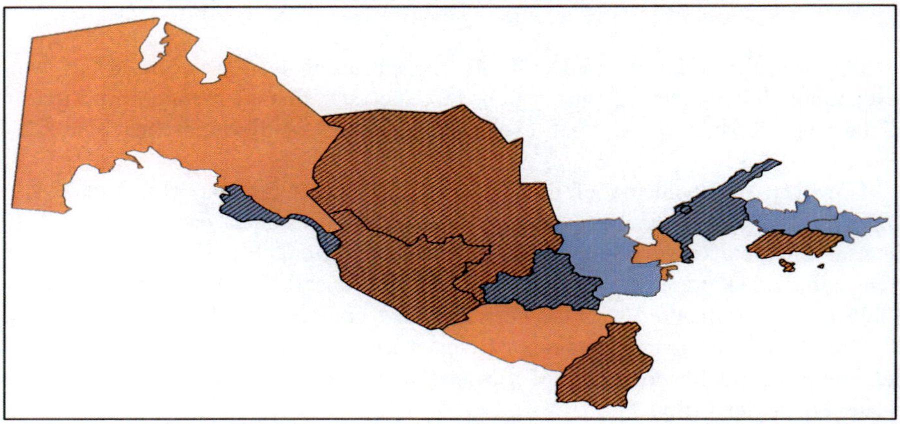

Fig. 7. Results of the final combination grouping of the regions of the Republic of Uzbekistan by the type of economic dynamics of industrial production. *Source* Author's calculation based on data for www.stat.uz

by 3.5 and 0.4% points. It follows from this that the situation of the coronavirus crisis coincided with a drop in industrial production in the regions of both groups due to the regular cyclicality.

Figure 7 shows the results of the final combination grouping of the regions of the Republic of Uzbekistan by the type of economic dynamics of industrial production.

5 Discussion of Results

Based on the results of evaluating the parameters of the dynamic models presented above, four types of regions have been identified that have the specifics of long-term dynamics of industrial production, determined by the features of the main trend (trend), the form of cyclicality, and a combination of these components of dynamics. The composition of these regional distinguished types is presented below: 1st type: Samarkand, Tashkent, Khorezm, Tashkent; 2nd type: Andijan, Jizzak, Namangan, 3rd type: Bukhara, Navoi, Surkhandarya, 4th type: Republic of Karakalpakstan, Kashkadarya, Syrdarya, Fergana.

According to data on the growth rates of industrial production by type of economic activity for 2010–2018. it was found that the general situation of industrial growth in the regions significantly affects the situation in the production of textiles. The influencing factors, in this case, are the general technical and technological development, improvement of infrastructure, improvement of the quality of labor, changes in the market environment, due to general market factors of supply and demand for industrial products. Calculations of the paired correlation coefficients of the annual growth rates of industrial production by type of economic activity show that the greatest response of the volume of textile production to industrial growth in the region occurs with a lag of 4 years (Fig. 8).

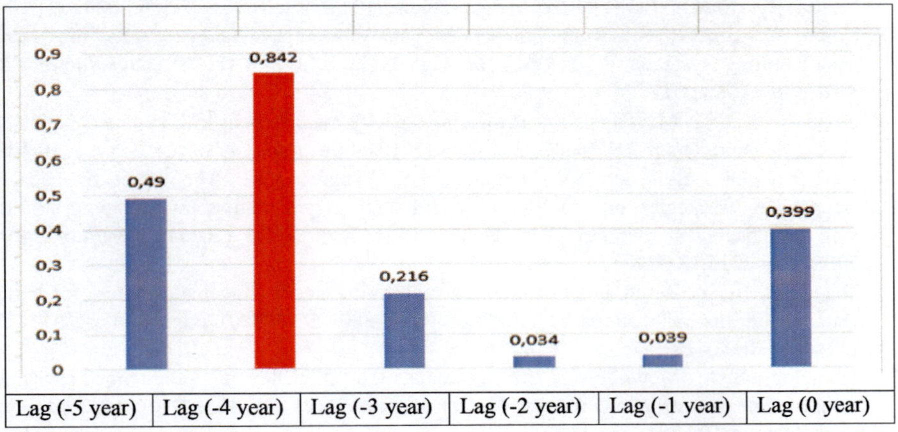

Fig. 8. Paired correlation coefficients of the growth rates of industrial production in general and the production of textiles with lags of lagging influence, according to the data of the Republic of Uzbekistan for 2010–2018. *Source* Author's calculation based on data for www.stat.uz

6 Conclusions

The performed calculations confirm the initial hypothesis about the need to take into account the spatial and dynamic features of industrial production in general, which are determined by both general industrial factors of technological progress and specific regional factors of cyclicality in the dynamics of the ratio of supply and demand for industrial products, in studies of the financial stability of textile enterprises in the regions of the Republic of Uzbekistan.

References

1. Bidzhoyan, D.S., Bogdanova, T.K.: Modelling the financial stability of an enterprise taking into account macroeconomic indicators. Bus. Inform. **3**(37), 30–36 (2016)
2. Burkhanov, A., Tursunov, B.O.: Main indicators of textile enterprises' financial security assessment. Vlakna a Textil **27**(3), 35–40 (2020). http://vat.ft.tul.cz/Archive/VaT_2020_3.html
3. Chien, C., Dou, R., Fu, W.: Strategic capacity planning for smart production: decision modeling under demand uncertain. Appl. Soft Comput. **68**, 900–909 (2018). https://doi.org/10.1016/j.asoc.2017.06.001
4. Davis, M., Dempster, M., Sethi, S., Vermes, D.: Optimal capacity expansion under uncertainty. Adv. Appl. Probab. **19**(01), 156–176 (1987). https://doi.org/10.2307/1427378
5. Huang, D., Lin, Z., Wei, W.: Optimal production planning with capacity reservation and convex capacity costs. Adv. Product. Eng. Manage. **13**(1), 31–43 (2018). https://doi.org/10.14743/apem2018.1.271
6. Kalandarovna, A.G., Gaibnazarovich, G.S., Turgunovna, S.N., Shuxratovna, F.D., Ortikmirzaevich, T.B.: Methodical aspects of establishing a control system over compliance with principles of decent work and social security in textile enterprises. J. Adv. Res. Dyn. Control Syst. **12**(5), 73–81 (2020). https://doi.org/10.5373/JARDCS/V12I5/20201691
7. Koltai, T., Stecke, K.: Route-independent analysis of available capacity in flexible manufacturing systems. Prod. Oper. Manag. **17**(2), 211–223 (2008). https://doi.org/10.3401/poms.1080.0017
8. Levin, Y., Nediak, M., Topaloglu, H.: Cargo capacity management with allotments and spot market demand. Oper. Res. **60**(2), 351–365 (2012). https://doi.org/10.1287/opre.1110.1023
9. Sebastiano, A., Belvedere, V., Grando, A., Giangreco, A.: The effect of capacity management strategies on employees' well-being: a quantitative investigation into the long-term healthcare industry. Eur. Manag. J. **35**(4), 563–573 (2017). https://doi.org/10.1016/j.emj.2016.12.001
10. Shao, J.F., Li, Y.: Multi-agent production monitoring and management system for textile materials and its applications. J. Ind. Text. **40**(4), 380–399 (2010). https://doi.org/10.1177/1528083710380428
11. Tursunov, B.O.: Methodology for assessing the efficiency of production capacities management at textile enterprises. Vlakna Textil, **26**(2), 74–81 (2019). https://doi.org/https://doi.org/10.5281/zenodo.3756262
12. Tursunov, B.O.: The mechanism for determining optimal management of the use of production capacity at the textile enterprises. Vlakna Textil, **27**(1), 99–106 (2020). https://doi.org/ https://doi.org/10.5281/zenodo.3787291
13. Zarova, E.: Statistical methodology for evaluating business cycles with the conditions of their synchronization and harmonization. Stat. Stat. Grow. Data Sets Growing Demand Stat. IntechOpen, 27–48 (2018)

Integration Mechanism of Educational Process Management Based on Value Chain and Life Cycle Model

Sergey B. Sorokin[1(✉)], Elena L. Neznahina[1], Alla G. Zinchenko[1], Dmitrii A. Kornilov[1], and Elena S. Lykova[2]

[1] Nizhny Novgorod State Technical University n.a. R.E. Alekseev, Nizhny Novgorod, Russia
sorokin@nntu.ru, alla@zinchenko.com

[2] Volga State University of Water Transport, Nizhny Novgorod, Russia

Abstract. Staff training affects the results of the entire organization. More organizations are looking for ways to improve learning effectiveness. The main goal of the educational process is to ensure a given level of the employee's competence throughout his life cycle. The effectiveness of the educational process depends on the degree of integration of all its participants. The goal of the research is to develop an integration mechanism for managing the educational process based on the value chain and life cycle model to ensure its competitiveness.

The educational process should be managed through a value chain. The main processes are defined as: the preparation stage, the training stage and the application stage. To ensure the integration of participants in the educational process, their target guidelines are formulated and role functions are defined. The absence of a mechanism for selecting employees for training and assessing their behavior after training is the main reason for the discrepancy among the actual and target results of the educational process.

An integration mechanism for managing the educational process based on the value chain and life cycle model has been developed to increase its competitiveness. The use of the life cycle model allows the manager to assess the level of competence of the employee before and after training.

The integration mechanism for educational process management is a set of actions aimed at strengthening coordination and interaction of the main participants in the educational process and ensuring the achievement of their goals.

Keywords: Education and economic development · Human capital · Learning · Value chain · Life cycle

JEL Codes: I25 · O15 · P46

E. G. Popkova (Ed.): ISC 2020, LNNS 368, pp. 611–619, 2022.
https://doi.org/10.1007/978-3-030-93244-2_67

1 Introduction

Currently, the competitiveness of any organization is ensured by the presence of competent and motivated personnel. Increasing the dynamism and variability of the external environment requires workers who can quickly and efficiently implement new values. Therefore, the leaders of modern organizations are increasingly paying attention to the problems of corporate education. According to statistics, the education level of the population in Russia is quite high [1], however, the problem of the education effectiveness is quite urgent. It is important the problem of training efficiency is not fragmented, but is systemic [2].

As part of the survey [3], employees rated corporate training in the company: 37% of employees consider corporate training to be completely ineffective, 30% consider training efficiency satisfactory and 33% of employees consider training to be really effective.

The main reasons for the inefficiency of corporate training are:

- Inability to determine actual training needs due to lack of communication among strategic goals of the organization and goals of personnel management units;
- Uncoordinated training format (dates, place and time), etc.;
- Imperfect structure of educational program: proportions of theory and practice, choice of priorities in favor of hardskills, or vice versa, softskills;
- Lack of motivation on the part of training workers: they do not know what or why they should be trained [4]. There is no link between the level of competence and wages or the possibility of career growth. Therefore, it is impossible to broadcast to the employee the goals of his training, and, therefore, the procedure for assessing the effectiveness of training becomes impossible.

The educational process is a collection of numerous interconnected processes which many participants take part in. Taking into account the peculiarities of the modern situation, it is advisable to apply the principles of the process approach in the development and implementation of the educational process.

Stages of the educational process can be distinguished on the basis of D. Kirpatrik's model [5]. The levels of learning effectiveness assessment applied in the D. Kirpatrik's model can be considered as interrelated stages of the educational process.

The stage of preparation is the preparation of the profile of the student of the educational process. According to the results of research by the Higher School of Economics, in Russia only 25% of managers determine what competencies its employees lack before starting training. The result of the process is an assessment of the primary response to learning, expectations of the learning process and the degree of engagement of potential students. A positive response does not ensure the successful development of new competencies; however, a negative response to learning significantly reduces the efficiency of learning.

Training stage determines the content of the educational process and the format of its implementation. The criterion for the effectiveness of this process is the degree of assimilation of the obtained competences by the participants in the educational process and the compliance of the received knowledge with the expected ones.

The stage of application (implementation) consists in ensuring the possibility of applying the obtained competencies in the workplace or in the process of personal activity. The criterion for the effectiveness of this process is the degree of competencies' implementation acquired by participants in the training process at the workplace. The effectiveness of staff training is 50% dependent on post-training work with participants and only 24% depends on the training process itself and, 26% - on what knowledge a person has come with [6].

The efficiency of the educational process is ensured by the efficiency of the implementation of each of the stages. Application of principles of process approach in development and implementation of educational process covers all elements of chain of creation of educational process' values [7, 8].

Highlighting key processes and focusing on their implementation will ensure the effectiveness of the process as a whole. The analysis of the stages of the educational process in terms of their position in the value chain allows seeing non-obvious opportunities and making more effective decisions [9].

2 Methodology

By the chain of creating the value of the educational process, it is proposed to mean a set of role functions of the main participants in the process which ensure its competitiveness.

An analysis of the educational process based on the value chain allows identifying the main participants in the process:

1. The head of the organization or any department, in fact, the customer of the process;
2. The organizer of the educational process (performer), usually it is any organization providing educational services, however, there may also be internal organizational units;
3. Trainee – an employee directed to training.

Each of the participants in the educational process has its own target guidelines and implements a certain set of role functions to achieve them.

The head as the customer of the educational process forms the requirements for the results of the educational process and its organization. The goal of the head of the educational process is: to increase the employee's competence in accordance with the strategic goals of the organization.

The organizer of the educational process (performer) ensures the production of the result, in accordance with the requirements of the customer. The goal of the organizer of the educational process is to ensure the trainee learns the required competencies in a given training format.

The trainee is an employee of the organization aimed at training within the framework of a specific educational program. The trainee can be called a passive participant, since he cannot have a managerial impact on the educational process. The purpose of the trainee is to increase the professional and personal level of competence. Lack of competencies, not only does not allow the employee to effectively realize the tasks assigned to him, but also reduces his level of motivation [10].

The peculiarity of the chain of creating the value of the educational process is that the main participants in the process are simultaneously its consumers. Achieving the goal of one of the participants in the educational process becomes impossible if the other participants do not achieve the goals.

For example, the formation of employee targets is the result of managerial impact. With inefficient management, the employee's goals may not meet organizational goals. For example, an employee may consider it his main goal to increase the volume of production and not associate its achievement with training.

In the training group, workers who are negatively related to the training process find themselves quite often. This situation may arise for the following reasons:

- the trainee does not have training needs because he does not understand what competencies he lacks to effectively perform current duties or what new competencies he may need in the future;
- the trainee does not want to learn because of personal circumstances or skills.

The value chain is broken, as the manager's goals do not correspond to the goals of the trained employee. The manager will not be able to achieve the goal, since in this situation the organization will not receive an employee with the necessary competencies.

A similar situation arises when the head of the organization sends employees for training, not taking into account his actual competence, the physical possibility of the employee's participation in the educational process and the degree of motivation.

As for the performer, i.e. the educational organization, in this situation it will also not be able to achieve its objectives - maximum customer satisfaction, since the requirements of the manager do not correspond to the goals of the trained employee. As a result, the educational program most often does not correspond to the level of education, the possible time and place of its implementation are not taken into account, and the lack of motivation of the employee does not allow him to master the educational material as much as possible.

The following basic conditions for the effectiveness of corporate training can be distinguished:

- The definition of training needs should be based on the strategic objectives of the organization.
- Transmission to the employee of the training goals from the point of view of achieving the strategic goals of the organization and the expected personal effect (salary increase, career growth, participation in a new project, etc.).
- Coordination of choice of training format, terms, time and place of classes by all participants of educational process.
- Analysis of training results.
- Organization of post-training work.

To ensure the transparency of the flow of value and the effective interaction of all participants in the value chain, it is necessary to define role functions (responsibilities) for each participant in the educational process.

Role functions (responsibilities) of the educational project customer:

1. Define the purpose of the training. It is necessary to focus on training workers whose level of competence is not sufficient for the implementation of job responsibilities or for future professional development in accordance with organizational goals. The formulation of the objective is necessary to define standards and criteria for assessing the effectiveness of the curriculum.
2. Personnel certification before and after training. Certification before training allows you to determine the starting level of competencies, the employee. This allows you to draw up a targeted training program that allows you to master the necessary competencies. Certification after training allows you to evaluate changes in the behavior of a worker after completing training.
3. Formulation of organizational conditions for the educational process (place, time, form and terms of training).
4. Development of personal motivators for employees sent for training. It is important to determine the degree of motivation of employees sent for training and, if necessary, to develop additional motivators for them.
5. Ensure that the competencies obtained can be applied at the workplace after training Develop adaptation events and appropriate tools for selecting and evaluating parameters that may have been subject to change after training.

Often managers do not take into account the importance of creating conditions for changes in employee behavior after training. The absence of changes in participants' behavior does not always mean that training was ineffective. However, if such conditions are created, a staff member who has learned the programme better will show more change in targeted behavior.

Role functions (responsibilities) of the educational project organizer:

1. Development of the educational program structure according to the customer's needs.
2. Provision of organizational conditions for the educational process (place, time, form and terms of training) in accordance with the customer's needs.
3. Selection of teaching staff ensuring the best mastery of the required competencies within the framework of the educational program.
4. Organization of feedback. Conducting questionnaires of trainees and managers - customers of the educational process.

Role functions (responsibilities) of the employee aimed at training within the framework of a specific educational project (with effective interaction of all participants in the value chain):

1. Maximum assimilation of educational material presented in the educational program.
2. Adaptation of acquired competencies to production activities.
3. Formulation of requirements to the management of the organization for realization of the obtained competencies in production activities.

3 Results

On the basis of certain role functions it is possible to allocate main units of process of creation of value to the educational process and to define their sequence: formation of value, creation of value, implementation of value (Fig. 1).

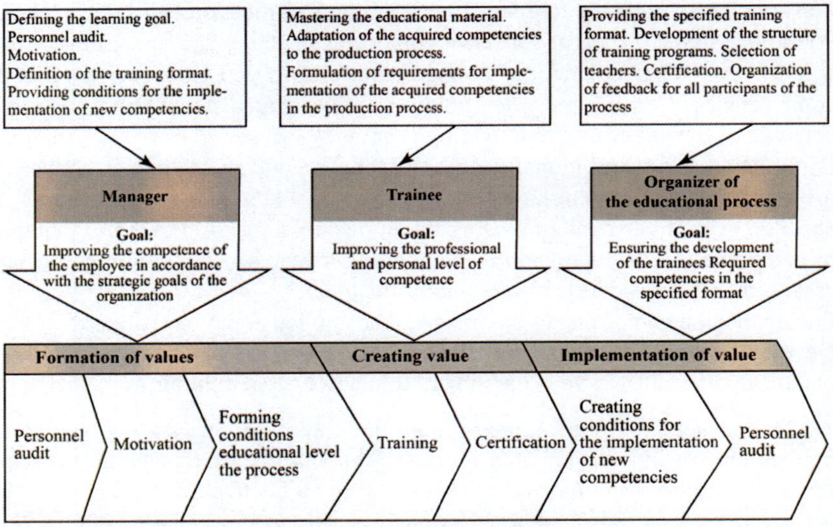

Fig. 1. Integration mechanism of management of educational process on the basis of a chain of creation of value. Source Compiled by the authors

The analysis of a chain of creation of educational process' value allows revealing the main discrepancies between valid and targeting results of processes – gaps:

1. Discrepancy among the purposes of the head and the purposes of the trained worker.
2. Discrepancy among the purposes of the organizer of the educational process and the head's purposes.
3. Discrepancy among the purposes of the trained worker and conditions of educational process.

Inefficient performance of the key role function of the head – the customer of the educational project «Personnel Audit» is the reason for certain discrepancies. As it was already noted, at selection of workers on training it is important to consider three parameters: professional competence, physical capacities and degree of motivation. In spite of the fact that the trainee is a passive participant, his characteristics are basic data for the organization and management of the educational process.

For effective realization of the role function «Personnel Audit» it is expedient to use the model of life cycle of the labor unit [11]. The model visualizes the current state of employees of the organization or division in general, and each worker separately on the basis of parameters of professional competence, physical capacities and degree of motivation and is created by a possibility of the forecast.

Here is the next description of model parameters:

- physical potential – characterizes the physical capacities of the worker caused, first of all, by age;
- motivation – characterizes desire of the worker to develop in the professional activity;
- professional competence – is characterized by education level, experience, professional qualification.

The presentation of the model allows timely making management decisions for formation of necessary purposes for the worker and development of the operating motivators.

Parameters of characteristics change in time and the speed of change of each component isn't identical. For creation of a model of life cycle of labor unit everyone a component is considered as separate process in which the main criteria of the levels of development are defined.

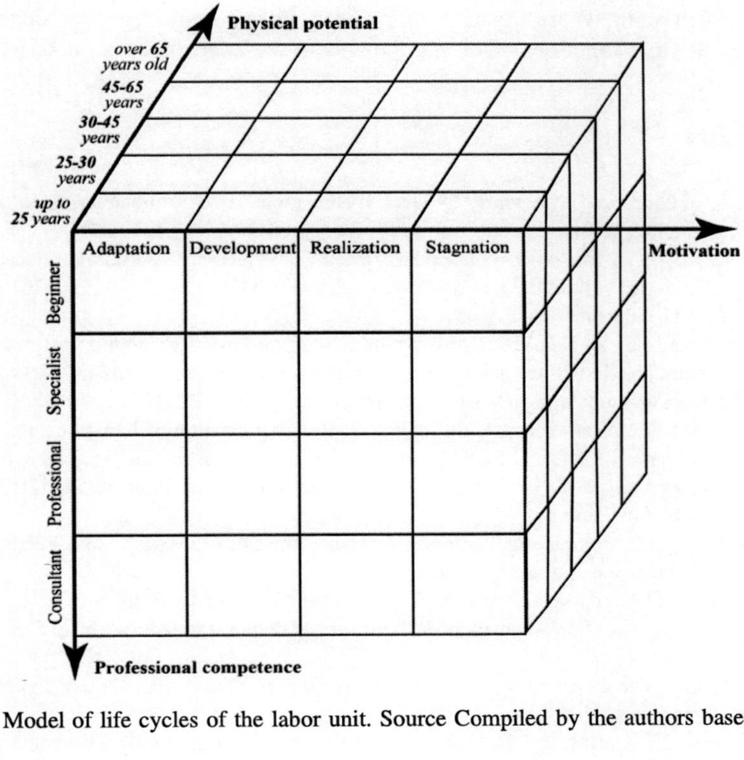

Fig. 2. Model of life cycles of the labor unit. Source Compiled by the authors based on [10]

Defining an employee-specific status criterion for each process determines the profile of the work unit. The model reflects a set of criteria for the characteristics of both the individual employee and the organization or its subdivision (Fig. 2).

In the value chain of the educational process, the role function «Personnel Audit» is implemented when selecting employees for training and to determine changes in the behavior of the employee after completing training.

The life cycle model of the labor unit allows observing these changes and analyzing their nature to ensure the effectiveness of the educational process.

4 Conclusion

The developed integration mechanism for controlling the educational process clearly demonstrates the logic of the educational process. Structuring and transparency of the process makes it possible to increase coordination and interaction of the main participants in the educational process. The analysis of employee profile characteristics based on the life cycle model of the labor unit before and after training makes it possible to more clearly formulate the goals of the educational process for all the main participants.

Within the framework of the integration mechanism of educational process management based on the value chain, recommendations and tools have been developed to implement the role functions of all participants in the educational process.

References

1. Ranking of countries of the world by level of education (2020). Humanitarian portal. https://gmarket.ru/education-index/education-index-info. Accessed 20 Feb 2021
2. Zavyalova, E.K., Ardishvili, A.: Modern problems of Russian corporate education: is it possible to study without development? Russ. Manag. Mag. **17**(4), 499–516 (2020). https://doi.org/10.21638/spbu18.2019.405
3. Corporate training in Russia 2020: expectations of employers vs opinion of professionals. HR in Russian. https://hr-elearning.ru/korporativnoe-obuchenie-v-rossii-2020-ozhidanija-rabotodatelej-vs-mnenie-professionalov/. Accessed 20 Feb 2021
4. Maranova, N.V., Neznakhina, E.L.: Approach to the evaluation of human capital based on the determination of the value (contribution) of employees of various professional competencies and the degree of motivation. Curr. Probl. Sci. Educ. **4**(54) (2014). http://www.science-education.ru
5. Kirkpatrick, D.L.: Evaluating Training Programs: The Four Levels, p. 229. Berrett-Koehler Publishers, Inc., San Francisco (1998)
6. Kirkpatrick, D.L., Kirkpatrick, J.D.: Transferring Learning to Behavior. Using the Four Levels to Improve Perfomance, p. 220. Berrett-Koehler Publishers, Inc., San-Francisco (2005)
7. Porter, M.E.: The Competitive Advantage: Creating and Sustaining Superior Performance, p. 526. Free Press, New York (1985)
8. Kuznetsov, V., Garina, E., Bardakov, A., Kornilov, D., Lapaev, D.: Re-engineering the business process of sales in view of life cycle stage and the company organisational structure. Int. J. Trade Glob. Mark. **12**(3/4), 412–423 (2019). https://doi.org/10.1504/IJTGM.2019.101554

9. Rackham, N.: Major Account Sales Strategy, p. 218. McGraw-Hill Inc., New York (1989)
10. Sorokin, S.A., Neznakhina, E.L., Firsov, A.S.: Analysis of staff training efficiency problems. Econ. Entrepreneurship **10**(99) (**12**(10)), 1095–1098 (2019)
11. Maranova, N.V., Neznakhina, E.L.: Multidimensional model of interaction of life cycle types that make up the labor unit profile. Innov. Invest. **4**, 187–197 (2014)

Assessment of Budgetary Stresses for Socio-economic Development of Regions

Nadezhda I. Yashina[1], Oksana I. Kashina[1], Nataliya N. Pronchatova-Rubtsova[1], Sergei N. Yashin[1], and Victor P. Kuznetsov[2(✉)]

[1] Lobachevsky State University of Nizhny Novgorod, Nizhny Novgorod, Russia
[2] Minin Nizhny Novgorod State Pedagogical University, Nizhny Novgorod, Russia

Abstract. Purpose: The main aim of the paper is to develop methodology and techniques of implementation of budgetary stress assessment and financial response mechanisms to social stability risks in Russian regions in the pandemic condition by using of the example of the Nizhny Novgorod region.

Design/Methodology/Approach: A current research is a methodological approach to assessing regional budgetary stresses in the crisis conditions of reforming and developing the government budgetary system. The application of this approach allows to solve the discussion problems of assessment of budgetary stresses and financial mechanisms for responding to threats to social stability Russian regions during the pandemic. Object of research - local and regional budgets of Nizhny Novgorod region; subject – a set of theoretical and methodological questions of budgetary stress assessment and financial mechanisms of response to threats to social stability in the regions of Russia during the pandemic and in the post-pandemic period.

Findings: As a result, the research allow us to estimate the stability of public finances during a pandemic, and to specify actions for the short and long-term fiscal policy in the region in such crises.

Originality/Value: The research result is the development of the methodology for evaluation of budgetary stresses and financial mechanisms for responding to threats to social stability Russian regions in the pandemic condition, which allows to study, detail, and present a list of recommendations for anti-crisis management of public finances.

Keywords: Budget execution · Budget assessment · Risk · Public finance · Budget stress · Standardized ratio

JEL Codes: H7 · H21 · H61 · H63 · H68

1 Introduction

Outbreaks of a new coronavirus infection showed the thin spots of budgets of all levels and helped only assume long-term financial problems. Public finance in 2020 was aimed at solving short-term problems, however, along with short-term measures. It is important to develop directions for increasing the sustainability of public finance in the

E. G. Popkova (Ed.): ISC 2020, LNNS 368, pp. 620–631, 2022.
https://doi.org/10.1007/978-3-030-93244-2_68

long or medium term and, most importantly, their effective use to solve social and economic problems in the country after the pandemic. Primarily, they should maintain investment activity and restructuring economies.

Assessment of budgetary stresses and financial mechanisms for responding to threats to social stability in the Russian regions allows to assess the risk of financing liabilities within the framework for the implementation of national projects and government programs in the regions in the short and long term. It is proposed to understand budgetary stress as tense budget execution, which violates its ability to maintain stability, which is determined by the indicators provided for in the initially approved budget for the next financial year.

Three stages should be noted in budget stress:

- mobilization of own budget revenues to finance priority and socially significant budget expenditures;
- impossibility (decrease) of collection of own budget revenues and, as a result, temporary problems with financing of priority and socially significant budget expenditures;
- a drop in the collection of own budget revenues. As a result, serious problems with financing priority and socially significant budget expenditures, and the impossibility of ensuring a decent standard of living for the population.

2 Methodology

The need for an urgent assessment of the impact of COVID-19 on society and the global environment has been emphasized in their works by many researchers Chakraborty and Maity [1], Husain [3], Lucchese and Pianta [8]. In addition to the importance of assessing the economic consequences, the problem of developing possible actions that can lead to a more stable and sustainable trajectory of development of territories is urgent, as noted by Gamukin [2], Ivanov and Korobova [4], Kolesov [6], Kireeva [5], Yashina and Poyuscheva [11], Yashina and Emelyanova [10], Kuryleva et al. [7].

Assessment of the budgetary stress is proposed to carry out on the basis of monthly dynamics formed a system of indicators of local and regional budgets of RF revenue mobilization and financing costs. These actions are needed in order to bring them to a comparable form and the proposed methodology for assessing stressful periods based on of the definition of a standardized indicator. Complex estimation involves the conversion of coefficients by standardization based on the requirements for indicators (minimization or maximization).

For standardization, depending on the economic meaning of the indicator, when minimizing (growth is undesirable) and maximizing (growth is desirable), the following formulas (1) and (2) are used, respectively.

$$K_{ij}^{*} = \frac{K_{ij} - K_{imin}}{K_{imax} - K_{imin}}, \tag{1}$$

$$K_{ij}^{*} = \frac{K_{imax} - K_{ij}}{K_{imax} - K_{imin}}, \tag{2}$$

where K_{ij}^{*} is the standardized indicator I of the proposed indicator of the system for estimating budgetary stresses in the j month, K_{ij} is the calculated the value of the i proposed indicator of the system for estimating budgetary stresses in the j month, K_{imax} is the highest estimated value I index in the analyzed period, K_{imin} is the lowest calculated value of the I index in the analyzed period.

Now the chosen method of standardization leads to transformation of the indicators' value in the range from 0 to 1.

Standardized indicators of budget stress evaluation systems are summarized each month in order to determine the total standardized coefficient. The smaller value of the total cost standardized coefficient estimation section stresses in periods, then the greater the ability to maintain of current execution of the budget stability. Monthly dynamics of the generated system of indicators of the budgets of the Nizhny Novgorod region is analyzed based on monthly budget execution reports for the years 2019–2020.

Execution risks for the income and expenses of the budget are determined taking into account the budget or with a revised budget plan. Standard deviation is used to determine the risk of collecting one's own income and the risk of financing expenses. The larger the indicator of standard deviation associated with great budgetary risk.

The system of the indicators that are determined budget stresses includes:

- total receipts of budget own revenues and expenditure of budgetary funds per month in 2019 and 2020 and their horizontal structural, comparative, factorial analysis;
- financial independence, provision of own funds for financing industrial sectors, investment in human capital, interest expenses, debt burden, total debt burden, tax debt in 2019 and 2020;
- performance rates of the budget and the specified plan;
- coefficients of variation in the performance of the budget and the specified plan and their comparison in 2019 and 2020;
- growth rates (decrease) of budget revenues and expenditures compared to the same period last year;
- the number of amendments to the budget law in 2019–2020;
- optimization of cash flows taking into account the synchronization method based on the correlation coefficients between receipts and financing of expenses in 2019 and 2020.

The set of indicators can be expanded. The current values of the assessment indicators are determined based on data on budget execution.

3 Results

Let us calculate the level of budgetary stress in the Nizhny Novgorod region for 2019–2020 according to the developed methodology for a comprehensive assessment of budgetary stress on the basis of indicators: coefficients of budget execution, the risk of budget execution or a revised plan, growth rates in comparison with the same time period of last year (Tables 1, 2, 3, 4, 5, 6, 7 and 8).

Analyzing the current values of some indicators for assessing the budgetary stress of revenue mobilization, it can be seen (Table 1) that in 2019 the budget was revised 11 times (except for March), and in 2020 it was adjusted monthly. The data of budget execution illustrate that in 2020 the mobilization of own revenues amounted to 102.3% of the revised plan, and in 2019 - 101.6%. Significant threats of execution of income side of the budget were observed already from June to September 2020 comparison with the same time period of last year. Own revenues in 2020 amounted to 182.579 billion rubles or 2.3% that more than in 2019 (RUB 181.971 billion).

Table 1. Analysis of the execution of consolidated budget's tax and own incomes (non-tax revenues) in 2019–2020 (fragment)

Month	The execution of consolidated budget's tax and own incomes (non-tax revenues)							
	2019				2020			
	Revised plan (thousand rubles)	Execution (thousand rubles)	Execution rate (%)	Variation coefficient (%)	Revised plan (thousand rubles)	Execution (thousand rubles)	Execution rate (%)	Variation coefficient (%)
January	170841316.3	9947023.9	5.8	22	189866045.3	10675503.4	5.6	19
February	170887979.1	20076304.4	11.7	12	189949257.2	21754033.5	12.8	11
March	170887979.1	39570890.2	23.2	12	189856186.2	43182738.1	22.7	9
…								
May	171578759.4	75254885.4	43.9	10	190992080.1	71183134.3	37.3	7
June	171701546.0	84228078.0	49.1	10	183531371.6	82155317.3	44.8	13
July	174303144.8	104817169.9	60.1	10	182359629.7	102069156.1	56	20
…								
September	176958218.6	128728870.9	72.7	8	182765459.7	127068423.3	69.5	11
October	177056713.8	148905179.7	84.1	7	182913515.1	145050821.4	79.3	8
November	178484225.1	163905036.6	91.8	6	183037561.7	161377218.7	88.2	7
…								

Source Compiled by the authors based on [9]

Table 2 presents a comparative analysis of monthly execution, execution risk and growth rates of consolidated budget's own incomes in 2019–2020. Assessment of budgetary stress level can be carried out on those indicators that help determine the response of public finances to stress and allows us to find ways to deal with it in the early stages.

Table 2. Comparative analysis of monthly execution, execution risk, growth rates of consolidated budget's tax and own incomes (non-tax revenues) in 2019–2020 (fragment)

Name of the month, indicators	Tax and own incomes (non-tax revenues)					
	2019			2020		
	Execution rate (%)	Variation coefficient (%)	The growth rate that fixed in the previous year (%)	Execution rate (%)	Variation coefficient (%)	The growth rate that fixed in the previous year (%)
January	5.8	22	119.2	5.6	19	107.3
February	11.7	12	116.2	12.8	11	109.0
March	23.2	12	112.3	22.7	9	109.1
…						
May	43.9	10	113.3	37.3	7	94.6
June	49.1	10	109.7	44.8	13	97.5
July	60.1	10	111.7	56	20	97.4
…						
September	72.7	8	111.6	69.5	11	98.7
October	84.1	7	107.9	79.3	8	97.4
November	91.8	6	107.1	88.2	7	98.5
December	101.6	12	107.7	102.3	5	100.3
Standard	54.0	10.6	112.3	51.1	11.0	100.4
Max. value	101.6	22.0	120.1	102.3	20.0	109.1
Min. value	5.8	6.1	107.1	5.6	4.7	94.6
The scale of variation	95.8	15.9	12.9	96.7	15.3	14.5

Source Compiled by the authors based on [9]

The presented analysis gives an objective view of the functional disorders of the regional budget system in response to the stress caused by the new coronavirus infection.

The results of a comprehensive evaluation by cumulative stress budget standardized coefficient in 2019 year – Table 3.

Table 3. Standardized indicators for assessing budgetary stress based on monthly execution of own revenues, execution risk and growth rates of consolidated budget's tax and own incomes (non-tax revenues) in 2019 (fragment)

Month	Standardized income indicators, 2019			
	Standardized Execution Ratio	Standardized coefficient of variation	Standardized growth rate that fixed in the previous year	Complex standardized indicator
January	1.000	1.000	0.068	2.068
February	0.938	0.371	0.297	1.606
March	0.818	0.371	0.602	1.791

(*continued*)

Table 3. (*continued*)

Month	Standardized income indicators, 2019			
	Standardized Execution Ratio	Standardized coefficient of variation	Standardized growth rate that fixed in the previous year	Complex standardized indicator
…				
May	0.602	0.245	0.525	1.372
June	0.548	0.245	0.805	1.598
July	0.433	0.245	0.651	1.330
…				
September	0.302	0.119	0.658	1.079
October	0.183	0.044	0.940	1.167
November	0.102	0.000	1.000	1.102
December	0.000	0.358	0.955	1.314
Standard	0.497	0.280	0.600	1.378

Source Compiled by the authors based on [9]

An analysis of the indicators in Table 3 makes it possible to understand that, based on the results of the assessment of budgetary stress. The least stressful budget execution, characterized as a result of relative stability (in particular, a low level of risk) of budget indicators, was recorded in April, September, October, and November. The results of a comprehensive assessment of budgetary stress using the cumulative standardized ratio in 2020 are presented in Table 4.

Table 4. Standardized indicators for assessing budgetary stress based on monthly execution of own revenues, execution risk and growth rates of consolidated budget's tax and own incomes (non-tax revenues) in 2020 (fragment)

Month	Standardized income indicators, 2020			
	Standardized Execution Ratio	Standardized coefficient of variation	Standardized growth rate that fixed in the previous year	Complex standardized indicator
January	1.000	0.935	0.124	2.059
February	0.926	0.412	0.011	1.349
March	0.823	0.282	0.000	1.105
…				
May	0.672	0.151	1.000	1.823
June	0.594	0.543	0.797	1.934
July	0.479	1.000	0.808	2.287
…				
October	0.238	0.208	0.806	1.252
November	0.146	0.131	0.734	1.010
December	0.000	0.000	0.610	0.610
Standard	0.530	0.414	0.601	1.544

Source Compiled by the authors based on [9]

The results of budgetary stress evaluation based on the analysis of the aggregate standardized coefficient (Table 4) show that the least stressful budget execution was in March, October, November, and December. A new coronavirus infection affected the level of budget stress standard, which became higher in 2020. Analysis of the current values of some indicators for assessing the budgetary stress of financing expenditures showed (Table 5) that in 2019–2020 the budget was revised on a monthly basis.

Table 5. Analysis of the execution of expenses of the consolidated budget in 2019–2020 (fragment)

Month	The performance of expenditure							
	2019				2020			
	Revised plan (thousand rubles)	Execution (thousand rubles)	Execution rate (%)	Variation coefficient (%)	Revised plan (thousand rubles)	Execution (thousand rubles)	Execution rate (%)	Variation coefficient (%)
January	201671811.8	12585371.7	6.2	12	233628863.5	14418280.2	6.2	11
February	202606144.3	26863784.2	13.3	10	234627913.1	30422195.6	12.9	16
March	207473976.0	39672290.0	19.1	17	241693047.0	45819101.6	19.0	19
…								
May	216431398.1	72597227.7	33.5	25	252262092.5	85237555.3	33.8	25
June	216652546.7	88964233.0	41.5	24	256676920.7	103790907.9	40.4	26
July	220374321.0	106944755.0	48.5	23	258455149.8	127427558.8	49.3	25
…								
September	223945720.4	138640994.2	61.9	24	267412607.5	164479411.2	61.5	27
October	224457661.4	161636713.9	72.0	8	270517997.6	188113649.7	69.5	8
November	226908192.3	178979858.6	78.9	7	274285951.2	209455132.6	76.4	8
December	225884425.5	215247100.5	95.3	2	271855672.6	261591222.3	96.2	3

Source Compiled by the authors based on [9]

The budget execution data show that in 2020, financing costs amounted to 96.2% and in 2019 - 95.3%, which is certainly better, but not significantly. The risks of financing budget expenditures were relatively strong over the year in August and September 2020. The total sum of appropriations in 2020 was RUB 261.591 billion or 21.5% higher than in 2019 (RUB 215.247 billion).

Table 6 presents a comparative analysis of monthly execution, execution risk and growth rates of consolidated budget expenditures in 2019 and 2020.

Table 6. Comparative analysis of monthly execution, execution risk and growth rates of expenses of Nizhny Novgorod region's consolidated budget in 2019–2020 (fragment)

Name of the month, indicators	Budget expenditures					
	2019			2020		
	The coefficient of performance	The coefficient of variation	The growth rate that fixed in the previous year	The coefficient of performance	The coefficient of variation	The growth rate that fixed in the previous year
January	6.2	12	108.4	6.2	11	114.6
February	11.7	10	104.8	12.9	16	113.2
March	19.1	17	102.4	19.0	19	115.5

(*continued*)

Table 6. (*continued*)

Name of the month, indicators	Budget expenditures					
	2019			2020		
	The coefficient of performance	The coefficient of variation	The growth rate that fixed in the previous year	The coefficient of performance	The coefficient of variation	The growth rate that fixed in the previous year
…						
May	33.5	25	105.0	33.8	25	117.4
June	41.5	24	103.6	40.4	26	116.7
July	48.5	23	105.6	49.3	25	119.2
…						
September	61.9	24	109.9	61.5	27	118.6
October	72.0	8	110.9	69.5	8	116.4
November	78.9	7	111.3	76.4	8	117.0
December	95.3	2	115.6	96.2	3	121.5
Standard	46.0	16	107.7	45.7	18	117.2
Max. value	95.3	25	115.6	96.2	28	121.5
Min. value	6.2	2	102.4	6.2	3	113.2
The scale of variation	89.1	23	13.2	90.0	25	8.3

Source Compiled by the authors based on [9]

Table 7. Standardized indicators for assessing budgetary stress of financing expenditures based on monthly execution, execution risk and growth rate of consolidated budget expenditures in 2019 (fragment)

Period	Standardized expenditure rates, 2019			
	Standardized execution ratio	Standardized coefficient of variation	Standardized growth rate that fixed in the previous year	Complex standardized indicator
January	1.000	0.423	0.547	1.970
February	0.938	0.335	0.820	2.092
March	0.855	0.645	1.000	2.500
…				
May	0.694	1.000	0.802	2.495
June	0.604	0.956	0.909	2.469
July	0.525	0.911	0.760	2.197
…				
October	0.261	0.260	0.354	0.875
November	0.184	0.197	0.328	0.710
December	0.000	0.000	0.000	0.000
Standard	0.554	0.614	0.599	1.767

Source Compiled by the authors based on [9]

Results of a comprehensive assessment of fiscal stress using the cumulative standardized ratio in 2019 are presented in Table 7.

Analysis of indicators in Table 7 makes it possible to understand that the evaluation of budgetary stress less tense budget expenditure recorded from September to December. Results of a comprehensive assessment of budgetary stress using the cumulative standardized coefficient in 2020 are shown in Table 8.

Table 8. Standardized indicators for budgetary stress assessment of expenditure funding based on monthly performance, execution risk and cost growth rates of consolidated budget in 2020

Period	Standardized expenditure rates, 2020			
	standardized execution ratio	Standardized coefficient of variation	Standardized growth rate that fixed in the previous year	Complex standardized indicator
January	1.000	0.325	0.841	2.166
February	0.926	0.524	1.000	2.449
March	0.858	0.643	0.729	2.229
…				
May	0.693	0.881	0.497	2.072
June	0.620	0.921	0.587	2.128
July	0.521	0.881	0.287	1.689
…				
October	0.296	0.225	0.622	1.143
November	0.221	0.203	0.544	0.967
December	0.000	0.000	0.000	0.000
Standard	0.561	0.610	0.525	1.696

Source Compiled by the authors based on [9]

In accordance with the results of the budget stress assessment based on the values of the total standardized coefficient (Table 8), budget execution was the least stressful in the second half of the year. At the end of the assessment of budgetary stresses, it is proposed to determine the degree of synchronization of budget's execution based on

Table 9. Indicators of budgetary stress for the synchronization of revenue mobilization and financing of expenditures based on monthly execution of the consolidated budget in 2019–2020

Name of month, indicator	Income performance ratio in 2019	Expense execution ratio in 2019	Income performance ratio in 2020	Expense execution ratio in 2020
January	5.8	6.2	5.6	6.2
February	11.7	13.3	12.8	12.9
March	23.2	19.1	22.7	19

(*continued*)

Table 9. (*continued*)

Name of month, indicator	Income performance ratio in 2019	Expense execution ratio in 2019	Income performance ratio in 2020	Expense execution ratio in 2020
…				
May	43.9	33.5	37.3	33.8
June	49.1	41.5	44.8	40.4
July	60.1	48.5	56	49.3
…				
October	84.1	72.0	79.3	69.5
November	91.8	78.9	88.2	76.4
December	101.6	95.3	102.3	96.2
Correlation coefficient	0.993		0.998	

Source Compiled by the authors based on [9]

the correlation coefficients between the receipt of incomes and expenses in 2019–2020 (Table 9).

Synchronization is based on the covariance of revenue mobilization and financing of expenses. Results of the method are evaluated using a correlation coefficient, which in the optimization process must strive to the value "+1". The correlation coefficient was equal 0.993 in 2019 and it was equal 0.998 in 2020.

The analysis of budgetary stress in the Nizhny Novgorod region showed that the budget stability of the region remains high (for example, the coefficient of financial independence, characterized by the share of own revenues in the total budget revenues, is more than 70%). The detailed analysis of the reports on budget execution for 2019–2020 showed that Nizhny Novgorod region didn't have budgetary stress in 2020.

4 Conclusion

This study made it possible to formulate the following directions of "anti-stress" budget management.

- It is necessary to focus on Cash-taxes, not just current tax revenues based on accruals. It is important to develop a strict policy for the collection of tax arrears from enterprises that are unpaid budgets of all levels.
- Improve the efficiency of control over costs: reduce costs to a minimum, to oblige the heads of ministries and departments, government and municipal budget and autonomous accountable for spending; focus on reducing indirect spending.
- Review investment plans and work out borrowing opportunities.

Public finance in the era of the coronavirus has once again shown its importance. Evaluating funding for health, education, and culture spending in terms of benefits does not stand up to any criticism. Sometimes new economic methods proposed by ministries and organizations are made dependent on profitability. Then free medical care, that is, treatment of a person, his vaccination also depends on profitability.

Currently, the population of the Nizhny Novgorod region is decreasing, mortality in many areas overlaps the birth rate, and, most importantly, the value of human health is in the last place due to insufficient or unprofitable funding.

A decline in GDP, the worsening economic situation and public finance instability affect the health of the population as directly or indirectly. Consequently, in order to implement measures to stabilize and improve the state health care system, it is important to reform the system of its financing, including through the possible liquidation of the Federal Fund for Mandatory Medical Insurance and the accumulation of health cares' funds in budgets of the constituent entities of the Russian Federation.

In order to increase funding, it is important to increase the tax incomes of the budget, and the following financial mechanisms are proposed:

- selection of qualified personnel of tax and customs authorities;
- strengthening work with taxpayers, their audit taking into account risk factors to strengthen tax discipline;
- the use of information super masses with the help of IT data systems to strengthen tax discipline and fight corruption.

Thus, using the monitoring local budgets based on the methodological tools that take into account the budgetary stresses the authorities of the constituent entities of the Russian Federation and local authorities can to make effective management decisions at all stages of the budget process. Important areas of countering budgetary stresses are the formation of budget revenues taking into account "stressful" scenarios and optimizing budget expenditures taking into account timely and full amount of funding for sections affecting the population livelihood.

Acknowledgments. The study was carried out within the framework of the basic part of the state assignment of the Ministry of Education and Science of the Russian Federation, project 0729-2020-0056 «Modern methods and models for diagnosing, monitoring, preventing and overcoming crisis phenomena in the economy in the context of digitalization as a way to ensure the economic security of the Russian Federation».

References

1. Chakraborty, I., Maity, P.: COVID-19 outbreak: migration, effects on society, global environment and prevention. Sci. Total Environ. **728** (2020). https://doi.org/10.1016/j.scitotenv.2020.138882
2. Gamukin, V.B.: Budgetary risks: options for trajectories. Questions Econ. Regul. **6**(4), 120–130 (2015)
3. Husain, A.: Coronavirus Pandemic: Effects, Prevention and Management, 153 p. The Readers Paradise (2020)
4. Ivanov, V.V., Korobova, A.N.: Municipal Management. INFRA-M, p. 718 (2012)
5. Kireeva, E.V.: Improving the efficiency of regional financial management. Region. Econ. Manag. Electron. Sci. J. **2–1**(50), 30–34 (2017)

6. Kolesov, A.S.: On the integrated assessment of the financial position of budget financing objects. Finance **6**, 9–11 (2000)
7. Kuryleva, O.I., Yashina, N.I., Bogomolov, S.V.: Improving the Methodology for Determining the Budget Sustainability of Territories and Assessing the Effectiveness of Budget Expenditures, p. 276. SEI HPE Volzhsky State Engineering University, Nizhny Novgorod (2009)
8. Lucchese, V.M., Pianta, M.: The coming coronavirus crisis: what can we learn?. Intereconomics **55**(2), 98–104 (2020)
9. Ministry of Finance of Nizhny Novgorod Region (2019, 2020). http://mf.nnov.ru/. Accessed 07 Jan 2021
10. Yashina, N.I., Emelyanova, O.V.: Methodology for assessing the consolidated budgets of constituent entities of the Russian Federation based on the definition of a consolidated standardized indicator. Finance Credit **40**(328), 22–35 (2008)
11. Yashina, N.I., Poyushcheva, E.V.: Improvement of the methodology for assessing the tax potential areas of risk. Region. Econ. Theory Pract. **6**, 132–138 (2007)

Accounting and Calculation Support for Reducing Losses of Material Resources not Used in the Production Process

Igor E. Mizikovsky[1(✉)], Victor P. Kuznetsov[2], Ekaterina P. Garina[2], Natalia S. Andryashina[2], and Elena V. Romanovskaya[2]

[1] Lobachevsky State University of Nizhny Novgorod, Nizhny Novgorod, Russia
[2] Minin Nizhny Novgorod State Pedagogical University, Nizhny Novgorod, Russia

Abstract. Today rational use of internal resources, improvement of techniques, and methods of material flow management, are the main factors in the growth of the economic potential at an enterprise. Effective management of enterprises in the real sector of the economy involves not only prompt identification and elimination of emerging losses but also the development of productive solutions to prevent them.

Our study aims to develop information support for making management decisions to identify and identify material losses arising in the production process.

The systematization of the movement of materials not used in production and the methodology for calculating their cost characteristics are done at each stage of this process. This has created an environment that minimizes unproductive costs.

Institutionalization of calculation developed by the accounting tools and their incorporation into the management accounting systems or manufacturing plant management, will significantly improve the results of their activities to ensure the efficiency of production costs.

Keywords: Losses · Production · Management tools · Material resources · Returns · Sales · Warehouse · Disposal · Leave for production · Equipment repair

JEL Codes: D24 · M21 · M41 · O33

1 Introduction

The economic development of business entities in the processing industries is undergoing significant changes associated, first of all, with the growth of competition in the sales markets, the complication of the used production technologies, and a noticeable increase in requirements for the quality and production time of products from the end consumers. In these conditions, it is necessary to constantly build up and improve the set of management tools that ensure the development of effective strategies and the

E. G. Popkova (Ed.): ISC 2020, LNNS 368, pp. 632–639, 2022.
https://doi.org/10.1007/978-3-030-93244-2_69

adoption of productive decisions to advance towards the intended financial and economic goals.

An important component of these tools is the management accounting tool aimed at solving the problems of rational use of internal resources, primarily by systematically reducing production losses. Our study carried out by the author in the period from 2016 to 2019 at a number of enterprises in the Nizhny Novgorod region, whose usual activity is the repair and adjustment of industrial electrical equipment. This made it possible to reveal a trend of the growth of production losses ahead of an increase in production volumes (Fig. 1). Of course, this situation is a significant problem for the economic life in the organization.

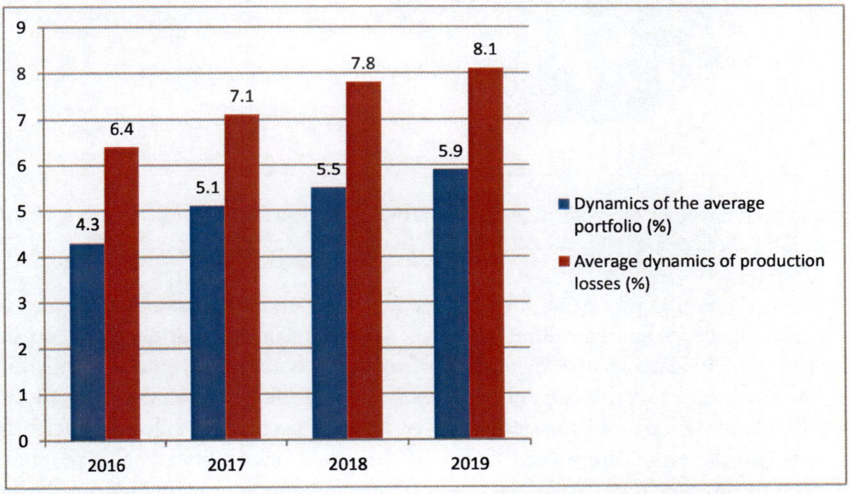

Fig. 1. Dynamics of losses and production on the test plants in the period from 2016 to 2019. *Source* Developed and compiled by the authors.

Structural analysis of production losses study enterprises showed that the largest share of loss of basic material resources released from the warehouse to the production, but not used in production (see Fig. 2). The genesis of this type of loss has a number of objective reasons due to the peculiarities of the individual type of production; the specifics of the preparation and repair of complex, usually seriously worn out and non-standard equipment design. Often, the production processes under study have the character of experimental work with a low level of unification of operations and value creation flow. It is no coincidence, on the test plants likelihood of re-use of material in the production of an average ranging from 8 to 15% (e.g., for some bronze castings is less than 1%).

In these conditions, it is objectively difficult to plan the procurement of material resources for use in the production process. For certain types of repairs, some of the basic materials (cable products; insulating, rubber-repair and technical materials; asbestos-rubber products) do not even have consumption indicators. They are purchased as needed in the "manual" mode during the execution of the order.

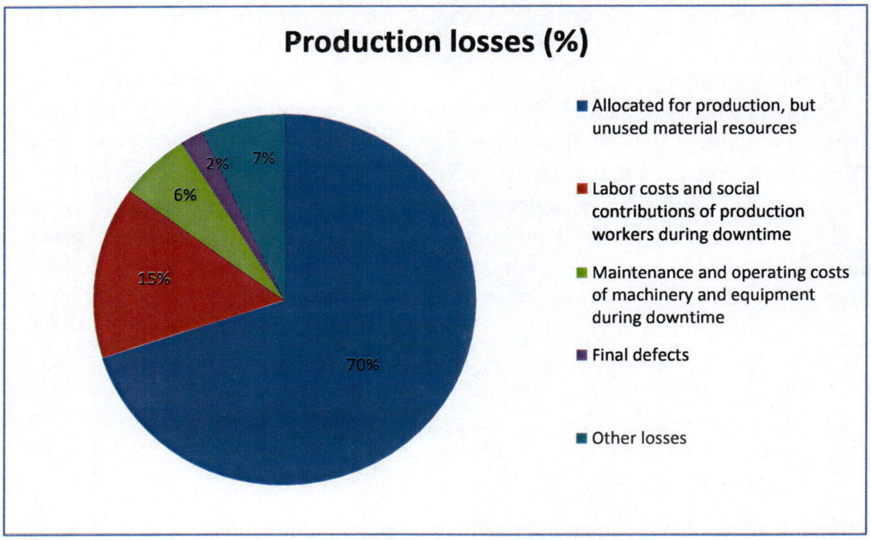

Fig. 2. Specific weight of production losses at the investigated enterprises in the period from 2016 to 2019. *Source* Developed and compiled by the authors.

It does not require proof that the absence of well-planned indicators of procurement of material resources has a negative effect on the purchasing and pricing policies of the enterprise. So, it leads to the formation of excess stocks of inventories and illiquid assets in warehouses. Replacements for materials that are not available for sale or are unacceptable at the cost of procurement, provided for by the technology of performing repairs at positions that are absent in it, but there is a possibility of their purchase or manufacture on their own. Also, an unreasonable release of materials from the warehouse to production leads to their accumulation in the repair zones, unproductive costs for the return of these resources, and the irrational use of logistics facilities, etc.

Almost all of its researchers see the problem of reducing production losses in the implementation of a lean production system, considering as them all operations and processes that do not create value [1–8]. The most significant group of losses, most authors position the loss of useful time of work of personnel, machinery, and equipment. [9–13]. At the enterprise's understudy, this type of loss, as follows from Fig. 2, has a significant volume. But, at the same time, is seriously inferior to the losses arising from the formation of residues of materials that have not found their application in production. This fact determined the purpose of the study presented in this article, which is the development of accounting and calculation support for decision-making. In order to reduce it or, if possible, eliminate the loss of material resources. The achievement of this goal is associated with solving problems of structuring the information field of the movement of materials released but not used in production; with the development of a methodology for calculating their cost characteristics at the stages of emergence and liquidation; on the institutionalization and incorporation of the developed methodology into the system of management accounting and production management operating at the enterprise, including the implementation of loss prevention strategies.

2 Methodology

Application of observation, measurement, formalization and generalization methods in research is forming an information base. Classification methods and decomposition - differentiate loss of production enterprises by type; structural and comparative analyzes - to match the dynamics of the studied parameters; graphical tools - to visualize them. It should be noted that the collection of information for the presented study was carried out over several annual periods based on primary documents, management reporting, surveys, questionnaires of executives and leading employees of enterprises.

3 Results

This study carried out by the author showed that at all the studied enterprises the cost of release from the warehouse for production, but unused material resources is calculated by the formula (1):

$$\mathrm{MP} = \sum\nolimits_{n}^{1} (Pri * \mathrm{K}i) \tag{1}$$

where:

Pr - an accounting price of the *i*-material;

K - quantity of *i* -material.

Of course, the advantage (1) is the efficiency and simplicity of accounting and calculation support of the process of identification and reflection in the accounts of losses of this type. Also here is an effective visualization of the final result, which significantly increases the speed of production by the subject of management on the control object. At the same time, based on the above definitions of the concept, it is obvious that it is necessary to attribute to the cost of the considered type of material residues of *all* technological operations. Studies have shown that such operations include: moving residues from places of unused materials at the warehouse, unloading their delivery vehicles, stacking, possibly packing and stacking, further storage, etc.

A calculation of the MP indicator, taking into account the value-added, which is arising from the above chain of operations, is presented in the formula (2):

$$With\ \mathrm{MP} = \sum\nolimits_{n}^{1} (Pri * \mathrm{K}i) + \mathrm{P}i + \mathrm{C}i \tag{2}$$

where:

P *i* - costs of moving *i* - material in the storage location;

C *i* - costs of warehouse operations.

Effective management requires immediate measures to eliminate the MP resulting in losses in the form of non-liquids and excess stocks in warehouses. These measures include operations involving the release of material items returned from production or again to production sites use in other orders. If this alternative is not possible, items sent to recycling. Unfortunately, for most of the investigated companies, there is no directive established local acts about the implementation of these actions. Internal reporting is not compiled, where data on the movement of the released, but not used material resources are systematized. The absence of regulated accounting and calculating support of the managerial procedures required in this situation. In addition, the mapping of their conduct leads to low control of this sphere and creates conditions for the growth of production losses.

The author systematized the stages of movement in the form of an enlarged block diagram of this process (Fig. 3).

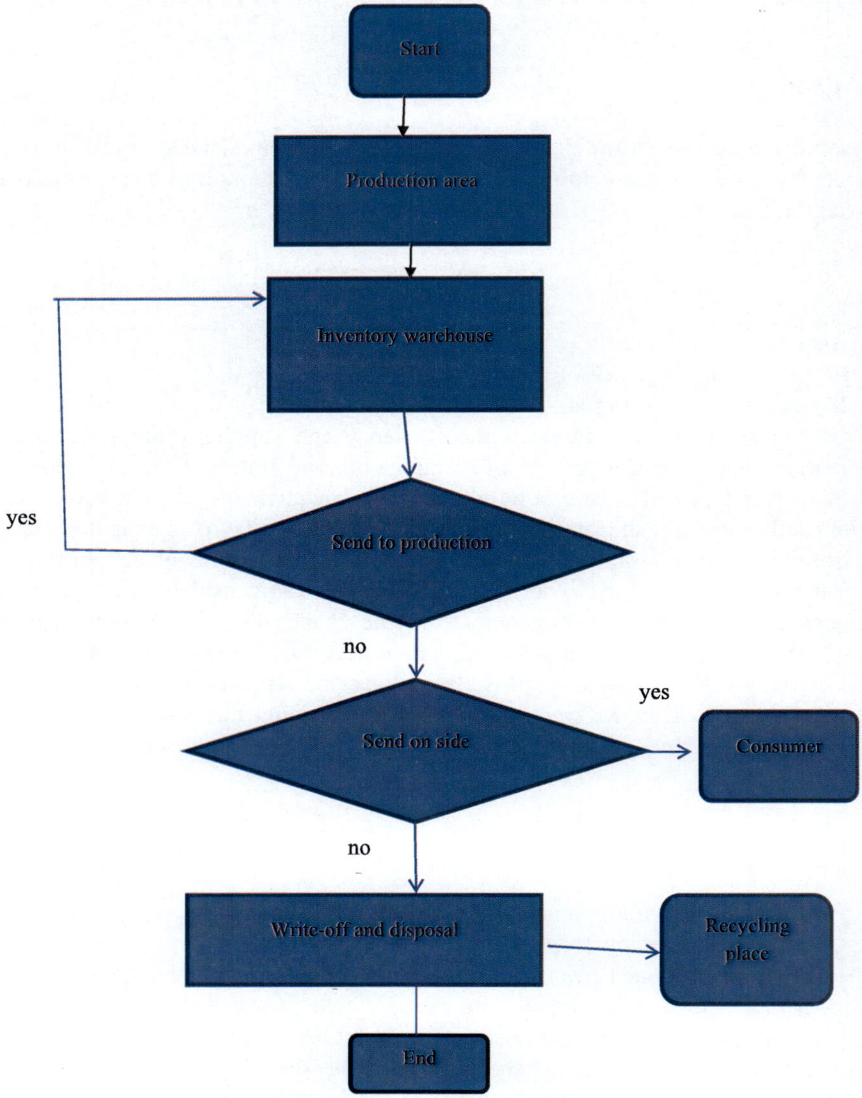

Fig. 3. An enlarged block diagram of the movement of materials released for production, but not used in this process. *Source* Developed and compiled by the authors.

It should be noted that the indicator MP *c* forms the cost of materials returned to the warehouse and again released for production, and also forms the basis of MP *m* for pricing in case of outsourcing:

$$\mathrm{MP}\ t = \sum\nolimits_{n}^{1} (Pri * \mathrm{K}i) + \mathrm{P}i + \mathrm{M}i + \mathrm{C}i + \mathrm{SC}i \tag{3}$$

where:
SCi - selling costs.

In a situation where been sale costs, but the transaction did not take place and this type of material resources recognized as unmarketable intended for recycling. The cost should be defined by the formula (4):

$$\mathrm{MP}\ p = \sum\nolimits_{n}^{1} (Pri * \mathrm{K}i) + \mathrm{P}i + \mathrm{M}i + \mathrm{C}i + \mathrm{CS}i + \mathrm{Y}i, \tag{4}$$

where:
Y*i*—disposal costs, including delivery to the disposal site.

In that general case, the cost of a resource, taking into account utilization, is proposed to be calculated using the formula:

$$\mathrm{Mpy} = \sum\nolimits_{n}^{1} (Pri * \mathrm{K}i) + \mathrm{P}i + \mathrm{C}i + \ + \mathrm{Y}i \tag{5}$$

A systematic representation of the process of movement of materials released but not used in production, according to the authors, should be reflected in the corresponding accounting registers. The author's register of the movement of unused materials in production, a fragment of which is presented in Table 1. It serves as an important information resource for control and accounting work, as well as for making management decisions, and is a necessary element of internal reporting.

The institutionalization of proposed tools by the author requires appropriate adjustments to the local acts of an economic entity, including: in the mapping of the value flow creation in part of the reflection of material flows used in production processes and operations; the regulation on management accounting of production costs; in the document flow scheme; to the accounting registers, as well as the formation of instructions for the use of tools in practice and training of personnel.

Incorporation of the tools proposed by the author into the complex fabric of management will significantly increase the level of awareness of those who make management decisions and develop strategies to prevent the loss of material resources, to ensure the rational use of internal resources in the value stream, to find and identify effective ways to use existing production reserves.

Table 1. Statement of accounting of unused materials for September 2019 (*fragment*) (RUB).

№ p/n	Resource name	Code	Not used in production	Intra-plant movements	Warehouse operations	Total return to a warehouse at the beginning of the month line 7 = line 4 =	Send to production	Send on side	Selling costs	Total sales line 11 = line 9 = line 10	Disposal costs	Total recycling line 13 = line 7 line 10 + line 12	Stock balance at the end of the month line 14 = line 7 - line 8 - line 11- line 13
1	2	3	4	5	6	7	8	9	10	11	12	13	14
1	Motor oil	0551	2890	60	980	3930	-	1445	50	1495	-	-	1395
2	Technical fabric	8976	1330	60	20	1410	1410	-	-	-	-	-	-
Total			4220	120	1000	5340	1410	1445	50	1495	-	-	1395

Source Developed and compiled by the authors.

Further studies, according to the authors, should be aimed at improving the material and technical supply of production which is based on a comprehensive analysis of the applicability of specific groups and individual nomenclature items in various orders, the use of resource consumption forecasts, or consolidated norms that summarize data on the use of homogeneous materials in certain types of repair works.

4 Conclusion

Of course, the concept of lean manufacturing, by definition, is not an algorithmic fund that is instantly tuned to a specific problem of a particular enterprise. At the same time, it does not require proof that its ideology is a catalyst for the development of diverse, effective methods of economic calculations, including management accounting. It is an important source for the formation of high-quality information space for decision-making and strategy development. Algorithmization of the movement of unused material resources in production, accounting, calculation, and information support of this process helps to see the connections between the elements of this complex process, timely localize emerging problems or losses, generate management actions aimed at reducing and preventing unproductive costs.

References

1. Balashova, E.S., Gromova, E.A.: Genesis and systematization of the results of implementation of lean production in various sectors of the Russian economy. Bull. Perm Univ. **3**(26), 61–68 (2015)
2. Uraev, N.N.: Classification of production losses and recipients of the economic effect of their elimination. Bull. Econ. Law Sociol. **4**, 93–96 (2014)
3. Akhmadeeva, G.Ch., Antonova, I.I., Antonov, S.A.: Approaches to reducing production losses at mechanical engineering enterprises. Bus. Educ. Law. Bull. Volgograd Bus. Inst. **3** (32), 47–52 (2015)
4. Marchvinski, C., Shuk, D.: Illustrated Lean Glossary, 123 p. Alpina Publisher (2005)
5. Levinson, W., Rerik, R.: Lean Manufacturing: A Synergetic Approach to Reducing Waste, 272 p. Publishing House: Standards and Quality (2007)
6. Wumek, J.P., Jones, D.T.: Lean Manufacturing: How to Get Rid of Waste and Make Your Company Thrive, 478 p. Alpina Business Books (2005)
7. Mann, D.: Lean Management of Lean Production, 208 p. Publishing House: Standards and Quality (2009)
8. Safiullin, N.Z., Kushimov, A.T.: Economic analysis of losses in the organization of machine-building production. Bull. KSTU n.a. A.N Tupolev **2**, 333–335 (2012)
9. Myshovskaya, L.P.: Determination of the main types of production losses during the reproduction of real estate by redistribution and at the stages of the life cycle. Constr. Arch. **1**, 95–102 (2008)
10. Grodzensky, S.Y., Emanakov, I.V., Ovchinnikov, S.A., Grudzinsky, P.V.: Methodology for identifying losses in the production system. Innov. Inf. Commun. Technol. **1**, 94–98 (2017)
11. Potsebneva, I.V., Lepeshki, V.I., Aleksandrov, D.O., Alaa, Sh.: Analysis of lean production systems aimed at eliminating losses when working with production equipment. In: Collection of scientific articles of the 2nd International Scientific Conference of Advanced Developments of Young Scientists, vol. 5, pp. 139–143 (2017)
12. Uraev, N.N.: Classification of production losses and recipients of obtaining economic effect from their elimination. Bull. Econ. Law Sociol. **4**, 93–96 (2014)
13. Lashmanova, Yu.Yu.: The system of indicators for assessing the production block of the production potential of the enterprise. Russ. Entrepreneurship **18**(7), 1105–1126 (2017)

Logistic Solutions for the Efficiency of Urban Transport in Nizhny Novgorod

Zhanna V. Smirnova(✉), Marina L. Gruzdeva, Zhanna V. Chaykina, Svetlana N. Kuznetsova, and Marina V. Artemyeva

Minin Nizhny Novgorod State Pedagogical University, Nizhny Novgorod, Russia
{z.v.smirnova,gruzdeva_ml}@mininuniver.ru

Abstract. The article discusses the efficiency of logistics solutions for the operation of urban transport routes. For practical implementation, the approaches of logistic solutions to the efficiency of route transport are proposed, a study of the organization of passenger traffic in Nizhny Novgorod was carried out according to the author's methodology, which consists in determining the technical and operational indicators of the rolling stock: technical and operational indicators of bus routes; determination of possible passenger traffic; calculation of a productive program for the operation of rolling stock; indicators for assessing the quality of passenger transportation. The main elements of the activity of a motor transport enterprise, which have an impact on market competitiveness, economic efficiency, and the quality of transport services for the population, are highlighted. The market and the demand of the population for passenger transportation have been analyzed. Logistic approaches and methods are used in the organization of urban transport, which make it possible to increase the efficiency of movement of the population at the right time with minimal costs of all types. The developed author's technique of technological calculation with an economic justification can be used in the practice of developing the urban transport system in Nizhny Novgorod.

Keywords: Logistics solution · Transport routes · Technical and economic indicators · Volume of transportation

JEL Codes: R01 · F22 · F63 · J15 · J61 · O15

1 Introduction

Nizhny Novgorod occupies a geographically advantageous area and has a fairly extensive transport infrastructure. One of the socially significant industries is the urban passenger transport services. Currently, intracity passenger transportation is carried out by land electric transport, road transport, metro, as well as water transport on the territory of the city [1]. The stable functioning of the transport system ensures the normal city life, the performance of enterprises, organizations and institutions, the

E. G. Popkova (Ed.): ISC 2020, LNNS 368, pp. 640–648, 2022.
https://doi.org/10.1007/978-3-030-93244-2_70

opportunity for citizens to receive goods and services necessary to maintain life, health, intellectual and spiritual level.

Large volume of passengers' transportation using buses requires competent planning and organization of the transport logistics process on the routes.

Since July 2017 [4], the complex transport scheme (CTS), which is the division of public transport routes into mainline and supply, has begun to be introduced in Nizhny Novgorod. Large and extra-large capacity buses operate on the mainline routes passing through the main transport routes of the city that allows to cope with a large volume of traffic, usually municipal transportation. The supply routes are required to ensure the delivery of passengers from remote areas of the city or suburb to stopping points through which mainline routes pass.

The CTS also provides for transfers for passengers within a certain time (60–90 min), which allows using buses for the cost of one ticket.

The low efficiency of buses on the route and the comparable level of transport services for the population determines the relevance of the selected research topic.

2 Methodology

In order to achieve the objective, it is necessary to analyze methods of increasing the efficiency of buses, foreign experience in the organization of land transport, to determine the current technical and operational performance of the rolling stock, to determine the current passenger traffic and draw up its distribution diagrams by hours of the day, days of the week and months of the year, to develop a program for the operation of buses, to draw up a bus schedule and work schedules for drivers, to determine the final technical and economic indicators of the route, to consider the issues of dispatch control and ensuring the safety of the transport.

Using the most appropriate methods to improve the efficiency of bus use is reflected in the studies of domestic scientists in the field of road transport. The quality of transport services for the population is an important element of the activity of a motor-transport enterprise which has an impact on market competitiveness and economic efficiency [3]. Among the many such methods, the most successful ones should be noted, in our opinion: flexible combined routes are organized in the evening; method of routes on duty - at the end of the evening rush hours, some routes are closed, and on the rest ones there is a high frequency of buses; method of mixed routes - buses of various capacities, medium and small buses (route taxis) are sent on the routes, giving the right to choose a passenger [7].

General methods of analysis and comparative analysis aimed at ensuring the effective use of all organizational, economic, technological and social resources of the organization of operational control transport were used as the methodological basis of the study. Achievement of this goal allows us to meet the needs of the population in transportation in a high-quality, timely manner and in full [5]. For the practical implementation of the proposed approach of logistics solutions to the efficiency of route transport a study of the organization of passenger traffic in Nizhny Novgorod was carried out according to the author's methodology, which consists in determining the technical and operational indicators of the rolling stock: technical and operational indicators of bus routes; determination of possible passenger traffic; calculation of a productive program for the operation of rolling stock; indicators for assessing the quality of passenger transportation.

3 Results

The study of the process of logistics solutions in passenger transport also provides for the introduction of ways to improve the efficiency of buses on the route №210 «Tunnel of Moskovsky railway station - Berezovaya poyma village» in Nizhny Novgorod. The following is a route indicators table (Table 1).

Table 1. Technical and operational indicators of the bus route

Indicator	Value
Length of the route «Tunnel of Moskovsky railway station - Berezovaya poyma village», km	24,18
Length of the route « Berezovaya poyma village - Tunnel of Moskovsky railway station», km	24,29
Length of the circle, km	48,47
Trip time, min	59,25
Circulating trip time, min	118,5
Average technical speed, km/h	36,35
Communication speed, km/h	31,03
Operational speed, km/h	26,78
Mileage utilization rate	0,9
Capacity utilization rate	0,67
Average travel distance of one passenger, km	15,6

Source Developed and compiled by the authors

The investigated route is a pendulum one and it complies with all the rules and regulations for the organization of passenger transportation. The route passes through the Moskovsky and Kanavinsky districts of the city of Nizhny Novgorod, and also, in part, through the territory of the urban district of the city of Dzerzhinsk.

Names of streets and highways in the forward direction:

N. Novgorod: Moscow highway; highway Volga Moscow - Vladimir - Nizhny Novgorod - Kazan - Ufa; Berezovaya poyma village: Luchistaya str.

Names of streets and highways in the backward direction:

Berezovaya poyma village: Luchistaya str.; N. Novgorod: highway Volga Moscow - Vladimir - Nizhny Novgorod - Kazan - Ufa; Moscow highway.

The transport mobility of the population in the city leads to the formation of passenger flows with different power and direction.

The study of passenger traffic allows identifying their direction, time distribution, route length [10].

Rational and efficient use of rolling stock is impossible without a systematic study of passenger traffic on the route.

It is possible to select the kind and type of transport, as well as the required number of rolling stock units, only if you have data on the direction and size of passenger traffic.

Passenger flows are characterized by [2, 6]:

- Capacity, i.e. the number of passengers passing at a certain time through a specific section of the route or the entire transport network of the locality in one direction (Q_{max}).
- The tension in individual sections of the route or in general along its length, as well as the number of passengers transported for each section of the route per unit of time, in the forward and backward directions of bus traffic.
- The volume of traffic, i.e. the number of passengers transported in total along the route or route network per unit of time in the forward and backward directions [8].

Basically, the following methods are used in passenger traffic exploration:

- questionnaire method (survey questionnaires which are filled in in process of working on the route are built);
- coupon method (specially prepared multi-colored coupons are issued to accountants);
- visual or eye method (visually determining the fullness of the cabin according to the point system);
- survey method (the accountant interrogates passengers in the cabin to determine their correspondence);
- the reporting and statistical method is based on ticket-registration sheets and the number of tickets sold [9];
- tabular method (prepared tables are filled in by the accountants located inside the bus near each door).

The analysis of operational data on the number of passengers carried (reporting and statistical method) has been conducted, as well as a survey method has been applied among the population to determine the volume of traffic on the route in this study [9].

As a result of these events, it was revealed: the number of passengers transported per day (Sunday) - 4772, passengers per month (July 2019) - 119375.

The maximum passenger traffic capacity is $Q_{max} = 453$ passengers per hour, $Q_{maxд} = 4472$ passengers per day, $Q_{maxм} = 119{,}4$ thousand passengers per month.

Table 2 shows the distribution of passenger traffic by hours, then Table 3 shows the coefficient of unevenness by hours.

Table 2. Division of passenger traffic by hours

Hours of the day	Number of passengers passengers/h	Hours of the day	Number of passengers passengers/h
05–06	130	14–15	130
06–07	365	15–16	157
07–08	453	16–17	193
08–09	411	17–18	287
09–10	385	18–19	365
10–11	289	19–20	379
11–12	193	20–21	396
12–13	164	21–22	273
13–14	97	22–23	104
Total			4472

Source Developed and compiled by the authors

Table 3. Coefficient of unevenness by hours

Hours of the day	$K_{н}$	Hours of the day	$K_{н}$
05–06	0,29	14–15	0,29
06–07	0,81	15–16	0,35
07–08	1,00	16–17	0,43
08–09	0,91	17–18	0,63
09–10	0,85	18–19	0,81
10–11	0,64	19–20	0,84
11–12	0,43	20–21	0,87
12–13	0,36	21–22	0,60
13–14	0,21	22–23	0,23

Source Developed and compiled by the authors

Passenger traffic diagrams are shown in Fig. 1.

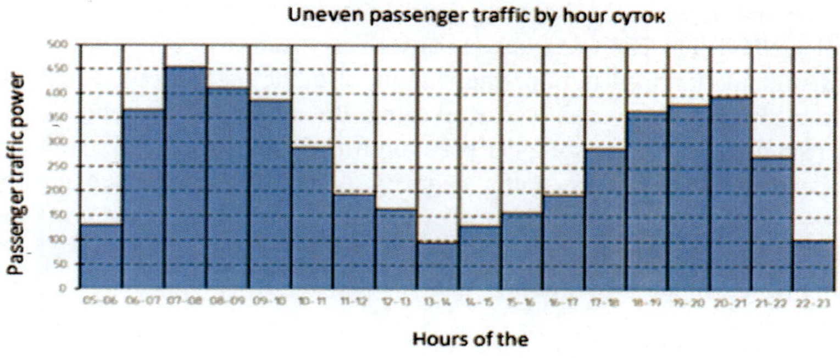

Fig. 1. Unevenness of passenger traffic by hours of the day. *Source* Developed and compiled by the authors

From the Figs. 2, 3, 4 of passenger flows it can be seen that the greatest need for the carriage of passengers arises:

1. On weekends, as well as the day preceding and following them.
2. In the summer months, as well as the month preceding and following them.

Despite the fact that the route is suburban by definition, in essence it is radial, i.e. connects the outskirts of the city (Berezovaya poyma village belongs to the Moskovsky district of the city of N. Novgorod) with a large hub of the city (Moskovsky railway station) [2].

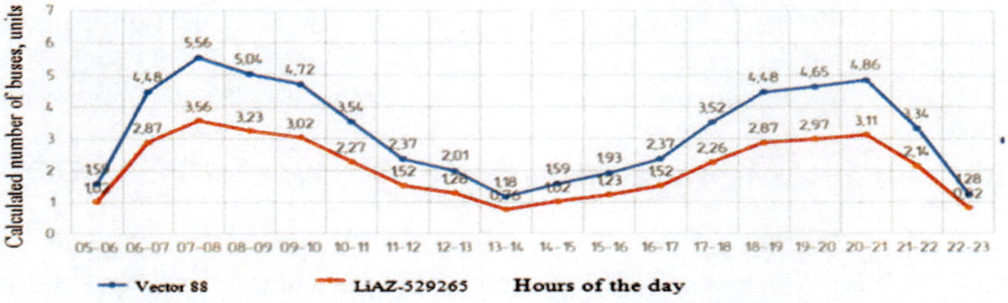

Fig. 2. Calculated number of buses. *Source* Developed and compiled by the authors

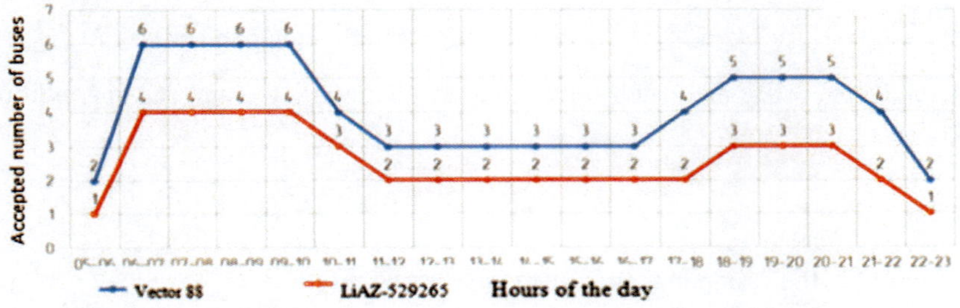

Fig. 3. Accepted number of buses. *Source* Developed and compiled by the authors

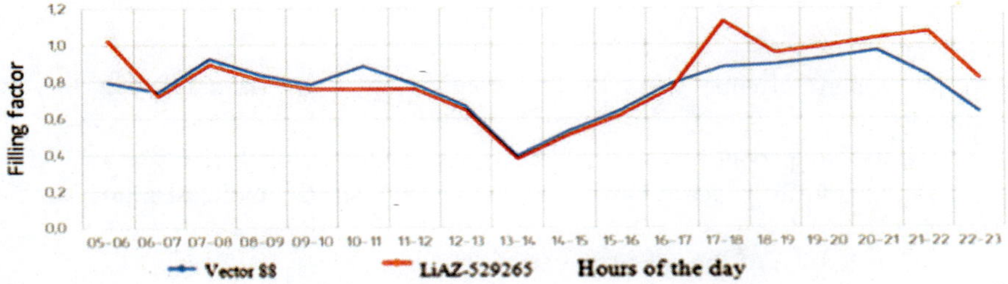

Fig. 4. Filling factor. *Source* Developed and compiled by the authors

Passenger transportation is carried out by various types of buses differing in a whole mass of parameters, and it is the choice of the most suitable rolling stock for a specific route that allows effective providing the population with transport services at minimal costs.

The choice of the kind and type of buses depends on:

А) Passenger traffic capacity in one direction;
Б) Uneven distribution of passengers by hours of the day;
В) A reasonable interval for buses by the hour of the day;
Г) Cost of bus transportation;
Д) Road conditions for bus traffic and street capacity.

It is possible to start choosing a rational type of bus actual data on the study of passenger traffic on the route.

A suburban bus should be chosen to serve passengers because the optimized route is suburban.

Let's carry out a comparative analysis between buses PAZ-320414 «VEKTOR 8.8» and LiAZ-5292.65.

As a result of computational study, it was revealed: one of the quality indicators is the coefficient $K_к$, which is defined as the ratio of the amount of time spent on the trip under the given theoretically favorable travel conditions t^3_{per} to the actual time spent on the trip in real conditions t^f_{per}:

$$K_k = t^3_{per} \div t^f_{per}$$

The average daily estimated travel time in theoretically comfortable conditions is determined by the formula:

$$t^3_{per} = 12,25 + \frac{3}{1,2} + 0,17\sqrt{C}$$

where C is the residential (built-up) area of the city.

It is recommended to determine the indicator of the quality of transport services in cities according to the formula:

$$K_K = \frac{t_n}{t_f} \times \frac{y_n}{y_f} \times R$$

where:

t_n – standard of time spent by a passenger on a trip, min (40 min for these conditions);

t_f – actual time spent on the trip, min;

y_n – standard filling factor. On average, no more than 0.4, and at rush hour no more than 0.8.

y_f – actual value of the filling factor;

R – indicator of regularity of the movement (for suburban routes it was adopted at 0.98).

The regularity and accuracy of traffic is a defining indicator of the quality of passenger service, it is impossible to be limited to just one indicator.

Let's calculate the quality indicators of passenger transportation on the investigated route.

Let's calculate the quality factor $K_к$ by rush hour:

$K_к = (40/33,16) \times (0,7/0,9) \times 0,98 = 0,91$ (Exemplary).

Let's calculate the quality factor $K_к$ on average per day:

$K_к = (40/30,16) \times (0,4/0,81) \times 0,98 = 0,64$ (Good).

Based on the data obtained from the calculations of the study, let's build radar charts showing the level of quality of passenger service (Figs. 5 and 6).

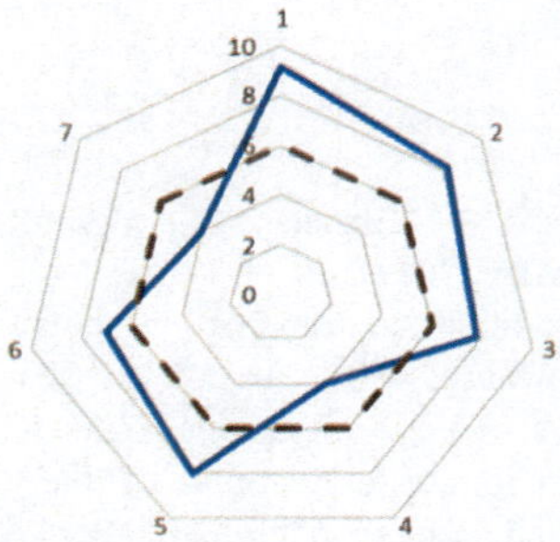

1. Technical speed 2. Communication speed 3. Operating speed 4. Movement interval 5. Travel time 6. Technical readiness factor 7. The degree of unevenness of passenger traffic

Fig. 5. Radar chart №1. *Source* Developed and compiled by the authors

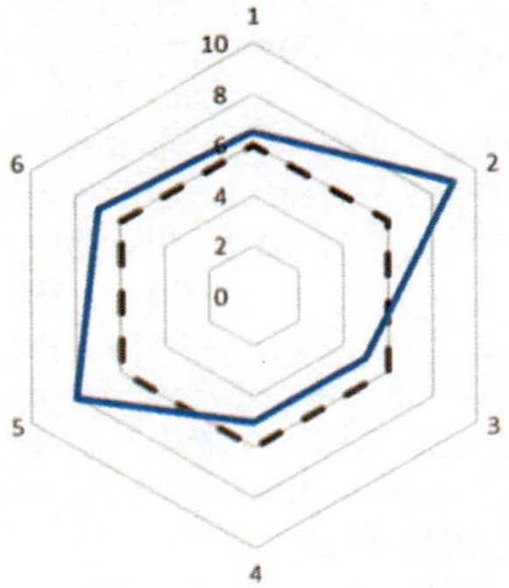

1. Filling factor 2. Straightness of the route 3. Cost of transportation 4. Profitability 5. Quality factor at rush hour 6. Quality coefficient on average per day

Fig. 6. Radar chart №2. *Source* Developed and compiled by the authors

4 Conclusion

Clear calculated planned logistics solutions, market analysis and population demand for passenger transportation are needed for the further development of passenger transport in Nizhny Novgorod.

The use of logistic approaches and methods in organizing urban transport makes it possible to increase the efficiency of movement of the population at the right time with minimal costs of all kinds.

The developed author's methodology of technological calculation with an economic justification can be used in the practice of developing the urban transport system in Nizhny Novgorod.

References

1. Aleksakhin, C.B., Nikolaev, A.B., Kuznetsov, I.A.: Automated Control Systems for Road Transport, 288 p. ACADEMIA (2011)
2. Bochkarev, A.A.: Logistics of urban transport systems: textbook. In: Bochkarev, A.A., Bochkarev, P.A. (eds.) Allowance, 150 p. Yurayt, Moscow (2018)
3. Garina, E.P., Romanovskaya, E.V., Andryashina, N.S., Kuznetsov, V.P., Shpilevskaya, E. V.: Organizational and economic foundations of the management of the investment programs at the stage of their implementation. In: Popkova, E.G., Sergi, B.S. (eds.) ISC 2019. LNNS, vol. 91, pp. 163–169. Springer, Cham (2020). https://doi.org/10.1007/978-3-030-32015-7_20
4. Gerami, V.D.: Management of Transport Systems. In: Gerami, V.D., Kolik, A.V. (eds.) Transport Provision of Logistics: Textbook and Workshop, Higher School of Economics - National Research University, 438 p. Yurayt, Moscow (2018)
5. Isimova, N.O.: Transport logistics. In: Isimova, N.O., Shepelin, G.I. (eds.) Science Through the Prism of Time, vol. 6, no. 15, pp 72–74 (2018)
6. Minko, R.N.: Technology of Transport Processes: Textbook (Ed. By, R.N. Minko, A.I. Shaposhnikov), 120 p. Direct-Media, Moscow (2016)
7. Smirnova, Z.V., Vaganova, O.I., Chanchina, A.V., Koldina, M.I., Kutepov, M.M.: Development of research activity of future economists in the university. In: Popkova, E. G. (ed.) Growth Poles of the Global Economy: Emergence, Changes and Future Perspectives. LNNS, vol. 73, pp. 371–379. Springer, Cham (2020). https://doi.org/10.1007/978-3-030-15160-7_38
8. Smirnova, Z.V., Kuznetsova, E.A., Koldina, M.I., Dyudyakova, S.V., Smirnov, A.B.: Organization of an inclusive educational environment in a professional educational institution. In: Popkova, E. (ed.) Growth Poles of the Global Economy: Emergence, Changes and Future Perspectives. LNNS, vol. 73, pp. 1065–1072. Springer, Cham (2020). https://doi.org/10.1007/978-3-030-15160-7_107
9. Smirnova, Z.V., Kamenez, N.V., Vaganova, O.I., Kutepova, L.I., Vezetiu, E.V.: The experience of using the webinar in the preparation of engineering specialists. Amazonia Investiga **8**(18), 279–287 (2019)
10. Yakunina, N.V.: Transportation of passengers by road. In: Yakunina, N.V., Yakunin, N.N. (eds.) Workshop: Tutorial, 126 p. Ministry of Education and Science of the Russian Federation, Orenburg State University: OSU, Orenburg (2017)

Hybrid Quantum-Classical Information Security in Video Surveillance Systems

Elena I. Shevaldina(✉)

Ufa State Petroleum Technological University, Ufa, Russia

Abstract. The article deals with the relationship between economic and information security, ensuring the protection of information in state and municipal communication channels and studying the possibility of using the method of quantum-classical information security for solving practical problems. This study is the result of testing in the existing telecommunications infrastructure of the Ufa Public Security Center of the prototype of the Data Transmission Complex with hybrid quantum-classical protection (hereinafter referred to as the Complex), developed at the Federal State Autonomous Educational Institution of Higher Education "St. National Research University of Information Technologies, Mechanics and Optics" (ITMO University) with the Kazan National Research Technical University named after A.N. Tupolev - KAI (hereinafter the "Developer"). The tests are carried out by the Developer and the Ufa Public Security Center on an initiative basis in order to study the functional and technical characteristics of the Complex for ensuring information security in municipal information systems while passing large amounts of information through communication channels. An online video stream from video cameras operating on the territory of Ufa, included in the video surveillance subsystem of the hardware and software complex (hereinafter referred to as the APC) "Safe City", are used as the transmitted data.

Keywords: APK "Safe City" · Information protection in municipal information systems · Quantum channel · Quantum signal · Data transmission complex with hybrid quantum-classical protection

JEL Codes: C93 · H56 · L63 · L86

1 Introduction

Information becomes the main resource of a developed state in a modern post-industrial society [1]. The provision of information capabilities has become the most important condition for the existence of the state as a whole and individual sector of the economy, the functioning of state and municipal authorities, individual enterprises, and institutions. The formation of the modern digital economy is inextricably linked with ensuring its safe functioning in the conditions of increasing threats of unauthorized impact on the digital environment, management processes, databases, and other digital reality. Everything related to the receipt, accumulation, storage, processing, and use of information may be subject to encroachments from unfair competitors with the aim of stealing, using, changing, or destroying information [2]. Therefore, ensuring

E. G. Popkova (Ed.): ISC 2020, LNNS 368, pp. 649–657, 2022.
https://doi.org/10.1007/978-3-030-93244-2_71

information security has become not just a new aspect of economic security, but the most important condition for the existence of the economy [3].

The dissemination of information occurs in the information environment, which, among other things, consists of hardware and software for information transmission. Communication channels have become the most important component of the information environment; they are also the most vulnerable from the point of view of information protection [4]. The hybrid quantum-classical method of cryptography can significantly contribute to the protection of information in communication channels, especially in the channels of government and the banking sector [5]. Its approbation in practice in the context of the need to encrypt and decrypt significant amounts of information was facilitated by the deployment of the Complex on the infrastructure of the agro-industrial complex “Safe City” in the city of Ufa [6].

APC “Safe City” contributes to solving complex problems of ensuring public and economic security, law, and order and safety of the environment at the municipal, regional, and federal levels. However, its normal functioning is impossible without solving complex problems of information protection, because the digital streams circulating in its subsystems are information of limited access. The need to protect the information in telecommunication networks of the APC “Safe City” is determined by a number of federal and regional regulatory documents [7].

2 Methodology

2.1 Basic Abbreviations and Concepts

QBER is the quantum bit error rate [8];

QCSF - quantum cryptography at side frequencies - a method of sending quantum encryption keys with coding at side frequencies of modulated radiation [9];

SQC is a system of quantum communication.

Sender block is a hardware component of the SQC that generates and sends quantum signals over a quantum channel in an optical network.

The receiver block is a hardware component of the SQC that performs registration of quantum signals transmitted over a quantum channel in an optical network.

Complex - a data transmission complex with hybrid quantum-classical protection.

A quantum channel is a physical medium for the transmission of quantum signals. The channel must not contain any other optical signals. The channel shouldn’t be amplified or otherwise processed by the hardware.

A quantum key is a random binary sequence, which is generated and processed in the sender and receiver blocks of the SQC using secure quantum communication methods.

A quantum signal is a weak optical pulse (1550 nm) with average energy in time counting not more than the energy of a single photon, used as an information carrier by quantum communication systems as part of the Complex.

Quantum bit error rate is the number of errors that occur in the process of quantum communication. It’s used to control access to the quantum channel.

A sync channel is a physical transmission medium for a sync signal.

A synchronization signal is a strong optical signal (1550 nm), transmitted between the blocks of the complex, intended for clock synchronization of the operation of the system blocks.

2.2 Test Object

The test object is a prototype of a data transmission complex with a hybrid quantum-classical protection. The complex is a network gateway designed to organize a transparent tunnel encrypted with AES-256 keys, which is generated by the quantum communication system [10].

A prototype of the Complex that is presented for testing, consisting of:

1. SQC at the side frequencies of the modulated radiation (1 set);
2. SQC software preinstalled in the sender and recipient blocks;
3. Software for creating an encrypted network tunnel, preinstalled in the sender and receiver blocks.

Completeness of the SCC:

- The sender unit is a PC running Linux (Nau Linux) operating system in a 19-in. 3-unit case 550 mm deep and built-in hardware designed to implement the distribution of quantum keys using the QCSF method. The unit contains two network interfaces (Gigabit Ethernet) for receiving and transmitting encrypted and decrypted information and two fiber-optic outputs (FC/APC) for transmitting quantum and synchronization signals over a standard telecommunication fiber passing in the fiber-optic ring of the "Public Security Center of Ufa", to the side of the recipient's block.
- The recipient unit is a PC running Linux (Nau Linux) operating system in a 19-in. 3-unit case 550 mm deep, built-in hardware designed to implement the distribution of quantum keys using the QCSF method. The receiver unit includes a separate module for a single photon detector (with dimensions 380 * 255 * 170 mm), designed to register a quantum signal. The unit includes two network interfaces (Gigabit Ethernet), two fiber-optic inputs (FC/APC) for connecting quantum and synchronization channels, and one fiber-optic output for connecting a single-photon detector.
- Fiber-optic filter FWDM-TV1550-P(1520-1600)-R(1260-1500)-733, 2 pcs. Since a fiber-optic channel is applied to deliver the synchronization signal, which has already been used for data transmission at wavelengths of 1270 and 1330nm, SQCs are equipped with a pair of spectral fiber filters that implement the function of multiplexing and demultiplexing signals.
- The units are equipped with input/output means (monitors, mice, keyboards) required for carrying out commissioning activities.

2.3 Test Site

The network infrastructure at the disposal of the Ufa Public Security Center is a network with a ring topology, which includes 10 nodes. The nodes of the ring are connected by two fibers, one of which is currently used for data transmission at wavelengths of 1,270 nm and 1,330 nm, the second "dark" fiber isn't used for data transmission. Dynamic routing in the network is organized using the OSPF protocol. The main source of traffic in the network is the cameras of the video surveillance subsystem of the APC "Safe City", mounted on the branches of the nodes and using both TCP and UDP protocols at the transport level. The total network traffic at the time of testing didn't exceed 100 Mbit/s.

For the installation, there were selected nodes, which are located at the following addresses:

- 1 Pobedy St., Ufa (hereinafter Node No. 1)
- 54 Ulyanovs St., Ufa (hereinafter Node No. 2)

The choice of nodes is due to the need to have a direct optical connection without re-switching and repeaters and the ability to organize access to server nodes for installation.

The distance between the nodes is 1.24 km. and the optical loss is 0.5 dB.

2.4 Connection Diagram

A "dark" optical fiber is used as a quantum channel. The synchronization signal of the Complex units is brought into the data fiber using a WDM filter and analogically branched off on the side of the receiving unit. The diagram is shown in Fig. 1.

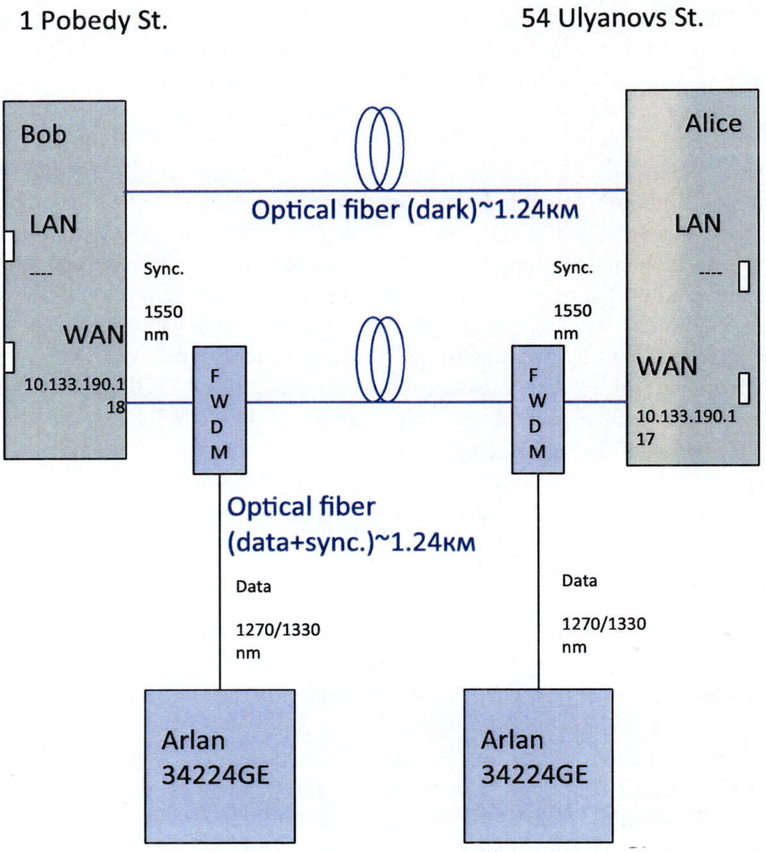

Fig. 1. Optics connection diagram. Source: compiled by the author

The network cards of the units are connected to the network switches of the nodes using a UTP cable with RJ-45 connectors.

The managed switches of the Arlan-3424GE nodes are configured according to the scheme, which is shown in Fig. 2. With such a network organization in the nodes, all incoming data passes through the Complex, where they are encrypted or decrypted using quantum keys, depending on the direction of packet movement.

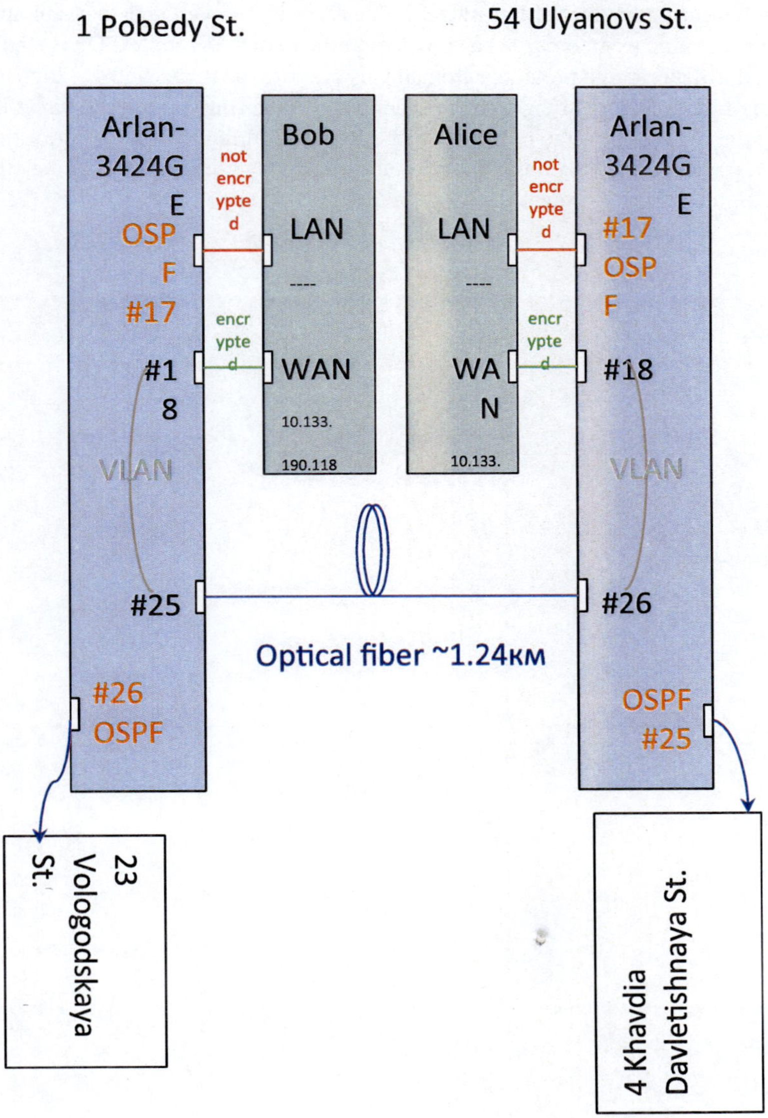

Fig. 2. Diagram of network connections. Source: compiled by the author

3 Results

The distribution of quantum keys between the nodes was launched during the installation. The average number of photons per modulation clock according to the protocol is 0.2/tact. The average value of the rate of generation of quantum keys was 7 Kbit/s with an average value of the quantum error rate of no more than 3.5% with a loss in the quantum channel of 0.5 dB and a distance of 1.24 km. This generation rate allows the 256-bit encryption key to be changed up to 27 times per second.

After reconfiguration of the switches in accordance with the diagram in Fig. 2, insignificant traffic went through the network interfaces of the Complex. Due to the fact that the dynamic data routing protocol automatically chooses the routes with the highest speed, the bulk of the traffic went through the second leg of the ring. In order to direct all traffic through the interfaces of the complex, without reconfiguring the routing priorities in the network, it was decided to physically break the second leg of the ring. After that, the entire volume of traffic went through the encrypted section and totaled 80 Mbit/sec (Fig. 3).

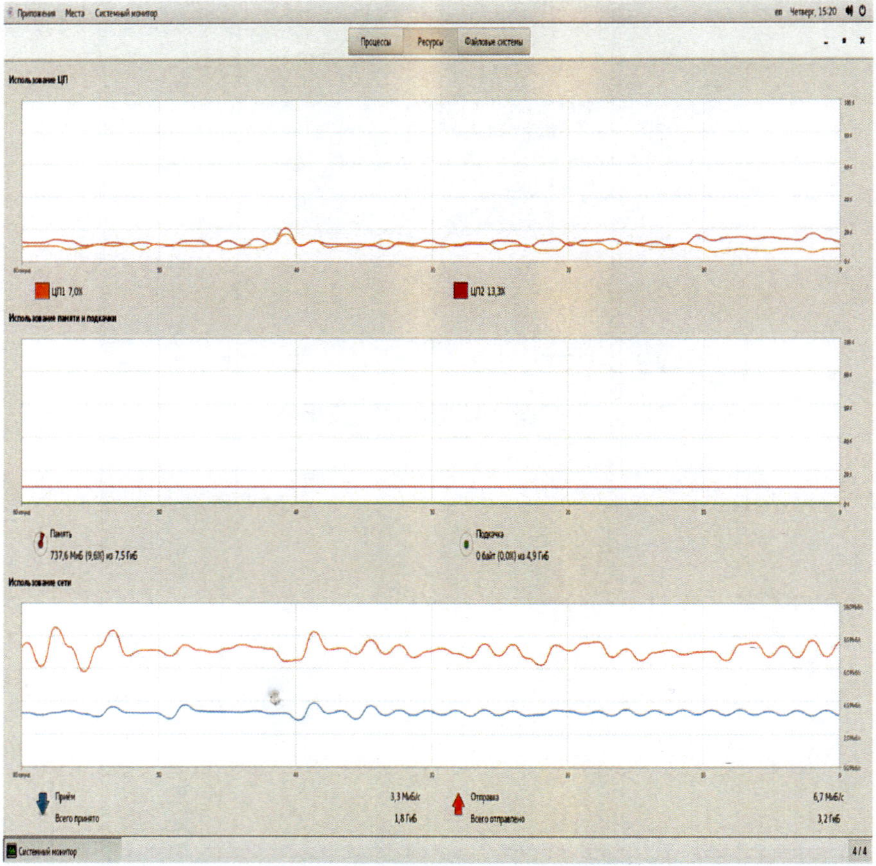

Fig. 3. Screenshots of the system resource monitor in load mode. Source: compiled by the author

At the same time, the work of the network wasn't disrupted; the data from the cameras successfully reached the addressees. With this amount of traffic, each 256-bit key block is used to encrypt 2963 Mbit of data and then deleted it.

When the quantum channel is broken, the quantum error rate exceeds the control value of 11% and is set at an average value of 50%. The generation of quantum keys and their sending to the software to create the encrypted network tunnel is stopped. When the supply of keys is exhausted, data transmission through the network tunnel is suspended or encryption is continued with the last key, depending on the setting. The main indicators obtained as a result of the tests are presented in Table 1.

Table 1. Key characteristics based on measurement results

Parameter, units	Value
Line length, km	1.24
Line loss, dB	0.5
Average number of photons per modulation clock	0.2
Quantum key generation rate, kbps	7
Quantum error rate, not more than %	3.5
Traffic, Mbps	80

Source: compiled by the authors

4 Conclusion

The studies carried out led to the following conclusions:

1. A software reconfiguration of the node switches is required to route traffic through the blocks of the Complex to embed the Complex into the infrastructure of the municipal information subsystem of video surveillance of the APC "Safe City" in the territory of Ufa.
2. To organize the quantum and synchronization channels, the Complex requires the presence of two fibers between the nodes of the protected section, "dark" fiber, and fiber with a free WDM-window at a wavelength of 1550 nm. In this regard, an additional fiber-optic channel is required to deploy the Complex at a number of infrastructure sections.
3. The possibility of building secure communication channels in the operating information subsystem of video surveillance of the APC "Safe City" on the territory of Ufa in optical cables using hybrid quantum-classical protection was shown. A hybrid system with a quantum key distribution module, in contrast to the classical one, allows you to change encryption keys dozens or more times per second, while controlling compromise in the process of key distribution. Due to the fact that the cryptographic strength of the cipher is entirely determined by the secrecy of the key (the Kerkhoffs Principle); this approach allows achieving exceptional resistance to cracking. If an attacker attempts to infiltrate the quantum key distribution channel, or if there is a problem with the quantum distribution module, the system switches to the classic encryption mode without changing the keys. Today, it's required to

work out a mechanism for notifying the information system control center about changes in the operating modes of the distribution module.

4. In the process of testing the Complex of violations of the functioning of the information subsystem of video surveillance of the APC "Safe City" wasn't revealed. According to the test results, it was confirmed that the actual operating modes correspond to the parameters of the Complex declared in the documentation.
5. It was confirmed that the Developer has a scientific and technical groundwork in the area of secure quantum communications, implemented in the form of a prototype network gateway designed to organize a transparent tunnel encrypted with AES-256 keys, which are generated by a quantum communication system at the side frequencies of modulated radiation. This technology has the potential to become an effective information security tool to ensure the confidentiality of transmitted information in public communication networks (PCN).
6. The expediency of further work on the development of quantum methods and data transmission complexes with hybrid quantum-classical protection, the importance of further development of this technology and devices based on it have been confirmed.
7. With regard to the existing video surveillance system of the APC "Safe City", operated by the "Public Security Center of Ufa", the Complex can't be considered as equipment that protects information at this moment, for the following reasons:
   - High cost of equipment;
   - Impossibility to protect information channels from video cameras to fiber-optic ring nodes due to the absence of a second "dark" optical fiber;
   - The presence of 10 switches in the fiber-optic ring and the absence of a direct optical connection along the length of the entire ring without re-switching and repeaters, which is critical for the functioning of the Complex.

References

1. Information and legal portal "Consultant Plus". Federal Law "On Information, Information Technologies and Information Protection" dated July 27, 2006 No. 149-FZ. http://www.consultant.ru/document/cons_doc_LAW_61798/. Accessed 22 Feb 2021
2. Skoudis, E.: Opposition to Hackers (Ed. by, Skoudis), 512 p. DMK Press (2003)
3. Yasenev, V.N.: Information Security in Economic Systems: Textbook. Publishing House of NNSU, Novgorod (2006)
4. The portal of the Associate Professor of the Department of Information Technologies and Systems of the Far Eastern State University of Railways V.V. Anisimov. https://www.sites.google.com/site/anisimovkhv/home. Accessed 22 Feb 2021
5. Internet portal Tadviser.ru. Article "Data transmission complex with hybrid quantum-classical protection in a telecommunication network". https://www.tadviser.ru/index.php/. Accessed 22 Feb 2021
6. Information security portal "SecurityLab". https://www.securitylab.ru/. Accessed 22 Feb 2021

7. Information and legal portal "Garant.ru". Order of the Government of the Russian Federation of December 3, 2014 No. 2446-r «On the Concept of Russian Federation "About the Concept of construction and development of the hardware and software complex "Safe City". https://www.garant.ru/products/ipo/prime/doc/70717448/. Accessed 22. Feb 2021
8. Internet portal "N + 1". Article "Quantum technologies. Module 5". https://nplus1.ru/material/2020/02/06/course-quantum-technology-chapter5. Accessed 22 Feb 2021
9. Kozubov, A.V., et al.: Fundamentals of Quantum Communication: Part 1.- SPb, 85 p. ITMO University (2019)
10. Egorov, V.I., Gleim, A.V., Rupasov, A.V.: System of quantum key distribution at subcarriers of modulated radiation with compensation for signal distortions. Uchenye zapiski KFU. Phys. Math. **155**(1) (2013)

Modeling Employment and Labor Productivity Estimates from a Sustainability Perspective

Guzel A. Salimova[1(✉)], Natalya A. Shaikhutdinova[2], Yuliya A. Pavlova[2], Yuliya A. Zapolskikh[1], and Gulnara S. Muchametshina[3]

[1] Bashkir State Agrarian University, Ufa, Russia
[2] Ufa State Petroleum Technological University, Ufa, Russia
[3] Birsk Branch of the Bashkir State University, Birsk, Russia

Abstract. The promotion of full, efficient, and high-productivity employment and decent work for everyone is one of the most urgent goals of the sustainable development agenda developed by the United Nations. Nowadays, in the conditions of new challenges, the provision of decent work is an urgent task in all countries worldwide. This study identifies indicators and builds models characterizing the sustainable development of the country and regions. The authors analyzed the structure of the country's economy in terms of employment, wage fund, and productivity. The authors constructed models of dependence of the employment, labor productivity, and the share of labor remuneration in the population's income (these indicators are consistently interrelated from the perspective of sustainable development goals) on the factors in the Russian regions. The authors analyzed the data of the Federal State Statistics Service of the Russian Federation. The analyzed data characterize the dynamics and intensity of various processes in the labor market and production. The conclusions on the models are consistent and do not contradict economic laws. Currently, it is necessary to increase employment and labor productivity and improve the quality of employment, taking into account the preservation of public health. Therefore, special measures are needed to improve working conditions and account for costs for these purposes. The authors developed recommendations for the further development of the country and regions and defined the directions for developing research methodology.

Keywords: Labor economics · Salary · Structure · Personal income · Econometric model · Industry · Investments

JEL Codes: J2 · J3 · M2 · R1

1 Introduction

The UN Sustainable Development Goals [20] include promoting full and productive employment and decent work for all people (Goal 8). Nowadays, humanity faces new challenges. Therefore, providing a decent job is an urgent task for governments and businesses of all countries. Scholars point out that employment is an important

E. G. Popkova (Ed.): ISC 2020, LNNS 368, pp. 658–669, 2022.
https://doi.org/10.1007/978-3-030-93244-2_72

production factor of economic development. Scholars conduct studies on the empirical relationship between economic growth, labor, capital, education, health, and ecology, linking various goals of sustainable development. In close connection with employment, it is noted that labor productivity is the primary factor determining the country's long-term development [15].

Measuring sustainable development requires a multidimensional approach since this process is marked with complex relationships between the social, economic, and environmental spheres. National economic development can be expressed through various indicators [11]. Thus, the competitiveness of countries in the world market is evaluated based on the analysis of 108 statistical indicators describing the 17 Sustainable Development Goals (SDGs) [13]. A study conducted by B. Seyoum based on data from 180 countries examines the impact of social cohesion on human development [17]. The research touches on a variety of areas of social life and, in particular, employment. Thus, it explores the relationship between the population and the environment, which affects sustainable development. In this connection, the research widely touched upon environmental issues. Scholars point out that sustainable development means protecting the environment and reducing overdependence on dwindling natural resources (including primary raw materials) in the economic sector [21]. The bioeconomy helps restore the environment, stimulates economic growth, and supports jobs [8]. It also deals with industrial development from the perspective of sustainable development. The concepts of sustainable industrial development are explained [19], including investment support [3].

Turning from the various studies of sustainable development specifically to labor, productivity, and employment, it is necessary to note the research on the transition from consumption to the development of human resources by incorporating the concept of sustainability [5]. Of particular interest are the studies on gender equality as a condition for the formation of human capital, increased productivity, job creation, poverty reduction, and overall socio-economic and human development [14].

Many studies examine the aging of the population in relation to health care costs and the balance of the workforce by age [4]. The issues of motivating employees to productive work are also widely studied [20]. The relationship of labor and capital to scientific results is explored in the study by E. Erfanian, N. Ferreira, and B. Amir [7]. The analysis of the interaction between the SDGs is a growing research area. However, there is not enough research on the dependence of SDGs [6]. Therefore, we consider some goals and models and evaluate sustainable development indicators using the example of the Russian Federation.

The paper aims to identify indicators and build models characterizing the sustainable development of the country and regions. The research tasks are as follows:

- To analyze the structure of the country's economy in terms of employment, wages, and productivity;
- To build models of the dependence of the level of employment, labor productivity, and the share of wages in the population's income on factors in the Russian regions;
- Develop recommendations for further sustainable development.

2 Materials and Methods

We will assess structural changes in the economy using the Ryabtsev index of structural differences. The Ryabtsev index is calculated according to the following formula:

$$I_R = \frac{\sqrt{\sum_{i=1}^{n} (d_i^1 - d_i^0)^2}}{\sum_{i=1}^{n} (d_i^1 + d_i^0)^2} \tag{1}$$

where d_i^1, d_i^0 is a specific weight (share) of the analyzed attributes in the studied populations for the reporting and baseline periods; i – the number of gradations in the studied structures [16].

The index can be applied to any studied data set. The available scale for assessing the significance of structural differences allows us to explain the results without comparative analysis.

From the study of structure, we move to the study of regional differences with the construction of models of multiple correlations and regression analysis. These models allow us to assess the closeness of the relationship between the result and the factors. The dependency models were built using data from a population of 76 regions from all federal districts of the Russian Federation for 2015–2019 (federal cities and regions with incomplete data were excluded). We used spatiotemporal sampling [1]. The final models include four-five most significant factors. The requirements of statistical science for a load of observations on the factors are met.

Since all attributes are presented quantitatively, the degree of strength of the linear relationship between the indicators in the proposed article is determined by using various kinds of multiple and pair correlation coefficients. The directions and the degree of mutual influence of each factor on the results were estimated using regression, elasticity, and beta coefficients. The selection of factors in the correlation and regression model was made based on substantive qualitative theoretical economic analysis and based on the construction and analysis of the correlation matrix (Table 1).

Table 1. The list of resultant and factor attributes of correlation and regression models

Designation	Attribute	Theoretical justification for the inclusion of a factor trait in the model
Model of dependence of the level of employment in the Russian regions (y – employment rate, %)		
x_1	Average monthly salary (average for the previous five years), thousand $	The level of wage is an incentive for employment (the incentive function of wages is implemented). The choice of five-year aggregate periods is caused by a long-term interest in the field of employment and the possibility of obtaining an appropriate education

(*continued*)

Table 1. (*continued*)

Designation	Attribute	Theoretical justification for the inclusion of a factor trait in the model
x_2	Amount and level of investment in fixed capital (average level for the previous five years), thousand $ per capita	Investments help create new jobs and increase employment. The choice of five-year aggregate periods is conditioned by the consideration of the payback period of investments and obtaining the effect
x_3	Share of urban population, %	It is usually assumed that there are more job opportunities in cities due to the attractiveness of infrastructure and the concentration of the population in urban areas (the average share of the urban population in Russia was 74.7% in 2019)
x_4	Proportion of the working-age population, %	The composition of the population by age usually has a direct and strong influence on employment. Simultaneously, part of the working-age population may not need to be employed
Model of dependence of the level of labor productivity in the Russian regions (y – labor productivity, GRP per person employed in the economy, $)		
x_1	Average monthly wage, $	Labor productivity and wages are factors of the same level, affecting each other. They are often considered in correlation and regression analysis models and econometric equation systems
x_2	Amount and level of investment in fixed capital (average level for the previous five years), thousand $ per capita	Investments increase the ability to develop production. Therefore, they allow obtaining a more significant amount of gross value added and gross product and contribute to the better technical equipment of production
x_3	Share of industrial production in GVA, %	It allows considering the impact of output in different types of economic activity
x_4	Level of innovation activity of organizations, %	Innovative component of production and obtaining the product. Innovative solutions usually contribute to the development of production and growth of productivity

(*continued*)

Table 1. *(continued)*

Designation	Attribute	Theoretical justification for the inclusion of a factor trait in the model
x_5	Value of internal R&D costs (average level for the previous five years), $ per person employed	While being expenses of an organization, R&D expenditures also contribute to the development of enterprises
Model of the dependence of the share of wages and salaries in the total income of the population in the Russian regions (y –level of wages and salaries in the income of the population, %)		
x_1	Employment rate, %	The resulting indicator in one model is included as a factor in this model since we analyze the impact on the share of wages in the income structure of the country and regions. Wages and salaries are income from employment
x_2	GRP per person employed in the economy (labor productivity), thousand $	Labor productivity should contribute to higher wages, hence its higher share in the population's income
x_3	Value of turnover of small businesses, thousands $ per capita	One of the types of income is the income from entrepreneurial activity. Consequently, it is necessary to estimate the impact of self-employment on the structure of income
x_4	Level of demographic burden, per 1,000 people	It indirectly shows whether social payments or salary form the population's income and affect the income structure. The health of the nation, the payment of certain social benefits, the availability of free time for the employed population, and, therefore, the general well-being of the society also depend on the level of the workload in the future

Source Compiled by the authors

The significance of partial and pair correlation coefficients was checked by Student's t-test. The significance of multiple correlation coefficients and determination coefficients was checked by F-criterion [1]. The data of the Federal State Statistics Service of the Russian Federation for 2010–2019 served as the source of information at all research stages [9].

3 Results

In this research, we study the level of employment, labor productivity, and the share of wages in the population's income as indicators that are consistently interrelated from the perspective of SDGs. Let us assess the structural differences in the Russian economy by type of economic activity according to selected indicators for 2015–2019 (the dynamics since the year of the creation of the UN development goals) (Table 2).

Table 2. Values of the Ryabtsev index of structural differences in the Russian economy

Years	By gross value added		By number of people employed		By payroll	
	Index value	Characterization of the measure of difference	Index value	Characterization of the measure of difference	Index value	Characterization of the measure of difference
2015	0.031	Very low level	0.005	Equality	0.015	Equality
2016	0.030	Equality	0.010	Equality	0.016	Equality
2017	0.029	Equality	0.017	Equality	0.032	Very low level
2018	0.048	Very low level	0.007	Equality	0.019	Equality
2019	0.015	Equality	0.010	Equality	0.013	Equality
2019 to 2014	0.091	Low level	0.028	Equality	0.059	Very low level

Source Compiled by the authors

The data presented in Table 2 allow us to conclude that there were insignificant changes in the structure of the economy according to the highlighted indicators from year to year. While the structure of the economy is completely identical in terms of the number of employees, there are some differences in gross value added and the wage fund in some years. There are interdependencies between output and wages. Differences in the structure of the economy by type of economic activity in one year led to differences in wages in another year. Conversely, differences in payroll led to differences in gross value added in another year. Thus, industrial ties are stronger than social ties.

From the structure of the economy, let us move on to modeling the level of employment (characterizing the labor market), labor productivity (reflecting the level of production), and the share of wages in the income of the population (expressing the social component of the labor market and the profitability of society). Tables 3, 4, and 5 show the results of the constructed models of multiple correlations and regression analysis.

Table 3. Statistical characteristics of the obtained regression equations and estimates of the relative impact of factors on the level of employment in the dynamics

Indicators	2015	2016	2017	2018	2019
a_1 – regression coefficient with factor x_1	−0.172	−1.311	4.890	4.615	4.076
a_2 – regression coefficient with factor x_2	0.394	0.469	0.311	0.347	0.397
a_3 – regression coefficient with factor x_3	0.210	0.215	0.121	0.116	0.148
a_4 – regression coefficient with factor x_4	0.079	0.140	0.420	0.432	0.586
F_{fact} – calculated value of Fisher criterion	17.08	20.37	25.29	28.43	25.56
R – multiple correlation coefficient	0.700	0.731	0.767	0.785	0.768
R^2 – multiple determination coefficient	0.490	0.534	0.588	0.616	0.590
$Э(x_1)$ – elasticity coefficient y from factor x_1	−0.002	−0.014	0.056	0.050	0.042
$Э(x_2)$ – elasticity coefficient y from factor x_2	0.015	0.018	0.012	0.012	0.013
$Э(x_3)$ – elasticity coefficient y from factor x_3	0.227	0.233	0.145	0.139	0.179
$Э(x_4)$ – elasticity coefficient y from factor x_4	0.071	0.123	0.403	0.409	0.570
$\beta(x_1)$ – beta coefficient of factor x_1	−0.010	−0.075	0.274	0.264	0.193
$\beta(x_2)$ – beta coefficient of factor x_2	0.184	0.226	0.144	0.162	0.153
$\beta(x_3)$ – beta coefficient of factor x_3	0.617	0.662	0.380	0.385	0.427
$\beta(x_4)$ – beta coefficient of factor x_4	0.032	0.060	0.192	0.214	0.248

Source Compiled by the authors

Table 3 shows that all factors positively affect employment, except wage levels in the models of 2015 and 2016. This was expected and consistent with the laws of economics. The closeness of the relationship between factors and outcome increases over time, with a correlation coefficient of 0.768 in 2019. The employment rate is 59.0% dependent on the factors included in the model.

Table 4. Statistical characteristics of the obtained regression equations and estimates of the relative impact of factors on labor productivity in the dynamics

Indicators	2015	2016	2017	2018	2019
a_1 – regression coefficient with factor x_1	9.134	9.606	8.941	7.831	8.212
a_2 – regression coefficient with factor x_2	3.990	2.576	1.982	2,855	4.568
a_3 – regression coefficient with factor x_3	196.383	97.212	63.293	99.302	154.901
a_4 – regression coefficient with factor x_4	−167.547	−44.457	15.028	1.217	−134.672
a_5 – regression coefficient with factor x_5	0.209	0.672	0,591	1.579	2.537
F_{fact} – calculated value of Fisher criterion	147.18	176.63	245.23	229.02	174.30
R – multiple correlation coefficient	0.956	0.963	0.973	0.971	0.962
R^2 – multiple determination coefficient	0.913	0.927	0.946	0.942	0.926
$Э(x_1)$ – elasticity coefficient y from factor x_1	0.230	0.306	0.444	0.353	0.275
$Э(x_2)$ – elasticity coefficient y from factor x_2	0.490	0.470	0.400	0.409	0.492
$Э(x_3)$ – elasticity coefficient y from factor x_3	0.286	0.226	0.187	0.234	0.315
$Э(x_4)$ – elasticity coefficient y from factor x_4	−0.071	−0.026	0.018	0.001	−0.068

(continued)

Table 4. (*continued*)

Indicators	2015	2016	2017	2018	2019
$Э(x_5)$ – elasticity coefficient y from factor x_5	0.001	0.007	0.007	0.013	0.016
$\beta(x_1)$ – beta coefficient of factor x_1	0.140	0.190	0.290	0.229	0.151
$\beta(x_2)$ – beta coefficient of factor x_2	0.701	0.695	0.634	0.662	0.690
$\beta(x_3)$ – beta coefficient of factor x_3	0.223	0.176	0.146	0.182	0.224
$\beta(x_4)$ – beta coefficient of factor x_4	−0.061	−0.028	0.015	0.001	−0.054
$\beta(x_5)$ – beta coefficient of factor x_5	0.003	0.015	0.015	0.030	0.032

Source Compiled by the authors

Table 4 shows that all factors positively affect labor productivity levels, except for the level of innovation activity of organizations in the models of 2015, 2016, and 2019. Thus, during the initial introduction of innovations, it is possible that they worsen the production process and do not contribute to the growth of labor productivity. In general, the conclusions of the models are consistent and do not contradict economic laws. The closeness of the relationship between factors and the result in this model is the highest among other models, reflecting the closest relationship between production indicators than between the social ones. In 2019, the correlation coefficient was 0.962. The level of labor productivity is 92.6% dependent on the factors included in the model.

Table 5. Statistical characteristics of the obtained regression equations and estimates of the relative impact of factors on the share of wages in income in the dynamics

Indicators	2015	2016	2017	2018	2019
a_1 – regression coefficient with factor x_1	0.873	0.803	0.835	0.799	0.748
a_2 – regression coefficient with factor x_2	0.363	0.659	0.848	0.663	0.407
a_3 – regression coefficient with factor x_3	0.368	0.341	0.768	0.760	0.836
a_4 – regression coefficient with factor x_4	−0.012	0.005	0.026	0.033	0.031
$F_{факт}$ – calculated value of Fisher criterion	13.37	12.31	14.58	13.42	11.54
R – multiple correlation coefficient	0.656	0.640	0.672	0.656	0.628
R^2 – multiple determination coefficient	0.430	0.410	0.451	0.431	0.394
$Э(x_1)$ – elasticity coefficient y from factor фактора x_1	1.105	0.997	0.927	0.851	0.784
$Э(x_2)$ – elasticity coefficient y from factor фактора x_2	0.147	0.170	0.175	0.170	0.125
$Э(x_3)$ – elasticity coefficient y from factor фактора x_3	0.027	0.019	0.051	0.056	0.052
$Э(x_4)$ – elasticity coefficient y from factor фактора x_4	−0.187	0.072	0.399	0.495	0.449
$\beta(x_1)$ – beta coefficient of factor x_1	0.866	0.797	0.832	0.797	0.746
$\beta(x_2)$ – beta coefficient of factor x_2	0.018	0.021	0.022	0.022	0.020
$\beta(x_3)$ – beta coefficient of factor x_3	0.003	0.002	0.005	0.006	0.006
$\beta(x_4)$ – beta coefficient of factor x_4	−0.003	0.003	0.017	0.022	0.021

Source Compiled by the authors

The data in Table 5 show that all factors positively affect the level of wages and salaries in the amount of total income of the population, except for the coefficient of a demographic burden on the working-age population in the model of 2015. In general, the results were expected, except for the influence of the turnover of small businesses. With an increase in this indicator, the share of wages also increases, indicating either the problems in the small business when income from the entrepreneurial activity does not change or the outstripping growth in wages from employment. The closeness of the relationship between factors and outcome decreases over time, with a correlation coefficient of 0.628 in 2019. The share of labor compensation in the population's income by 39.4% depends on the factors included in the model.

Figure 1 shows the value of the multiple correlation coefficient for the developed models.

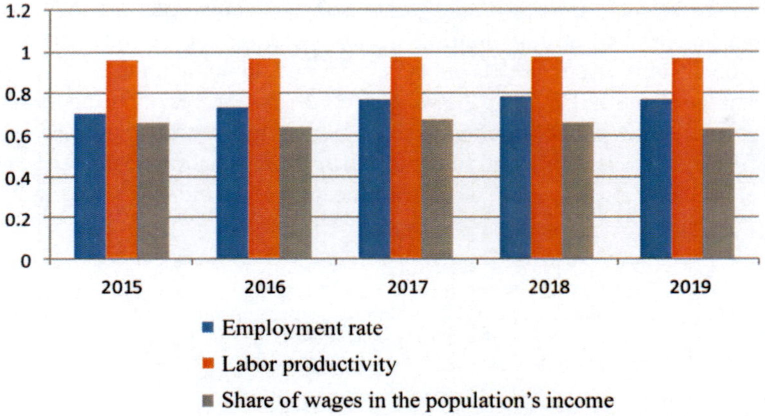

Fig. 1. Dynamics of multiple correlation coefficients R^2 for the output traits of the models. *Source* Compiled by the authors

The values of the correlation coefficient, presented graphically, clearly illustrate the decrease in the degree of closeness of correlation between the resultant indicators with the factor indicators during the transition from direct production to the distribution of the public product and decent employment (sustainable development goal No. 8).

The results obtained are consistent with the results of research by international scientists and complement them. Thus, modeling shows that during 2000–2019, nearly 80% of the targets of SDGs had significant interactions, both positive (synergies) and negative (trade-offs). The identification of numerical dependencies can help countries reach their goals [6].

Investment, a favorable business environment, and a legal environment contribute to economic development [2]. Low labor productivity and insufficient qualification of personnel that does not meet the labor market requirements slow down economic development [11]. We also analyze indicators of investment and labor productivity in relation to the level of employment and sustainable development of the country as a whole. One of the studies [3] notes that an investment of 4.3 thousand euros returns in the first year and increases net employment per person after four years.

It is emphasized that sustainable industrial development should promote a competitive economy, create jobs, and protect the environment through the efficient use of resources [19]. In our study, industrial production also has a positive effect on labor productivity. The service sector has the greatest impact on employment [5].

Researchers point to factors of high productivity, including business scale, investment in fixed and human capital, the use of modern digital technology, export activities, and employee training [18]. Direct foreign investment and the quality of human capital have a direct positive impact on long-term productivity and labor efficiency [15]. The high level of sustainable development can be explained by high labor productivity and efficient sectoral structure of employment [10]. This is also the context of our study. Speaking of employment, it is essential to note current trends such as social cohesion [17], social responsibility [12], and flexible employment [5].

Scholars link the ecological aspects of economic development, population distribution over the territory, urban infrastructure, population migration to the most attractive areas, employment [21], labor productivity, living standards, the health of the population, and sustainable economic growth. These aspects of sustainable development, which are of interest to scientists, are also relevant for Russia. This is also confirmed in our study.

4 Conclusion

The conducted research allowed us to draw the following conclusions:

1. There were insignificant annual changes in the economic structure regarding gross value added, the number of employees, and the number of wage funds. We revealed interdependencies between output and labor remuneration. It was also determined that industrial ties are stronger than social ones.
2. In general, the results obtained by the models of correlation and regression analysis are consistent and not contrary to economic laws. The closeness of the relationship between factors and the result in the labor productivity model is the highest among other models. This shows the closest relationship between the production indicators than between the social ones.
3. The obtained results are consistent with the results of studies by international scholars. Similar to international scholars, we analyzed indicators of investment and labor productivity in correlation with the level of employment and sustainable development of the country as a whole.
4. In the new epidemiological conditions, it is necessary to increase employment and labor productivity, as well as to improve the quality of employment, taking into account the preservation of public health. Therefore, it is necessary to implement special measures to improve working conditions, taking into account the costs for these purposes.

The system of modeling methods proposed in the article is of applied value. This system allows us to consider and analyze the attainment of SDGs in relation to each other and achieve integrated socio-economic development of the country. Further development of the research methodology is seen in improving the methodology for measuring the indicators of SDGs, multidimensional modeling of indicators, and the construction of forecast estimates of indicators up to 2030 based on the analysis of panel data.

References

1. Abdrakhmanova, G.I., et al.: Encyclopedia of Statistical Terms: Instrumental Methods of Statistics, vol. 2. Federal State Statistics Service, Moscow (2011). http://www.gks.ru/free_doc/new_site/rosstat/stbook11/tom2.pdf
2. Battalova, A.A., Burenina, I.V.: Formation of a comprehensive indicators system to assess the potential for sustainable development. In: European Proceedings of Social and Behavioural Sciences EpSBS, vol. 79. https://doi.org/10.15405/epsbs.2020.03.25
3. Cosculluela-Martinez, C.: Sustainable knowledge investment increases employment and GDP in the Spanish agricultural sector more than other investments. Sustainability **12**(8), 3127 (2020). https://doi.org/10.3390/su12083127
4. Cristea, M., Noja, G., Stefea, P.: The impact of population aging and public health support on EU labor markets. Int. J. Environ. Res. Publ. Health **17**(4), 1439 (2020). https://doi.org/10.3390/ijerph17041439
5. Davidescu, A.A., Apostu, S.-A., Paul, A.: Work flexibility, job satisfaction, and job performance among Romanian employees-implications for sustainable human resource management. Sustainability **12**(15), 6086 (2020). https://doi.org/10.3390/su12156086
6. de Miguel Ramos, C., Laurenti, R.: Synergies and trade-offs among sustainable development goals: the case of Spain. Sustainability **12**(24), 10506 (2020). https://doi:org/10.3390/su122410506
7. Erfanian, E., Ferreira, N., Amir, B.: Scientific output: labor or capital intensive? An analysis for selected countries. Scientometrics **112**(1), 461–482 (2017). https://doi.org/10.1007/s11192-017-2369-z
8. Fava, F., Gardossi, L., Brigidi, P.: The bioeconomy in Italy and the new national strategy for a more competitive and sustainable country. New Biotechnol. **61**, 124–136 (2021). https://doi.org/10.1016/j.nbt.2020.11.009
9. Federal State Statistics Service of the Russian Federation (n.d.). Official website: https://rosstat.gov.ru/
10. Georgescu, M.-A., Herman, E.: Productive employment for inclusive and sustainable development in European Union countries: a multivariate analysis. Sustainability **11**(6), 1771 (2019). https://doi.org/10.3390/su11061771
11. Ginevicius, R., Kliestik, T., Stasiukynas, A.: The impact of national economic development on the shadow economy. J. Competitiveness **12**(4), 39–55 (2020). https://doi.org/10.7441/joc.2020.04.03
12. Katz, M.: The need for socially responsible university educated professionals in the extractive industries. Extr. Ind. Soc. Int. J. **7**(4), 1351–1353 (2020)
13. Kiselakova, D., Stec, M., Grzebyk, M.: A multidimensional evaluation of the sustainable development of European Union countries-an empirical study. J. Competitiveness **12**(4), 56–73 (2020). https://doi.org/10.7441/joc.2020.04.04
14. Mishra, P.K., Mishra, S.K., Sarangi, M.K.: Do women's advancement and gender parity promote economic growth? Evidence from 30 Asian countries. Millennial Asia **11**(1), 5–26 (2020). https://doi.org/10.1177/0976399619900603

15. Nguyen, H.L., Luong Vinh, Q.D., Bui, H.N.: Effects of foreign direct investment and human capital on labor productivity: evidence from Vietnam. J. Asian Finance Econ. Bus. **6**, 123–130 (2019). https://doi.org/10.13106/jafeb.2019.vol6.no3.123
16. Ryabtsev, V.M., Chudilin, G.I.: Regional Statistics. MID, Moscow (2001)
17. Seyoum, B.: State fragility and human development: a study with special emphasis on social cohesion. Int. J. Soc. Econ. **48**(1), 39–61 (2021). https://doi.org/10.1108/IJSE-07-2020-0445
18. Simachev, Y., Kuzyk, M.G., Fedyunina, A.A.: Labor productivity in Russian companies: how to foster sustainable growth. Zhurnal Novaya Ekonomicheskaya Assotsiatsiya [J. New Econ. Assoc.] **4**, 205–217 (2020). https://doi.org/10.31737/2221-2264-2020-48-4-10
19. Ulbrych, M.: Progress in achieving sustainable industrial development – the case of the Czech Republic and Poland. Comp. Econ. Res. Cent. East. Eur. **23**(4), 109–128 (2020). https://doi.org/10.18778/1508-2008.23.30
20. UN General Assembly: Resolution A/RES/70/1 "Transforming Our World: The 2030 Agenda for Sustainable Development" (September 25, 2015 No. A/RES/70/1). UN, New York (2015). https://www.un.org/ga/search/view_doc.asp?symbol=A/RES/70/1&Lang=E
21. Wysokinska, Z.: A review of transnational regulations in environmental protection and the circular economy. Comp. Econ. Res. Cent. East. Eur. **23**(4), 149–168 (2020). https://doi.org/10.18778/1508-2008.23.32

Assessment of Corporate Performance via Balanced Scorecard as an Element of Economic Security

Svetlana F. Kochetkova(✉), Natalia N. Luneva, and Tatiana M. Levina

Ufa State Petroleum Technological University, Branch of the University in the City of Salavat, Salavat, Russia

Abstract. This study focuses on the development of balanced scorecard indicators, the optimal values of which guarantee the reliable functioning of companies in the oil and gas complex (OGC). We analyzed the approaches to the term "economic security" and highlighted its criteria regarding the specifics of the OGC enterprises. We suggest continuously assessing corporate activities via the developed balanced scorecard system to ensure national economic security. Moreover, we identified the key performance indicators of an OGC enterprise according to the strategic goal chart of the industry. Monitoring these indicators allows detecting the obstacles that obstruct the development and lower the economic security of OGC enterprises. To ensure the rapid assessment, we developed the "Balanced Scorecard" software; it can be used to assess the activities of different companies. The results were tested at an OGC enterprise – "Construction Materials and Structures Factory" LLC. As a result, we drew several conclusions and proposed measures to improve the sustainability of the company via the created software for monitoring the balanced scorecard.

Keywords: Key performance indicators · Oil and gas complex · Balanced scorecard · Strategic goal chart · Economic security · Software

JEL Codes: L12 · O1 · D24

1 Introduction

The oil and gas complex (OGC) consists of high-tech organizations that perform various stages of a single technological process. The ultimate goal of the OGC is to meet the needs for energy resources for industries and the general public. The OGC constituent enterprises operate in an uncertain and risky market environment. Therefore, maintaining the operating condition of all OGC components is vital to national economic security since they provide for the interests of the public and the state [8]. Recently, economic security at all its levels (international, national, regional, corporate) became an essential issue to be addressed at the highest level of the national economic management. The problem of preserving and developing OGC enterprises is an important area of research and various activities [5].

E. G. Popkova (Ed.): ISC 2020, LNNS 368, pp. 670–678, 2022.
https://doi.org/10.1007/978-3-030-93244-2_73

OGC supports and helps develop all areas of the national economy. In this case, the production unit that performs a stage of the technological process is an OGC company. Its success determines the condition of national economic security, affecting the industrial, budget, financial, innovation, and investment policies of the state. The OGC economic security is affected by the development factors of the oil sector. These factors may threaten the sustainable economic growth of the sector or, on the contrary, greatly increase its protection against this threat [3].

Ensuring corporate economic security requires continuous monitoring of business processes. The ultimate aim of the monitoring is to identify factors that impede the growth of the company and developing a development strategy that ensures a reliable and cost-effective economic cycle. Balanced scorecard is an especially important tool in economic security.

There are several different interpretations of the "balanced scorecard" concept. Its distinctive feature is the assessment of corporate performance through financial and non-financial indicators [2].

2 Materials and Methods

Assessment of corporate performance is a complex and iterative procedure necessary for developing the strategy of corporate economic security. Scholars recommend performing the evaluation via the balanced scorecard with the following indicators: finances, production, competitors, and sales [7]. We used the balanced scorecard concept to create a strategic goal chart for the "Construction Materials and Structures Factory" LLC ("ZSMiK" LLC) (Fig. 1).

The main objective of the strategic development of OGC enterprises is ensuring economic security. The financial component of economic security is important for determining the level of effective resource use and sustainable economic development. The element refers to the financial stability, solvency, independence, and liquidity of the corporate working capital. This element is also fundamental since it (same as the financial capacity of the company) determines whether the company can achieve its strategic business goals [6]. In examining the financial element, we concluded that the budget revenue from state investment programs and taxation is rapidly declining.

To excel both within and outside the industry, an enterprise should match its indicators to the optimal values. Financial solvency requires the enterprise to be independent of external sources of financing and to have a sufficient level of solvency. This can be accomplished by increasing profits and the profitability of products. The increase in positive financial results stems from the growth of sales revenue. The main indicators of sales growth are the increase in the number of customers and the expansion of sales markets.

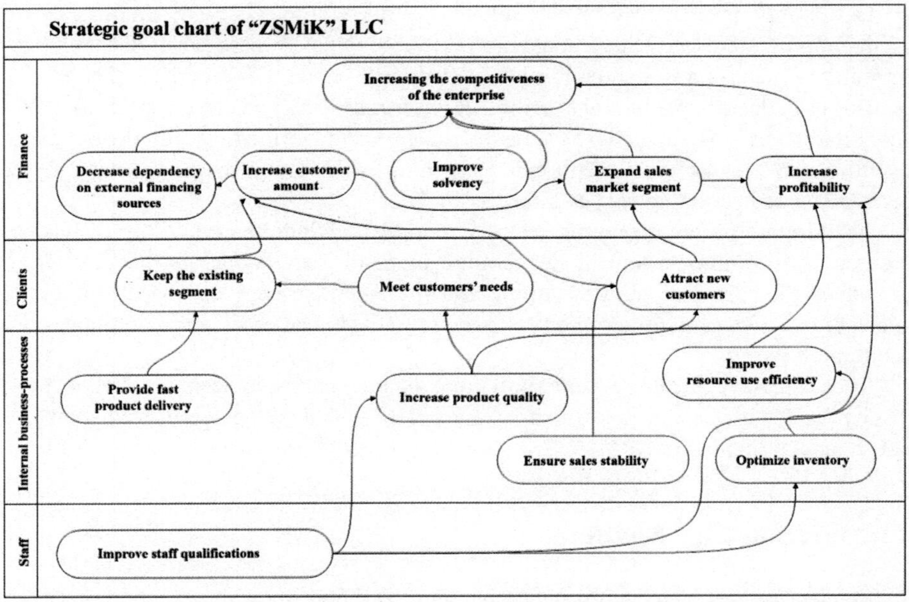

Fig. 1. Strategic goal chart of "ZSMiK" LLC. *Source* Compiled by the authors

Product sales can be increased by working with customers according to market segmentation research. The positive impact of these factors is made possible by retaining existing customers, meeting customer needs, and attracting new customers.

In turn, the company needs to improve its internal business processes: ensuring timely delivery of products and stability of sales activities, raising product quality and resource use efficiency. Reducing acquisition costs of raw materials and supplies will boost profitability. Inventory turnover depends on its optimal availability.

Human resource policy should ensure sufficient personnel qualifications. This factor directly affects the product quality, the efficiency of resource utilization, and optimal inventory volume. Moreover, it is necessary to improve the incentive system for employees that use resources sparingly.

Within each prospective area of economic security, we defined the key indicators to be assessed (Fig. 2). In this case, the autonomy coefficient serves as an indicator of independence from external financial sources. The normative value of the indicator points to the sufficient share of own corporate capital funds and, accordingly, the return on equity.

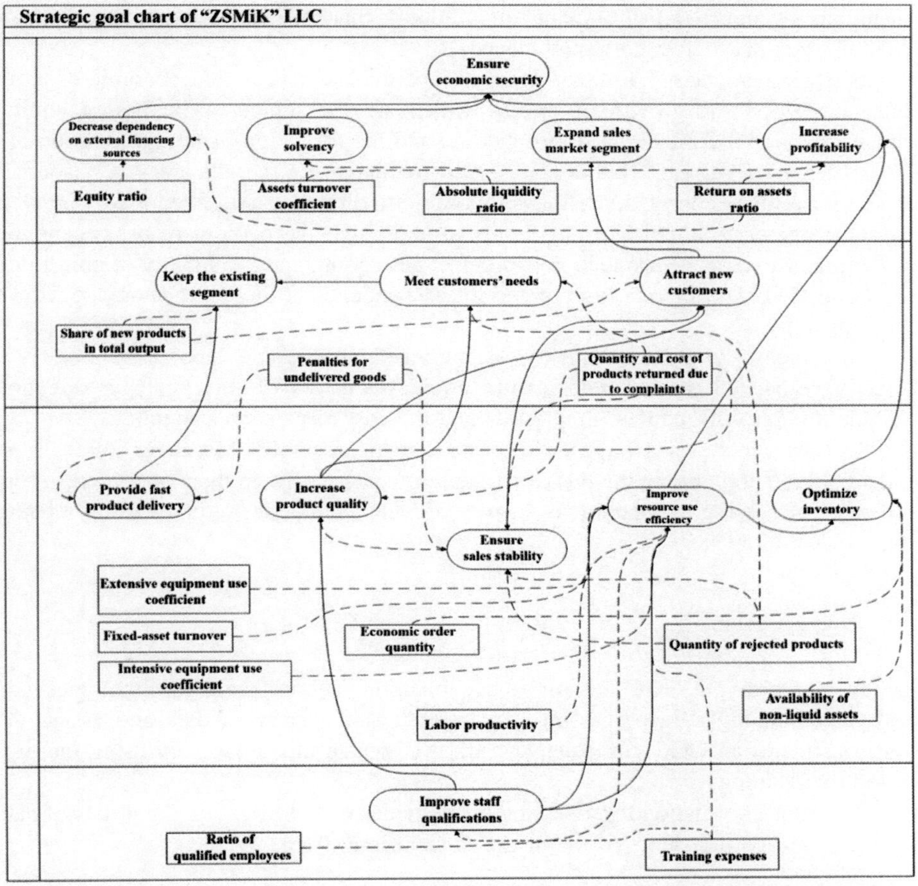

Fig. 2. Key indicators of economic security at "ZSMiK" LLC in the framework of the balanced scorecard. *Source* Compiled by the authors

Solvency depends on such indicators as liquidity ratio and working capital provision.

The profitability of "ZSMiK" LLC is reflected in the efficiency of sales and equity capital use. These indicators show the net profit ratio in equity and in sales revenue. Expansion of the market segment allows increasing profitability of sales.

One indicator reflecting the work with customers is the increase in the ratio of new products in the total output. New products shall be understood as new or upgraded products that meet the competitiveness criteria. This indicator determines the opportunities to enter new markets and attract new customers.

Timely delivery and stability of sales are what maintains the existing market conditions. Indicators of these strategic directions are penalties for undelivered products, as well as the quantity and cost of advertised products. One can also attract new customers by improving the quality of products; this indicator is comprised of production costs and defect correction costs [10].

Increasing operating profitability is one of the main tasks of economic security since it reflects the growth of profits, which are the source of increasing equity, expansion and modernization of production, and the formation of budgetary and non-budgetary resources [9]. This task is solved by increasing the efficiency of resources use and inventory optimization. Indicators of these directions are: fixed-asset turnover, labor productivity, coefficients of intensive and extensive equipment use, economic order quantity, costs of production and defect correction, and availability of non-liquid assets. In the chart, the relationship and interdependence of these indicators are shown by arrows [4].

An increase in product quality, resource efficiency, and labor productivity is directly connected to personnel qualification. Indicators of this strategic direction include the ratio of qualified employees and the cost of personnel training.

The chart includes a large array of data to be assessed. Moreover, there is no functional relation among the pieces of data, as well as between them and the resulting indicator. Assessing the economic security of OGC enterprises requires using modern digital technologies.

3 Results

Today, not a single OGC enterprise can function without digital technologies and automation of big data processing. To processing the array of data and assess the performance of an OGC company, one must automate this process by using the proposed methodology.

The input information for assessing the activities of "ZSMiK" LLC consists of data on the proposed indicators of the balanced scorecard (Table 1).

Table 1. Input data for key indicators of the balanced scorecard at "ZSMiK" LLC as of January 1, 2021

Indicator	Source	Symbol	Unit of measurement	Value
Equity	Balance sheet, row 1300	EQ	thou. RUB	5052600
Balance sheet total	Balance sheet, row 1700	BST	thou. RUB	26892546
Non-current assets	Balance sheet, row 1100	NCA	thou. RUB	181328
Current assets	Balance sheet, row 1200	CA	thou. RUB	26711218
Revenue	P&L, row 2110	RV	thou. RUB	12848189
Cash assets	Balance sheet, row 1250	CASH	thou. RUB	9939856
Short-term financial investments	Balance sheet, row 1240	SFI	thou. RUB	0
Short-term liabilities	Balance sheet, row 1500	SL	thou. RUB	21780926
Net income	P&L, row 2400	NI	thou. RUB	2317419
Refinancing rate	Central Bank of Russia	REF	unit fraction	0.0425
Marketable output, total	Analytical data	MO	thou. RUB	12956147
New product output	Analytical data	MO_{new}	thou. RUB	1248940
Penalties for undelivered goods	Analytical data	P_{und}	thou. RUB	27079

(*continued*)

Table 1. (*continued*)

Indicator	Source	Symbol	Unit of measurement	Value
Quantity and cost of products returned due to complaints	Analytical data	RET	thou. RUB	589
Loss from manufacturing defective goods	Analytical data	L_{def}	thou. RUB	839
Expenses for defective products	Analytical data	C_{def}	thou. RUB	356
Expenses incurred to correct defective products	Analytical data	EXP_{def}	thou. RUB	254
Cost of scrap at second use price	Analytical data	C_{use}	thou. RUB	547
Deductions made by those responsible for defective products	Analytical data	C_{ded}	thou. RUB	0
Production costs	P&L, row 2120	C	thou. RUB	8965104
Storage cost per unit of materials	Analytical data	COS_{unit}	thou. RUB	1200
Shipping and handling costs per order	Analytical data	COS_{ship}	thou. RUB	890
Annual materials demand	Analytical data	DEM_{anu}	Units	6163638
Value of fixed assets at the beginning of the evaluation period	Notes, row 5200	FA_{beg}	thou. RUB	168308
Value of fixed assets at the end of the evaluation period	Notes, row 5200	FA_{end}	thou. RUB	137022
Average annual cost of fixed assets	$(FA_{beg} + FA_{end})/2$	FA_{anu}	thou. RUB	152665
Production capacity	Analytical data	PC	thou. RUB	13000000
Actual working time of equipment	Analytical data	T_{act}	hours	3997
Effective working time of equipment	Analytical data	T_{eff}	hours	4096
Availability of non-liquid assets	Analytical data	A_{nlq}	thou. RUB	45
Number of qualified employees	Analytical data	ST_q	people	956
Average number of employees	Analytical data	ST	people	1012
Training expenses	Analytical data	TRN	thou. RUB	510000

Note P&L – profits and loss statement; Notes – notes to the balance sheet and profits and loss statement; Analytical data – information generated by management accounting and constituting a trade secret. *Source* Compiled by the authors

The input data form contains the values of "ZSMiK" LLC performance indicators as of January 1, 2021. The output form (Table) containing the automatically generated list and values of balanced scorecard indicators. Then, we compare the calculated values with the recommended values and identify discrepancies. Values that do not meet the requirements are automatically highlighted in red, thus indicating the need for intervention to ensure the stability of the balanced scorecard.

The recommended values for financial indicators are established proceeding from the "Methodological recommendations for analyzing financial-economic activities of organizations" document (approved by the State Statistics Committee of the Russian Federation on November 28, 2002) [1]. To determine the criteria of non-financial indicators, we involved experts from the middle management of business processes: corporate security and economic security department, financial-economic department, legal department, procurement department, chief engineer department, chief mechanic department, etc.

Table 2. Output form: the assessment of balanced scorecard indicators that ensure economic security

Indicator	Formula	Symbol	Unit of measurement	Recommended value	Actual value
1. Finance					
Equity ratio	EQ/BST	RA_{eq}	unit fraction	>0.6	0.188
Assets turnover coefficient	RV/BST	CF_{at}	unit fraction	>(EQ-CA)/EQ	**0.182**
Absolute liquidity ratio	(CASH+SFT)/SL	RA_{abslq}	unit fraction	0.1–0.3	**0.456**
Return on assets ratio	NI/EQ	RA_{roa}	unit fraction	REF	**0.459**
Return on sales ratio	NI/RV	RA_{ros}	unit fraction	0.19	**0.18**
2. Clients					
Share of new products	MO_{new}/MO	RA_{new}	unit fraction	0.3–0.5	0.096
Penalties for undelivered goods	P_{und}	P_{und}	thou. RUB	0	27079
Quantity and cost of products returned due to complaints	RET	RET	thou. RUB	0	**74.363**
Quantity of rejected products (production costs and defect correction)	$Q_{rej} = (C_{def} + EXP_{def} - C_{use} - C_{ded}) * (RA_{ros} + 1)$	Q_{rej}	thou. RUB	0	**589**
Indicator	Formula	Symbol	Unit of measurement	Recommended value	Actual value
3. Production					
Economic order quantity	$EOQ = \sqrt{\frac{2 * COS_{ship} * DEM_{anu}}{COS_{unit}}}$	EOQ	units	$<RV/DEM_{anu}$	3023.694
Fixed-asset turnover	RV/FA_{anu}	FA_t	thou. RUB	>0	**84159**
Intensive equipment use coefficient	MO/PC	CF_{ext}	thou. RUB	0.9–1.0	**0.997**
Extensive equipment use coefficient	T_{act}/T_{eff}	CF_{ext}	thou. RUB	0.9–1.0	**0.976**
Availability of non-liquid assets	A_{nlq}	A_{nlq}	thou. RUB	0	**45**
Labor productivity	RV/ST	LP	thou. RUB per person	>500	**12695839**
4. Staff					
Ratio of qualified employees	ST_q/ST	R_{aq}	units	0.9–1.0	**0.945**
Training expenses per employee	TRN/ST	TRN_{per}	thou. RUB	500–1000	95.85

Source Compiled by the authors

Table 2 shows the output form of the software. The data shows that the following indicators values do not meet the recommended values: the equity ratio, the share of new products, penalties for undelivered goods, economic order quantity, and training expenses per employee. To ensure economic security, "ZSMiK" LLC needs to increase the equity ratio. We advise providing for the growth of retained earnings as a part of the equity capital. This can be achieved by increasing sales volumes, which depend directly on the market segments, the number of new customers, and the elimination of undelivered goods. Optimizing order quantity entails reducing the cost of inventory storage. The enterprise also should increase employee training expenditures since the training cost per employee is less than the recommended value.

4 Conclusion

In this study, we drew the following conclusions:

1. Preservation and development of OGC depend on the economic security of each constituent company. The obstacles to the development of OGC companies are best identified via continuous monitoring of balanced scorecard indicators.
2. Balanced scorecard methodology increases the economic security of the enterprise and allows to reveal the between discrepancy recommended and actual values of key performance indicators. Moreover, it helps to identify the available reserves for improving economic solvency. These reserves act as a basis for stable and independent development – the main strategic goal of any company.
3. Since the assessment of key performance indicators is a continuous process, we developed the "Balanced Scorecard" software that helps to automate the procedure. The software allows:
   - Speed up and simplify data processing;
   - Monitor the compliance of indicators with the recommended values;
   - Promptly identify the factors that lower economic security;
   - Produce management decisions;
   - Store information for all evaluation periods.
4. Testing results indicate that the proposed balanced scorecard methodology is viable and can be used by other companies of various economic branches for creating their own development strategy and improving economic security.

References

1. Goskomstat of Russia: Methodological recommendations for the analysis of financial and economic activities of organizations (2002). http://www.consultant.ru/document/cons_doc_LAW_142116/. Accessed 1 Feb 2021
2. Blagodarov, K.V., Eliseev, D.M.: Technique of perfection of strategic planning of competitiveness on the basis of system of the balanced indicators. Bull. Saratov State Soc. Econ. Univ. **1**(35), 145–147 (2011)

3. Glotova, I.I., Tomilina, E.P.: Threats to economic security and directions of their neutralization in the system of economic security of the enterprise. In: Collection of Scientific Papers of the International Scientific and Practical Conference "Economic security: Legal, Economic, Environmental Aspects" , pp. 29–33. Publishing house CJSC "University book", Kursk, Russia (2016)
4. Kochetkova, S.F., Romanova, O.V.: Features evaluating the performance of scientific-production associations industry. Kazan Sci. **12**, 46–48 (2015)
5. Kochetkova, S.F., Luneva, N.N., Romanova, O.V., Khabibullina, L.R.: Improving the corporate security component. Innov. Invest. **10**, 109–113 (2019)
6. Leybert, T.B., Khalikova, E.A.: Organizational-economic aspects of controlling the financial condition of the company. Audit Financial Anal. **1**, 201–207 (2016)
7. Leybert, T.B., Reznova, V.P.: Formation of the management decision-making system in the enterprises of the oil industry on the basis of controlling. Audit Financial Anal. **3**, 225–231 (2016)
8. Rakhmatullina, S.F.: The mechanism of adaptation of enterprises of the military-industrial complex to the market using outsourcing (Dissertation of Candidate of Economic Sciences). Ufa State Aviation Technical University, Ufa, Russia (2009)
9. Sahiti, A., Ahmeti, S., Sahiti, A., Aliu, M.: The impact of balanced scorecard on improving the performance and profitability of the implementing companies. Mediterr. J. Soc. Sci. **7**(4), 60–65 (2016). https://doi.org/10.5901/mjss.2016.v7n4p60
10. Stoenoiu, C.-E., Cristea, C.: Comparative analysis for estimating production costs. In: MATEC Web of Conferences, vol. 184, p. 04004 (2018). https://doi.org/10.1051/matecconf/201818404004

Methodology for Assessing Information Security Risks at Oil Refining Enterprises

Natalia N. Luneva(✉), Tatiana M. Levina, and Natalia G. Evdokimova

Ufa State Petroleum Technological University, Branch of the University in the City of Salavat, Salavat, Russia

Abstract. The basis of Russia's economic security is the effectiveness of financial management of oil refining enterprises in the context of global digitalization and the transition of enterprises to a new technological order. Moreover, the current high level of automation of business processes of oil refining enterprises leads to the adoption of development strategies, taking into account the risks from compliance with information security.

Discussion of law enforcement practice by oil refineries, of the Federal Law dated July 27, 2006 No. 149 Federal-Law "On information, information technology, and information protection", issued to replace the Federal Law of February 20, 1995 No. 24 Federal Law "On Information, Informatization, and Information Protection", which are held with the support and participation of the Federal Security Service of the Russian Federation, Ministries of Internal Affairs of the Russian Federation and Defense of Russia, Federal Service for Technical and Export Control of Russia, Federal Security Service of Russia, Ministry (Emergency Situations) for Civil Defense, Emergencies and Elimination of Consequences of Natural Disasters of the Russian Federation, Ministry of Digital Development of the Russian Federation, Roskomnadzor (FS for Supervision in the Area of Communications), Ministry of Economic Development of the Russian Federation, Ministry of Energy of the Russian Federation, Ministry of Transport of the Russian Federation, Ministry of Finance of the Russian Federation, The Central Bank of the Russian Federation, and many of the remaining federal ministries and departments of the Russian Federation is posed as a particularly important task of the "digital" economy of Russia in terms of the importance of identifying, accounting, and reducing threats from information security risks.

The main purpose of this article is to develop a methodology for assessing the information security risks of enterprises in the oil refining industry, taking into account the delineation of areas of responsibility in the event of threats to information security from the security services and information technologies of the enterprise. Justification of the calculation of information security risks, taking into account the transition of oil refineries to the conditions of the "digital" economy set out on the example of the standards of the general process

E. G. Popkova (Ed.): ISC 2020, LNNS 368, pp. 679–690, 2022.
https://doi.org/10.1007/978-3-030-93244-2_74

"Risk Management" and regulatory documents: ISO 20000 series; ISO 27000 series; ISO 22000 series; Interstate Standard R 53647 series; ISO 31000 series.

The proposed methodology for assessing information security risks of oil refining enterprises was implemented using the example of ISO 31000, the choice of indicators of direct and indirect types of threats in assessing the risk category was determined according to the Interstate Standard R ISO/IEC 13335-3-2007 security methods and means. A set of methods for the internal process of oil refineries, risk assessment, and, as a result, damage prevention, a mathematical apparatus was used for statistical data processing.

A redistributed map of business processes of a refinery was obtained, which allows assessing possible threats, identify channels of information leakage, and take prompt measures to counter information security risks at enterprises of the oil refining industry as a way to improve the efficiency of an enterprise in a digital economy in the area of economic security of PJSC "GAZPROM".

Keywords: Economic security · Information security · "Digital" economy · Methodology · Assessment · Risks · Risk models · Regulatory documents · Categories · Risk probability · Severity of consequences · Business processes

JEL Codes: R1 · J5 · C8 · D2 · D7 · L60 · L71

1 Introduction

The digital economy poses unusual tasks to the oil refining industry. Global digitalization is transforming the approaches of enterprises to technological, economic, and personnel structures [10]. And also the fact that every competitive enterprise in the oil refining industry has a high level of business process automation makes it necessary to develop development strategies, taking into account the risks from compliance with information security.

Methods in relation to the result of financial management in conditions of compliance with economic activities for the safety of oil refining enterprises are considered in relevant articles and publications [9].

The annual forums of representatives of the oil refining industry business community are the most important events to discuss the law enforcement practice of the Federal Law dated July 27, 2006 No. 149 Federal-Law "On information, information technology, and information protection", issued to replace the Federal Law of February 20, 1995 No. 24 Federal Law "On Information, Informatization, and Information Protection", which are held with the support and participation of the Federal Security Service of the Russian Federation, Ministries of Internal Affairs of the Russian

Federation and Defense of Russia, Federal Service for Technical and Export Control of Russia, Federal Security Service of Russia, Ministry (Emergency Situations) for Civil Defense, Emergencies and Elimination of Consequences of Natural Disasters of the Russian Federation, Ministry of Digital Development of the Russian Federation, Roskomnadzor (FS for Supervision in the Area of Communications), Ministry of Economic Development of the Russian Federation, Ministry of Energy of the Russian Federation, Ministry of Transport of the Russian Federation, Ministry of Finance of the Russian Federation, The Central Bank of the Russian Federation, and many of the remaining federal ministries and departments of the Russian Federation. That is why the assessment of information security risks of oil refineries is very important for modern science and practice.

2 Materials and Methods

The need to assess possible threats, identify channels of information leakage, and take measures to counter information security risks at enterprises of the oil refining industry is emphasized in the works of Alexandrovich et al. [4], Aydinyan et al. [5, 6], and Balanovskaya [7]. At the same time, research at the empirical level of methods for assessing information security risks, in particular, a practical solution that allows updating and flexibility of decision-making at the strategic level of enterprise management is very rarely found in scientific papers and messages. Therefore, it also requires further research, taking into account the new technological order.

The transition of refineries to Industry 4.0 as an industry of a new technological order is considered from the point of view of defining the indisputable identity of the “digital” economy. There are some conclusions about the undeniable identity of the “digital” economy. The first approach is “classical”, where the “digital” economy is an economy, which is based on DT (digital technologies) and this is sufficient when characterizing the identity exclusively in the area of electronic goods and services. It’s also confirmed by the concept of the World Bank: “Digital” economy is the structure of economic, social, and cultural interactions based on the use of computer information and communication technologies (Fig. 1) [11].

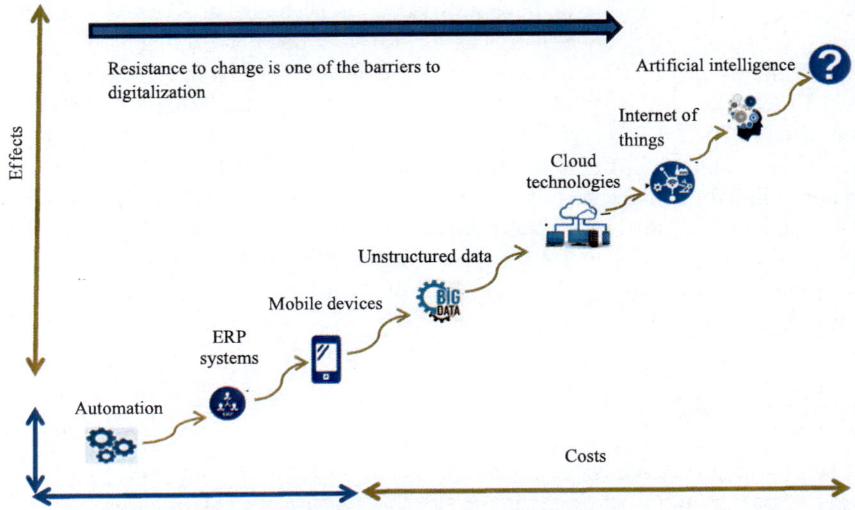

Automation of individual processes	**Creation of an integrated production system**
Delayed collection of information for management personnel	**Receive information in real-time**
Limited development options	**Multivariate development of an asset**

Fig. 1. Paradigm shift of information security management functions at a refinery in Industry 4.0. *Source* Oil refineries and petrochemical enterprises start Industry 4.0 [11]

The second approach is an in-depth one, where the "digital" economy is, in an economic sense, the connection of production with the use of digital technologies. This is confirmed by Rumyantsev, Director General of the Digital Development Fund of the Digital Platform Economy, who points out that identifying the digital economy in the world of economic relations doesn't have a stable, pleasant concept. He said: "Our task is to understand the terminological apparatus, the conceptual framework to facilitate the work of the participants of the digital market" [8]. The problems of digitalization of economic relations are aimed at increasing its flexibility, the speed of interaction between its subjects are used to draw up a performance map, during setting up business processes and, as a result, to increase its key performance indicators.

Let's consider the most relevant risk - information security risk assessment models in the distribution of exchange and consumption of information flow at oil refineries:

1) RiskWatch (RiskWatch Inc);
2) Cramm (the UK Analysis and Management Method Government Risk);
3) Octave (Operationally Critical Threat, Asset &Vulnerability).

The first risk model is based on the selection of qualitative criteria ROI (return on investment) and ALE (annual loss expectancy).

The application of the second risk - the CRAMM model at an oil refinery requires the management of the enterprise to control the software and hardware complex, i.e. it isn't possible to expose IT and communications management units for audit.

The application of the third risk - the Octave model implies the connection to the information security risk analysis process only of the internal services of the enterprise without the participation of external organizations and has been used to represent the Octave model.

Risk analysis according to the Octave methodology must be carried out and permanently once a year [12].

The considered risk models for assessing information security risks allow making prompt decisions on emerging threats on time, but don't provide an opportunity to cut off threats that have lost their relevance.

The process of identifying information security risks that have lost their relevance is possible only with the division of responsibility for identifying between the security services and information technologies of an oil refinery (Fig. 2).

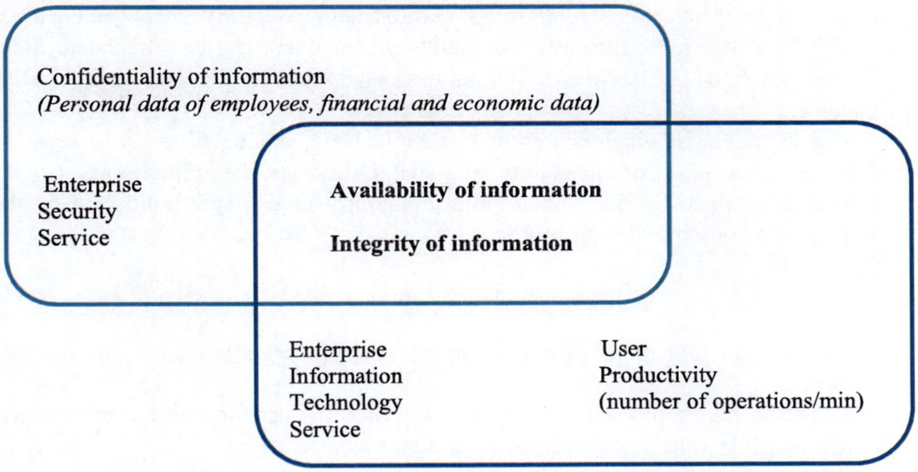

Fig. 2. Requirements of business processes for security and information technology services of an enterprise. *Source* compiled by the authors

The work of security and information technology services at an oil refinery is responsible for ensuring information security, which is also obvious from the risk models: RiskWatch, CRAMM, and Octave - regulated in Interstate Standard, ISO, OS, and GD.

So, the responsibility for ensuring information security of information technology services includes the following types:

- Responsibility for the operation of the process "Ensuring information security", the degree of responsibility of the heads of the unit operating the protected objects of information infrastructure;
- Responsibility for ensuring, monitoring, compliance of the settings of subsystems and information protection tools with the requirements of the documents of the process "Ensuring information security", participation in the formation of requirements, registration, and investigation of incidents, as well as the degree of responsibility of the information security administrator;
- Responsibility for the installation, configuration, updating, operation, and administration of information protection tools and subsystems, participation in the formation of requirements, identification, registration, and investigation of incidents, the degree of responsibility of the administrator of information protection tools and subsystems.

Responsibility for ensuring information security for security services at a refinery can be attributed to compliance and ensuring unified standards for managing the quality of information security services, as part of the process of ensuring business continuity from the side of the business digitalization of the enterprise. Ensuring information security at a refinery is an integral part of the overall Risk Management process and is regulated by the ISO series: ISO 20000 series - Information Technology Service Quality Management; ISO 27000 series - Information Security Management System; ISO 22000 series - Business Continuity Management System; Interstate Standard R series 53647 - Business continuity management; ISO 31000 series - Risk management.

Using the oneness, which is presented in ISO 31000, as a basis, it can be revealed that risk (R) is the effect of uncertainty on a goal. Otherwise, the ability to assess risks is always determined by the value of the probability of damage, multiplied by the severity of the consequences from it.

$$R = p * S, \tag{1}$$

Where p is the probability of risk occurrence; S is the severity of the consequences of damage.

Calculation of the probability of risk occurrence (p) depends on the probability-weighted possible values of the discrete random value M (X):

$$M(X) = \sum p_i \cdot u_i, \tag{2}$$

Where N is the number of discrete values of possible damages (of the same type and the same dimension) or the groups that combine them; p_i is the objective way of their occurrence; u_i - the resulting value of the loss of the enterprise.

If the damage u is a continuous random variable having a probability distribution density F (u), then the risk is calculated by the next formula:

$$R = \int uf(u)du. \tag{3}$$

The integral is taken over the entire range of damage u.

The characteristics of random numbers, including the values of probability and damage, are usually determined from a representative sample limited in size and time. In this case, formula (1) takes the following form:

$$R = \sum_{i=1}^{N} p_i \cdot u_i, \tag{4}$$

Where R* - risk assessments available for individual cases; p_i^* - in all values of the measure of occurrence u_i of damage.

In the basic concept of risk (formula 3), it's necessary to enter the parameters responsible for information security:

$$R = p_{vul} * p_{exp} * u_i, \tag{5}$$

Where p_{vul} is the probability of vulnerability in the information systems of a refinery; p_{exp} is the probability, which consists in the fact that vulnerability will be exploited by an attacker or lead to losses as a result of random circumstances.

Even though within the framework of the requirements of the Interstate Standard, ISO series of standards, explicit analytical expressions aren't analyzed for calculating information security risks, taking into account the transition of oil refineries to the conditions of a "digital" economy, it can be noted that, according to formula (4), the security services and information technology enterprises can indirectly influence all three factors, and the reduction of the first two factors must be calculated on practical examples.

3 Results

Let's consider, following State Standards [1–3], the assessment of information security risk through taking into account direct and indirect types of threats.

Indirect threats allow assessing risks concerning the result, which has a causal staging with risks and choose the deviation of the observed conditions from incorrect conditions. Preference can be determined according to the Interstate Standard R ISO/IEC 13335-3-2007 "Methods and means of ensuring safety. Part 3" and highlighting the risks by the following degrees of threats:

- Degree one (I), all special actions have information technology assets;
- Degree two (II), there are all events realized by employees that can accidentally damage the information technology fund;
- Degree three (III), there are special actions, which aren't based on events, in full action committed by employees.

The authors represent the degrees of threats, taking into account the different categories (Tables 1, 2, 3 and 4).

Table 1. List of general degrees of information security threats depending on natural phenomena

Event	Degree		
	I	II	III
Earthquake	–	–	+
Flooding	+	+	+
Hurricane	–	–	+
Lightning strikes	–	–	+
Max and min values of T (temperature), as well as water vapour value	+	+	+
Dust	–	–	+

Source compiled by the authors

Table 2. List of general degrees of information security threats depending on technical defects/malfunctions

Event	Degree		
	I	II	III
Damage in electrical communications	+	–	–
Breakdowns/defects in water communications	+	–	–
Defects in air communications	+	+	–
Technical failures	+	–	–
Amplitude voltage changes	+	+	–
Electromagnetic field	+	+	+
Statically charged particles	–	–	+
Deteriorating storage conditions	–	–	+
Hardware-software gaps	+	+	–
Hardware malfunctions of telecommunication components	+	–	–
Data gaps	+	–	–
Damaged connections	+	+	–
Maximum loaded data stream	+	+	–
Problems in communication services	+	+	–

Source compiled by the authors

Table 3. List of degrees of threats to information security from employees of the enterprise

Event	Degree		
	I	II	III
Employee protests	+	+	–
Not simple damage	–	+	–
Theft	–	+	–
Use of storage devices without permission	–	+	–
Incorrect work with the listener's software solution	+	+	–
The result of the assessment during the maintenance of systems	+	+	–
Application of software by a non-permissive employee	+	+	–
Using the software in a non-permissive way	+	+	–
The infiltration of an illegal listener under the guise of a legal one	–	+	–
Illegal use of software	+	+	–
Acquired/developed software	–	+	–
Incorrect actions from an employee	+	+	–
Access for unauthorized users to the network	–	+	–
Using network facilities in an unauthorized way	–	+	–
Data interception	–	+	–
Unauthorized access to communication facilities	–	+	–
Data flow analysis	–	+	–
Directing data to the wrong address	+	–	–
Changed text file route	–	+	–
Distortion about data	–	+	–
Staff shortage	+	–	–
Employee errors	+	+	–
Improper use of resources	+	+	–
Damage to hardware and software devices	–	+	–
Activities that violate the law	–	+	–

Source compiled by the authors

Table 4. List of general degrees of information security threats depending on other factors

Event	Degree		
	I	II	III
Shelling from different types of weapons	+	+	–
Use of weapon	+	+	–
Combustion	+	+	–
Malicious software	+	+	–

Source compiled by the authors

The proposed methodology for assessing the information security risks of oil refineries uses a mathematical apparatus for statistical data processing.

The standard error S_c of the sample for the mean risk for all three degrees (I, II, III) is: $S_c = 0,1633$

The normal error of the intermediate indicates how the mean over a segment of 0.6 differs from the mean of the initially defined aggregate.

The largest (max) sample is $\varepsilon = t_{kp}\, s_c = 2.685 * 0.163 = 0.438$.

The unspoken interval is 0.6 – 0.438; 0.6 + 0.438) = (0.162; 1.038).

Let's distribute the threat categories from the average value by 0.49 ≈ 50%, so, the next data were received: types of information security threats that have all three categories - 100% of risk probability; types of information security threats that have no more than two categories - 70% of risk probability; types of information security threats that have no more than one category - 30% of risk probability.

The degree of the severity of the consequences on the part of the security service of the enterprise (personal data of employees, financial, and economic data), on the part of the information technology service of the enterprise (number of operations/min.), is characterized by the parameters given in Table 5, regulated in the detection method, assessment, and minimization of risks in the area of economic security of PJSC GAZPROM.

Table 5. The degree of increasing probability of the severity of the consequences (damage through information transmission channels, leakage of personal data of employees, as well as data on the financial and economic activities of the enterprise)

Difficulty in identifying the severity of the consequences	Feature
1 - Not particularly significant	– Non-academic excluded; – Obeying instructions; – Countless breakdowns/failures/errors are needed
2 - Significant	– It is not natural, but it can happen; – Depends on the obedience of following the instructions (administrative document); – Countless breakdowns/failures/errors are needed
3 - Medium significant	– From time to time; – Level of training (qualification) of personnel; – A single occurrence may cause an accident/incident/accident
4 - Highly significant	– Subordination to the case from the possibility of solutions; – Frequently repeated fact; – Time-dependent observable event
5 - Very high significant	– Will happen – Empirically determined what will happen – Regularly observed event

Source compiled by the authors

4 Conclusion

Thus, this methodology includes:

- Risk models RiskWatch, Cramm, and Octave;
- Regulatory documents Interstate Standard, ISO, OS, GD;
- Distribution of responsibility of security and information technology services at an oil refinery;
- Accounting the formula for calculating the probability of information security risk;

– Methodology for identifying, assessing, and minimizing risks in the area of economic security of PJSC GAZPROM.

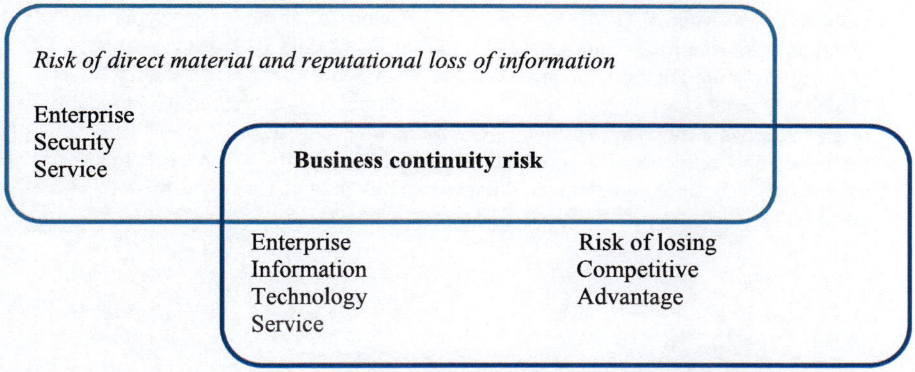

Fig. 3. Redistribution of possible information security risks of an oil refinery to the security and information technology services of the enterprise. *Source* compiled by the authors

This methodology also provides obtaining a redistributed map of business processes of an oil refinery (Fig. 3), which allow assessing possible threats, identifying channels of information leakage, and taking prompt actions to counter information security risks at enterprises of the oil refining industry. It will lead to an increase in the key performance indicators of the enterprise in the context of the "digital" economy.

References

1. Interstate Standard R ISO/IEC 20000-1-2013: Information technology (IT). Service management. Part 1. Requirements for the service management system. https://docs.cntd.ru/document/1200109304. Accessed 20 Jan 2021
2. Interstate Standard R ISO/IEC 27005-2010 Information technology (IT): Methods and means of ensuring safety. Information security risk management. https://docs.cntd.ru/document/1200084141. Accessed 20 Jan 2021
3. Interstate Standard R ISO 31000-2019 Risk management: Principles and guidelines. https://docs.cntd.ru/document/1200170125. Accessed 20 Jan 2021
4. Alexandrovich, G.Ya., Nesterov, S.I., Petrenko, S.A.: Automation of the assessment of information risks of a company. Inf. Secur. Confide. Inf. Method. J. **2**(5), 78–81 (2003)
5. Aydinyan, A.R., Tsvetkova, O.L., Kikot, I.R., Kazantsev, A.V., Kaplun, V.V.: On the approach to assessing the information security of an enterprise. In: System analysis, Management and Information Processing: A Collection of Articles of 5th International Scientific Seminar in Don State Technical University, pp. 109–111 (2014)
6. Aydinyan, A.R., Tsvetkova, O.L.: An approach to assessing DLP systems using fuzzy logic tools. Eng. Bull. Don **4** (2017). ivdon.ru/ru/magazine/archive/n4y2017/4432. Accessed 25 Dec 2017

7. Balanovskaya, A.B.: Ensuring information security of industrial enterprises. Bull. Samara Univ. **3**(84), 72–79
8. Digital economy: how experts understand this term. https://ria.ru/20170616/1496663946.html. Accessed 20 Jan 2021
9. Leibert, T.B., Khalikova, E.A., Sakaeva, D.R.: Factors - indicators of economic security of an oil and gas enterprise. Eurasian Law J. **8**(135), 359–361 (2019)
10. National Program: Digital Economy of the Russian Federation (2018). Russia, 90
11. Pugin, V.V., Gubareva, O.Yu.: Review of methods for analyzing information security risks of an enterprise information system. T_Comm **6**, 54–57 (2012)
12. Refineries and petrochemical enterprises are starting Industry 4.0. https://chemtech.ru/neftepererabatyvajushhie-i-neftehimicheskie-predprijatija-pristupajut-k-industrii-4-0/. Accessed 20 Jan 2021

Methodology for Assessing Economic Security of the Oil Industry Enterprise

Marina M. Gayfullina(✉)

Ufa State Petroleum Technological University, Ufa, Russia

Abstract. The paper offers a methodical approach to assessing the economic security of the enterprises of the oil industry. This approach allows to identify possible threats and hazards to these enterprises and make timely management decisions on their neutralization. This methodology is based on calculating the integral indicator, taking into account the system of micro and macro indicators. Micro indicators show the internal capabilities and performance of the enterprise. Macro indicators consider the parameters of external factors affecting the company. Considering the specifics of the oil industry, the author offers a system of economic security indicators for the enterprises in the oil industry. Within each indicator, the author proposes threshold values to determine the nature of the enterprise's economic security at four levels of economic security: high, warning, dangerous, and critical. The author uses the example of Russian oil companies Rosneft Oil Company PJSC and Gazprom Neft PJSC to try out the proposed approach. The research allowed the author to assess the level of economic security of oil enterprises. The author draws several conclusions and identifies the main factors that negatively affected the economic security of oil companies in 2020. The proposed methodology allows one to determine the direction and nature of the management of the micro indicators to improve the economic security of the oil enterprise.

Keywords: Indicator · Indicator of economic security · Integral index · Oil industry · Enterprise · Oil company · Threat to economic security · Economic security

JEL Codes: D20 · L10 · L71

1 Introduction

The economic security of an enterprise characterizes a state of effective use of the enterprise's resources, which allows it to prevent external and internal threats.

The relevance of this research lies in the increasing importance of the economic security of enterprises in terms of escalating the threats of crisis phenomena in the external environment, which affect the company's performance.

Simultaneously, the level of economic security is largely influenced by the set of indicators considered in the assessment. An essential point in assessing economic security is the consideration of the enterprise's industry specifics.

E. G. Popkova (Ed.): ISC 2020, LNNS 368, pp. 691–701, 2022.
https://doi.org/10.1007/978-3-030-93244-2_75

The paper aims to determine the indicators of economic security of the oil industry enterprise and offers a methodological approach to assessing the economic security of the oil industry enterprise considering industry specifics.

The research subject is the methods for assessing the economic security of the oil industry enterprise.

The research hypothesis is the assumption that the set of tools for regulating economic security depends on the industry specialization of the enterprise and is generally similar for enterprises of the same industry.

The author assesses the economic security of different oil companies to test the hypothesis using the proposed methodology.

2 Materials and Methods

The methodological basis of this research consists of methods of logical analysis, grouping, comparison, and generalization.

Many authors discuss the problems of assessing and ensuring the economic security of entities at various levels of management. In particular, several works consider and study the problem of assessing economic security at the level of the country and the region [13, 15], as well as at the level of particular enterprises [11]. When considering economic security at the level of certain enterprises, the authors consider the management of economic security at the enterprise level [16], the categories of security, and the place of economic security in the system of enterprise management [23]. Secondary literature also analyzes the strategic potential of companies, identifies the threats and risks [12] associated with management decisions in the field of economic security and balanced development [3], and assesses the impact of the crisis on the economic security of the enterprise [17].

The most common approaches to assessing the level of economic security of an enterprise are as follows:

1. Calculation of the economic security of the enterprise through a system of indicators. The boundaries of permissible deviations (limits (thresholds)) are defined for each indicator;
2. Resource and functional approach, which involves structuring the economic security of the enterprise by functional components and determining in these components indicators that characterize the effectiveness of the use of resources.

There are also other methods and approaches to determine the level of economic security of the enterprise, which are much less common in scientific research [21].

Researchers also pay due attention to the sectoral aspects of assessing the economic security of enterprises, including construction [6], machine-building [4], and other enterprises.

Thus, in the modern economical literature [14], a significant number of alternative methods for assessing the economic security of the enterprise are discussed. Each of these methods is debatable and cannot be used to determine the level of economic security of oil industry enterprises without some adaptation.

The need to develop such a methodology is due to the specific threats experienced by oil industry enterprises, especially considering the fact that the set of these threats changes over time.

It is proposed to evaluate the level of economic security of the oil industry enterprise through the integral indicator calculated as the arithmetic mean value of point estimates of the indicators of micro-and macro-levels:

$$I = \frac{E_1 + E_2}{2}, \tag{1}$$

where

E_1 – total score of micro-level indicators;

E_2– total score of macro-level indicators.

An important issue in the development of the methodology was the justification of the indicators of economic security considering the specifics of the oil industry.

The study of a considerable number of internal factors and external threats to the economic security of the enterprise, as well as the objects and directions of their impact and potential consequences for the activities of the enterprise, involves significant time and labor costs.

However, to carry out a successful activity, each enterprise should identify (predict) the most significant (dangerous) internal and external sources of risk based on the current situation and form a set of measures for timely detection, prevention, or mitigation of these risks.

Internal sources of risks to the economic security of enterprises are formed directly in the enterprise's production sphere.

Problems of research of internal factors and threats to the economic security of oil companies were not left without the attention of researchers. It is necessary to highlight the following factors considered by researchers in relation to the oil complex:

- Factors of industrial competition [9];
- Production and resource factors [5];
- Technological and investment factors;
- State regulation of the oil complex;
- Global oil prices [10];
- Social and governance factors [7, 18];
- Factors affecting the financial security of the company [2] and others.

Having analyzed the existing research on internal factors and external threats to the economic security of oil companies, taking into account the principles of sufficiency and objectivity, the author selected and formed a set of indicators of micro-and macro-levels of economic security of the oil company (Tables 1 and 2).

The indicators of micro-level reflect the influence of internal factors on the economic security of the oil company. The indicators of macro-level reflect the influence of external macroeconomic factors.

At the initial stage, the author calculated single indicators of economic security of the oil enterprise at micro-and macro-levels.

Then, the authors determined the level of economic security for each indicator according to the threshold values of the indicators proposed in Table 1 and Table 2.

Table 1. The micro-level indicators of the economic security of an oil enterprise

Factors	Indicator	CR	DS	WG	HH
Production	Coefficient of change in the volumes of oil production (X_1)	$X_1 < 0.9$	$0.9 \leq X_1 < 1.0$	$1.0 \leq X_1 < 1.1$	$X_1 \geq 1.1$
	Coefficient of change in the level of proven reserves (X_2)	$X_2 < 0.9$	$0.9 \leq X_2 < 1.0$	$1.0 \leq X_2 < 1.1$	$X_2 \geq 1.1$
	Coefficient of oil reserves replacement (X_3)	$X_3 < 0.8$	$0.8 \leq X_3 < 1.0$	$1.0 \leq X_3 < 1.2$	$X_3 \geq 1.2$
	Coefficient of change in oil refining volumes (X_4)	$X_4 < 0.9$	$0.9 \leq X_4 < 1.0$	$1.0 \leq X_4 < 1.1$	$X_4 \geq 1.1$
	Depth of oil refining, % (X_5)	$X_5 < 70$	$70 \leq X_5 < 80$	$80 \leq X_5 < 90$	$X_5 \geq 90$
	Coefficient of change in the value of fixed assets, % (X_6)	$X_6 < 0.9$	$0.9 \leq X_6 < 1.0$	$1.0 \leq X_6 < 1.1$	$X_6 \geq 1.1$
Financial condition	Current liquidity ratio (X_7)	$X_7 < 1.0$	$1.0 \leq X_7 < 1.5$	$1.5 \leq X_7 < 2.0$	$X_7 \geq 2.0$
	Solvency ratio (X_8)	$X_8 < 0.5$	$0.5 \leq X_8 < 0.6$	$0.6 \leq X_8 < 0.7$	$X_8 \geq 0.7$
	Return on sales, % (X_9)	$X_9 < 5$	$5 \leq X_9 < 10$	$10 \leq X_9 < 15$	$X_9 \geq 15$
Human resources	Share of employees with higher education, % (X_{10})	$X_{10} < 30$	$30 \leq X_{10} < 45$	$45 \leq X_{10} < 60$	$X_{10} \geq 60$
	Staff turnover, % (X_{11})	$X_{11} > 10$	$7 < X_{11} \leq 10$	$5 < X_{11} \leq 7$	$X_{11} \leq 5$
	Coefficient of change in the average wage, % (X_{12})	$X_{12} < 0.9$	$0.9 \leq X_{12} < 1.0$	$1.0 \leq X_{12} < 1.1$	$X_{12} \geq 1.1$

(continued)

Table 1. (*continued*)

Factors	Indicator	CR	DS	WG	HH
Investment	Coefficient of capital investment change, % (X_{13})	$X_{13} < 0.9$	$0.9 \leq X_{13} < 1.0$	$1.0 \leq X_{13} < 1.1$	$X_{13} \geq 1.1$
	Investment coefficient (X_{14})	$X_{14} < 0.2$	$0.2 \leq X_{14} < 0.3$	$0.3 \leq X_{14} < 0.4$	$X_{14} \geq 0.4$
	Coefficient of investment activity (X_{15})	$X_{15} < 0.2$	$0.2 \leq X_{15} < 0.4$	$0.4 \leq X_{15} < 0.6$	$X_{15} \geq 0.6$
	Depreciation coefficient (X_{16})	$X_{16} > 0.85$	$0.7 < X_{16} \leq 0.85$	$0.5 < X_{16} \leq 0.7$	$X_{16} \leq 0.5$
Safety	Coefficient of change in environmental protection costs, % (X_{17})	$X_{17} < 0.9$	$0.9 \leq X_{17} < 1.0$	$1.0 \leq X_{17} < 1.1$	$X_{17} \geq 1.1$
	Coefficient of change in industrial safety costs, % (X_{18})	$X_{18} < 0.9$	$0.9 \leq X_{18} < 1.0$	1. $0 \leq X_{18} < 1.1$	$X_{18} \geq 1.1$
	Coefficient of change in labor protection costs, % (X_{19})	$X_{19} < 0.9$	$0.9 \leq X_{19} < 1.0$	$1.0 \leq X_{19} < 1.1$	$X_{19} \geq 1.1$

Source Compiled by the authors

The following levels of economic security of the oil enterprise are proposed:

- high level (HH);
- warning level (WG);
- dangerous level (DS);
- critical level (CR).

When developing the threshold values of indicators, the recommendations of researchers [19, 22] and the average values of the indicators for oil enterprises were considered.

Further, the calculated values of the unit indicators of economic security are translated into a point estimate. The grades (B_{Xi}, B_{Yj}) are given as follows:

- High level (HH) – 0 point;
- Warning level (WG) – 5 points;
- Dangerous level (DS) – 10 points;

Table 2. The macro-level indicators of the economic security of an oil enterprise

Factors	Indicator	CR	DS	WG	HH
Social	Coefficient of change in wages in the economy (Y_1)	$Y_1 < 0.9$	$0.9 \leq Y_1 < 1.0$	$1.0 \leq Y_1 < 1.1$	$Y_1 \geq 1.1$
	National employment rate, % (Y_2)	$Y_2 < 50$	$50 \leq Y_2 < 60$	$60 \leq Y_2 < 80$	$Y_2 \geq 80$
	Staff turnover by industry, % (Y_3)	$Y_3 > 25$	$20 < Y_3 \leq 25$	$15 < Y_3 \leq 20$	$Y_3 \leq 15$
Economic	Inflation rate for the year, % (Y_4)	$Y_4 > 15$	$10 < Y_4 \leq 15$	$5 < Y_4 \leq 10$	$Y_4 \leq 5$
	Key rate of the Bank of Russia, % (Y_5)	$Y_5 > 12$	$10 < Y_5 \leq 12$	$8 < Y_5 \leq 10$	$Y_5 \leq 8$
	Coefficient of change in the annual average ruble-dollar exchange rate (Y_6)	$Y_6 > 1.3$	$1.2 < Y_6 \leq 1.3$	$1.05 < Y_6 \leq 1.2$	$Y_6 \leq 1.05$
	Price level for Urals oil, USD/barrel (Y_7)	$Y_7 < 50$	$50 \leq Y_7 < 70$	$70 \leq Y_7 < 100$	$Y_7 \geq 100$
	Coefficient of change in mineral extraction tax (oil) (Y_8)	$Y_8 > 1.2$	$1.1 < Y_8 \leq 1.2$	$1.05 < Y_8 \leq 1.1$	$Y_8 \leq 1.05$
	Coefficient of change in customs duty on oil exports (Y_9)	$Y_9 > 1.2$	$1.1 < Y_9 \leq 1.2$	$1.05 < Y_9 \leq 1.1$	$Y_9 \leq 1.05$
Technological	R&D expenditures in GDP, % (Y_{10})	$Y_{10} < 1.0$	$1.0 \leq Y_{10} < 1.5$	$1.5 \leq Y_{10} < 2.0$	$Y_{10} \geq 2.0$
	Coefficient of change in the output of	$Y_{11} < 0.9$	$0.9 \leq Y_{11} < 1.0$	$1.0 \leq Y_{11} < 1.1$	$Y_{11} \geq 1.1$

(*continued*)

Table 2. (*continued*)

Factors	Indicator	CR	DS	WG	HH
	innovative products in the country (Y_{11})				
	Level of innovation activity, % (Y_{12})	$Y_{12} < 10$	$20 \le Y_{12} < 10$	$20 \le Y_{12} < 30$	$Y_{12} \ge 30$
	Coefficient of change in the cost of innovation in the country (Y_{13})	$Y_{13} < 0.9$	$0.9 \le Y_{13} < 1.0$	$1.0 \le Y_{13} < 1.1$	$Y_{13} \ge 1.1$

Source Compiled by the authors

- Critical level (CR) – 15 points.

The total score of the micro-level indicators is found according to the following formula:

$$E_1 = \sum_{i=1}^{n} B_{X_i}, \quad (2)$$

where

B_{Xi} – evaluation of the micro-level indicator *Xi* in points;
n – number of micro-level indicators.

The total score of the macro-level indicators is calculated as follows:

$$E_2 = \sum_{j=1}^{m} B_{Y_j}, \quad (3)$$

where

$B_{Y\,j}$ – score of the macro-level indicator *Yj* in points;
m – number of macro-level indicators.

After calculating the integral indicator of economic security, let us determine the nature of economic security in accordance with the scale (Table 3).

Table 3. Scale of the assessment of economic security of an oil enterprise

Nature of economic security	Value of the integral
High level (HH)	$0 \leq I \leq 5$
Warning level (WG)	$5 < I \leq 10$
Dangerous level (DS)	$10 < I \leq 15$
Critical level (CR)	$I \leq 15$

Source Compiled by the authors

3 Results

The developed methodology was tested on the example of Rosneft Oil Company PJSC and Gazprom Neft PJSC, which are the largest oil companies in Russia.

Table 4 shows the grades of micro-and macro-level indicators.

Table 4. Assessment of the level of economic security of companies

Indicator, company	2018	2019	2020
Total score of micro-level indicators (E_1)			
Rosneft Oil Company PJSC	5.0	5.8	6.0
Gazprom Neft PJSC	4.5	3.5	5.8
Total score of macro-level indicators (E_2)	6.4	5.0	4.6
Integral index (I)			
Rosneft Oil Company PJSC	5.7	5.4	5.3
Gazprom Neft PJSC	5.5	4.3	5.2
Level interpretation			
Rosneft Oil Company PJSC	WG	WG	WG
Gazprom Neft PJSC	WG	HH	WG

Source Compiled by the authors based on [8, 20]

In general, it is possible to note a different level and dynamics of economic security in the considered oil companies. While the economic security of Rosneft Oil Company PJSC slightly decreased in 2020, the economic security of Gazprom Neft PJSC slightly increased. The integral level of economic security in the considered companies is characterized as pre-crisis. It can be noted that the evaluated companies managed to maintain their economic security in 2020 at a relatively high level, despite the record oil prices in the world and the decline in demand for petroleum products.

At the same time, the factors that negatively affected the economic security of the examined oil companies are mostly similar (Table 5).

Thus, the calculations confirm the hypothesis that the set of tools for regulating economic security depends on the industry specialization of the enterprise and is generally similar for enterprises of the same industry. For the examined oil companies, four out of five reasons that influenced the reduction of their economic security coincided.

Table 5. Factors that negatively affected the economic security of companies in 2020

Company	Factors that negatively affected the economic security
Rosneft Oil Company PJSC	1. Deterioration of financial results: decrease in revenue by 29%, decrease in profit on sales by 58%, and decreased sales profitability
	2. Decrease in the efficiency of fixed assets (return on assets) by more than 33%
	3. Decrease in the volume of oil refining (by 5.6%)
	4. Decrease in oil production (by 11.4%)
	5. High employee turnover
Gazprom Neft PJSC	1. Deterioration of financial results – decrease in sales revenue by 16.4%, negative profit on sales
	2. Decrease in the efficiency of fixed assets (productivity of funds) by more than two times
	3. Decrease in oil refining (by 2.7%)
	4. Decrease in capital investments
	5. High employee turnover

Source Compiled by the authors based on the calculations

4 Conclusion

In contemporary economic science, there is no generally accepted approach to assessing the economic security of enterprises. Various concepts, typologies, and approaches are presented in the secondary literature.

However, the known methods for assessing economic security do not sufficiently consider and reflect the industry characteristics of oil enterprises, taking into account the influence of specific (industry) internal and external factors.

The paper systematized the available research and analyzed the average statistical values of the indicators for the industry to form a set of indicators of economic security in relation to the oil companies and determine the boundaries of their threshold values by level.

The relevance of this study is also determined by the fact that different sets of macro-level and micro-level indicators differently affect certain significant aspects of the economic security of the oil company.

For example, company size, market shares, and the degree of vertical integration significantly affect profit margins in oil and gas production, as well as in its refining and marketing [1], which affects the value of energy security indicators at the micro-level.

The main scientific and practical results of this research are as follows:

1. The use of the proposed methodology allows to determine the current level of economic security of the oil enterprise and assess the directions and levels of impact of threats affecting it;
2. In determining the level of economic security of the oil enterprise, it is proposed to use indicators reflecting the industry specifics of the oil enterprise (e.g., the world price of oil, the volume of production and refining of oil in the company, etc.).

The application of the proposed tools increases the scientific validity and quality of management decisions in terms of the development of particular business segments of the oil company (e.g., Upstream, Downstream), as well as the company as a whole.

The results of this study may be helpful for:

- Enterprises – when assessing the current level of economic security and identifying potential threats to activities;
- Investors – when assessing the attractiveness of enterprises for investment.

References

1. Andersen, A.T., Moody, C.E., Rasmussen, J.A.: Strategy, structure and performance of major U.S. Oil companles: evidence from line of business data. Rev. Ind. Org. **1**, 290–306 (1984). https://doi.org/10.1007/BF02354128. Accessed 31 Jan 2021
2. Biliomistniy, O., Bilomistna, I., Galushko, Y.: Influence external and internal factors to financial security of enterprise. Financial Credit Activity Probl. Theory Pract. **1**(22) (2017). https://doi.org/10.18371/fcaptp.v1i22.109935. Accessed 25 Jan 2021
3. Biryukova, V.V.: Production system management based on a balanced development model. IOP Conf. Ser. Mater. Sci. Eng. **753**, 062014 (2020). https://iopscience.iop.org/article/10.1088/1757-899X/753/6/062014/meta. Accessed 30 Jan 2021
4. Bublyk, M., Koval, V., Redkva, O.: Analysis impact of the structural competition preconditions for ensuring economic security of the machine building complex. Mark. Innov. Manag. **4** (2017). https://doi.org/10.2139/ssrn.3184338. Accessed 20 Jan 2021
5. Burenina, I.V., Evtushenko, E.V., Kotov, D.V., Battalova, A.A., Gaifullina, M.M., Gamilova, D.A.: Integral assessment of the development of Russia's chemical industry. J. Environ. Manag. Tour. **8**(5), 1075–1108 (2018)
6. Forrester, S.V., Ustinova, G.H., Popok, L.E.: Economic security of construction industry enterprises. In: Ashmarina, S.I., Mantulenko, V.V. (eds.) IPM 2020. LNNS, vol. 161, pp. 675–683. Springer, Cham (2021). https://doi.org/10.1007/978-3-030-60926-9_84
7. Gayfullina, M.M., Nazimova, G.Z., Makov, V.M.: Formation of strategy of effective management of human resources of an oil company. Oil Ind. **4**, 8–11 (2018)
8. Gazprom Neft PJSC: Official website (n.d.). https://www.gazprom-neft.ru/
9. Gong, B.: Total-factor spillovers, similarities, and competitions in the petroleum industry. In: Gong, B. (ed.) Shale Energy Revolution, pp. 151–217. Springer, Singapore (2020). https://doi.org/10.1007/978-981-15-4855-0_8. Accessed 25 Jan 2021

10. Hall, S.G., Kenjegaliev, A.: The effect of oil price changes on the price of Russian and Chinese oil shares. Empir. Econ. **53**, 1639–1656 (2017). https://doi.org/10.1007/s00181-016-1176-3. Accessed 25 Jan 2021
11. Ianioglo, A., Polajeva, T.: The essence and phases of the comprehensive system of ensuring the economic security of enterprise. Int. J. Learn. Change **9**(1), 59–74 (2017). https://doi.org/10.1504/IJLC.2017.10005203. Accessed 30 Jan 2021
12. Karanina, E., Ryazanova, O., Timin, A.: The risk-based approach to rating the competitiveness of transport enterprises in the system of economic security. In: Murgul, V., Popovic, Z. (eds.) EMMFT 2017. AISC, vol. 692, pp. 318–325. Springer, Cham (2018). https://doi.org/10.1007/978-3-319-70987-1_34
13. Kazakova, N.A., Bolvachev, A.I., Gendon, A.L., Golubeva, G.F.: Monitoring economic security in the region based on indicators of sustainable development. Stud. Rus. Econ. Dev. **27**, 638–648 (2016). https://doi.org/10.1134/S1075700716060058. Accessed 19 Jan 2021
14. Klimuk, V.V., Piskunov, V.A., Pecherskaya, E.P., Tarasova, T.M.: Methodical approaches to assessing the economic security. Eur. Proc. Soc. Behav. Sci. EpSBS **57**, 283–290 (2019). https://doi.org/10.15405/epsbs.2019.03.29. Accessed 30 Jan 2021
15. Kolesnikov, S.I., Dolzhenko, L.M.: Methodology for assessing the level of the territory's economic security. In: Pinelas, S., Kim, A., Vlasov, V. (eds.) CONCORD-90 2018. Springer Proceedings in Mathematics & Statistics, vol. 318, pp. 359–364. Springer, Cham (2020). https://doi.org/10.1007/978-3-030-42176-2_35. Accessed 25 Jan 2021
16. Korzh, M.: Formation of organizational support for the management of the economic security of engineering enterprises: methodical and practical aspects. Bus. Theory Pract. **20**, 317–328 (2019). https://doi.org/10.3846/btp.2019.30. Accessed 21 Jan 2021
17. Melnyk, S., Shuprudko, N., Kolosovska, I. Berest, I., Pasichnyk, M.: Anti-crisis personnel management in the process of ensuring the economic security of the enterprise. Bus. Theory Pract. **21**(1), 272–281 (2020). https://doi.org/10.3846/btp.2020.11438. Accessed 25 Jan 2021
18. Nurgali, A.: The impact of economic crisis on personnel management. Eur. Sci. **4**(26), 54–57 (2017)
19. Orazalin, N., Mahmood, M., Narbaev, T.: The impact of sustainability performance indicators on financial stability: evidence from the Russian oil and gas industry. Environ. Sci. Pollut. Res. **26**, 8157–8168 (2019). https://doi.org/10.1007/s11356-019-04325-9. Accessed 31 Jan 2021
20. Rosneft Oil Company: Official website (n.d.). https://www.rosneft.ru/
21. Shynkar, S., Gontar, Z., Dubyn, M.: Assessment of economic security of enterprises: theoretical and methodological aspects. Bus. Theory Pract. **21**(1), 261–271 (2020). https://doi.org/10.3846/btp.2020.11573. Accessed 25 Jan 2021
22. Uktamov, K.: Problems of evaluation and procuring economic security at enterprises. Asian J. Technol. Manag. Res. **10**(1), 123–129 (2020). http://ajtmr.com/papers/Vol10Issue1/Vol10Iss1_P18.pdf. Accessed 25 Jan 2021
23. Wu, Y., Meng, F.: Categorizing security for security management and information resource management. J. Strateg. Secur. **11**(4), 72–84 (2019). https://doi.org/10.5038/1944-0472.11.4.1694. Accessed 21 Jan 2021

Accounting and Analytical System of Added Value from the Position of Controlling

Lyudmila V. Popova[1(✉)], Irina A. Maslova[1], Yulia A. Kotlova[2], Mikhail V. Selyukov[3], and Zoya V. Mkrtchyan[1]

[1] Orel State University named after I.S. Turgenev, Orel, Russia
[2] Khabarovsk State University of Economics and Law, Khabarovsk, Russia
[3] Surgut State University, Surgut, Russia

Abstract. Purpose: The purpose of the article is to determine the criteria for a systematic approach to accounting and analytical support of value-added management, to consider it from the standpoint of controlling tools and methods.

Design/methodology/approach: The authors highlight two key theses in the accounting and analytical provision of added value: the purpose of the functioning of an economic entity is reduced to the achievement of resistance through strategically directed accounting and analytical influences; in the process of controlling, there is a purposeful impact of the subject on the object, through certain methods, principles and management functions.

Findings: It is shown that modern economic subjects need tools for flexible management of their functioning parameters. The problem of ensuring the resistance of an economic entity lies in the underdevelopment of scientific and methodological support of the management process and added value, depending on the context and current needs of the economic entity.

Originality and value: A promising method for solving this problem is the use of controlling tools to assess and regulate based on data provided by the accounting and analytical system. Integration of the information clusters included in it allows generating information flows that provide a flexible response to the achieved values of the added value of an economic entity, thereby allowing it to be in a state of resistance.

Keywords: Added value · Controlling · Resistance · Cost approach · Accounting · And analytical system · Financial stability

JEL Code: M10 · M20 · M40

1 Introduction

In modern times, there is a need to control the parameters that arise in the external and internal environment. These parameters, as management tools, are necessary for the transparency of the strategy, for organizing accounting and reporting, for planning and forecasting, for generating goal-setting processes. For example, a tool such as sensitivity analysis is used when it is necessary to establish how adequately the management

E. G. Popkova (Ed.): ISC 2020, LNNS 368, pp. 702–710, 2022.
https://doi.org/10.1007/978-3-030-93244-2_76

decision-making model reacts to factors leading to changes or growth in structural costs, to the return on net operating capital, to a critical success factor, to constant, systematic and continuous improvement. And here the activity of the enterprise is aimed at setting goals, developing rules in the field of planning, regulation, control, testing, monitoring strategies. Controlling is such a management tool.

2 Methodology

The need to form integrated accounting and analytical systems, including for managing the cost of an organization, is emphasized in the works of Melnik, Popova, Maslova, Vasilyeva and Savkina, Sorokina) and others [7, 8, 13, 14]. At the same time, even though the formation and management of value-added have long been a subject of discussion among economists, it continues to cause controversy among economists and determines the ambiguity of the interpretation of this category. Moreover, as the study has shown, there is still no single methodological approach to defining the role of the accounting and analytical system in generating added value from the standpoint of controlling, which confirms the relevance of further studying this problem and finding ways to solve it.

There are two specific aspects to note:

- added value, as an economic category, is the main indicator of the effectiveness of the functioning of an economic entity;
- the processes of formation and assessment of added value within the organization are based on the principles of integration of the accounting and analytical information space formed by the accounting, reporting, analytical and control, and monitoring clusters.

The flexibility of managing the cost of an economic entity can be ensured by monitoring the tendencies of its change within the framework of controlling, based on the integration of the accounting and analytical information space [3].

3 Results

Controlling closely, through a system of indicators, interacts with accounting, analysis, reporting, that is, the main components of the accounting and analytical system. The use of a systematic approach makes it possible for controlling to combine indicators using both direct and feedback. Rybak O.V. writes that "the performance of the functions of analysis and control is possible subject to the formation of the optimal number of indicators, with the help of which it is possible to obtain information about the quality of management" [11]. Controlling is designed to provide an information system of service management functions for making effective and timely management decisions that are implemented through the accounting and analytical system.

Each system has its characteristics, which are classified by a plurality of elements, integrity, unity, and the presence of a certain structure. Also, each system has certain properties that allow it to focus on the final results, it acquires character and gives feedback, while it can be open or closed.

Applying a systematic approach, accounting and analytical information interconnected with controlling represents:

- purposeful influence of the subject on the object;
- resource movement management;
- long-term profit maximization [10].

Resistance management involves the implementation of a process of management influences. It is primarily aimed at achieving a normal state of financial stability of the enterprise. And we have the right to designate a goal that boils down to achieving resistance through strategically directed accounting and analytical influences (Fig. 1).

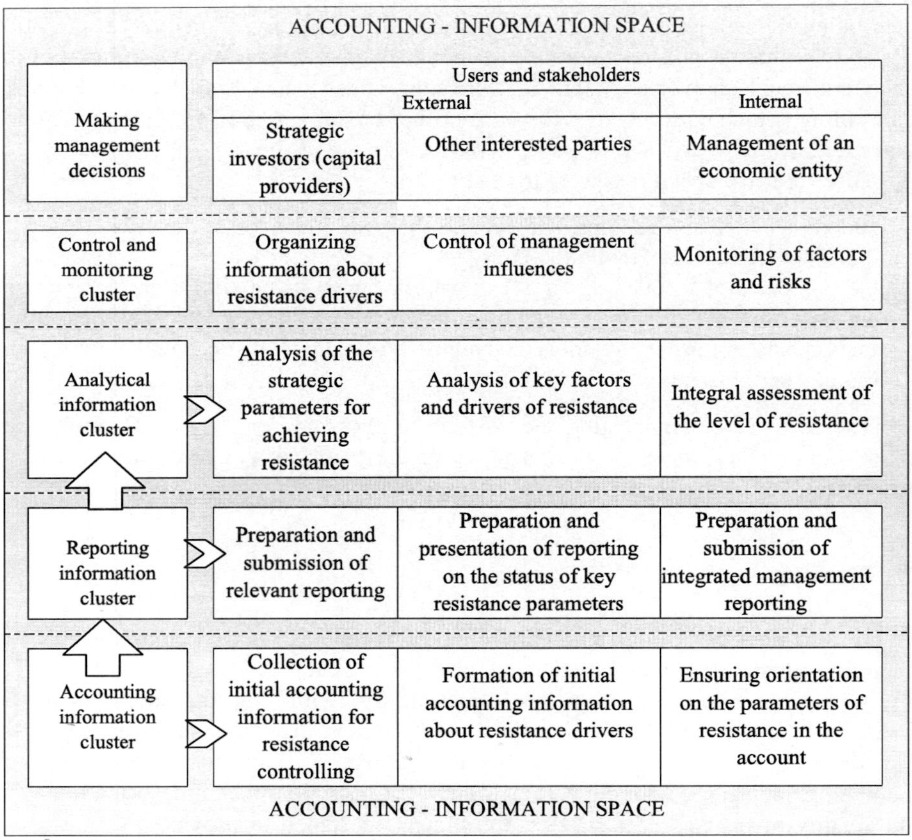

Fig. 1. The architecture of the system of accounting and analytical support for controlling the resistance of an economic entity. *Source* Compiled by the authors

For the goal to be achieved, it is necessary to determine the tasks related to financial and economic stability, growth in business value, profitability, cash flow growth, budgeting, and monitoring of indicators of the economic entity's resistance.

The history of the creation of the accounting and analytical system is based on three main vectors: operational, tactical, strategic, which were first described in detail by the scientist Popova L.V. in 1993–1995, and subsequently, we observe how the knowledge in this area of research was growing (Fig. 2).

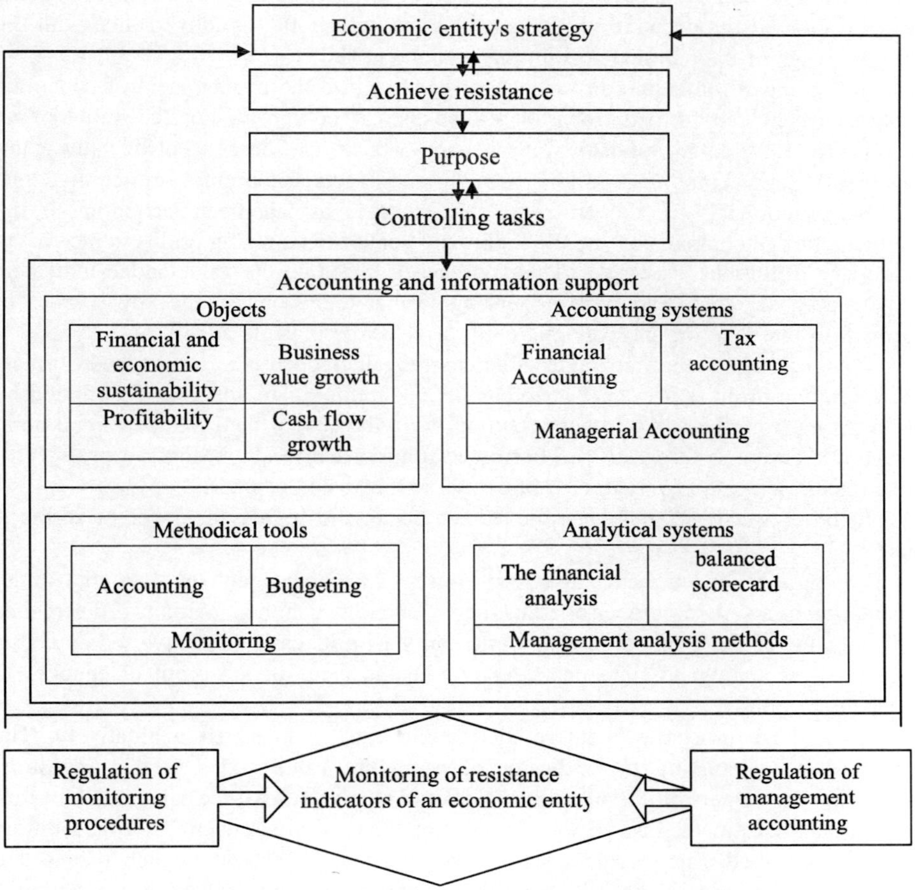

Fig. 2. Conceptual model of accounting and analytical support for controlling the resistance of an economic entity. *Source* Compiled by the authors

From the point of view of applying a systematic approach, the accounting and information space of controlling is an integrated system that combines certain elements and the relationship between them [9]. In the process of controlling, there is a purposeful impact of the subject on the object, through certain methods, principles, and management functions. It is this managerial impact, which is a management process, that is the "core" of the enterprise resistance management system.

In turn, one of the key system-forming criteria for the efficiency and sustainability of an enterprise is its value, the measurement of which makes it possible to qualify the business as a system (value creation), assess the management impacts on it (value assessment) and directly organize the process of managing it (value management).

It should be noted that over the past decade, the concept of "added value" has been actively used by economists, researchers, and practitioners. Such interest in this category is determined by "the new requirements of the domestic analytical complex for the organization of the accounting process of value indicators, its convergence with the requirements of the national accounting system" [13].

In this regard, in economic science, an approach to the management of economic entities have been formed, focused on the use of not profit, but the value of an economic entity as an indicator for assessing a business (the concept of value management). So, Sokolov Ya.V. in his work "Management accounting" argued that "one of the prerequisites for the emergence of strategic management accounting is the reorientation of companies from the traditional goal of maximizing profits to increasing the value of the business" [12]. This approach is based on the value-added indicator.

Modern scientists and practitioners agree that today one of the key sources of growth in the value of an economic entity as a whole is value-added.

It can be argued that the value of economic value-added is a transformed estimate of economic profit, since, as a general rule, economic value added is determined by subtracting from the adjusted net operating profit the cost of the invested capital of an economic entity (both equity and borrowed funds) used to obtain this net profit (product of the weighted average cost of capital and invested capital).

In other words, "the higher the added value, the higher the level of resource use" [4].

The application of the value-added concept in management involves the implementation of a certain number of adjustments, due to which the maximum convergence of the value of the profit of an economic entity with its cash flow, invested capital is achieved, excluding possible interference with the value of the profit of changes in accounting policy.

Only those processes that are measurable can be managed qualitatively. The meaning of calculating the indicator of economic value added is that investors, including the owners of the organization, should receive at least the same rate of return as similar investment risks in the capital market. Otherwise, even if the accounting profit is received, there may be no real profit from the activity, which means that investors do not receive benefits from the investment activities of the organization.

In continuation of the above, we note that the value of added value can well be determined not only for the organization as a whole but also separately for each business area and even for the product (work, service) produced.

So, in the accounting and analytical system of the organization, in the course of managing its value, it is possible to calculate several value indicators, each of which claims to be the main criterion of value: MVA (added market value), EVA (added economic value), CVA (net added value), SVA (shareholder value-added), CFROI (cash return on investment) [2].

Thus, most economists agree that the indicator of economic value added is a universal cost criterion of business efficiency, and, as a consequence, a criterion for making managerial decisions in the course of managing the value of an organization. Along with indicators of the financial and economic stability of an economic entity, it can be successfully used in assessing its resistance. This means that all management tools must obey the task of maximizing the added value of an economic entity.

Moreover, adherents of the concept of value management T. Copeland, T. Koller, D. Murrin in their famous publication "Company Value: Assessment and Management" interpret value management as "an integrating process aimed at qualitatively improving strategic and operational decisions at all levels of the organization" [1].

The system of accounting and analytical support of each enterprise is unique, and its structure is determined by the peculiarities of the "information needs of managers, the current regulatory framework (Fig. 3).

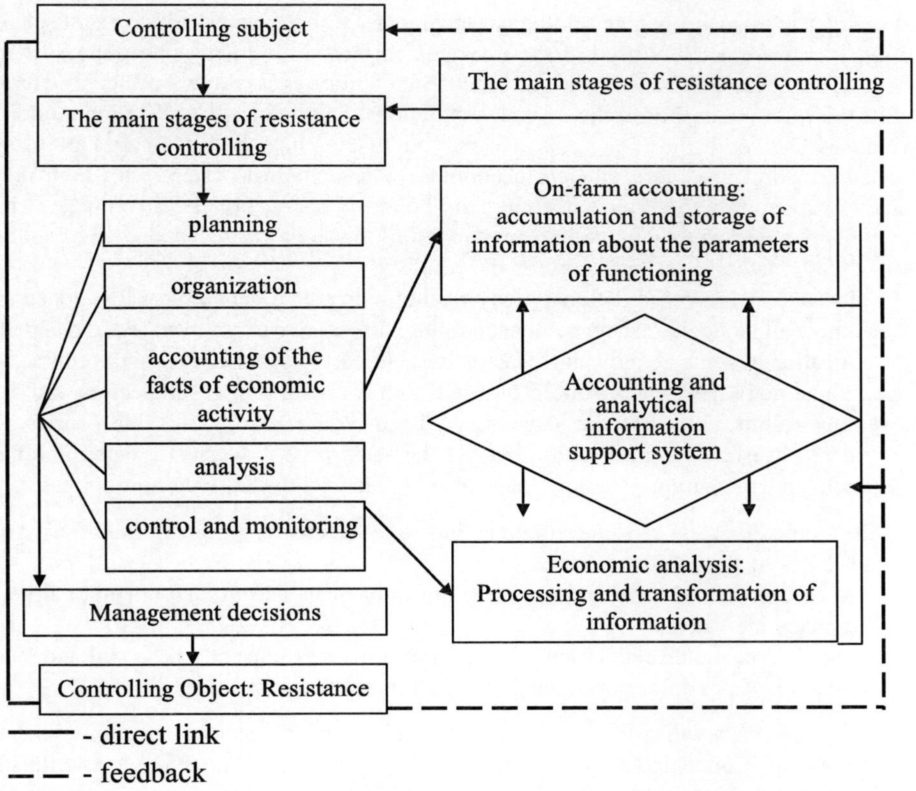

Fig. 3. Accounting and analytical support for controlling the resistance of an economic entity. *Source* Compiled by the authors

Currently, there is no unified concept of controlling the resistance of an economic entity and its information support. This problem is aggravated by the fact that many subjects, depending on the scale and industry of operation, have their characteristics that affect the formation of the information field. Information related to the resistance controlling process should be classified according to the information cluster.

In this regard, it is obvious that the most efficient way to carry out the cost management process is possible within the framework of controlling, which, as mentioned above, interacts with the accounting and analytical system of the organization through a system of indicators [6]. A holistic system of controlling an economic entity should give an objective picture of reality. Management of various levels, selecting part of the relevant information, should be able to obtain all the necessary information about his condition, fully and objectively reflecting reality.

One of the ways to improve the quality of the information in the accounting and analytical system of an economic entity is the integration of indicators of its added value with a balanced system of indicators. Both of these systems are equally well consistent with such his methodological techniques as budgeting and long-term planning, which are actively used in the management accounting and controlling system both at the level of management of an economic entity as a whole, and its structural units – centers of responsibility. Such integration of the concept of economic added value with other controlling tools allows you to combine the advantages of these systems to obtain a more efficient accounting and analytical system with high predictive capabilities [5]. So, concerning the balanced scorecard, the advantage is its integrity and consistency in the implementation of the strategy, in the concept of added value - the mathematical accuracy of the results.

Moreover, to increase the analyticity of the indicator of economic value added in the contour of strategic management accounting, it is allowed, and even recommended, the allocation of separate indicators (factors or value drivers) controlled in the course of operational activities, which should be used both for setting target indicators and for assessing results work. At the same time, the drivers (factors) of value cannot be considered in isolation from each other. At the same time, to correctly determine the factors of value creation, it is important to be guided by the following principles:

a) the factors (drivers) of value should be linked to the creation of value and brought to all levels of the economic entity;
b) cost factors should be measured using not only financial but also operational performance indicators;
c) value drivers should reflect both the level of efficiency currently achieved and take into account possible long-term prospects for further growth [3].

This approach will allow ensuring the relationship of each cost factor with the indicators based on which management decisions are made, and therefore to coordinate management decisions with their consequences for the cost of an economic entity. All these elements together form the so-called value thinking, without which it is impossible to implement the concept of resistance controlling and genuine cost management.

At the same time, we note that the determination and constant monitoring of the values of added value makes sense only if the management of the economic entity pays due attention to the allocation of centers of responsibility, as well as to the system of motivating performers for the level or amount of personal contribution of each of them to the achievement of the overall result.

4 Conclusion

Thus, the accounting and analytical system of an economic entity accumulate all possible data streams, through which it is possible to assess the features, nature, and results of the functioning of an economic entity, and, as a consequence, the circulation of its added value. It is in the controlling system that an integrated accounting and analytical system that combines three types of accounting: financial, management and tax, in conjunction with analytical procedures, implements the possibility of consolidating information flows of data reflecting the state of an economic entity, its capabilities, and prospects for achieving resistance, and therefore and further successful development and functioning.

References

1. Copeland, T., Koller, T., Murrin, J.: Valuation: Measuring and Managing the Value of Companies, 3rd edn., p. 576. McKinsey & Company, New York (2000)
2. Danko, T.P., Golubev M.P.: Value-Oriented Management and Marketing: Textbook, 94 p. (2011)
3. Kotlova, Yu.A.: Conceptual foundations of economic value added in the management accounting system. Bull. Khabarovsk State Univ. Econ. Law (1), 47–55 (2018)
4. Kuzmin, D.A.: Application of controlling tools in the process of managing the company's value. Econ. Reg. (2), 219–224 (2008)
5. Maslova, I.A.: Formation and distribution of value in the accounting and analytical system during the transition to IFRS: monograph, 135 p. (2005)
6. Maslova, I.A.: Theoretical and methodological aspects of the accounting and analytical system for the formation of production costs in the industry. Manag. Account. (7), 89–100 (2009)
7. Melnik, M.V.: Modernization of accounting, control and analytical processes in the digital economy. Account. Anal. Audit (5), 129–130 (2018)
8. Popova, L.V.: An Integrated System of Accounting, Analysis, and Audit in the Economic Mechanism of Commercial Activity, p. 364. Russia, Moscow (1996)
9. Popova, L.V., Maslova, I.A., Bondareva, A.Yu.: Conditions for using the fair value in practice. Audit Statements (1–2), pp. 76–84 (2017)
10. Popova, L.V.: Financial management and its role in the modern management system. Econ. Humanit. Sci. (8(235)), 87–91 (2011)
11. Rybak, O.V.: Theoretical features of determining the controlling indicators of diagnostics of the productive activity of the enterprise. Manag. Account. (12), 36–43 (2013)
12. Sokolov, A.Yu.: Management Accounting (Advanced Level). Lecture Notes. KSU, Kazan, 70 p. (2014)

13. Sorokina, M.S.: Element-Wise Assessment of Added Value Based on the Principles of Formation of a Single Accounting Space. Oryol, Russia, 198 p. (2014)
14. Vasilieva, M.V., Savkina E.S.: Methodology for assessing the added value in the conditions of the functioning of the complex of accounting and analytical support for social and sectoral budgetary support. Econ. Humanit. Sci. (10), 55–62 (2012)

Labor Market and Employment Problems: Analysis of Long-Term Dynamics and Prospects of Development in Russian Regions

Inessa G. Nazarova[1], Gulnaz F. Galieva[2,3], Ekaterina V. Sazanova[4], Eleonora M. Chernenko[5], and Evgeniya K. Karpunina[6(✉)]

[1] Ukhta State Technical University, Ukhta, Russia
[2] Financial Research Institute of the Ministry of Finance of the Russian Federation, Moscow, Russia
[3] Ufa State Petroleum Technological University, Ufa, Russia
[4] Northern (Arctic) Federal University named after M. V. Lomonosov, Arkhangelsk, Russia
e.sazanova@narfu.ru
[5] Kuban State Technological University, Krasnodar, Russia
[6] G.R. Derzhavin Tambov State University, Tambov, Russia

Abstract. The study was aimed at conducting a comprehensive analysis of the long-term dynamics of the labor market in the Russian regions, as well as evaluating the effectiveness of the implemented state policy to maintain employment and reduce unemployment during the COVID-19 pandemic. The analysis of long-term trends in the labor market allowed the authors to identify and assess the "tension" points of the current model of the Russian labor market including a decrease in the number of employed in 2010–2019, a decline in the level of employment in the regions, a reduced average annual number of employed in the system-forming sectors of the economy (agriculture, manufacturing, education, and healthcare), the existence of a discrepancy between the level of official and hidden unemployment. The authors calculated the level of tension in the labor market and formed a relevant rating of Russian regions. The authors assessed the negative impact of the COVID-19 pandemic on the long-term dynamics of the Russian labor market at the federal district level. The mass transfer of employees to remote work mode required the implementation of emergency measures to stimulate employment in the regions. The authors propose a set of measures to overcome the reduction in the level of employment and the growth of unemployment in the long term to form an effective model of the Russian labor market.

Keywords: Unemployment · Employment · Region · Labor market · Labor force · Hidden unemployment · COVID-19 pandemic · Russia

JEL Codes: J21 · J64 · R23 · R28

E. G. Popkova (Ed.): ISC 2020, LNNS 368, pp. 711–722, 2022.
https://doi.org/10.1007/978-3-030-93244-2_77

1 Introduction

The state of the labor market is a direct reflection of the laws of social development and the division of labor. The influence of external factors of turbulence, intra-country contradictions, and regional development imbalances determine the emergence of problem areas in the field of employment and the actualization of issues of regulating unemployment.

The development of the Russian labor market can be divided into several stages. The first stage (1991–1998) is associated with its creation in the context of the breakdown of the Soviet planned system of economic management and the global transformation caused by the transition of the Russian economy to a market system of management. At the second stage (1999–2008), the labor market was formed against the background of a positive economic situation in the country, including due to the growth of incomes of enterprises and households [21]. The third stage (2008–2015) is associated with overcoming the global crisis when the state developed effective measures to maintain employment and prevent a sharp increase in unemployment against the background of the global crisis [20]. The post-crisis stage (2015–2019) is determined by the development of labor relations in the context of mass automation, robotization, and the implementation of digital technologies [11, 13]. Computerization and digitalization have provoked the emergence of new types of vacancies in the labor market and an increase in their number while reducing the demand for "traditional" working specialties. The coronavirus pandemic, which broke out in 2020, gave an impetus to the transition to a new stage in the development of the labor market - the "coronavirus stage".

The specifics of the Russian labor market during the passage of the first two stages are sufficiently covered in scientific sources. However, the patterns of its development in crisis periods, from 2010 to the present, are poorly understood. In particular, there is practically no description of the regional trends in the development of the Russian labor market during this period. Since the regional development of Russia is extremely polarized, timely diagnostics of emerging employment problems at the level of individual territories is required.

2 Literature Review

The problems of unemployment and employment were studied in detail by the founders of economic science (A. Smith, D. Ricardo, J.-B. Say, A. Marshall, A. Pigou, K. Marx, J. M. Keynes). Currently, the study of the problems of employment and unemployment in the Russian Federation is devoted to the works of economists Ramazanova [17], Gimpelson et al. [9], Chizhova et al. [5]. The analysis of the materials of scientific publications allows the authors to identify the following global trends in the development of the Russian labor market. First, the trend of a decline in production, a fall in labor productivity, and the corresponding adaptation of the labor market by reducing the cost of labor, downsizing working hours, and changes in the level of employment. This adaptation has led to the widespread use by employers of such mechanisms as part-time work, forced leave, delayed wages, shadow payments, and in-kind payment, a decrease in the variable part of wages, which makes up a significant share of wages The analysis of the materials of scientific publications allows the authors to identify the following global trends in the development of the Russian labor market. Firstly, the trend of a decline in production, a fall in labor productivity, and the adaptation of the labor market by reducing the cost of labor, lowering working hours, and changes in the level of employment. This adaptation has led to the widespread use by employers of such mechanisms as part-time work, forced leave, delayed wages, shadow payments, and in-kind payment, a reduction in the variable part of wages, which makes up a significant share of wages [5, 9]. Secondly, the active use by employers during periods of economic downturn and global economic crises of the mechanism of "informal" employment in the labor market with a shadow wage system. The mechanism of hidden unemployment only formally means that there is a job with actual unemployment [3, 8, 18, 19, 26, 27]. Thirdly, the trend of forming a "modern" portrait of the labor market: low official unemployment, stable employment; the uncontrolled gap between general and registered unemployment, the flexibility of working hours and wages; unregistered employment and mass shadow wages [5, 23, 25]. Fourthly, the low actual "insensitivity" of employment to changes in the economic environment against the background of the successful application of the developed measures of anti-crisis employment management in Russia [9, 16]. Fifthly, expert and predictive testing of previously developed practical methods for studying the Russian labor market in stable and crisis periods [2, 5, 14, 15, 24]. Sixth, the coincidence of the Russian way of developing the labor market with the experience of developed countries against the background of falling employment and rising unemployment [4, 9, 10, 14]. Seventhly, the negative impact of the 2014–2015 crisis on the labor market, as well as the implementation of a low-unemployment model and a sharp decline in real wages [9, 12, 14]. Eighthly, the presence of long-term negative trends in the Russian labor market, such as the lag of the educational services market from the demands of the economy, the excess of labor supply over demanding with a growing discrepancy between the structure of labor demand and the structure of its supply, the presence of illegal labor migration, low competitiveness of the labor force in the world labor market [22]. Ninthly, the "coronavirus" impact on employment under quarantine and economic measures, as well as the emergence of risk of deterioration in the quality of human capital. Tenthly, the heterogeneous structure of the Russian labor market with the formation at the regional level of segments, patterns, sectors, and spheres with pronounced territorial features [1].

3 Methodology

The purpose of the study is to conduct a comprehensive analysis of the long-term dynamics of the labor market in the Russian regions, as well as to assess the implemented state policy to maintain employment and reduce unemployment during the COVID-19 pandemic.

Research objectives: 1) conducting a comprehensive analysis of long-term regional trends in the Russian labor market, as well as identifying employment problems and increasing unemployment in the regions; 2) assessment of the negative impact of the

COVID-19 pandemic on the long-term dynamics of the labor market at the federal district level; 3) identification of measures to stabilize the situation in the labor market in the Russian regions and overcome negative long-term trends in it.

Research methods: the theoretical analysis method, the method of comparison, the graphical method, the method of systematization, the economic analysis method, the system approach.

The information base of the research is the materials of the Federal State Statistics Service of the Russian Federation on the situation on the labor market and in the field of employment, including by regions of Russia, as well as statistical materials of regional authorities and materials of scientific publications.

4 Results

To identify long-term trends in the Russian labor market for the period 2010–2019, we will begin by analyzing the changes in the country's population by federal districts (Table 1).

Table 1. Dynamics of the average annual population (thousands of people), 2010–2019

Federal district	2010	2019	2019/2010 * 100
Southern Federal District	13,852.5	16,460.3	118.83
North Caucasus Federal District	9,396.1	9,898.9	105.35
North-Western Federal District	13,615.0	13,977.0	102.66
Central Federal District	38,390.3	39,405.8	102.65
Ural Federal District	12,087	12,355.4	102.22
Siberian Federal District	17,191	17,145.9	99.74
Volga Federal District	29,937	29,342.5	98.02
Far Eastern Federal District	8,380.6	8,178.9	97.59
Russian Federation	142,849.5	146,764.7	102.74

Source Compiled by the authors based on [6]

In Russia, the average annual population in 2010–2019 increased by 102.74%, and only in two federal districts (the Southern Federal district, the North Caucasus Federal district), the average annual population growth was higher than the national average. In most of the districts (North-Western, Central, Ural, Siberian, Volga, and Far Eastern), a negative trend of population decline can be noted. The negative trend of the lack of population growth in the country affected the dynamics of the average annual number of employees in 2010–2019 (Fig. 1).

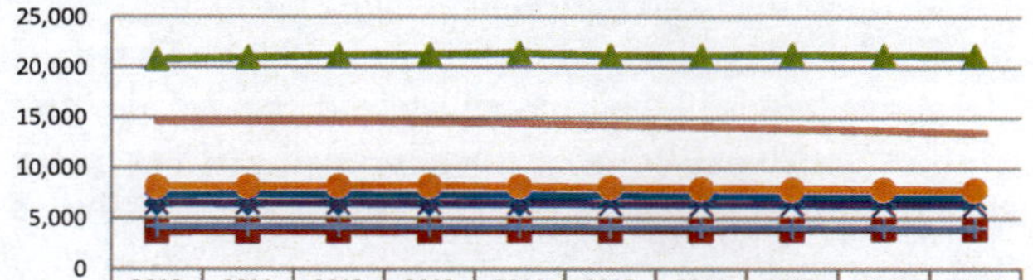

	2010	2011	2012	2013	2014	2015	2016	2017	2018	2019
Southern Federal District:2019 to 2010,% -114.5	6,478.6	6,470.1	6,505.4	6,493.1	6,436.1	7,442.5	7,402.8	7,455.0	7,449.8	7,417.3
North Caucasus Federal District:2019 to 2010,% -106.9	3,650.4	3,683.3	3,694.6	3,711.8	3,742.4	3,745.8	3,778.9	3,839.6	3,913.6	3,904.1
Central Federal District:2019 to 2010,% -101.9	20,782.3	20,956.7	21,184.5	21,254.4	21,390.2	21,178.0	21,181.9	21,259.7	21,198.5	21,171.2
Ural Federal District:2019 to 2010,% -97.9	6,458.3	6,497.9	6,467.9	6,482.7	6,462.1	6,425.9	6,347.1	6,366.7	6,353.0	6,322.1
North-Western Federal District:2019 to 2010,% -96.9	7,286.3	7,396.1	7,367.2	7,363.2	7,330.6	7,296.8	7,251.1	7,161.5	7,080.4	7,065.3
Siberian Federal District:2019 to 2010,% -95.9	8,121.2	8,150.6	8,206.3	8,217.4	8,166.0	8,061.3	7,918.2	7,865.6	7,852.8	7,795.2
Far Eastern Federal District:2019 to 2010,% -95.6	4,136.5	4,168.6	4,158.1	4,142.7	4,103.8	4,053.7	4,069.0	4,039.7	4,030.6	3,954.7
Volga Federal District:2019 to 2010,% -92.1	14,579.5	14,577.5	14,613.7	14,511.2	14,449.4	14,221.0	14,116.2	13,854.8	13,683.0	13,434.6

Fig. 1. Average annual number of employees in the regions of Russia, thousands of people, 2010–2019. *Source* Compiled by the authors based on [6]

The negative trend of reducing the population, including the working-age population, affected the average annual number of employees. In Russia, in 2010–2019, it decreased by 0.6% or 428.6 thousand people. The average annual number of employees in the analyzed period increased in the Southern (+938.7 thousand people or 14.5%), North Caucasus (+253.7 thousand people or 6.9%), and Central Federal district (+388.9 thousand people or 1.9%). The decrease in employment was observed in the Ural (−136.2 thousand people or 2.1%), the North-Western (−221 thousand people or

	North Caucasus Federal District	Far Eastern Federal District	Central Federal District	Ural Federal District	Southern Federal District	Russian Federation	Siberian Federal District	Volga Federal District	North-Western Federal District
Change in the employment rate of the population, in%, for the period from 2010 to 2019 (2019-2010)	1.0	-1.2	-2.8	-2.9	-3.2	-3.3	-4.0	-4.6	-5.0

Fig. 2. The level of employment of the population in the regions of Russia, % of the total population, 2010–2019. *Source* Compiled by the authors based on [6]

3%), the Siberian (−326 thousand people or 4%), the Far East (−181.8 thousand people or 4.4%), the Volga Federal district (−1144.9 thousand people or 7.9%).

Negative trends in the Russian labor market are also confirmed by the decline in the level of employment in the regions in 2010–2019 (Fig. 2).

Conducting an industry analysis of changes in the share of the average annual number of employees will help to structure the problems of the Russian labor market while maintaining the negative trend of declining employment (Fig. 3).

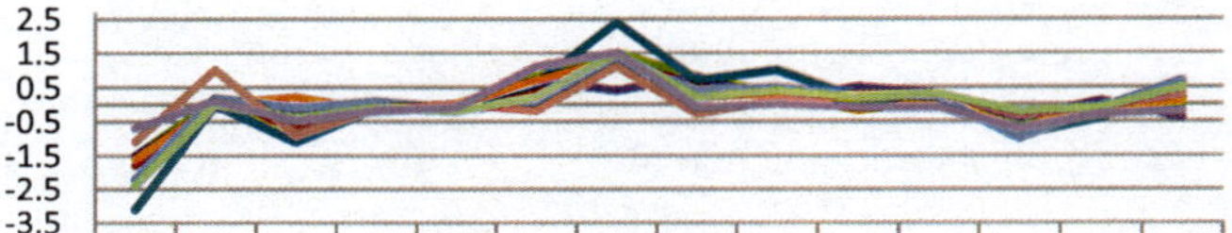

	Agricultural industry	Mining operations	Processing plants	Providing resources	Water supply, waste collection and disposal	Construction	Wholesale and retail trade	Transportation and storage	Public catering and hotels	Information and communication	Realty	Education	Health and social services	Other activities
Russian Federation	-1.8	0.1	-0.7	-0.1	0.0	0.4	1.5	0.5	0.4	0.2	0.2	-0.6	-0.2	0.2
Central Federal District	-1.5	0.1	-1.1	-0.1	-0.1	0.8	1.5	0.8	0.3	0.1	0.1	-0.6	-0.3	0.0
North-Western Federal District	-1.5	-0.1	-0.3	-0.1	-0.1	0.7	0.4	0.8	0.1	0.5	0.3	-0.4	0.1	-0.4
South federal district	-3.1	0.0	-1.1	-0.1	0.0	0.6	2.4	0.7	1.0	0.2	0.2	-0.9	-0.4	0.6
North Caucasus Federal District	-1.6	0.0	0.2	-0.2	0.0	0.7	1.3	0.2	0.3	-0.2	0.1	-0.7	-0.1	0.1
Volga Federal District	-2.2	0.2	-0.1	0.1	-0.1	-0.1	1.2	0.4	0.5	0.3	0.2	-1.0	-0.1	0.7
Ural Federal District	-1.1	1.0	-0.9	-0.1	0.0	-0.2	1.1	-0.3	0.1	0.3	0.3	-0.3	-0.1	0.3
Siberian Federal District	-2.4	0.0	-0.4	-0.1	-0.2	0.2	1.5	0.2	0.4	0.2	0.3	-0.2	-0.1	0.5
Far Eastern Federal District	-0.7	0.1	-0.5	-0.2	-0.1	1.1	1.5	-0.1	0.0	-0.1	-0.1	-0.7	-0.3	-0.2

Fig. 3. The change in the share of the average annual number of employees by economic sector in the regions of Russia, 2019 relative to 2010. *Source* Compiled by the authors based on [6]

The above graph allows us to identify several problem areas of reduction in the average annual number of employees: agriculture (a decrease in the share of the total structure of employment in 2019 compared to 2010 in all federal districts); manufacturing (a decrease in the country as a whole and all regions, except for the North Caucasus Federal district). The share of people employed in education and healthcare has decreased. Significant growth is typical only for the retail and wholesale trade sectors. This raises concerns, as the share of trade continues to grow in the overall structure of the economy, and the share of the "backbone" sectors of the Russian economy - manufacturing, mining, agriculture, health, and education - tends to decline.

The existing problems of unemployment inherent in the Russian labor market are compounded by the low level of official unemployment, as well as the presence of shadow and hidden unemployment (Fig. 4).

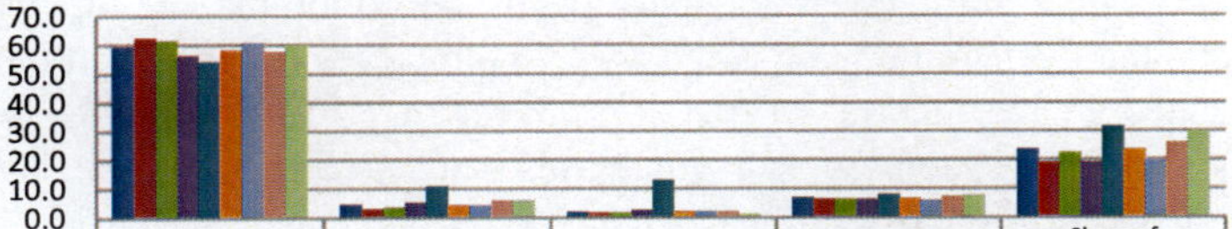

	Employment rate of the population aged 15 years and older, percent	Unemployment rate, percent	Tension coefficient	Average time for unemployed people to find a job months	Share of unemployed people looking for work 12 months or more, percent
Russian Federation	59.4	4.6	2.1	6.9	23.8
Central Federal District	62.3	2.9	1.6	6.4	19.0
North-Western Federal District	61.6	3.6	1.7	6.4	22.4
South federal district	56.5	5.3	2.6	6.5	19.2
North Caucasus Federal District	54.3	11.0	13.0	8.2	31.7
Volga Federal District	58.1	4.2	2.1	6.6	23.5
Uralsky District federal district	60.7	4.3	2.2	6.1	19.6
Siberian Federal District	57.5	5.9	2.0	7.1	25.7
Far Eastern Federal District	60.0	6.0	1.1	7.7	29.8

Fig. 4. Selected indicators of labor market tension in the regions of Russia, 2019. *Source* Compiled by the authors based on [6]

The data in Fig. 4 allows us to calculate the weighted average rating of the federal districts of Russia by the level of "tension" in the labor market in 2019 (Table 2).

Table 2. Rating of tension in the Russian labor market in 2019

Federal district	Employment rate	Unemployment rate	Tension coefficient	Average time for unemployed people to find a job	Share of unemployed people looking for work 12 months or more	Summary rating on the level of tension n 2019
Central Federal District	1	1	2	3	1	1.6
North-Western Federal District	2	2	3	2	4	2.6
Ural Federal District	3	4	6	1	3	3.4
Volga Federal District	5	3	5	5	5	4.6
Southern Federal District	7	5	7	4	2	5
Far Eastern Federal District	4	7	1	7	7	5.2
Siberian Federal District	6	6	4	6	6	5.6
North Caucasus Federal District	8	8	8	8	8	8

Source Compiled by the authors

The minimum level of “tension” of the labor market in 2019 (calculated as a weighted average number based on five indicators of tension in the labor market) was reached in the Central (1.6), North-Western (2.6), Ural (3.4), Volga (4.6) Federal districts. The Southern (5), Far Eastern (5.2), Siberian (5.6), and North Caucasus (8) Federal districts demonstrate significant labor market tension. We will conduct a more detailed analysis of the unemployment rate in 2010–2019 (Fig. 5).

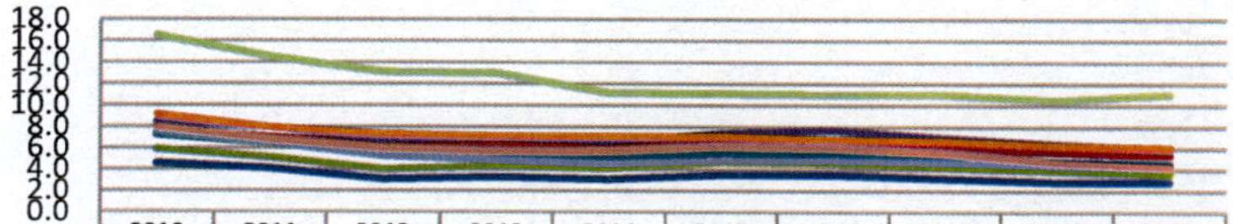

	2010	2011	2012	2013	2014	2015	2016	2017	2018	2019
Central Federal District	4.6	4.1	3.1	3.3	3.1	3.5	3.5	3.2	2.9	2.9
Southern Federal District	7.6	7.0	6.2	6.5	6.2	6.7	6.4	6.0	5.6	5.3
North-Western Federal District	5.9	5.1	4.0	4.3	4.1	4.7	4.6	4.2	3.9	3.6
Siberian Federal District	8.5	7.9	6.8	7.0	6.7	7.4	7.7	7.0	6.4	5.9
Russian Federation	7.3	6.5	5.5	5.5	5.2	5.6	5.5	5.2	4.8	4.6
Far Eastern Federal District	9.1	7.9	7.3	7.1	7.1	7.1	6.8	6.7	6.3	6.0
Volga Federal District	7.6	6.5	5.3	4.9	4.5	4.8	4.8	4.7	4.4	4.2
Ural Federal District	8.0	6.8	6.0	5.7	5.8	6.2	6.1	5.6	4.7	4.3
North Caucasus Federal District	16.5	14.5	13.1	13.0	11.2	11.1	11.0	11.0	10.5	11.0

Fig. 5. Unemployment rate in Russian regions, %, 2010–2019. *Source* Compiled by the authors based on [6]

In 2010–2019, the official unemployment rate decreased in all federal districts of Russia. Short-term growth was recorded during the crisis of 2015 when the unemployment rate in the whole country increased from 5.2% to 5.6%. The largest decrease in the unemployment rate in 2010–2019 occurred in the North Caucasus (−5.5%), the Ural (−3.7%), the Volga (−3.4%), and the Far Eastern Federal District (−3.1).

Assessing long-term trends in the labor market requires identifying discrepancies between the level of official and hidden unemployment (as the ratio of the economy's need for workers to the level of registered unemployed) (Fig. 6).

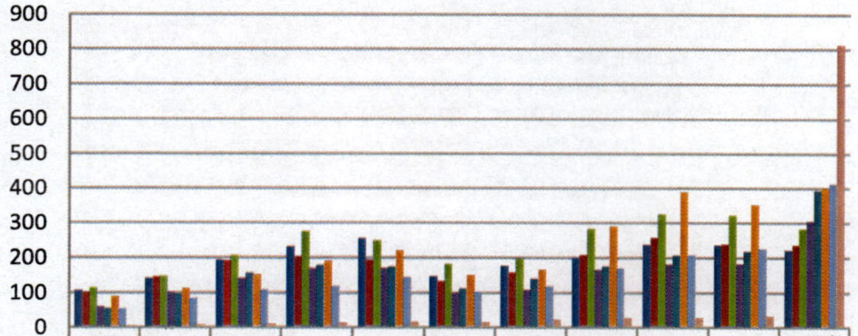

	2 010	2 011	2 012	2 013	2 014	2 015	2 016	2 017	2 018	2 019	2019/2 010*10 0,%
North-Western Federal District	106	142	194	232	256	147	178	200	237	236	222
Southern Federal District	102	145	191	202	193	133	157	207	257	239	235
Central Federal District	114	147	206	274	248	181	197	282	325	322	283
Ural Federal District	60	103	141	170	171	99	108	166	181	183	305
Volga Federal District	55	98	157	180	175	113	141	177	208	218	394
Far Eastern Federal District	88	112	151	190	219	149	165	289	388	351	400
Siberian Federal District	55	85	109	120	145	103	120	171	209	227	413
North Caucasus Federal District	4	6	9	12	16	14	22	27	28	32	812

Fig. 6. The discrepancy between the level of official and hidden unemployment, %, 2010–2019
Source Compiled by the authors based on [6]

The figure shows that there is a clear trend of divergence between the need for workers and the number of registered unemployed. In 2010, the maximum value of this indicator was recorded in the Central Federal district (114). In the future, there was a constant increase in the ratio of the number of required workers to the level of registered unemployed. In 2019, in the Central Federal district, this indicator was 322. The growth of this indicator (the level of 2019 to the level of 2010 in %) indicates a decrease in the need for the unemployed to register with state bodies.

The coronavirus pandemic has had a negative impact on the long-term trends in the Russian labor market. During the pandemic, there were significant changes in the labor market, in particular, the mass transfer of employees to remote work. This required the development of new measures to support employment, including incentives for employers to maintain the level of employment of the population and to seek other forms of cost reduction. However, the implementation of the above measures has not yet proved its effectiveness (Fig. 7).

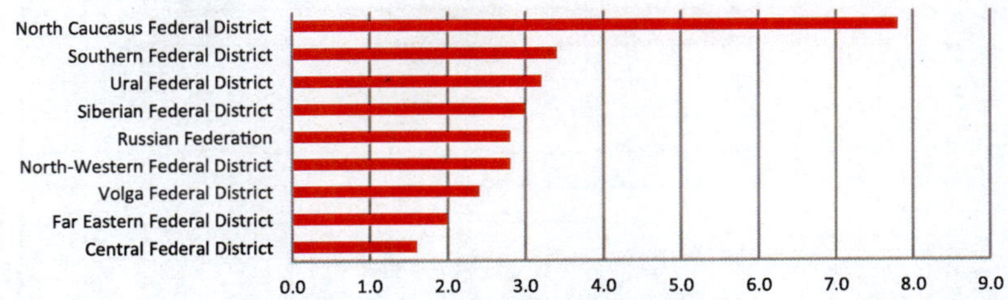

	Central Federal District	Far Eastern Federal District	Volga Federal District	North-Western Federal District	Russian Federation	Siberian Federal District	Ural Federal District	Southern Federal District	North Caucasus Federal District
■ Change in the registered unemployment rate in December 2020 to December 2019, 2020-2019 %	1.6	2.0	2.4	2.8	2.8	3.0	3.2	3.4	7.8

Fig. 7. Change in the official unemployment rate, %, December 2020 to December 2019 *Source* Compiled by the authors based on [7]

In all federal districts, the unemployment rate has increased. In some federal districts (Central, Far Eastern, Volga, North-Western), the dynamics of changes in the unemployment rate is better than the national average. In the Siberian, Ural, Southern, and North Caucasus Federal districts, it is more critical. The problem of rising unemployment at the "coronavirus" stage requires urgent measures to support employment and develop an anti-crisis program.

The formation of an effective model of the Russian labor market requires the implementation of new measures to ensure the growth of employment and reduce the unemployment rate in the regions:

- developing and implementing a model for assessing the discrepancy between the level of official and hidden employment and unemployment, based on the assessment of declared vacancies or labor needs;
- stimulating the flow of labor from the wholesale and retail sectors to agriculture and manufacturing;
- ensuring the inflow of skilled labor into the problematic and system-forming sectors of the Russian economy (introduction of a mass system of personnel re-profiling with the allocation of a target and priority segment of the unemployed population of the country - young people aged 15 years and up to 30 years);
- developing and implementing common criteria and approaches to the adjustment of state programs to increase employment and reduce unemployment to avoid the risk of a long-term and stagnant crisis in the labor market (ensuring a high and protected level of wages, maximizing the use of effective mechanisms to protect against layoffs, developing a system of support for the unemployed, increasing the role of trade unions in defending the interests of employees);
- increasing the role of the digital factor in priority and system-forming industries to create new types of vacancies on the labor market, while simultaneously reducing the demand for "traditional" working specialties;

- ensuring the migration flow of skilled labor by simplifying the procedure for registering migrants and increasing social state payments, for example, reducing taxes and fees for qualified foreign workers.

5 Conclusion

The analysis of the employment situation in 2010–2020 allowed the authors to identify the "tension" points of the current model of the Russian labor market, as well as to identify measures that can ensure employment growth and reduce the unemployment rate in the regions of Russia. Firstly, the authors identified the trend of a decline in the number of employees in 2010–2019 and a decline in the level of employment in the regions of the country, identified areas of reduction in the average annual number of employees in agriculture, manufacturing, education, and healthcare. Secondly, the authors highlighted the problem of low official unemployment, the presence of shadow and hidden unemployment, and the existence of a discrepancy between the level of official and hidden unemployment. The authors calculated the level of tension in the labor market and formed a rating of Russian regions. Thirdly, the authors pointed to the negative impact of the pandemic on the labor market and employment in the Russian regions. The mass transfer of employees to remote work mode required the implementation of emergency measures to stimulate employment in the regions. Fourth, the authors formulated proposals to ensure the growth of employment and reduce the level of unemployment in the regions as part of the formation of an effective model of the Russian labor market.

References

1. Akhapkin, N.: Structural changes in the Russian labor market (regional aspect). Bull. Inst. Econ. Rus. Acad. Sci. **6**, 26–41 (2019)
2. Astafyeva, E., et al.: Some Approaches to Modeling Individual Macroeconomic Indicators. Scientific Works, Moscow (2019)
3. Bobkov, V., Lok'tyukhina, N., Rozhkov, V., Chernyshova, M.: Informal employment: trends of decline and research priorities. Manag. Megalop. **1**(43), 29–43 (2015)
4. Burjalov, F., Gontmakhera, E.: Labor Market: Reaction to the Crisis (Based on Materials from Foreign Countries). IMEMO RAS, Moscow (2011)
5. Chizhova, L., Sadovaya, E., Kuzmin, V.: Innovative Economy: Employment, Labor Motivation, Labor Efficiency. Institute of Macroeconomic Research, Moscow (2011)
6. Federal State Statistics Service of the Russian Federation: Statistical Collection "Russian Regions. Socio-economic indicators 2020 (2021). https://rosstat.gov.ru/folder/210/document/13204. Accessed 12 May 2021
7. Federal State Statistics Service of the Russian Federation: Report "Socio-economic situation of Russia" (2021). https://rosstat.gov.ru/compendium/document/50801. Accessed 12 May 2021
8. Gimpelson, V., Kapelyushnikov, R.: In the Shadow of Regulation: Informality in the Russian Labor Market. HSE House, Moscow (2014)

9. Gimpelson, V., Kapelyushnikov, R., Roshchin, S.: Russian Labor Market: Trends, Institutions, Structural Changes. Center for Strategic Research, Moscow (2017)
10. Kapelyushnikov, R., Gimpelson, V.: The Russian model of the labor market: the test of the crisis. J. New Econ. Assoc. **2**(26), 249–254 (2015)
11. Karpunina, E., Isaeva, E., Abalakina, T., Shilova, I., Kulikova, M.: The development of human capital as a factor of economic security in the digital age. In: Proceedings of the 2nd International Conference CEED 2019, Beijing, China, 26–27 October 2019, pp. 51–56 (2019)
12. Maleva, T.: The Economic Crisis is a Social Dimension. Information and Analytical Bulletin. RANEPA, Moscow (2015)
13. Molchan, A., Karpunina, E., Kochyan, G., Petrov, I., Velikanova, L.: Effects of digitalization: new challenges for economic security systems. In: Proceedings of the 34th IBIMA Conference, Madrid, Spain, 13–14 November 2019, pp. 6631–6639 (2019)
14. Mukhina, I.: Socio-economic problems of employment and the labor market in Russia. In: Proceedings of the International Scientific Conference Lomonosov Readings, Faculty of Economics of Lomonosov Moscow State University, pp. 1011–1018 (2016)
15. Mukhina, I.: Parameters of the social and labor sphere in the system of macroeconomic forecasting. In: Proceedings of the First Efimov Readings. VAVT of the Ministry of Economic Development of Russia, pp. 146–151 (2019)
16. Podvoysky, G.: Labor activity of workers in the conditions of the crisis. Pers. Manag. **3**, 76–80 (2016)
17. Ramazanova, A.: Analysis of the problems of employment and unemployment of the Russian Federation. In: Proceedings of International Student Scientific and Practical Conference, vol. 1, no. 41, pp. 344–346 (2019)
18. Semenov, A.: From accumulating hidden unemployment to stimulating employment. Man Labor **6**, 13–18 (1994)
19. Sindyashkina, E.: Employment in the informal sector of the economy. Economist **6**, 51–56 (1998)
20. Sindyashkina, E.: Unemployment in the run-up to the crisis. Factors influencing the dynamics and scale of unemployment in Russia. In: Collection of Scientific Works of IMEI, vol. 1, pp. 162–193 (2015)
21. Sinyavskaya, O.: Informal employment in Russia: measurement, scale, dynamics. Econ. Sociol. **6**(2), 12–28 (2005)
22. Smolenskaya, S.: Problems and prospects of development of the labor market in the Russian federation. Bull. Ulyanovsk State Tech. Univ. **4**, 56–59 (2018)
23. Strelchenko, E., Marchenko, Y.: The modern model of the labor market: institutional dynamics and features in Russia. J. Econ. Regul. **10**(3), 37–49 (2019)
24. Sukhanova, E., Shirnaeva, S.: Forecasting the unemployment rate: an econometric approach. In: Science of the XXI Century: Current Directions of Development, vol. 1–2, 226–231 (2016)
25. Timacheva, E.: Russian labor market: trends and prospects. Labor Soc. Relat. **3**, 68–81 (2015)
26. Volgin, N., Rakitskiy, B., Shcherbakov, A.: Hidden Unemployment: Phenomenon, Analysis, Consequences, Moscow (1998)
27. Yamshchikova, T., Vnukova, A.: Informal employment: between unemployment and additional opportunities. Econ. Humanit. Sci. **4**(315), 106–118 (2018)

Cluster Development as a Factor of Sustainable Economic Development: Scientific Analytics and Management Prospects

Anastasia A. Sozinova(✉) and Olesya A. Meteleva

Vyatka State University, Kirov, Russia

Abstract. The purpose of this research is to test the hypothesis and scientific analytics of cluster formation as a factor in sustainable economic development and to identify prospects for managing this factor in order to maintain sustainability. The authors have carried out a statistical analysis of depending indicators of cluster development in the country and existing manifestations of sustainable development. Global Competitiveness Index 4.0, Economic growth rate, GDP per capita, Sustainable Development Index were chosen as manifestations of sustainable development.

The methodological apparatus of the research is based on methods of economic statistics (econometrics): the method of correlation analysis, the method of regression analysis, and the analysis of variance. The authors also use such methods of scientific research as classification, problem, and comparative analysis, synthesis, method of analysis of cause-and-effect relationships, hypothetical-deductive method.

The article proves that there are two forms of manifestation of sustainable development, which are influenced by the cluster development - Global Competitiveness Index 4.0 and Sustainable Development Index. The results of developing clusters in countries are an increase in the country's rating in terms of key manifestations of sustainable development.

Keywords: Cluster · Formation of clusters · Factor · Sustainable development · Sustainable economic development · Economic development · Scientific analytics · Governance perspectives · Pandemic · COVID-2019 · Post-COVID economy conditions · Risks · Innovation

JEL Codes: C31 · C33 · C38 · F63 · Q01 · Q56 · O11 · O12 · O31 · O32 · O33 · O38 · P25 · P48 · R11 · R13 · R58

1 Introduction

The COVID 2019 pandemic and conditions of the post-COVID economy is a new milestone in the history of sustainable development. The global economic crisis, caused by a sharp decline in business activity, is a major obstacle to sustainable development.

This research hypothesizes that the formation of clusters is an effective factor in sustainable economic development, independent of the economic crisis. This fact is proved by the results of forming clusters in countries and regions with different income levels.

E. G. Popkova (Ed.): ISC 2020, LNNS 368, pp. 723–733, 2022.
https://doi.org/10.1007/978-3-030-93244-2_78

At present, we have to admit that only some constituent entities of the Russian Federation are implementing a policy to ensure sustainable development. At the same time, the transition to the sustainable development of the country as a whole is possible only if sustainable development of all regions is ensured.

From the point of view of sustainable development, differences in the economic development of regions can have a pronounced impact on the balance of regions in the economic system of the Russian Federation, on the state structure, structure, and efficiency of the economy in general. In addition, differences in the economic development of regions can become a significant factor of additional risks that hinder innovation.

The study aims to test the hypothesis and scientific analytics of forming clusters as a factor of sustainable economic development and to identify the prospects for managing this factor in order to maintain sustainability.

2 Materials and Methods

The following authors consider the topic of manifestations and factors of sustainable development of the modern economy: [1, 6–8, 10–12, 15–23, 26, 36–38].

Prospects for improving the practice of making management decisions in conditions of economic crises are considered by [2, 4, 9, 13, 14, 24, 27, 28, 32]. Development of the cluster approach in economics is discussed by [3, 12, 25, 29–31].

Nevertheless, the content analysis has shown that existing sources do not allow forming a systemic vision of the cluster formation process as a factor of sustainable economic development. Therefore, additional scientific research is required.

The study uses the example of the leaders (countries from the top 20) of the World Bank global ranking for 2020 according to the value of the doing business index for 2020.

Using the method of regression analysis, firstly, the authors have determined the dependence of manifestations of sustainable development, highlighted in the content analysis of sources, on the cluster development in the country (indicator 5.2.2 State of cluster development in The Global Innovation Index (GII) 2020) [35] by plotting regression curves. Secondly, a multifactorial model of the regression dependence of cluster development in the country on the selected potential factors was built (Table 1).

Table 1. Statistics for correlation and regression analyzes in the top 20 countries

Position in the ranking doing business 2020	Country	State of cluster development points 1–100	Manifestations of sustainable development				
			Global Competitiveness Index 4.0, points 1–100	Economic growth rate, %	GDP per capita 2020, PPP$	GDP per capita 2019, PPP$	Sustainable Development Index, points 1–100
		x	y1	y2	y3	For reference, for calculating x2	y4
Developed countries (top 10)	New Zealand	49.5	76.7	−10.94	35,744.00	40,135.40	43.6
	Singapore	69.2	84.8	−10.23	90,080.20	100,344.70	7.9
	Denmark	67.1	81.2	−9.75	47,040.40	52,120.50	35.5
	The Republic of Korea	60	79.6	−5.54	39,059.70	41,350.60	27.4
	USA	74.8	83.7	−9.20	56,844.30	62,605.60	18.1

(continued)

Table 1. (*continued*)

Position in the ranking doing business 2020	Country	State of cluster development points 1–100	Manifestations of sustainable development				
			Global Competitiveness Index 4.0, points 1–100	Economic growth rate, %	GDP per capita 2020, PPP$	GDP per capita 2019, PPP$	Sustainable Development Index, points 1–100
		x	y1	y2	y3	For reference, for calculating x2	y4
	UK	65.9	81.2	−10.55	40,881.30	45,704.60	39.1
Developing countries (top 20)	Malaysia	69.8	74.6	−6.98	28,705.90	30,859.90	46.6
	UAE	69.5	75	−12.63	60,618.60	69,381.70	11
	Thailand	51.4	68.1	−8.72	17,778.80	19,476.50	73.7
	Russia	40.3	66.7	−11.58	25,878.70	29,266.90	68.5
	China	59.6	73.9	−5.98	17,027.50	18,109.80	51.7
	Turkey	47.5	62.1	−11.73	24,675.50	27,956.10	67.5
Average	–	60.38	75.63	−9.49	40,361.24	44,776.03	40.88

Source Compiled by the authors from [5, 33–35]

3 Results

Based on the data from Table 1, the following results of correlation analysis were obtained (Fig. 1). At the same time, the indicators of the State of cluster development and the Sustainable Development Index are measured in points but are reciprocal. Since, in accordance with the methodology, the State of cluster development varies from 1 to 100. The closer to 100, the better the development of clusters in the country. In accordance with the methodology for calculating the Sustainable Development Index, the closer the country is to 1, the better.

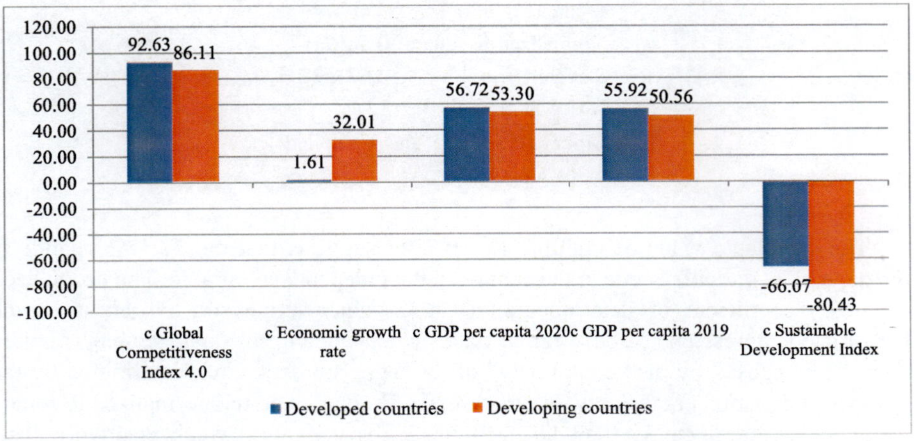

Fig. 1. Correlation analysis of the dependence of manifestations of sustainable development on cluster development. *Source* Compiled by the authors

As can be seen from Fig. 1, in the top 10 countries, cluster development affects global competitiveness to a greater extent (92.63%). At the same time, a fairly high connection between the development of clusters and the index of sustainable development (−66.07%) and GDP (56.72% in 2020 and 55.92% in 2019) has been proved.

A high level of correlation is among the studied indicators in the top 20 countries. Also, the highest dependence on cluster development is shown in the global competitiveness of countries. At the same time, in the top 20 countries, the development of clusters has a greater impact on the sustainable development index (−80.43%) than in the top 10 countries.

As you can see, the correlation analysis has partially confirmed the scientific hypothesis about the dependence of the manifestations of sustainable development on cluster development, identified in the content analysis. Three of the four manifestations of sustainable development depend on cluster development: Global Competitiveness Index 4.0, GDP per capita, and Sustainable Development Index. The economic growth rate has shown a weak relationship with cluster development, and therefore we exclude it from further regression analysis.

Firstly, let us consider the regression statistics of the correlation of the dependent variable y_1 (Global Competitiveness Index 4.0) on the factor x (State of cluster development). By the method of analysis it has been determined (by the coefficient of determination) that the dependence is better described by the linear function $y_1(x) = a_1 + b_1 * x$. The regression statistics of this equation are shown in Table 2.

Table 2. Regression statistics

Regression statistics	
Multiple R	0.777653
R-square	0.604744
Normalized R-square	0.565219
Standard error	4.671522
Observations	12

Source Developed and compiled by the authors

The calculated value of multiple R = 0.7777 suggests that the 77.77% change in the dependent variable is due to the change in the independent variable. The normalized (reduced) coefficient of determination (0.5652) characterizes the closeness of the constructed regression to the original data, which contain an "undesirable" random component caused by the "complexity" of the regression equation, determined by the number of equation coefficients. In our case, the R-square and the normalized R-square do not differ significantly (0.6047 and 0.5652, respectively), which positively characterizes the constructed model. The results of the analysis of variance are shown in Table 3.

Table 3. Analysis of variance

Analysis of variance				
	df	*SS*	*MS*	*F*
Regression	1	333.8955	333.8955	15.30008
The remainder	10	218.2312	21.82312	
Total	11	552.1267		

Source Developed and compiled by the authors

Table 3 shows that the observed value of Fisher's F-criterion: $F_{obs} = 15.3$. The tabular value of F-Fisher's test at the significance level $\alpha = 0.05$, $k_1 = m = 1$ and $k_2 = n - m - 1 = 12 - 1 - 1 = 10$ is $F_{tabl} = 2.156$. Since $F_{obs} > F_{tabl}$ ($15.3 > 2.156$), the equation is considered statistically significant. The calculated coefficients of the regression equation are given in Table 4.

Table 4. Calculated coefficients

	Coefficients	Standard error	t-statistics	P-value	Low 95%	High 95%
a_1	45.05639	7.932595	5.679905	0.000204	27.38146	62.73131
b_1	0.506381	0.129458	3.911532	0.002906	0.217929	0.794832

Source Developed and compiled by the authors

Table 4 allows concluding that a model is obtained: $y_1 = 45.0563 + 0.5064 * x$. The calculated p-value does not exceed 0.05 and is 0.003 for b_1, therefore, the model is statistically significant at the $\alpha = 0.05$ level. To assess the statistical significance of the regression coefficient, let us analyze the observed value of the Student's t-test: 3.912 (for the coefficient b_1). For a given level of significance ($\alpha = 0.05$) and $n - 2 = 12 - 2 = 10$ degrees of freedom, the tabular value of the t-criterion: $t_{tabl} = 2.228$. Since t_{obs} of each regression coefficient is greater than t_{tabl} ($3.912 > 2.228$), then we recognize the statistical significance of the regression coefficients.

Thus, the regression analysis has shown that the growth of cluster development by 1 point contributes to the growth of the Global Competitiveness Index 4.0 0.51 points.

Secondly, let us consider the regression statistics of the correlation of the dependent variable y_3 (GDP per capita 2020) on the factor x (State of cluster development). By the method of analysis it has been determined (by the coefficient of determination) that the dependence is better described by the linear function $y_3 (x) = a_3 + b_3 * x$. The regression statistics of this equation are shown in Table 5.

Table 5. Regression statistics

Regression statistics	
Multiple R	0.620996106
R-square	0.385636164
Normalized R-square	0.324199781
Standard error	17,293.68041
Observations	12

Source Developed and compiled by the authors

The calculated value of multiple R = 0.621 suggests that the 62.1% change in the dependent variable is due to the change in the independent variable. The normalized (reduced) coefficient of determination (0.3242) characterizes the weak proximity of the constructed regression to the original data. In this regard, further analysis is not advisable.

Thirdly, let us consider the regression statistics of the correlation of the dependent variable y_4 (Sustainable Development Index) on the factor x (State of cluster development). By the method of analysis it has been determined (by the coefficient of determination) that the dependence is better described by the linear function $y_4(x) = a_4 + b_4 * x$. The regression statistics of this equation are shown in Table 6.

Table 6. Regression statistics

Regression statistics	
Multiple R	0.793968
R-square	0.630385
Normalized R-square	0.593423
Standard error	14.15265
Observations	12

Source Developed and compiled by the authors

The calculated value of multiple R = 0.794 suggests that the 79.4% change in the dependent variable is due to the change in the independent variable. The normalized (reduced) coefficient of determination (0.5934) characterizes the closeness of the constructed regression to the original data, which contain an "undesirable" random component caused by the "complexity" of the regression equation, determined by the number of equation coefficients. In our case, the R-square and the normalized R-square do not differ significantly (0.6304 and 0.5934, respectively), which positively characterizes the constructed model. The results of the analysis of variance are shown in Table 7.

Table 7. Analysis of variance

Analysis of variance	*df*	*SS*	*MS*	*F*
Regression	1	3,416.103	3,416.103	17.05515
The remainder	10	2,002.974	200.2974	
Total	11	5,419.077		

Source Developed and compiled by the authors

Table 7 shows that the observed value of the Fisher's F-criterion: $F_{obs} = 17.06$. The tabular value of F-Fisher's test at the significance level $\alpha = 0.05$ and $k1 = m = 1$ and $k2 = n - m - 1 = 12 - 1 - 1 = 10$ is $F_{tabl} = 2.156$. Since $F_{obs} > F_{tabl}$ (17.06 > 2.156), the equation is considered statistically significant. The calculated coefficients of the regression equation are given in Table 8.

Table 8. Calculated coefficients

	Coefficients	Standard error	t-statistics	P-value	Low 95%	High 95%
a_4	138.6868	24.03226	5.77086	0.00018	85.13959	192.234
b_4	−1.61971	0.392202	−4.12979	0.002045	−2.49359	−0.74583

Source Developed and compiled by the authors

Table 8 allows concluding that a model is: $y_4 = 138.687 - 1.6197 * x$. The calculated p-value does not exceed 0.05 and is 0.002 for b_4, therefore, the model is statistically significant at the $\alpha = 0.05$ level. To assess the statistical significance of the regression coefficient, let us analyze the observed value of the Student's t-test: 4.129 (for the coefficient b_4). For a given level of significance ($\alpha = 0.05$) and $n - 2 = 12 - 2 = 10$ degrees of freedom, the tabular value of the t-criterion: $t_{tabl} = 2.228$. Since t_{obs} of each regression coefficient is greater than t_{tabl} ($4.129 > 2.228$), then we recognize the statistical significance of the regression coefficients.

Thus, the regression analysis has shown that the growth of cluster development by 1 point contributes to the growth of the Sustainable Development Index 4.0 by almost 1.62 points.

Based on the presented calculations, the prospects for managing the development of clusters have been identified. The results are given in Table 9.

Table 9. Prospects for managing cluster development to maintain sustainability

Indicators	Manifestations of sustainable development			
	Global Competitiveness Index 4.0, y_1	Economic growth rate, y_2	GDP per capita, y_3	Sustainable development Index, y_4
Independent variable	State of cluster development, x			
The relationship between variables	Direct	Direct	Direct	Reverse
Assessment of depending manifestations of sustainable development on cluster development	High (correlation 86.11%–92.63%)	Weak (correlation 1.61–32.01)	Quite high (correlation 56.72%–53.30%)	High (correlation 66.07%–80.43%)
Regression dependence	$y_1 = 45.0563 + 0.5064 * x$	–	The weak closeness of the constructed regression to the original data	$y_4 = 138.687 -1.6197 * x$

(continued)

Table 9. (*continued*)

Indicators	Manifestations of sustainable development			
	Global Competitiveness Index 4.0, y_1	Economic growth rate, y_2	GDP per capita, y_3	Sustainable development Index, y_4
Consequences of the rise/fall of the independent variable	If the independent variable changes by 1%, the dependent variable improves/worsens its performance by 0.51	–	–	If the independent variable changes by 1%, the dependent variable improves/worsens its performance by 1.62
Conclusion on the prospects for management	Cluster development is an important element in driving global competitiveness	It is not possible to assess the prospects due to the weak relationship between the indicators	It is not possible to assess the prospects due to the weak relationship between the indicators	Cluster development is an important element in stimulating sustainable development

Source Compiled by the authors

As you can see from Table 9, the hypothesis that cluster development stimulates sustainable development is proven.

4 Discussion

A series of statistical studies have allowed identifying two forms of manifestation of sustainable development, which are influenced by the development of clusters - the Global Competitiveness Index 4.0 and the Sustainable Development Index. The direct results of the development of clusters in countries are an increase in the country's rating in terms of key indicators characterizing sustainable development.

The debatable issue in the context of this study is the impact of cluster development on such manifestations of sustainable development as Economic growth rate and GDP per capita, which will be the focus of further scientific research.

5 Conclusion

The study is of theoretical significance as its results and authors' conclusions contribute to the development of the concept of sustainable development. The scientific value and novelty is an assessment of the prospects for managing cluster development in the context of sustainable development methodology, results of a statistical analysis of depending manifestations of sustainable development on cluster development.

References

1. Andronova, I.V., Chernova, V.Y., Starostin, V.S., Degtereva, E.A.: Study of sector-specific innovation efforts: the case from the Russian economy. Entrepreneurship Sustain. Issues **7**(1), 540–552 (2019)
2. Bogoviz, A.V., Sozinova, A.A., Ostrovskaya, V.V.: Approaches to managing economic growth of socio-economic systems. In: Endovitsky, D.A., Popkova, E.G. (eds.) Management of Changes in Socio-Economic Systems. SSDC, vol. 135, pp. 131–137. Springer, Cham (2018). https://doi.org/10.1007/978-3-319-72613-7_11
3. Burtseva, T.A., Sozinova, A.A., Fokina, O.V., Kalinin, P.A., Grabar, A.A.: Information technologies as a tool for increasing clustering effectiveness. Espacios **39**(28), 6 (2018)
4. Fufacheva, L.A., Lepeshkin, S.M., Fokina, O.V., Sozinova, A.A., Shchinova, R.A.: Improvement of the mechanism of provision of food security of russia within management of risk system of entrepreneurship. In: Popkova, E.G. (ed.) Overcoming Uncertainty of Institutional Environment as a Tool of Global Crisis Management. CE, pp. 9–14. Springer, Cham (2017). https://doi.org/10.1007/978-3-319-60696-5_2
5. Global Competitiveness Report 2019: How to end a lost decade of productivity growth. www.weforum.org/reports/how-to-end-a-decade-of-lost-productivity-growth. Accessed 08 May 2021
6. Goyal, S., Sergi, B.S.: Social entrepreneurship and sustainability – understanding the context and key characteristics. J. Secur. Sustain. Issues **4**(3), 269–278 (2015)
7. Haabazoka, L.: Project finance for Africa's construction sector: can stabilization funds work? In: Popkova, E.G. (ed.) ISC 2018. LNNS, vol. 57, pp. 32–60. Springer, Cham (2019). ISBN 978-3-030-00102-5. https://doi.org/10.1007/978-3-030-00102-5_4
8. Inshakov, O.V., Bogachkova, L.Y., Popkova, E.G.: The Transformation of the global energy markets and the problem of ensuring the sustainability of their development. In: Inshakov, O., Inshakova, A., Popkova, E. (eds.) Energy Sector: A Systemic Analysis of Economy, Foreign Trade and Legal Regulations. LNNS, vol. 44, pp. 135–148. Springer, Cham (2019). https://doi.org/10.1007/978-3-319-90966-0_10
9. Kataeva, N.N., Marakulina, I.V., Sanovich, M.A., Sozinova, A.A., Vasilyuk, N.: Transformation of approach to market segmentation within crisis management of global entrepreneurship. In: Popkova, E.G. (ed.) Overcoming Uncertainty of Institutional Environment as a Tool of Global Crisis Management. CE, pp. 491–496. Springer, Cham (2017). https://doi.org/10.1007/978-3-319-60696-5_62
10. Morozova, I.A., Popkova, E.G., Litvinova, T.N.: Sustainable development of global entrepreneurship: infrastructure and perspectives. Int. Entrepreneurship Manag. J. **15**(2), 589–597 (2018). https://doi.org/10.1007/s11365-018-0522-7
11. Petrenko, E., Pritvorova, T., Dzhazykbaeva, B.: Sustainable development processes: service sector in a post-industrial economy. J. Secur. Sustain. Issues **7**(4), 781–791 (2018).https://jssidoi.org/jssi/papers/papers/view/309
12. Popkova, E.G., et al.: Clustering as a growth point of modern russian business. In: Popkova, E.G., Sukhova, V.E., Rogachev, A.F., Tyurina, Y.G., Boris, O.A., Parakhina, V.N. (eds.) Integration and Clustering for Sustainable Economic Growth. CE, pp. 55–63. Springer, Cham (2017). https://doi.org/10.1007/978-3-319-45462-7_7
13. Popkova, E.G., Saveleva, N.K., Sozinova, A.A.: A new quality of economic growth in "smart" economy: advantages for developing countries. In: Popkova, E.G., Sergi, B.S. (eds.) ISC 2020. LNNS, vol. 155, pp. 426–433. Springer, Cham (2021). https://doi.org/10.1007/978-3-030-59126-7_48

14. Popkova, E.G., Savelyeva, N.K., Sozinova, A.A.: Smart technologies in entrepreneurship: launching a new business cycle or a countercyclical instrument for regulating the economic situation. In: Popkova, E.G., Sergi, B.S. (eds.) ISC 2020. LNNS, vol. 155, pp. 1722–1730. Springer, Cham (2021). https://doi.org/10.1007/978-3-030-59126-7_188
15. Popkova, E.G., Dubova, Y.I., Yakovleva, E.A., Azarova, N.A., Titova, E.V.: Role of ecological marketing in the formation and development of the ecological cluster. Asian Soc. Sci. J. **10**(23), 1–8 (2014)
16. Popkova, E.G., Poluyufta, L., Beshanova, Y., Popova, L.V., Kolesnikova, E.: Innovations as a basis for marketing strategies of Russian oil companies in the conditions of oil prices reduction. Contrib. Econ. **9783319606958**, 449–455 (2017)
17. Popkova, E.G., Shakhovskaya, L.S., Abramov, S.A., Natsubidze, A.S.: Ecological clusters as a tool for improving environmental safety in developing countries. Environ. Dev. Sustain. **18**(4), 1049–1057 (2016)
18. Pozdnyakova, U.A., Popkova, E.G., Kuzlaeva, I.M., Lisova, O.M., Saveleva, N.A.: Strategic management of clustering policy during provision of sustainable development. Contrib. Econ. **9783319454610**, 413–421 (2017)
19. Ragulina, Y.V., Alekseev, A.N., Strizhkina, I.V., Tumanov, A.I.: Methodology of criterial evaluation of consequences of the industrial revolution of the 21st century. Stud. Syst. Decis. Control **169**, 235–244 (2019a)
20. Ragulina, Y.V., Semenova, E.I., Avkopashvili, P.T., Dmitrieva, E.A., Cherepukhin, T.Y.: Top priority directions of implementing new internet technologies on the territories of rapid economic development. Adv. Intell. Syst. Comput. **726**, 182–188 (2019b)
21. Sergi, B.S. (ed.): Exploring the Future of Russia's Economy and Markets: Towards Sustainable Economic Development. Emerald Publishing, Bingley (2018)
22. Sergi, B.S., Popkova, E.G., Borzenko, K.V., Przhedetskaya, N.V.: Public-private partnerships as a mechanism of financing sustainable development. In: Ziolo, M., Sergi, B.S. (eds.) Financing Sustainable Development: Key Challenges and Prospects, pp. 313–339. Palgrave Macmillan (2019a)
23. Sergi, B.S., Popkova, E.G., Sozinova, A.A., Fetisova, O.V.: Modeling Russian industrial, tech, and financial cooperation with the Asia-Pacific region. In: Sergi, B.S. (ed.) Tech, Smart Cities, and Regional Development in Contemporary Russia, pp. 195–223. Emerald Publishing Limited (2019b). https://doi.org/10.1108/978-1-78973-881-020191012
24. Soboleva, O., Sozinova, A., Spengler, A., Fokina, O., Savelyeva, N.: Mechanisms of regulation of economic processes in a region. In: Popkova, E.G. (ed.) Overcoming Uncertainty of Institutional Environment as a Tool of Global Crisis Management. CE, pp. 403–415. Springer, Cham (2017). https://doi.org/10.1007/978-3-319-60696-5_51
25. Sozinova, A.A., Androsova, I.V., Abramov, V.L., Tikhomirov, E.A., Redin, D.V., Oganesyan, T.L.: Formation of the new forms of management systems spatially localized economies in the paradigm of the cluster approach. Int. Rev. Manag. Mark. **6**(S1), 250–254 (2016b)
26. Sozinova, A.A., Burtseva, T.A., Fokina, O.V., Grabar, A.A., Tufyakova, E.S.: The concept of industrial enterprises' economic development amid the global financial and economic crisis. In: Popkova, E.G. (ed.) ISC 2018. LNNS, vol. 57, pp. 1085–1093. Springer, Cham (2019). https://doi.org/10.1007/978-3-030-00102-5_114
27. Sozinova, A.A., Fokina, O.V., Fufacheva, L.A.: Reorganization of entrepreneurial structures within global crisis management: problems and perspectives. In: Popkova, E.G. (ed.) Overcoming Uncertainty of Institutional Environment as a Tool of Global Crisis Management. CE, pp. 3–8. Springer, Cham (2017). https://doi.org/10.1007/978-3-319-60696-5_1

28. Sozinova, A.A., Kosyakova, I.V., Kuznetsova, I.G., Stolyarov, N.O.: Corporate social responsibility in the context of the 2020 economic crisis and its contribution to sustainable development. In: Popkova, E.G., Sergi, B.S. (eds.) ISC 2019. LNNS, vol. 198, pp. 83–90. Springer, Cham (2021b). https://doi.org/10.1007/978-3-030-69415-9_10
29. Sozinova, A.A., Novikov, S.V., Kocnikov, S.N., Nemchenko, G.I., Alenina, E.E.: Peculiarities of Isolated Clusters Operation. Int. J. Econ. Financial Issues **6**(8), 19–23 (2016a)
30. Sozinova, A.A., Okhimenko, O.I., Goloshchapova, L.V., Kolpak, E.P., Golovanova, N.B., Tikhomirov, E.A.: Industrial and innovation clusters: development in Russia. Int. J. Appl. Bus. Econ. Res. **15**(11), 111–118 (2017)
31. Sozinova, A.A., Sofiina, E.V., Safargaliyev, M.F., Varlamov, A.V.: Pandemic as a new factor in sustainable economic development in 2020: scientific analytics and management prospects. In: Popkova, E.G., Sergi, B.S. (eds.) ISC 2019. LNNS, vol. 198, pp. 756–763. Springer, Cham (2021a). https://doi.org/10.1007/978-3-030-69415-9_86
32. Sozinova, A.A.: Dataset of a fight against the viral threat based on management and marketing during the COVID-19 economic crisis. Mendeley Data, 1 (2020). https://doi.org/10.17632/g6p5kjz657.1
33. Sustainable development index. www.sustainabledevelopmentindex.org/time-series. Accessed 08 May 2021
34. The Global Innovation Index (GII) 2019. www.wipo.int/global_innovation_index/en/2019/. Accessed 08 May 2021
35. The Global Innovation Index (GII) 2020. www.globalinnovationindex.org/home. Accessed 08 May 2021
36. Tripathi, D., Priyadarshi, P., Kumar, P., Kumar, S.: Micro-foundations for sustainable development: leadership and employee performance. Int. J. Organ. Anal. **28**(1), 92–108 (2020). https://doi.org/10.1108/IJOA-01-2019-1622
37. Zavyalova, E.B., Studenikin, N.V., Starikova, E.A.: Business participation in the implementation of socially oriented sustainable development goals in countries of Central Asia and the Caucasus Region. Cent. Asia Cauc. **19**(2), 56–63 (2018)
38. Zhang, D., Wang, C., Zheng, D., Yu, X.: Process of innovation knowledge increase in supply chain network from the perspective of sustainable development. Ind. Manag. Data Syst. **118**(4), 873–888 (2018). https://doi.org/10.1108/IMDS-06-2017-0243

Methodology for Risk Analysis of Industrial Enterprises Based on the Probability Distribution

Yaroslav S. Potashnik(✉), Victor P. Kuznetsov, Ekaterina P. Garina, Svetlana N. Kuznetsova, and Elena V. Romanovskaya

Minin Nizhny Novgorod State Pedagogical University, Nizhny Novgorod, Russia

Abstract. Risk analysis of industrial enterprises is often based on averaged single estimates of damage/gain and their probabilities. At the same time, risks are random values, the consequences of which are a certain interval, where the individual probability corresponds to each value. In this regard, the current practice of analysis makes it difficult for specialists to develop the most adequate measures for risk management and reduces the efficiency of activities and the sustainability of industrial enterprises. One of the reasons for this situation, in our opinion, is the insufficient methodological elaboration of risk analysis based on the probability distribution. The present study is devoted to solving this problem. The article proposes a methodology based on predicting statistical indicators of the consequences of risk (mathematical expectation, variance, standard deviation), plotting a distribution curve, determining the risk characteristics required for making further decisions on responding. The methodology is based on statistical analysis, tools of probability theory and expert assessments.

Keywords: Risk · Risk analysis · Probability distribution · Innovative project

JEL Code: G320

1 Introduction

Cross-country economic integration is a trigger for the growth of requirements for the quality of management decisions taken by business entities. The volatility of the external environment contributes to the growth of uncertainty about the conditions and final results of the planned implementation. Numerous risks arise, the lack of attention to which or an inadequate assessment of the consequences are sometimes fraught with unacceptable losses. In this regard, the development of the theory and practice of risk management is becoming one of the mainstreams of modern management science.

E. G. Popkova (Ed.): ISC 2020, LNNS 368, pp. 734–740, 2022.
https://doi.org/10.1007/978-3-030-93244-2_79

The works of many specialists, including I.R. Balabanov, P.G. Grabovoi, M.V. Lapusta, B. Hedges, N. Taleb, D. Hubbard and others are devoted to the issues of risk management. However, in our opinion, certain methodological aspects, reflected in the publications not fully enough, still exist. In particular, the methodology for analyzing the risks of enterprises based on the probability distribution needs to be clarified. This article is devoted to the study of this issue.

2 Methodology

Risk is an uncertain event or condition, the occurrence of which negatively or positively affects the achievement of goals [5, 8].

Risk management is a continuous iterative process, which is based on the identification and analysis of risks, the development and implementation of responses to risks, monitoring the effectiveness of the response and making necessary adjustments. The most important component of risk analysis is determination of consequences of risks and their likelihood [7]. The data obtained form the basis for decisions on priorities, goals and ways of responding to risks.

Traditionally, conclusions about the probabilities and consequences of risks are formulated on the basis of their averaged, single values. So, for example, if it is predicted that the possible damage in the event of a risk event may amount to 300 000 rubles or more up to 700 000 rubles, then the average value (arithmetic mean) of 500 000 rubles is included in the calculations, an average probability of occurrence is assigned to it (for example, 0,4). Sometimes averaged quantitative values are translated into linguistic values (for example, high, medium, low) or points (for example, from 0 to 1). Risks are displayed on the map (risk map) in accordance with the average estimates, its basic form is presented in the form of Table 1, on the basis of which a plan of further actions in relation to each specific risk (acceptance, reduction, transfer, avoidance, etc.) is developed.

Table 1. Risk map layout

Probability		Damage/opportunities				
		(0; 0,2]	[0;2; 0,4]	[0,4; 0,6]	[0,6; 0,8]	[0,8; 1,0)
		VL	L	M	H	VH
VL	[0,8; 1,0)					
L	[0,6; 0,8]					
M	[0,4; 0,6]					
H	[0;2; 0,4]					
VH	(0; 0,2]					

Abbreviations: VL – very low; L – low; M – medium; H – high; VH – very high

Source Developed and compiled by the authors

This approach is quite convenient for implementation; it does not require a large amount of preliminary research and calculations. However, at the same time, all other possible outcomes of a risk event, except for the averaged one, and the corresponding probabilities remain outside the focus of analysis, which significantly reduces the accuracy of estimates and conclusions [9]. In our opinion, risk analysis should be based not on average values, but on the probability distribution.

Risks are random variables that can take on a particular value in case of implementation, and it is not known in advance what value [2, 6]. Risks can be divided into 2 groups as random variables:

1) discrete – taking strictly defined predetermined known values in case of occurrence; for example, if the values of the possible price depending on market conditions are fixed in the contract for the supply of a material;
2) continuous – taking any values from a certain interval in case of occurrence; for example, inflationary risk, foreign exchange risk, etc. [10].

The values of the mathematical expectation, variance and standard deviation are determined to characterize the probability distribution of random variables.

The mathematical expectation of a random variable is the sum of multiplications of all possible values of a random variable by the probabilities of these values [11]. It characterizes the average expected value for multiple repetitions of tests. In general, the mathematical expectation is calculated by the formula [2]:

$$M(X) = \sum_{i=1}^{n} XiPi \tag{1}$$

where M (X) – mathematical expectation; i – number of possible values of a random variable; n – number of possible values of a random variable; Xi – i-th value of a random variable; Pi – the probability of the i-th value of a random variable.

The variance of a random variable is a measure of the spread of the values of a random variable relative to its mathematical expectation. The larger the variance, the more scattered the values. In general, the variance (D (X)) is calculated by the formula [2]:

$$D(X) = M[(Xi - M(X))^2] \tag{2}$$

Standard deviation is an additional (in addition to variance) indicator of the dispersion of the values of a random variable relative to its mathematical expectation. The standard deviation values are closer to the values of the random variable, as opposed to the variance, which facilitates subsequent analysis. The standard deviation is generally calculated by the formula [2]:

$$\sigma = \sqrt{D[X]} \tag{3}$$

The smaller the values of the variance and standard deviation, the smaller the scatter of the values of the random variable relative to its average expected value, the more predictable the possible result is and the lower the risk.

Formulas (1)–(3) can be directly used in relation to risks – discrete values. If the risks are continuous values, then the formulas presented in [2] are applied to them.

It is important to know the form of the probability distribution to characterize the risks related to continuous random variables. As practice shows, the most common are uniform, beta, triangular and normal distribution [3, 5]. The type of distribution corresponding to a specific case is determined based on the collection of preliminary statistical data and expert estimates.

The following methods were used in the study: a systematic approach, logical analysis and synthesis, abstraction, statistical analysis, the method of expert assessments.

3 Results

To characterize the influence of risk in the form of a probability distribution, the authors propose a methodology implemented in the following sequence:

1. The interval of values of damage/gain in the event of a risk is predicted on the basis of the collection and analysis of statistical information and expert assessments.
2. The type of random variable to which the considered risk belongs (discrete or continuous) is determined. If the risk refers to discrete random variables, all its possible values and probabilities of their occurrence are determined. If the risk refers to continuous random variables, then an auxiliary (approximate) probability distribution is calculated, including several possible values of damage/gain from risk and the probability of their occurrence.
3. The mathematical expectation, variance and standard deviation are determined for a given risk; the graph of the probability distribution is built if necessary.
4. Conclusions, including on the possible values of damage/gain in the event of a risk, their probabilities, priority and options for responding to risk are formulated.

A principal example of the application of the proposed methodology is presented below.

Risk analyzed: change in the purchase price of material Z. On the basis of statistical information and expert assessments, it was determined that this risk refers to continuous random variables, while the price in the period of interest can take any value in the range from 500 rubles up to 700 rubles. The predicted auxiliary probability distribution is presented in Table 2.

Table 2. Predicted probability distribution

Price, rub	500	550	600	650	700	750	800
Probability of occurrence	0,05	0,1	0,2	0,3	0,2	0,1	0,05

Source Developed and compiled by the authors

According to Table 2, it is calculated:

- mathematical expectation (rub.): M (X) = 550 * 0,05 + 550 * 0,1 + 600 * 0,2 + 650 * 0,3 + 700 * 0.2 + 750 * 0,1 + 800 * 0.05 = 650;
- variance (rub.): D (X) = (500 – 650) * 0,05 + (550–650) * 0,1 + (600–650) * 0,2 + (650–650) * 0,3 + (700–650) * 0,2 + (750–650) * 0,1 + (800–650) * 0,05 = 5250;
- standard deviation (rub.): $\sigma = \sqrt{5250} \approx 72,5$

This risk is assumed to have a standard probability distribution. The density function is plotted (Fig. 1). Software products including Microsoft Excel can be used for this.

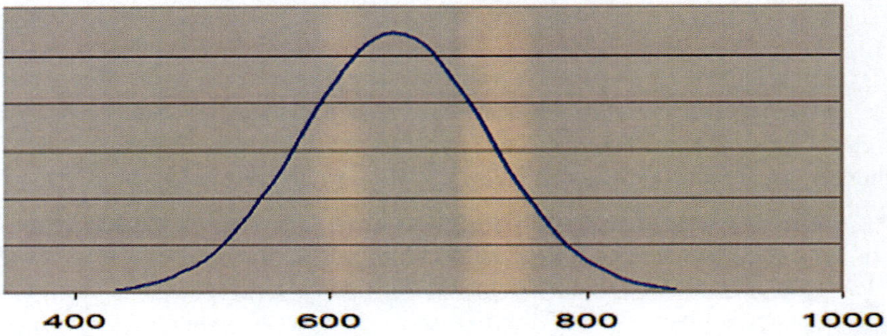

Fig. 1. The density function of the probability distribution of the considered risk. *Source* Developed and compiled by the authors

The conclusions necessary for making decisions on responding to this risk are formulated on the basis of the calculations and the schedule. For example, let's determine the probability that the price of material Z will be higher than 570 rubles. To do this, we calculate how many standard deviation values are 570 rubles behind the mathematical expectation according to the formula:

$$A = \frac{X - M(X)}{\sigma} \tag{4}$$

$$A = \frac{570 - 650}{72,5} = -1,1$$

The obtained result shows that the value of the price of 570 rubles lags by 1,1 standard deviations from the expected value of the probability distribution of possible values of the resource price (negative value indicates that 570 rubles are on the left side of the value of the mathematical expectation).

We find that the probability that some observation will lag by 1,1 standard deviations to the left of the mathematical expectation will be 0,1357, using the tables of the normal distribution of probabilities [1, 4]. That is, the price of the material will not exceed 570 rubles with a probability of 14% and the price will exceed this value with a probability of 86%.

Let's also define a range of values for the price of material Z, corresponding to 90% probability. The number of standard deviations from the mathematical expectation, cutting off the sides of the normal distribution curve of the area corresponding to the value of 0,05, is calculated for this (because 0,05 * 2 = 0,1). We find that A = 1.65 from the tables of the normal distribution of probabilities. This means that the price value with a 90% probability will be in the range from $M\ (X) - 1{,}65 * \sigma$ to $M\ (X) + 1{,}65 * \sigma$, that is from 530 rub. to 770 rub.

4 Conclusion

The method of risk analysis (the consequences of its occurrence and their probabilities) is presented in the form of a probability distribution. The methodology is based on predicting statistical indicators (mathematical expectation, variance, standard deviation), plotting a distribution curve, determining the risk characteristics required for making further decisions. The practical implementation of the proposed approach will contribute to improving the quality of management decisions made at industrial enterprises in conditions of uncertainty.

References

1. Van Horn, D.K.: Fundamentals of Financial Management, 12th edn. Translated from English./Van Horn, D.K., Vakhovich, D.M. (Jr.), – M.: Publishing house «Williams», 1232 p. (2008)
2. Wentzel, E.S.: Probability theory: Textbook for universities/E.S. Wentzel. – 10th edn. Stereotyped – M.: Higher school, 375 p. (2006)
3. Marketing in an innovative economy: Cherchenko, N.V., [and others]; under total. Cherchenko, N.V., Marmashova, S.P., (eds.) GIUST BSU. Department of Marketing. – Minsk, GIUST BSU, 188 p. (2017)
4. Knight F.H.: Risk, Uncertainty and Profit/Translated from English. – M.: Delo, 360 p. (2003)
5. Guide to the Project Management Body of Knowledge (PMBOK Guide). – 5th edn. Project Management Institute, Inc., 586 p. (2013)
6. Risk management standards. Federation of European Associations of Risk Managers, 16 p. (2008)
7. Taleb, N.: Black Swan. Under the sign of unpredictability. – M.: KoLibri, 736 p. (2016)
8. Organizational risk management. Integrated model: Committee of Sponsoring Organizations of the Treadway Commission (COSO), 111 p. (2014)
9. Garina, E.P., Romanovskaya, E.V., Andryashina, N.S., Kuznetsov, V.P., Shpilevskaya, E. V.: Organizational and economic foundations of the management of the investment programs at the stage of their implementation. In: Popkova, E.G., Sergi, B.S. (eds.) ISC 2019. LNNS, vol. 91, pp. 163–169. Springer, Cham (2020). https://doi.org/10.1007/978-3-030-32015-7_20

10. Potashnik, Y.S., Kozlova, E.P., Kuznetsova, S.N., Tsymbalov, S.D., Chelnokova, E.A.: Methodology of risk management assessment at an industrial enterprise. In: Popkova, E.G., Sergi, B.S. (eds.) ISC 2019. LNNS, vol. 129, pp. 265–271. Springer, Cham (2019). https://doi.org/10.1007/978-3-030-47945-9_29
11. Potashnik, Y.S., Kuznetsov, V.P., Garina, E.P., Romanovskaya, E.V., Andryashina, N.S.: About the place and role of risk management in the adoption of management decisions at the enterprise. In: Kolmykova, T., Kharchenko, E.V. (eds.) Digital Future Economic Growth, Social Adaptation, and Technological Perspectives. LNNS, vol. 111, pp. 3–9. Springer, Cham (2020). https://doi.org/10.1007/978-3-030-39797-5_1

System of Success Indicators for Older Employees in Routine Labor Conditions

Irina B. Durakova, Ekaterina V. Mayer, and Ivan V. Grigorov(✉)

Voronezh State University, Voronezh, Russia
durakova@econ.vsu.ru

Abstract. The paper aims to represent the success of older employees in a routine work environment. The authors show the features, advantages, and disadvantages of routine labor. Moreover, they provide the interpretation of routine labor in the narrow and broad senses. The authors build a system of success indicators of older employees using the model of determinants of success.

Keywords: Routine · Routine labor · Success determinants · Success indicators · Older employees

JEL codes: J14 · J20 · J28

1 Introduction

An aging population and a correspondingly aging workforce are a problem of recent history. Signs of difference demographically "old" countries include the high level of economic development and the resulting achievements in health care, new standards of quality of life and working conditions, and culture of a health-oriented balanced diet. The cumulative effect of these advances is the provision of a longer lifespan of generations compared to their predecessors.

As life expectancy increases, each successive generation becomes less numerous than the previous one. Statistics explain these realities of recent history by the downward trend in the birth rate. A shortage of labor led to a decrease in labor quality [8].

The digital economy is becoming an impetus to accelerate the transformation of the competencies required by the employee. The problems of a timely response to the current challenges include ensuring an effective policy of age diversity related to the prerequisite of an increase in the structure of employed older employees and the development of indicators of their success in performing routine and innovative work.

2 Materials and Methods

The formation of a system of success indicators of older employees includes three stages.

E. G. Popkova (Ed.): ISC 2020, LNNS 368, pp. 741–749, 2022.
https://doi.org/10.1007/978-3-030-93244-2_80

The first step is determining the specifics of routine work, its advantages, and disadvantages for older employees.

The definition of routine as a conservative regime and work method and slavish adherence to an established pattern that has become a mechanical habit [7] shows two characteristics of this category. On the one hand, routine is compulsory, brought to the mechanical pursuit of the existing algorithm of activity. This approach characterizes a necessary attribute of framework conditions of work. On the other hand, the negative definition of "slavish following" indicates the impossibility of introducing elements of creativity and novelty into the work routine, meaning that the activity remains unchanged.

According to the definition, the negative attitude towards routine work is not always justified. Organizations usually operate so that processes are repetitive and need to be managed quickly, safely, reliably, and with low friction losses and costs [10]. These processes are called routine [5]. Organizations only make sense if they provide a long-term guarantee that all repetitive work will be done according to clear competencies and established routines [10].

Distinguishing between individual and organizational routine, the authors understand the individual routine as the stability of processes and their simplification by the employee. Organizational routine is not a direct and primary attribute of the organization but an attribute of its workers [9]. In the narrow sense, routine is a system (complex, set) of highly automated behaviors characterizing a high degree of repeatability of information processing [3]. It is a repetitive and recognizable pattern of interdependent action involving multiple participants [5], a designation of the result of a rational division of labor [1].

In a broader sense, denoted as "rules of thumb," a routine is an initial approximation or relatively simple decision-making rules for quantitative questions consciously formed and requiring little information processing.

The advantages and disadvantages of routine work can be presented in a classification of three groups:

1. Organizational and process working conditions: Results and prospects;
2. Social guarantees, health, motivation, and adaptation;
3. Prerequisites and results of labor.

The second step is determining the process of achievement and determinants of professional success of older employees under conditions of employment in routine work.

According to research, jobs in low- and middle-income countries are more routine than those in high-income countries within the same occupation. As a consequence of the stability of routine labor, the low level of technological development in less developed countries justifies their classification as donors to the labor markets of other nations experiencing labor shortages.

Ensuring the success of routine workers in developed and less economically developed countries is equally important. In developed countries, it keeps routine workers in their jobs. In less economically developed countries, it helps to keep routine workers in the country.

The transformation of the labor market due to changes in the values of young people, the desire of young people for creative activities resulting in a service rather than a product leads to the fact that older employees are increasingly finding themselves in routine jobs.

The achievement and evaluation of work results as successful are possible through a logical chain: “purpose → action → result → evaluation” [4].

A purpose is a condition for determining the employee’s behavior, activity, sense of self-worth, and psychological well-being. Success-oriented behavior is determined by dispositions and motives for achievement, attitudinal readiness, and potential.

First, the need for the result of a particular action must be a prerequisite for forming a success-oriented attitude. It can be differentiated into three types of work: task-driven, context-driven (cooperation, coordination, and interaction), and adaptive work (adaptation to change, innovation).

According to the target orientation, each of these types can be represented by one (several) of the following positions:

- Flexing work – employees adapt (apply) existing, interdependent actions in a manner familiar to all members of the workgroup;
- Stretching work – employees adapt (apply) the existing interaction procedure with other participants not familiar to them.
- Inventing work – an employee, not relying on the existing order, builds a new pattern of action not familiar to the members of the working group.

The internal grouping of functions in routine work shows that the same participants must be adapted to new conditions changing the routine from time to time. The relationship between old and new routines creates a sustainable and long-lasting opportunity to change, select, and maintain new practices and patterns of action within routines. This allows routines to generate a wide range of results, from apparent stability to significant change. The existing rules still guide actions in unprogrammed situations marked with an innovative nature.

Second, with age, there is a decrease in motivation to work, especially in routine activities. The motive for achieving success can be the realization of one’s experience, years of polished techniques in performing the job, proven ability to solve a specific production problem (fundamental and expert competencies), mentoring, and the desire to stay in line [2].

Third, employee’s attitudinal readiness differs from other types of work. Subjectively, readiness is defined by the employees’ sense of their resources, allowing them to implement the work functions without assistance. Older employees need regular breaks since their bodies are poorly prepared for the physical work ahead, and it is not easy to work up and recover quickly. An age-related decrease in physical strength and endurance can interfere with the work requiring physical ability. In turn, an age-related decrease in physical strength has little or no effect on mental workers. An age-related decrease in the body’s ability to maintain homeostasis can be a problem for successful activities in areas that require rapid adaptation to different working conditions, such as the works requiring employees to withstand fluctuations in ambient temperature. Thus, employees of older employees often freeze and get cold in a poorly heated room. In combination with low mobility, overcooling develops even at positive temperatures, which can lead to illness. At the same time, these employees have difficulty tolerating high temperatures, especially high humidity [6].

Objective readiness for successful routine work is determined by the feedback from product or service consumers, supervisors, co-workers, technical supervisors, the media, wages and bonuses, and individual performance ratings.

The third stage is the development of success indicators of older employees in accordance with the work performed. According to the target orientation, routine work can be "enlivened" to varying degrees by choosing one of its three positions: flexing work, stretching work, and inventing work.

3 Results

First, routine work is an obligatory attribute of framework conditions of work marked with repetitive actions in accordance with clear competencies and established order. Its benefits to the organization are the possibility of fast, safe, and low-cost management with minimal intra-organizational friction.

Second, routine is viewed in terms of narrow and broad scientific approaches. In a narrow sense, routine is a system of highly automated behaviors resulting in a rational division of labor through a recognizable pattern of interdependent actions involving multiple participants. In a broader sense, which is referred to as "rules of thumb," routine is the initial approximate answer (relatively simple decision-making rules) to the issues that are quantitative, consciously formed, and require a low level of information processing.

Third, the identified advantages and disadvantages of routine work are divided into three groups (Table 1):

Table 1. Advantages and disadvantages of routine activities for older employees

Advantages	Disadvantages
Group 1. Organizational and process working conditions: realities and prospects	
1. Possibility of effective collective work organization due to clear structuring of processes and routines;	1. Errors and accidents on the job due to loss of vigilance associated with fatigue;
2. Ensuring the individual ability of workers to perceive and interpret messages from other members of the team or collective, as well as to check their operational knowledge and skills against prospective recurring tasks of the organization;	2. Lack of something new in perspective (the future repeats the retrospective);
3. Ability to repeat activities due to knowledge of specific working procedures;	3. In comparison with innovative work, a rarer focus on achievement;

(*continued*)

Table 1. (*continued*)

Advantages	Disadvantages
4. Possibility of threefold autonomous control of (1) one's actions and their rationalization, (2) social and physical environment (reflexive control), and (3) one's body (joints, spine, sensory organs, etc.);	4. Reduction of the emotional component in the perception of a repeated result of labor, orienting the desire to repeat it in the future;
5. With optimal management, there is the possibility of development and horizontal career – from Flexing work to Stretching and Inventing work	5. Problems with unforeseen circumstances in work due to lack of flexibility
Group 2. Social guarantees, health, motivation, and adaptation	
1. Maintaining social order in the organization;	1. Health problems due to the one-sided strain of physical and mental work;
2. Ensuring the safety, stability, security, and confidence of workers. Routines drive a wedge between the explosive content of the unconscious and the reflexive control of actions [4];	2. Possible decrease in the intelligence quotient;
3. High probability of minimizing conflicts and their resolution in a diverse workforce, including through the adoption of work schemes by all team members;	3. Decreased motivation to perform the job;
4. Rotation, freedom of function of the related department, and horizontal development	4. Difficulties in adapting to new conditions due to the loss of skills to find alternatives in dealing with issues
Group 3. Prerequisites and results of the labor process	
1. Saving time and energy due to the lack of constant need for new solutions;	1. One-sided career guidance (including the division of high school education into fields) narrows professional choices, and the employee finds himself or herself in the wrong place;
2. Guaranteed quality of work;	2. Emotional burnout due to monotonous work is a prerequisite for the deterioration of the image of the profession or position and the employer's brand on the labor market;
3. Maintaining a form of professional memory (crystalline intelligence), storage of specific operational knowledge and skills;	3. Heavy and strenuous work provokes fatigue, which affects the speed and quality of customer service (hotels, restaurants, banks, etc.);
4. Possibility of repeating a proven mechanism of action in solving a new problem, opening the way to innovation	4. Smoothness of the monotonous business process dulls the professional outlook

Source Compiled by the authors

Fourth, according to the target orientation, routine work can be "enlivened" to varying degrees by choosing one of its three positions: flexing work, stretching work, and inventing work.

Fifth, the success of older employees can be presented according to the success determinants (Table 2).

Table 2. Indicators for evaluating the success of older employees according to the success determinants

Success determinants	Success indicators	Commentary
Fully accomplished work	1. Presenting the completed work within the designated deadline; 2. Availability of methodological and process competencies; 3. Individual absenteeism rate (absenteeism); 4. Individual prezentheism indicator; 5. Costs of employee absence from the workplace; 6. Dynamics of the employee's ranking place in their total list	- **Factors of physical risk:** low level of injuries and accidents in the workplace, subjective evaluation of the employee's work by the manager as safe, taking into account the number of sick leaves to the total number of working days; - **Physical requirements: intensity**: medical examination; psychological evaluation; - **Emotional pressure**: emotional lability, proneness to conflict
Timing and quality of execution	1. Ability to make decisions according to the situation; 2. Ability to cooperate; 3. Ability to communicate; 4. Ratio of professional and qualification characteristics of workers over 45 years old to job requirements; 5. Share of employees over 45 years old who requested skills and competency assessments; 6. Share of employees over 45 years old participating in multi-age teams; 7. Number of employees over 45 years old who successfully passed the probationary period (certification); 8. Number of working pensioners who successfully passed the probationary period (certification); 9. Availability of materials on conflict prevention in the organization; 10. Number of jobs to be optimized due to technological changes	- **Autonomy** – the ability to influence the performance of one's immediate work (choose methods, the pace of work, and standards); - **Clarity of tasks** and consistency of results – the extent to which work tasks are clearly defined, workers receive feedback, and the extent to which the work performed is consistent with those tasks; - **Discrimination** – the degree to which an employer treats some groups of employees more favorably than others; - **Managerial practices** – the extent to which managers implement best practices in work organization, conflict resolution, etc.; - **Social support and good relationships at work** – supportive people and good relationships in the workplace; - **Organizational involvement and voice in the workplace** – the degree to which employees can apply their ideas to the workplace and use their creativity in the process

(*continued*)

Table 2. (*continued*)

Success determinants	Success indicators	Commentary
Innovation and novelty in the content or methods of performing a job (conquest, creativity)	1. Number of submitted and implemented rationalization proposals; 2. Readiness for achievements (competence and engagement); 3. Ability to learn; 4. Willingness to learn	- **Trainability** (ability to learn); - **Career growth** (career opportunities); - **Self-fulfillment** (opportunity for self-fulfillment)
Labor costs	1. Willingness to achieve work-life balance; 2. Share of jobs where work can be transferred to a part-time, flexible schedule, and flexible workplace; 3. Share of workers over 45 years of age preferring part-time or flexible work schedules	- **Work schedule** – a need to work at non-standard hours (e.g., evenings, nights, weekends, and holidays); - **Flexibility of working hours** – whether employees can choose or influence their working hours; - **Work-life balance** – the number of vacations, days off, compensatory time off, absences, and family leave to the total number of working hours
Scope and expected duration of use of the result achieved	1. Number of students prepared to cooperate and continue activities; 2. Number of rationalization proposals accepted for implementation	- **Motivation** – the degree of value of the work performed for clients (society) and the share of target settings
Compliance with instructions and regulations	1. Responsiveness; 2. Mental stability; 3. Tenacity; 4. Number of incidents, conflicts, and accidents; 5. Information processing speed; 6. Fatigue rate; 7. Fluctuations in performance throughout the day	- **Routine work** – the employee's willingness to perform repetitive actions at a given pace without loss of quality; - **Work stress**: the assessment of stress tolerance is appropriate if an employee has to work under constant pressure in an environment where there is always a mismatch between the knowledge and skills available and the demands of the situation
Emotional feelings of employees, which form their professional perspective of achievements	1. Level of emotional intelligence of the employee; 2. Level of emotional burnout of the employee; 3. Job satisfaction; 4. Resilience of an employee; 5. Engagement Ratio (ER)	- **Subjective insecurity** – the degree of confidence in the possibility of losing a job within a year; - **Job satisfaction** – the affective aspect (how happy the employee is at work) and the cognitive aspect (evaluation of work, salary, hours, and job content)

Source Compiled by the authors

4 Conclusion

The results of research on the success of older employees are contradictory. Scholars constantly develop new lists of myths and realities. It is more difficult to achieve success and experience a sense of victory (which will cause further achievements) in the conditions of routine labor than in innovative work. Success controlling is an important function of success evaluation. Its result is appropriate to determine the contribution of personnel management to the success of the organization. Gathering information to develop a system of success indicators is somewhat difficult. The first problem is different methods for calculating indicators and causes of staff turnover, including without differentiating them by age group. The second problem is the avoidance of employers to identify causal relationships (e.g., between the motivation of older employees and the organization's success in the labor market). The third problem is the bureaucratic fear and protection of information from public scrutiny, which, to some extent, blocks the possibility of registering achievements.

Our research advances the scientific notion of measuring the professional success of older employees. The novelty of this research lies in the fact that success indicators are differentiated according to the success determinants in a familiar work environment. The obtained results can be used in developing the theory of slowing down personnel aging, improving the system for controlling success in organizations, and in the practice of retention of age personnel through the current policy of motivation and development.

Acknowledgments. This article was supported by the RFBR under the scientific project 20-010-00263.

References

1. Becker, M.C.: Organizational routines: a review of the literature. Ind. Corp. Chang. **13**(4), 643–678 (2004). https://doi.org/10.1093/icc/dth026
2. Calzavara, M., Battini, D., Bogataj, D., Sgarbossa, F., Zennaro, I.: Aging workforce management in manufacturing systems: state of the art and future research agenda. Int. J. Prod. Res. **58**(3), 729–747 (2020). https://doi.org/10.1080/00207543.2019.1600759
3. Cohen, M., Bacdayan, P.: Organizational routines are stored as procedural memory: evidence from a laboratory study. Organ. Sci. **5**(4), 554–568 (1994). https://doi.org/10.1287/orsc.5.4.554
4. Durakova, I.B., Holyavka, M.G.: Older workers success: biological functions and managerial nudging balance. In: Ashmarina, S.I., Mantulenko, V.V. (eds.) IPM 2020. LNNS, vol. 161, pp. 566–575. Springer, Cham (2021). https://doi.org/10.1007/978-3-030-60926-9_72
5. Feldman, M.S., Pentland, B.T., D'Adderio, L., Lazaric, N.: Beyond routines as things: introduction to the special issue on routine dynamics. Organ. Sci. **27**(3), 505–513 (2016). https://doi.org/10.1287/orsc.2016.1070
6. Hedge, J.W., Borman, W.C., Lammlein, S.E.: The Aging Workforce: Realities, Myths, and Implications for Organizations. American Psychological Association, Washington, DC (2006). https://doi.org/10.1037/11325-000

7. Ozhegov, S.I., Shvedova, N.Y.: Explanatory Dictionary of the Russian Language. ITI Technologies, Moscow, Russia (2015)
8. Scheibe, S.: Employee age moderates within-person associations of daily negative work events with emotion regulation, attention, and well-being. Eur. J. Work Organ. Psychol. (2021). https://doi.org/10.1080/1359432X.2021.1873772
9. Schreyogg, G., Koch, J., Sydow, J.: Routines and path dependency. In: Schreyogg, G., Werder, A.V. (eds.) Pocket Dictionary of Business Management and Organization, pp. 1296–1304. Schaeffer Poeschel, Stuttgart, Germany (2004)
10. Yi, S., Knudsen, T., Becker, M.C.: Inertia in routines: a hidden source of organizational variation. Organ. Sci. **27**(3), 782–800 (2016). https://doi.org/10.1287/orsc.2016.1059

Risks of the Shared Consumption Economy in Russia and the European Union Countries

Svetlana V. Plyasova[1], Natalia A. Bondareva[1], Elena V. Titova[2], Zarema N. Stash[3], and Evgeniya K. Karpunina[4](✉)

[1] Moscow University for Industry and Finance "Synergy", Moscow, Russia
splyasova@synergy.ru
[2] Voronezh State Forestry Engineering University named after G.F. Morozov, Voronezh, Russia
[3] Kuban State Technological University, Krasnodar, Russia
[4] G.R. Derzhavin Tambov State University, Tambov, Russia

Abstract. The research is aimed at revealing the specifics of the risks that arise in the process of developing the shared consumption economy in different countries, as well as at designing tools for levelling them. The objects of the study were the sectors of the shared consumption economy of the leading EU-28 countries (France, Great Britain, Poland, and Spain), as well as Russia. The authors identified the features of the development of the shared consumption economy in the leading countries of the EU and Russia, conducted a comparative analysis of their sectoral structure. The authors systematized the risks of the development of the shared consumption economy faced by the EU-28 countries, as well as analyzed the measures implemented by the states to level the emerging risks. The article defines the specifics of the spatial distribution of the sectors of the shared consumption economy in Russia and identifies new trends in its development. The authors identified specific risks of the development of the shared consumption economy in Russia and identified strategic directions for levelling them (creating institutional foundations for the functioning of the shared consumption economy; maintaining information security of digital platforms of the shared consumption economy; spreading standards in the field of standardization of the quality of goods and services to the sectors of the shared consumption economy).

Keywords: Shared consumption economy · Risk · Sharing services · Rent · EU · Russia · Risk levelling

JEL Codes: E21 · O11 · O33

1 Introduction

The shared consumption economy as a new form of organization of economic relations arose as a result of changes in the perception of private property, a shift in the paradigm of commodity relations, excessive development of the consumer society, and, of course, digitalization [5, 23].

E. G. Popkova (Ed.): ISC 2020, LNNS 368, pp. 750–761, 2022.
https://doi.org/10.1007/978-3-030-93244-2_81

The joint use of goods and services, exchange and various forms of rent are becoming not only a tool for the transition from a model of overconsumption to lean consumption but also a source of GDP growth in many countries. Thus, by 2025, the global volume of the shared consumption economy will amount to 282.7 billion Euros. At the same time, in some countries, in particular, in China, this market segment has reached 1.67% of the country's GDP [10, 32].

However, on a global scale, the distribution of sectors of the shared consumption economy is very uneven both in space and in time. This applies to the development of its various areas (from renting things, finding transport, renting premises for work to sharing agile offices and knowledge), as well as state support for ongoing sharing projects [34].

In developed countries, the formation of the economy of shared consumption occurred at an earlier time [25]. In emerging economies, these innovations came much later and are currently only gaining their power. Here, the construction of a shared consumption economy plays a special role, since it is associated with the desire to overcome the lag behind the world leaders [2, 38]. And its formation is associated with a lot of risks.

2 Literature Review

The specifics of the implementation of shared consumption models in different sectors of the economy are described in the works of [5, 11, 14, 16, 18, 19, 21–23, 35, 36]. The authors identify the reasons for the development of the shared consumption economy, in particular, the interest of subjects in obtaining economic benefits, reducing transaction costs, realizing social needs for mutual assistance, the widespread use of digital platforms and mobile applications.

The development of the shared consumption economy in different countries has significant differences. Access to the Internet, the formation of digital infrastructure, the peculiarities of consumer demand in the regions, state regulation, the existence of risks of the development of sharing models affect the development of the shared consumption economy [6, 12, 14, 19, 35].

Some countries achieve significant benefits from the development of the economy of shared consumption, consisting in the creation of additional added value by subjects and an increase in GDP, in additional employment of the population, in reducing the burden on the environment due to the reuse of resources, in lowering transaction costs associated with the marketing of free assets [4, 13, 15, 39].

Other countries face risks in the process of their development. In particular, [19] point to the risk of a reduction in production and a slowdown in GDP growth due to a decrease in consumer demand for goods and services of traditional sectors of the national economy, the risk of undermining the conditions of normal competition, the risk of not receiving tax fees.

3 Methodology

The purpose of the study is to reveal the specifics of the risks that arise in the process of developing the shared consumption economy in Russia and the EU countries, as well as to develop tools for levelling them.

Research objectives:

(1) to highlight the features of the development of the shared consumption economy in Russia and some EU countries;
(2) to systematize the risks of developing a shared consumption economy;
(3) to form a tool for levelling the risks of the development of the shared consumption economy in Russia.

Methods used in the study: theoretical analysis and synthesis, comparative analysis, systematization method, system approach, graphical method.

4 Results

The economy of shared consumption is developing dynamically. According to PwC's forecast, the global revenue of companies in the field of the sharing economy will reach 282.7 billion Euros by 2025[40].

In 2018, the value of the shared consumption economy in the EU-28 amounted to 22.4 billion Euros (about 0.17% of the total GDP of the countries). This market segment is most developed in France (25% of the total volume of the shared consumption economy in the EU-28), the United Kingdom (17%), Poland (10%), and Spain (10%), followed by Germany, Italy, and Denmark.

In general, the seven largest markets of the joint consumption economy in the European Union account for about 80% of the total joint income of the EU-28 in 2018 (Fig. 1).

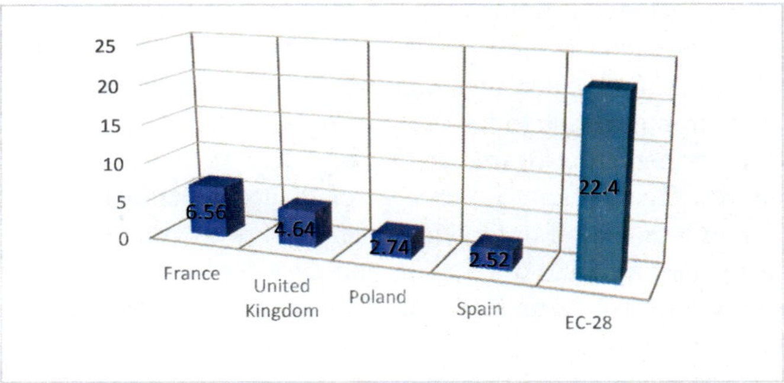

Fig. 1. The volume of the shared consumption economy in Russia and the EU countries, billion Euros, 2018 *Source* Compiled by the authors based on [24]

The majority of activities are carried out in four sectors: housing (0.83% of the sectoral GDP), the financial sector (0.03%), transport (0.03%), online skills (0.01%) [13]. The shared consumption economy provides approximately 394,000 jobs across the EU, which is about 0.15% of the total number of people employed in the 28 EU countries [13].

France is the undisputed leader among the EU countries in the development of the shared consumption economy. The contribution of the shared consumption economy to the country's GDP in 2018 was 0.38%, in other words, 2.5 times higher than the EU average. During this period, 99 sharing platforms operating in the online skills sector (35 platforms), finance (27), transport (21) and housing (16) were identified in France. The largest employment was achieved in the transport sector – 32,386 employees, 19,000 employees in the accommodation sector, 14,300 employees in the finance sector, and 9,000 employees in the online skills sector. Despite this distribution of employees, the financial sector and the placement sector provided the largest revenues to the French economy – 2.2 billion Euros each, the online skills sector-more than 1 billion Euros. The volume of investments in the transport sector amounted to 47.5 million Euros, in the placement sector-40 million Euros. The reasons for the intensive development of the shared consumption economy in France are the large stock of unused items, a growing interest in using the product rather than property, a synergistic effect between traditional companies and platforms, state support for the sector and measures implemented to level risks.

In the UK, the total market volume of the shared consumption economy has reached 4.6 billion Euros (0.2% of the country's GDP), and the number of people employed in its sectors is more than 69,431 people. The transport sector (represented mainly by the international platforms BlaBlaCar and Uber and the domestic platform JustPark) is the largest in the UK's shared consumption economy, it provides maximum income and employment, as well as continues to grow dynamically. The financial sector receives an annual income of 1.8 billion Euros and employs an average of 9,936 employees. The country is dominated by domestic digital platforms with sharing services that meet the requirements for debt and equity financing. The placement sector has attracted investments worth more than 140 million Euros in recent years and currently provides a fifth of the total income from the UK's shared consumption economy with average revenue of 828 million Euros per year. Most of the platforms in this sector are international (for example, Airbnb) and are focused on providing housing rental services. The online skills sector in the UK remains underdeveloped and mainly provides on-demand consumer services [13]. The development of the shared consumption economy in the UK is facilitated by the general support of sharing services at the national level within the framework of the formation of open markets, as well as the government's desire to become the country's centre for innovative solutions.

There are 36 platforms of the shared consumption economy operating in Poland, and the average annual revenue from its operation is 2.7 billion Euros or 0.66% of GDP. The transport sector is formed by both national and international platforms (BlaBlaCar, Taxify, and Uber), which operate according to the ridesharing doctrine. The co-hosting platforms are also divided into national (Stancja) and international (Airbnb), providing 2,509 jobs and 81.3 million Euros per year. The financial sector of the shared consumption economy has the greatest potential for development, where 15 platforms of domestic origin and three international ones operate. The sector employs 5,300 jobs and an income of 285 million Euros per year [13]. The online skills sector, which offers on-demand household services, is the largest in the Polish shared consumption economy, it employs about 50,000 employees and a guaranteed income of 2.3 billion Euros per year. Another direction of the Polish shared consumption economy is "Foodsharing Warsaw" – a public grassroots initiative aimed at limiting food waste [27]. This can be achieved through education, active actions of food rescuers, volunteers who provide delivery of unsold products to special points with a kitchen cabinet and a refrigerator for further use. The success of the shared consumption economy in Poland is due to the effective coordination of investment efforts and a strict regulatory framework [26].

The sectors of the economy of shared consumption in Spain have demonstrated intensive development after 2016 when its volume exceeded the traditional for the country in the amount of 0.24% of GDP. The most developed sectors are housing and shared services. In Spain, there are 81 platforms for the sharing economy, mainly of domestic origin. Most of them serve the financial sector and the transport and accommodation sectors. The online skills sector provides the maximum employment of 16,293 people and generates the highest income – more than 1 billion Euros per year. In terms of revenue, this sector is comparable to the placement sector [3]. The financial and transport sectors provide the country with 454 million Euros and 118 million Euros per year, respectively. The factors of the growth of the economy of shared consumption in Spain are the development of telecommunications networks and the increase in the use of mobile phones, which contribute to the spread of digital platforms; the economic crisis, which promotes the development of alternative credit channels; changes in the consumer culture focused on access to services, rather than ownership [8].

In Russia, the dynamics of the development of the shared consumption economy is characterized by constancy and continuous growth. According to the RAEC, in 2018, the volume of transactions on the main platforms of the sharing economy in Russia amounted to 5.78 billion Euros (compared to 2017, the volume of transactions increased by 30%), in 2019, the volume of this segment was already about 8.7 billion Euros [29]. In 2020, the volume of the shared consumption economy reached the level of 12.1 billion Euros (Fig. 2).

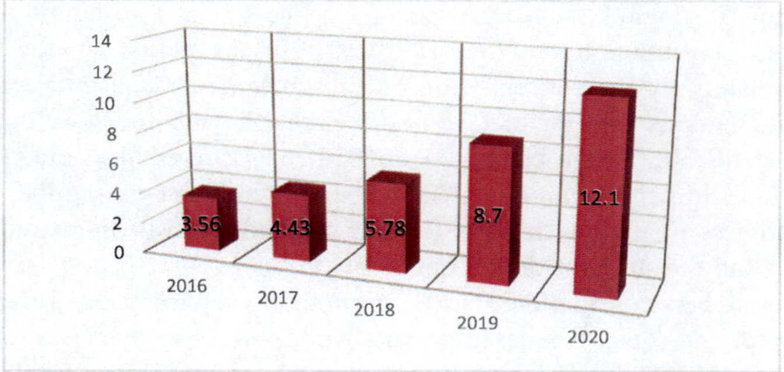

Fig. 2. Dynamics of the development of the shared consumption economy in Russia, 2017–2020 *Source* Compiled by the authors based on [30]

The development of the shared consumption economy in Russia has its sectoral specifics. In 2020, 78.3% of the total volume of all transactions accounted for C2C sales of things, 17.1% provided the services of private freelancers (P2P services), 2.06% was made by the carsharing sector, 1.1% was accounted for carpooling, the share of office sharing was 0.57% [28] (Fig. 3).

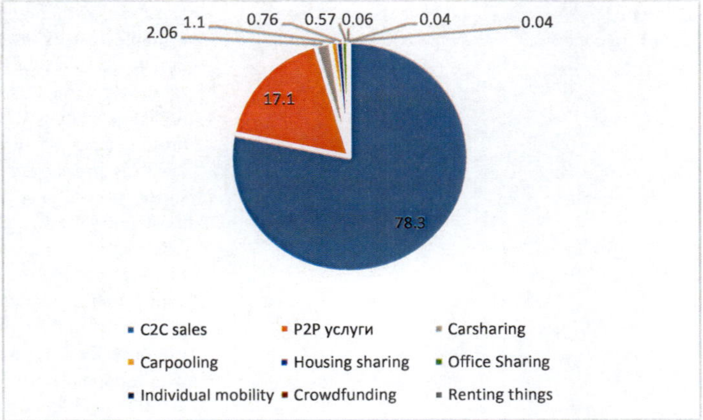

Fig. 3. The sectoral structure of the shared consumption economy in Russia, %, 2020 *Source* Compiled by the authors based on [30]

In addition, it should be noted the specifics of the spatial distribution of the sectors of the shared consumption economy in Russia. In the period up to 2017, it was determined mainly by the concentration of consumer demand in megacities, uneven access to the Internet and the unfavourable situation with the development of the transport system [30]. In the period 2018–2020, new trends have emerged, such as regionalization and penetration of sellers into the regions, increasing the indicators of network readiness of Russian regions [6] and intensification of infrastructure support and ensuring the operation of digital online platforms [17].

Some researchers consider the shared consumption economy as a social process of exchange, implying social ties based on such values as trust, openness, equality, participation, care [37]. However, the lack of such basic values in society, as well as the state's inattention to the regulation of this sector of the economy can become a deterrent to the development of the sharing sectors and a source of risks for the economy and society (Table 1).

Table 1. Risks of the development of the shared consumption economy in the countries of the European Union and the tools for levelling them

Prerequisites	Risk	Risk levelling measures
France		
Lack of transparency of platforms, the working legal status of employees or the financial regime of transactions.	The development of unfair competition and the lack of support for the economic development of the sector.	Adoption of a set of rules concerning transport, placement and the crowdfunding sector. For example, drivers providing private transport must be registered at the national level, drivers providing VTC must be professionals. There are three main laws in the accommodation sector: peer-to-peer providers must notify the city administration when they rent out secondary housing; permission and compensation are required in case of changes in the conditions of use of residential premises; cities with a population of more than 200,000 people must request permission from the host country to rent out their housing, regardless of the duration and category of residence.

(*continued*)

Table 1. (*continued*)

Prerequisites	Risk	Risk levelling measures
United Kingdom		
The lack of an institutional framework for regulating employment in the shared consumption economy	Reducing the level of user confidence in the platforms of the shared consumption economy. The imbalance of the labour market and employment. Political instability in the sectors of the shared consumption economy (tariff setting policy, licensing of transport transportation, planning of short-term lease, taxation of property and profits	General support of sharing services at the national level for the formation of open markets. Development of a conceptual framework to support the collection and dissemination of statistical data on activities in the field of the sharing economy. Creation of an industry association of platforms for the economy of shared consumption using the seal of trust (a self-regulation tool). Limitations of the legislative framework to support the peer-to-peer housing rental sector, the implementation of measures to update and introduce some new rules that are adequate for the cohabitation sector
Poland		
The lack of a regulatory framework regulating activities in the sectors of the shared consumption economy	Shortfall in tax payments to the state budget, GDP reduction, stagnation	Effective coordination of investment efforts and the formation of a strict regulatory framework, primarily in the field of taxation. Creation of the Development Fund (FIU) as an umbrella institution for a group of institutions involved in the implementation of development projects
Spain		
Fragmented regulatory framework for regulating transactions in the sectors of the shared consumption economy at the local level. Suspension of the issuance of licenses until the lease of their property. Banning the P2P version of the platform for calling Uber passengers. Tax restrictions and social security obligations for employees of the online skills sector	The lack of conditions for further growth of the sectors of the economy of shared consumption. Violation of the conditions of normal competition	Identification of obstacles to the development of the shared consumption economy and the formation of recommendations for further growth. Ministerial consultations aimed at developing a strategy for the digital economy and the shared consumption economy. Government initiatives to introduce special tax provisions for cooperation platforms and colleagues, including at the regional and local levels

Source: Compiled by the authors.

As for Russia, the country entered the path of development of the sectors of the economy of shared consumption much later than European countries, which contributes to the emergence of some specific risks of its intensive development. In particular, the researchers identify the following risks that are characteristic of the Russian reality:

- reducing the quality of services provided in the absence of institutional responsibility for the owner of resources [9, 19];
- lack of a systematic approach to pricing and, as a result, overestimation of prices for goods and services provided [1];
- information attacks and cybercrime actions due to the vulnerability of digital platforms [9];
- fraudulent transactions and loss of property [31];
- reduced confidence and reduced consumer demand due to poor quality of services [7, 20];
- slowing of GDP growth due to a reduction in consumer demand for goods and services of traditional sectors of the national economy [19];
- violation of the conditions of normal competition [33];
- the emergence of new forms of inequality and polarization in property relations [12, 33];
- shortfall in tax revenues due to the lack of institutional frameworks for regulating the sectors of the shared consumption economy [1].

Thus, the risks of developing a shared consumption economy in Russia may arise concerning all its subjects-consumers, resource owners, the state. It is important to take this into account when developing risk levelling tools (Fig. 4).

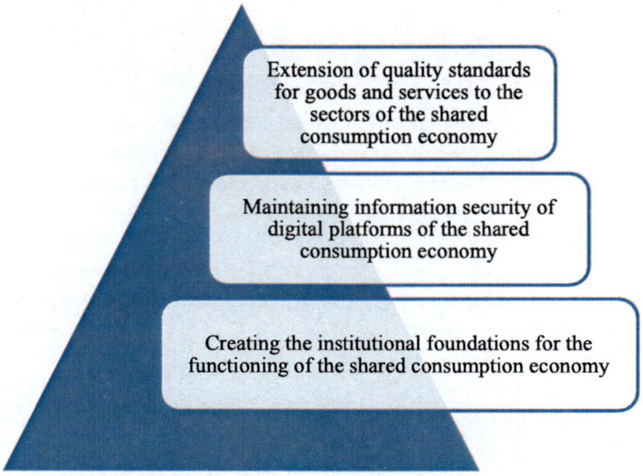

Fig. 4. Tools for levelling the risks of the development of the shared consumption economy in Russia *Source* Compiled by the authors

5 Conclusion

Firstly, the authors identified the features of the development of the shared consumption economy in the EU and Russia. It is revealed that the countries with the largest segment of the shared consumption economy in the EU are France, United Kingdom, Poland, and Spain. The specific features of the development of the shared consumption economy in these countries, as well as its sectoral structure, are determined. It is revealed that in Russia, the dynamics of the development of the shared consumption economy is characterized by constant and continuous growth, especially in the period 2017–2020. The specifics of the spatial distribution of the sectors of the shared consumption economy in Russia are determined. New trends in the development of the shared consumption economy in Russia have been identified (penetration of sellers into the regions, increasing the indicators of network readiness of Russian regions, intensification of infrastructure support and ensuring the operation of digital online platforms). Secondly, the authors systematized the prerequisites for the emergence and risks of the development of the shared consumption economy in the EU countries, studied the practice of levelling them in France, the United Kingdom, Poland and Spain. The specific risks of the development of the shared consumption economy in Russia are systematized. Thirdly, the analysis of effective experience in overcoming the risks of the leaders' countries allowed the authors to form tools for levelling the risks of the development of the shared consumption economy in Russia.

References

1. Analytical Center under the Government of the Russian Federation: The shared consumption economy as a new economic model. https://ac.gov.ru/files/publication/a/23715.pdf (2019). Accessed 12 June 2021
2. Anokhov, I.: Mass and lean production. Soc. Econ. **7**, 51–64 (2018)
3. Barcelonacat: Special Tourism Accommodation Plan (PEUAT) 2017. http://ajuntament.barcelona.cat/pla-allotjaments-turistics/en/ (2017). Accessed 12 June 2021
4. Blablacar: Trips of your choice at the lowest prices. https://www.blablacar.ru/ (2021). Accessed 12 June 2021
5. Botsman, R., Rogers, R.: What's Mine is Yours: How Collaborative Consumption is Changing the Way we Live. HarperCollins Publishers, New York, NY (2010)
6. Bychkova, N., Tavbulatova, Z., Ruzhanskaya, N., Tamov, R., Karpunina, E.: Digital Readiness of Russian Regions. In: Proceeding of the 36th IBIMA conference, 4–5 November 2020, pp. 2442–2461. Granada, Spain (2020)
7. Chica, M., Chiong, R., Adam, M.: An evolutionary trust game for the sharing economy. In: Proceedings of the IEEE Congress on Evolutionary Computation, CEC, pp. 2510–2517 (2017)
8. CNMC: Press release "Las app para vender or alquilar productos de segunda mano, las más utilizasas por los internautas". https://www.cnmc.es/node/365389 (2017). Accessed 12 June 2021
9. Cohen, M., Sundararajan, A.: Self-regulation and innovation in the peer-to-peer sharing economy. Univ. Chicago Law Rev. Dialogue **8**, 116–120 (2015)

10. Daxueconsulting: "Sharing Economy" only as one part of exploding Collaborative Economy in China. https://daxueconsulting.com/exploding-collaborative-economy-in-china/ (2017). Accessed 12 June 2021
11. Dillahunt, T., Malone, A.: The promise of the sharing economy among disadvantaged communities. In: Human Factors in Computing Systems proceedings of the International Conference, pp. 2285–2294 (2015)
12. Dostmohammad, S., Long, J.: Regulating the Sharing Economy: Applying the Process for Creative Destruction. Dalhousie University, Nova Scotia (2015)
13. European Commission: Study to monitor the economic development of the collaborative economy at sector level in the 28 EU Member States. https://publications.europa.eu/en/publication-detail/-/publication/0cc9aab6-7501-11e8-9483-01aa75ed71a1/language-en/format-PDF/source-72448580 (2018). Accessed 12 June 2021
14. Filimonova, N., Kapustina, N., Bezdenezhnykh, V., Kobiashvili, N.: Trends in the sharing economy: a bibliometric analysis. In: Popkova, E., Sergi, B. (eds.) Digital Economy: Complexity and Variety vs. Rationality, vol. 87, pp. 145–154. Lecture Notes in Networks and Systems (2016)
15. Fremstad, A.: Essays on information, income, and the sharing economy: doctoral dissertation. https://scholarworks.umass.edu/dissertations_2/470/ (2015). Accessed 12 June 2021
16. Frenken, K., Meelen, T., Arets, M., Van de Glind, P.: Smarter regulation for the sharing economy. The Guardian, 2015. https://www.theguardian.com/science/political-science/2015/may/20/smarter-regulation-for-the-sharing-economy (2015). Accessed 12 June 2021
17. Galeeva, G., Ishtiryakova, L.: Investments in the sharing economy and their impact on the employment structural changes. Finance Theory Pract. **24**(5), 128–148 (2020)
18. Glukhov, V., Glukhova, Z.: The economy of joint consumption as a new form of economic relations. Waste Resour. **4**, 10–19 (2019)
19. Gusev, V.V., Magomedbekov, G.U., Galieva, G.F., Gundorova, M.A., Shadrina, Z.A.: 11 Sharing economy: how digital technologies have changed economic reality. In: Alpidovskaya, M.L., Karaseva, L.A., Mamagulashvili, D.I., Bogoviz, A.V., Krivtsov, A. (eds.) Industry 4.0: Implications for Management, Economics and Law, pp. 95–106. De Gruyter (2021). https://doi.org/10.1515/9783110654486-011
20. Hawlitschek, F., Teubner, T., Adam, M.: Trust in the sharing economy: an experimental framework. In: Proceedings of the International Conference on Information Systems, ICIS, pp. 1–14. Dublin, Ireland (2016)
21. Heinrichs, H.: Sharing economy: a potential new pathway to sustainability. GAIA **22**(4), 228–231 (2013)
22. Kiryak, S.: Sharing economy – a new model of consumption. Beneficiary **58**, 29–31 (2019)
23. Lymar, E.: The economy of joint consumption in modern Russia. Bulletin of the Chelyabinsk State University. Econ. Sci. **12**(422), 63, 67–72 (2018)
24. Marketologi: The sharing economy will grow 10 times. https://www.marketologi.ru/img/news/2019/06/13/3.pdf (2019). Accessed 12 June 2021
25. PwC: Sharing or paring? Growth of the sharing economy. https://www.pwc.com/hu/en/kiadvanyok/assets/pdf/sharing-economy-en.pdf (2015). Accessed 12 June 2021
26. PwC. This section has drawn extensively on the analysis carried out by PwC (2016) (Współ) dziel i rządź!, Prawno-podatkowe aspekty ekonomii współdzielenia w Polsce. https://www.pwc.pl/pl/pdf/ekonomia-wspoldzielenia-raport-2-pwc.pdf (2016). Accessed 12 June 2021
27. Radiopolsha: Foodsharing Warsaw – a way to save food from the trash. http://archiwum.radiopolsha.pl/6/249/Artykul/426757 (2019). Accessed 12 June 2021
28. RAEC: The Economy of shared consumption in Russia 2018. https://tiarcenter.com/sharing-economy-research/ (2019). Accessed 12 June 2021

29. RAEC: The shared consumption economy in Russia-2019. https://tiarcenter.com/wp-content/uploads/2020/03/RAEC_Sharing-economyin-Russia-2019_March-2020.pdf (2020). Accessed 12 June 2021
30. RAEC: The Economy of shared consumption in Russia 2020. https://raec.ru/upload/files/raec-sharing-economy-2020.pdf (2021). Accessed 12 June 2021
31. Rbc: We have reached the revenue: the profitability of the sharing economy-a myth or a reality. https://www.rbc.ru/trends/sharing/5de4e5f19a7947f2c0f0d0a7 (2018). Accessed 12 June 2021
32. Rbc: The Economics of Sharing in 30 figures and facts Sources: Brookings Institute, Global Market Insights, NYSE, NYU, Pew Research, PwC, WEF, RAEC. https://www.rbc.ru/trends/sharing/5ddbb3279a7947b01be74c19 (2019). Accessed 12 June 2021
33. Richardson, L.: Performing the sharing economy. Geoforum **67**, 121–129 (2015)
34. Shaikomalova, K.: New in real estate: office sharing. Why is it profitable? https://vc.ru/flood/33970-novoe-v-nedvizhimosti-shering-ofisov-pochemu-eto-vygodno (2018). Accessed 12 June 2021
35. Shmeleva, A., Bezdelov, S., Rybakov, M.: Prospects for the development of the sharing economy in Russia. Competency (Russia) **7**, 4–10 (2020)
36. Steuer, E.: Robin Chase on the (financial) value of sharing. www.good.is/post/robin-chase-on-thefinancial-value-of-sharing/ (2009). Accessed 12 June 2021
37. Tagarov, B.: Problems of the development of the economy of joint consumption in Russia. Russian Entrepreneurship **20**(2), 593–606 (2019)
38. Tagarov, B.: Specifics of the economy of joint consumption and the conditions of its development. ECO **7**, 140–155 (2019)
39. Vedomosti: Marriott creates a home rental business to compete with Airbnb. https://www.vedomosti.ru/business/news/2019/04/29/800468-marriott-konkurentsii-s-airbnb (2019). Accessed 12 June 2021
40. Weforum: How much is the sharing economy worth to GDP? https://www.weforum.org/agenda/2016/10/what-sthe-sharing-economy-doing-to-gdp-numbers/ (2016). Accessed 12 June 2021

Analysis of the Russian Banking System Institutional Saturation Over the Past Decade

Maxim V. Milenin, Olga A. Grishina, Tatyana M. Rogova(✉), Yuliya A. Vlasova, and Dinara R. Tutaeva

Plekhanov Russian University of Economics, Moscow, Russia
{milenin.mv, grishina.oa, rogova.tm, vlasova.ya, tutaeva.dr}@rea.ru

Abstract. This article analyses the development of the Russian banking system over the past 10 years and studies its tendencies. The leading tendencies revealed are (a) the annual decrease of the number of banks and (b) redistribution of assets between existing groups of banks leading to concentration and centralization of the banking system. The author forecasts the number of banks in Russia in the next few years in case the mentioned tendencies persist. The article analyses the regions, in which the number of banks decreased most significantly. The example of these regions shows what negative consequences the liquidation of local banks brings to the regions and why big banks are not always able to solve such problems. The author outlines possible ways to preserve the diversity of financial institutions, which is necessary for the harmonious development of individual regions' economics in Russia.

Keywords: Banking in Russia · Banking system saturation · Central Bank of the Russian Federation · Regional banking systems · Banking system development

JEL Codes: G21 · E58 · P25

1 Introduction

Russian banking system current situation analysis

One of the most significant trends of the last two decades in the Russian economy is the decline in the number of banks and other financial institutions. Starting from 2003 there was no year in which the number of financial institutions increased. Moreover starting from 2015 annual decrease started overlapping 10%.

During the last two years, this trend has slightly slowed down to 9% and 8% of players leaving this market in the years 2019 and 2020 accordingly. However, these figures are still significant.

This study pursues the following goals, which are relevant for the current situation in the Russian financial sphere:

- To outline the consequences of the permanent decline in the number of banks and other financial institutions;

E. G. Popkova (Ed.): ISC 2020, LNNS 368, pp. 762–771, 2022.
https://doi.org/10.1007/978-3-030-93244-2_82

- To forecast the number of financial institutions in the Russian economy in the next few years;
- To estimate the number of banks necessary and sufficient for the Russian economy;
- To work out the recommendations on how to prevent the Russian financial system from the insufficiency of credit institutions.

The main reason for the closure of credit institutions in recent years in Russia was the license revocation by the Central Bank. Starting from 2013, the Regulator proclaimed and consistently carried out the program to improve the banking system. This program has included tracking the ineffective and unscrupulous banks and introducing appropriate measures to such banks up to their liquidation.

The attitude of the banking and academic communities to the increased regulatory pressure on the banking sector by the Central Bank has split. The experts have outlined the following negative consequences of such a clean-up:

1. regional financial systems deterioration [1]
2. system-based stability risks [13],
3. the lack of financial institutions and services in some regions [17],
4. the nationalization of the banking business [16] and,
5. as a result – monopolization of the banking services market [14].

On the other hand, some experts suppose that the consequences stated above are temporary, and as time passes, the banking system will become stable and competitive and it will provide a sufficient level of services to businesses and the population [14]. Some experts also consider the creation of systemically important banks[1] cluster to provide an additional advantage to the banking system [5].

Methodology

The research carried out in this article is based on statistical data analysis published by the Central Bank of the Russian Federation. Contrary to previous studies [13, 15, 17], where the regional banking systems of Russia were considered only at the level of the Federal Districts, this article analyzes the indicators of banking systems at province (in Russian – oblast') and republic levels. This makes analysis more detailed and accurate and allows to confirm the hypotheses made earlier by [1] and shows how the closure of local banks negatively affects the institutional development of regional banking systems. Detailed analysis also has allowed describing the advantages of local banks over federal banks in the conditions of regional markets.

To forecast the number of banks in Russia in the nearest future, the current rate of decline in the number of banks was extrapolated to coming years.

The author also studied the experience of the countries with a similar complex federal system and comparable GDP to formulate recommendations for maintaining the efficiency of the Russian banking system.

Additionally, the article uses expert assessments, analysis, deduction, and hypothesis as the methods of scientific research.

[1] According to the Central Bank of the Russian Federation [6] annually approves a list of systemically important credit institutions.

2 Results

Despite the reduction in the number of financial institutions in recent years (Fig. 1), the total assets of the banking sector have shown steady growth (Fig. 2), comparable to the growth rate of Russia's GDP, sometimes even overlapping it.

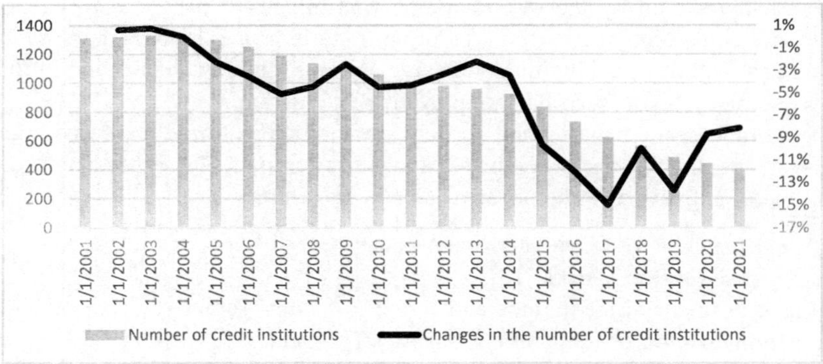

Fig. 1. Number of credit institutions in Russia. *Source* The diagram compiled by the author based on materials from the Central Bank of Russia [7–10]

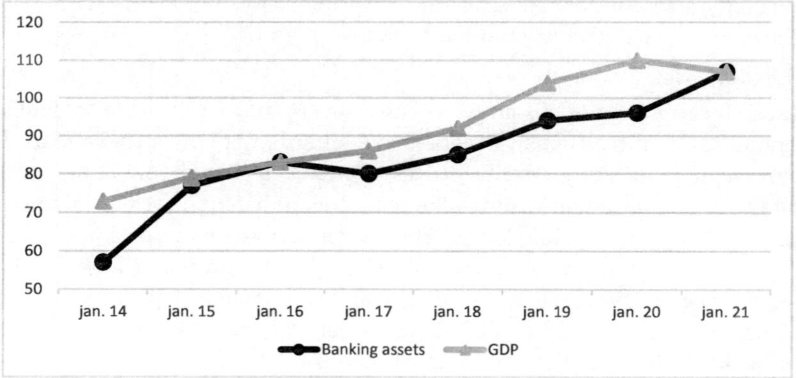

Fig. 2. GDP of Russia vs overall assets of the Russian banking system (trillion rub.). *Source* The diagram compiled by the author based on materials from the Central Bank of Russia [7–10]

It means that the constant disposal of banks, mostly small ones, was compensated by the consolidation of the leading Russian banks, including consolidation through the acquisition of the former by the latter (Fig. 3).

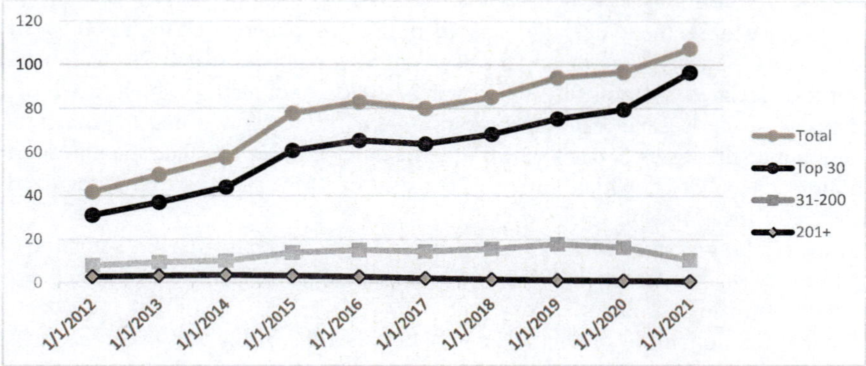

Fig. 3. The banking system of Russia assets concentration (trillion roubles). *Source* The diagram compiled by the author based on materials from the Central Bank of Russia [7–10]

Figure 3 shows that from 2012 to 2019, the assets growth rate of the top-30 Russian banks coincided closely with the total banking system figures[2]. However, according to the results of the last two years, the assets growth rate of the top-30 banks exceeded the total indicator of the industry, and the total assets of banks in the group "31–200" decreased significantly, which may be the beginning of a new trend for this group. As for the banks outside the first two hundred, their total assets have been steadily falling since 2015 (Table 1).

Table 1. The assets of the banks 201+ rated by the number of assets (billion rub.)

Date	1.1.12	1.1.13	1.1.14	1.1.15	1.1.16	1.1.17	1.1.18	1.1.19	1.1.20	1.1.21
Overall assets	2,734	3,171	3,520	3,145	2,725	1,974	1,462	1,056	835	540
Average assets	3.51	4.19	4.87	4.96	5.11	4.67	4.05	3.72	3.45	2.62

Source The table compiled by the author based on materials from the Central Bank of Russia [7–10]

[2] Grouping of banks to 1–30, 30–200 and 201+ is made by their assets amounts according to the reports of the Central Bank of the Russia Federation [11].

Moreover, as the analysis shows, this diminishing has been caused not only by reducing the number of banks outside the first two hundred. Since 2016, the average size of the assets of the banks from this group has been steadily decreasing. Thus, the weight of the banks outside the first two hundred diminishes markedly.

How the situation may evolve?

As of January 2021, the total share of assets of this group amounted to only 0.5% of all Russian banks' assets. To compare: the assets of Sovcombank, which Russian financial authorities declared systemically important in 2020, amounted to about 1.5% of the total assets of the Russian banking system. That is, if we assume that the indicator of the amount of the assets is decisive for the systemic significance, then the entire group of bottom-rated banks, which currently consists of 166 banks can be recognized as systemically insignificant.

If the banking system will continue shrinking by an average of 10% annually, then in the next two years less than 300 banks will remain in Russia, and by 2027, the third hundred may vanish at all.

Today it is difficult to predict accurately how fast the banks will leave the Russian economy. More important is that, taking into account the trends discussed above, the closure of any number of banks from the current third and fourth hundred in the coming years will not have a significant macroeconomic impact on the performance of the Russian banking system as a whole.

At the same time, it is impossible to imagine that disappearance of several dozen banks will have no consequences at all. Any bank, regardless of size, has its clients, provides them with a range of services, and performs significant functions within the local market. That is why the experts suppose [1] that the process of small regional banks vanishing has a negative impact.

With the chances high, it does not affect Moscow, where several dozen leading banks are fully represented. In Moscow, the provision of the population and business with banking institutions is almost one and a half times ahead of the average in Russia. However, in other regions, this may have a serious impact on the local banking market.

The consequences of current trend progression

In recent years, the Central Bank of the Russian Federation [7] has already indicated in its reports that the saturation of local markets with banking services is decreasing (Fig. 4). In its reports, the Regulator has named the withdrawal from the market of regional banks, as the main reason for this process [7–10].

[3] As can be seen, the North-Caucasian region's saturation differs a lot from all other Russian regions. The banking development tendencies of this region are studied by [15].

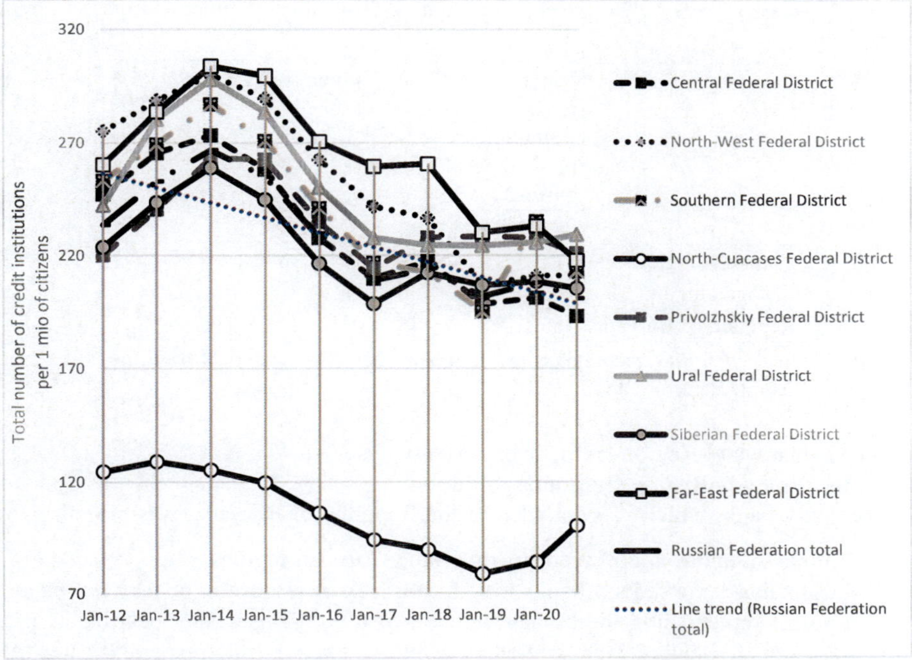

Fig. 4. Russian banking system saturation with credit institutions, incl. banks, their branches and offices. *Source* The diagram compiled by the author based on materials from the Central Bank of Russia [7–10][3]

In general, we can suppose that the decrease in the number of branches and offices has been caused by the digitalization of many banking services, and the physical presence of the client in the bank office is often not required. However, inside the federal districts, the decrease in the provision of the population with banking institutions is very uneven. The leaders in this indicator's fall over the past 10 years are present in Table 2.

As can be seen from the figures presented, the institutional saturation of the banking system in these regions is falling at a rate that is more than 2 times ahead of the all-Russian indicators. Suppose that the main reason for this decline is the closure of local banks. Indeed, in these regions, this indicator is very significant. The number of local banks registered in the listed regions for the specified period decreased as follows:

[3] As can be seen, the North-Caucasian region's saturation differs a lot from all other Russian regions. The banking development tendencies of this region are studied by [15].

Table 2. Russian regions in which saturation with credit institutions decreased most significantly

Region	Number of credit institutions per 1 mil of citizens		Change in number	Change in %
	2020	2011		
Yaroslavl region	185	304	−119	−39%
Vologda region	196	332	−136	−41%
Republic of Buryatia	165	281	−116	−41%

Source The table compiled by the author based on materials from the Central Bank of Russia [7–10]

- In the Yaroslavl Region from seven to one
- In the Vologda Region from nine to three
- The only bank, which was founded in the Republic of Buryatia, was closed.

A similar situation is observed in some other Russian regions.

It means that large federal banks may face a lack of resources, or may not have a hankering for replenishing the branch network of the region, where local banks previously operated. Banks headquartered in a different region may experience the following difficulties in opening and managing a local branch network:

- Insufficient knowledge of the local market
- Insufficiently established relations with the local administration
- The need to coordinate decisions with the head office.

All the facts stated above lead to the deterioration in the provision of banking services to the local population and businesses. That is, if the Central Bank of the Russian Federation [10] says that competition between banks and the saturation of the sector with services remain at the same level, it is true only for the average indicators of the state in general. As can be seen from the figures presented above, the situation in the country is very heterogeneous, and certain regions may face problems due to the massive closure of banks, especially regional ones.

According to the research made by [18] for Banki.ru, “the so-called cleaning of the banking sector from unscrupulous players was completed in 2019”. At the same time, statistics of recent years show that the pace of bank closures has hardly slowed down. In his research, [18] “expects 30–35 small credit institutions to leave the market in 2021”. Without a doubt, the Regulator’s campaign to remove unscrupulous players from the market was necessary and successful. However, the analysis carried out in this article shows, that in the current market conditions, the further withdrawal of bona fide but relatively weak regional players from the market under the pressure of strict Central Bank’s regulations and requirements will most likely have negative consequences for regional economies.

How to preserve credit institutions diversity

If the regional markets are developed sufficiently, the benefits of the local banks' operation are obvious. Taking into account that (a) some Russian regions have just started to face the lack of financial institutions in last years and (b) the Russian Central Bank's campaign against massive violations in the banking system comes to an end, it can be assumed that Russian banking system is still close to its equilibrium. The lack of financial institutions has not become dramatic yet thus it will not require an extraordinary effort from financial authorities to preserve credit institutions diversity. The Russian banking system needs several hundred financial institutions to fill all the economic niches in all the regions.

International experience also shows that countries with a comparable population and GDP, as well as a complex federal structure, have hundreds and sometimes thousands of financial institutions in their banking systems. The experience of various foreign systems can be studied to preserve the institutional variety of the Russian financial system and to slow down the trend of increasing centralization and concentration. For example, a transition to a three-tier banking system is possible, as in Germany, where local banks do not compete with large commercial banks at the federal level and provide only the regional financial needs. In Germany, only 282 of 1,783 financial institutions are large universal commercial banks, 386 are savings banks, and 878 are credit cooperatives operating at local levels [12]. It is also possible to consider the Brazilian experience, where only 137 of 1631 financial institutions are universal commercial banks, and the rest are highly specialized banks, credit and financial institutions [16].

The basis of the modern banking system in Russia was laid in the conditions of the liberal economy proclaimed about three decades ago. At present, the centralization and concentration of the economic system as a whole are increasing. In order to preserve the diversity of financial institutions, regulatory protection of their medium and small representatives is required. This aim may be reached through their special functional and/or regional status assignment.

3 Conclusion

The analysis carried out in this article shows that at the macro level, the current withdrawal of many banks from the economy does not have a serious negative impact. However, the article proves that the local banks' closure negatively influences the economies of the Russian regions. Taking into account that Central Bank's program on banking system clean-up has succeeded, further diminishing of the banking system is not needed and will be harmful. Moreover, the current rate of banks liquidation may lead the Russian banking system to critical institutional saturation insufficiency in the next 5–7 years.

In this regard, the Central Bank of Russia, having completed the program of closing unscrupulous banks, should consider a program for the development and support of banks that meet the established standards. The Russian banking system has already made the first steps to a 3-tier banking system, such as the creation of a systemically important banks cluster, which differs significantly from all other banks. Now the aim is not to destroy all other banks. Though they are not systemically important for the whole country, many studies including this one prove that they are important for the successful development of local economies [2, 3].

In future present research may be proceeded and developed in the following directions:

- How to transform the Russian banking system into a three-tier one,
- Creation of the methodology and regulations to support, preserve and transform the Russian banking system.

References

1. Afanasyeva, O.N.: Directions of the institutional transformation of the Russian banking system. Banking **4**, 35–37 (2016)
2. Anopchenko, T., Gorbaneva, O., Lazareva, E., Ougolnitsky, G., Roshchina, E.: Dynamic model of macroregion subjects' interests coordination. In: Proceedings – 2019 of the 1st International Conference on Control Systems, Mathematical Modelling, Automation and Energy Efficiency, SUMMA 2019, pp. 141–146 (2019). https://doi.org/10.1109/SUMMA48161.2019.8947516
3. Anopchenko, T.Y., Gorbaneva, O.I., Lazareva, E.I., Murzin, A.D., Ougolnitsky, G.A.: Advanced solow model as a tool for coordination of interests of spatial economic systems' development (on the materials of the South Russian macro-region). Adv. Syst. Sci. Appl. **19** (4), 1–13 (2019)
4. BCB (Banco Central do Brasil): Chart 01 – Quantity of authorized institutions by type. National financial system evolution information – Annual publication. https://www.bcb.gov.br/en/statistics/annualreportsnfs (2020). Accessed 13 July 2021
5. Blokhin, A.A.: The Russian banking system's development: institutional differences and institutional rent. Finance Credit **25**(1), 15–25 (2019)
6. CBR (The Central Bank of the Russian Federation): Ordinance No. 3737-U of July 22, 2015. On the Methodology for Determining Systemically Important Credit Institutions. https://base.garant.ru/71167794/ (2015). Accessed 13 July 2021
7. CBR (The Central Bank of the Russian Federation): The report on banking sector development and banking supervision in 2015. http://www.cbr.ru/collection/collection/file/24206/bsr_2015.pdf (2016). Accessed 13 July 2021
8. CBR (The Central Bank of the Russian Federation): The report on banking sector development and banking supervision in 2016, p. 14. http://www.cbr.ru/collection/collection/file/24205/bsr_2016.pdf (2017). Accessed 13 July 2021
9. CBR (The Central Bank of the Russian Federation): The report on banking sector development and banking supervision in 2017, p. 35. http://www.cbr.ru/collection/collection/file/24204/bsr_2017.pdf (2018). Accessed 13 July 2021
10. CBR (The Central Bank of the Russian Federation): The report on banking sector development and banking supervision in 2018, p. 37. http://www.cbr.ru/collection/collection/file/24203/bsr_2018.pdf (2019). Accessed 13 July 2021
11. CBR (The Central Bank of the Russian Federation): Aggregate balance sheet for 30 largest credit organizations of the Russian Federation as of 01 September 2020. https://cbr.ru/Collection/Collection/File/29332/svst01092020.pdf (2020). Accessed 13 July 2021

12. DBB (Deutsche Bundesbank): Banking statistics February 2020. Statistical Supplement 1 to the Monthly Report, p. 104. https://www.bundesbank.de/resource/blob/825952/8a9a20d5cebfbc45d3b4859f88d0c6a4/mL/2020-02-bankenstatistik-data.pdf (2020). Accessed 13 July 2021
13. Krylova, L.V., Krylov, S.V.: The institutional structure of the Russian banking system and the directions of its transformation. Financ. Anal. Sci. Experience **7**(40), 2–10 (2014)
14. Ponomarenko, A., Sinyakov, A.: The impact of strengthening banking supervision on the structure of the banking system: conclusions based on agent-based modelling. Money Credit **77**(1), 26–50 (2018)
15. Tashtamirov, M., Mazayeva, L., Kuriev, Z.: A retrospective analysis of the provision of regions with banking services. Eurasian Sci. J. (2019). https://doi.org/10.15862/35ECVN519
16. Tashtamirov, M.R., Varaev, A.A.: Systemically important banks: current state and threat of systemic risk. Sci. Notes Crimean Eng. Pedagogical Univ. **2**(60), 178–184 (2018)
17. Tavbulatova, Z.K., Tashtamirov, M.R.: Institutional features of the banking structure and Russian banking system stability problems at present. Russian Economical Internet Journal, 1. http://www.e-rej.ru/upload/iblock/e28/e28f6fe40cd676c19555ef60b8769b55.pdf (2017). Accessed 13 July 2021
18. Tikhonov, V.: Banking sector in 2020. Banki.ru. March 19. https://www.banki.ru/news/research/?id=10943188 (2021). Accessed 13 July 2021

Company Risk Management in Export Activities

Elena A. Bratukhina(✉), Eleonora V. Nagovitsyna, and Dmitry S. Tusin

Vyatka State University, Kirov, Russia
{ea_bratuhina, ev_nagovicina}@vyatsu.ru

Abstract. The article deals with the ability to manage risks that an enterprise may face when doing business internationally. It informs what you need to know about risks to identify, assess, and manage them.

While studying the risks of exporting products, the authors have used economic-statistical, abstract-logical methods, comparative analysis, systemic and institutional approaches. They allow assessing the impact of risks on the export of products, identifying the trend of risk changes and predicting the most probable and dangerous risks in the future. Participants need to competently build a system for managing risks as their structure is constantly changing, depending on the direction of foreign economic activity.

According to the research, the authors have identified the main risks that enterprises face in export activities. They have worked out a risk management system for export companies and proposed a risk management matrix, which can be used as an instruction for managing or mitigating risks.

The majority of small and medium-sized enterprises enter foreign markets taking into account existing opportunities. They deal with only limited management of associated risks since they have neither the resources nor the necessary experience. This research proposes the main tools that can support companies with limited resources in managing export risks.

Success in foreign trade needs hard work. The article shows what tools can help to carry out risk management in foreign markets in order for enterprises to obtain maximum profits from exports at minimum costs.

Keywords: Currency risks · Foreign markets · Foreign economic activity · Political risks · Risk assessment · Risk insurance · Risk management · Financial risks · Export activities

1 Introduction

Due to the COVID-19 pandemic and the global financial and economic crisis [16–18], many Russian small and medium-sized enterprises have realized that there is a risk in the export business. Currency losses reduce sales margins. Economic downturns weaken income in foreign markets. Riots and strikes hinder the distribution of food. Foreign competitors copy patented technologies. Floods and storms delay transport. These are typical examples of the risks faced by Russian companies in export transactions. This article deals with the problem of how to cope with such risks in order to

E. G. Popkova (Ed.): ISC 2020, LNNS 368, pp. 772–783, 2022.
https://doi.org/10.1007/978-3-030-93244-2_83

keep them under control and avoid threats, which in turn seriously affect the financial stability of an enterprise. Risk management is currently becoming one of the most important tasks of enterprise management [5, p. 159]. Practice shows that export risk management is not about risk elimination. There is no risk-free business, especially when dealing with business partners and clients in other countries. On the contrary, risk management means taking measures aimed to inform the company what risks it faces and how much of this risk can be reduced [17]. An interconnected set of methods and tools, including administrative, economic, organizational and financial, can help enterprises focus on both individual risks and the entire portfolio of export risks they face. [4, p. 33].

2 Materials and Methods

Risk management in export activities in the context of the transformation of export risks, existing methods and mechanisms of a management cease to be fully adequate to new realities of globalization and informatization [20]. Therefore, when organizing risk management at export enterprises, it is important to rely on advanced scientific and methodological developments and international standards in the field of risk assessment and management.

The main methodological basis in this research is the economic-statistical, abstract-logical methods, comparative analysis, systemic and institutional approaches. The theoretical and information base is the works of Russian and foreign scientists, statistical reports and forecasts, regulatory legal acts, Internet resources of the state and public organizations.

3 Results and Discussion

Nowadays exports are an important macroeconomic indicator. It is a way of creating new jobs, developing infrastructure and increasing welfare in society. Due to the high uncertainty of the external environment and instability in the macrosystem, management has to assess risks in export activities in order to effectively manage them [11].

Risk assessment methods are different for different sectors and organizations, but the method adopted should best suit the organizational process. Risk assessment is the definition of a quantitative or qualitative assessment associated with a well-defined situation and a recognized threat (hazard). Quantifying risk requires calculating two risk components (R): the magnitude of potential losses (L) and the likelihood (p) that a loss will occur. Thus, there are always five basic steps used to assess risks:

1. Identify the risk;
2. Decide who or what can cause harm;
3. Assess the risk;
4. Make a record of results.

Working in an unfamiliar, foreign market always means new challenges, no matter how developed the company is in the domestic market. Risk assessment methods may differ depending on the industry, direction of export, the volume of products supplied, and other factors [1]. However, there are four general risks that all Russian small and medium-sized enterprises should be aware of when doing business in international markets. They are shown in Table 1.

Table 1. Types of risks

Types of risks
Political risks. An uncertain political environment can impede export operations for several reasons. A trade embargo can affect the delivery of goods, civil war or political violence can affect the safety of employees and partners. Political instability can lead to defaults on payments, confiscation of property and assets, and blocking the transfer of income [2].
Legal risks. Legal requirements and processes can vary significantly in different markets, therefore it is necessary to conduct research and obtain legal advice in order to understand the legal position. The company should know contractual terms, patent registrations and intellectual property requirements, dispute resolution processes, and health and safety laws [2].
Operational risks. Exporters need to become familiar with the operating environment of new markets as it can be very different from the operating environment in Russia. Some important things are labour relations policies and practices, various regulations and import requirements [2].
Currency risks. Unfavourable fluctuations in foreign exchange rates represent a risk of doing business abroad and can result in loss of income or profits. You can protect yourself by opening a foreign exchange office that allows locking in exchange rates and hedge foreign exchange risks [14, 15].

Source: Compiled by the authors.

As practice shows, there are no universal ways to completely avoid risks when exporting. As with any other business risk, you should adhere to a sound risk management process so that your business is in the most powerful position when you face them. Thus, Australian analysts propose four main steps that need to be taken to manage export risk:

1. It is necessary to identify potential sources of risk in the chosen export market.
2. Consult bankers, lawyers, insurers and accountants about the risks that a company may face in a particular foreign market. Also, you should find other participants in export activities who have previously worked or are working in the country where you plan to have business in order to get useful information from their experience.
3. While it is not possible to predict the occurrence of specific risks, it is possible to make a reasonable assessment of the likelihood and impact of most scenarios that could affect a particular business. The enterprise must also rank the likelihood and importance of various risks in order to intelligently prioritize and direct efforts to those areas that need protection and support.
4. Identify steps or measures that can be taken to avoid, reduce or minimize the risk. At the same time, the risk management strategy should be as clear and simple as possible, and each employee in the enterprise should know what he needs to do [9].

Experienced USA analysts and executives also point out that a company seeking international expansion needs to be familiar with the market it chooses to penetrate and its risks. For example, exporters need to understand barriers and risks associated with personnel, distribution integration, financial transactions, currency fluctuations, changes in foreign investment laws, and economic and social unrest. Exporters should be aware of the risks they may face. American managers divide risks into pre-export and post-export. Pre-export risks, similar to those before shipment in the domestic market, are primarily related to the seller's performance. The main challenge is to accurately assess the manufacturer's ability to manufacture the product, meet all required letters of credit specifications and ship within the agreed time frame. To successfully complete an overseas transaction, the exporter must also be aware of several other pre-export risks. They include the ability of the exporter to comply with all regulations of the importing country and accurately account for duties and taxes. If there is some problem that prevents the exporter from completing these tasks, the deal may be cancelled.

Post-export risk includes commercial and political risks. Commercial risks are often viewed in terms of the strength of the transaction, with an emphasis on receivables. They include the buyer's ability to pay for the item, terms of payment, and the buyer's bank trust. Political risks are often assessed in terms of the stability of a country. Commercial and political risks are often blurred and confused as one influences the other. A very important rule of prudent exporters and lenders is "to know the customer". It is necessary to understand as accurately as possible the buyer's capabilities. It is also important that the exporter allocates the resources necessary to develop strong and long-term relationships, which can ultimately lead to extended loan repayments. However, the longer the duration of the credit conditions, the higher the risk, since it is difficult to determine whether a foreign client has the financial ability to make payment in six or 12 months. To determine the likelihood of timely payment of debt obligations, it is necessary to analyze some questions and answer them. For example, whether a foreign customer is a solvent, whether there are problems with the flow of funds that may delay payment, what the reputation of this customer is, how long a potential customer has been in business, etc. It is worth paying attention to the bank report, which is an important tool for facilitating the sale abroad from the point of financiers' view. Banks around the world, with the buyer's permission, are ready to provide exporters with a credit report. Also, credit reporting agencies specialize in collecting valuable data on companies around the world [7]. One of the largest credit reporting agencies is Dun & Bradstreet [6]. Another important point for the exporter is to assess the ability of the importer's bank to meet its payment obligations. In order to identify risks that may lead to non-payment, it is important to carefully study the reputation of a foreign bank. It depends on various factors, including history, ownership, capital, competence and its legal address. When studying the history of a bank, one should take into account the level of its experience. In a global environment where new banks appear and disappear every day, track records are one of the most valuable intangible assets. Bank ownership is also an important factor in determining risk. Thus, if a small bank in a developing country has an unfamiliar name, but is partially owned by a larger and well-known institution, it may be assigned to represent the interests. The location of the bank must be taken into account. Prudent exporters need to analyze the stability of the country where the bank is located.

In addition to these risks, exporters need to be well aware of the country where they plan to work. That is, political and commercial conditions of the importing country need to be considered, as they may adversely affect the foreign client or his ability and willingness to pay debt obligations. In many developing countries, leadership changes are often of instability. If social instability engulfs the country, the cessation of activities could lead to the fact that the importer goes bankrupt. In the event of a military coup, the new government could change economic policies that could prevent the exporter from receiving money. Therefore, the exporter needs to know about the political relations of his country and the countries with which he intends to cooperate. In addition to the risks associated with late or non-payment, adverse fluctuations in exchange rates as a result of political or economic instability can undermine the exporter's profits or lead to losses. The factors affecting currency trends are complex and depend on an independent activity, rhetoric and very smooth capital flow, as well as macroeconomic factors. Thus, it is very difficult and extremely risky to predict whether a currency will increase or decrease in value. Many economists, traders and investors did not foresee the chain reaction and did not identify the complex factors that led to the devaluation of the Asian currency that began in 1997.

In today's global economy, it is important to plan for future receivables, payables and capital flows, making it necessary to manage foreign exchange risk. This task requires an understanding and assessment of many factors and conditions that affect currency volatility, including the following three important indicators:

1. Macroeconomic fundamentals (economic growth, inflation, unemployment and trade balance);
2. Political considerations (upcoming elections, politics and the level of confidence in the government);
3. Social considerations (labour unrest and violence).

In theory, if the economy of the importing country is stagnant and unemployment is high, there is a chance that the government will lower interest rates and/or increase spending. In this case, the value of the importing country's currency is likely to fall. On the other hand, if the economy grows rapidly, the currency will likely become stronger. However, in today's heavily traded markets, the number of considerations and potential responses are endless. As a result, managing foreign exchange risk may require more than research-based forecasts – this may require a prudent hedging strategy. By using spot, forward and option contracts, the exporter can isolate his business and minimize risks of losses due to currency fluctuations. Spot contracts convert the buyer's currency at current rates for settlement within two business days. This simple method allows the exporter to take advantage of favourable rates immediately but does not provide insurance against future rate changes. Forward contracts allow the exporter to fix the exchange rate for periods from three days to 12 months. This popular hedging method protects against future losses. Options contracts offer the exporter the ability to fix the rate for approximately one month [7].

Exports continue to represent a good growth opportunity for industrial companies. But often the majority of export businesses fail to enter foreign markets. Researchers estimate the failure rate for all international extensions are between 70% and 90%. Common causes include supply chain disruptions, market saturation, and poor strategic decisions – all linked to basic market research and business planning.

Before being international, smart industrial enterprises must plan carefully, understand the market, staff, and accept that the enterprise can fail. Businesses need to be realistic about external expansion, plan to minimize losses, and calculate how much the business can afford to lose if expansion fails. You need to detail the exit strategy your business will use, develop metrics to track performance and set goals and be prepared to shut down quickly to minimize impact. In 2015, US analysts conducted a study that found that 95% of businesses lost 95% of global customers because they did not enjoy the benefits of exports. Today, more and more companies are updating their growth plan strategies, revising and increasing their online presence. Thanks to careful planning and strategy, many manufacturers have successfully entered new markets – and many businesses are improving their positions and reducing their risks without resorting to high costs [1, 9].

Export risk management is the process of systematically considering all possible unwanted outcomes before they occur, and then developing procedures that help either avoid or minimize these risks or help cope with their impact [8, 13]. The heads of Russian exporting enterprises need to establish their effective management system for foreign economic activity that will meet all modern requirements and conditions, take into account the dynamics of the international and domestic markets, based on the principles of adaptability and the ability to make decisions flexibly and quickly. The enterprise must systematize risk management to neutralize risk factors, plan, monitor, and influence them. For this, we have developed a risk management system. It is shown in Fig. 1.

Risk management in the foreign economic activity of an enterprise must be carried out within the framework of a systemic and process approach, taking into account the specifics of the industry, using modern effective management methods and production organizations, and risk management tools. For effective risk management of an enterprise, an integrated approach is required to assess the maximum number of risk factors for an enterprise's activities carried out in a dynamic economic environment. This is the aim of the risk management system for export companies [10].

In 2020, the wave of the COVID-19 pandemic covered the world and currently many countries have not recovered from it. Many exporters find it difficult to assess the situation in a particular country. The globalEDGE website can be used to assess the risks of the country to which shipments are planned. For example, a Russian company plans to supply its products to the Belgian market and checks information about the country on this website. So, according to the site, today the country risk rating of this country is A3. Changes in the generally favourable but somewhat volatile political and economic environment can affect the solvent behaviour of corporations. However, a generally secure business environment can be challenging for companies from time to time. The likelihood of corporate default is, on average, quite acceptable. The business climate rating is A1, that is, the business environment is very good. Corporate financial information is accessible and reliable. Debt collection is effective. The institutional quality is very good. Intercompany transactions run smoothly.

Strengths of Belgium:

- The optimum location between UK, Germany and France
- Presence of European institutions, international organizations and global groups
- Ports of Antwerp (the second largest in Europe) and Zeebrugge, canals, motorways
- Well-trained workforce through vocational education, multilingualism.

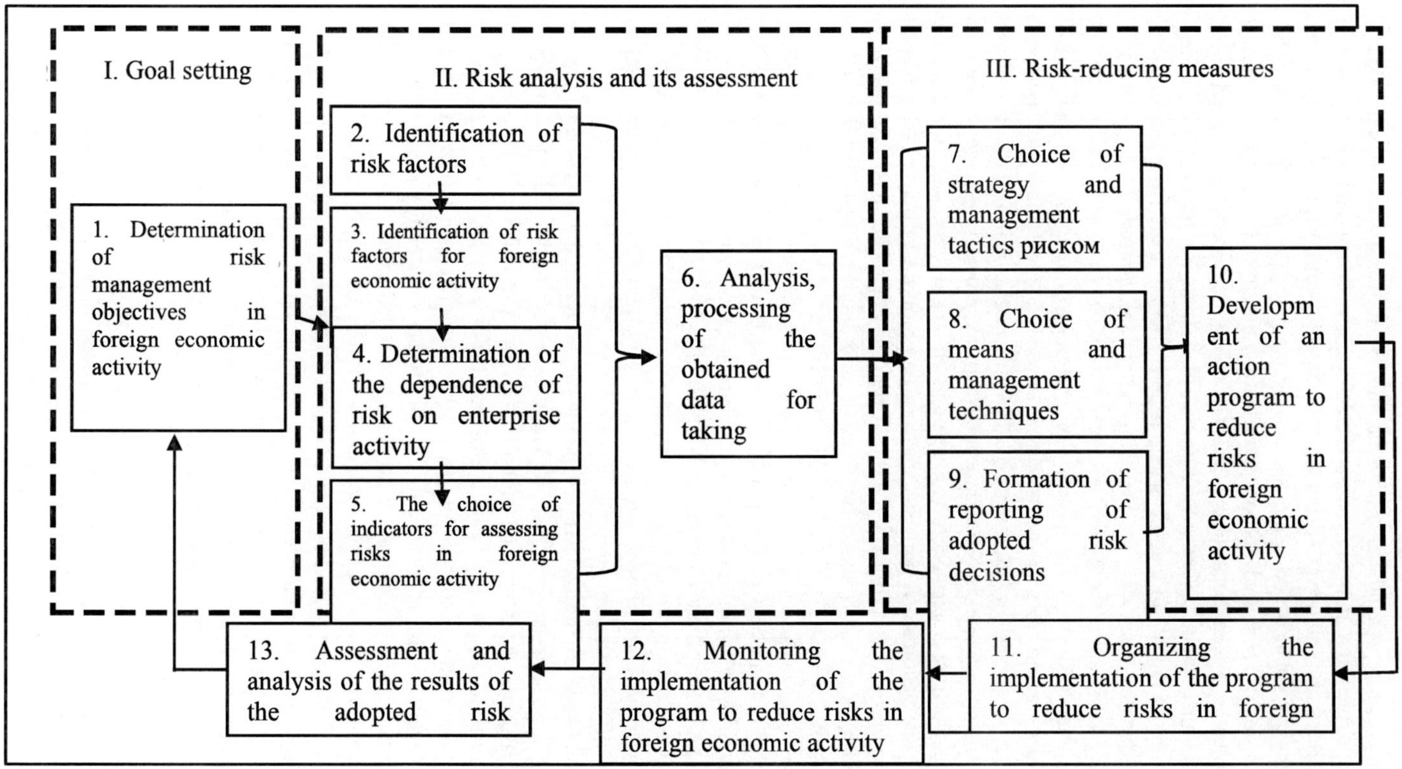

Fig. 1. Risk management system for an export company. *Source*: Compiled by the authors

Disadvantages:

- Political and financial tensions between Flanders and Wallonia
- Complex institutional structure and multiple administrative levels
- Strong dependence on the economy of Western Europe (exports of goods and services = 107% of GDP, of which more than 70% to the rest of the EU)
- Export focused on intermediate products
- High structural unemployment
- High government debt
- Dense housing market
- Saturated transport infrastructure

Current trends

The recovery is limited by a lag in world trade. Belgium will return to growth in 2021 after becoming one of the countries hardest hit by the pandemic in 2020, both in terms of health (highest deaths per capita at the end of 2020) and economics. To limit the spread of the virus, the country's government was forced to quarantine and close nonessential businesses in the spring and fall of 2020, resulting in an unprecedented decline in the activity. As the health situation gradually improves, domestic demand is expected to rebound sharply in 2021. As household consumption is dictated by the uncertainty surrounding health status, recovery should accelerate in the second half of the year. Companies are also expected to increase their investments after receiving significant government support (government-guaranteed loans, deferred social security payments and tax deadlines). However, the resumption of investment will be limited by economic uncertainty, less fluid trade between the United Kingdom and the European Union (EU) – despite the signing of an agreement in December 2020 – and a gradual reorganization of production chains. Highly dependent on regional trade – although 40% of exports to the rest of the EU are the imports in transit – the economy will depend on a trade lag. Moreover, Government accounts are still in deficit to support activities. The government deficit will remain very large in 2021, after a sharp deterioration in 2020 due to the costs incurred to contain the economic and health impacts of the epidemic (total cost was 3.9% of GDP in 2020), combined with a drop in related income. As most of the support measures have been extended until early 2021, the modest deficit reduction will be mainly attributable to a rebound in income from economic growth, as well as an increase in tobacco excise taxes. After a sharp rise in 2020, as in other eurozone countries, public debt should stabilize, albeit at a very high level.

The current account is expected to remain relatively balanced in 2021. Due to the country's integration into European production chains and its status as a regional centre, imports move in line with exports, so the balance of goods and services has recently been fairly balanced. At the same time, the income balance had a slight deficit (−0.4% of GDP in the first half of 2020), with dividends from foreign investment in the country exceeding dividends from Belgian investment abroad. Despite the very large external debt (268% of GDP, three-quarters of which corresponds to private liabilities), it is much lower than significant assets held abroad, as a result of which the country had a net external surplus of 45% of GDP at the end of June 2020. Political instability is likely to persist in 2021 [3].

Having analyzed the country where the export is planned, we propose to develop a risk management matrix. The risk management matrix will not only clarify risks that an enterprise may face in a given country but will also provide guidance to work with when managing or mitigating risks. This document must be attached to the export plan of the enterprise.

The risk management matrix should start by creating a simple list of potential risks. A risk management matrix is not only a good way to determine the likelihood of risks occurring and their consequences but also helps to prioritize issues that cannot be ignored. The risk management matrix helps focus on critical areas and mitigate them before they escalate into a crisis.

For example, if your entire export business is tied to one client in Malaysia and that company becomes insolvent, the result could be disastrous. But if the likelihood of insolvency is low, your risk rating for this event will be more moderate, although it may still require monitoring.

The enterprise is encouraged to regularly review its risk management matrix to ensure that any changes in emerging markets are taken into account. Table 2 provides an example of what might be included in a risk matrix.

Table 2. Risk management matrix

Type of a risk	Rating (low, medium, high)	Consequences	Steps to manage or reduce this risk
Exchange rate risk (change in RUB/USD rate)	High	Payments in USD will not cover the invoice amount that would result in a loss or decrease in profit.	The company's policy is to hedge all export orders accepted in US dollars at the time the order is received
Legal risk: US Product Responsibility Laws	High	Lawsuits can be brought against our company, followed by financial claims from clients.	Get up-to-date legal advice and insurance on US liability laws
Non-payment	Medium	Debtors are difficult to pursue abroad, financial losses	Checking the creditworthiness of clients to whom we provide credit conditions, checking the list of debtors before sending any orders. The first order needs prepayment or a letter of credit

Source: Compiled by the authors.

The company independently determines forms, methods and volumes of work in the international market, and is also fully responsible for the results of its activities. A significant share of risks is formed under the influence of economic, socio-political and even geographical features of the location of a particular country for export-oriented companies [12, p. 414]. Export management is considered successful when, as a result of such management, the enterprise achieves its goals. The main components of the success of an enterprise in a market economy are determined by the survival, efficiency and effectiveness of the enterprise, its productivity and the practical implementation of the decisions made [19].

4 Conclusion

The study shows how important it is to assess risks in the implementation of activities, to identify causes of risks and try to minimize them. With high uncertainty in the external environment and instability in the macrosystem, management needs to effectively manage risks. A strong approach to export risk management enables companies to do business with a large number and variety of international business partners. A company with a strong management team can more confidently provide loans and favourable payment terms to accelerate growth and strengthen important business relationships, as well as make contracts with importers with whom other enterprises cannot contract due to weak management. When assessing risks, the policy of companies must be adjusted for the constantly changing macro-environment, otherwise, the company may face a new risk or insufficiently assess current risks, which will lead to an incorrect strategy to eliminate or minimize them. Such mistakes entail serious costs for the company and can lead to its complete liquidation. Businesses need to build a risk management matrix and are encouraged to revise it regularly to ensure that any changes in emerging markets are taken into account. It helps to structure export knowledge, identify main actors, relationships between them, consequences and methods that can be applied to amplify the impact of positive factors and eliminate negative ones. This process also helps to show where and at what stage it is possible to face certain risks when entering the international market and to promptly influence them.

References

1. Methods of risk assessment you should know – HSEWatch [Electronic resource]. https://hsewatch.com/methods-of-risk-assessment. Accessed 02 Mar 2021
2. Risks of Exporting Manufactured Goods – And How to Avoid Them [Electronic resource]. https://blog.thomasnet.com/risks-in-exporting-manufactured-goods-and-how-to-avoid-them. Accessed 04 Mar 2021
3. Belgium: Risk Assessment >> globalEDGE: Your source for Global Business Knowledge [Electronic resource]. https://globaledge.msu.edu/countries/belgium/risk. Accessed 03 Mar 2021

4. Borodushko, I.V.: Risks of entrepreneurial activity in the international business space. Scientific notes of the St. Petersburg branch of the Russian Customs Academy named after V.B. Bobkov, **3**(67), 33–36 (2018)
5. Dautkhadzhieva, M.K., Mulkaev, Y.M.: Risks of economic activity of an enterprise and their impact on economic stability. Fundamental and applied scientific research: topical issues, achievements and innovations. Collection of articles of the XVII International scientific and practical conferences. 2 parts, 158–161 (2018)
6. Dun & Bradstreet – Accelerate Growth and Improve Business Performance [Electronic resource]. https://www.dnb.com/. Accessed 03 Mar 2021
7. Export Risk Assessment [Electronic resource]. http://www.manzellareport.com/index.php/trade-finance/405-risk-assessment/. Accessed 03 Mar 2021
8. Export Risk Management. Euler Hermes USA [Electronic resource]. https://www.eulerhermes.com/en_US/insights/export-risk-management.html. Accessed 02 Mar 2021
9. Four risks for exporters – and how to overcome them. Export Finance Australia [Electronic resource]. https://www.exportfinance.gov.au/resources-news/news-events/latest-news/2020/february/four-risks-for-exporters-and-how-to-overcome-them/. Accessed 02 Mar 2021
10. Nagovitsyna, E.V., Tusin, D.S.: Assessment of the company's risks in the implementation of foreign economic activity. Bull. NGIEI **8**(99), 95–104 (2019)
11. Nagovitsyna, E.V., Tusin, D.S.: Russian and foreign experience in improving the efficiency of export activities. Bull. NGIEI **9**(112), 112–125 (2020)
12. Polonik, S.S., Khorobrykh, E.V., Litvinchuk, A.A.: On the formation of factors contributing to the prevention of risks in foreign economic activity. Multifactor challenges and risks in the context of the implementation of the strategy of scientific, technological and economic development of the macro-region "north-west". In: Materials of the All-Russian Scientific and Practical Conference, pp. 413–420 (2018)
13. Risk management – Austrade [Electronic resource]. https://www.austrade.gov.au/Australian/Export/Guide-to-exporting/Risk-management. Accessed 04 Mar 2021
14. Saveleva, N.K.: Transnational capital in banking market: making investment decisions with the help of ai in the conditions of competition. In: Popkova, E.G., Sergi, B.S. (eds.) ISC 2020. LNNS, vol. 155, pp. 1714–1721. Springer, Cham (2021). https://doi.org/10.1007/978-3-030-59126-7_187
15. Savelyeva, N.K., Timkina, T.A.: "Smart technologies" as a mechanism for regulating bank competition in cross-border markets. In: Popkova, E.G., Sergi, B.S. (eds.) ISC 2020. LNNS, vol. 155, pp. 1322–1330. Springer, Cham (2021). https://doi.org/10.1007/978-3-030-59126-7_145
16. Sozinova, A.A., Kosyakova, I.V., Kuznetsova, I.G., Stolyarov, N.O.: Corporate social responsibility in the context of the 2020 economic crisis and its contribution to sustainable development. In: Popkova, E.G., Sergi, B.S. (eds.) ISC 2019. LNNS, vol. 198, pp. 83–90. Springer, Cham (2021). https://doi.org/10.1007/978-3-030-69415-9_10
17. Sozinova, A.A., Savelyeva, N. and Alpidovskaya, M.: Post-COVID Marketing 2019: Launching a New Cycle of Digital Development. In: de la Poza, E., Barykin, S.E., (eds.) Global Challenges of Digital Transformation of Markets. Series: Economic Issues, Problems and Perspectives. Novapress (2021)
18. Sozinova, A.A., Sofiina, E.V., Safargaliyev, M.F., Varlamov, A.V.: Pandemic as a new factor in sustainable economic development in 2020: scientific analytics and management prospects. In: Popkova, E.G., Sergi, B.S. (eds.) ISC 2019. LNNS, vol. 198, pp. 756–763. Springer, Cham (2021). https://doi.org/10.1007/978-3-030-69415-9_86

19. Successful enterprise management. Success factors. General approaches to management aimed at success. [Electronic resource]. https://studopedia.ru/6_22949_uspeshnoe-upravlenie-predpriyatiem-faktori-uspeha-obshchie-podhodi-k-upravleniyu-napravlennomu-na-uspeh.html. Accessed 04 Mar 2021
20. Zirchenko, L.A.: Improving the Risk Management System in the Development of Entrepreneurial Structures. Ph.D. Thesis. St. Petersburg, 165 P (2019)

Demographic Problems of Russian Regions: Analysis of Long-Term Dynamics and Tools for Overcoming

Evgeniya S. Tishchenko[1], Evgeniya K. Karpunina[2](✉), Tatiana D. Strelnikova[3], Marina A. Gundorova[4], and Marieta K. Gonezhuk[1]

[1] Kuban State Technological University, Krasnodar, Russia
[2] G.R. Derzhavin Tambov State University, Tambov, Russia
[3] Lipetsk Branch of the Financial University under the Government of the Russian Federation, Lipetsk, Russia
[4] Vladimir State University named after Alexander and Nikolay Stoletovs, Vladimir, Russia

Abstract. The research aims at identifying the factors that determine the occurrence of demographic problems in the Russian regions, as well as defining the guidelines of state policy for their solution. The novelty of the study consists in conducting a comprehensive economic and statistical analysis of changes in the demographic situation of Russian regions through the prism of the situation in the health system and environmental safety of territories. The authors evaluate the natural population growth in the Russian regions and analyze the long-term dynamics of the indicator of life expectancy at birth. The authors evaluate the dynamics of indicators reflecting the degree of readiness and ability of the healthcare system to solve the problem of slowing down population growth in Russian regions. The article examines the factors of environmental safety of the regions, reveals their impact on the demographic situation. The analysis allowed the authors to identify the "bottlenecks" of public administration in the federal districts of Russia, as well as to justify the guidelines of state policy aimed at solving demographic problems in Russia.

Keywords: Region · Demography · Population · Population growth · Life expectancy · Health care system · Environmental safety of the region

JEL classification: J11 · R23

1 Introduction

Preserving the territorial integrity and socio-economic identity of the country requires solving the most important demographic problems [3, 9, 10]. The decline in the population of Russia, the increase in morbidity and the reduction in life expectancy of the population reflect an unfavorable trend in social development.

Assessing this situation, Rosstat forms several demographic forecasts [17]. Rosstat's demographic forecast in its optimistic version is focused on positive changes in

E. G. Popkova (Ed.): ISC 2020, LNNS 368, pp. 784–797, 2022.
https://doi.org/10.1007/978-3-030-93244-2_84

the migration and natural movement of the population. According to this forecast, it is assumed that the total population at the beginning of 2020 is 146.798 million people and will increase to 150.126 million people by 2036. According to the pessimistic version of the Rosstat forecast, the birth rate in the country will continue to fall, the growth rate of life expectancy will slow down, the migration influx will decrease, thus, the total population will decline to 134.277 million people by 2036 [17].

The implementation of the forecast according to the negative option may lead to a decrease in the socio-demographic security of national and regional development [2]. The positive fulfilment of the forecast can be ensured by such factors of socio-demographic security as improving the quality of life of the population, improving medical care, optimizing the health protection system, maintaining a favorable environmental situation [21].

The consideration of demographic problems in terms of the healthcare system development and environmental safety is especially relevant since the regional situation in Russia differs significantly both in its economic basis and in its social and cultural characteristics.

2 Literature Review

Researchers associate the causes of demographic problems of modern society with the transition to demographic ageing of the population, as well as with the action of some socio-economic factors, in particular, a drop in income, a decrease in the standard of living of the population, the transformation of the institution of marriage and family [20, 22]. Meanwhile, stimulating population growth is possible due to the systematic development of the health system, as well as maintaining environmental safety. The attention of scientists is attracted by the analysis of the dysfunctionality of regional health systems. As a rule, it is performed by the method of interregional comparison based on an assessment of the level of performance of the functions of the regional health system [6, 8, 11, 15]. For Russia, the problem of the environmental safety is especially relevant, since the regions are distinguished by a variety of environmental conditions and the lifestyle of the population. For example, insufficient provision of high-quality drinking water to the population, a decrease in the quality characteristics of the atmosphere are becoming key factors in the deterioration of public health and the growth of regional instability [3–5]. In addition, the main environmental problems in the regions include intensive pollution of ground and surface waters [13].

The demographic situation in the regions and the health of the population are also determined by the effectiveness of public administration and the formation of regional institutions of socio-economic regulation [16].

3 Methodology

The purpose of the study is to determine the factors that provoke the emergence of demographic problems in Russia, as well as identify the directions of stimulating population growth at the regional level.

The study period is 2010–2019. Research objectives are 1) to assess the natural population growth in the Russian regions and the long-term dynamics of the coefficient of life expectancy at birth; 2) to assess the situation in the healthcare system of the regions; 3) to analyze the factors and dynamics of environmental safety of the regions, to estimate their impact on the demographic situation; 4) to determine the guidelines of state policy aimed at solving the demographic problems of the Russian regions.

4 Results

Let's start the study by analyzing the dynamics of the natural population growth rate in the regions of Russia for the period from 2010 to 2019 (Table 1).

Table 1. The coefficient of natural population growth in the regions of Russia, 2010–2019

Region	Natural population growth rate per 1,000 people in 2010	Natural population growth rate per 1,000 people in 2019	Change on the rate of natural population growth per 1,000 people of the population for the period from 2010 to 2019. (2019–2010)	Change on the rate of natural population growth per 1,000 people of the population for the period from 2010 to 2019. (2019–2010) * 100
Siberian Federal District	−0.1	−2.5	−2.4	2,500.0
Far Eastern Federal District	−0.6	−1.1	−0.5	183.3
Southern Federal District	−2.3	−3.0	−0.7	130.4
Russian Federation	−1.7	−2.2	−0.5	129.4
Volga Federal District	−2.6	−3.3	−0.7	126.9
Northwestern Federal District	−3.5	−2.8	0.7	80.0
Central Federal District	−4.5	−3.3	1.2	73.3
North Caucasus Federal District	8.7	6.3	−2.4	72.4
Ural Federal District	1.1	−0.8	−1.9	−72.7

Source Compiled by the authors based on [18, 19]

In the period 2010–2019 in Russia, the rate of natural population growth decreased by 0.5% and amounted to −2.2%. Natural population growth is observed in the North Caucasus Federal District, and the rest of the federal districts are characterized by a natural population decline.

The relevance of the research is increasing during the coronavirus pandemic[7, 12, 14]. The pandemic has exposed the problem of population decline (per 1,000 people): in 2020, the population decline in Russia increased to 4.7 against 2.2 in 2019 (Fig. 1).

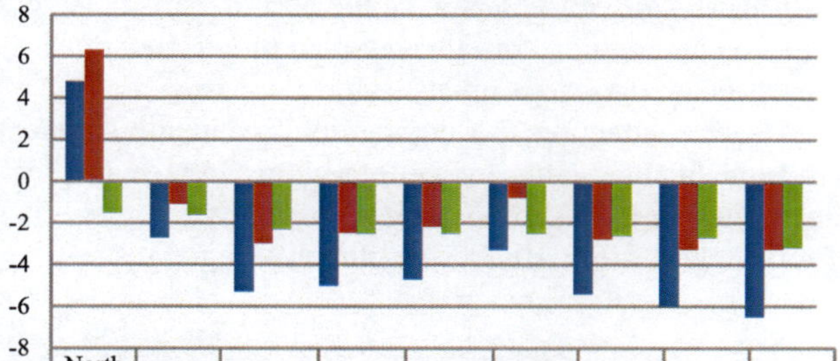

	North Caucasus Federal District	Far Eastern Federal District	Southern Federal District	Siberian Federal District	Russian Federation	Ural Federal District	North-Western Federal District	Central Federal District	Volga Federal District
Natural population growth (decline) per 1000 people, in 2020	4.8	-2.7	-5.3	-5	-4.7	-3.3	-5.4	-6	-6.5
Natural population growth (decrease) per 1000 people, in 2019	6.3	-1.1	-3	-2.5	-2.2	-0.8	-2.8	-3.3	-3.3
Change in natural population growth (loss) per 1000 people of the population, in 2020 to 2019, (2020-2019)	-1.5	-1.6	-2.3	-2.5	-2.5	-2.5	-2.6	-2.7	-3.2

Fig. 1. Natural growth per 1,000 people of the population, in 2020 relative to 2019 *Source* Compiled by the authors based on [18]

Population decline has increased in all federal districts, which requires the attention of state authorities in terms of stimulating population growth in the regions of the country.

An important indicator reflecting the demographic situation in the country is the indicator of life expectancy at birth. The analysis of this indicator for the period 2010–2019 is shown in Fig. 2.

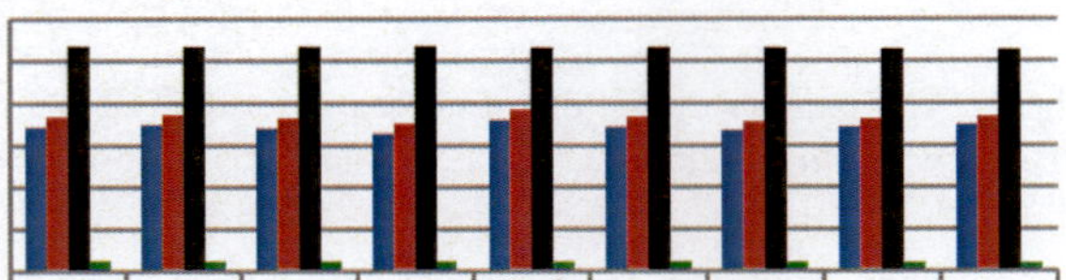

	North-Western Federal District	Central Federal District	Volga Federal District	Far Eastern Federal District	North Caucasus Federal District	Russian Federation	Siberian Federal District	Ural Federal District	Southern Federal District
Life expectancy at birth of all categories of citizens, number of years, in 2010	68.9	69.9	68.4	65.8	72.2	68.9	67.1	68.8	70.1
Life expectancy at birth of all categories of citizens, number of years, in 2019	73.7	74.5	72.9	70.2	76.6	73.3	71.1	72.5	73.7
Change in life expectancy at birth of all categories of citizens, number of years, in 2019 to 2010, (2019/2010)*100,%	107.0	106.6	106.7	106.8	106.2	106.4	106.0	105.4	105.1
(2019-2010), years	4.8	4.6	4.6	4.5	4.5	4.4	4.0	3.7	3.6

Fig. 2. Life expectancy of citizens at birth in the regions of Russia, 2010–2019 *Source* Compiled by the authors based on [18, 19]

In Russia life expectancy for the period from 2010 to 2019 increased by 4.5 years to 73.3 years in 2019 for all categories of citizens. Such dynamics is the characteristic of each Federal District under study.

A factor affecting the demographic situation in the country and regions is the provision of the healthcare system with qualified medical personnel – doctors of all categories. To assess this factor, we will analyze the indicator "the number of doctors of all categories per 10,000 population" (Fig. 3).

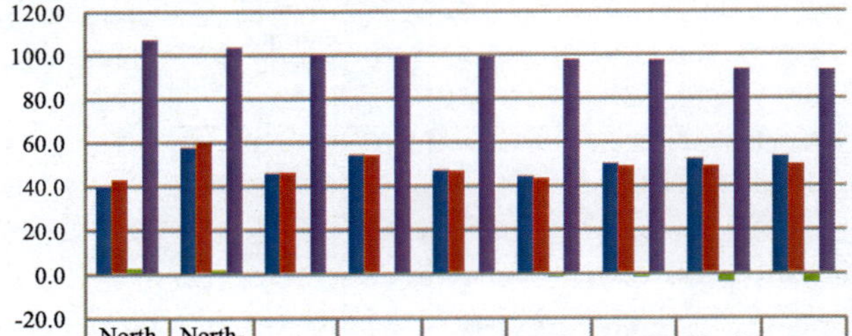

	North Caucasus Federal District	North-Western Federal District	Ural Federal District	Far Eastern Federal District	Volga Federal District	Southern Federal District	Russian Federation	Siberian Federal District	Central Federal District
Number of doctors of all specialties in 2010, per 10,000 population, people	40.1	57.8	46.0	54.2	47.2	44.4	50.1	52.4	53.5
Number of doctors of all specialties in 2019, per 10,000 population, people	42.9	59.8	46.0	53.9	46.7	43.3	48.7	48.8	49.6
Change in the number of doctors of all specialties in 2019 to 2010, per 10,000 people of the population, people, (2019-2010)	2.8	2.0	0.0	-0.3	-0.5	-1.1	-1.4	-3.6	-3.9
Change in the number of doctors of all specialties in 2019 to 2010, per 10,000 people of the population, people, (2019/2010)*100	107.0	103.5	100.0	99.4	98.9	97.5	97.2	93.1	92.7

Fig. 3. The number of doctors of all categories per 10,000 population in the regions of Russia, 2010–2019 *Source* Compiled by the authors based on [18, 19]

In general, in Russia, we can state a negative situation with the provision of doctors in the healthcare system. During the period from 2010 to 2019, the number of doctors of all categories per 10,000 population decreased from 50.1 doctors in 2010 to 48.7 doctors in 2019. The number of doctors per 10,000 population during the analyzed period increased in the North Caucasus Federal District and the Northwestern Federal District (the maximum value of the indicator) and did not change in the Ural Federal District.

The analysis of the availability of hospital beds in the Russian healthcare system is presented in Fig. 4.

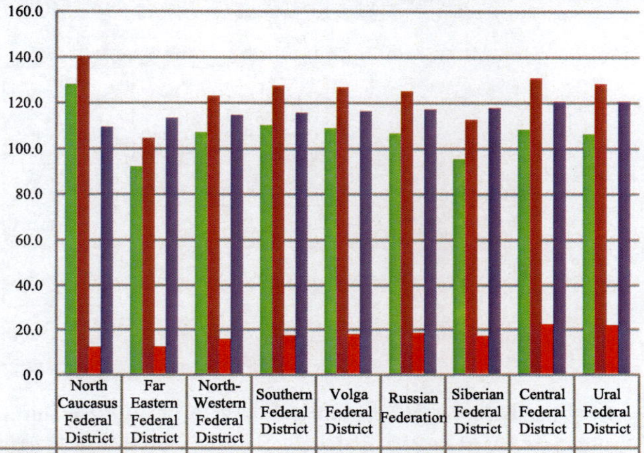

	North Caucasus Federal District	Far Eastern Federal District	North-Western Federal District	Southern Federal District	Volga Federal District	Russian Federation	Siberian Federal District	Central Federal District	Ural Federal District
Population per hospital bed, year-end, 2010	128.2	92.2	107.2	110.2	108.9	106.7	95.6	108.4	106.4
Population per hospital bed, year-end, people in 2019	140.3	104.5	122.9	127.4	126.7	125.0	112.5	130.6	128.2
Population change per hospital bed, year-end, people in 2019 to 2010, (2019-2010)	12.1	12.3	15.7	17.2	17.8	18.3	16.9	22.2	21.8
Population change per hospital bed, year-end, people in 2019 to 2010, (2019/2010)*100,%	109.4	113.3	114.6	115.6	116.3	117.2	117.7	120.5	120.5

Fig. 4. Population per hospital bed in the regions of Russia, 2010–2019 *Source* Compiled by the authors based on [18, 19]

The provision of the population with hospital beds in the whole country in 2010–2019 increased from 106.7 people per hospital bed in 2010 to 125 people per hospital bed in 2019.

We will analyze the capacity of outpatient polyclinic organizations in the Russian healthcare system (Fig. 5).

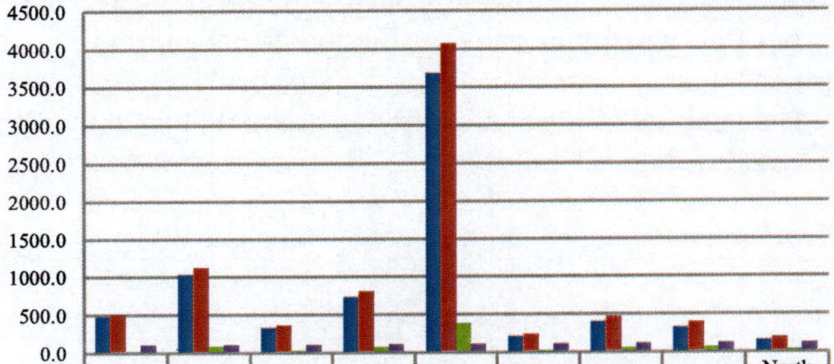

	Siberian Federal District	Central Federal District	Ural Federal District	Volga Federal District	Russian Federation	Far Eastern Federal District	North-Western Federal District	Southern Federal District	North Caucasus Federal District
Capacity of outpatient clinics, vthis year, thousand visits per shift, in 2010	482.7	1033.5	332.8	732.0	3685.1	213.7	404.7	326.4	159.4
Capacity of outpatient clinics, vthis year, thousand visits per shift, in 2019	502.2	1117.4	361.9	804.0	4072.4	236.3	462.3	394.0	194.2
Change in the capacity of outpatient clinics, inthis year, thousand visits per shift, in 2019 to 2010, (2019-2010)	19.5	83.9	29.1	72.0	387.3	22.6	57.6	67.6	34.8
Change in the capacity of outpatient clinics, inthis year, thousand visits per shift, in 2019 to 2010, (2019/2010*100),%	104.04	108.12	108.74	109.84	110.51	110.58	114.23	120.71	121.83

Fig. 5. Capacity of outpatient clinics, thousands of visits per shift, 2010–2019 *Source* Compiled by the authors based on [18, 19]

The capacity of outpatient clinics in the Russian healthcare system increased by 110.6% from 2010 to 2019, which indirectly indicates a deterioration in the morbidity situation in Russia. Moreover, the burden on medical institutions has increased in all federal districts.

We will conduct an additional analysis of the registered cases of diseases in patients with a diagnosis established for the first time in their lives per 10,000 people of the population (Fig. 6).

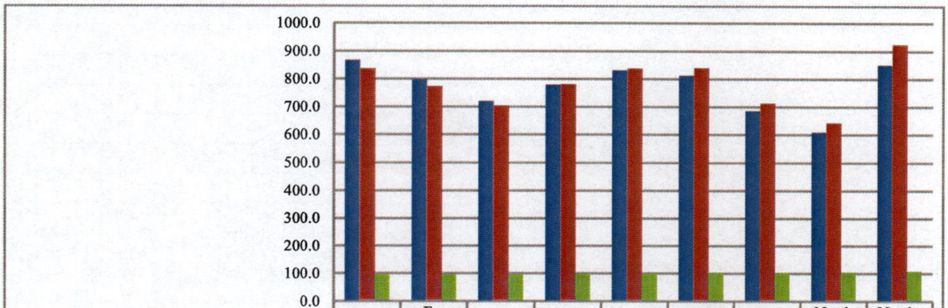

	Volga Federal District	Far Eastern Federal District	Central Federal District	Russian Federati on	Siberian Federal District	Ural Federal District	Southern Federal District	North Caucasu s Federal District	North-Western Federal District
■ Morbidity rate in patients diagnosed for the first time in life, per 1000 people of the population in 2010	867.2	799.7	720.2	780.0	831.4	813.3	685.6	609.4	850.8
■ Morbidity rate in patients diagnosed for the first time in their life, per 1000 people of the population in 2019	836.2	772.1	703.0	780.2	837.5	839.9	711.7	641.6	924.5
■ Change in morbidity in patients diagnosed for the first time in life, per 1000 people of the population in 2019 to 2010, (2019/2010)*100	96.4	96.5	97.6	100.0	100.7	103.3	103.8	105.3	108.7

Fig. 6. Registered cases of diseases in patients with a diagnosis established for the first time in life per 10,000 people of the population, 2010–2019 *Source* Compiled by the authors based on [18, 19]

Figure 5 shows that the level of morbidity in patients with a diagnosis established for the first time in their lives remained unchanged during the study period. In other words, the increase in the burden on the country's medical institutions is ensured by patients with repeated cases of diseases. This also indirectly indicates the inefficiency of the Russian healthcare system.

The demographic situation in the regions is influenced by the state of the environment, and in particular, environmental safety [1]. Let's analyze the dynamics of the volume of emissions of pollutants into the atmospheric air coming from stationary sources (Fig. 7).

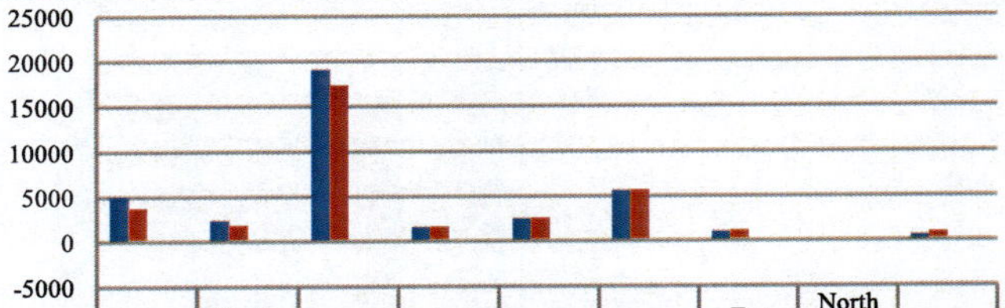

	Ural Federal District	North-Western Federal District	Russian Federation	Central Federal District	Volga Federal District	Siberian Federal District	Far Eastern Federal District	North Caucasus Federal District	South federal district
Volume of emissions of pollutants into the atmosphere from stationary sources,in 2010, thousand tons	5105	2389	19100.0	1618	2513	5634	1070	138	648
Volume of emissions of pollutants into the atmosphere from stationary sources,in 2019, thousand tons	3655	1748	17300.0	1603	2509	5632	1099	165	885
Changes in the volume of air pollutants emitted from stationary sources, ((2019-2010) / 2010))*100, thousand tons, in %	-28.4	-26.8	-9.4	-0.9	-0.2	0.0	2.7	19.6	36.6
Changes in the volume of air pollutants emitted from stationary sources, ((2019/2010))*100, thousand tons, in %	71.6	73.2	90.6	99.1	99.8	100.0	102.7	119.6	136.6

Fig. 7. The volume of emissions of pollutants into the atmospheric air coming from stationary sources, thousand tons, 2010–2019 *Source* Compiled by the authors based on [18, 19]

The volume of emissions into the atmosphere from stationary sources decreased by 9.4% for the period from 2010 to 2019, and in the Ural Federal District and the Northwestern Federal District, the volume of emissions decreased significantly – by 28.4% and 26.8%, respectively, the volume of emissions in the Siberian Federal District remained unchanged. In the Far Eastern, North Caucasus and Southern Federal Districts, the volume of emissions into the atmosphere has increased, which allows us to conclude that there is insufficient attention to the problems of environmental safety on the part of state authorities in these federal districts.

An additional effect in maintaining environmental safety and improving the demographic situation can be provided by increasing the capture of air pollutants coming from stationary sources (Fig. 8).

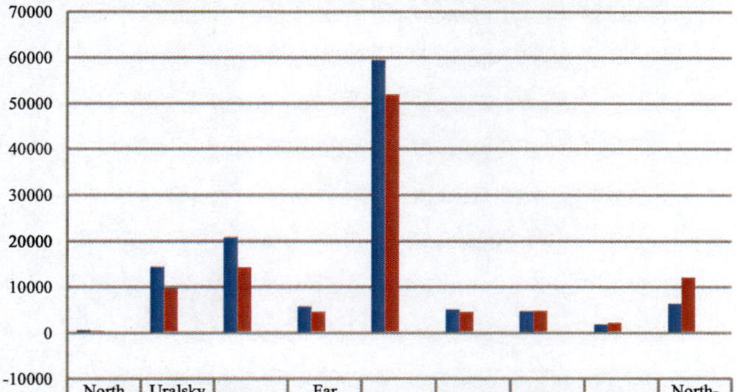

	North Caucasus Federal District	Uralsky District federal district	Siberian Federal District	Far Eastern Federal District	Russian Federation	Central Federal District	Volga Federal District	Southern Federal District	North-Western Federal District
Volumes of capture of air pollutants from stationary sources in 2010, thousand tons	590	14381	20838	5722	59500.0	5061	4724	1817	6386
Volumes of capture of air pollutants from stationary sources in 2019, thousand tons	368	9637	14206	4507	52000.0	4449	4722	2138	11987
Change in the volume of capture of air pollutants from stationary sources in 2019 compared to 2010, thousand tons, in %, ((2019-2010)/2010)*100	-37.6	-33.0	-31.8	-21.2	-12.6	-12.1	0.0	17.7	87.7
Change in the volume of capture of air pollutants from stationary sources in 2019 to 2010, thousand tons, in%, (2019/2010)*100	62.4	67.0	68.2	78.8	87.4	87.9	100.0	117.7	187.7

Fig. 8. Capture of air pollutants from stationary sources in the regions of Russia, 2010–2019 *Source* Compiled by the authors based on [18, 19]

During the analyzed period, the capture of emissions in Russia decreased by 12.6% or by 7.5 million tons. In the Volga Federal District, the volumes of capture in 2019 remained unchanged compared to 2010, while they increased in the Southern and Northwestern Federal Districts.

An increase in the volume of use of fresh and high-quality drinking water by the population directly affects the health of the population and increase life expectancy. The analysis of changes in the volume of fresh drinking water use in 2010–2019 is shown in Fig. 9.

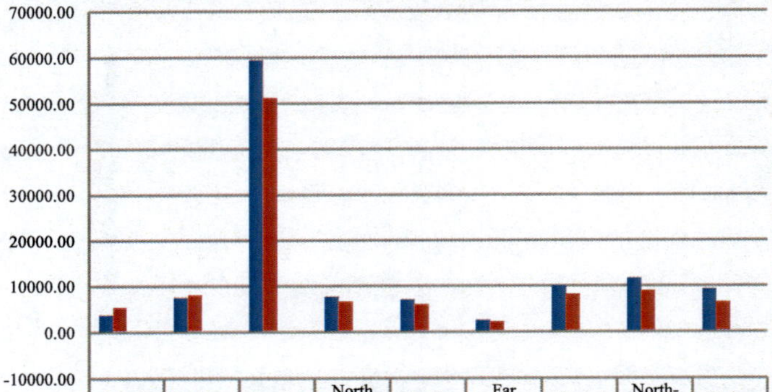

	Ural Federal District	Southern Federal District	Russian Federation	North Caucasus Federal District	Siberian Federal District	Far Eastern Federal District	Central Federal District	North-Western Federal District	Volga Federal District
Fresh water use in 2010, (million cubic meters)	3745.00	7465.00	59455.00	7635.00	7005.00	2617.00	10062.00	11671.00	9254.00
Fresh water use in 2019, (million cubic meters)	5267.00	7942.00	51163.00	6554.00	5886.00	2179.00	8084.00	8860.00	6392.00
Change in fresh water use in 2019 to 2010, (million cubic meters), (2019/2010)*100	140.64	106.39	86.05	85.84	84.03	83.26	80.34	75.91	69.07
Change in fresh water use in 2019 to 2010, (million cubic meters), (2019-2010)/2010 * 100	40.64	6.39	-13.95	-14.16	-15.97	-16.74	-19.66	-24.09	-30.93

Fig. 9. The volume of fresh drinking water use in the regions of Russia, 2010–2019 *Source* Compiled by the authors based on [18, 19]

In the whole country, the volume of use of freshwater by the population in 2019 compared to 2010 decreased by 14%, and only in the Ural and Southern Federal Districts, the volume of use of freshwater increased. There is an obvious negative trend, which is increasing in the conditions of growing emissions into the atmosphere due to a decrease in the volume of capture.

Environmental safety is affected by a reduction in the volume of discharge of contaminated wastewater into surface water bodies (Fig. 10).

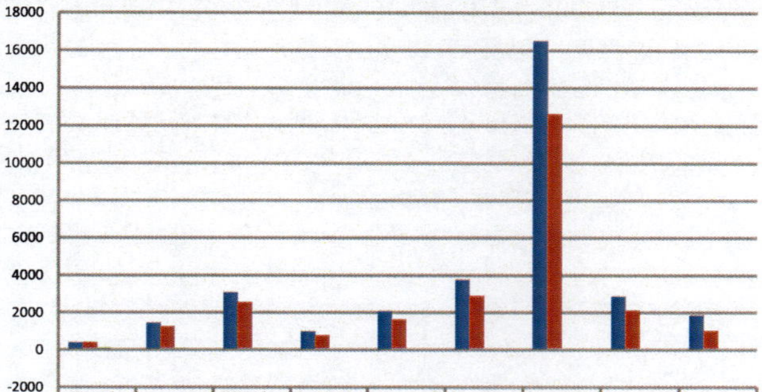

	North Caucasus Federal District	Southern Federal District	North-Western Federal District	Far Eastern Federal District	Siberian Federal District	Central Federal District	Russian Federation	Volga Federal District	Ural Federal District
Volume of contaminated wastewater discharged into surface water bodies in 2010, (million cubic meters)	390	1446	3088	989	2099	3761	16516	2883	1860
Volume of contaminated wastewater discharged to surface water bodies in 2019, (million cubic meters)	399	1239	2548	767	1625	2879	12602	2120	1025
Change in the volume of contaminated wastewater discharged to surface water bodies in 2019 to 2010, (million cubic meters), in%, (2019/2010)*100	102.3	85.7	82.5	77.6	77.4	76.5	76.3	73.5	55.1
Change in the volume of contaminated wastewater discharged to surface water bodies in 2019 by 2010, (million cubic meters), in %, ((2019-2010)/2010)*100	2.3	-14.3	-17.5	-22.4	-22.6	-23.5	-23.7	-26.5	-44.9

Fig. 10. Volumes of discharge of polluted wastewater into surface water bodies in the regions of Russia, 2010–2019 *Source* Compiled by the authors based on [18, 19]

In 2019, in comparison with 2010, the volume of wastewater discharge into surface water bodies decreased in the whole country by 23.7%, and the amount of discharge decreased in all Federal districts, except for the North Caucasus Federal district (+2.3%).

The analysis allowed identifying the demographic problems of the Russian regions, which have become the guidelines for public authorities for the implementation of regional policy (Table 2).

Table 2. Guidelines of the state policy aimed at solving the demographic problems of the regions of Russia

The direction of state policy in Region	North-Western Federal District	Far Eastern Federal District	Volga Federal District	Ural Federal District	Siberian Federal District	Central Federal District	North Caucasus Federal District	Southern Federal District
Increase in the number of doctors and hospital beds per capita	+	+	+	+	+	+	+	+
Reducing the burden on outpatient organizations	+	+	+	+	+	+	+	+
Reducing the morbidity of the population	+			+			+	+
Increasing the use of freshwater	+	+	+		+	+	+	
Reducing the volume of emissions into the atmosphere and increasing their capture		+	+	+	+	+		+
Reduction of the volume of polluted wastewater discharge							+	

Source Compiled by the authors

5 Conclusions

Firstly, the authors identified and evaluated the key factors affecting the population decline in the Russian Federation; put forward a hypothesis about the need for a systematic enhancement of the demographic situation by improving the health care system and maintaining environmental safety. Secondly, the authors assessed the state of the healthcare system as the basis for smoothing demographic problems, conducted a comparative analysis in terms of the regions of Russia. Thirdly, the authors investigated the factors and dynamics of environmental safety of the regions, assessed its impact on the demographic situation. Fourthly, the authors proved that an effective state policy aimed at radically improving the demographic situation in the regions should be based not only on economic measures but also on systemic ones. In particular, the implementation of measures aimed at improving the health system and improving environmental safety in the regions.

References

1. Averina, O., Gridnev, Y., Lavrikova, N., Smirnova, E., Karpunina, E.: Sustainable development during the COVID-19 pandemic: opportunities for ecological innovation in Russian regions. In: Proceeding of the 37th IBIMA Conference, 1–2 April 2021, Cordoba, Spain, pp. 1118–1129 (2021)
2. Belousova, S.: Social and demographic security of national and regional development. Law Secur. **1–2**, 113–117 (2013)

3. Egorov, V.: Medico-social aspects of information activity in the field of ecology (based on the work of industrial enterprises). The Dissertation of Ph.D. in Medical Sciences, Volgograd (2006)
4. Eliseeva, I.: Demography and Population Statistics. Finance and Statistics, Moscow (2014)
5. Golubchikov, Y.: Demographic troubles of Russia. Energy Econ. Technol. Ecol. **10**, 124–130 (1999)
6. Karpunina, E., Molchan, A., Lapushinskaya, G., Ryabova, T., Jusufov, A.: Mechanisms for managing changes in the socio-economic development of lagging regions. In: Proceedings of the 35th IBIMA Conference, 1–2 April 2020, Seville, Spain, pp. 6538–6549 (2020)
7. Karpunina, E., Butova, L., Sobolevskaya, T., Badokina, E., Pliusnina, O.: The impact of the COVID-19 pandemic on the development of russian national economy sectors: analysis of dynamics and search for stabilization measures. In: Proceeding of the 37th IBIMA Conference, 1–2 April 2021, Cordoba, Spain, pp.1213–1226 (2021)
8. Karpunina, E., Petrov, I., Klimentova, E., Sozaeva, J., Korkishko, I.: Mechanisms of self-development of subsidized regions. In: Proceedings of the 35th IBIMA Conference, 1–2 April 2020. Seville, Spain, pp. 2282–2293 (2020)
9. Kuklin, A., Shipitsyna, S., Naslunga, K.: Comparison of the effectiveness of budget financing and social security of the region. Economy Reg. **12**(3), 638–653 (2016)
10. Kuklin, A., Shorikov, A., Tyulyukin, V., Cherepanova, A., Vasilyeva, E., Nekrasova, E.: Diagnostics and modeling of the effectiveness of healthcare system management for ensuring socio-demographic security of Russian regions. Spatial Econ. **4**, 72–92 (2011)
11. Kuznetsov, V., Rybakovsky, L.: The Strategy of Demographic Development of Russia. TSP, Moscow (2015)
12. Lisova, E., Petrov, I., Koryakina, T., Kalombo Mulamba, V., Karpunina, K.: The well-being of Russian regions under threat: economic and social impact assessment of COVID-19. In: Proceeding of the 36th IBIMA Conference, 4–5 November 2020, Granada, Spain, pp. 7603–7616 (2020)
13. Mamyrbayev, A., Umarova, G.: Modern aspects of the state of public health (literature review). Med. News Georgia **5**(254), 61–64 (2016)
14. Mejokh, Z., Korolyuk, E., Sozaeva, D., Pilipchuk, N., Karpunina, E.: Economic security of russian regions: risk factors and consequences of the COVID-19 pandemic. In: Proceeding of the 36th IBIMA Conference, 4–5 November 2020, Granada, Spain, pp. 8197–8205 (2020)
15. Pepelyaeva, A.: Dysfunctions of regional health care as a threat to the social security of the region. Bull. Prikamsky Soc. Inst. **1**(76), 109–117 (2017)
16. Rimashevskaya, N., Rusanova, N.: The health of the Russian population in the conditions of socio-economic modernization. Population **4**, 96–105 (2015)
17. Rosstat: Rosstat Has Presented an Updated Demographic Forecast Until 2036. https://rosstat.gov.ru/folder/313/document/72529 (2019). Accessed 14 July 2021
18. Rosstat: Statistical Collection "Regions of Russia, Socio-Economic Indicators, 2020". https://rosstat.gov.ru/folder/210/document/13204 (2021). Accessed 14 July 2021
19. Rosstat: The Report "The Socio-Economic Situation of Russia". https://rosstat.gov.ru/compendium/document/50801 (2021). Accessed 14 July 2021
20. Ryazantsev, S.: Practical Demography. CSP, Moscow (2015)
21. Sadykov, R.: The level and quality of life of the population as a factor of ensuring the social security of territorial entities. Fundam. Res. **11–1**, 201–205 (2014)
22. Sinelnikov, A.: Is it possible to immediately stop the decline of the population in Russia? Demogr. Res. **7**. https://istina.fnkcrr.ru/journals/5218880/ (2008). Accessed 14 July 2021

Development of Risk Management Tools for Innovation-Oriented Industrial Enterprises

Anna A. Ter-Grigor'yants[1], Natalia V. Lazareva[1](✉), Tatiana A. Kulagovskaya[1], Elena A. Drofa[2], and Liliy V. Semenova[2]

[1] North Caucasus Federal University, Stavropol, Russia
nlazareva@ncfu.ru
[2] Don State Technical University, Rostov-on-Don, Russia

Abstract. The chemical industry is the basic segment of the Russian industry. It is of great importance for the innovative development of the country and its regions. In current global markets, innovative chemical enterprises increasingly face the problems of adapting to the situation, which makes the identification, assessment, and management of risks especially relevant. The paper aims to develop theoretical and methodological provisions and practical solutions to improve the risk management system of innovation-oriented chemical enterprises. The methodological basis of the research includes general scientific and specific cognitive methods. The research problems were solved using economic and statistical methods, such as comparison, grouping, systemic analysis, systematicity, integrity, logical modeling, etc. The practical significance of the research lies in the fact that the results can be used to develop risk management programs for innovation-oriented organizations, including the chemical industry organizations.

Keywords: Industry and its structure · Innovative enterprises · Chemical complex · Risk management · Preventing risks and threats

JEL classification: D81 · D84 · F63

1 Introduction

Constantly changing factors of the environment of innovative industrial enterprises, determining the uncertainty and riskiness of economic activity, and the evolution of the understanding of risks and threats, as well as the understanding of their types, cause the need to improve the theory and practice of risk management in the context of developing specific industries. The chemical industry in Russia and the Stavropol Territory is one of the most promising industries. Simultaneously, this industry is lagging behind the world leaders. For the last five years, investments in chemical production have exceeded 2 trillion rubles. Nevertheless, despite such considerable financial influence, the industry still faces several macroeconomic, transformational, financial, and external economic risks and threats, significantly affecting development dynamics. During the COVID-19 crisis in Russia, we can observe a lesser decline in chemical production

E. G. Popkova (Ed.): ISC 2020, LNNS 368, pp. 798–807, 2022.
https://doi.org/10.1007/978-3-030-93244-2_85

compared to developed countries. In the current reality, only large and innovative companies in the chemical industry use a comprehensive approach to risk management in practice [6]. We believe that the assessment and analysis of real risks and threats, as well as the development of an adequate risk management strategy, are complex tasks in an unstable environment and the lack of domestic experience and technology. The solution to these tasks is linked to the development of mechanisms and adaptive tools for risk management.

The lack of adequate methodological approaches to risk management in the chemical industry has determined the relevance of this research.

The analysis of secondary literature shows that certain issues of nature, theory, and methodology of risk have been addressed in the works of A. P. Algin [1], S. N. Vorobyov and K. V. Baldin [8], E. V. Ioda [4], V. S. Stupakov and G. S. Tokarenko [7], T. Barton, W. Shenkir, and P. Walker [2], et al.

Russian and foreign theory and practice of risk management have a considerable number of techniques. Nevertheless, these techniques are marked with the ambiguity of approaches to risk management. Nevertheless, in our opinion, the issues of forming a risk-oriented culture of innovative chemical enterprises, applied aspects of determining the ratio of risks and opportunities at the industry level, exposure of industry enterprises to risks, and the introduction of the elements of risk management in the strategic planning process have not been studied sufficiently. These unresolved problems and several other issues force scholars and analysts to return to this topic.

2 Methodology

The methodological basis of the research consists of standards of risk management, materials of scientific and practical conferences and seminars at various levels, and fundamental developments of Russian and foreign scholars in the field of risk management and management of innovative industrial enterprises. The research is based on system analysis and a dialectical approach to studying patterns of formation and development of economic systems.

The research utilizes a systematic approach to studying the mechanism of risk management, the principle of diversity of forms of research and comprehensive evaluations, the principle of information support of the research, and the principle of the practical feasibility of the research results [3, 4].

The research tools include general scientific methods of knowledge and systematic, comparative, and economic-statistical methods of analysis.

The practical significance of further research lies in the fact that the proposed conceptual provisions on improving the risk management system and the program on preventing threats can serve as a basis for improving the risk management of innovative enterprises, especially of enterprises of the chemical industry.

3 Results

The year 2020 was challenging for almost all businesses due to the effects of the COVID-19 pandemic. This trend was also observed in the North Caucasus Federal District.

The revenues of major companies in the North Caucasus Federal District showed a negative trend for the first time in a long period. We believe that the reason for this situation lies in the behavioral aspect of leaders, reorganization, competition, and the ongoing power supply collapse.

Only six industries showed positive dynamics (Table 1) [5].

Table 1. Dynamics of indicators of industries in the North Caucasus Federal District

Industry	Revenue, million rubles		Change, million rubles	Share of industry revenues in the total, %	
	2018	2019	2019–2018	2018	2019
Wholesale industry	401264.00	285739.00	−115525.00	43.27	37.05
Energy industry	144120.00	137831.00	−6289.00	15.54	17.87
Chemical industry	113616.00	110642.00	−2974.00	12.25	14.35
Food industry	60215.00	63111.00	2896.00	6.49	8.18
Construction	48819.00	53954.00	5135.00	5.26	7.00
Transportation and communication industry	62416.00	45405.00	−17011.00	6.73	5.89
Agro-industrial complex	17508.00	15748.00	−1760.00	1.89	2.04
Oil and gas industry	14865.00	15126.00	261.00	1.60	1.96
Construction materials industry	9471.00	10974.00	1503.00	1.02	1.42
Retail	20039.00	8575.00	−11464.00	2.16	1.11
Glass industry	3225.00	6883.00	3658.00	0.35	0.89
Iron and steel industry	12146.00	6496.00	−5650.00	1.31	0.84
Instrumentation	5798.00	5524.00	−274.00	0.63	0.72
Non-ferrous metal industry	0.00	2648.00	2648.00	0.00	0.34
Housing and public utilities	2505.00	2552.00	47.00	0.27	0.33
Mechanical engineering and metalworking	7830.00	0.00	−7830.00	0.84	0.00
Service companies	3446.00	0.00	−3446.00	0.37	0.00
Total	927283.00	771208.00	−156075.00	100.00	100.00

Source [5]

The obtained data clearly shows that the top three industries remain unchanged – the wholesale industry, the energy industry (represented by OGK-2, the largest company in the district), and the chemical industry.

In our opinion, the state of industrial enterprises is determined by interrelated and interdependent factors, including the economic and political situation in the country. However, the individual features of the enterprises play a decisive role (e.g., the activity management model, organizational and management structure, and industry specifics). The chemical industry is the most promising in the North Caucasus Federal District. Nevertheless, this industry is marked with several risks. At the current stage, the enterprises of the chemical industry face not only external challenges but also challenges related to the peculiarities of the economic, political, and demographic situation in the country and the region. These challenges are associated with innovative development. To develop theoretical and methodological issues and practical recommendations for creating a risk management system, the authors chose one of the innovation-oriented enterprises of the chemical industry, "Monocrystal" JSC, whose products are exported to many countries around the world. During the COVID-19 pandemic in 2020, the market for ultraviolet LEDs for air, water, and surface disinfection received an unprecedented boost. In their technology, the LEDs are based on sapphire wafers manufactured by "Monocrystal" JSC.

The factors affecting the functioning of "Monocrystal" JSC, the authors identified and structured strengths and weaknesses, as well as opportunities and threats (Table 2).

Table 2. SWOT analysis of "Monocrystal" JSC

Strengths	Weaknesses
– Increase in the volumes of production and sales; – Use of advanced technologies for sapphire cultivation; – Application of ISO 14001 environmental management system; – Developed service units; – Advantageous location; – Prominence in the market (as evidenced by its weighty market share); – Financial stability; – Proactive leadership and loyal employees	– Impact of international factors; – Dependence on the political situation; – Constant adaptation to the requirements of the global market
Opportunities	**Threats**
– Expansion of service projects in Russia and abroad; – Improvement of the efficiency of the development of innovative products; – Development of innovative technologies for growing sapphires; – Implementation of innovative environmentally friendly technologies	– Increased influence from the government in the future; – Loss of competitive position; – Emergence of alternative products; – Unstable price situation on the energy market; – Price fluctuations on the world commodity and financial markets

Source Compiled by the authors

Table 2 shows that the strengths of the financial factor outweigh the weaknesses. "Monocrystal" JSC has its own risk management system; no third-party specialized firms are involved. The main risks are the threats associated with the implementation of foreign economic activities, particularly currency risks associated with the political situation and currency fluctuations. Trying to identify and assess threats and develop measures to minimize them, the authors conducted a risk analysis using the method of riskiness assessment, the so-called "risk rose" or "risk star." A 10-point riskiness rating system was used for the analysis. The results are presented in Fig. 1.

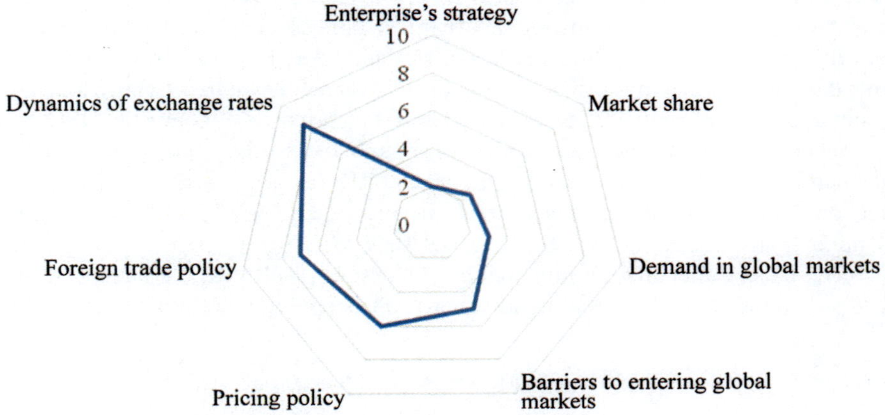

Fig. 1. "Risk star" at "Monocrystal" JSC. *Source* Compiled by the authors

The conducted analysis confirmed that the foreign trade policy and the variability of exchange rate dynamics significantly impact foreign economic activity.

To minimize currency risks, "Monocrystal" JSC uses diversification of currency assets, restrictions on the use of certain currencies and export transactions in a certain currency by contract, and hedging using currency futures.

Innovative "Monocrystal" JSC operates in a changing competitive environment, with different internal environments, staffing, level of production capacity, etc. Consequently, there are risks and threats associated with the specifics of foreign economic, technological, manufacturing, commercial, financial, and other activities.

The authors believe that it is extremely important to identify risks and threats promptly, determine the probability and timing of their occurrence, and predict the possible damage. It is impossible to eliminate all threats, which determines the need to anticipate, systematize, and assess risks, as well as to find risk management strategies.

Over the last five years, "Monocrystal" JSC has increased its synthetic sapphire production ninefold, having invested over $220 million in capacity expansion and application of innovative technologies.

High-tech production and unique specifications of many products made "Monocrystal" JSC quite competitive compared with other chemical enterprises, including foreign ones. The structure of sales in non-CIS markets is shown in Fig. 2.

In our opinion, creating an effective risk management system should include a risk management program covering the internal and external environment of an industrial enterprise.

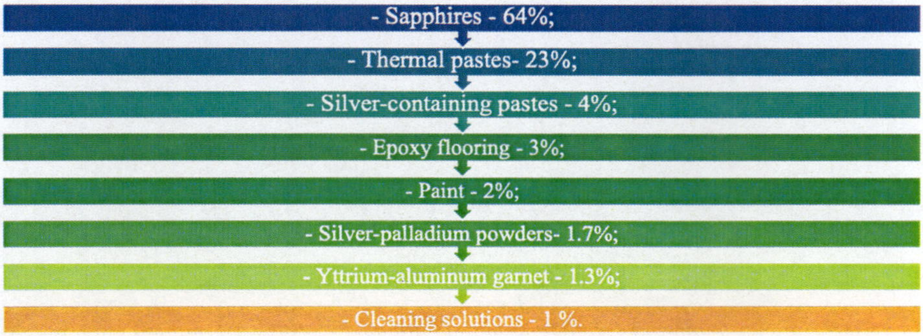

Fig. 2. Structure of sales of "Monocrystal" JSC on non-CIS markets. *Source* Compiled by the authors

The proposed risk management program, including the program aimed at reducing the risk of reducing export volumes of "Monocrystal" JSC, should contain identification (detection), analysis, planning, response, monitoring, and management.

Since each type of threat has individual features affecting the choice of its management, the program's formation should consider the possibility of occurrence of a particular threat, the strength of its impact, and the possibility of management. It is advisable to use a tool to determine the parameters of integral assessments of general and specific risks of foreign economic activity of "Monocrystal" JSC based on a combination of statistical, analytical, and expert assessments. The analysis allowed us to identify the most possible general and specific risks (Table 3).

Table 3. Typology of general and specific risks of foreign economic activity of "Monocrystal" JSC

General risks	Specific risks
1. Risks associated with the progressive movement of the contract	1. Risks associated with the peculiarities of customs clearance; 2. Investment risks; 3. Logistics risks; 4. Product certification risks; 5. Risks of misrepresentation of the necessary information
2. Risks associated with territorial dislocation	1. Risks arising in the domestic market; 2. Risks arising at the border; 3. Risks arising in the foreign market
3. External risks	1. Partner reliability risks; 2. Risks of political instability; 3. Macroeconomic risks; 4. Risks of legal support; 5. Risks associated with the marketing campaign and market research
4. Risks associated with the contractual process	1. Commodity risks; 2. Risks of delivery terms; 3. Risks associated with the contract terms; 4. Risks of contract termination and failure to fulfill contractual obligations

Source Compiled by the authors

To obtain quantitative indicators of risk assessment, the authors propose to use the following indicators:

$$R = \sum_{i=1}^{n} W_i \cdot V_i, \tag{1}$$

$$W_i = W_k \cdot \frac{(k-i) \cdot f + i - 1}{k-1}, \tag{2}$$

$$\sum_{i=1}^{k} W_i = 1, \tag{3}$$

$$f = \frac{W_1}{W_k}, \tag{4}$$

$$W_k = \frac{2}{k(f+1)}, \tag{5}$$

where:
R – integral index of the assessment of the occurrence of risk event;
Wi – risk weighting by groups, i = 1, 2,…,k;
k – number of groups;
f – ratio of multiplicity;
Wk – weight of the group with the lowest priority;
Vi – probability of the occurrence of risk event.

The results of the assessment of the risks of foreign economic activity of the organization based on the considered methodological tools are summarized in Table 4.

Table 4. Estimated indicators of external risks of foreign economic activity of the organization based on the results of activities in 2018–2020.

General risks	Weight, W_i
1. Risks associated with the progressive movement of the contract	0.221
2. Risks associated with territorial dislocation	0.205
3. External risks	0.176
4. Risks associated with the contractual process	0.103

Source Compiled by the authors

According to the results, the authors conclude that the most significant risks are related to territorial dislocation.

Within the framework of the presented general risks, the authors identified specific risks, their weights, priorities, and probabilities of their occurrence (Table 5).

Table 5. Assessment of specific and general risks of foreign economic activity of innovative organization based on the results of activities in 2018–2020

Types of risks	Priority, P_i	Weight, W_i	Probability, V_i	Integral evaluation, $W_i * V_i$	Strength of impact
External risks					
1. Macroeconomic risks	0.354	0.084	90	7.56	0.145
2. Risks of political instability	0.273	0.069	75	5.175	0.102
3. Partner reliability risks	0.161	0.037	70	2.59	0.085
4. Risks of legal support	0.113	0.019	35	6.65	0.057
5. Risks associated with the marketing campaign and market research	0.072	0.012	15	0.18	0.021
Total for the group	–	0.221	–	22.155	0.39
Risks associated with the contractual process					
1. Risks associated with the contract terms	0.221	0.092	55	5.06	0.135
2. Risks of delivery terms	0.177	0.072	30	2.16	0.09
3. Commodity risks	0.134	0.03	20	0.6	0.04
4. Risks of contract termination and failure to fulfill contractual obligations	0.098	0.011	10	0.11	0.01
Total for the group	–	0.205	–	7.93	0.275
Risks associated with the progressive movement of the contract					
1. Risks associated with the peculiarities of customs clearance	0.265	0.061	60	3.66	0.061
2. Investment risks	0.201	0.05	30	1.5	0.035
3. Product certification risks	0.187	0.024	18	0.432	0.026
4. Logistics risks	0.154	0.032	2.5	0.08	0.018
5. Risks of misrepresentation of the necessary information	0.051	0.009	15	0.135	0.01
Total for the group	–	0.176	–	5.807	0.132
Risks associated with territorial dislocation					
1. Risks arising in the domestic market	0.491	0.062	50	3.1	0.031
2. Risks arising at the border	0.315	0.024	40	0.96	0.024
3. Risks arising in the foreign market	0.207	0.017	25	0.425	0.012
Total for the group	–	0.103	–	4.485	0.067

0 – low risk;
25 – risk is unlikely to be realized;
50 – uncertain risk;
75 – risk is likely to be realized;
100 – risk is most probable

Source Compiled by the authors

According to the obtained estimates, the highest probability of occurrence is observed in macroeconomic risk (90 points out of 100). The group with the lowest probability of occurrence includes logistics risks, risks of contract termination, default on contractual obligations, and risks of distortion of the necessary information.

To identify the most significant specific risks of foreign economic activity of the innovation-oriented enterprise, the authors singled out the risks with the integral assessment values close to 10 (macroeconomic risks [7.56 points]).

According to the logic of our research, the construction of the risk matrix is an essential stage of risk assessment to make informed managerial decisions. The basis of this matrix is the positioning of the innovative chemical enterprise, depending on the level of influence of private and general risks on its activities. This representation of risks is demonstrated by the graph, where risks are arranged according to their probability and strength of impact (Table 6).

Table 6. Risk matrix of the innovative chemical enterprise "Monocrystal" JSC

Risks	Significance			
	I	II	III	IV
I	Risks of legal support	Partner reliability risks	Risks of political instability	Macroeconomic risks
II	Risks of contract termination and failure to fulfill contractual obligations	Commodity risks	Risks of delivery terms	Risks associated with the contract terms
III	Risks arising in the domestic market	Risks arising in the foreign market	Risks arising at the border	-
IV	Logistics risks	Product certification risks	Investment risks	Risks associated with the peculiarities of customs clearance

Note Quadrant I – the strength of risks is minimal; Quadrant II – average strength of risks; Quadrant III – significant strength of risks; Quadrant IV– maximum strength of risks. *Source* Compiled by the authors

Thee in the red sector of the matrix deserve special attention from the perspective of adapting to risks. The presented calculations show that the risks that deserve special attention are external risks of an innovative organization since most of the specific risks included in this group are marked with a high probability of occurrence (more than 75%) and significant impact power.

4 Conclusions

The developed methodical approach of risk assessment of foreign economic activity of innovation-oriented chemical enterprises allowed identifying the private and general risks of its functioning, highlighting the significant risks, and analyzing the strength of their impact. The authors believe that solving the problems of effective risk management in the field of foreign economic activity of innovative chemical enterprises should be based on reliable and complete analysis of the processes occurring in the world market. The foreign economic risk of innovation-oriented enterprises of the chemical industry reflects the financial losses arising during activities in the world market. The peculiarities of their management predetermine the specifics of the activity of innovative industrial enterprises. In the context of the global crisis, enterprises of the chemical industry are experiencing several problems related to the depreciation of production and technology and macroeconomic trends. The solution to these problems lies in improving the risk management system, which implies the formation of an adequate risk management program. In the course of developing such a program, it is necessary to consider the industry characteristics of such enterprises since they provide a choice of factors and sources of value formation of chemical organizations.

Thus, the risk management of foreign economic activity of innovative chemical enterprises becomes an objective necessity in the current unstable international market economy. The development of effective measures to adapt to internal and external risk factors will create the necessary conditions for doing business on an expanded basis, which is fundamental to the sustainable and dynamic development of the organization's economy at the international level.

References

1. Algin, A.P.: Risk and Its Role in Public Life. Mysl., Moscow, USSR (1989)
2. Barton, T., Shenkir, W., Walker, P.: Making Enterprise Risk Management Pay Off. Williams, Moscow, Russia (Original Work Published 2002) (2003)
3. Gorshenina, E.V.: Methodological principles, levels, and tasks of research. Econ. Stud. **2**, 25–29 (2011)
4. Ioda, E.V.: Enterprise risk management. Retrieved from https://cyberleninka.ru/article/n/upravlenie-riskami-predpriyatiya (2003). Accessed 12 September 2021
5. Rakul, E.: Big Business SCFD 2020: minus 20% in general and growing construction projects in particular. Expert South. Retrieved from https://expertsouth.ru/articles/krupnyy-biznes-skfo-2020-minus-20-v-obshchem-i-rastushchie-stroyki-v-chastnosti/ (2020). Accessed 12 September 2021
6. Sharova, A.A.: Formation of the corporate risk management system in the enterprises of the chemical industry. Dissertation of Candidate of Economics, Nizhny Novgorod State Technical University, Nizhny Novgorod, Russia. Retrieved from http://dissers.ru/1ekonomika/formirovanie-korporativnoy-sistemi-upravleniya-riskami-na-predpriyatiyah-himicheskoy-promishlennosti-specialnost-08-00-05-ekonomika.php (2012). Accessed 12 September 2021
7. Stupakov, V.S., Tokarenko, G.S.: Risk Management: Handbook. Finance and Statistics, Moscow, Russia (2005)
8. Vorobyov, S.N., Baldin, K.V.: Risk Management in Entrepreneurship. Dashkov and Co., Moscow, Russia (2013)

Ensuring the Sustainable Development of Entrepreneurial Structures in Unstable Economic Realities

Innara R. Lyapina[1(✉)], Olesya A. Stroeva[2], Aleksandr V. Zakharov[3,4], Elena N. Tokmakova[1], and Elena A. Sotnikova[5]

[1] Orel State University, Orel, Russia
[2] LLC "Transformation", Orel, Russia
[3] Moscow University for Industry and Finance "Synergy", Moscow, Russia
[4] Plekhanov Russian University of Economics, Moscow, Russia
[5] Orel State University of Economics and Trade, Orel, Russia

Abstract. The instability of economic processes in modern realities leads to barriers and associated risks related to the development of entrepreneurial structures. Entrepreneurial structures are a rather complex category. The highlighted condition is associated with the lack of a clear declaration of the concept of entrepreneurial structure in Russian legislation. Additionally, transformational processes manifested in the emergence of post-pandemic consequences lead to the elimination of entrepreneurial structures. This condition violates the integrity of the entrepreneurial environment and requires the formation of a new system of functioning and implementation of entrepreneurial structures in unstable economic realities. The paper aims to form a model to ensure the sustainable development of entrepreneurial structures in unstable economic realities. The authors consider the main elements of the development of entrepreneurial structures in Russia. Moreover, they discuss the relevance of the national project "Small and medium entrepreneurship and support for individual entrepreneurial initiatives" in an unstable economic development. A model for the sustainable development of business structures in unstable economic realities was formed. The instrumental basis of the research includes the methods of classification, data typology, normative and programmatic nature, analytical perception, highlighting and clarifying problems, simulation, and comparative analysis.

Keywords: Entrepreneurial structure · Small and medium businesses · National project · Model · Entrepreneurial activity · Entrepreneurial actions · Development · Instability

JEL codes: L26 · M2 · Q01

E. G. Popkova (Ed.): ISC 2020, LNNS 368, pp. 808–816, 2022.
https://doi.org/10.1007/978-3-030-93244-2_86

1 Introduction

The processes associated with the introduction of digital technologies and the innovative agenda of the technology sector have destabilized the economic component that affects the various spheres and structures of the national economy. Ensuring sustainable development is a fairly important criterion for long-term and efficient functioning in unstable economic realities. The sustainable development of entrepreneurial structures in unstable economic realities is dictated by the regeneration of public support and the foundations of the formation of the national economy. Entrepreneurial structures are a set of organizational units, which have legal forms necessary to make profits and participate in market relations [4]. According to this concept, entrepreneurial structures appear as a wide layer of the national economy independently engaged in economic activities to make profits from using property, selling goods, and providing services. The highlighted condition indicates a wide range of directions of market relations and organizational and economic characteristics. Even though legal norms secure the development of entrepreneurial structures, unstable economic realities produce several problems that lead to systematic shifts reducing business competitiveness.

The first problem is the lack of a single algorithm for supporting entrepreneurial structures in Russia (the principle of systemic support). Currently, the foundations for the development of entrepreneurial structures are declared through disparate regulations, which shift and decompose this category into small components:

- Federal law “On the development of small and medium entrepreneurship in the Russian Federation” (July 24, 2007 No. 209-FZ) examines the structure of firms;
- Federal law “On banks and banking” (December 2, 1990 No. 395-1) focuses on the banking structures;
- Law of the Russian Federation “On the organization of insurance business in the Russian Federation” (November 27, 1992 No. 4015-1) describes insurance structures;
- Federal law “On investment activities in the Russian Federation, carried out in the form of capital investment” (February 25, 1999 No. 39-FZ) describes investment structures.

This legal regulation raises the question of the unity of entrepreneurial structures under this type of differentiation [13]. Otherwise, entrepreneurial structures are considered in the context of identification with the concept of business structures. This condition also has a disparate regulatory component focusing on the aspects of the development of small and medium-sized enterprises (e.g., Federal law “On the development of small and medium entrepreneurship in the Russian Federation” [July 24, 2007 No. 209-FZ] and national project “Small and medium entrepreneurship and support for individual entrepreneurial initiatives”). The conditions highlighted above do not allow implementing the principle of systemic support for entrepreneurial structures.

The second problem lies in denying the importance of the partnership assistance of the state and business structures (the principle of exploitation). In theoretical studies and practical realities, the government perceives entrepreneurial structures as a source of budget replenishment [8], denying the existence of partnerships in the formation of long-term conditions for business development. Public-private partnership as a mechanism to implement the principle of interaction has replaced several social and territorial problems using entrepreneurial structures [11]. Simultaneously, this interaction was implemented based on the principle of exploitation – the possibility of additional use of free funds at a lower cost to the government [6].

The third problem lies in the emergence of difficulties identified during the assessment of the state of business and the effectiveness of support measures provided for business entities (the principle of instability of the current situation). The principle of instability of the current situation means the emergence of problems caused by the transformation of the current economic state [9]. According to the RBC monitoring of the state of business and measures on the effectiveness of public support, we can highlight the following problems inherent in entrepreneurial structures:

- Reduced demand for goods and services (59.8%);
- Lack of money to pay property taxes and rent payments (27.3%);
- Lack of funds to pay salaries and wages (25.3%);
- Emergence of non-payments by counterparties for delivered products (22.3%) etc. [7].

In the short term, the highlighted problems of entrepreneurial structures in unstable economic realities project a decrease in company turnovers [3], the bankruptcy of business structures, and the weakening of competitive advantages in the international market [15].

The formulated problems allowed to highlight the importance and sufficient relevance of the chosen research topic. The paper aims to form a model for ensuring the sustainable development of entrepreneurial structures in unstable economic realities. To implement the highlighted research purpose, it is necessary to do the following:

- Consider aspects of the development of entrepreneurial structures in modern Russia;
- Determine the relevance of the national project "Small and medium entrepreneurship and support for individual entrepreneurial initiatives" in the context of unstable economic development;
- Form a model to ensure the sustainable development of entrepreneurial structures in unstable economic realities.

2 Materials and Methods

The methodological basis of the research is based on the use of materials selected based on the highlighted research tasks. The consideration of aspects of the development of entrepreneurial structures in Russia is associated with the application of the method of classification and data typology [10] and the semantic model of whole-part [14]. The relevance of the national project "Small and medium entrepreneurship and support for individual entrepreneurial initiatives" in an unstable economic development is determined using the methods of regulatory and programmatic nature, analytical perception, and highlighting the problems [1]. The formation of a model to ensure the sustainable development of business structures in unstable economic realities focuses on the implementation of the methodological apparatus of simulation, comparative analysis, and the method of clarification of problems.

3 Discussion

Let us consider the fundamentals of the entrepreneurial structure in modern Russia. The essence of the entrepreneurial structure is due to the definition of the nature of the activity and the specific features of the subjects included in its structure. An entrepreneurial structure is a set of units engaged in entrepreneurial activities to make profits [2]. The highlighted concept rather conventionally reveals the category of entrepreneurial structure. On the one hand, making a profit is the main goal in the activities of commercial organizations in accordance with Article 50 of the Civil Code of the Russian Federation. In this case, an entrepreneurial structure is perceived as a commercial organization. Compared to a commercial organization, an entrepreneurial structure is more general-declarative. For example, an entrepreneurial structure is an organizational and economic unit, the functioning of which is based on the production of goods or services with the ultimate goal – profit [16]. According to Article 50 of the Civil Code of the Russian Federation, the commercial organization is a legal entity pursuing profit as the primary purpose of its activities. Additionally, an entrepreneurial structure is defined as a form of business entities engaged in entrepreneurial activity. [5]. In general, this definition allows us to conclude that the entrepreneurial structure contains the elements of a commercial organization.

To consider the foundations of entrepreneurial structures, the constituent parts, which allow determining the categorical meaning of this definition and additional elements revealing the functional nature of the highlighted concept, are distinguished. The constituent elements characterizing entrepreneurial structures are conditioned by the availability of own capital for business development with differentiation of types of entrepreneurial activity. The highlighted basis allows us to conclude that the critical feature of entrepreneurial structures is (1) capital used to implement (2) types of entrepreneurial activities (3) aimed at providing services, goods, and works.

The key problems of small and medium-sized enterprises (SMEs) identified in 2020 include a sharp decline in demand in the SME sector and the growth of debts of SMEs to the state, followed by the liquidation of firms. As part of the national project "Small and medium entrepreneurship and support for individual entrepreneurial initiatives," the key factors in the implementation of the processes of development and improvement in 2021 are as follows:

- Increasing the volume of purchases made by large customers from SMEs (about 3.8 trillion rubles);
- Increasing the number of people employed in the economy to 21.6 million;
- Increasing the share of SMEs in GDP by 25% (9.25% of subjects from non-resource exports).

Increasing negative trends in the deterioration of SMEs in the conditions of post-COVID economic development do not allow to correct the indicated problems in the context of the national project "Small and medium entrepreneurship and support for individual entrepreneurial initiatives." Under conditions of unstable economic development, there is a decrease in the growth index of SMEs, the lack of fair distribution of public support (assistance within the Russian National Classifier of Types of Economic Activity [OKVED]) among entrepreneurial structures, the abandonment of business development towards liquidation, and the increase in the production costs in the abandonment of planned business development.

The national project "Small and medium entrepreneurship and support for individual entrepreneurial initiatives" acts as a strategic document forming a long-term view on the development of SMEs in Russia. When considering SMEs as part of the entrepreneurial structure, it is necessary to note the role of entrepreneurial activity of subjects in the system of unstable economic development. In this case, the problems encountered in the current activities of SMEs hinder the development of entrepreneurial structures and reinforce trends towards a weakening of competitiveness in the Russian and foreign markets.

The functioning of entrepreneurial structures in an unstable economy depends on government policies and existing barriers that slow down development and reduce the competitiveness of business entities. Within the framework of the analysis presented by the Institute for Emerging Market Studies of Skolkovo Business School, the barriers to the development of entrepreneurial structures are based on administrative pressure of government agencies (24.3%) [12]. In general, the main barriers slowing down the development of entrepreneurial structures in an unstable economy are of a state-financial nature.

The key factors indicated as barriers to entrepreneurial structures are as follows:

1. Lack of a fair tax system (18.6%). This barrier is due to the presence of unequal conditions for entrepreneurs in the proportionality of the tax burden, which is due to (1) the priority of doing business for the government and (2) the presence of OKVED, corresponding to the support measures provided during the COVID-19 pandemic. The highlighted barrier provokes a shadow economy associated with shifting tax regimes and reducing officially registered employees of entrepreneurial structures. The orientation of entrepreneurial structures aims to choose the optimal taxation regime, with limitations in scaling the business and entering new markets.
2. Lack of motivation for development (18.3%). Under the constraints that emerged due to the COVID-19 pandemic, lower profits and higher risks lead to the reorganization of debt-ridden entrepreneurial structures. The scheme, when a new company opens instead of the old one, destabilizes the entrepreneurial sphere, forming unstable conditions of economic development.
3. Excessive reporting (16.4%). The availability of quarterly and monthly reporting creates extra work, the substitution of which will optimize business development.
4. Financial burden (14.2%). Within this barrier, business taxes paid on behalf of employees are an essential issue. The payroll tax system projects conditions associated with shadow employment when it becomes unprofitable to hire people officially. Additionally, the financial burden on business structures can be attributed to the state accounting systems (e.g., Platon – the system of transportation payments for vehicles over 12 tons, Mercury – cargo certification system, etc.).

5. Other (8.2%). Other barriers to the development of entrepreneurial structures include the lack of motivation for the development of entrepreneurial structures among municipal and state authorities, neglect of the initiatives of entrepreneurial structures, blocking of accounts of entrepreneurial structures in any (even minor) debts, and increasing the costs of entrepreneurial structures after the government initiatives.

It is worth noting that entrepreneurial structures are a multifaceted concept with a clear vertical and horizontal basis. In this case, legislative fixation and regulation of the development of entrepreneurial structures in unstable economic conditions lead to a shift in priorities. Thus, the national project "Small and medium entrepreneurship and support for individual entrepreneurial initiatives" allows defining strategic goals of development of a part of a horizontal entrepreneurial structure. Nevertheless, it does not focus on solving current issues requiring activities aimed at creating a harmonized business environment. The identified condition is stated by the highlighted barriers to the development of entrepreneurial structures. Thus, it is necessary to form a model to ensure the sustainable development of entrepreneurial structures in unstable economic realities.

4 Results

The model to ensure the sustainable development of entrepreneurial structures in unstable economic realities is aimed at identifying conditions that will minimize the risks associated with losses. Figure 1 shows a three-stage model algorithm to ensure the sustainable development of entrepreneurial structures in unstable economic realities. The first stage of the model involves defining a single entrepreneurial structure in the overall business system at the legislative level. As noted earlier, entrepreneurial structures are not enshrined at the legislative level in the context of a regulatory act. The interpretation of the category "entrepreneurial structure" comes from the author's vision and understanding of academic researchers.

The second stage is conditioned by the design of programs to develop entrepreneurial structures in unstable economic conditions. This tool is concentrated on territorial levels – municipal, regional, and federal. The key task in the context of the formation of this program is to ensure the interest in the development of business structures of state and municipal authorities. In the highlighted aspect, it is worth noting that local government and public authorities are interested in the development of entrepreneurial structures indirectly. To solve this problem, it is necessary to form a bonus KPI. The basis of the bonus KPI will consider several indicators such as (1) the number of registered business structures and (2) the number of provided government (municipal) support measures, allowing to increase the profit or productivity of each entrepreneurial structure.

The third stage of the model is forming a pool of focal points for the long-term, medium-term, and short-term periods. The allocation of a pool of focal points is necessary to identify activities to solve problems of current and strategic nature. Short-term activities in the focal point pool are driven by quick decisions. Short-term measures include preferential refinancing of leases and loans, stabilization (moratorium) of the cadastral value, the extension of the moratorium on inspections with subsequent reduction of the significance of this process, and freezing the growth of tariffs. Medium-term actions imply conscious decisions. Conscious decisions mean the formation of a pool of measures designed to solve medium-term problems based on the already created algorithm for the implementation of quick decisions. Medium-term actions involve reducing the tax burden, limiting the growth of monopoly tariffs, reducing the administrative burden, and developing the purchasing power of the population in small towns.

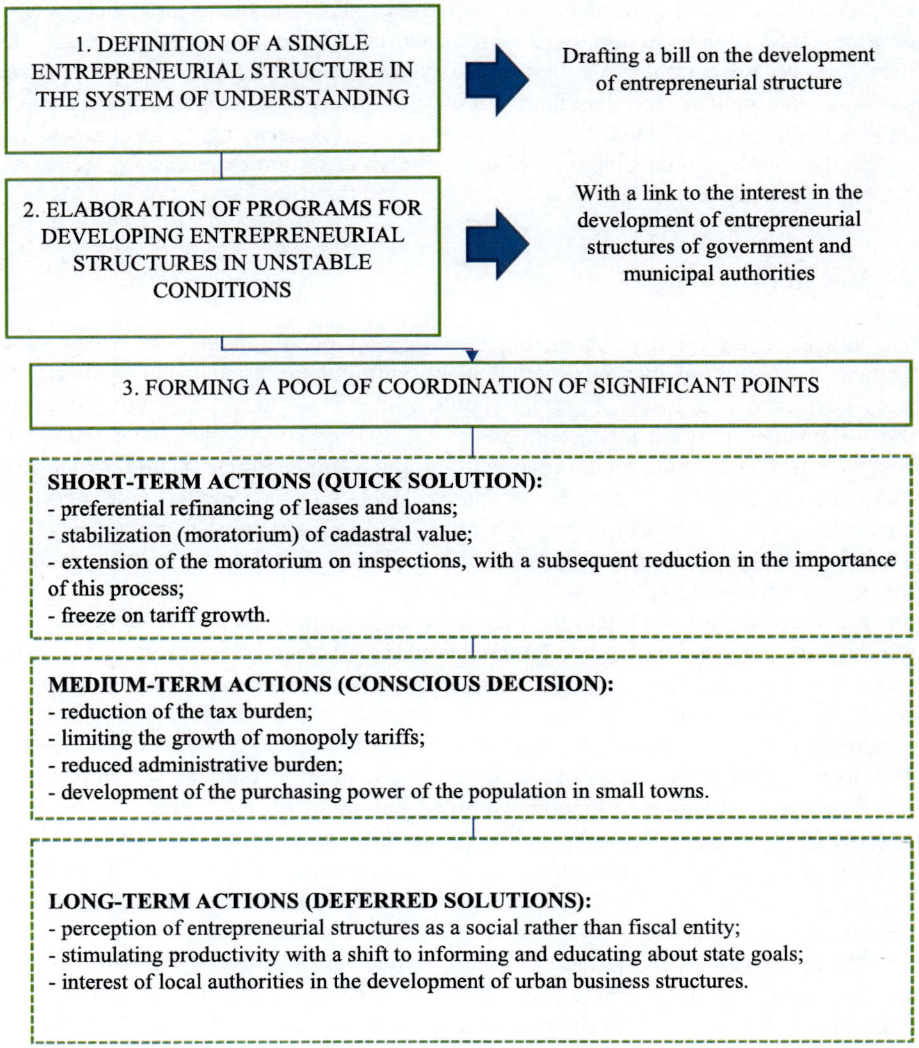

Fig. 1. Model to ensure the sustainable development of entrepreneurial structures in unstable economic realities. *Source* Compiled by the authors

Long-term actions are based on deferred strategic solutions to ensure the sustainable development of entrepreneurial structures in unstable economic realities. Long-term actions involve implementing systemic measures that allow forming an effective system of functioning of entrepreneurial structures for a period of 5–8 years. Long-term actions are based on the perception of entrepreneurial structures as a social rather than fiscal entity, the stimulation of productivity with a shift to informing and educating about state goals, and the interest of the local government in the development of urban entrepreneurial structures.

5 Conclusion

The conducted research on the sustainable development of entrepreneurial structures in unstable economic realities allowed us to formulate the following conclusions:

1. The examined aspects of the development of entrepreneurial structures in modern Russia have established that this category is not enshrined in the form of a normative-legal act with clear differentiation. Entrepreneurial structures are regarded as commercial or business structures. The most general concept of entrepreneurial structure is an organizational and economic unit, the functioning of which is based on the production of goods or services with the ultimate goal of making profits. Entrepreneurial structures have a horizontal and vertical classification. The vertical classification includes the division of entrepreneurial structures according to the elements-agents of microeconomics: firms, banking structures, insurance companies, and investment companies. Horizontal classification is made based on the division into large, medium, and small enterprises.
2. The relevance of the national project "Small and medium entrepreneurship and support for individual entrepreneurial initiatives" in the context of unstable economic development is long-term and strategic in nature. The national project affects part of the horizontal classification of entrepreneurial structures (small and medium-sized enterprises), replacing current problems with long-term elements of development. Thus, the current problems of entrepreneurial structures are formed as administrative barriers.
3. The model formed to ensure the sustainable development of entrepreneurial structures in unstable economic realities is implemented based on stages to create conditions for developing entrepreneurial structures in the short, medium, and long term. The model to ensure the sustainable development of entrepreneurial structures in unstable economic realities includes (1) the definition of a single entrepreneurial structure within the legal framework, (2) designing programs for the development of entrepreneurial structures in unstable conditions, and (3) the formation of a pool of coordination of significant points for developing entrepreneurial structures in the short, medium, and long term.

References

1. Bakirov, M.I., Koch, X., Amanova, E.A.: Development of small and medium enterprises in the region. Euras. Sci. Assoc. **5–3**(75), 174–176 (2021)
2. Battakhov, P.P.: Development of legal regulation of entrepreneurship in advanced foreign countries and in Russia. Law State Theory Pract. **3**(195), 84–88 (2021). https://doi.org/10.47643/1815-1337_2021_3_84
3. Blokhin, E.A.: The development of entrepreneurship in Russia. Regul. Issues FEC Reg. Feder. **3**, 12–15 (2019)
4. Goncharenko, A.A., Isaeva, E.V.: Cluster entrepreneurship structures in the system supporting the development of programs of import substitution in the Russian Federation. Bull. Adyghe State Univ. Ser. 5: Econ. **4**(230), 221–228 (2018)
5. Khusenov, S.: Development of entrepreneurship in the emerging post-crisis economy. Econ. Yesterday Today Tomorrow **3**, 85–90 (2021)
6. Li, J.: The role of small and medium-sized businesses in Russia and China. Compon. Sci. Technol. Prog. **3**(57), 9–12 (2021)
7. Peremitin, G., Tkachev, I.: Monitoring "Assessment of the state of business and effectiveness of state support measures." RBC News. Retrieved from https://www.rbc.ru/newspaper/2021/02/25/603664ba9a79472b2daabe02 (2021, February 25)
8. Piskun, E.I.: Corporate entrepreneurship: mechanisms for coordination of innovation and investment activity. Drukerovskij Vestnik **5**(31), 90–97 (2019). https://doi.org/10.17213/2312-6469-2019-5-90-97
9. Pyataev, D.A., Martynova, O.: Evolutionary development of entrepreneurship. World Sci. **4**(49), 204–206 (2021)
10. Sahiti, F.: Institutions and entrepreneurial activity: a comparative analysis of Kosovo and other economies. J. Entrepreneur. Public Policy **10**(1), 98–119 (2021). https://doi.org/10.1108/JEPP-08-2020-0056
11. Satybekova, E., Maulenkulova, G., Shevchenko, I., Aliyeva, E.: Business structures in the agro-industrial complex as a factor of increasing competitiveness. Stat. Acc. Audit **1**(80), 92–97 (2021)
12. SKOLKOVO Institute for Emerging Market Studies (IEMS): Sustainable Development in Russia: A Guide for Transnational Corporations. Retrieved from https://iems.skolkovo.ru/downloads/documents/SKOLKOVO_IEMS/Research_Reports/SKOLKOVO_IEMS_Sustainable_Business_Lab_Research_2016-07-13_ru.pdf (2016)
13. Skorobogatova, T.N., Marakhovskaya, I.Y.: Knowledge-intensive business structures as economic systems: parameters, equifinality and competitiveness. Bull. Adyghe State Univ. Ser. 5: Econ. **2**(240), 106–113 (2019)
14. Sokolsky, I.A.: Development of forms and mechanisms of green entrepreneurship. Innov. Invest. **3**, 145–148 (2021)
15. Tursunov, I.E., Rakhimova, D.O.: Development of virtual entrepreneurship based on the digital economy. Economics **1**(48), 14–18 (2021). https://doi.org/10.24411/2410-289X-2021-10102
16. Venâncio, A., Pinto, I.: Type of entrepreneurial activity and sustainable development goals. Sustainability **12**(22), 1–25 (2020). https://doi.org/10.3390/su12229368

On the Potential Danger of Using Exoskeletons to Commit Crimes

Vadim A. Chukreev(✉)

Prosecutor's Office of the Sverdlovsk Region, Ekaterinburg, Russia

Abstract. Currently, the development of exoskeletons for civilian and military use is rapidly developing all over the world. The purpose of this work is to determine the level of danger to society from the uncontrolled use of exoskeletons by the state, including for illegal purposes. The study reveals that, in general, the threats to social relations from the unlawful use of exoskeletons by criminals can range from the creation of mass disturbances in public places, violent crimes and threats to citizens to the emergence of more sophisticated ways to steal, burglary and art thefts. The original conclusion is that a government strategy is needed to recognize and manage the risks in misusing exoskeletons while encouraging their responsible use for the public good, as well as creating a legal liability mechanism as a preventive measure to prevent misuse of exoskeletons.

Keywords: Exoskeleton · Dual purpose · Crime · Law · Regulation · Liability

JEL Codes: K14 · K16 · K33

1 Introduction

Modern high-tech, complex devices invented and implemented in various spheres of our life activity, in addition to the obvious useful properties, may contain potential risks to the interests of the state, society and the individual when used for illegal purposes.

At present, the development of exoskeletons and exosuit technologies is rapidly developing all over the world for civilian use, in particular for medical, industrial and defence purposes. The first exoskeleton was made at the Pentagon and was called the Hardiman. It was a machine with a big arm in which you had to insert your hand to operate it, its lifting capacity was 110 kg, but Hardiman was not of any serious use because it was big and heavy (680 kg).

In 2014, the global market for exoskeletons was $68 million [6], and in 2016 it was already $98 million [21] (mostly for medical purposes). The exoskeleton market is expected to grow by 43% [6] annually and is projected by scientists to reach $1.8 to 8.3 billion USD by 2025 [2, 23].

Unfortunately, the number of illegal exoskeleton owners will increase as prices will decrease and competition from manufacturers increases, for example, SuitX has almost halved its prices to $40,000 [25] for the Phoenix exoskeleton compared to the ReWalk, with a starting price of about $75,000 to $80,000 [13]. It is predicted that by 2027,

E. G. Popkova (Ed.): ISC 2020, LNNS 368, pp. 817–824, 2022.
https://doi.org/10.1007/978-3-030-93244-2_87

exoskeletons will be widely used and available in a greater variety to serve multiple user purposes.

In modern dictionaries – exoskeleton (from Greek ἐξω – external and σκελετος – skeleton) – a device designed to replenish lost functions, increase human muscle strength, expand the range of motion through the external frame and driving parts, transfer load when carrying loads through the external frame to the support area of the exoskeleton foot[1].

A wide variety of exoskeletons such as power, passive, pseudo-passive or active have been developed and operated around the world and serve a variety of purposes, from rehabilitation care assistance [10, 17] to industrial production [12] and combat operations [18, 24, 28]. Exoskeletons can be worn on the human body (or part of the body), carried around [10], there are exoskeletons and exoskeletons for arms, legs or back [17].

The different types of exoskeletons available serve multiple purposes, so they are used for rehabilitation purposes and are usually powered or active. This means that they require a power source or motor, arm or hydraulics [11]. When the power source is a battery pack, hydraulic unit, or pneumatic unit attached to the exoskeleton (and usually placed in a backpack with the user), the exoskeleton is considered autonomous [18] in the sense that it is not tethered. Tethered versions, on the other hand, use electrical wires to connect to an external outlet. Commercially available active rehabilitation exoskeletons such as ReWalk [14] and The Phoenix [18] are autonomous and assist in outpatient rehabilitation of patients with spinal cord injuries, for example [19].

Autonomous exosuits with partial body power are also under development for rehabilitation and there is a soft robotic exosuit from the Wiss Institute at Harvard University for lower limb mobility.

Unlike electrically powered exoskeletons and exosuits, passive versions do not require a power source; they use human energy, are autonomous and lighter, but are less mobile and fast than tethered versions [26]. Passive exoskeletons and exosuits are usually used for weight distribution or shock absorption, such as in industrial production, to facilitate lifting, moving loads [6, 12, 14].

In 2014, workers at South Korea's Daewoo Shipbuilding and Marine Engineering were already lifting industrial loads weighing up to 30 kg en masse in special exoskeletons [28] developed by their company; now they are talking about plans to develop exoskeletons capable of lifting loads up to 100 kg [14]. In the automotive industry, the use of exoskeletons has grown from a few dozen to thousands of devices in the creation of BMW, Ford, Honda, Nissan, Toyota and Volkswagen cars [23]. Exosuits in the sports and recreation industry [16] are also widely used and can be repurposed for industrial or military use [27].

For example, the SuitX passive modular flexible exoskeleton (MAX) made by the University of California, which combines removable leg, back and shoulder exoskeletons (for workers) sells for less than $5,000. While the retail price of the SuitX-powered Phoenix exoskeleton is eight times higher at $40,000 [2] and the ReWalk powered exoskeleton is about $75,000 [25].

---

[1] See: Wikipedia.org, section "exoskeleton".

In addition to this significant price difference, a key difference is that both Phoenix and ReWalk are intended for medical use in rehabilitation care and will be available to an undetermined number of individuals. In contrast, passive exoskeletons and upper or lower limb exoskeletons, such as the SuitX MAX, are not only more affordable but are intended for use by workers in the entire spectrum of the public and private sectors, construction, material handling, automotive, airports, etc.

In the Pentagon, the DARPA Advanced Research Projects Agency is currently working on this and similar projects are underway in China, Japan, Israel, and European countries.

Exoskeletons for military combat appear to be particularly dangerous for uncontrolled use, which optimizes the strength capabilities of soldiers; they range from those that increase a person's carrying capacity and fighting ability for long periods to those that increase resistance to various forms of attack [2].

The ExoHiker exoskeleton makes it easier to carry heavy backpacks during a hike, weighs 14 kg, but relieves a load of up to 70 kg. On a single charge without a solar battery connection, a soldier can walk 65 km with it at a speed of about 4 km/h [15]. The suit monitors vital signs and even stops bleeding.

Vorobyov proposed the following classification of exoskeletons: by energy source and drive principle (passive, active); by application point (localization) – (upper limb exoskeleton; lower limb exoskeleton; suit exoskeleton); by cost (low – $1,000–10,000; medium – $100,000–50,000; high – over $500,000; by application field (military; medical; industrial; space); by weight of design (light – up to 5 kg; medium – from 5 to 30 kg; heavy – over 30 kg); by the number of functions (simple purpose; dual purpose; with extended functions); by patient's mobility (mobile; fixed (stationary)) [31].

Legally exoskeletons are already in use and are of undeniable benefit. In the International Classification of Goods and Services for the Registration of Marks (ICS-11), adopted in Nice 15.06.1957 in the list of names for goods and services included exoskeletons robotic, other than used for medical purposes (robotic exoskeleton suits, other than for medical purposes).

2 Methodology

In order to obtain reliable results, the author used the materials of official sources containing current and developing regulations, reporting materials and statistical data in her research. In addition, we analyzed data from independent Internet portals and media resources. We have specially developed evaluation criteria to determine the degree of danger to the civil turnover of exoskeletons. The results of the study were obtained based on the method of comparative data analysis.

3 Results

The Russian Roadmap for the Development of "End-to-end" Digital Technology "Components of Robotics and Sensorics" The Roadmap (RD) for "End-to-end" Digital Technology (ETS) "Components of Robotics and Sensorics" is one of the fundamental documents in implementing the Federal Project "Digital Technologies" of the National Program "Digital Economy of the Russian Federation". The document contains the technological task of developing and implementing multimodal human-machine interaction systems for exoskeletons and prosthetics for people with musculoskeletal problems.

In the Order of the Government of the Russian Federation dated 06.06.2020 No. 1512-r "On approval of the Consolidated Development Strategy of the Manufacturing Industry of the Russian Federation until 2024 and for the period until 2035" among the most demanded and promising long-term technological trends of the Russian market until 2035 is highlighted, in particular such trend as the use of biomechanics, including the creation of exoskeletons, robotization, organ printing on 3D printers to treat cancer and cardiovascular disease.

The use of exoskeletons generates controversy about terminology and their legal status. The authors say that not all robots can be characterized as things in the traditional sense at the stages of their use (for example, nanorobots, roboprotheses and exoskeletons), which will require clarification of rules depending on the specific circumstances of their use [1].

Filipova considers it necessary to establish the specifics of the status for workers who use exoskeletons and neuroprosthetics (body parts with artificial intelligence). Indeed, workers who use such devices can increase their striking properties by several times if they wish to inflict a blow or other harm.

European law is analyzing whether exoskeletons and exosuits can be recognized as "dual-use items" [8] to help control their trade, transfer and use within European borders. According to the European Union export control policy, the term "dual-use" means "goods, software and technology that can be used for both civilian and military purposes" and the unfair use of these items is classified as "improper use" [9].

On the one hand, dual has the advantage of recognizing the dual "civilian and military uses" and the dangers arising from the misuse of autonomous exoskeletons and exosuits. However, this designation of "dual-use items" would result in exoskeletons being subject to Council Regulation (ECR) 428/2009 "to control the export, transfer, brokering and transit of dual-use items" (ECR, 2009) from Europe to non-EU countries. This will have no direct impact on the trade, transfer and use of autonomous exoskeletons and exosuits in Europe.

The European Firearms Council Directive 91/477 EEC (hereinafter CFD), which defines common minimum rules for controlling the acquisition and possession of firearms in the EU and the transfer of firearms to another EU country, was reformed in 2017 to improve citizen safety. It aims to prevent the misuse of exoskeletons and exosuits in Europe [7]. The 2017 revision of the Firearms Directive made the rules for acquiring high-capacity weapons more complicated and allows for better tracking of firearms to reduce the risk of diversion to illegal markets [30].

The European Parliament is considering new protective measures for firearms, in particular stricter conditions for purchasing firearms, parts or ammunition over the Internet [22] and expresses excitement about increased import and export controls on exoskeletons and exosuits from countries with less stringent security standards than in Europe.

Recognition of exoskeletons and exosuits as accessories to firearms has several advantages for combating possible threats from the misuse of exoskeletons.

4 Conclusion

Thus, in the context of an overall regulatory system for the ownership and trade of exoskeletons and exosuits for internal law enforcement in Europe, CFD considers it necessary to tighten legal requirements for their acquisition based on authorization and licensing, tighten conditions for online purchase, introduce ownership tracking using existing "information exchange" networks and cooperation between EU countries, prevent entry into illegal markets, create a set of rights and responsibilities for owners of hospitals and enterprises [4].

The Russian Army needs three types of exoskeletons: a marching one for rapid movement during combat, an engineering one for low-mobility tasks, and a cargo one for moving cargo and performing other work. Russia has tested (including in combat conditions) a passive exoskeleton for the needs of the Armed Forces developed by Rostec State Corporation; at 6 kg the device can carry combat equipment and weapons weighing up to 60 kg [32]. These products are now supplied to the engineering forces and used in hospitals to rehabilitate patients. Russian companies are also working to improve the active exoskeleton, the introduction of these devices into the troops will noticeably increase soldiers' physical capabilities on the battlefield, since the exoskeleton is hidden under their equipment, does not hinder movement and relieves the musculoskeletal system, and joints can be blocked to stay in a certain posture for a long time.

The Russian Armed Forces should be widely used for passive exoskeletons in the form of lever-hinged mechanical devices that replicate human joints. An important characteristic of which is the simplicity of manufacture and operation, as well as the absence of electronic components. This type of exoskeleton could be indispensable for soldiers during long marches and assault operations since it allows relieving a soldier's musculoskeletal system. The active exoskeleton is a more complex type of army ammunition with its advantages and disadvantages. It has an increased payload which is over 100 kg and the soldier experiences virtually no fatigue and no loss of manoeuvrability on the battlefield [32]. An exoskeleton is a complex and expensive specimen consisting of many auxiliary structures, a battery, electric drives and other units; such a product can be controlled by a neural network [32].

Exoskeletons designed to reduce arm weight by 50–70% when carrying welding or painting guns [5] would also improve an intruder's ability to wield firearms; making it easier to do more damage to public places and endanger lives.

We suggest that commercially adapted versions of exoskeletons and exosuits, originally designed for industrial, medical or military use, pose a threat to public order and citizen safety in the hands of abusers. The only example we have found so far to use exoskeletons for illegal purposes in Russia is the case of Anton Maltsev, a former GRU and SOBR officer, who was accused of illegal arms trafficking by the Pushkin court in the Moscow Region. He used weapons in disputes with debt collectors of "Rosbank" to whom he owed $118 thousand for a mortgage. The judge found that the collectors drove Maltsev to mental disease (paranoia) and because of them he began to arm himself; based on the diagnosis the ex-special force's officer was completely released from criminal responsibility [3].

The photo of the exoskeleton used by Maltsev is of interest to us. The Special Forces soldier welded a metal structure from titanium and rebar, attached a helmet on top, and made provisions for the insertion of a bulletproof vest. A fact was documented when the exoskeleton was used in a chase around the yard for collectors with a hunting carbine [3]. Maltsev's exoskeleton was destroyed by court order.

In general, this device does not fall under the norms of the current legislation, it cannot be recognized as an object used as a weapon. However, exoskeletons considerably enhance the defeating properties of the person himself which is not yet regulated by the law in any way.

Thus, exoskeletons or exosuits enhance positive human capabilities and are intended for the useful purposes of healthcare, industry and national defence [9] but these characteristics of human enhancement can have negative, harmful effects if used incorrectly [8, 20, 29].

We are sure that society does not agree with the uncontrolled existence of artificially enhanced citizens which can cause significant harm to the inertia of the state, society and the individual.

For example, Sarcos Robotics' Guardian XO, which produces a "supersoldier" exoskeleton for the US military, is already aware of the threat to public law and order, and therefore actively includes design features to make illegal use more difficult [20]. However, they will make misuse of the exoskeleton more difficult, but not prevent it.

In general, threats to public relations from the misuse of exoskeletons by criminals can range from creating mass disturbances in public places, committing violent crimes and threatening citizens to more sophisticated methods of theft, burglary and art theft. Exoskeletons will make it possible to break into locked doors or open padlocks with a single movement of the mechanically reinforced wrist, to open safes, and it is also possible to predict fights between workers in exosuits in certain industries.

Take the example of Stephen Paddock, who in 2017 shot 58 people to death and injured 869 from his room window at the Mandalay Bay Hotel in Las Vegas. He had a firearm accessory, a "bump stock" which added lethality to his semiautomatic rifle [30]. Since that time, exosuits, such as the universal "MAX" in the USA, have come into use which is not only comfortable to wear all day but can also reduce muscle strength by 60% to increase the "lethality" of shooting [20].

Body-tethered exoskeletons or exosuits are unlikely to be a favourite weapon of rogues because they limit quick escape; make it easier to locate; and are virtually impossible to adapt into autonomous versions. On the contrary, autonomous, untethered, flexible versions are more likely to be misused and law enforcement should pay attention to their illicit trafficking.

The focus needs to be on threats to public order from the misuse of exoskeletons and exosuits by criminals.

Public policies are therefore needed to recognize and manage the risks of exoskeleton misuse while encouraging responsible use for the public good, such as disability rehabilitation, industrial productivity and worker health.

Right now, exoskeletons have not massively entered the global market but we foresee the need to develop and adopt regulations. No country in the world has established legal regulations to counteract illegal exoskeleton trafficking yet. Perhaps Russia needs to get ahead of the world community and develop a set of such measures. In our opinion, Article 63 of the Criminal Code "Aggravating Circumstances" could be amended to include paragraph "c" as follows: committing a crime using exoskeletons, exosuits and other devices that enhance human defeat.

The goal of states is to safeguard society from the misuse of promising emerging technology without limiting its usefulness or expanding the existing framework of technological innovation.

References

1. Arkhipov, V.V., et al.: Regulation of robotics: an introduction to "Robo Law". In: Neznamova, A.V. (Ed). Legal Aspects of the Development of Robotics and Artificial Intelligence Technologies, Infotropik Media, Moscow, 232 p. (2018)
2. Bickford, A.: "Kill-proofing" the soldier: environmental threats, anticipation, and US military biomedical armour programs. Curr. Anthropol. **60**(S19), S39–S48 (2019)
3. Blinov, M.: A Special Forces Soldier Who Fought with Collectors in an Exoskeleton was Released. https://life.ru/p/1082779 (2018). Accessed 20 September 2021
4. Burton, S.D.: Responsible use of exoskeletons and exosuits: ensuring internal security in the European context. Paladyn J. Behav. Robot. **11**(1), 370–378 (2020)
5. Butler, T.R.: Exoskeleton technology: making workers safer and more productive. Prof. Saf. **61**(9), 32–36 (2016)
6. Coren, M.J.: Robot Exoskeletons are Finally Here, and They're Nothing Like the Suits from Iron Man". https://qz.com/971741/robot-exoskeletons-are-finally-here-and-theyre-nothing-like-the-suits-from-iron-man (2017). Accessed 20 September 2021
7. Council: "Council's Firearms Directive of 18 June 1991 on control of the acquisition and possession of weapons," 91/477/EEC. Off. J. Eur. Commun. L256/51–58 (1991)
8. Dugan, C., Baloi, A.: Practical and ethical issues of persistent behavioural and neurophysiological monitorization for human performance optimization. In: International Conference RCIS'18, Redefining Community in Intercultural Context, Bucharest, Romania, pp. 17–19 (2018)
9. ECR: "Council Regulation (EC) No 428/2009 of 5 May 2009 setting up a communist regime for the control of exports, transfer, brokering and transit of dual-use items (Recast). Off. J. Eur. Union. https://eur-lex.europa.eu/legal-content/EN/TXT/?qid=1527179601283&uri=CELEX:02009R0428-20171216115221 (2009). Accessed 20 September 2021
10. Fisahn, C., Aach, M., Jansen, O., Moisi, M., Mayadev, A., Pagarigan, K.T., et al.: The effectiveness and safety of exoskeletons as assistive and rehabilitation devices in the treatment of neurologic gait disorders in patients with spinal cord injury: a systematic review. Global Spine J. **6**(8), 822–841 (2016)
11. Gorgey, A.S.: Robotic exoskeletons: the current pros and cons. World J. Orthop. **9**(9), 112–119 (2018)
12. Heary, J.: Top 5 DARPA Projects of 2010. http://www.networkworld.com/community/node/57992 (2010). Accessed 20 September 2021

13. Hitch, J.: Uncanny Exosuit: The Industrial Exoskeleton Made for Real Work, Not Movies. https://www.newequipment.com/research-and-development/article/22059105/uncanny-exosuit-the-industrial-exoskeleton-made-for-real-work-not-movies (2017). Accessed 20 September 2021
14. Hodson, H.: The Robotic Suit Gives Shipyard Workers Super Strength. https://www.newscientist.com/article/mg22329803-900-robotic-suit-gives-shipyard-workers-super-strength/ (2014). Accessed 20 September 2021
15. Investlab: Exoskeletons in the Service of Man: In Medicine, Factories, and the Army. https://invlab.ru/texnologii/ekzoskelety-na-sluzhbe-u-cheloveka/ (2019). Accessed 20 September 2021
16. Jarrass´e, N.: Contributions on the use of active exoskeletons for neuromotor re-education. Ph.D. Thesis, Institute of Intelligent Systems and Robotics, Pierre and Marie Curie University, Paris (2010)
17. Kandilakis, C., Sasso-Lance, E.: Exoskeletons for personal use after spinal cord injury. Arch. Phys. Med. Rehabil. 1–7 (2019)
18. Lin, P.: Ethical blowback from emerging technologies. J. Milit. Ethics **9**(4), 313–331 (2010)
19. Louie, D.R., Eng, J.J.: Powered robotic exoskeletons in post-stroke rehabilitation of gait: a scoping review. J. Neuroeng. Rehabitat. **13**, 53 (2016)
20. Manaugh, G.: Exoskeletons are the New Weapon of Choice for Ambitious Criminals. https://www.wired.co.uk/article/exoskeletons-crime-industry (2019). Accessed 20 September 2021
21. Manning, J., de Cameron, S., Sharkey, N.M., Shenkman, G.: Health, humanity and justice: emerging technologies and health policy in the 21st century. Health **11**, 1–30 (2020)
22. Marcello, V.E.: Dual-use in the 21st century: emerging risks and global governance. Swiss Med. Wkly. **148**, w14688 (2018)
23. Marinov, B.: Passive Exoskeletons Establish a Foothold in Automotive Manufacturing. https://www.forbes.com/sites/borislavmarinov/2019/05/15/passive-exoskeletons-establish-a-foothold-in-automotive-manufacturing/ (2019). Accessed 20 September 2021
24. Moreno, J.D.: The emerging life sciences and the national security state. Strateg. Stud. Q. **10** (3), 9–15 (2016)
25. Peck, M.: "Real-Life Iron Man? China is Developing Military Exoskeletons". National Interest. https://nationalinterest.org/blog/buzz/real-life-iron-man-china-developing-military-exoskeletons-115221 (2020). Accessed 20 September 2021
26. Sawicki, G.S., Beck, O.N., Kang, I., Young, A.J.: The exoskeleton expansion: improving walking and running economy. J. Neuroeng. Rehabitat. **17**, 25 (2020)
27. Sylla, N., Bonnet, V.: Ergonomic contribution of the ABLE exoskeleton in the automotive industry. Int. J. Ind. Ergon. **44**(4), 475–481 (2014)
28. Taraska, P.A.: How can the use of human enhancement (HE) technologies in the military be ethically assessed? Ph.D. Thesis, Duquesne University, Duquesne, USA (2017)
29. Tennison, M.N., Moreno, J.D.: Neuroscience, ethics, and national security: the state of the art. PLoS Biol. **10**(3), e1001289 (2012)
30. The Telegraph: Bump-Stock' Devices: How Las Vegas Shooter Stephen Paddock Used 'Souped-Up Semi-Automatic' Weapons. https://www.telegraph.co.uk/news/2017/10/03/bump-stock-devices-las-vegas-shooter-stephen-paddock-could-have/.115221 (2017). Accessed 20 September 2021
31. Vorobyov, A.A.: Terminology and Classification of Exoskeletons. http://www.volgmed.ru/uploads/journals/articles/1476165386-vestnik-2015-3-2459.pdf (2015). Accessed 20 September 2021
32. Zakvasin, A., Komarova E.: Under the Equipment: How the Latest Exoskeletons Can Change the Appearance of the Russian Army – RT (in Russian). https://russian.rt.com/russia/article/787558-ekzoskelet-rosteh-armiya115221 (2020). Accessed 20 September 2021

To the Question of Digital Transformation of Strategic Planning in the Field of Environmental Protection and Rational Use of Natural Resources in the Russian Federation with Consideration of Integration with the EAEU

Kira V. Kolesnikova[1(✉)], Liliya V. Postulga[1], Daria U. Usckova[1], Nadezhda V. Miroshnichenko[2], and Yulia V. Labovskaya[2]

[1] North Caucasian Federal University, Stavropol, Russia
[2] Stavropol State Agrarian University, Stavropol, Russia

Abstract. The paper aims to identify the main problems of strategic planning of socio-economic (and environmental) development of the Russian Federation and the reasons for their emergence. Moreover, the paper develops the ways to solve these problems in a fully automated process in managing the socio-economic sphere of social relations. The conducted legal analysis found that the low level of effectiveness of strategic planning (particularly in the field of environmental protection and environmental management) is due to two aspects. On the one hand, it is due to the framework nature of the provisions of the law on strategic planning. On the other hand, the low level of effectiveness of strategic planning is caused by its "overloading" with strategic documents, which, in the absence of proper methodological support, leads to misunderstanding of the order of actions by participants in strategic planning, and, as a consequence, to the lack of involvement in the planning process, focusing only on solving individual problems, and abstracting from the achievement of common priorities and goals. During legal research, the authors use formal-legal analysis, synthesis, generalization, and comparative-legal scientific methods. The authors develop proposals to improve the effectiveness of state strategic planning through its digital transformation using the accumulated national and international experience and considering integration with the EAEU. A legal analysis of the main risks that can negatively impact the effectiveness of digital transformation in this area has been conducted. The research may be of practical interest to organizations and authorities of all levels involved in digital transformation.

Keywords: Public administration · Strategic planning · Environmental protection · Digital economy · Digital transformation · Eurasian Economic Union

JEL codes: K32 · O21 · Q-58

E. G. Popkova (Ed.): ISC 2020, LNNS 368, pp. 825–831, 2022.
https://doi.org/10.1007/978-3-030-93244-2_88

1 Introduction

Nowadays, the system of state strategic planning in the Russian Federation is regulated by Federal Law "On strategic planning in the Russian Federation" (the law) [3]. The adoption of the law was perceived as an important stage (new round) in the development of the system of state planning. However, the provisions of the law have failed to ensure the effective functioning of the strategic planning system and have not solved the problem of ensuring a breakthrough socio-economic development of Russia.

In our opinion, the low effectiveness of strategic planning can be explained by the framework nature of the law's provisions (its norms are too general, there is no regulation of the order and sequence of actions of participants of strategic planning). On the other hand, the low effectiveness of the law is due to its "overloading" with strategic documents (it provides for 16 types of documents for the federal level and 12 documents for regional and municipal levels; during the implementation of the law, over 55 thousand documents of strategic planning have been developed and approved). In the absence of proper methodological support, these facts lead to misunderstanding of the order of actions by participants in strategic planning, and, as a consequence, to the lack of involvement in the planning process, focusing only on solving individual problems and abstracting from the achievement of common priorities and goals. As a result, there is a mismatch of goals, objectives, and activities to the established goals and objectives, as well as the inconsistency of interrelated strategic documents, which, in turn, leads to poor results of their implementation throughout the vertical.

2 Materials and Methods

Within the framework of this research, the authors analyzed the indicated problems using formal-legal analysis, synthesis, generalization, and comparative-legal methods, which allowed developing possible ways to solve them.

3 Results

We believe that one of the ways to solve the problem of low efficiency of strategic planning in the Russian Federation, along with the need to update and improve the regulatory framework (including methodological), is the digital transformation of the strategic planning, including in the field of nature conservation and sustainable use.

Various aspects of digitalization and digital transformation of the strategic planning system are considered in economic [5, 9, 13, 15] and legal scientific literature [6, 8, 12, 14, 16, 17].

This research considers digital transformation as a profound reorganization of processes in public administration of the socio-economic sphere of social relations with the active use of a wide range of digital tools, which significantly improves the characteristics of existing processes. These processes include the reduction of the time to perform an action (e.g., generating reports), the elimination of some processes (e.g., eliminating paper-based workflows), the reduction of the resources spent on some processes, and the creation of new processes, digital models, and digital assets.

The digital transformation of the economy and the entire system of state management of socio-economic development at the national and supranational (interstate) level is particularly important. In 2018, EAEU member countries discussed the creation of a unified strategy and program of digital transformation as the most important task within the framework of the EAEU Digital Agenda. The "Guidelines for the implementation of the EAEU Digital Agenda" (the guidelines) note that without the development of the digital economy, the country remains within the traditional processes (i.e., deprived of new opportunities), while there is a depreciation of traditional assets. However, it should be noted that EAEU member countries use an excessively "cautious" approach to implementing digitalization and digital transformation projects at the union level, which was also highlighted by the Russian Prime Minister at the forum Digital Almaty 2021. The strategy for the development of the EAEU digital space was adopted only in December 2020. The overall digital transformation program has not yet been adopted, which, in our opinion, may become an obstacle to achieving significant results in the development of the digital economy of EAEU member countries and the formation of a common digital space.

Currently, all EAEU member countries have adopted national programs of digital transformation at the domestic level. The analysis of the content of these programs allows us to conclude that they were developed without uniform standards and approaches. Therefore, the goals and objectives of these programs are similar to some extent but not entirely the same.

Decrees of the President of the Russian Federation [1, 2] defined digital transformation as a national development goal, one of whose targets is the "digital maturity" of key sectors of the economy and, among other things, public administration, which should be achieved through the introduction of digital technologies and platform solutions in the relevant areas.

The Federal Information System for Strategic Planning (FIS SP) was created as part of the federal project "Digital Government," which is part of the Digital Economy national program. In 2020–2022, based on FIS SP, it is planned to create a unified digital platform for strategic planning and support of public administration in the Russian Federation. In our opinion, it is driven by the need to introduce a tool to provide a new level of cooperation and interaction between strategic planners, which can lead to the creation of new networks and global chains of digital interaction patterns and digital dividends. The project plans to make the transition from a "document environment" to a "digital data environment," eliminate the principle of "document for document's sake" (i.e., when strategic documents are adopted, but the targets they set are not met), and achieve "digital maturity" of public administration. Nevertheless, the implementation of the considered federal project has some difficulties, and its shortcomings have been analyzed in the legal literature [17].

Even though strategic planning documents assign a vital role to digital technologies in implementing public administration in the socio-economic (and environmental) development of the Russian Federation, digital transformation is also associated with certain risks and difficulties. These risks include legal and regulatory risks, in particular the framework regulation of this process, the lack of a more detailed legal elaboration of the issue of introducing digital technologies in sectoral legislation (e.g., water, air, forest, etc.), the absence of these tasks in the strategic development documents in the field of nature protection and its rational use, and the inconsistency of interrelated strategic acts. There are also technical risks (lack of national software for the platforms developed) and human resources risks (lack of qualified personnel, their outflow from the country, and inability to provide competitive salaries). Another problem is the lack of competencies in the field of digital development (to accumulate such competencies, EAEU organizes expert platforms with the participation of representatives of government agencies, business, and science). Among the factors that slow down the digital development of EAEU member countries, scientific literature highlights the inertia of government agencies in addressing issues of informatization, the lack of motivation to change business processes, insufficient funding from government and business, and low interest in opportunities for public-private partnerships [7].

The possible risks of the digital transformation project also include the lack of independent and objective monitoring of the implementation of planned indicators. An example of the realization of this risk is the results of the federal target program "Electronic Russia (2002–2010)," several measures of which (with a deadline in 2003) have been relevant at the time of the development of the considered federal project. These measures include creating an integrated information system integrating statistical data, the implementation of reporting systems, and the analysis of the performance of government agencies [5]. Another risk lies in the lack of a full volume of reliable data in all related government information systems to be used by the FIS SP, which may make the creation of a digital ecosystem meaningless. One of the examples is the use of the GIS to account for waste from the use of goods when it does not receive full information about payers of the environmental levy. Another example is the GIS on solid municipal waste management and the GIS of public control in the field of environmental protection and rational use of natural resources "Our Nature," for the creation of which more than 119 billion rubles have been spent since 2013, but the system is still not functioning [4].

To ensure the effectiveness of the digital transformation of the strategic planning system, it is essential to develop a detailed national and supranational regulatory framework, including the formation of joint solutions, the reduction of risks, and the use of a regulatory sandbox mechanism to test new projects, digital models, and processes. The scientific literature points to the fact that the EAEU must create a uniform and unified system of regulation of relations related to digital transformation, the basis of which can be fixed, for example, in the form of a separate international treaty [11]. The EAEU documents do not highlight issues of integration cooperation, environmental protection, and rational use of natural resources as an independent area of integration interaction. Accordingly, the digital transformation of strategic planning in this area is considered in the context of the digital transformation of the economy. It seems necessary to harmonize the national legislation of EAEU member countries in this area and develop a unified regulatory framework at the union level [10]. Moreover, to ensure the effectiveness of the digital transformation of the strategic planning system, it is necessary to ensure the interoperability of data (i.e., the ability of systems to interact based on information and communication technologies) in developed information systems domestically and at the EAEU level. Additionally, the effectiveness of the digital transformation of the strategic planning system should be ensured by using common principles, approaches, and technologies to ensure the compatibility of national digital platforms, which will prepare and ensure the transition to interstate integration. Using elements of artificial intelligence, a single digital platform should provide systematization and analysis of big data (i.e., compare goals, objectives, and indicators from the entire set of related strategic documents). Next, based on this information, this platform should assess the effectiveness and efficiency of strategic documents and strategic planning participants, form forecasts and plans, and collect information in real-time to plan activities in the considered area. The digital transformation should focus not just on the automation of existing processes but also on creating fundamentally new models of activity, which include a wide exchange of data between participants in strategic planning – a comprehensive digital interagency interaction forming a digital ecosystem.

During the integration cooperation of the EAEU member countries, they must unite efforts to develop and implement a common program of digital transformation aimed at the creation of a single digital space and a single digital platform for the interaction of the EAEU member countries and the development of a digital ecosystem of the EAEU based on common principles, standards, and technologies. Such a program should include specific activities, including the following:

- Changes in the procedures for the preparation of the EAEU regulatory framework (the introduction of a mandatory pre-modeling of processes and the development of rules for the use of common digital platforms, procedures, and standards for project management);
- Implementation of unified digital accounting systems of identification of citizens and businesses, recognition of digital electronic signature;
- Development of procedures for the exchange of personal data and their protection and mechanisms to ensure the development and implementation of projects within the framework of the Digital Agenda.

It is critical to conduct this integration taking into account the principle of national interest.

4 Conclusion

Thus, the implementation of the project of digital transformation of the system of strategic planning is of paramount importance for improving the effectiveness of strategic planning, including in the field of nature conservation and environmental management. However, it is possible to achieve a positive result only if the possible risks are assessed, and relevant accumulated national and international experience is considered. The successful implementation of the digital transformation project at the national level will further contribute to integration, the achievement of high rates of economic development, global competitiveness, and a high level of welfare of the EAEU countries. The problems of protecting the natural environment and its rational use are the problems of the entire global community. Therefore, a single integrated digital platform in the considered area will ensure their effective solution, ultimately acting as an important tool for improving the global level of environmental quality and, consequently, the quality of human life.

The research may be of practical interest to government agencies and organizations involved in digital transformation.

Acknowledgments. This research was supported by the Russian Foundation for Basic Research – project 20-511-00015 Bel_a "Legal Problems of Forming a Common Environmental Safety Space of Eurasian Economic Union Member Countries."

References

1. President of the Russian Federation: Decree "On National Development Goals of the Russian Federation" (May 7, 2018 No. 204). Moscow, Russia. Retrieved from https://base.garant.ru/71937200/ (2018). Accessed 7 Sep 2021
2. President of the Russian Federation: Decree "On the national development Goals of the Russian Federation for the period up to 2030" (July 21, 2020 No. 474). Moscow, Russia. Retrieved from http://www.kremlin.ru/acts/bank/45726(2020). Accessed 7 Sep 2021
3. Russian Federation: Federal Law "On strategic planning in the Russian Federation" (June 28, 2014 No. 172-FZ). Moscow, Russia. Retrieved from https://base.garant.ru/70684666/(2014). Accessed 7 Sep 2021
4. Accounting Chamber of the Russian Federation: Environmental state information system. Bulletin of the Accounting Chamber of the Russian Federation, **12**(277), 7–8. Retrieved from https://ach.gov.ru/upload/iblock/fdf/fdfa306ca7498d6522d5100d4a76c670.pdf (2020). Accessed 7 Sep 2021
5. Dvinskikh, D.Yu., et al.: Digital transformation of public administration: myths and reality. In: Dmitrieva, N.E. (ed.) Publishing House of the Higher School of Economics, Moscow, Russia (2019)
6. Koroleva, A.N.: Transformation of legal regulation of strategic planning of socio-economic development of Russia and foreign countries in the digital economy. Actual Problems of Jurisprudence **1**(65), 7–14 (2020)
7. Kovalev, M.M., Golovenchik, G.G.: Digital Economy is a Chance for Belarus. Publishing House of the BSU Center, Minsk, Belarus (2018)
8. Kuznetsov, P.U.: Digital transformation of public administration as a stage of informatization development in Russia. Bulletin of the South Ural State University **21**(1), 84–95 (2021). https://doi.org/10.14529/law210113
9. Lipina, S.A., Belyaevskaya-Plotnik, L.A., Sorokina, N.: Approaches to automatization of data collection and presentation in order to monitoring the implementation of strategic planning documents. Scientific Research and Development Economy **6**(1), 12–19 (2018). https://doi.org/10.12737/article_5a8d40cea87b67.44080109
10. Makarova, T.I.: Environmental law of the eurasian economic union: problems of formation. In: Makarova, T.I. (ed.) Influence of Interstate Integration Processes on the Development of Agrarian, Environmental, Natural Resource and Energy Law, pp. 55–58. BSU Publishing House, Minsk, Belarus (2018)
11. Mikhaleva, T.N.: The digital agenda of the EAEU: legal standards and prospects. In: Mikhaleva, T.N. (ed.) States and Law in the XXI Century: Materials of the International Scientific and Practical Conference Dedicated to the 95th Anniversary of the Faculty of Law of the Belarusian State University, pp. 42–47. BSU Publishing House, Minsk, Belarus (2021)

12. Peretelet, R.A.: Ecological aspects of the digital economy. The World of the New Economy **12**(4), 39–45 (2018). https://doi.org/10.26794/2220-6469-2018-12-4-39-45
13. Pisareva, O.M.: Reforming of state strategic planning system and actual problems of improving methodological support for the development of strategic decisions in the conditions of the formation of the digital economy. In: Kleiner, G.B. (ed.) Strategic Planning and Enterprise Development: Materials of the Nineteenth All-Russian Symposium, pp. 633–636. CEMI RAN, Moscow, Russia (2018)
14. Sukhova, E.A., Abanina, E.N.: Legal problems of digital transformation of the environmental management system as a mechanism for ensuring environmental security. Russian Justice **8**, 17–20 (2020)
15. Zaversky, S.M., Kiseleva, E.S., Kononova, V., Plekhanov, D.A., Churkina, N.M.: Strategic planning for economic development: international experience and lessons for Russia. Bulletin of the Institute of Economics **2**, 22–39 (2016)
16. Zhavoronkova, N.G., Shpakovskiy, Y.: Digitalization in the field of environmental security: administrative and legal aspects. Lawyer **4**, 14–19 (2019). https://doi.org/10.18572/1812-3929-2019-4-14-19
17. Zubarev, S.M.: Legal risks of digitalization of public administration. Actual Problems of Russian Law **15**(6), 23–32 (2020). https://doi.org/10.17803/1994-1471.2020.115.6.023-032

Conclusion

Perspectives of the Practical Use of the Imitation Models of the Digital Economy Markets

Imitation modeling – as a scientific and methodological approach to the research of the digital economy markets – has proved its effectiveness. As this book has shown, the perspectives of the practical application of imitation models of the digital economy's models are very wide. Thus, the influence of globalization and digitalization on the modern economy and sectorial markets has been studied systemically. The presented imitation models allow forecasting the change of economic processes and systems depending on the direction and intensity of globalization and digitalization, as well as determining their optimal level to achieve various market priorities.

The results obtained in this book have also demonstrated that the use of the leading technologies in the digital economy markets is active, but not always profitable. The compiled imitation models allow determining the level of digitalization that is most profitable for economic subjects and systems. As this book has shown, business management and business integration in the digital economy received a new impulse for development, but they require flexible management – it could be provided by imitation models and Game theory.

This book has also shown that markets and companies in the digital economy face increased risks and threats to economic security and sustainable development. The offered imitation models allow for the full determination of these risks and the development of their highly effective management. The proposed game approach to managing the markets of the digital economy allows balancing the interests of technological modernization and sustainable development.

Despite the filling of the scientific and methodological gaps, imitation modeling – as an approach to studying the markets of the digital economy – has demonstrated new gaps in science, which include the complexity of choice between private (based on the experience of certain companies) and general (reflecting the

E. G. Popkova (Ed.): ISC 2020, LNNS 368, pp. 835–836, 2022.
https://doi.org/10.1007/978-3-030-93244-2

experience of whole markets or economies) imitation models, each of which is useful in its own way in certain cases.

They also include the narrowness of the directions of using imitation models (similarly to conceptual models). The models reflect either the issues of economic effectiveness and leading technologies, or the issues of economic security, risk management, and sustainability. Due to this, the complex consideration of these issues during the research of the same economic process or economic system requires the use of several imitation models at the same time. Therefore, it is necessary to enlarge the imitation models and achieve their comprehensive coverage.

The above gaps in science require further scientific elaboration, which should be performed in future studies. However, despite certain limitations of imitation modeling, this book has demonstrated its advantage (as compared to conceptual modeling) and perspectives of use during the study of the digital economy markets.

Author Index

E. G. Popkova (Ed.): ISC 2020, LNNS 368, pp. 837–840, 2022.
https://doi.org/10.1007/978-3-030-93244-2